Biophysics

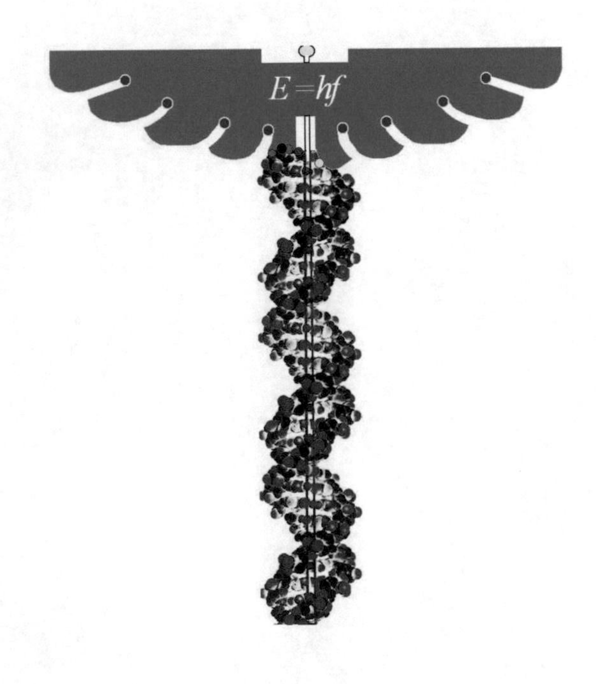
E=hf

William C. Parke

Biophysics

A Student's Guide to the Physics of the Life Sciences and Medicine

William C. Parke
Professor of Physics Emeritus
George Washington University
Washington, WA, USA

ISBN 978-3-030-44148-7 ISBN 978-3-030-44146-3 (eBook)
https://doi.org/10.1007/978-3-030-44146-3

Cover image: © Dkosig / Getty Images / iStock

This Springer imprint is published by the registered company Springer Nature Switzerland AG
The registered company address is: Gewerbestrasse 11, 6330 Cham, Switzerland

Preface

"Begin at the beginning," the King said gravely, "and go on 'till you come to the end. Then stop."

—Lewis Carroll, *Alice in Wonderland*

This is a student's guide to biophysics, following a course given for many years to biophysics, premedical, and biomedical engineering students. The full range of physical principles underlying biological processes are introduced and applied to biophysics and medicine. By starting with the foundations of natural processes and then showing how to arrive at some consequences in biosystems, students may develop their own thinking in applying our understanding of nature. These creative endeavors are useful to future applications of the field, facilitating progress in biophysical and biomedical technology.

The level of the course is intended for advanced undergraduates who have already had a first year physics course and an introduction to calculus and for first-year graduate students. Although the presentations sometimes contain models whose mathematical tactics are new to some students, the use of such techniques is not to teach mathematical physics, but rather to show that the ideas behind definitive statements in the subject are logically firm. The art of solving the more sophisticated models for particular examples in nature is left to more advanced courses.

I wish to especially thank Dr. Thanhcuong Nguyen, who shared her extensive course notes, and Dr. Craig Futterman, for his continuous and lively encouragement through the years. With discussions and exchanges, Prof. Leonard C. Maximon was particularly helpful in his ability to zero in on the essence of a question, as was Prof. Ali Eskandarian for his insightful ideas.

Please notify the author (wparke@gwu.edu) if errors are found. Unattributed figures were created by the author or are in the public domain. No liability is assumed for losses or damages due to the information provided.

Solutions to the exercises are available to instructors for their use, but not for distribution, from https://www.springer.com/us/book/9783030441456.

Washington, DC, USA William C. Parke

Contents

Chapter 1
Introduction: The Nature of Biophysics

Nature's imagination is so much greater than man's, she's never going to let us relax.

—Richard Feynman

Summary This chapter describes how biophysics fits into the sciences. Included is a discussion on the character and limitations of models and theories of nature, and introduces the idea that complex systems develop their own emergent principles not originally part of the underlying physics.

1.1 The Study of Life as a Physical System

Biophysics is the study of life enlightened by physical principles. Several aspects make the subject particularly interesting, not the least being the fact that we are developing a fundamental understanding of ourselves, a life form with the ability to make sense of our own observations. After several million years of hominid thinking, we are just now realizing what constitutes life, and that life processes follow natural laws common to the sequence of all events in our universe. We now know that 'living organisms' are complex self-reproductive systems in an open environment containing the elements, interactions, and energy flow necessary for activity and the creation of order. We can construe the continuing evolution of life as a natural process and clarify the detailed mechanisms used by life on Earth with the same physical laws already successfully applied to the behavior of planets and atoms. Because of the evolving order in biological systems toward optimizing systems and behavior in a given environment, new organizing systematics and biological principles develop. It is the aim of this presentation to describe this ongoing 'natural science', a process of studying, organizing, and developing our ideas about nature.

In biophysics, as in any science, our models are designed to match observation, as far as measurement allows. Concepts in our models should be defined unam-

W. C. Parke, *Biophysics*, https://doi.org/10.1007/978-3-030-44146-3_1

biguously and must be logically self-consistent. The 'best models' cover the widest range of natural events, with the least number of propositions.

Scientifically answering certain questions may be difficult, such as the end of a string of whys. In addition, because our physical instruments are finite, there may always be gaps in our knowledge. And, because logical systems with an infinite number of elements may have unprovable propositions (a form of Gödel's Theorem), the 'laws of nature' may themselves be unlimited and even evolving. If so, happily, there will always be more to do in our quest to understand and to develop.

1.2 Conceptual Models of Nature

Everything should be made as simple as possible, but not simpler!

— Albert Einstein

We are a species driven, in part, by 'curiosity'. Curiosity leads to exploration, exploration to a form of understanding, understanding to an ability to effectively anticipate, manage, and develop our environment. The elaboration of such intellectual processes seems quite natural in evolution, as such abilities give a distinct advantage over those life forms with lesser talents in this direction.

Central to the process of understanding is the creation of a 'model' to reflect a set of related observations. A conceptual model of nature synthesizes a set of observations into a mental structure which incorporates a predictive scheme. The relationships in a physical model are required not to be contradictory, and to form an economical way of consolidating and connecting facts.

If a model is based on an underlying logical and causal connections between observed events, then the model becomes a 'theory'. Theories of nature give the user a framework on which to evaluate and plan ahead. The Babylonians, Greeks, and Mayans were able to predict lunar eclipses using careful records showing periodicity. As impressive and useful as these predictions were, the scheme was not yet a theory of lunar eclipses. No underlying causal connections were given. Those who imagined that a lunar eclipse could be generated by the shadow of a giant bird who periodically intercepted the Moon's light had a theory, albeit a flighty and easily falsifiable one. Aristarchus, who thought of the Earth as a spherical body blocking sunlight as the Sun, Earth, and Moon moved on determined paths, had a better theory, not only because it agreed with observations, but it could also make correct predictions of events other than lunar eclipses.

We judge theories based on the agreement of their predictions with observations and their relative simplicity. If a competing theory makes the same predictions in the realm of initial application but has fewer constructs and/or wider applicability, we say it is a more 'refined theory'. We are attracted to a more refined theory, as it has the advantage of being easier to keep in mind, and it might give us more understanding of what we see, while generating insights into what might be not yet seen.

A theory which is capable of not only matching known facts, but contains new predictions, achieves a special stature when those predictions are later verified. The probability of making a correct detailed numerical prediction matching a measurement, by chance or by an incorrect theory, is usually quite small. There are many instances in the development of our understanding of nature when a new theory has made significant predictions later found true. In physics, perhaps the most famous of these are Maxwell's prediction (1864) of radio waves, Einstein's prediction (1905) of the energy equivalence of mass, and Dirac's prediction (1931) of antimatter. In biology, Darwin (1859) predicted how species evolve under selective environmental pressure, and Mendel's rules (1865) described the behavior of inherited units of trait later discovered to be genes on our DNA. In chemistry, Dalton's law (1803) of multiple proportions in chemical reactions preceded the demonstration that atoms exist, and Mendeleev predicted (1871) the properties of as yet undiscovered elements from the pattern of periodic behavior of known elements.

Predictions arising from good theories often have profound influence on the development of technology, including medical applications.

Theories successfully describing observations give us the ability to think clearly about the possible as well as the improbable.

In contrast to the search for a logical basis for our observations, there has also been a tendency in the mind of a multitude of humans to submit to the solace and comfort of supernatural explanations, often promulgated by so-called higher authorities in societies. In many cases, those higher authorities were motivated by a desire to maintain their status rather than to seek the truth. Throughout history, individuals have said we should give up an exploration at the frontiers of knowledge, with declarations equivalent to "God made it that way." This capitulation seems to go against our natural curiosity. But the desire for a childlike feeling of security based on an inscrutable and benevolent higher intelligence in the universe can be stronger. As genuine as this feeling is, keeping an open mind willing to explore has led our species to see deeper and deeper logical underpinnings to natural events. With such a frame of mind and the resulting knowledge, we have the ability to judge when a story is based on false or unverifiable notions. Science organizes our description of nature's logic. The success of science in finding answers about nature has expanded the domain of the understandable. There is no evidence that the domain of the knowable has limits.

Logically, conceptual models contain 'primitive elements' whose nature is defined only by relationships with other elements. So it is in both mathematics and science. Ultimately, there will be, in each conceptual model of nature, ideas which have no antecedent, such as space and time in Newtonian theory. These ideas take their significance from their connection to other ideas. In newer formulations, it may be possible to find deeper significance to a set of ideas, or to give alternate interpretations to observations. These can be useful in stimulating new ideas. Selecting between inequivalent models may require more precise measurements, or a fuller exploration of the model's realm. If there are no other compelling reasons to embrace one model over another, we apply 'Occam's Razor': Do not introduce complication if a simpler explanation suffices! The best models over a given realm of observation use the least number of assumptions to predict what is measured.

Many early conceptual models of the world developed by the human species were contained in stories and lessons based on an attempt to make sense of their experiences. Inevitable to the evolution of world descriptions is confrontation with observation. A natural reaction to a discovery of discrepancy with a favored model is dismissal or denial of the observation. Models with a degree of success are not easily erased from the mind, or from the lore of a society, particularly if there are no ready replacements. We all have a resistance to changing our world view, since our deep-seated views can be stabilizing factors in our lives.

But there is a chance that individuals or groups will be able to formulate a better model of the world, or the universe. The chances for an improved model increase with the number of smart individuals freely thinking about related concepts and problems, and may increase geometrically with the number of such individuals communicating through the free exchange of ideas. Thus, skilled individuals engaged in careful observation, in the creative process, and in the sharing of ideas, are catalytic elements in conceptual model building.

1.3 Conceptual Models and Natural Sciences

God made the integers, all else is the work of man.

——Leopold Kronecker

If we restrict a study to conceptual models which conform to exacting (quantitative) observations, then that study becomes a science. Those concepts which are imprecisely introduced or ill-defined are not refined enough to be part of a science, although they may make good poetry and be stimulating to thought. They might even be precursors to a measurable entity in an exact science. In contrast, if a concept is connected to a reproducible measurement, then that concept becomes an 'observable', whose behavior can be scientifically studied in relation to other observables. Beauty is not yet a scientific concept. So far, there is no accepted physiological, psychological or neurological definition of beauty which can be consistently measured. On the other hand, time is an observable: We simply define time to be what is measured by a clock. In turn, a clock is a device which passes through a finite sequence of states repeatedly, with each smallest repetition determining a single period of the clock. Comparing the number of such periods to the progress of other sequences of events, such as the beating of a heart, constitutes a measurement of the time interval for those other events.

Pinning a concept to a measurement circumvents much of the ambiguity in which argument thrives. As confirmation of the predictions of a model relies on the scope and precision of measuring instruments, verification of a theory is limited by these same instruments. Part of the effort of science involves the refinement of measurement and the expansion of the realms over which observation is possible.

Often, the well-tested and accepted statements in a natural theory are designated as physical laws, but the phrase gives a false impression that the propositions of the theory cannot be violated. Inversely, a common misconception is that a physical theory is "just a theory," implying that no tests have been made. Statements within a theory should not be assumed absolute or even independent of the human mind. None of our instruments which measure a physical quantity (as opposed to pure numbers) can have infinite precision or make determinations over all space or all time. Therefore, it does not make sense to claim that a scientific theory giving relationships between such physical quantities describes an absolute truth. Only records of discrete integers can be considered absolute.

Note that the language of a theory contains preconceived notions of what is to be taken as primitive and fundamental. Alternatives always exist.[1] We know that a logical theory can be mathematically transformed into an equivalent one with a completely different language but identical predictions. So we should recognize that when we talk about certain 'concepts', such as particles or electric fields, we are not talking about nature herself, but rather about elements of our models of nature.

All of our conceptual models must be considered tentative. They catch observations no better than a fisherman's net catches objects in the ocean. Some will get away. Our physical laws about observables with physical dimensions are never absolute. They are known to hold only in the realm where they have been tested, and there only to within a certain precision.

The process which has been utilized that favors the development of a science is the scientific method. The essence of this process is observation, model construction, and the design of a careful set of tests to check if there is a degree of agreement of the model's prediction with measurements. Observations or experiments must be sufficiently precise and reproducible to make the tests definitive and convincing. We define 'precision' by the smallness of the variability of a measurement after repeated measurements, while 'accuracy' is determined by how close a measurement comes to a prediction.

In the process of checking a hypothesis or world view, Nature may reveal yet another mystery to be solved. Moreover, complex models often show their own limits, at the boundaries of their realm. This is true for some of our best theories: Newton's, Maxwell's, Einstein's and Dirac's.

The scientific method has had remarkable success in the last few centuries for several reasons, not least among these are: A climate of independent thinking; rapid distribution of the news of a new discovery and of ideas; the use of impartial testing

[1] It is possible to transform Maxwell's theory of electromagnetism to an 'action-at-a-distance' theory, with no electric or magnetic fields. The new formulation is completely equivalent to Maxwell's theory in its predictions. It is also possible to make a 'non-local' quantum theory without any reference to wave functions, completely equivalent to conventional quantum theory.

of conceptual models through careful observation; and the simple realization that we are capable of successfully expressing natural behavior in logical form.

Amazingly, our minds have proven of sufficient intellect to unravel some logic behind Nature, at least on her surface. This includes models describing the structure of the whole Universe as well as being able to predict observations at particle separations down to much smaller than a proton diameter.

In observing the natural world, we find it possible to diminish external influences onto certain 'localizable systems'. Natural processes appear isolatable in space and reproducible in time. It therefore becomes possible to study systems under controlled conditions. If the behavior of such systems depends predominantly on only changes in a few observables, those systems are said to be simple. The motion of two or three point-like masses under gravity is relatively simple, but so is the macroscopic behavior of 10^{23} particles in an ideal gas if that gas is near thermodynamic equilibrium, for then the intrinsic macroscopic properties of the gas depend simply on its density, specific heat, temperature, and pressure.

Of the exact sciences, 'classical mechanics' was successful over 300 years ago because it could be applied to simple isolatable systems, such as planetary systems. The theory of mechanics given to us by Isaac Newton in 1687 is still used by NASA to plot the motion of spacecraft. The phenomenal success of classical mechanics is still a source of wonder, and a stimulant to the development of other sciences. Its descendant, quantum mechanics, is capable at once of accurately tracking planets, baseballs, molecules, and electrons within a theory built on just a few logical statements. (NASA does not use quantum theory to plot spacecraft motion because Newton's theory is a very good approximation for objects much more massive than molecules, and moving much slower than light speed, and because Newton's theory is much simpler to handle than quantum theory.)

A distinction must be made between the number of independent observables within a system and the complexity of its ramifications. As Henri Poincaré recognized in the nineteenth century, simple mechanical systems can show chaotic behavior, manifest in how small differences in the initial conditions can evolve exponentially into large differences in the resultant paths of motion. Given the finite resolution of our instruments, we are incapable of determining the long-term evolution of certain systems with the same precision as our knowledge of their present state. Even so, definitive statements often can be made about the probabilities of a class of future paths, and about their possible confinement.

A collection of interacting fundamental elements will be a simple system if its dynamics can be followed by using only a few relationships between selected system observables. If a large system (containing many elements, also called a macrosystem) requires a large amount of information to specify its dynamics, it is called a 'disordered system'. Those systems whose long-time behavior becomes intractable exponentially with time because of inevitable information loss are called 'chaotic systems'. Those systems whose behavior depends on the development of transformable substructures with evolved relationships between those substructures, are said to be 'complex systems'.

Natural ordering within open macrosystems may lead correspondingly to simple behavior when disturbances are weak. The science of macrosystems began with the study of simple systems with a large number of elements. Thermodynamics is a classic example. It reveals that new macroscopic observables which have no meaning for individual elements may follow simple relationships, such as the ideal gas law, when the system dynamics has quenched fluctuations in those observables.

A unification of principles occurs when macroscopic laws are derivable from microscopic laws and collective statistical behavior. Thermodynamics, initially developed from macroscopic observation, is now supported by statistical mechanics. The laws of evolution, the behavior of neural networks, and the rules of genetics from genotype to phenotype are gaining similar support.

Life systems are intrinsically complex. In order to exist and thrive, they take advantage of a variety of interacting elements, a spectrum of possible metastable states, and a flow of energy, to support life-enhancing algorithms. Even so, as we shall see, subsystems in an organism may be simple.

For complex systems, internal natural ordering and optimization makes it possible for there to exist general organizing principles, rules, behaviors and relationships among participating 'agents'. The endeavor to discover these 'emergent principles' and relationships is a central frontier of biophysics.

1.4 Biophysical Studies

We must trust to nothing but facts: These are presented to us by Nature, and cannot deceive. We ought, in every instance, to submit our reasoning to the test of experiment, and never to search for truth but by the natural road of experiment and observation.— Antoine Lavoisier (1790) Elements of Chemistry

Biophysics covers both macroscopic and microscopic applications of physical principles. Historically, our focus was on macroscopic properties of life systems. These were closer to observations and easier to measure. Isaac Newton was pleased to find a relationship between the weight of land-dwelling animals and their required skeletal structure. Several centuries ago, a physical basis was ascribed to the optics of eyes and the acoustics of ears. Other internal systems were also studied using physical laws. For example, fluid mechanics was used to study blood pressure and flow. With the successes of microscopic physics, atomic chemistry and molecular biology, new 'nanoscopic' and 'mesoscopic' approaches became possible, with the intent to find the machinery of life at the deepest level, and to discover the full spectrum of organizing principles leading to the behavior of life systems.

1.5 Focus of This Study

This text is written for those who already have some background in physics (such as an introductory college course) and wish to further explore how to apply physical laws and principles to the understanding of life systems and to instruments used in biology, medicine, and diagnostics. Biology and pre-medical-doctor students, pre-medical-engineering students, medical technologists, and other science majors might benefit from the practical aspect of such knowledge while taking part in the enjoyment that accompanies such an exploration. Medical researchers need to know such things if they are to make better instruments and cures, and doctors should be expected to know the important workings, capabilities, limitations, and dangers of the devices they use for diagnostics and treatment of the ill.

1.6 Organization of Study

In this study, we will identify particular phenomena in nature which arise from basic physical interactions. For each natural effect, we will attempt to give the following, when applicable:

- Defining Concepts;
- Nature of the Phenomenon;
- Relation to other Phenomena;
- Generation, Detection, and Measurement;
- Bioreceptors, Biogeneration, and Functional Organs;
- Uses in Research, Diagnosis, and Treatment;
- Dosimetry and Safety.

The topics treated are taken from each of the principal areas of study in physics which developed in the historical order: Mechanics, Electricity and Magnetism, Thermodynamics, Statistical Mechanics, and Quantum Theory. This sequence has the advantage that the focus of study starts predominantly macroscopically, with the familiar, and ending predominantly microscopically, i.e. with the less familiar to us humans.

For ready reference while doing chapter problems, some relevant mathematical, physical, biophysical constants, notation, and conversions are given in Tables Tables J.1, J.2, J.3, J.4, and J.5. Additionally, data for nerves, hearing, balance, taste, smell, vision, feeling, and touch are collected in Tables J.6, J.7, J.8, J.9, J.10, J.11, J.12, and J.13.

Problems

1.1 Suppose you are attempting to answer a friend who asks a "why" question. Formulate such a question, and then generate answers to a series of at least three "why"s from your friend following your answers.

1.2 Challenge: Argue that in a single universe, the logic that applies to its dynamics must also be singular. For example, the laws that govern how gravity operates cannot be unlinked to how electromagnetism operates.

1.3 Challenge: Create two different models to explain how DNA might direct an embryo to have a heart on the left side (rather than the right). If possible, apply Occam's razor to these two models.

1.4 Explain: A Newtonian description of the gravitational interaction of stars in our galaxy, taken as point masses, no longer applies if a pair of masses collide. How does this kind of boundary exist for the electric and magnetic interaction of point-like charges?

1.5 In brief, what more recent fundamental ideas now support Darwin's theory since his discovery?

1.6 Give an example of an emergent principle in the crystallization process in inorganic systems.

1.7 Argue for and then against the proposition that the advancement of science required a series of geniuses.

1.8 Argue for and then against the proposition that all life forms eventually evolve into an intelligence capable of science.

1.9 What actions in a biological system might go against Darwinian evolution?

1.10 Give an example of an emergent principle in the social realm.

Chapter 2
The Kinds of Ordinary Materials

Have not the small particles of bodies certain powers, virtues or forces, by which they act at a distance ... ?

—Isaac Newton

Summary Life systems have evolved to use a wide variety of substances, for both structures and activities. Biological structures are built from molecules and mixtures of molecules in a number of different states and phases. The activity of life relies on protein systems acting as builders, renewers, dismantlers, catalysts, shepherds, facilitators, inhibitors, nano pumps and motors. This chapter reviews the underlying nature of atoms and molecules in materials important to life on Earth.

2.1 Nature of Matter

Common materials on Earth, and much of the ordinary matter in the universe, is made from protons, electrons, and neutrons. Protons and electrons are stable (over astronomical times), but free neutrons are not: A free neutron will transform into a proton (p), an electron (e), and an antineutrino ($\bar{\nu}_e$) with about half of any collection of free neutrons transforming every 10.3 min. Such a transformation causing a loss of a certain material is called a decay. In this case, the products of the decay include energetic subatomic particles. Such a process is called a 'radioactive decay'. However, neutrons can be stabilized when combined with protons. A deuteron (D) has one neutron and one proton bound together, and is stable. The neutron bound within cannot decay because the system has insufficient energy to make a free proton, electron, and neutrino.[1]

Deuterons were created about 13.82 ± 0.12 billion years ago in the first 3 min of the Universe's life, and can now be found in deuterium, or 'heavy hydrogen', which makes up about one atom in 6420 of the hydrogen in ocean water. 'Heavy water'

[1]The binding energy of the deuteron, 2.224 MeV, is greater than the mass energy of a neutron minus that of hydrogen: $(m_n - m_H)c^2 = 0.796\,\text{MeV}$.

W. C. Parke, *Biophysics*, https://doi.org/10.1007/978-3-030-44146-3_2

has molecules with two deuterium atoms bound to one oxygen atom.[2] Binding two neutrons with one proton makes tritium (T), but this nucleus is not stable, decaying to helium-3 ($^3_2$He), an electron, and an antineutrino, with a half-life of 12.3 years. Almost all the heavier elements beyond hydrogen, deuterium, lithium, helium and beryllium came from the supernovae debris of the explosions of dying individual stars and the explosions resulting from dense-star merges. Elements up to iron in atomic weight result from cooking in stars over millions of years before exploding, while those heavier than iron are made in the seconds after the explosion.

The planets of the solar system arose from stellar debris: rocks, dust, and gases, released from early supernovae and stellar merges. Within the gas, dust, and rocks were the chemicals needed for life. Because of thermal gradients, the chemicals were naturally differentiated in regions around the Sun, in the molten interiors of planets, and in the liquids which formed on the surface of planets after cooling. Today, the predominant gases in the Earth's atmosphere at sea level are nitrogen (78%), oxygen (21%), water vapor (~1%), argon (0.93%), carbon dioxide (0.04%), and trace amounts of neon, helium, krypton, methane, hydrogen, radon and volatile organics.[3] Elements in the Earth's crust, found mostly as compounds, include oxygen, silicon, aluminum, iron, calcium, sodium, potassium, magnesium, titanium, hydrogen, phosphorus, manganese, fluorine, barium, carbon, strontium, and sulfur, these given in order of abundance.

To understand the differentiation of elements and compounds on Earth, and the various phases possible for the resulting material, we need to understand the forces between atoms and the natural processes of differentiation at prevailing temperatures and differences of temperatures across the space occupied by the materials.

Atoms have a positively-charged nucleus (made from protons and neutrons) surrounded by negatively-charged electrons, with a total charge of zero. The electrons most distant from the nucleus take part in chemical bonding of atoms, and are called 'valence electrons'. These electrons near the 'surface' of the atom are held with the least average force, and require typically about 0.1 to 10 electron volts of energy to release one of them.[4] If an electron is shared between two or more atoms, causing the atoms to hold together over distances in the tenths of a nanometer, the resulting chemical bond is called 'covalent'. If the sharing is unequal, with the center of charge for the shared electron not centered halfway between the atoms, then the

[2]Curiously, if you drink a bottle of pure heavy water, you are likely to die. This observation came via a private communication in 1975 from Prof. Robert Corbin Vincent, Chemistry Dept., GWU, while he held up a pint of liquid deuterium oxide, and explained that the lethality comes from the fact that the D^+ ion moves about 30% slower than its cousin p^+, hindering critical biochemical reaction rates.

[3]When the Earth was formed about 4.54 ± 0.05 billion years ago, the Earth's atmosphere had little oxygen. Our present atmospheric oxygen was produced by the activity of cyanobacteria and then plants, starting about 2.5 ± 0.2 billion years ago.

[4]In comparison, the average kinetic energy of a small atom or molecule in your body is about 0.04 eV, due to its thermal motion.

covalent bond is call ‘polar’. If one atom holds one or more extra electrons to itself or releases one or more electrons to its neighbors, the atom has become ionized. The coupling holding oppositely charged ions is called an ‘ionic bond’, and the force ‘Coulombic’.

Now consider a large number of atoms each with some kinetic energy and found in some region of space. After some time, collisions between atoms will distribute their kinetic energies. If the resulting average kinetic energy is less than the atomic binding energy, then molecules can form with some permanence. Residual weak electromagnetic forces between atoms and molecules, insufficient to cause lasting bonds at the prevailing temperature, are called ‘van der Waals forces’.[5] Van der Waals forces include weak hydrogen bonding, permanent and induced charge multipolar interactions, and dispersion forces, i.e. forces due to the quantum fluctuation of charges. (See Sect. 11.4 for more on molecular forces.) Van der Waals forces between molecules diminish dramatically as the molecules are pulled apart, dropping at least as the inverse seventh power of the separation distance.

A material will be a gas if the weak intermolecular forces cannot keep the material localized against thermal outward pressure. Otherwise, the material will be a liquid or a solid. Liquids are distinguished from solids by flowing under shearing force.[6] Layers of a fluid move continuously in response to a shear. Solids can sustain a shear statically.

If, in a solid or liquid, some fraction of the valence electrons are not captured and held by neighboring atoms, but rather form a delocalized ‘sea’ around the positive atoms, the material is a metal. Compared to non-metals, metals are good conductors of electricity, and, when polished, good reflectors, all because of those loose electrons. The electrons also make metals good conductors of heat, but some non-metals are also good thermal conductors, such as diamonds, because the atoms within are so tightly coupled. If the valence electrons in a solid are localized at low temperatures, but some significant fraction of them become delocalized at room temperatures or by a weak electric field (typically 0.2 V/mm), then that solid is called a semi-conductor. If some atoms or molecules in a material hold a positive or negative charge (but the net charge within volumes larger than the atoms or molecules approaches zero as the volume increases), then the material is called ionic. If the material is in a gas phase with delocalized electrons as a sea around positive ions, the material is called a plasma. The solar wind is a plasma.

[5]Johannes Diderik van der Waals, Ph.D. Thesis (1873).

[6]A shearing force will exist if a segment of a material initially in the shape of a cube has a tangential force on one of the surfaces of the cube, with an oppositely directed force on the opposite side of the cube. You can shear your textbook by putting one of your hands on the top of a book with fingers toward the binding, the other hand on the bottom cover and then moving your hands sideways but in opposite direction, causing the textbook pages to slip a little, one to the next, limited by the book binding.

Ordinary materials, whether gas, liquid, or solid, contain interacting atoms or molecules. A common property of ordinary material particles is the repulsive nature of the force between them when they are brought in proximity. For atoms in such materials, this repulsion is due to the behavior of the valence electrons in atoms, which experience mutual electrical repulsion, and a quantum mechanical repulsion between electrons.

Quantum theory predicts that no two fermions may occupy the same quantum state at the same time. All particles which have a spin quantum number of $1/2,\ 3/2,\ 5/2,\ldots$ are fermions. Those with integer spin quantum number are called bosons. Electrons and protons are fermions. If two electrons are brought together and are forced to stay nearby over a period of time, with their spins aligned, they will be close to a single quantum state. Quantum theory predicts that this configuration becomes more and more unlikely as the electrons approach one another. This property was first proposed by Wolfgang Pauli in 1925, to account for atomic structure, before the formal quantum argument was realized. The behavior is now referred to as the 'Pauli exclusion principle'. The available quantum states for electrons in the lowest energy states within atoms and molecules are few. For atoms with atomic number greater than hydrogen, the lowest states are filled with electrons, and cannot accept any more. Any added electrons must go into quantum states further from the atomic nucleus, and have higher energy than the inner filled states. Without the exclusion effect, the electrons would all fall to the lowest energy state, and atoms would collapse, obliterating all of chemistry and biology!

Collision forces act on gas molecules to oppose compression. These momentary forces come from the same atomic repulsion experienced constantly by molecules in solids and liquids and also account for the repulsive force from a container wall, opposing gas expansion and liquid encroachment. If the material is confined, there are also forces present to keep the constituents together. In the case of solids, bonding between atoms and molecules is due to an attraction coming from electrical effects and from quantum properties. Atoms and molecules in liquids feel a longer-range but weak attractive force, combining electrical and quantum-fluctuation effects. As described in Sect. 11.7.7, the quantum-fluctuation forces, or dispersion forces, come about from the induced-electric-dipolar attraction caused by quantum jiggling of the electronic states in atoms and molecules. In general, magnetic forces have a much less significant role in the binding of ordinary non-ferrous materials.[7]

A system will be self confined and in mechanical equilibrium if internal forces oppose attempts to expand or compress that system. Adding energy to a system

[7]The work of P.A.M. Dirac in 1931 implies that if particles with only north or only south magnetic poles existed, the strength of the smallest magnetic charge g would be related to the electric charge e by $g = \hbar c/(2e)$, so that north- and south- pole particles would have attracted with a force 4692 times greater than electrons and protons. Their primordial binding would explain why we observe the remaining magnetic effects in nature as relatively weak.

in mechanical equilibrium can cause internal oscillations, provided that the added energy is insufficient to deconfine the system, and does not quickly cause particles to thermalize or escape. Parts of the system in proximity can transfer distortional energy to the next part in sequence, producing a wave. Such a wave in a material substance is sound. Sound waves have become important to life systems, and will be treated in Chap. 5. More generally, a 'wave' is a disturbance that self-propagates across space. Waves exist in materials and in fields.

Fields are defined for each of the known interactions in nature, often divided into the categories of gravitational, weak, electromagnetic, and strong. Each such field can have waves by jiggling the source of the field. Electromagnetic waves, made by jiggling charges, are of particular importance in biological systems due to their strength and range in macrosystems, and so they will be extensively treated later. Material waves are distinct from field waves in that material waves consist of a series of coupled oscillations in a substance; field waves do not require a background material. Sunlight passes through the vacuum of outer space.

The bonding of atoms in a molecule comes about through the effects of the electromagnetic forces and the quantum behavior of the outer atomic electrons. When atoms or molecules are pressed together, their electron clouds come closer together, resulting in electrical repulsion, but also repulsion due to the Pauli exclusion principle, a quantum effect that keeps electrons from occupying a single quantum state. Pressed even further, electrical repulsion between nearby nuclei becomes important.

Life relies on the existence of a variety of complex structures and on activity. Carbon, silicon, nitrogen, phosphorus, sulfur and boron are capable of linking up to form large molecules. Carbon is much better at such feats than the other candidates. Life on Earth uses carbon-based compounds in liquid water, where they can be active. Water is a relatively abundant compound with several properties advantageous for life. Because the water molecule has a relatively large dipole moment, pointing from the oxygen to the middle of the pair of hydrogen atoms, liquid water makes a good solvent for other polar molecules and ions (Fig. 2.1). Unlike most materials, when water freezes, it expands (between 4 to 0 $^{\circ}$C), so that ice is less dense than liquid water. The fact that ice floats makes ponds and lakes

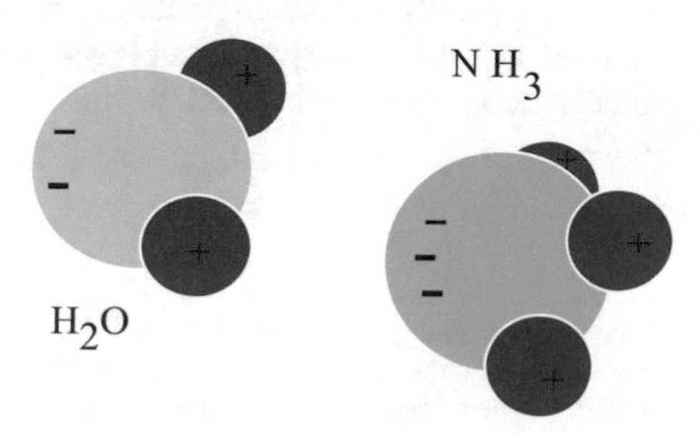

Fig. 2.1 Good solvent molecules

freeze from the top down, leaving aquatic life to feed on the bottom.[8] If this rather unusual property of life's solvent did not hold, ice ages over the Earth would have made life much more difficult.

Earth's surface average temperature is currently 15 °C above the freezing point of water.[9] On planets at an average surface temperature below −33 °C, liquid ammonia (that is anhydrogenous NH_3, not NH_4OH) would serve as a good solvent, since the ammonia molecule, like the water molecule, is strongly polar.

2.2 Mixtures, Solutions, Colloids, Sols, and Gels

Experiment is the sole judge of scientific 'truth'.
— Richard Feynman

Mixtures of materials can have the properties of the three ordinary phases of matter, solid, liquid, or gas, depending on the conditions. Some of these mixtures are used by life systems. Homogeneous fluid mixtures with the mixed particles comparable in size and density to the background fluid material are called 'solutions'. Note that we use the term fluid to include both liquids and gases. The mixed particles are the 'solute', and the background fluid the 'solvent'. If the interactions of a fluid with another material results in a solution, we say the material has 'dissolved'. If the mixture material in a fluid is also a fluid, we say that the two fluids are 'miscible'. If the size of the mixed particles is much less than four thousand ångströms, a clear fluid will remain clear to visible light after the particles are added and the fluid is stirred to make it homogeneous. A homogeneous mixture containing a dispersion of particles much larger than the background fluid particles but small enough to remain dispersed over times much larger than typical observation times is called a 'colloid'. Colloids may contain within the fluid a dispersion of microstructures each with thousands to billions of atoms. The colloidal particles typically make the colloid appear cloudy even if the background fluid was clear. A mixture in a fluid with dispersed particles large enough to settle under the influence of the Earth's gravity is called a 'suspension' if settling is incomplete. In this case, typically, the mixed particles are a few tens of thousands ångströms in size. Mixtures with the mixed particles bigger than a few hundred thousand ångströms are called heterogeneous or bulk mixtures.

[8] However, under arctic and antarctic surface ice, 'brinicles' can form from concentrated and dense brine dropping to the ocean floor, making structures looking like stalactites with frozen water at the interface of the brine with the liquid ocean, even creating a layer of ice over the brine that spread on the ocean floor, killing most of the covered life forms.

[9] Without some greenhouse effect in our atmosphere, water would be in the solid phase over the entire Earth.

To prevent coalescence of the embedded particles in a fluid mixture, the particles must have a net repulsive force between neighboring pairs, or sufficient kinetic energy to overcome any weak attractive force. In general, mixed particles will have electrical interactions with the background medium, and with other such particles, just as atoms do. Over atomic scales, these interactions are dominated by charge and quantum effects: ionic forces and charge sharing interactions. In addition, there are van der Waals forces coming from charge-dipole, dipole-dipole, dipole-induced dipole forces, and induced-dipole induced-dipole (dispersion) forces. Also, the net interaction between mixed particles must include the effect of any media molecules adsorbed onto the surface of the particles. For example, gold nanoparticles can be put in water suspension if the gold particles acquire a charge through reagents such as citrate ions or thiol that attach to the gold surface, making the gold particles repel each other.

All nearby atoms and molecules will have short-range electrical interactions due to the dispersion force. This force is always attractive, and is larger between similar molecules than dissimilar molecules of the same molecular weight, accounting for why identical macromolecules in solution, even with no charge or dipole moments, tend to clump together and 'congeal'. The dispersion force, even in the presence of permanent dipole interaction, can dominate the attractive interaction of nearby molecules, because the dispersion force grows with the size of the molecules, their proximity (flat molecules can get closer together, making the average distance between atoms in adjacent molecules smaller), the ease with which the electrons in the molecule can shift position due to an external electric field (polarizability), and the number of matching electron resonant frequencies.[10] One result is that homogeneous mixtures of fluids, each with neutral molecules having no fixed dipole interactions, still separate from each other over time, making the fluids largely 'immiscible'. In contrast, the immiscibility of oil and water comes from the fact that water molecules are polar, so that the water molecules attract each other with a force larger than that between a water molecule and a non-polar fat molecule. We say that oil is 'hydrophobic'.

Note that soaps have molecules with a hydrophilic head by being negatively charged, and a fatty hydrophobic tail. The tail will have van der Waals forces with the oil. This means that soap molecules can form a water-attractive capsule around oils, with the soap molecule tail immersed into the oil droplets. Van der Waals forces between the fatty ends and other hydrocarbon chains will have enhanced attraction because of the common frequencies of induced-dipole resonances. Small clusters of lipid molecules in water with their hydrophobic ends pointing inward are called 'micelles', and the process of their formation is called 'emulsification'.

The separation between mixed particles may be maintained by thermal fluctuations that tend to scatter the particles. These dispersive effects may act against local interactions or any long-range forces such as gravity or external electromagnetic fields. The thermal scattering and diffusion of the mixed particles in low concen-

[10]Resonance is defined in Sect. 5.12.

Table 2.1 Colloid phase possibilities

		Dispersed particles		
		Gas	Liquid	Solid
Medium	Gas	[Vapor] e.g. Moist air	Droplet aerosol e.g. Fog, mist	Particulate aerosol e.g. Smoke
	Liquid	Fluid foam e.g. Whipped cream	Emulsion e.g. Milk, blood, pharmaceutical creams	Liquid sol e.g. Paint, ink, sol-cytoplasm
	Solid	Rigid foam e.g. Aerogels, styrofoam	Rigid gel e.g. Gelatin, gelcytoplasm	Rigid sol e.g. Bone, ruby glass

Vapor here is not strictly a colloid, but rather a gas-gas solution
A froth is a fluid foam stabilized by solid particles

trations occurs through collisions of those particles with the background molecules, a process first properly described by Einstein in 1905.[11] We will return to a more detailed description of the diffusion of substances such as perfume in air or oxygen from a blood cell in Sect. 9.20.

Table 2.1 shows the possible phase states of colloids. As indicated, a rigid 'gel' is a colloid that has a solid matrix with an intervening dispersed liquid. Such a material holds its shape under mild stresses. In contrast, a liquid 'sol', a colloid with a liquid medium with a dispersed solid, does not hold its shape. The protoplasm within living cells can convert between these two kinds of colloids.

A xerogel is a dried out open solid foam left after the medium of a gel is removed. A cryogel is a xerogel dried in a partial vacuum after freezing. An aerogel is solid foam with a tenuous solid matrix produced by replacing a liquid medium in a gel formed from a network of structures with a gas. Aerogels have been made with densities as low as 330 times less than water. The matrix of an aerogel is made of structures smaller than the wavelength of visible light, making an aerogel translucent blue from 'Rayleigh scattering', i.e. scattering from particles much smaller than the wavelength of the light scattered.

A 'lyophilic' substance is a solid material which forms a liquid sol easily when its dry-powdered form is mixed with a specified background liquid medium. Starches and proteins are lyophilic in water.

A 'lyophobic' material does not easily form a liquid sol with a specified liquid. Iron hydroxide is lyophobic in water.

A 'thixotropic' gel will maintain its shape until subject to a critical shearing force, at which point it turns to a sol. The behavior is reversible when the sol is allowed to stand. "Quick Clay", which is thixotropic, has caused landslides after mild earthquakes. Thixotropic gels are an example of 'network gels', in which the dispersed material occurs as interconnected filaments or plates.

[11] A. Einstein, *Investigations on the theory of Brownian movement*, Ann der Phys **17**, 549 (1905).

Multiple colloids, with more than two kinds of particles dispersed in a background material, also are evident, such as oil-water-air in porous rock.

When two metals are mixed homogeneously, the combination is called an 'amalgamation'. Dental 'amalgam' is an amalgamation of mercury, a liquid at room temperature, with silver, a solid at room temperature. The result is a liquid. Mercury readily forms amalgams with copper, silver, gold and even with neutral ammonium NH_4, which acts like a metal in the amalgam.

If the structure of one substance in another exists at a 'lower dimension' (filaments or surfaces), then the material is a matrix. If the matrix consists of long thin threads, the material is fibrous. If the fibers are, to some degree, aligned, the matrix is 'ordered'.

2.2.1 Mixtures in Biological Structures

Live systems on Earth use mixtures of carbon-based compounds in salty water to generate the complexity and activity need for the tasks of sustaining life.

The material structure elements of biological systems must be able to keep individual life forms localized within the organism, and be stable enough to maintain the individual life form for a certain replacement lifetime. Additionally, repair mechanisms and the ability to respond to a variety of environmental pressures are distinct advantages.

The simplest of life-forms, viral particles, preserve and safeguard their structure through intermolecular forces within their protein capsid. The much larger and far more sophisticated cell structures became successful by the fact that important water-based metabolic processes can be achieved through the highly sophisticated and ordered mixtures of substances contained in porous cells.

Biologic cells maintain intracellular structural stability by using microtubules, a polymer of two globular proteins, and by using microfilaments, a linear polymer of the protein actin. Cellular cytoplasm shows both a sol and a gel behavior by how the microfilaments make viscoelastic gel-sol transitions. Amoebae pseudopodia use a sol-gel transformation and contractile microfilaments to ambulate. A cell nucleus, having embedded macromolecular coils of DNA is a colloid.

In addition, plant cells maintain shape using cellulose, a long chain of sugar molecules. Wood is produced as cellulose bound in a complex polymer of cross-linked phenols called lignin.

Extracellular auxiliary structures are sometimes created by multicellular life forms to gain quasi-rigidity. These structures can then hold specialized cells in place relative to other specialized cells, or relative to gravity. Among the components in exoskeletons are secretions with proteins and calcium carbonate in the case of mollusks, and chitin (an N-acetylglucosamine polymer) in the case of insects.

Collagen is an extracellular structural protein, often in the form of elongated fibrils, affording some rigidity. Collagen consists of long chains of a triple helix molecule, giving collagen significant tensile strength compared to other organic molecular tissue.

Vertebrates create and maintain bone as an endoskeletal structure. Bone is a calcium phosphate in the chemical arrangement termed calcium hydroxyapatite, $Ca_{10}(PO_4)_6(OH)_2$, embedded in collagen.

2.2.2 *Detecting Particle Size in Colloids*

Dynamic Light Scattering

Consider a colloid of small round particles undergoing Brownian motion in a fluid. By measuring the diffusion rate of the mixed particles, the effective diameter 'd' of the particles can be found using the Stokes-Einstein Relation:

$$d = \frac{k_B T}{3\pi \eta D} \tag{2.1}$$

where D is the diffusion constant for the suspended particles. The viscosity of the fluid, η, and its temperature, T, are assumed known.

The diffusion constant D can be measured by employing the technique of 'dynamic light scattering' from the particles in the fluid. If laser light, of wavelength λ selected to be larger than the average interparticle distances, is passed through the colloid, the light scattered from neighboring mixed particles will add coherently. However, because the interparticle distances vary through Brownian motion, the relative phase of the scattered light from individual particles in a neighboring cluster will vary over time, typically fractions of a microsecond. This will cause the scattered light intensity to show a variation when measured over time, as shown in Fig. 2.2.

The variation of the scattered light intensity $I(t)$ due to Brownian motion can be characterized by the behavior of the normalized 'auto-correlation' function $G(\tau)$, defined by

$$G(\tau) \equiv \frac{\langle I(t) \cdot I(t+\tau)\rangle}{\langle I(t) \cdot I(t)\rangle}.$$

For short times (typically in the microseconds), there should be near perfect overlap of intensity $I(t)$ with itself, while for very long-time displacements τ (typically in the hundreds of milliseconds), there will be a much smaller correlation between various phases of light scattered from the particles as they are shifted by 'random' motion, so that $G(\tau)$ drops down as τ increases. One can show that for particles of uniform size,

$$G(\tau) = (1 + \beta e^{-2\Gamma\tau}),$$

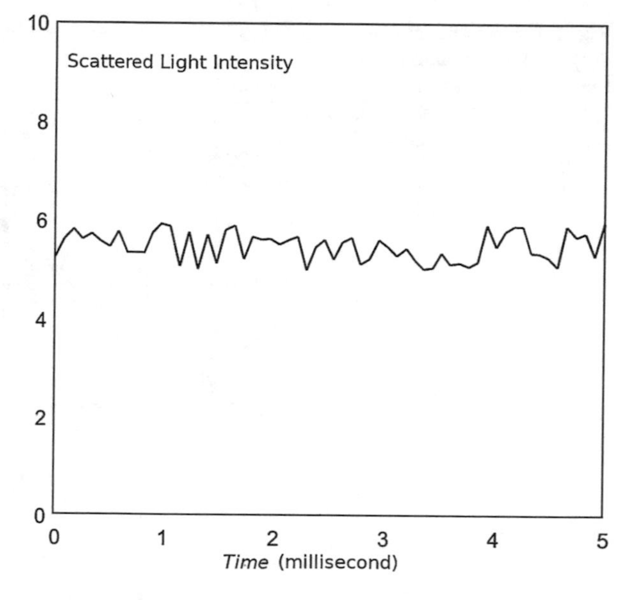

Fig. 2.2 Representative scattered-light intensity variation due to Brownian motion of colloidal particles

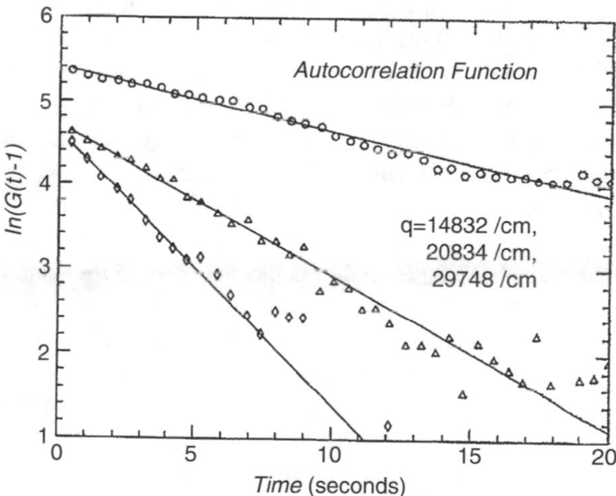

Fig. 2.3 An example plot of the scattered-light-intensity autocorrelation function $G(\tau)$, expressed as $\ln(G(\tau) - 1)$, versus the time τ. The value q is the momentum transfer to the particle by the light. (From Appollo Wong & P. Wiltzius, *Dynamic Light Scattering with CCD Camera*, Rev Sci Inst **64, 9**, 2547–2549 (Sept. 1993))

where the relative rate of drop in $G(\tau)$, 2Γ, is connected to the particle diffusion by

$$\Gamma = Dq^2 = D((4\pi nf/c)\sin(\theta/2))^2 . \tag{2.2}$$

In the above equation, n is the refractive index of the dispersant, q is the magnitude of the momentum transfer of the laser photon to a colloidal particle divided by $\hbar$, f is the frequency of the laser light, c is the speed of light, and θ is the light scattering angle. A plot such as that shown in Fig. 2.3 is used to determine Γ, and from Γ the diffusion constant D for the colloid particles using a plot such as Fig. 2.4.

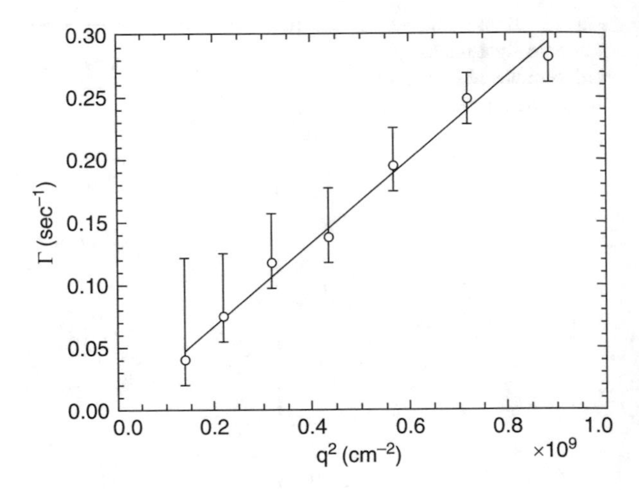

Fig. 2.4 Fractional rate of drop of autocorrelation as a function of the square scattering wave number, q^2. (From Wong & Wiltzius as in Fig. 2.3)

Optical Effects

If the particles in a solution are comparable in size (or larger) to the wavelength of the light illuminating them, they will become visible. For our eyes, this means greater than or equal to about 4000 ångströms. The 'Tyndall effect' is the observation of a bluish tinge to light scattering by a colloidal suspension of particles whose size is comparable to the wavelength of the scattered light. As in Rayleigh scattering, the intensity of the scattered light is greater for blue than red. However, the effect is not as strong as in Rayleigh scattering, in that the particles are not far smaller than the wavelength of the light, nor are they much larger than the wavelength, which would be the realm of 'Mie scattering', for which the light of all colors coherently produce no color in the reflected light, such as the light we see from the surface of a droplet of water. The Tyndall effect can be used to estimate the size of colloidal particles suspended in a clear liquid.

Blue eye color is due to the Tyndall effect, in that light is scattered from a turbid translucent layer in the iris, without much melanin present to cause a brown color. The blue shade to very fine smoke particles in air is also a Tyndall effect, as is the blue ting to clouds of water droplets as the droplets evaporate.

2.2.3 Separating Materials

Any fluid mixture in an open system able to expel heat into the environment will have a tendency to separate, by natural processes, into bulk amounts of the embedded materials. Witness ore deposits in geologic material initially fluid at relatively high temperature and then allowed to slowly cool, or the crystallization of salts from a brine solution. Natural processes that can cause separation include:

- cohesive attraction of like substances in solutions,
- repulsion of dissimilar molecules,
- adsorption of molecules to surfaces,
- differential diffusion in bulk material,
- osmotic differential diffusion through membranes,
- gravitational separation due to varying buoyancies,
- filtration and sieving,
- fractional vaporization,
- fractional freezing,
- precipitation,
- sedimentation,
- solute vaporization or sublimation,
- differential fluid drag separation and natural winnowing.

Besides taking advantage of natural processes, life systems have a plethora of techniques using active processes for separating atoms, molecules, and other particles. Organisms employ such organizational processes to create, transport and concentrate complex molecular materials and to maintain the activity of life.

Our species has developed methods to direct, control and enhance the separation of mixtures into products, for use in food production, industry or for the analysis of individual components, including medical analysis. We give below a few examples of devices and technologies used in bioengineering and medicine.

- Gravitational separation takes advantage of the varying densities of components in a mixture. In a fluid, buoyant forces act against gravity to separate materials of differing density.
- The centrifuge is a device which can separate various-sized nano- and micro-scale particles in solutions, using acceleration forces much larger than the acceleration of gravity, provided the intrinsic particle density is greater or less than the background fluid density. We will describe the centrifuge in more detail in Chap. 3.
- Evaporation concentrates low vapor pressure dissolved materials from a volatile solute. Some dissolved materials, when so concentrated, will form pure crystals. Condensation of the vapor back into a fluid can be used to capture the purified fluid.
- Water distillation and desalination, solar evaporation, partial freezing, and reverse osmotic pressure devices, are all used to concentrate potable water.
- Chemical precipitation, filtration, and bacterial conversion of hazardous materials helps generate potable and palatable water. Filtration to remove toxins from blood is the basis of kidney dialysis.
- Electrodialysis uses a semi-permeable membrane in an electric field to transport selected ions across the membrane. A major application of electrodialysis is in the desalination of brackish water.
- Chromatography is an analysis tool that separates substances by diffusion according to their affinity for a particular solvent. Often, this is done by putting the mixture on filter paper and then letting the solvent soak in from one side of the paper.

- Electrostatic precipitators can be used to remove colloidal particles such as dust, smoke, or oil mist from air. They work by creating a net charge on the colloid particles and then an electric field to sweep them away. This process begins by negatively charging wires until they emit corona discharge electrons into the passing air mixture. The electric field near the wires accelerate the electrons, which collide with air molecules (and colloidal particles), knocking off electrons from those molecules, making positive ions near the wires. But further away from the wires, the electrons are not moving as fast, and tend to be captured by air molecules. So, a negative set of ions are created, which feel a force moving them away from the wire and toward the collection plates. Some of these negative ions make contact with and adhere to the colloidal particles, imparting to them a negative charge. As the colloidal particles are much larger than the air molecules, they can accumulate a number of such charges. As the air moves through the device, the negatively-charged colloid particles are attracted to and adhere to collection plates made positively charged relative to the coronal wires. The plates are periodically cleaned with a fluid, or by mechanical 'rapping'.
- Electrophoresis uses differential diffusion in a fluid to separate molecules with varying mobilities. The process is the following: The molecules are first given a charge (if they had no net charge initially) by attaching anions or cations to their surface. They are then pulled through the background fluid or gel by an electric field. The fluid will exert a viscous force which depends on the size and shape of the base molecule together with any adsorbed molecules on its surface and on the tangential forces of the fluid on this molecular complex. After being dragged a certain distance by the external electric field, the location of differing molecules along the field direction will vary.
- Advantage can be taken of the adhesion of molecules to both natural and specially coated surfaces. Oil can be separated from water using treated sponges and other porous materials. The effectiveness of such an open matrix within the material is proportional to the surface area of the network boundaries within the matrix as well as the adhesive properties of the molecules to the fiber surfaces.
- If one or more components of a mixture is ferromagnetic, then magnets can be used to effect separation of those substances. Magnetic fields with strong gradients can be used to separate materials of varying paramagnetism and diamagnetism.
- A 'mass spectrometer' is capable of separating atoms and molecules according to their mass. However, the device requires that the material be first vaporized in a vacuum, ionized, and then accelerated by an electric field into a magnetic field. The magnetic field will deflect the particles into a curve whose radius depends on the particle's mass to charge ratio.[12] Collection and counting of the particles according to their separated beams is then possible.

[12]Newton's 2nd law gives a radius of curvature of $R = p/(qB)$, where q is the charge of the deflected particle, p its momentum, and B the magnetic field strength. This formula for R works even for relativistic particle speeds v if one takes $p = mv/\sqrt{1-(v/c)^2}$.

- An impressive application in biological research is the instrument put together by Prof. Akos Vertes (Chemistry Dept., GWU). With his device, he can analyze the proteins taking part in some action in a live cell. First, he knocks the molecules out of a particular cell region or organelle with a highly focused infrared laser. Then, he collects the vaporized material in a micro-vacuum cleaner hovering over the cell, and sends the material to a mass spectrometer. Knowing the masses of various proteins, he can tell which ones were present in the cell region at the time of vaporization.

As we have indicated, by expending energy (and releasing heat), life systems also have dynamical processes for substance separation and transport. These processes effect separation against opposing forces and diffusion. The pumping of sodium out of a biologic cell (via 'active transport') is a good example. The transport by motor proteins of molecules ('cargo') along intracellular filaments of the cytoskeleton is another.

Problems

2.1 Based on the mechanisms for geological differentiation and the solubility of metals in metal, argue that at the center of the Earth's core, the iron may contain more than 50% gold.

2.2 Explain why natural diamonds could not have been formed in the Earth by the compression of plant carbon deposits. Describe the current theory on natural diamond formation, and how we are able to find diamonds near the surface of the Earth.

2.3 In short, what is the current scientific view on how life started on Earth?

2.4 Describe the microscopic changes that cause a homogeneous mixture of dispersed microspheres of oil in water to naturally separate.

2.5 In the measurements of Wong and Wiltzius, a large number of microscopic particles of uniform size were mixed into glycerol (index of refraction of 1.473, temperature of 51 °C, viscosity 137 cp) to create a semi-transparent colloidal suspension. The mixture was then put into a 1 mL cuvette sample holder. Light from a Helium-Neon laser beam was scattered from the sample. The scattered light was then focused with a lens onto a CCD detector containing pixels spread over an area. The direction of the light rays to a given pixel determined the scattering wave number $q = (4\pi n f/c)\sin(\theta/2)$. From the intensity of the light measured over time, the autocorrelation function $G(\tau)$ was calculated, and then fitted to $1 + \beta\exp(-2\Gamma\tau)$ to find Γ for several values of q^2. Assuming $\Gamma = Dq^2$, a value for the diffusion constant for the Brownian motion of the particles in glycerol can be found. Using the graphs in the text and the Stokes-Einstein Relation, find the diameter of the colloidal particles.

2.6 Argue the feasibility of using the principles behind a mass spectrometer to separate garbage and other waste material into usable elements.

Chapter 3
Mechanical Aspects of Biosystems

...those who rely simply on the weight of authority to make an assertion, without searching out the arguments to support it, act absurdly.

—Vincenzo Galilei, Galileo's father

Summary The materials we are made from follow the same 'laws' of mechanics that inorganic materials do. From the sizes of macromolecules to the sizes of whales, these laws are those discovered by Isaac Newton back in 1665. We can understand much of our construction, operations, and our physical limitations by studying these Newtonian principles. The subject of the study is called mechanics, and the ideas in this subject describe relationships between forces, accelerations, stress, strain, and viscous flow. These connections will be applied to a number of biophysical systems.

Mechanics applied to macroscopic systems was the first successful science, and not surprisingly, is close to everyday observation. Archimedes studied the behavior of fluids, discovering a basic principle of buoyancy. Leonardo da Vinci invented many new mechanical devices, and made a careful study of the human musculature, blood flow, and anatomy. Galileo studied mechanical motion, realizing that massive bodies retain their motion unless acted on. In the hands of Isaac Newton, our understanding of the dynamics of particle systems extended far beyond planetary systems, to any system of masses, including the matter of which we are made.

In what follows, we will see examples of the vast territory over which Newton's ideas can take us. However, when our vision reaches to the smallest scales of biological interest, the atoms that make us, we will find that Newton's ideas are subsumed by those of quantum theory, or when we wish to become relativistic space travelers or apply geosynchronous satellites for positioning, then we use the ideas of Einstein to expand those of Newton.

W. C. Parke, *Biophysics*, https://doi.org/10.1007/978-3-030-44146-3_3

3.1 Newtonian Principles in Biostatics and Biodynamics

Never trust an experimental result until it has been confirmed by theory.

—Sir Arthur Stanley Eddington

Newtonian principles enter biology from the scale of macromolecules to that of the largest animals. Life systems intrinsically involve forces. On the microscopic level, nuclear and electric forces allow atomic structures, and therefore chemistry. Weak interactions have practically an unmeasurable force outside the nucleus of atoms, but are the cause for radioactive decay, a factor in genetic mutation. With the known forces and particles, biochemistry and the evolution of life become understandable as a natural process at the atomic and molecular level. On a nanometer scale, the activity around biostructures can be formulated in terms of predominantly electrical forces. For more massive biosystems evolving on a planet, organisms must also cope with gravitational forces. The dominance of electrical and gravitational forces in the interactions within biological systems comes from the long-range behavior of these forces, which is in contrast to nuclear and weak forces, which are short range. These latter two drop exponentially outside a range the size of the nuclei of atoms.[1]

Currently, to accurately predict the behavior of atomic and molecular systems, quantum mechanics is applied, a topic we will reserve until our study of molecular biophysics. As yet, no known life systems have traveled close to the speed of light relative to us, so that for understanding biosystems, we can postpone explicit consideration of Einsteinian mechanics. For the dynamics of life systems above the molecular scale, Newtonian physics and a little statistics will be sufficient.

3.2 Newton's Laws Applied to a Biosystem

Few things are harder to put up with than the annoyance of a good example.

—Mark Twain

3.2.1 *Basic Concepts*

Before discussing the effects of forces in biosystems, we will review some physically defined concepts.

---

[1]The exponential drop in the force can be associated in quantum theory with the exchange of particles with mass, making the force proportional to $\exp(-\mu c\, r/\hbar)/r^2$, where r is the distance from the charge center, $\hbar$ is Planck's constant over 2π, μ the mass of the quantum in the field causing the force, and c is the speed of light. The quanta of the electromagnetic interaction, called photons, appear to have no mass ($\mu < 10^{-14}\mathrm{eV/c^2}$), so the exponential factor becomes one, characteristic of long-range forces.

The Concept of Time and of Space

Time is a count of the number of repetitions which occur in a 'periodic system'. A system is periodic if it deviates from a given state, returns, and then repeats this same sequence of changes. The 'period' of a system is the smallest unit of repeat. A periodic system may serve as a 'clock'. A good clock is one for which many copies, put under varying but ordinary conditions, all agree in count with the original, to within an acceptable tolerance. Since nature gives us a number of good examples of a periodic system, time is correspondingly given a unit according to which system is being used as a clock, and how larger periods are put together from the smallest.

In modern use, the heart is not a good clock. But it served Galileo, while sitting in church watching a swinging lamp, when he noticed, using his own heartbeat, that a pendulum's period appears to be independent of its amplitude.

Atomic clocks can presently be constructed which would agree with each other to within a second in a few 100 million years.

Bodies occupy and are separated by what we call space, which we define through a measure of separation called 'distance'. A spatial separation exists between two bodies if light takes time to move between them. The calibration of distance is called 'length'. Recognizing that the speed of light, c, is a universal and fixed number, length is defined by the time it takes for light to travel the length multiplied by the speed of light.[2] In our Universe, determining the unique position of bodies from a give one requires at least three independent distances. Giving the extent of these three distances gives the 'location' of the body relative to the first.

Neural networks making a brain capable of thinking are easier to construct in at least three spatial dimensions.

Starting from a given origin and set time, the measured location and the time of observation of a localized body or entity determines an 'event', often communicated by four numbers, (ct, x, y, z). Because of conservation laws, it follows that each atom within you follows a contiguous sequence of events.[3] The track of a body in space-time is called a 'world line'.

The Concept of Mass

Mass is a measure of the inertia of a body, i.e. the body's resistance to change of motion. Note that mass is a concept defined independent of weight. Astronauts have inertia even far from the gravitational pull of planets.

In 'classical' descriptions (i.e. with Newton's laws combined with Maxwell's equations), the mass, length, and time are independently defined. Standard unit sizes

[2]The speed of light is set to be $c = 299,792,458$ m/s, thus defining the meter in terms of a time measurement.

[3]In this case, local baryon and lepton number conservation insures that isolated atoms detected in one region of space will be found a short time later near that same region.

are selected for each. All other quantities with physical units can be defined in terms of the units for each of these three. When there is no compelling reason to select otherwise, we will be using the 'cgs' system of units, with centimeter-gram-second unit sizes, or, secondarily, the 'mks' system, meter-kilogram-second.

For systems moving fast compared to us, such as life forms moving in relativistic spaceships, we apply the extension of Newton's laws given by Einstein. In Einstein's Special Theory of Relativity, a generalization of Newtonian mechanics to bodies with relative motions close to the speed of light, the concept of space cannot be separated from the concept of time. That idea leaves only two independently defined units, often taken to be length and mass. Time units are converted to length units by multiplying the time measure by the speed of light, c. The units of energy can be expressed in mass units, using Einstein's famous discovery in the analysis of Special Relativity that $E = mc^2$.[4]

Since Special Relativity limits the speed of ordinary particles and quanta of interaction to be no greater than a universal constant c, Einstein knew that Newton's Theory of Gravity also had to be corrected, since Newtonian gravity acts instantaneously over long distances. By thinking about rides in falling and accelerated elevators, he could explain why inertial mass was the same as gravitational mass with the proposal that the local effects of acceleration are indistinguishable from gravity. Finally, using his Special Theory to compare the circumference to the radius of a fixed platform compared to a rotating platform, he concluded that space-time must be curved in an accelerated frame. In this way, he knew that non-uniform gravity must be associated with space-time curvature. His General Theory of Relativity is the mathematical embodiment of these propositions.

In Einstein's General Theory of Relativity, which associates gravity with a curvature of space-time due to nearby masses, the concept of mass is not independent of length (mass produces gravitationally-curved space, with curvature defined by the inverse of a length). Moreover, gravitational mass in Einstein's theory is identical to inertial mass.[5] Within Einstein's theory, we can convert all physical units to a length. For example, using the universal gravitational constant: $G = 6.67408 \times 10^{-11}\,\mathrm{N\,m^2/kg^2}$, we can multiply mass by $2\,\mathrm{G/c^2}$ to get a length. If a mass were to be compressed into a ball of this radius, it would be a black hole. For the Sun, $2\,\mathrm{GM/c^2} = 2.953\,\mathrm{km}$.

So, in the present state of our theories, there remains only one scale with units, conventionally taken as length units.

[4]Einstein was the first to see, in 1905, that mass holds a very large quantity of energy. A chemical reaction which releases energy causes an almost imperceptible loss of mass. Nuclear reactions can produce much more energy from mass than chemical reactions because much more binding energy is involved.

[5]Inertial and gravitational mass have been shown equivalent in a number of very precise measurements. See, e.g., P.G. Roll, R. Krotkov, and R.H. Dicke, *The Equivalence of Inertial and Passive Gravitational Mass,* Ann Phys **26**, 442–517 (1964).

The Concept of a Force

Force characterizes the effect on one body due to other nearby bodies, causing a change in the momentum of the affected body if the net force is non-zero. The Newtonian momentum of a body is its mass times its velocity. Einstein's Special Relativity Theory makes momentum the mass times velocity divided by the square root of one minus the square of the velocity over the speed of light. For slow speeds compared to the speed of light, Einstein's theory reduces to Newton's.

Observations indicate that in Nature, no more than four types of forces are needed to explain all interactions. As we have noted, two of these (the 'Nuclear force' and the 'Weak force') have ranges which do not measurably extend much beyond the diameter of a proton. The other two, the 'Gravitational force' and the 'Electromagnetic force', act over distances we can see. They are called 'long-range' forces.

Our successful mathematical descriptions of the known interactions uses the concept of a 'field'. A field is any quantity defined at each point in space. In our models of all interactions, we say that bodies with mass or charge 'create a field' around them. Other masses or charges in the vicinity are then affected by this field.

The force of one body on another due to gravity or electromagnetism diminishes with increasing separation r between the bodies, with a power law $1/r^2$ for bodies with mass or electric charge. In Einstein's theory, gravity is always attractive and, in Newtonian theory, is proportion to the mass of bodies, while electromagnetic forces can be attractive or repulsive, and, in Maxwell's theory, are proportional to the size of the electric charge on each of the interacting bodies.

As there is no negative mass in our best models of nature, we cannot 'cancel' the gravitational field outside a mass.[6] However, because there are both positive and negative electric charges, we can shield bodies from the effect of an electric field. For example, to protect sensitive electronics or people, one can surround a system with a 'Faraday cage', which will exclude any external electric field. 'Mu-metal' (made from a nickel-iron alloy) can shield against an external magnetic field. A superconducting material can also prevent a magnetic field from penetrating. (This is called the Meissner effect.)

The local interactions from one molecule to another in life systems can be explained in terms of electromagnetism and quantum behavior. Of course, longer

[6]This inability to cancel the Earth's gravitational field over a finite region is not just a matter of not finding negative mass: The nature of the gravitational interaction seems to not allow repulsive gravity. In Einstein's General Theory, gravity must be attractive, as the gravitational acceleration of a body is due to the distortion of space-time, and not any internal property of the body. In quantum theory, a supposed 'graviton' carries a spin of two units times Planck's constant over two pi. A spin-2 interaction is sufficient to explain the purely attractive nature of gravity, to be contrasted with a spin of one unit for the photon, the particle that carries the electromagnetic interaction. An odd-integer spin particle exchange allows for both attractive and repulsive interaction. For details on this topic, see Anthony Zee's book, **Quantum Field Theory in a Nutshell**, Princeton Univ Press (2003).

range interactions such as gravity play a significant role for life systems that have developed near a big mass, like a planet, or which regularly experience significant accelerations.[7]

Our best present theory of how matter behaves uses quantum mechanics rather than Newtonian mechanics, with forces being replaced by changes in field-interaction energies across space. Even so, Newtonian mechanics has proven to work well from solar-system sizes down to mesoscopic scales.

Newton's Laws

Newton's 1st Law: There exists frames of reference, called inertial frames,[8] in which isolated bodies move uniformly or remain at rest.

Newton's 2nd Law[9]: When a body is acted on by a net unbalanced force determined by interactions with other nearby objects, the body no longer moves uniformly, but rather accelerates in proportional to that net force and inversely proportional to the inertia of the body measured by its mass: $\sum \mathbf{F} = m\,\mathbf{a}$.

Newton's 3rd Law[10]: If body A acts on body B with a force $\mathbf{F}_{AB}$, then body B must act back on body A with an equal and opposite force $\mathbf{F}_{BA}$: $\mathbf{F}_{BA} = -\mathbf{F}_{AB}$.

Newton's Law of Linear Superposition[11]: The Newtonian proposition that forces add is the following: If one finds that body 2 acts on body 1 with a force $\mathbf{F}_{12}$ when the two are isolated, and body 3 acts on body 1 with a force $\mathbf{F}_{13}$ when the two are isolated, then when all three are present but isolated from others, the force on 1 due to both bodies 2 and 3 will be $\mathbf{F}_{12} + \mathbf{F}_{13}$.

Newton's Law of Gravity[12]: An attractive gravitational force exists on each of a pair of masses, in proportional to each of their masses and inversely proportional to the square of their distance of separation: $F_G = G m_1 m_2 / r^2$.

---

[7] As we have noted, Einstein showed how to make an equivalence between acceleration and gravity.

[8] A frame of reference separated from local bodies and moving at constant velocity relative to the average motion of the distant stars is a good inertial frame.

[9] The alternative form of Newton's 2nd law, $\sum \mathbf{F} = d\mathbf{p}/dt$, where $\mathbf{p}$ is the particle momentum, actually still works in Einstein's Special Theory.

[10] Note: Newton's 3rd law should NOT be written as 'for every action there is an equal and opposite reaction'. The writer of this textbook found this phrasing of Newton's 3rd law in a sociology textbook, whose author, with apparent seriousness, claimed it justified the statement that for every social action there must be an opposing action. Newton would have rolled his eyes.

[11] This 'Law' is found to hold, to a very good approximation, for electromagnetic forces, but is violated for very strong gravitational forces, such as nearby black holes. In such cases, we say that the gravitational effects are acting non-linearly.

[12] The number G is Newton's universal gravitational constant, first directly measured by Henry Cavendish in 1797. We note that very strong gravitational fields exist near very massive and dense bodies, such as neutron stars and black holes. In these cases, Newton's Laws no longer work well, but Einstein's General Theory of Relativity successfully (so far) predicts the gravitational interaction between two or more massive bodies. Near neutron stars and massive black holes, no simple force law can be used, nor do strong gravitational effects simply add when several bodies are considered, as they do for Newtonian gravity.

Newton knew these laws in 1666, and formally published them in 1687, in his book entitled **Principia**. A complete description of electromagnetic fields and forces came in 1865 with Maxwell's paper on the subject. A new theory of gravity, encompassing Newton's for weak gravitational fields, was presented by Einstein in 1915. With quantum theory interpretations, no observational deviations have been found from Maxwell's or Einstein's theory in over one hundred years.

3.2.2 Application of Newtonian Ideas

We can use Newtonian laws of dynamics to describe mechanical equilibrium, kinetics, and dynamics within life forms and between life forms and their environment. We will examine several illustrative examples.

According to Newton, the external forces on every mass m within a system determines its motion through the relation $\mathbf{F} = m\mathbf{a}$, where $\mathbf{a}$ is the acceleration of the mass. By applying the left- and right-hand sides of Newton's 2nd Law to each element in a system of masses, and then adding, each element labeled by the number 'i', the right-hand side can be written $d^2 \sum m_i \mathbf{r}_i / dt^2$. This expression can be recast into the form $(\sum m_i) d^2\{(\sum m_i \mathbf{r}_i)/(\sum m_i)\}/dt^2$, i.e., the mass of the whole system times the acceleration of a special location within called the center-of-mass, defined by $\mathbf{R} \equiv \sum m_i \mathbf{r}_i / (\sum m_i)$. Alternatively, the right-hand side is also the change with time of the total linear momentum of the system, $\mathbf{P} \equiv \sum m_i \mathbf{v}_i$. On the left-hand side is the sum of external and internal forces on each element of mass in the system. But because of Newton's Third Law, $\mathbf{F}_{ij} = -\mathbf{F}_{ji}$, the internal forces cancel in pairs, leaving the statement that only the external forces determine the motion of the center-of-mass of a system. We have

$$\sum \mathbf{F}^{ext} = \frac{d}{dt}\mathbf{P} . \tag{3.1}$$

One evident consequence: No internal forces in a bacterium can change the position of the center-of-mass of the bacterium! For humans, you cannot lift yourself up by your 'bootstraps'. (Boots once had straps![13])

Thus, if the external forces on the system add to zero, its momentum will be conserved, i.e. not changed in time, and the center-of-mass, if initially at rest, will remain at rest. However, relative motion within the system is still possible. Rotation is a subclass of any relative motion which does not affect the distance from a given system point to any of other system points. More generally, if the distances between mass elements change with time, then the shape of the system changes or is distorted. Life systems are of this kind.

[13]The phrase, 'booting your computer' comes from this old adage. The earliest computers had no static memory, so that an operator had to enter the code to start the machine. Figuratively, the computer lifted itself up. (With a little help from its creators!)

Just as the total external force determines the motion of the center-of-mass of the system, the net external torque determines the pure rotation of the system. Torque measures the effectiveness of a force in causing a rotation about an axis. It is a vector defined in size by the magnitude of force times a 'lever arm', the minimum distance from the line of the force to the line of the axis of rotation. Torque as a vector is directed along the axis in the 'right-hand' sense: Curl your fingers of your right hand in the direction that the body would rotate if the force acted alone. Then your thumb points in the direction of the torque. This idea is conveniently expressed by the vector cross-product of the force times the displacement from the axis to the point where the force is acting on the system: $\boldsymbol{\tau} \equiv \mathbf{r} \times \mathbf{F}$. The cross-product of two vectors is another vector pointing perpendicular to the vectors in the product, directed according to the way one's thumb points if one turns the fingers of one's right hand from the first to the second vector, and has a magnitude equal to the product of the length the two vectors times the sine of the angle between them.[14]

If we construct the net torque on a system of masses, we can apply Newton's Third Law again, collecting the internal-force terms into pairs of the form $(\mathbf{r}_i - \mathbf{r}_j) \times \mathbf{F}_{ij}$. These will vanish if the forces act along the line joining the mass elements, as do gravitational and electrical, leaving only the external torques contributing to the net torque.[15] From Newton's 2nd Law, this net torque is given by $\sum m_i \mathbf{r}_i \times \mathbf{a}_i = d/dt \sum (\mathbf{r}_i \times \mathbf{p}_i)$. The sum defines the net angular momentum of the system, $\mathbf{L}$, so that Newton's 2nd Law for rotation has the form

$$\boldsymbol{\tau}_{net}^{ext} = \frac{d}{dt}\mathbf{L} \,. \tag{3.2}$$

If the net external torque vanishes, the angular momentum of the system is conserved. This observation has become very useful in analyzing the behavior of systems with internal spin. For example, an astronaut in space who exerts a torque on a spacecraft will cause the spacecraft to spin, and the astronaut spins, but in the opposite direction and with a greater rotational velocity, so that the total vector angular momentum of the two, spacecraft and astronaut, still add to zero. As another example, the spin of an electron in hydrogen can flip by absorbing a photon, because the photon itself carries spin. The total angular momentum of atom plus photon is conserved.

For pure rotations about the origin, every coordinate of the mass elements changes with time according to

---

[14] For those sending an E-mail message to another life form on some other planet who want to know what is 'the right-hand rule', tell them if they turn the first vector toward the second as an anti-neutrino rotates, then the torque will be in the direction the anti-neutrino moves forward. No light message, audio, or visual, can communicate handedness, because photons are their own antiparticles. If the alien's screens show your image reversed, we could not tell just by light signals!

[15] Magnetic forces acting on moving charges do not act along the line joining the charges. There will be added torque coming from the electromagnetic field.

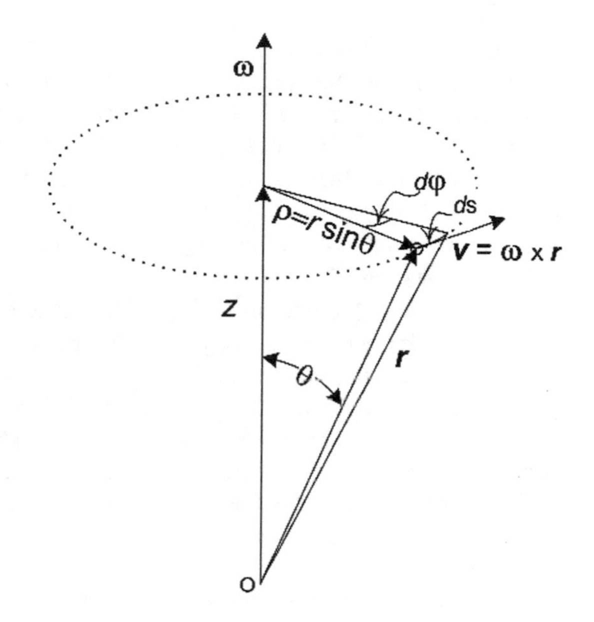

Fig. 3.1 Quantities used to characterize a mass rotating on the dotted circle, with coordinate origin at "O"

$$v_i = \frac{d\mathbf{r}_i}{dt} = \boldsymbol{\omega} \times \mathbf{r}_i \ , \tag{3.3}$$

where $\boldsymbol{\omega}$ is the angular velocity of the rotation with direction along the axis of rotation in the 'right-hand' sense. This relation follows from the definition of the angle and the cross product, and is depicted in Fig. 3.1. In turn, the acceleration of a coordinate vector under pure rotation becomes

$$\frac{d^2\mathbf{r}_i}{dt^2} = \boldsymbol{\omega} \times (\boldsymbol{\omega} \times \mathbf{r}_i) = -\left(\omega^2\mathbf{r}_i - (\boldsymbol{\omega} \cdot \mathbf{r}_i)\,\boldsymbol{\omega}\right) = -\,\omega^2\boldsymbol{\rho}_i \ . \tag{3.4}$$

The second equality in Eq. (3.4) follows from the "double cross" vector identity, while the third equality follows by re-expressing r_i as $z_i\widehat{\boldsymbol{\omega}}+\boldsymbol{\rho}_i$, i.e., the radial vector can always be expressed as a vector, $z_i\widehat{\boldsymbol{\omega}}$, along the axis of rotation and another one, $\boldsymbol{\rho}_i$, perpendicular to the axis. The vector geometry can be seen in Fig. 3.1. The result in Eq. (3.4) is the infamous centripetal acceleration, a vector pointing inward toward the axis of rotation of size $\omega^2\,\rho$, angular speed square times the radius of the circle of motion ρ. (The expression may be more familiar as speed squared over radius, which here is v^2/ρ.)

The angular momentum of a 'rigid' body is usefully written in terms of the angular velocity of the masses, because under rotation of a rigid body, the angular velocities for all the masses within are the same, and can be factored out of the summation:

$$\mathbf{L} = \sum m_i \left\{r_i^2\,\boldsymbol{\delta} - \mathbf{r}_i\,\mathbf{r}_i\right\} \cdot \boldsymbol{\omega} \ , \tag{3.5}$$

The displacements $\mathbf{r}_i$ to each mass m_i are calculated from the center of mass. The symbol $\boldsymbol{\delta}$ represents the 'identity tensor', defined so that when 'dotted' into a vector, it turns that vector into itself. The factored expression in the braces depends only on the geometric distribution of the masses, and is called the 'moment of inertia tensor'. For systems whose masses are distributed about the axis symmetrically, the angular momentum $\mathbf{L}$ expression simplifies to $I\boldsymbol{\omega}$, wherein the moment of inertia for the system becomes a number given by $I \equiv \sum m_i \rho_i^2$, where ρ_i is the distance from the axis of the mass m_i.[16]

An ice skater spinning on the tip of her skate has some frictional force acting on her skate, but with a relatively small lever arm (no bigger than the radius of the small circle the skate carves in the ice when she spins). This means the external torque on her will be relatively small, and the value of her angular momentum, $L = I\omega = \left(\sum m\rho^2\right)\omega$, will not change much over several rotations. If she pulls in her arms, some mass in her arms will have a smaller distance from the axis, making her moment of inertia smaller. From the conservation of angular momentum we conclude that as she pulls in her arms, her angular speed must increase, keeping $I\omega$ constant.[17]

Examples of spinning objects in biology include the tail of certain bacteria, ballerinas, and centrifuges to separate organics.

3.2.3 *Conserved Quantities*

Isaac Newton did not use energy conservation in solving for the motion of masses, since, in the problems he posed, the forces were known or could be found.[18] However, as we will describe, energy conservation is deeply seated in Nature and has wide practical purpose, even for dissipative processes if heat and microscopic kinetic motion are recognized.

In a brilliant paper in 1918, Emmy Nöther[19] showed that for every continuous symmetry of a system, there will be a corresponding conserved quantity, i.e. a function of the dynamical variables which will not change in time. With Nöther's theorem, momentum conservation is a consequence of the deeper idea of symmetry under translations. If the dynamics of a system does not change when the whole

[16]Some students learning integral calculus have fun finding $I = (M/V)\int \rho^2 dV$ for various simple uniform-density objects of mass M and volume V, such as a ball of radius a, for which $I_{ball} = \frac{2}{5}Ma^2$; a cylinder of radius a rotating about its axis, for which $I_{cyl} = \frac{1}{2}Ma^2$, or a stick of length l rotated about a perpendicular axis through its center, for which $I_{stick} = \frac{1}{12}Ml^2$.

[17]The increase in her kinetic energy comes from the work she did in pulling in her arms.

[18]Also, the energy concept was just evolving. Gottfried Leibniz used the phrase life-force ('vis-viva') for twice the kinetic energy of the particles. Adding the potential energy came in the work of Joseph-Louis Lagrange.

[19]E. Nöther, *Invariante Variationsprobleme*, Nachr. D. König. Gesellsch. D. Wiss. Zu Göttingen, Math-phys. Klasse, 235–257 (1918).

system is shifted in space, its momentum is conserved. Angular momentum of a system will be fixed if the system dynamics do not depend on the system's angular orientation. Energy conservation in a system follows if that system behaves the same way starting at any time, provided its initial conditions are reproduced.

Charge conservation results from a so-called 'gauge' symmetry, which is an invariance under any shift in the phase of the wave functions describing the particles coupled with the addition to the fields interacting with those particles by a gradient of that phase times the charge of the particle in the field.

3.3 Accelerated Motion and Life

Who studies petrified dinosaur droppings? Answer: A fecescist.

—wcp

Accelerations which affect humans are usefully expressed in terms of 'g', the gravitational acceleration of freely falling bodies near the surface of the Earth ($g = 9.80665$ m/s squared or about 32 feet per second squared). Since the dawn of aircraft, the human physiology during high and low acceleration has been of significant interest. Flying a plane into a steep dive and then pulling out can cause pilots to black out. Flying a large aircraft on the path and speed of a projectile thrown high into the air produces "zero g" for those floating inside. Table 3.1 indicates some effects on a person whose body is parallel with the direction of various accelerations.

3.3.1 *High-g*

A little kinematics shows that the minimum deceleration per unit 'g' after a sudden stop such as in a collision or after a fall can be found from $v^2/(2g\,x)$, where v is the initial speed and x the stopping distance. If a car hits a solid wall, the stopping distance may only be a little more than the thickness of an airbag, maybe three feet. At 60 mph (88 ft/s), the stop produces about 40 g's. A person will take only about 70 ms to stop from 88 ft/s to zero speed in 3 ft. Car air bags are designed to deploy before decelerations exceed 20 g's, and inflate in less than 30 ms.

A typical elevator will generate less than a 15% change in the force of the floor on the passengers, i.e. the effective 'g' is $(1.0\pm0.15)\,g$. Amusement rides are generally designed to keep riders below 3 g's, although some have reached 6.5 g's. The Apollo astronauts felt about 4 g's after liftoff. Test pilots encounter 20 g's lasting a few seconds. In 1954, Col. John Paul Stapp survived 46.2 g's for 1.4 s in a decelerating rocket sled.[20]

[20]Data from Nick T. Spark, *The Story of John Paul Stapp, the Fastest Man on Earth*, Wings and Airpower Magazine, p. 53, Republic Press, Cal. (Jul 2003).

Table 3.1 Unusual 'g' effects

Sustained acceleration	Effect
Positive (toward head)	
+3 to +5 g	Vascular pooling of blood in legs
	Muscle cramps
	Poor circulation
+5 to +9 g	Loss of vision
	Loss of hearing
	Blackout ('G-LOC) (g-induced loss of consciousness')
Negative (toward feet)	
−3 g	Pressure in the eye socket
	Headache
−3 to −5 g	Retina engorgement
	Loss of vision ("redout")
Above −5 g	Cerebral hemorrhages
Short-term zero g	Loss of a sense of balance
	Possible vertigo and anxiety
	Elation, feeling of strength
Long-term zero g	Loss of muscle tone
	Loss of heart strength
	Reduced red blood cell production
	Degeneration of labyrinth function
	Reduced bone strength

A concussion in a human will result from accelerations of over 90 g in less than a second, particularly if rotation is involved. Collisions between American footballers can generate 150 g's. Helmets may reduce the maximum expected brain acceleration to under 50 g's. But repeated lower-level collisions can create permanent damage to the human brain. Woodpecker brains, however, regularly undergo 500 g's without injury. Paracoccus denitrificans bacteria thrived living in a ultracentrifuge test tube experiencing over 400,000 g's.[21]

For aviators, the effects of acceleration can be reduced to some degree by wearing pressure suits which inflate when large g-forces are present, in order to reduce the pooling of blood, and by lying in a personally contoured seat, with the back perpendicular to the expected g-forces. This distributes the seat forces more evenly, and also minimizes the pooling of blood by minimizing the longitudinal separation between the high and low blood pressure regions in the body. Under these conditions, up to 9 g's can be tolerated for minutes, and 15 g's can be suffered for up to half a minute, after exercise and training (Fig. 3.2).

[21] S. Deguchi, et al., *Microbial growth at hyper-accelerations up to 403,627 g*, Proc Natl Acad Sci **108:19**, pp.7997–8002, NAS Press, Wash., DC (10 May 2011).

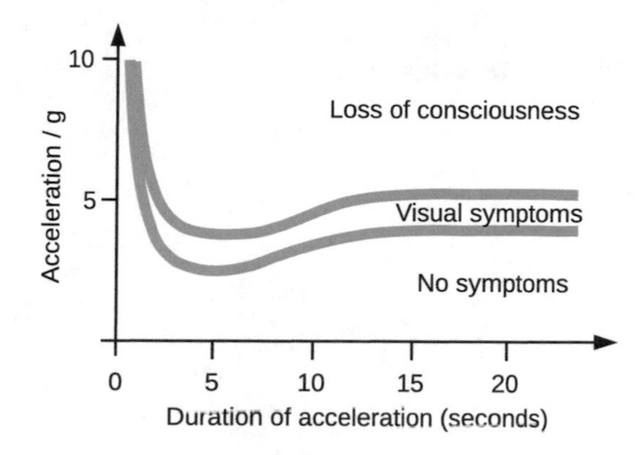

Fig. 3.2 Tolerance to high g's. Not shown is brain trauma and tissue tearing for accelerations above 50 g

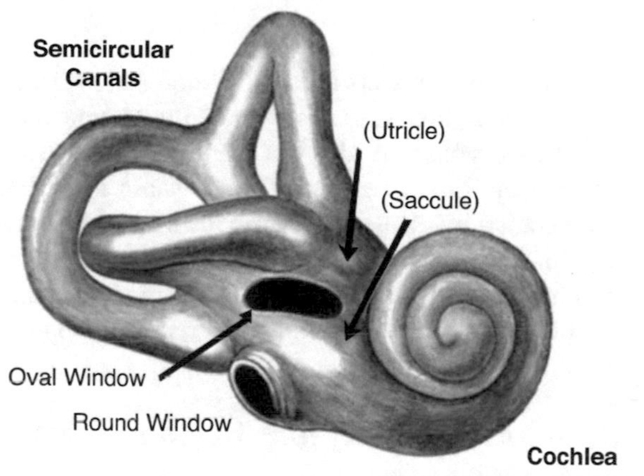

Fig. 3.3 Right inner ear

When the body undergoes acceleration, particularly when the acceleration is rapidly changing in magnitude and or direction, the brain can quickly get confusing signals from the inner ear, from muscle and joint proprioceptors, and from the eyes. Often, dizziness, disorientation, nausea and vertigo result. Experience, training, and mental attitude can be effective in reducing or even eliminating these effects.

To help us keep balance and orientation, and to help keep our eyes focused at one place even while your head moves, each inner ear has a vestibular system of three semicircular canals nearly at right angles (Fig. 3.3), holding a fluid (the endolymph), and two other organs, called otoliths, the utricle and the vestibule. The semicircular canals have nerve hairs cells which sense when endolymph fluid shifts when your head rotates. The otoliths sense the direction of gravity and linear acceleration. They contain tiny calcium stones (otoconia) which can move within a gel, and nerve hair cells to detect this movement. The action of the gel on the otoconia is analogous to the action of the springs on the mass shown in the Fig. 3.4. Within the linear elastic range of the springs, the greater the acceleration, the greater the displacement of the mass. Measuring this displacement is a measure of the acceleration, which can be calibrated in g's. Displacement of nerve hair cells in the gel in the utricle and saccule of our inner ear sends linear acceleration information to the brain.

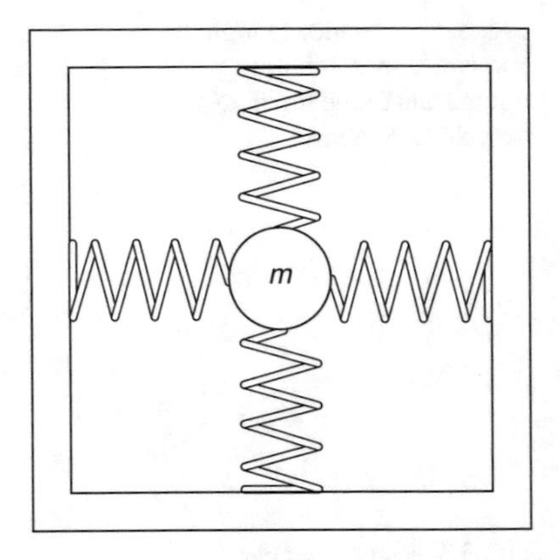

Fig. 3.4 A 'g-detector'. Two of six springs are not shown

Tiny accelerometers built on silicon and measuring only 0.5 mm across, such as those in cell phones, are constructed with a movable conductive layer with fingers interleaved with a fixed set of conductive fingers. The movable layer is suspended by thin elastic extensions. The movable layer has a mass which reacts to any acceleration by displacing, while the extensions limit the movement just as a spring would. Acceleration is detected by measuring the capacitance between the movable and the fixed fingers. Present devices in phones have an acceleration sensitivity of about $0.02\,g$ when $|a| \leq 2\,g$, and a range of $\pm 8\,g$.

3.3.2 *Microgravity and Weightlessness*

Near-zero gravity is sometimes referred to as 'microgravity'. If all masses within a body have the same force of gravity on them, and there is no opposing external force, all the mass points in that body will have a single acceleration. As long as this condition is sustained, then there will be no strain on the material within the body. If the body is a person, that person will experience 'zero gravity'. An astronaut 'floating' in orbit around the Earth is such a body. In fact, in Einstein's General Theory of Relativity, there is no difference locally between gravity and acceleration. A person in a free-falling elevator near the Earth will not perceive, from any local observation or measurement, any difference between that state and being free floating far from any gravitating mass. Nor could people distinguish their sense of gravity standing at rest on the Earth from what they would feel while accelerating in a rocket at $a = g$ in outer space.

A fish or a person in water with the same average density as the water may be able to move through the water seemingly weightless. In fact, astronauts go into tanks of water to practice 'space walks'. However, such astronauts still feel the gravitation of the Earth when in such tanks, because their flesh and organs are still hanging on their bones. The buoyant force acts on the surface of the body, not over the body volume, as does gravity. A person knows this because of proprioceptors which give information to the brain about the position of muscles.

As indicated earlier, it is possible to create real weightless conditions in so-called 'vomit comets', aircraft which move on a parabolic curve matching the speed of a free-falling rock which might be on the same curve. People inside experience 'zero-g' conditions for up to about 25 s.[22] Entering and exiting the parabolic path produces about 2.5 g on test subjects.

One can imitate zero-g conditions by opposing gravity at the molecular level with a magnetic force acting on each molecule. This is possible with materials that have a negative magnetic susceptibility if they are placed in a magnetic field having a gradient pointing in the direction of the gravitational force. The force per unit volume is given by $\mathbf{f} = (\chi/(2\mu_0))\nabla(B^2)$, where χ is the magnetic susceptibility, μ_0 is the magnetic permeability of free space, and B is the external magnetic field. For biological tissue, we can take advantage of the fact that cells contain a large fraction of water. Water is diamagnetic, i.e. $\chi < 0$, although the effect of this diamagnetism is quite small compared to ferromagnetic effects ($\chi_{water} = -9\times10^{-6}$ while $\chi_{iron} = 2 \times 10^5$). Live frogs and mice have been suspended in this way unharmed (see Sect. 6.5.8) in very strong magnetic field gradients. However, such fields are very dangerous if the tissue contains any ferromagnetic materials, such as a surgery pin!

Here are some observed effects of microgravity on humans and other organisms over a period of time:

- Disorientation: The brain must reprogram signals from the otoliths in the inner ear; The eyes become the main source of orientation information; On return to Earth, a new disorientation results which may take weeks to overcome.
- Euphoria is experienced by some after the initial disorientation;
- Bone density loss of 3% per month; calcium loss, kidney stones may develop from excess calcium expulsion;
- Muscle atrophy because of reduced demand;
- Reduced red blood cell production occurs, causing 'space anemia';
- Blood shifted to upper torso, brain senses excess fluid, kidney filtration rate increases, making dehydration;
- Heart reduced in strength because of lesser load;
- Immune system compromised by lack of exposure to pathogens;
- Bacteria acquire increased virulence when grown in zero-g, probably through reduced stress on cell structures.

Half of all astronauts are affected, with resulting nausea, headache, lethargy, and sweating. On return to Earth's gravity, the body recovers from this sickness in a few days, but balance, muscle tone and bone density restoration requires more time.

Zero gravity is experienced by astronauts in spacecraft orbiting the Earth, as long as air friction is negligible, and no propulsion is being used. In fact, occupants of any free non-rotating craft moving through space will have no weight. Because we have adapted to the Earth's gravity, physiology, pathology, and psychology are

[22]More than about 25 s and the airplane has trouble pulling out of the dive part of the parabola before hitting the ground!

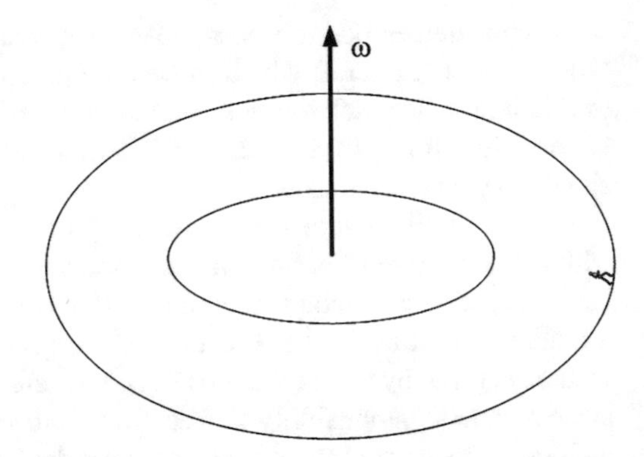

Fig. 3.5 Creating gravity in a spaceship. Shown is a person walking around inside a doughnut-shaped rotating spacecraft

all affected by micro or zero gravity. However, the effects are not life-threatening if zero-gravity exists for a limited time (weeks), or can be mollified with counteracting strategies, such as regiments of regular exercise.

Microgravity conditions occur in free-fall over the Earth with negligible air friction. Alan Eustace dropped (in a pressure suit) from 41.42 km (25.74 miles) above the Earth, where atmospheric drag on him was negligible for about the first 20 s, making him effectively weightless during that part of the drop. For shorter times, drop towers have been used. If an elevator drops with the acceleration of gravity, and the free-fall drop distance is 100 m, then the zero-g time would be about four and a half seconds ($t = \sqrt{2L/g}$).

3.3.3 *Rotating Spacecraft*

As we indicated in Sect. 3.3.2, humans have both physiological and psychological trouble under long-term micro and zero gravity. Long-term space travel will likely use rotational motion to introduce gravitational conditions for the travelers (Fig. 3.5).

If the inhabitants of a rotating spacecraft stand on the inner surface at its largest radius r, then they will experience the same gravitational effects of the Earth if the spacecraft is rotated at an angular speed of $\omega = \sqrt{g/r}$, where g is the acceleration of gravity near the Earth's surface. If the radius is 100 m, the period of rotation will be a gentle one-third of a minute ($2\pi\sqrt{r/g}$).

3.3.4 *The Centrifuge*

The effects of centripetal forces are used to advantage in medical procedures and biological research. A good example is the centrifuge. A centrifuge is a device for generating large rotational accelerations for the purpose of separating materials in suspensions and determining particle properties from the viscous drag force while the particle moves toward greater radii.

The ultracentrifuge that was developed by the Swedish chemist Theodor Svedberg in the early 1920's, with rotations of thousands per second, was capable of a million-g. He could separate and weigh large biological molecules. In the 1930's, Jesse Beams, using magnetically suspended rotors in a vacuum, made the rotors spin at more than a million revolutions per second, with rotational accelerations greater than a billion times the acceleration of gravity.[23]

Suspensions in a Centrifuge

Bodies in a liquid suspension undergoing accelerations much larger than free fall will separate from the liquid much faster than ordinary gravity causes. The rate of sedimentation depends on the fluid viscosity, the relative density of the particles, their size, and their shape.

Because fluids are necessarily compressible,[24] there will be a gradient of fluid density in a centrifuge.[25] Suspended particles of different densities can be made to accumulate at the location where the fluid density matches that of the suspended particle, called the 'isopycnic' point.

Centrifuges for biological and medical labs typically have a rotor with a seven centimeter diameter and rotate at up to sixty thousand revolutions per minute (making about 300,000 g's). These can be used to separate mixtures of proteins; different kinds of RNA; etc. At 10,000 g's, blood plasma, which normally makes up about $(55 \pm 5)\%$ by volume of blood, can be separated from blood cells in about 30 min.

Sedimentation Speed

Consider the forces on a body, such as a red-blood-cell, suspended in a liquid, such as blood plasma. In a centrifuge, several forces will act on the cell: Gravity (a body force), pressure forces (from the adjoining liquid, acting on the cell surface), and viscous forces if the cell moves through the liquid. To simplify this discussion, take the centrifuge axis vertical and the motion of its rotor and liquid containers horizontal. Suppose the rotor spins with angular speed ω. The net result of pressure

[23] Walter Gordy, *Jesse Wakefield Beams*, Biographical Memoirs **54**, [Nat Acad of Sci, Eng, Med] (1983).

[24] Special Relativity Theory forbids perfectly rigid bodies: Suppose such a body existed. Knock one side of the body. Then the other side must move instantaneously with the first side. One has made a signal from one end to the other move at an infinite speed. But Relativity limits signals to speeds no greater than c, which is found to be the speed of light.

[25] The average solution density itself can be adjusted by dissolving material which disperses at a molecular level, such as sugar or CsCl in water.

forces is the Archimedes buoyancy force. In the plane of rotation, this will be $B = \rho_F \, \omega^2 r V$, where ρ_F is the fluid density and $\omega^2 r$ is the centripetal acceleration of the cell when at radius r in the centrifuge. If the suspended cell is spherical (with radius a and density ρ_c), and its radial speed is sufficiently small, the viscous drag is well approximated by Stokes' law,[26]

$$F_\eta = 6\pi \eta a v \, , \tag{3.6}$$

where η is the fluid viscosity and v is the radial speed of the cell. (See Fig. 3.6.) For other shapes, the viscous force is still proportional to the speed, at least for relatively slow radial speeds typical of cells and macromolecules in a centrifuge. (The proportionality works well even for the viscous drag on a small particle moving in a gas, i.e. $F_v = b\,v$. In this case, Einstein proved that $b = k_B T/D$, with D the gas diffusion constant.)

The particles initially suspended in a liquid and then placed in a centrifuge may accelerate radially, but after a short time, will reach a 'terminal' velocity, when the forces on the particles balance. Newton's 2nd law in the horizontal plane then gives

$$6\pi \eta a v = (\rho_c - \rho_F)[(4/3)\pi a^3]\,\omega^2 r \, . \tag{3.7}$$

(The expression in the brackets is the volume of the cell $V = (4/3)\pi a^3$. In the vertical plane, $6\pi \eta a v_u = (\rho_c - \rho_F)[(4/3)\pi a^3]\, g$, where v_u is the upwardly directed velocity. Since $\omega^2 r >> g$, $v_u << v$.)

Optical centrifuges send focused light from the centrifuge chamber to the axis of rotation, and then out along the axis to an observer. In this way, sedimentation speeds can be directly observed and measured, as well as other static and dynamic reactions to high g.

3.3.5 Drag on Body

As we have indicated, for any object not moving radially outward too fast in the liquid, the viscous drag is proportional to the velocity, so the relation in Eq. (3.7) generalizes to

$$b\,v = (\rho_c - \rho_F) V \omega^2 r \, . \tag{3.8}$$

[26]Stokes' law works well here provided the Reynolds number associated with the fluid flow satisfies $\mathscr{R}_e \equiv \rho v/\eta < 0.2$. Empirical formulae for the range $0.2 < \mathscr{R}_e < 10^5$ exist. (See references in M.D. Mikhailov and A.P. Silva Freire, *The drag coefficient of a sphere*, Powder Technology **237**, 432–435 (2013).) An example given by D.J. Dunn is $F_\eta = [2/5 + 24/\mathscr{R}_e + 6/(1 + \mathscr{R}_e^{1/2})][\rho_f v^2/2 \cdot \pi a^2]$.

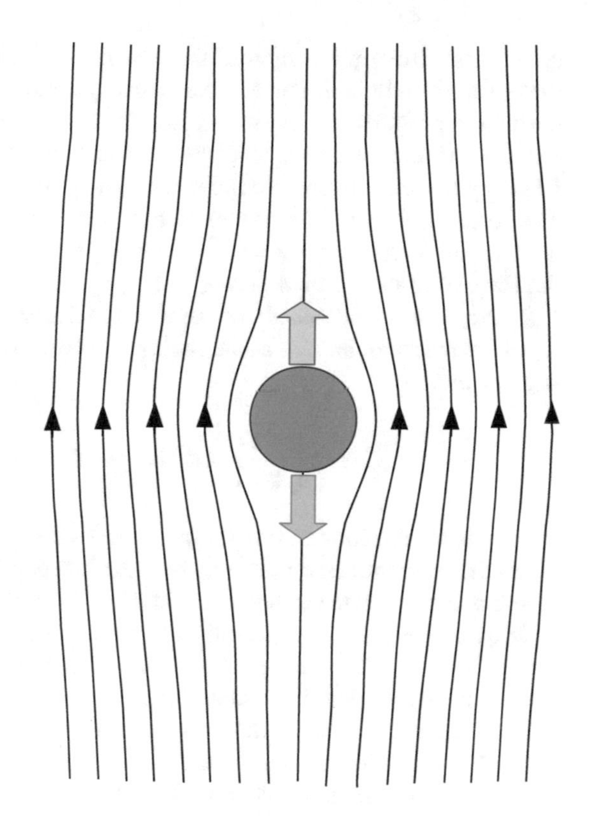

Fig. 3.6 A ball experiencing fluid drag forces, the upward force, while being pulled through fluid by the downward force. Fluid streamlines are shown in the frame of the ball. Stokes' law gives the drag force as $F_v = 6\pi\eta a v$. The downward force might come from gravity, $F_g = mg$, or the 'acceleration force' in a centrifuge, $F_a = ma$

The 'sedimentation time', defined by $S = v/(\omega^2 r) = (\rho_c - \rho_F)V/b$, can be measured from observation of the sedimentation speed v within a small range of radius r of the suspended bodies in the centrifuge. The relation for S is often used to get a value for the molecular weight of macromolecules. First, a separate determination of the molecule's volume V is needed. The frictional drag coefficient can be found by observing how the molecules diffuse over time in the same liquid not rotating. Einstein showed that $b = 6k_BT\,t/\langle r^2\rangle$, in which $\langle r^2\rangle$ is the average square distance the macromolecules disperse in the time t, in a liquid of temperature T. (k_B is Boltzmann's constant.) To calculate the macromolecule mass, we use

$$m = \rho_F\, V + b\, S\ . \tag{3.9}$$

Nanoparticles, such as macromolecules in suspension, will suffer thermal collisions with atoms in the liquid, just as Robert Brown noticed for pollen grains on the surface of water. These collisions tend to disperse the nanoparticles particles, with the less massive particles dispersing faster. The tendency to disperse is opposed by the centrifugal effects on the more dense particles. A gradient of concentration will be established after reaching equilibrium in a centrifuge. We can then apply the Boltzmann probability distribution to find the radial concentration of particles of a given mass. Boltzmann showed that when equilibrium is reached, the probability that a particle will be found to have an energy near ϵ in a system of such interacting particles is proportional to $\exp(-\epsilon/k_BT)$. The energy of a macromolecule at radius r in a centrifuge relative to its energy at some larger distance from the center can be found by the work needed to move that macromolecule from the larger radius to r. The force (in the rotating frame) on the macromolecule is the 'weight' mg' minus the buoyant force, where $g' = \omega^2 r$, so $\epsilon = \int (m\omega^2 r - \rho_F V\omega^2 r)dr$. (We can neglect the force from the Earth, mg, as $g' >> g$.)

Letting ρ_M be the density of the macromolecule, we find that the concentration, C, of the macromolecules at the radius r in terms of the concentration at radius r_0 is given by

$$C = C_o \exp\left[\frac{1}{2k_BT} m\omega^2 \left(r^2 - r_o^2\right)(1 - \rho_F/\rho_M)\right]. \tag{3.10}$$

The concentrations at various radii can be measured with an optical centrifuge. From the concentration gradient, and the independently measured density of the macromolecules, one can solve for the mass of an individual macromolecule.

If an object moves sufficiently fast in a fluid, the drag force will no longer be proportional to the velocity. The transition occurs as the fluid flow changes from streamline to turbulent flow around the object.[27] If the object must push fluid ahead as it moves, then the momentum transfer slows the object. The momentum loss of the object per unit time will be the object's velocity times the mass of fluid taking up momentum. The latter is proportional to the volume swept up by the object per unit time, which is the density of the fluid times the object's cross-sectional area A times the object's velocity. Thus, the net drag force has the form

$$f_{drag} = -bv - \frac{1}{2} C_d \rho A v^2 . \tag{3.11}$$

For a spherical ball, the quadratic drag coefficient can be approximated by $C_d \approx 24/\mathscr{R}_e$ for $\mathscr{R}_e < 60$ and $C_d \approx 0.4$ for $60 < \mathscr{R}_e < 2 \times 10^5$, where the unitless 'Reynolds' number $\mathscr{R}_e = 2\rho a v/\eta$ for a fluid of viscosity η moving around a ball with radius a.

3.4 Forces on Materials

And thus Nature will be very conformable to herself and very simple, performing all the great motions of the heavenly bodies by the attraction of gravity which intercedes those

[27] As we will describe in the chapter on fluid dynamics, this transition can be characterized by the value of the Reynolds number, which is $\mathscr{R}_e = 2\rho a v/\eta$ for a spherical body. As we will discuss, turbulence occurs for large values of the Reynolds number.

bodies, and almost all the small ones of their particles by some other attractive and repelling powers which intercede the particles.

—Isaac Newton

3.4.1 Systems in Mechanical Equilibrium

Systems in which no macroscopic accelerations occur are said to be in mechanical equilibrium, and their study is the subject of statics. Applications to cells, bodies, and biomaterials abound.

Having a 'rigid' system be static requires that the net force and torque on the system vanish. If the system is non-rigid, additional conditions must be applied. Mechanical equilibrium can be stable, neutral, or unstable. These are distinguished by slightly changing the body from its static position and then releasing. If the body forces act to restore the original position, the equilibrium was stable. If the forces still vanish, the equilibrium was neutral. If the body accelerates further away from its original position, the equilibrium was unstable. Standing a pen on its tip is an example of unstable equilibrium.

A fish with a fixed amount of air in a passive air sack will be in unstable equilibrium, because raising the fish slightly causes the air sack to expand under a reduced pressure, which increases the buoyant force on the fish. A live fish compensates. There is more on this buoyancy topic in Sect. 4.2.3.

3.4.2 Surface Tension

Identical molecules on the surface of a liquid interact with each other ('cohere') differently than when the same molecules are within the volume of the liquid. Without similar molecules attracting on one side of an interface, those on the surface tend to pack closer together than within the liquid, and tend to hold together. Polar molecules, such as water, will attract each other by electric dipole-dipole interactions, some atoms will have weak mutual electron orbital attraction, and even neutral identical molecules will attract by van der Waals forces. The van der Waals forces are effective in liquids when the molecules are separated by distances on the order of a nanometer or less.

If we imagine 'cutting' the surface of a liquid, the force needed to keep the two edges closed would depend linearly on the length of the cut. The force per unit length to hold a surface cut together defines the 'surface tension' σ (Table 3.2).

When the surface is stretched or expanded a relatively small amount, the surface molecules can cause an elastic restoring force, acting like a sheet of rubber.

Surface tension tends to decrease with increasing temperature. For water, the decrease is nearly linear from 20 °C to the boiling point.

Table 3.2 Surface tensions of certain liquids

Liquid (in air)	Surface tension.(dynes/cm)
Benzene	28.9
Carbon tetrachloride	27
Chloroform	27.1
Diethyl ether	72.8
Ethyl alcohol	22.3
Glycerol	63.4
Mercury	465
Water 20 °C	72.75
Water 100 °C	58.9
Typical mucus	~50
Alveolar surfactant	~3 to 15
Whole blood 37 °C	53
Urine, healthy 20 °C	66
Urine, w bile 20 °C	55

Fig. 3.7 Water strider (Image from wikispace.com)

Normal urine has a surface tension of about 66 dynes/cm. With jaundice (bile present in urine), the urine has a surface tension of about 55 dynes/cm. Normal urine can support sprinkled powdered sulfur (properly prepared) on its surface, but not with bile present.

Some insects have developed special feet with hydrophobic oils on their hair, enabling them to walk on water. (Figure 3.7 shows an image of a 'water strider'.) Surface tension is sufficient to support the weight of the insect. These insects also incorporate strategies for propulsion and for getting themselves to shore out of the surface depressions their feet make. Insects which can walk on water depress the surface with their hydrophobic feet and push on the angled walls of the depression for locomotion.

Some disinfectants carrying bactericides spread on bacteria cell walls better than water because they have a lower surface tension than water. One product has a surface tension of only about half that of water. Soaps (and detergents) lower the surface tension of water, making bubbles of air less well attached to the pores of cloth, and, as we described earlier, emulsify fats in water (See Sect. 2.2.)

Surface tension in water is due to 'cohesion' of water molecules with themselves. The contact angle of a liquid drop of water placed on a horizontal surface is called the 'wetting angle'. On hydrophobic surfaces, a contact angle is typically 70 to 90°. If the contact angle is even greater, the surfaces are called superhydrophobic (such as Teflon). Hydrophilic surfaces make the contact angle with surface drops of water approaches zero.

Surface tension is an important consideration in the inflation of the alveoli in our lungs. The surrounding mucous tissue fluid in an alveoli has a surface tension of about 50 dynes/cm. But this value is reduced by a factor from 2 to 15 by surfactants, which are phospholipids, predominantly dipalmitoyl lecithin. The surfactants are secreted into the fluid coating the interior of each alveolar sac, with their hydrophobic ends facing the air. With no surfactants, the alveoli would collapse.

3.4.3 Surfaces Curved by Differential Pressure

By balancing the pressure difference across any local region of the surface of a fluid with surface tension, there follows the Young-Laplace relation

$$\Delta p = \sigma \left(\frac{1}{R_a} + \frac{1}{R_b} \right) , \tag{3.12}$$

where σ is the surface tension and the $R's$ are the principal radii of curvature of the surface where the pressure is acting. For a cylindrical surface of radius R, one can take $R_a = R$ and $R_b = \infty$. For a spherical surface, $\Delta p = 2\sigma/R$.

The Young-Laplace relation can be applied to alveoli in our lung as we breathe and to the reaction of our arterial walls to increased pressure as our heart beats.

Surface tension holds a water droplet hanging onto horizontal objects until the drop grows large enough to make the weight of the drop overcome the force of tension. If the droplet is cylindrical at the contact line, then a balancing of forces produces Tate's law: $mg = 2\pi r\sigma$, where mg is the weight of the drop, r the contact cylindrical radius, and σ the surface tension.

Surface tension holds bubbles in a spherical shape. The Young-Laplace relation $\Delta p = 4\sigma/R$ can be used to find the difference in pressure between the inside and out of the bubble. (The extra factor of 2 occurs because a bubble has two liquid-gas interfaces creating surface tension.) The formation of microbubbles (cavitation) in cells due to intense ultrasonic waves will be discussed in Sect. 5.24.2.

Decompression sickness is caused by bubble formation from dissolved gases, principally nitrogen, in the body tissues and in blood when the outside pressure is quickly decreased. The bubbles can block blood flow in small vessels. (For more about nitrogen bubble formation in decompression sickness, see Sect. 9.18.) More details about bubbles is contained in the Appendix D.

The shape of a hanging drop of blood plasma, urine, or cerebrospinal fluid can be used to determine the fluid's surface tension. This value is of interest in medicine since the surface tension of these fluids is affected by various pathologies. The 'pendant hanging method' is the following: The drop is allowed to hang and form a pendant shape. Optical instruments determine the droplet's shape, including the radii of curvature at various levels in the droplet. Then a computer is used to solve for the surface tension, using two equations: (1) The Young-Laplace relation Eq. (3.12) and (2) The net zero force condition on a segment of the droplet from its base to a given height z:

$$2\pi r \sigma \cos\phi = \rho g V(z) + \pi r^2 p \,, \tag{3.13}$$

where r is the radius of the droplet at height z, ϕ is the angle between the surface tangent at height z and the vertical axis, ρ is the density of the fluid, and $V(z)$ is the volume of the droplet up to the height z.

3.4.4 Capillary Action

'Capillary action' is the movement of a liquid caused by adhesion to a hydrophilic contacting material. If the molecules of a liquid adhere to the surface of a material with an effective binding energy greater than that within the bulk of the liquid, the liquid 'wets' the material. This is the case for water and glass, because hydrogen bonding can occur between the water molecules and silicon dioxide, and because the water molecules can embed into the spaces between the molecules of the glass surface. Glass surfaces therefore are hydrophilic. In contrast, mercury does not wet glass.

If a solid material adheres to one fluid greater than another, and is placed across the boundary layer between these two immiscible fluids under the influence of gravity, a 'meniscus' (crescent shape) will form at the boundary between the fluids due to the contact forces between the fluids and the solid. Water placed in a vertical tube with a hydrophilic interior surface will rise in the tube until the weight of the column of water balances the force adhesion and surface tension σ holding the liquid at a given height. If we assume that the meniscus that forms has a contact angle of θ with the tube wall, then the two forces will balance when $2\pi r\sigma \cos\theta = \rho\, gh\pi\, r^2$, where r is the tube inner radius, and h the height of the liquid of density ρ. This gives $h = 2\sigma \cos\theta/\rho g r$.

Because water is the medium of life, the hydrophobic and hydrophilic behavior of biomolecular surfaces is often related to the function of the surface. Plants take advantage of capillary-like adhesive forces to raise water from the soil (which itself holds water by adhesive forces) against the force of gravity. An upwardly-directed gradient of water pressure at the top of a plant or tree is generated by

- capillary adhesive forces of water with the walls of the xylem and the elaborated microtubules at the top of the plant;
- water concentration gradients due to controlled evapotranspiration from foliage surfaces;
- osmotic pressure in the plant foliage due to the active transport of sugar solution down the phloem in the branches or trunk and active absorption of water from dew or mist settling on plant top foliage;
- upward osmotic pressure in the xylem from the roots of the plant.

If air leaks into the microtubules at the top of a tree, fluid cavitation can occur. The adhesive forces at the top level of the tree is enhanced by any branching a given xylem tube undergoes, which increases the ratio of water-contact surface area compared to the volume per length in the xylem tubule.

The tallest trees in the world are the California redwoods, standing as much as 112 m high. Calculations show that the height limitation does not come from physical constraints, such as resilience under gravity or wind bending. Rather, the limitation is likely to be dominated by hydraulic negative pressure causing cavitation at the top of the xylem fluid column (with negative pressures on the order of −2 MPa).

The interface between two immiscible fluids ('1' and '2') is called a capillary surface. Under static conditions, the surface is determined by a balancing of pressure forces and surface tension. Let $\widehat{\mathbf{n}}$ be the normal to the surface pointing in the direction toward fluid '2' from '1'. Then

$$p_2 - p_1 = \sigma \nabla \cdot \widehat{\mathbf{n}} \, . \tag{3.14}$$

This relation can be expressed as an equation for a capillary surface. More details can be found in Appendix D.

3.4.5 *Stress in Materials*

Humans, even at rest on the surface of the Earth, have stresses in their bones and elsewhere brought on by gravity.

Within material substances, such as bones or flesh, the forces on each small element of mass δm can be separated into two types: those which act throughout the volume of the mass, and those which act on the external surface areas of the mass. Newton's 2nd law can then be written: $\sum \mathbf{f_V} \delta V + \sum \mathbf{f_S} \delta A = \delta m \mathbf{a}$. For example, gravity generates body forces since it acts on every atom within the mass. In this case, $\mathbf{f_V} = \mathbf{f_G} \equiv \rho \mathbf{g}$ where ρ is the mass density, $\rho \equiv \delta m / \delta V$, and $\mathbf{g}$ is the local gravitational field. The surface force per unit area, $\mathbf{f_S}$, is referred to as the 'stress' acting on the surface. Stress is measured in newtons per square meter. One newton per square meter is defined as a 'pascal' (Pa).

In turn, surface forces can be divided into two types: tangential forces, also called 'shearing' forces, and forces normal to the surface, called 'pressure' forces. (Pressure forces will include negative ones, i.e. forces which pull out on a surface.)

Pressure Forces

A pressure is that part of a force per unit area applied perpendicular to the surface of a substance. Pressure can arise from adjacent layers within a material, or from other material interacting on a boundary layer. In the case of gases, a large part or all of the force on a macroscopic surface comes from the average impulse per unit time due to molecular collisions. Typically, at biological temperatures (i.e. from the freezing to the boiling temperature of water), the energy carried by the gas molecules (about $3k_BT/2 \approx 0.04$ electron volt at room temperature) is insufficient to cause excitation of the atoms or molecules in a material surface, so the collisions are largely elastic. (Even so, energy can be transferred from or to the gas from the surface.) A little Newtonian physics and statistics gives that the average pressure due to an ideal gas (one with average molecular separations much larger than the size of the molecules and with negligible long-range forces) to be proportional to the average kinetic energy of the molecules in the gas: $p = (1/3)\rho \langle v^2 \rangle$, where ρ is the mass density of the gas and $\langle v^2 \rangle$ the average speed squared of molecules in the gas.

Negative pressures can be created by pulling on a surface. Gases cannot do this, but you can glue a plate onto a surface and then pull on the plate. This uses either an attractive force between one layer of molecules and an adjacent layer, or an interlocking of material at the interface. Barnacles use a fibrous protein to give strength to their glue. Micro tentacles into surface cracks and holes is also effective.

For comparison, the pressure of the calm atmosphere at sea level is 1.01325×10^5 Pa, which is 14.7 pounds per square inch. This is called one atmosphere.[28] Often, pressure is communicated in units of the height of a column of mercury held up by air pressure. One atmosphere of pressure will support a column of mercury at 0 °C (where the mercury has a density of 13.595 gm/cm^3) to a height of 76.00 cm.

The pressure to support 1 mm of mercury is called a 'torr'. Another common unit of pressure used in meteorology is the 'bar', from the Greek word baros, for weight, and defined as 10^5 newton per square meter. This is close to the atmospheric pressure at sea level. Processes carried out at constant pressure are called 'isobaric'.

In physiology, the pressure needed to hold up a column of water rather than mercury is often used for the unit of pressure. Since the density of water is 1/13.5695 that of mercury (both at 4 °C), 1 mmHg is 1.36 cmH_2O.

As the atmospheric pressure does not vary quickly, short term pressure changes in our body are normally due to our own processes, such as our heart pumping blood

[28]Before 1982, one atmosphere was called the 'standard pressure'. In 1982, the International Union of Pure and Applied Chemistry recommended the 'standard pressure' be taken as exactly 10^5 Pa.

($\Delta p \approx 40$ mmHg), our diaphragm moving air in our lungs ($\Delta p \approx 10$ mmHg), or our alimentary canal peristaltically moving material in our digestive track ($\Delta p \approx 5$ mmHg). All of these process change internal pressures by less than 6%. Assuming regular metabolic processes occur both isobarically and isothermally is a good approximation.

Atmospheric lows and highs have a range at sea level from 635 to 813 mmHg, i.e. a 23% variation, but the changes occur over hours, not seconds. For those who live in the Andes (7 km high), the atmospheric pressure can be as low as 40% of sea level pressure. People who live at such heights develop more hemoglobin to better utilize oxygen, whose partial pressure has been reduced by 40%.

Measuring Pressure

Below is a list of some devices used to measure pressure using the elasticity of materials responding to pressure:

- A capacitor with an elastic dielectric will have a capacitance which increases with pressure acting to squeeze it. Very small such capacitors can be placed in the body and in the blood stream.
- A closed gas chamber with an elastic diaphragm acts as a barometer by placing a lever on the diaphragm.
- A column of mercury in a closed glass vertical tube with an open bottom end placed in a dish of mercury will have a height proportional to the atmospheric pressure. (Mercury is used because of its high density, and its low vapor pressure.) Sea-level atmospheric pressure will support a column 76.0 cm high.
- A sphygmomanometer measures blood pressure in the arm by determining the pressure in an inflatable cuff.
- Strain gauges measure pressure by the change of the electrical resistance in a stretched or compressed conductor, such as a metal foil, or carbon granules, placed in stress.
- A carbon microphone has a resistance which depends on how much the carbon granules are compressed by air pressure.
- An inductor with a core whose position varies with an external stress will have an inductance depending on that external pressure.
- A piezoelectric crystal, such as quartz, will produce an electric field when under strain. This strain can be caused by pressure. (Piezo means squeeze in Greek.)
- An ocular tonometer is a device to measure the intraocular pressure (normally 12–20 mmHg above atmospheric). The instrument invented by Hans Goldmann measures the force needed to flatten the cornea within a ring of 3.06 mm diameter. Assuming only very soft media between the cornea and the vitreous humor in the eye, that force divided by the area of the ring will be the intraocular pressure. The modern instrument corrects for tear surface tension and cornea elasticity.

Frictional Forces

Frictional forces are a type of shearing force of one layer of material against another, and are accounted for by how molecules interact. They can exist in the boundary between any pair of solids and liquids, and between gases and either solids or liquids.

Two solid layers of material can act on each other tangentially largely because their surfaces are never perfectly smooth, given that material surfaces are made from discrete molecules. On an atomic scale (or larger scale, if there are larger microscopic bumps), the molecules across the boundary can interleave and push sideways against each other. Since the interleaving becomes more intimate as the pressure acting to push the layers together increases, the frictional force available increases with this pressure.

With macroscopic materials under typical pressures, the frictional force will be proportional to the normal (perpendicular) force squeezing the layers together. This relationship is found to hold for most smooth surfaces, but is not a universal law; rather such relationships are called 'phenomenological laws'.

Because friction between two tangentially sliding solid surfaces can break molecules away from either surface, debris can be left between the surfaces, often rolling and acting like ball bearings. This reduces the frictional force needed to slide, for the given normal force. The induced vibration of surface molecules during slipping and the breaking of microbumps will generate heat, making friction a dissipative process. Rubbing our hands together can warm the skin.

If the frictional force is large during a macroscopic motion of one surface against another over a short enough time (so that the heat generated has insufficient time to spread much), then the solid material at the contact point can undergo thermal phase changes. Sparks made by striking flint is a good example.

Frictional forces caused by the motion of fluids acting on bodies are called viscous forces. We will deal with these in the section on fluid dynamics (4.3).

3.5 Response of Materials to Stress

The response of materials to unopposed force is to start moving. The initial motion will be linear acceleration of the center of mass of the material if the net forces on an element are unbalanced, and rotational acceleration about the center of mass if the net torques are unbalanced. Additional types of motion may occur if the materials are not acting rigidly. However, if mechanical equilibrium is re-established, the accelerations must vanish.

A lemma from Einstein's Special Theory of Relativity says there are no rigid bodies, i.e. bodies with a fixed separation between any pair of material points.[29]

[29]See footnote (24) in Sect. 3.3.4.

Of course, in nature, some bodies are 'softer' than others. We can sometimes approximate the hard ones as if they were acting rigidly. We use the term rigid only as an approximation, in cases where the internal relative displacements are not important to the problem being investigated. In this discussion, we are interested particularly in those displacements which distort a material.

Even in equilibrium, internal distortions of the material may generate forces to oppose the external surface and body forces acting on a given material element and cause internal forces of one material element on the next. For example, the response of a bone to a set of distorting forces is an opposing set of forces while the bone changes in shape. The response of an elastic band to forces at its ends is to stretch. Stretching produces internal forces on each piece of the rubber. The action of increased pressure on gas bubbles results in a decreased gas volume, thereby increasing the outward force on the agents causing the pressure. Confined sidewalks can explode from internal stress as the temperature of the concrete rises. The concrete attempts to expand, but may experience large opposing forces, and so the concrete might become compressed beyond its fracture limit.

To characterize the effect of forces in distorting a body, we can apply Newton's 2nd law[30] to each small element within that body, and Newton's 3rd law to follow the effects from one element to the next. The choice of the elements is arbitrary, and they need not be physical. However, for simplicity, the elements are usually taken sufficiently small that relevant quantities within do not vary appreciably (to some approximation acceptable in a calculation).

In the mathematical descriptions of the mechanical behavior of a system of entities, an enumeration of each entity is often given by specifying its spatial position in a selected coordinate system. Strategic choice of a coordinate system according to its symmetries can simplify calculations. For example, the elements in a bone with approximate cylindrical symmetry are best followed by their location in a set of appropriate cylindrical coordinates. As the system responds to forces, the elements may distort in shape as well as move. To be definite, we will circumscribe each element with an imaginary surface which follows all the atoms in the initial volume of the element. Keeping the total number and kind of atoms fixed also keeps the total mass of the element constant to within the precision of ordinary laboratory instruments.

In any set of 'curvilinear coordinates' (i.e. ones whose axes are locally perpendicular), small elements in a body can be selected to be initially approximate rectangular solids with surfaces perpendicular to coordinate lines. As we described, forces acting on such elements can be divided into two types: body forces and surface forces. Body forces act on all parts within the volume of the elements. The body forces acting on each element will be taken together as a single vector.

[30]A useful alternative to studying forces is to consider the energy changes in a body as it is distorted. One advantage of starting with energy is the fact that energy is a number, while force has both size and direction. However, our intuition is built first with the effect of forces, so we will continue with them.

Surface forces act only through the faces of the rectangular elements and come from interaction with adjacent elements.

The surface forces acting across each surface of a given small mass element will be summed into a single force acting on that surface. In the spirit of the approximation of small mass elements, the surface forces will not vary significantly across a single element's surface. This means that if any element's surface were cut in half, the force on that half surface should be half the force on the whole. Thus, the surface force per unit area should not be affected by the choice of size for the mass elements, as long as they are small enough. Similarly, the body force per unit volume should not depend on the choice of size of the mass element, as long as they are small enough. It is for these reasons that Newton's second law for the behavior of a distortable material is given a form that is independent of the choice of small mass elements.

Instead of writing Newton's 2nd law for the masses, we write a corresponding expression by dividing each side of Newton's 2nd law for a small mass element by its volume. The law is then expressed in terms of only 'intensive variables'[31]: body force per unit volume; the change across the material of surface forces per unit area; and the density of the material. We should expect that the laws of nature do not depend on sizes which we are able to pick arbitrarily, such as the size of the mass elements inside a body. Our choice of small mass elements is arbitrary, as long as their size is small enough that quantities which change across the body do not change significantly across the small mass elements, and large enough to make averages over atoms meaningful.

In Fig. 3.8, a pair of the surface forces denoted by $\mathbf{F_z}$ are shown acting on the surface of an approximate cubic element of material, with the two surfaces on which this pair of forces act lying perpendicular to the z-axis of the selected coordinate system. The unit vectors $\widehat{\mathbf{n}}$ are taken normal to the surface. Notice that $\mathbf{F_z}$ may have components along all three axes. The stress (force per area) on the 'top' surface in the figure is then $\mathbf{S_z} = \mathbf{F_z}/(dxdy) = \mathbf{F_3}/A_3$, where $A_3 = dx\,dy$ by using $(1, 2, 3)$ as indices for (x, y, z), and k can replace (i, j) as the 'missing' third index in (i, j, k). The components of this stress will be named $\mathbf{S}_3 \equiv \{S_{31}, S_{32}, S_{33}\}$; similarly for $\mathbf{S}_1$ and $\mathbf{S}_2$. Thus,

$$S_{ij} \equiv F_{ij}/A_i \; , \tag{3.15}$$

where F_{ij} is the jth component of the force $\mathbf{F}_i$, and A_i is the area element perpendicular to the i-th direction. Notice that the 'diagonal' components of the stress are pressures on each side of the cube: $S_{11} = S_{22} = S_{33} = -p$.

By adding the forces on all six sides of the distorted cube, one finds that the net force will be

$$\delta\mathbf{F} = -\nabla \cdot \mathbf{S}\,\delta V \; , \tag{3.16}$$

[31] Intensive variables do not depend on the size of selected elements; 'extensive variables' do.

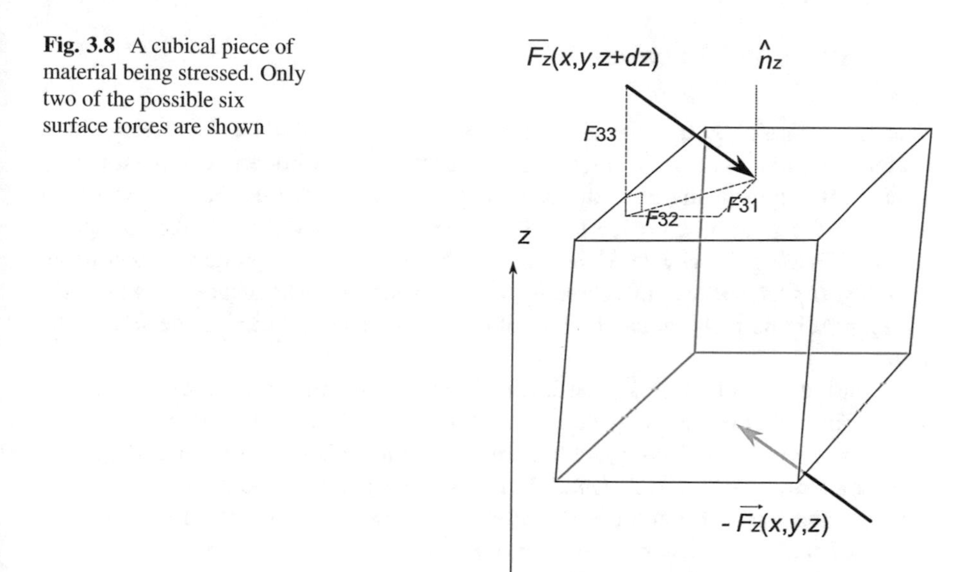

Fig. 3.8 A cubical piece of material being stressed. Only two of the possible six surface forces are shown

where δV is the volume of the cube. The entity $\mathbf{S}$ is a tensor, here made from the components S_{ij}. Static conditions (no net force) require $\nabla \cdot \mathbf{S} = 0$. By ensuring that the cube does not rotate under the stresses, i.e. that the net torque vanishes, the stress tensor components must be symmetric: $S_{ij} = S_{ji}$. This can be seen by finding the net torque on the cube due to the stress forces. The result is that net torque has components $\tau_k = -(S_{ij} - S_{ji})\delta V$, with i, j, k in cyclic order.

Given that ordinary materials have internal interactions between each of its elements, if we distort that material, we will have to exert a force to counteract internal forces. For example, solids have internal forces which bind the atoms and molecules together. Displacing any atom relative to its neighbor requires a force which initially increases with increasing displacement. Typically, the size of small distortions are in proportion to the forces causing them. For a macroscopic piece of material (i.e. one with lots of atoms), such as a ligament, a bone, an elastic band, or a glass fiber, many small atomic displacements can add to a large macroscopic displacement. Thus, whether solid, liquid, or gas, most materials will distort in proportion to the forces causing that distortion, if the relative displacement across the material is not too large.

3.6 Strain in Materials

The local 'strain' in a material is characterized as the fractional distortion of a material element as a result of stress. For example, if a bone is compressed, the change in length of the bone divided by the original length is the compressional strain. If the bone were twice the length, we should expect twice the change in length. Dividing the change in length by the original length produces an intrinsic quantity. In fact, strain is measured by a unitless number. The local effects of strain in uniform bone material can be described without having to know the size of the bone.

Shearing stress can cause one layer of material to shift relative to an adjacent layer. This relative shift is called a shearing strain. If your foot is attached to a ski, and you twist your body, each element of material in your femur undergoes a shearing strain. For a small material element rectangular in volume, a pure shear between the top and bottom of the volume causes the top to shift relative to the bottom, keeping the distance between top and bottom fixed. The shearing strain is then the size of the displacement of the top relative to the bottom, divided by the distance between the top and bottom. We will look at this case in more detail later.

We can define strain through the relative shift of material elements, but we should be aware that pure rotation produces relative motion of material elements without distortion. So, we can say that strain is that part of the relative shifts of material elements which causes distortion. Mathematically, if $\boldsymbol{\xi}(x, y, z)$ represents the vector displacement of a material element which started at position (x, y, z), then the relative shift in the material in the ith direction due to a displacement in the jth direction per unit length is given by $\partial\xi_j/\partial x_i$ which we will shorten to $\partial_i\xi_j$. We can always write: $\partial_i\xi_j \equiv (1/2)(\partial_i\xi_j + \partial_j\xi_i) + (1/2)(\partial_i\xi_j - \partial_j\xi_i)$. The second term can be shown to be pure rotation.[32] The strain in the material in the ith direction due to a displacement in the jth direction is defined as[33]

$$\sigma_{ij} \equiv (1/2)(\partial_i\xi_j + \partial_j\xi_i)\,. \tag{3.17}$$

There are six independent elements of the strain. Pure shearing strains are represented in the three independent off-diagonal elements of the three-by-three matrix formed from these terms, and the relative volume changes by the sum of

[32]For a small rotation through an angle $\delta\theta$, Eq. (3.3) or the Fig. 3.1 gives $\boldsymbol{\xi} = \delta\boldsymbol{\theta} \times (\mathbf{r} - \mathbf{r}_o)$, where $\mathbf{r}_o$ locates a point on the axis of rotation. Taking the vector curl of both sides shows that half the curl of the displacement field $\boldsymbol{\xi}(x, y, z, t)$ is a vector along the axis of rotation with size equal to the angle of rotation: $\delta\boldsymbol{\theta} = (1/2)\,\nabla \times \boldsymbol{\xi}$. The components of this curl are just the antisymmetric combination of the derivatives of the displacements we separated from the changes in displacement across the material to define the strain tensor. Inversely, if the change of the displacements across a small region of the body can be expressed as a curl, then this region is not being distorted, but rather it is being rotated.

[33]Caution: Engineers typically do not include the factor of (1/2) in their definition of strain.

the diagonal elements (called the 'trace' of a matrix).[34] The last two independent elements determine asymmetric strain.

Elastic and Inelastic Distortion

'Elastic' behavior is defined by the property that if the material is stressed, the energy needed to strain the material is all stored mechanically, so that no heat is produced. In contrast, 'inelastic' behavior is observed when some work done to deform a material is lost to heat or other forms of dissipation.

3.7 Connection Between Stress and Strain

Under stress, a material may deform in a variety of complicated ways, depending on the nature of the bonding between its molecules, the ordering of those molecules, the initial internal stresses, and the duration of the external stress.

For small stresses, we should expect there to be a proportionality between stress and strain. This is a form of Hooke's Law for materials:

$$\mathbf{S} = \mathbf{T} : \sigma \quad . \tag{3.18}$$

If this proportionality holds, we say the material is acting with linear elasticity.

Displacements of a material under stress do not have to be in the same direction as the stress causing them. When you push on a bone, the displacement this push causes will be preferentially in the direction that the atoms and molecules can more easily shift. Thus, the proportionality factor in Eq. (3.18) is not simply one constant, but rather a set of constants:

$$S_{ij} = \sum_{k,l=1}^{3} T_{ijkl}\sigma_{kl} \quad . \tag{3.19}$$

The set of coefficients T_{ijkl} are called 'tensor' components. Each of the labels on T_{ijkl}, such as i, is called an index, taking the values $1, 2,\ or\ 3 = x, y,\ or\ z$. With four indices, the values T_{ijkl} are said to make up the components of a fourth 'rank' tensor. Selecting a set of particular values for the indices (i, j, k, l), such as

[34]The sum of the diagonal elements becomes $\nabla \cdot \boldsymbol{\xi}$. That this represents a relative volume change can be seen as follows: Let $\Delta V = dxdydz$ be a small volume element in the material before it is stressed, and $\delta(\Delta V)$ be the change in that small volume after the material is stressed. If all three sides of the original volume are displaced in the direction of their normal, we will have $\delta(\Delta V) = \prod_i [\xi_{ii}(x_i + dx_i) - \xi_{ii}(x_i)]$, which is, to first order in the small dx_i, $(\sum_i \partial_i \xi_i)\, dxdydz = (\nabla \cdot \boldsymbol{\xi})\, \Delta V$.

1223, picks 'a component', in this case T_{1223}, of **T**. Each component of **T** is simply a number, with units of force over area. We will refer to **T** as the 'stress-strain' tensor.

This kind of object, **T**, is of general interest to mathematicians as well as physicists and engineers. In fact, the name tensor historically came from the relation expressed by Eq. (3.19)! These days, any object which behaves like a vector in each of its n indices is called a tensor of rank n.[35] (See Appendix G.) We have seen one good example already: The vector cross-product contains the operation which converts a vector into a vector. The entity **T** is more general because it relates one second rank tensor to another.

Symmetry reduces the number of independent components of **T**. Balancing torques on a cubical piece of material requires the stress tensor to be symmetric: $S_{ij} = S_{ji}$. The strain tensor is symmetric because it cannot have any part which is a pure rotation. These two conditions mean that the stress-strain tensor is symmetric in both its first and its second pair of indices. Thus **T** has no more than $6 \times 6 = 36$ independent components for any material. (That's still a lot of numbers to measure for a material!)

If the structure of the material looks the same in a mirror, then the stress-strain will have only 6 independent components left. But mirror symmetry will not apply to many biological materials, because amino acids and sugars have a left or right 'handedness'. Material with structural handedness may show more resistance to a left-handed twist than a right-handed twist. Twisting DNA to a tighter helix is harder than twisting it to a looser helix.

If the material is a cubic crystal (such as table salt), only 3 components of **T** are independent. And if the material is isotropic (the same in any direction), there are only 2 left![36] Typically, amorphous materials and fluids have such isotropy. The two independent components can be determined by the linear stretch and shear properties of the material, which are rather easily measured and tabulated.

For an isotropic and homogeneous material, the relationship between elastic stress and strain is often measured by how much the material stretches under stress and contracts at right angles to the stretch:

$$\frac{F}{A} = Y \frac{\Delta l}{l}, \tag{3.20}$$

$$\frac{\Delta w}{w} = \frac{\Delta h}{h} = -n_P \frac{\Delta l}{l}, \tag{3.21}$$

[35] Tensors are any entity which transform under coordinate changes $\boldsymbol{x}' = \boldsymbol{a} \cdot \boldsymbol{x}$ with the product of coordinate transformations and/or their inverses: $\mathbf{T}' = \boldsymbol{a} \cdots \boldsymbol{a} : \mathbf{T} : \boldsymbol{a}^{-1} \cdots \boldsymbol{a}^{-1}$.

[36] There are only two kinds of tensors in three dimensions which are unchanged by rotations. They are proportional to (1) The Kronecker delta: $\{\delta_{ij}\}$, which is 1 for $i = j$ and zero otherwise; and (2) the completely antisymmetric tensor $\{\epsilon_{ijk}\}$ in a three dimensional space, which is 1 when $ijk = 123$ or any even permutation, -1 for odd permutation, and 0 otherwise. To be invariant under rotation and symmetric in the first and the second pair of indices, the tensor T_{ijkl} must be of the form $a\, \delta_{ij}\delta_{kl} + b\,(\delta_{ik}\delta_{jl} + \delta_{jk}\delta_{il})$.

where l, w, h are the length, width, and height of a block of the material.

The constant Y is called 'Young's modulus'. The relative transverse contraction, n_P, is called the 'Poisson ratio'. For an isotropic material, one can show that the Poisson ratio must be less than $1/2$, and that $T_{xyxy} = Y/(1+n_P)$ and $T_{xxyy} = Yn_P/[(1-2n_P)(1+n_P)]$. The other components follow from symmetry.

Two important examples of the stress-strain relation are from bulk compression and pure shear. If an isotropic cube of material is put under uniform increase in pressure Δp from all sides, we expect it will compress in volume ΔV and density $\Delta\rho$ in proportion to the pressure change:

$$\Delta p = -B\frac{\Delta V}{V} = B\frac{\Delta\rho}{\rho}\,. \tag{3.22}$$

(The minus sign accounts for the fact that an increase in pressure results in a decrease in volume, so that measured values of B will always be positive.) In terms of our two constants Y and n_P, one can show that[37] the 'compression modulus' (also called the 'bulk modulus') B is given by $B = Y/(1-2n_P)$. The inverse of B is referred to as the 'compressibility' $C = 1/B$. You can see that if a material had $n_P = 1/2$, that material would be perfectly incompressible, an impossibility, so $n_P < 1/2$. If the cube is sheared with tangential forces F on opposite surfaces of area A, the cube tilts in the direction of the shear (just as a textbook shifts when its back is held and its cover is pushed tangent to the paper). The shearing angle θ is determined by

$$\frac{F}{A} = \frac{Y}{1+n_P}\,\theta = n_s\theta\,. \tag{3.23}$$

The quantity n_s is called the 'shear modulus'.

Stress in two-dimensional elastic membranes is often characterized by 'distensibility': If an elastic membrane surrounds a volume, then the change in that volume per unit volume per unit pressure increase is the distensibility D:

$$\frac{\delta V}{V} = D\delta p \tag{3.24}$$

Alternatively, the 'compliance' of the membrane is defined by the change in volume per unit pressure increase so that

$$\delta V = C\delta p\,. \tag{3.25}$$

Healthy lungs have a volume of about 6 liters, a compliance of about 100 mL per 1 cmH_2O, with an inspiration volume of air of 0.5 L. Attempts to fully expand the lung meets high resistance, because the alveoli are reaching their limiting volume

[37] See the **Feynman Lectures Vol. II**, [Addison-Wesley] (1964), p39.

before rupture. Thus, the compliance value decreases toward zero as the lung is forced to unusually large volumes. This behavior is similar to stretching a material beyond its linear region. The relation between the force and the stretch may become inelastic and possibly hysteretic, a property described below.

An elastic membrane of compliance C and surrounding a volume will store energy in the amount $\int p\delta V$ giving

$$\mathscr{E} = (C/(2)(p^2 - p_0^2) \tag{3.26}$$

when expanded by an increase of the internal pressure from p_0 to p.

3.8 Limits of Stress

As we know, we can take only a certain amount of stress(!). There are several possible reactions as stress is increased on a material. Table 3.3 shows the various material responses to stress.

If the stress on a material is relaxed and that material returns the energy it stored by distortion back into mechanical work with no heat generated, then the material is acting elastically and the process is called 'adiabatic'.[38] On an atomic level, an 'adiabatic stress' causes atoms to undergo a relative shift in position, but not to change their quantum states. Most solid materials will distort adiabatically if the

Table 3.3 Behavior of materials under stress

Behavior	Description
Linear elastic region	Strain is proportional to stress
Non-linear elastic region	Strain non-linearly related to stress
Inelastic region	Mechanical energy is lost during strain of the material
Material creep	Individual atoms within the material have relative movement in relief of long-duration stress
Material yield	Whole layers of atoms or molecules displace in relative position
Microfracturing	Microscopic two-dimensional regions of separation occur
Macroscopic fracture	One or more large two-dimensional regions separate but by a relatively small amount
Material failure	Fracture separation across the whole material occurs
Material burst	Some energy that is stored because of stress is released by material ejection

[38]Adiabatic is from the Greek *adiábatos* for 'not passable', in this case referring to insulating walls stopping heat flow.

stress is sufficiently small, since there is a minimum quantum energy that must be exceeded to excite a bond between atoms.

The stress at the transition from elastic to inelastic behavior is called the 'elastic limit'.

If a material does not return to its original shape when the stress is gone, but has been left deformed, then it shows mechanical 'hysteresis'.[39] (See Fig. 3.15.)

A material left in a deformation state, after the external forces on the material are brought back to zero, may still contain internal stresses. These stresses can be relieved by 'tempering', i.e. the application of heat until atoms or molecules in the material are able to move in response to the stress. Some unusual materials show hysteresis at low temperature, but will restore themselves to their original shape when heated. When cold, the material can hold internal stress, and in this way retains some 'memory' of its original shape.

If internal relative displacement occurs in a solid under shearing stress that carry atoms past each other, the material is said to undergo 'plastic flow'. 'Creeping' is plastic flow that occurs over relatively long times while the material suffers stress. If there are internal bands of material slipping in relative position, we say the material is yielding. The stress is likely to be dissipated into heat as atoms move past each other. A microscopic slip band which opens into a microscopic gap is a 'microfracture'.

Materials with a buildup of microfractures are called 'fatigued'. Bone can become fatigued, but bone can also repair itself, unlike the metal in airplane wings.

The time scale for measurable plastic flow under natural stress can be from nanoseconds to geological ages, depending on the material.

The stress at which a material macroscopically separates is called the 'rupture strength'. Rupture can occur when microfractures join to form a 'macrofracture'. The limiting 'stretching strength' before rupture is called 'tensile limit'; under compression, 'compressive limit'; and under shear, 'shear limit'. When a material under stress breaks after rupture, it is said to have undergone 'material failure'.

A material which has relatively large stress-strain tensor-components is called 'stiff'. If the material fractures after significant stress but relatively little strain, we say it is 'brittle'. Materials that yield easily but do not fracture when stressed are said to be 'malleable'. A material malleable under tension is call 'ductile'. While under tensile stress, a ductile material tends to 'neck', i.e. layers of material move to make the cross-section smaller. A material which is easily bent is call 'pliable'. A brittle material yields to stress by fracturing instead of having atomic creep or slippage. Glass at room temperature is an example. A material can be brittle at one temperature and ductile at a higher temperature or at a lower pressure.

[39]Hysteresis is Greek for deficiency.

3.9 Viscoelastic Materials

If the stress energy in a material is transformed into heat through plastic flow, the material shows viscous behavior. If the stress energy is stored as potential energy by the distortion of molecular bonds, the material is acting elastically. Some materials show both properties. Such a material is called viscoelastic. The transition from elastic to plastic behavior depends on the material, on the rate of buildup of stress, on temperature, and on the background pressure. Viscoelastic materials can be modeled by allowing the stress-strain relation to include temporal rates and time integrals of the stress or strain.

If a material has no static limit in response to a stress, the material is a fluid, for which the strain tensor depends on the relative displacement per unit time rather than just relative displacements alone.

The dissipative effects of macromolecules in biological systems are those that cause a loss of mechanical energy into heat. For example, a vibrating chain molecule might lose energy by bumping into adjacent molecules. We say dissipation ‘damps’ mechanical systems. ‘Dampers’ are devices which introduce forces which oppose any motion, while dissipating the work-energy of the forces causing the motion. Mechanically, they can be represented by ‘dashpots’, a cylindrical device holding a fluid with a piston at one end of the cylinder, but the piston allows a small amount of fluid to flow past to relieve any pressure in the fluid. On the suspension of cars, they are called ‘shock absorbers’. Magnetic dashpots use a magnet on an element which moves relative to a nearby good conductor. The induced Faraday currents oppose the motion of the magnet, and any object attached. In fluid motion, friction both on the surface of fluid elements and by squeezing the volume of those elements can cause viscous heating.

Even the protoplasm inside every living cell has viscoelastic behavior, since the microtubules and microfilaments (‘cytoskeleton’) within tend to give the material some elastic behavior for small stress over short times, but protoplasm can change from a gel like system to a sol when stress is exerted over longer times.

We can model viscoelastic materials with combinations of springs and dampers (as in Fig. 3.9). For such models, each spring has a restoring force in proportion to the stretch and compression of the spring, while each damper has an opposing force in proportion to the stretch or compressional speed of the damper.

3.9.1 Kelvin-Voigt Model

A ‘Kelvin-Voigt’ viscoelastic material acts as a spring in parallel with a damper, i.e. the material is elastic for slow changes in stress, and viscous for fast changes. By balancing forces, if the Kelvin material is stretched by δx, then there will be an opposing force given by $\delta F = -k\delta x - \eta d(\delta x)/dt$. We see that without the spring ($k = 0$), the material acts as a viscous fluid, and that without the damper ($\eta = 0$), the material acts as an elastic spring. As a stress/strain relation, we have, for the Kelvin-Voigt case,

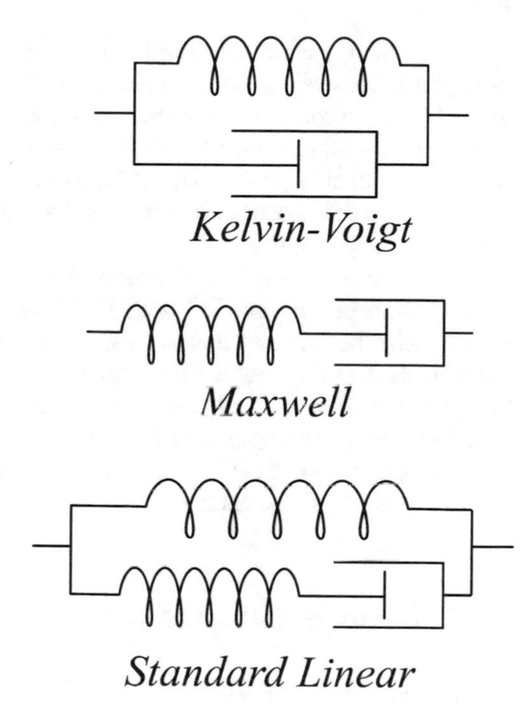

Fig. 3.9 Mechanical models for viscoelasticity. Top: A Kelvin-Voigt material; Middle: A Maxwell material; and Bottom: A 'Standard linear' viscoelastic material

$$S = Y\sigma + \eta \frac{d\sigma}{dt}\,, \tag{3.27}$$

where S is the stress and σ is the strain.

3.9.2 Maxwell Model

A 'Maxwell' viscoelastic material behaves as a spring in series with a damper, i.e. viscous for slow changes in stress but elastic for rapid changes. Using the fact that the displacements of the spring and the damper add to the displacement overall, δx, and that the force F between the spring and the damper must match with the external force F, the displacement of the material will satisfy $\delta x = (1/k)\delta F + (1/\eta)\int \delta F dt$. As a stress-strain relation, the Maxwell model has

$$\sigma = \frac{1}{Y} S + \frac{1}{\eta} \int S\, dt\,. \tag{3.28}$$

A Maxwell viscoelastic material is also called a 'rheid'. (Rheos means flow in Greek. Rheology is the study of material flow.) Maxwell viscoelastic materials can be brittle under a rapidly increasing stress, but plastic under a slowly increasing stress. Silly putty is a good example. It was originally made of a silicone polymer created in the effort to find rubber substitutes. Rock furnishes another example: Rock can flow under constant stress, but also it supports elastic sound waves. The bone around teeth respond to lateral forces by reforming. The bone around the inner ear is less viscous and more elastic than bone elsewhere, in order to help preserve the sound energy within.

At a molecular level, a viscoelastic material might consist of long chain molecules that become tangled, or that have weak side bonds if they are stretched out next to each other. A quick and impulsive stress will likely cause the entanglement or weak bonds to stretch, but not necessarily break. However, if the stress is sustained, the entangled chains can slip and the weak bonds break from thermal action on the stretched ones. After segments of the chains slip, they can sequentially re-establish new entanglements or new hydrogen bonding, which, under continued stress, is seen macroscopically as viscous flow.

3.9.3 Standard Linear Model

A 'Standard Linear model' for a material puts a Maxwell element in parallel with a spring, as shown in Fig. 3.9. For a 'standard linear' material, the relationship between stress and strain can be found from the constraints:

- The force of the material back on the external agent is the sum of the force due to the first spring, #1 (top one in Fig. 3.9), and the force of the second spring, #2 (bottom one in Fig. 3.9);
- The force of spring #1 is proportional to its compression;
- The force of spring #2 is proportional to its compression;
- the force of the dashpot is proportional to its compressional speed;
- the force of the second spring is the samc as that of the dashpot;
- the sum of the displacements of the second spring and the dashpot is the displacement of spring #1.

Thus, for the stresses and strains, we have

$$
\begin{aligned}
S &= S_1 + S_2 \\
S_1 &= Y_1\, \sigma \\
S_2 &= Y_2\, \sigma_2 \\
S_3 &= \eta\, d\sigma_3/dt \\
S_2 &= S_3 \\
\sigma &= \sigma_2 + \sigma_3
\end{aligned}
\tag{3.29}
$$

From these, the stress S is related to the strain σ by

$$S + \frac{\eta}{Y_2}\frac{dS}{dt} = Y_1\sigma + \eta\left(\frac{Y_1 + Y_2}{Y_2}\right)\frac{d\sigma}{dt} . \tag{3.30}$$

In this relation, there are two time scales. A shorter one given by

$$\tau_s = \frac{\eta}{Y_2} , \tag{3.31}$$

and a longer one given by

$$\tau_l = \frac{\eta}{Y_1} + \frac{\eta}{Y_2} . \tag{3.32}$$

These times determine how quickly a material responds if the stress is suddenly or slowly changed.

Note that the Standard Linear model contains both the Kelvin-Voigt model ($Y_2 \to \infty$) and the Maxwell model ($Y_1 \to 0$) as limits.

Neither the Kelvin-Voigt nor the Maxwell models work well when applied to the intervertebral discs of the human spine. Maxwell's model allows arbitrary large creep not characteristic of discs. The Kelvin-Voigt model does give a finite creep for a fixed stress, but does not well represent stress relaxation. The 'Standard Linear' model combines some features of both the Kelvin-Voigt and the Maxwell model, by adding a spring in series with a parallel spring-damper. This model can be used for spinal intervertebral discs, although good fits to data require a series of such combinations.

Researchers building more accurate mechanical models representing real biological materials such as bones, ligaments, films, and other quasi-rigid structures invoke a network of many springs and dampers in three dimensions. The models may also require a non-linear stress-strain relation if the strain grows too large. This follows from the fact that intermolecular forces become non-linear as the molecules are forced to separate or compress beyond the parabolic behavior of their interaction energy.

3.10 Bones, Ligaments, and Other Structures

To build organs optimized for certain functions, living systems have taken advantage of structural hierarchy, wherein at intermediate scales of size, each unit has substructure.

A major distinction with non-living materials is that living systems are adaptive by having active control over the behavior of their structures, and have repair mechanisms at the nanoscale. For example, bone under stress can be made stronger in the direction of the stress (to a limit), and purposefully yield (as in jaw bone under the side pressure of teeth). Our bridges and buildings do not yet have nanoscale repair, nor do they grow stronger at locations of greatest stress.

The strongest structures made from atoms have those atoms locked in atomic bonds. However, most ordinary solids have inhomogeneities due to being a mixture of differing materials or irregular arrays of crystals. Even single crystals grown from a single solute in solution tend to have nanoscale defects and atomic dislocations.

As a carbon atom can bond strongly to other carbons sharing one to four valence electrons with its neighbors, atomic structures built by linking many carbon atoms, such as in fullerenes and bucky tubes, are the basis of many of the strongest materials known.[40]

3.10.1 Stress and Strain in Bone

Bone is a composite material, made from collagen fibers embedded in an organic matrix containing crystals of calcium hydroxyapatite,[41] making bone reach a stiffness of $Y = 5$ to 30 giganewton per square meter, which is about 10 to 30 times the stiffness of collagen (See Table 3.4). Bone density is about 2 g per cubic centimeter. Increasing the mineral content of bone by 20% can increase its stiffness by 230%. A bone's strength roughly depends on its mineral density squared. However, this also makes the bone more vulnerable to breakage, because microfractures can more easily grow when mineral density is high. The petrous temporal surrounding the cochlea and middle ear ossicles is stiff in order to reflect and transmit sound without absorption, but is not as strong as leg bone.

Vertebrae fracture under compression near 200 meganewton per square meter; bones fail under tension at about 140 meganewton per square meter, and fracture under shear near 70 meganewton per square meter. Live bone, given sufficient time, can repair microfractures, such as those which might be produced in the lower limbs from vigorous running or jumping. Ground reaction forces on bones can reach 12 times body weight during jumping for hard ground. Live bone is an adaptive tissue capable of repair, regeneration, and reconfiguration according to its history of stress. With sufficient time, microdamage is repaired by osteoclasts (cells which reabsorb bone) and osteoblasts (cells which regenerate bone). After repeated stress, the bone material grows stronger, within limits, in the direction of the stress.

[40]Buckminster Fuller (1895–1983) was an architect, inventor, and philosopher who made innovative structures, including a geodesic dome built from triangles.

[41]$3Ca_3(PO_4)_2.Ca(OH)_2$.

Table 3.4 Stress and its limits for various materials

	Y	n_s	Tensile limit	Compr. limit	Shear limit
Steel	200	84	400	0.5	0.25
Glass	60	25	33	0.7	
Silicone elastomer	4		36		
Cartilage	12	8.3			
Cortical bone	5–30	3.2	140	200	
Tibia	12–21				140
Fibula					
Hair	0.2		200		
Erythrocyte cell wall		2.5 μ			
	Stress in megapascal except $\mu - 10^{-6}$ Pa				

Bone	Breaking torque	Twist breaking angle
Leg		
Femur	140 Nm	1.5°
Tibia	100 Nm	3.4°
Fibula	12 Nm	35.7°
Arm		
Humerus	60 Nm	5.9°
Radius	20 Nm	15.4°
Ulna	20 Nm	15.2°

3.10.2 Shearing a Femur

A long piece of blackboard chalk[42] can be used to model what happens to our leg bone when twisted too far (See Fig. 3.10). Breaking the chalk by a twist makes a fracture in the chalk which looks very similar to the X-ray image of a femur broken by a twist, such as that which might happen in a skiing accident.

Consider a long vertical cylindrical homogeneous tube, clamped at its lower end and twisted by a torque at its upper end. Figure 3.11 shows a very thin cylinder representing a part of a bone of radial thickness dr, with a torque acting on the cylinder produced by the small tangential forces on the top surface, with an equal but opposite set of tangential forces (not shown) on the bottom surface to keep the cylinder from rotating. We will assume the bottom surface is held fixed while the twist is being imposed. Suppose the cylinder undergoes an angular twist of Φ. A small quasi-cubical element of the cylinder is drawn, with edges of lengths dr, dz, and $rd\phi$. Adjacent material elements act on this cube to produce a shear, shifting its top in the direction of increasing ϕ relative to its bottom. Newton's 2nd and 3rd law tells us that, under static conditions, the same shearing torque acts on each slice of

[42] Yes, I still prefer chalk talk over power points, without referring to notes. During the creation of chalk expressions and drawings, both the students and the professor have some time to think.

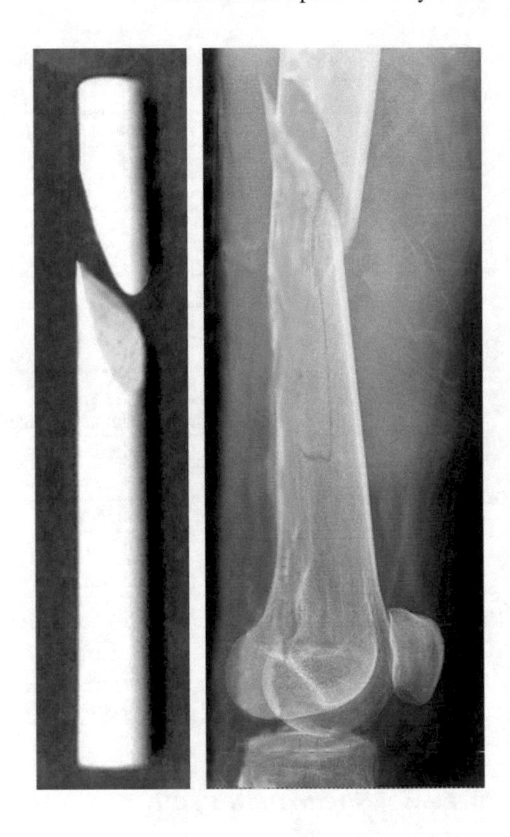

Fig. 3.10 The result of twisting a piece of chalk beyond the fracture limit compared to a femur spiral fracture

thickness dz throughout the length of the cylinder. The magnitude of the torque on this thin cylinder element will be the sum of the twisting forces times the radius out to the element. Suppose the quasi-cubical element selected has a tangential force d^2F acting on its top surface (and an opposite one on its bottom). The angle of shift is shown in the diagram and labeled θ. As can be seen in the diagram, $r\Phi/L = \theta$. By definition of the shear modulus n_s,

$$\frac{d^2F}{dr\,rd\phi} = n_s\,\theta = n_s\,r\Phi/L\ . \tag{3.33}$$

As a vector relation, this reads

$$d^2\mathbf{F} = n_s\frac{\Phi}{L}\,rdr\,d\phi\,\widehat{\mathbf{n}}\times\mathbf{r} \tag{3.34}$$

where $\widehat{\mathbf{n}}$ is a unit vector along the axis of the cylinder. We now can express the small torques on the infinitesimally thin cylinder as

$$d\boldsymbol{\tau} = n_s\frac{\Phi}{L}\,r^3dr\int_0^{2\pi}d\phi\,\widehat{\mathbf{n}}\ . \tag{3.35}$$

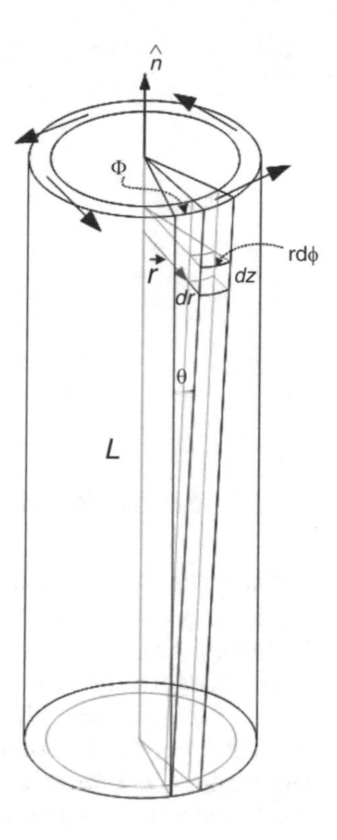

Fig. 3.11 A twisted cylindrical bone segment. Note small cubical element undergoing a shear

All lie along the cylinder's axis, and can be added in magnitude to find the net torque for a cylinder of finite thickness. If the outer radius is b and inner radius a, the result is

$$\tau = n_s \frac{\pi \Phi}{2L} (b^4 - a^4) . \tag{3.36}$$

As we can see from Eq. (3.36), a cylindrically-shaped bone with strength enhanced on the outer surface is far stronger under a twist than a corresponding one of equivalent average strength uniformly distributed over the radius of the bone. However, the extreme case of a hollow bone is not optimal, because once a fracture begins, the bone will collapse. For a given mass with limited strength, construction into a round cylinder with its highest density and strength at large radii and with a lighter but stiff material in the supporting core gives the greatest resilience to stress, not only for shearing strength, but also for bending, wherein the outer surfaces are compressed and stretched. For long bones, the cortical region contains dense osseous tissue, while less dense cancellous (spongy) bone forms near the center. This strategy is evident in the bones of flying birds, who particularly benefit by high strength to mass.

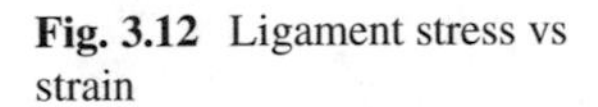
Fig. 3.12 Ligament stress vs strain

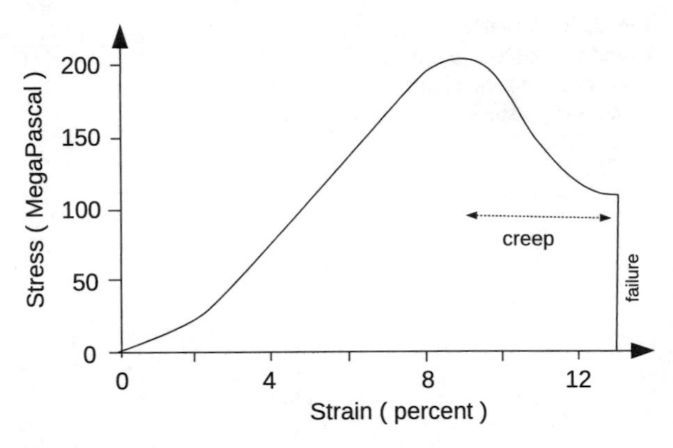

3.10.3 *Stress-Strain in Ligaments and Tendons*

Unlike bone, fasciae, ligaments, and tendons[43] all show an initial non-linear relationship between stress and strain under ordinary physiological conditions. Moreover, they show hysteretic and viscoelastic behavior. Figure 3.12 displays a graph of a typical response of a ligament to stress. The low slope in the curve at low stress (below 2% strain) indicates that the ligament stretches easily at first. This is due to the 'uncrimping' of molecular chains for low stress. When the chains become less crimped, the force needed to stretch them requires stretching intramolecular bonds, rather than acting against intramolecular bending forces, and therefore are significantly larger. At high stress, the bonds can no longer hold, and the ligament elongates by slippage (creep) between strands in the ligament. Eventually, the ligament fails.

Figure 3.13 shows the effect of Maxwellian viscoelastic creep in a ligament. Creep relieves stress, as shown in Fig. 3.14. The hysteresis of the ligament is evident in Fig. 3.15.

While walking, our Achilles tendon suffers a stress of four times our body weight, but can withstand a force of twelve times our weight or more. This follows from the fact that the ankle pivot point in the foot is typically four times the distance to where the ball of the foot makes contact with the ground compared to the distance from the pivot to the attachment of the tendon. As the tendon has a smallest diameter of about 6 mm, the stress in the tendon is about 70 million N/m^2 (70 MPa) for a 50 kg person during walking, and can fail at 210 million N/m^2. (The tendons of athletes become stronger principally by enlargement of the tendon's diameter.)

[43]Faciae, ligaments, and tendons are all made from collagen fibers. Faciae connect muscles to muscles, tendons connect muscles to bone, and ligaments connect bone to bone.

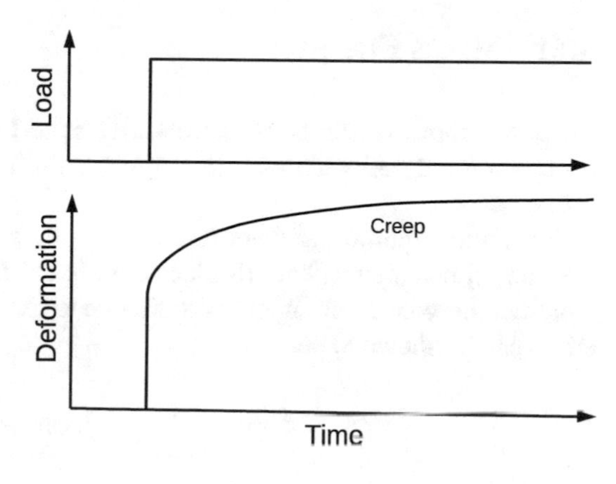

Fig. 3.13 Ligament creep

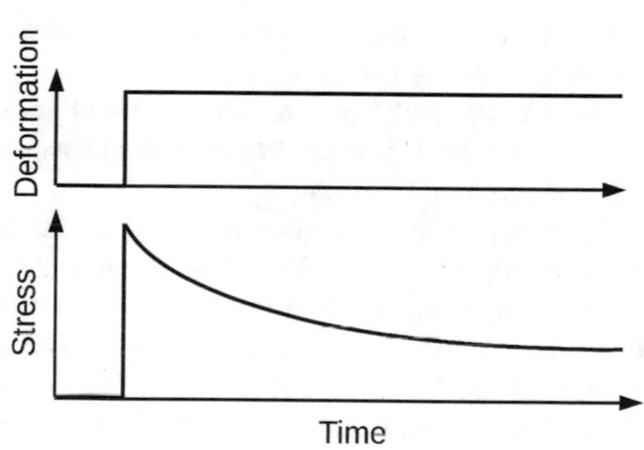

Fig. 3.14 Ligament relaxation

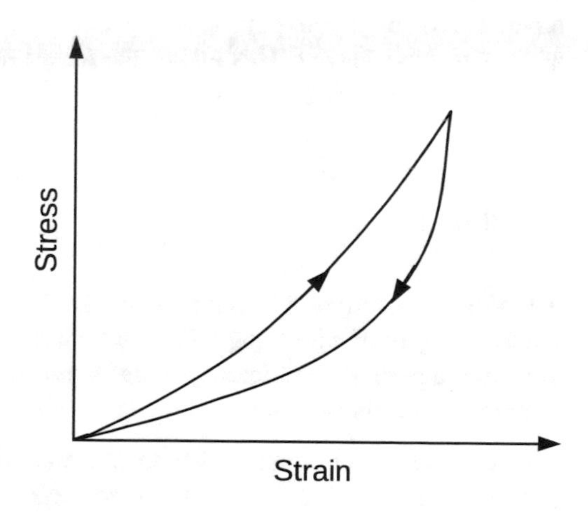

Fig. 3.15 Ligament hysteresis

3.11 Stress Energy

Materials under elastic stress necessarily store energy. Some of that energy can be subsequently released as work. Springs and elastic bands are famous for this property.

From the definition of work done on a body as the sum of the forces acting on the body times the resultant displacement in the direction of the force, the energy stored in the volume of an elastic solid due to the material of the solid being under stress can be shown to be

$$\mathcal{E} = \frac{1}{2}\int_{\text{body}} \sum T_{ijkl}\sigma_{ij}\sigma_{kl}\, dV \,. \tag{3.37}$$

This relation is the generalization of the more familiar expression for the energy stored in a spring, $(1/2)k\, x^2$.

Your bones and ligaments store a small amount of such 'spring' energy when compressed, bent, twisted, or put under tension. You get some of this energy back when the material relaxes.

However, if the compression of a material produces heat, the process will not act perfectly elastically. The stress-strain curve during loading over a finite time matches or is higher than the stress-strain curve during unloading back to the original no-strain condition, as in Fig. 3.15. The area between these two curves is the energy released per unit volume of material. The released energy is usually in the form of heat. Mammalian hair shows this behavior. It is composed largely of keratin, as are hooves and horns. The yielding behavior comes from a conversion of protein alpha helices to protein beta sheets as the keratin is stretched, and conversion back as the stress is relieved. This makes the material tough under stresses not ordinarily encountered by biological tissue.

Problems

3.1 What is the minimal amount of work a 55 kg person needs to climb the steps to the fourth floor of a building with 3 m high stories? If the person's muscles operate at an average of 20% efficiency, what minimal number of kilocalories would they require for the climb?

3.2 A typical adult human has a basal metabolism rate (BMR) of about 2200 kcal per day. Translate this figure into power expended in watts. Kleiber found that the BMR per unit mass of a wide variety of animals scales as their mass to the negative one fourth power. Estimate the BMR of a canary with a mass of 28 g.

3.3 Bacterial flagella are powered by an electrochemical motor. Why is such a motor not limited in efficiency by the thermodynamic maximal efficiency of engines? If the flagellum motor is powered by 7.8×10^4 protons per second passing through a potential difference of 140 mV, causing the flagellum to rotate at 70 times per second, what torque does the motor develop?

3.4 If a spacecraft in the shape of a doughnut, of radius 400 m, is spun around its natural axis, what angular speed is needed to make humans feel natural gravity standing on the inner shell of the doughnut?

3.5 Apply Stokes' Law in the form $F_v = b\,v$ to a person of mass 50 kg standing against the 44 m/s wind of a Hurricane. If the center-of-mass of that person is 1.2 m high and 0.4 m shifted horizontally from the pivot point at the feet, what is the air drag coefficient b for that person? (Assume that the effect of the wind on the person is the same as a force F_v on the center-of-mass.)

3.6 Determine the sedimentation speed for a hemoglobin molecule in a centrifuge with the molecule at a radius of 20 cm from the axis and a rotational acceleration of 100,000 g. Take the hemoglobin molecule as having a mass of 8000 Da and a radius of 0.0027 microns, and the plasma having a density 1.025 g per cubic centimeters and a viscosity of 0.015 poise.

3.7 The wind on Mars can reach 27 m/s (60 mph). The temperature of that wind in the midday Sun is about 20 °C (70 °F) but the atmosphere is only 1% as dense as at the surface of the Earth at the same temperature. For simplicity, suppose that a man's effective surface area of 0.5 m^2 stops all the momentum of the wind hitting it. Show that the force depends on the square of the speed of the wind (unlike the Stokes' drag force). Estimate the force of that Martian wind on a man standing against it.

3.8 An air bubble of diameter 0.45 mm is observed at a depth of 3.6 cm in blood plasma. Estimate the pressure in that bubble.

3.9 For a sitting person with a cuff pressure of 120/80 mmHg, what would you predict for the pressure in the carotid artery?

3.10 Amusement parks have rides which produce up to '4g' on the riders. If a 2 m high person's body is along the direction of the acceleration of the ride, what would be the difference in blood pressure from his head to toe?

3.11 If the compliance C of an aorta is measured to be 2.67×10^{-4} ml/mmHg, what change occurs in its volume as the heart creates a systolic gauge pressure of 120 mmHg compared to a diastolic gauge pressure of 70 mmHg. Estimate the extra elastic energy stored in the walls of the aorta at systole?

3.12 If a certain astronaut's lung has a compliance of 140 ml/mmHg, how much larger might his lung expand elastically if, during a space walk, his spacesuit loses pressure while he holds his breath?

3.13 Suppose the cervical region of the spine of a woman holds up her head, of mass 4.5 kg. The intervertebral discs make up about 25% of a length 11.5 cm for the cervical region, containing six such discs of cross-sectional area 145 mm^2.

Determine the stress on the discs when the woman stands. During the day, after 4 h, the strain of the cervical region has become 2%, which is 60% of its long-period strain. The spine expands back to its original length during the night when the body is prone. There is little compression of the vertebrae between the intervertebral discs.

Fit this data to a Kelvin-Voigt model of the intervertebral discs.

Plot the predicted strain as a function of time (in hours) during the day.

3.14 A 45 kg woman jumps from a height of 2.5 m to a solid floor. In an attempt to minimize the likelihood of injury, she lands on the ball of her feet (the pads just behind her toes) with her knees bent. What average force is needed on her feet to stop her fall if her center-of-gravity after touching the floor drops 80 cm before she stops? If the horizontal distance from the ball of her foot to her talus (where her tibia exerts its force) is 15 cm, and from talus back to the Achilles tendon is 3.6 cm, what average force does the Achilles tendon suffer during the stopping time?

3.15 The bone around your teeth behaves somewhat like a Maxwell viscoelastic material. When stressed by the pressure of a tooth, the bone, over time, moves to relieve the stress. Suppose we find that the stress on the bone next to a tooth drops linearly with the strain until it reaches zero after some time. Show that the stress drops exponentially with time, and the strain approaches a constant exponentially.

Chapter 4
Fluid Mechanics Applied to Biosystems

I keep the subject constantly before me and wait till the first dawnings open little by little into the full light.

—Isaac Newton

Summary Life on Earth could not be active without fluids. The behavior of ordinary fluids, whether in use by a life form, or otherwise in nature, follows Newtonian principles. Here we look at the consequences of mechanics applied to fluids, including the oceans, the atmosphere, and our own blood.

4.1 Nature of Fluids

A fluid is a material which undergoes macroscopic movement under stress or reaches a limiting boundary. Both liquids and gases are fluids. Apart from weak long-range forces such as gravity, gases expand by outward thermal pressure until they reach a bounding surface. Even without a container, liquids have an equilibrium volume produced by their own internal attractive molecular interactions. We can understand the distinction between a gas and a liquid by the interaction of the molecules therein.

For a gas, the motion of its molecules carries, on average, more kinetic energy for a pair than the energy which might be released by the binding of that pair of molecules. If the slower ones happen to stick together, the faster ones break them apart by collisions, thus maintaining the gaseous state at the given temperature. The collisions of molecules with the walls containing the gas produce the observed pressure.

By contrast, for a liquid, intermolecular forces tend to hold molecules as neighbors. Any shearing force acting on the liquid causes movement of layers of fluid against adjacent layers. A liquid at rest can exert pressure on its containing walls, this pressure coming from the repulsive forces acting between molecules when they are pushed against a wall and from molecules thermally jostling against the walls. However, because fluids move under stress, a liquid at rest cannot exert

W. C. Parke, *Biophysics*, https://doi.org/10.1007/978-3-030-44146-3_4

a shearing force internally on a bounding surface. When you are at rest under still water, the water can exert pressure on your skin, but will not push your skin sideways.

As we have described, a colloid, i.e. a mixture containing nanoscale structures having a significant surface area, can sometimes act as a solid, called a sol, and other times as a liquid, called a gel. The transition between a sol and a gel depends on temperature, pressure, and the rate at which a stress is imposed.

4.2 Laws for Fluids at Rest

A fluid may be sometimes modeled as a smoothed distribution of individual masses, each one able to move relative to others. It should be no surprise to find that the motion of fluids can be described by the same principles used to follow the behavior of interacting masses. To do so successfully, however, the segments of the fluid must be taken small enough that the velocities of components within these segments do not deviate much from the average over all the components, and large enough so that averages are meaningful. To use Newton's laws, the segments must not move relativistically, and their subparticle quantum waves must not significantly extend beyond the molecular level, so that quantum dynamics is well approximated by Newtonian dynamics for the fluid segments. For values of macroscopically measurable quantities such as pressure within a given segment not to deviate much from their average over the components of that segment, the number of molecules in the components must be large. Moreover, these molecules must interact sufficiently that the spatial averages within a segment approximate averages over short time spans.[1]

4.2.1 Pascal's Law for Pressure in a Fluid

Those of us who have dived into deep water likely know that water pressure increases with the depth of the water, based on how our ears react. As we go deeper, we experience pain in our middle ear from the pressure on our eardrum. We also may have observed that the extra pressure is felt irrespective of the angle of our head. A pressure gauge rotated under water reads the same value.

In 1646, Blaise Pascal investigated how pressure acts in fluids, and found that pressure at any one location in a fluid was independent of the orientation of the surface used to measure that pressure, and that if the pressure in a fluid is changed, that change is felt throughout the fluid by the time the fluid returns to rest. This discovery, known as Pascal's principle, is the result of the physical properties of a

[1] These 'short' time spans are the ones used to define time derivatives of the measurable quantities.

fluid under pressure. Wherever a local change of pressure occurs in the fluid, the nearby fluid must respond by moving, and each subsequent more distant layer of fluid must move by the extra pressure of the prior layer. If the fluid returns to static conditions, there will be new pressure forces on each element of fluid, but with no shearing forces.

Suppose a small triangular object, with sides of length a, b, and c, and some thickness d, were placed in the fluid. Under static conditions in a fluid, the forces on any surface of the object can only be pressure forces perpendicular to the object's surfaces. The three pressure forces on the sides must add to zero, so they also form a triangle, with the same angles as those of the physical triangle. The sides of these two similar triangles must be proportional: $F_a/a = F_b/b = F_c/c$. Since $F_a = p_a\, a\, d$, $F_b = p_b\, b\, d$, and $F_c = p_c\, c\, d$, the pressure $p_a = p_b = p_c$ is the same on all sides of the triangle, no matter its orientation. At a given depth of water, the pressure on any side of a fish will be nearly constant. However, the 'high' side of the fish will have less pressure than the 'low' side, being at a shallower depth. The result is buoyancy.

A balancing of forces on each fluid element throughout the fluid leads to Pascal's principle: If A and B are two locations where a change in pressure has occurred, then, after static conditions have returned,

$$\Delta p_A = \Delta p_B \,. \tag{4.1}$$

For a fluid which returns to rest, any change in pressure at one location will be reflected by the same pressure change elsewhere in that fluid.

For a fluid at rest held by gravity, such as our atmosphere and oceans, the pressure increases with depth. This is seen by balancing the forces on an element of the fluid at a given height. Taking a rectangular volume of height dz and area A, the pressure on the bottom of the volume must be greater than that on the top because of the weight of the material in the volume.[2]

Suppose the fluid has a density ρ at a given height z. Balancing forces on the fluid element gives

$$p(z)A - p(z + dz)A = \rho g dz A \,, \tag{4.2}$$

or

$$-\frac{dp}{dz} = \rho g \,. \tag{4.3}$$

[2]Gas does have weight, because the Earth's gravity accelerates gas molecules downward, and decelerates them upward. This makes the impulse of the molecules on the top of the volume less than that on the bottom. If the gas were held in a container and weighed, the extra impulse of the gas between the top and bottom surfaces causes the container to press further on the scales.

To integrate over height in the atmosphere, we need to know how air density depends on air pressure. For air modeled as an ideal gas at constant temperature, this would be $\rho_a = (m/k_B T)p$, where m is the average mass of a molecule in the air. To account for the decrease of the acceleration of gravity with height, we use Newton's Law of Gravity to express

$$g(z) = g_0 \frac{1}{(1 + z/R)^2} . \tag{4.4}$$

If the temperature were constant, then integration would give

$$p = p_o \exp\left(-\frac{m g_0}{k_B T} \frac{z}{1 + z/R}\right) , \tag{4.5}$$

where R is the radius of the Earth, z is the height above the Earth's surface, and p_o is the atmospheric pressure at sea level. A similar expression applies for the density of air. But the temperature is not constant. A plot of the temperature variation with height is shown in Fig. 16.1. As $T(z)$ is rather complicated, numerical integration of the relation dp/dz to find the pressure with height is used. Direct measure of pressure above the Earth can be performed by weather balloons up to 40 km (in the stratosphere). The Earth's atmosphere makes a transition to the solar environment at about 100 km. Above about 120 km, solar wind dominates the temperature, which can reach greater than 1000 °C above 150 km. However, the atmosphere is so thin there that little energy is available to transfer as heat to astronauts who leave their spaceship (as 'space walkers').

Our inner ear is sensitive to air pressure when the Eustachian tube is not open. Airplane travelers are quite aware of this sensitivity. It is caused by the fact that the atmospheric pressure diminishes with height. The air behind the ear drums are connected to the pharynx (upper throat) by the Eustachian tubes. When the tube is open, the pressure of the air in the inner ear is the same as atmospheric. However, the tube is normally closed until you strongly swallow. If you have not, then going up in the airplane makes a difference in pressure between the middle ear and the inner ear, causing the ear drum to distort outward. As a protective mechanism for the drum, nerves associated with the drum send a pain signals to your brain. Swallowing can open the Eustachian tubes from the nasal cavities to the middle ears, letting air move to equalize pressure on the eardrum.

For water in the oceans, the density ρ is nearly constant. (The increase of water density from the surface to a depth of 1000 m is only about 0.3%.) If taken to have a constant density, the Eq. (4.3) for the pressure a distance $|z|$ below the surface of a fluid integrates to

$$p = p_o + \rho g \, |z| . \tag{4.6}$$

Divers must take caution to let air out of their lungs as they rise from the depths, so that alveoli in the lungs are not ruptured.

4.2.2 Pascal Applied

The original application of Pascal's principle was to 'hydraulics', the study of the mechanical aspects of liquids.

Consider a force F_1 pushing against a piston of area A_1. In turn, suppose that piston pushes on a contained fluid and that the fluid is connected through a tube to act on larger piston of area A_2, held by an opposing force F_2. Pascal's Principle gives $F_1/A_1 = F_2/A_2$. The resultant force is magnified by A_2/A_1. There is no such principle as the conservation of force. But there is conservation of energy. If the first piston displaces a distance x_1 and the second x_2, then $F_1x_1 = F_2x_2$, so $x_2 = (A_1/A_2)x_1$, i.e. the displacement has been reduced when the force is magnified. Hydraulics is a useful replacement of levers to magnify forces, because fluid lines can be flexible, small in diameter, and a long distance from the originating source.

A number of biological systems use hydraulics to apply forces. Starfish (sea stars) use hydraulics to control their five radial feet and cups on those feet to attach to a clam. Hydraulic pressure is generated by muscles that squeeze water-filled ampoulae connected to canals extending through their feet and into their podia cups. With valves, the pressure can be maintained with little muscle effort. Sea stars can win the battle with clams, because the sea star does not have to expend much energy to keep an opening force on the two sides of a clam, but the clam requires energy to keep its shell muscles under tension. The clam eventually tires and succumbs.

A Venus fly trap uses hydraulics in the form of osmotic pressure to put its bimodal leaves under flexure tension, and to trigger the closure of those leaves onto a fly in only about 1/10th of a second. Amazingly, touching only one of the trigger hairs does not cause the reaction. Touching one twice within a second, or touching one, and then the second, will stimulate closure (Fig. 4.1).

The human arm from the skin surface down to the bone contains a large fraction of water held by elastic and pliable membranes. If the pressure on the arm is increased, that pressure increase is transmitted throughout the fluids of the arm and into the membranes that hold fluids. Among these fluids is the blood in arteries.

A sphygmomanometer is a device which uses a cuff to determine blood pressure. Whatever pressure is measured in the cuff air bag is the same as in the blood vessel. The cuff is inflated until a major artery is constricted to close. As the pressure is slowly released, there will be some pressure transmitted by the heart which pushes blood through the constriction, causing a turbulent sound which can be detected by listening through a stethoscope. During continued pressure release, the sound stops. At this lower pressure, the heart-generated pressure is sufficient to move blood through the artery and maintain its open condition. Then, far less noise is generated.

For a hypodermic needle, if the pressure is increased on the plunger, the pressure of the fluid in the needle is correspondingly increased. However, viscous drag within the needle results in a pressure near atmospheric when the fluid enters the body, since this is the pressure on the surface of the skin, and Pascal would tell us this pressure is communicated across all soft membranes.

Fig. 4.1 A Venus flytrap applying hydrolics to capture a fly

4.2.3 Archimedes' Principle and Buoyancy in a Fluid

The increase of pressure with depth in any fluid near the Earth accounts for the buoyant force on a volume in the fluid.

Consider a body at rest partially or fully submerged in a liquid, with air above. Pressure forces will exist as surface forces on the body, and gravity will act on all elements of mass within. For mechanical equilibrium, the net pressure force,

$$\mathbf{B} \equiv -\int p\,\widehat{\mathbf{n}}dA\ , \tag{4.7}$$

called the buoyant force, should balance the weight of the body. The negative sign in the buoyancy expression comes about because the pressure points inward on the surface, while the unit vector normal to the body surface, $\widehat{\mathbf{n}}$, is defined as pointing outward. In equilibrium, the net force on the body must vanish:

$$\mathbf{B} - W\widehat{\mathbf{k}} = \mathbf{0}\ , \tag{4.8}$$

where

$$W \equiv \int \rho g dV\ , \tag{4.9}$$

$\widehat{\mathbf{k}}$ is a unit vector directed upward, g is the acceleration of gravity, and ρ is the density of the body.

In addition, in order to prevent rotation, the net torque due to these forces must vanish. We know that if the net force vanishes, the torques around any axis must vanish, i.e.

$$-\oint \mathbf{r} \times p\,\widehat{\mathbf{n}}dA - \mathbf{R}_{cm} \times W\widehat{\mathbf{k}} = \widehat{\mathbf{0}} \tag{4.10}$$

for an arbitrary choice for the coordinate origin. The vector $\mathbf{R}_{cm}$ is the coordinate of the center of mass of the body.

We will construct a vector, $\mathbf{R}_B$, called the position of 'center of buoyancy' measured from the center of mass, by making the net torque due to buoyancy expressible as the torque of the single force $\mathbf{B}$:

$$-\oint \mathbf{r} \times p\,\widehat{\mathbf{n}}dA \equiv \mathbf{R}_B \times \mathbf{B}\,. \tag{4.11}$$

The surface integrals are taken over the whole body, including the part not in the liquid. Imagine dividing the surface of the body into two closed surfaces, by first cutting the volume along the plane of the liquid surface and then adding two fictitious horizontal surfaces within the cut, both infinitesimally close together at the level of the top of the liquid, with the lower one enclosing the submerged body and the top one enclosing the part of the body not in the liquid. On these surfaces we will imagine putting equal but opposite atmospheric pressure. The new pressures on the whole body cancel, so they will not affect the equilibrium conditions of the whole body. We will then have

$$\mathbf{B} = -\oint p_a\,\widehat{\mathbf{n}}dA - \oint p_L\,\widehat{\mathbf{n}}dA \tag{4.12}$$

and

$$\mathbf{R}_B \times \mathbf{B} = -\oint \mathbf{r} \times p_a\,\widehat{\mathbf{n}}dA - \oint \mathbf{r} \times p_L\,\widehat{\mathbf{n}}dA\,. \tag{4.13}$$

Inserting the dependence of the pressure in the liquid and the air on height, Eq. (4.6), the above relations give

$$\mathbf{B} = -p_o \oint_{out} \widehat{\mathbf{n}}dA + \rho_a g \oint_{out} z\,\widehat{\mathbf{n}}dA - p_o \oint_{in} \widehat{\mathbf{n}}dA + \rho_L g \oint_{in} z\,\widehat{\mathbf{n}}dA \tag{4.14}$$

and

$$\mathbf{R}_B \times \mathbf{B} = -\;p_o \oint_{out} \mathbf{r} \times \widehat{\mathbf{n}}dA + \rho_a g \oint_{out} z\mathbf{r} \times \widehat{\mathbf{n}}dA$$

$$- p_o \oint_{in} \mathbf{r} \times \widehat{\mathbf{n}} dA + \rho_L g \oint_{in} z\mathbf{r} \times \widehat{\mathbf{n}} dA \,. \tag{4.15}$$

The subscript on the integral signs, 'in' or 'out', refer to integrations over the surface of the body segments in the liquid or out of the liquid. The evaluations of the closed-surface integrals are simpler if we apply Gauss' expression $\oint \mathbf{B} \cdot d\mathbf{S} \equiv \oint \nabla \cdot \mathbf{B} \, dV$ to each, in the forms $\oint F\widehat{\mathbf{n}} dA \equiv \oint \nabla F \, dV$ and $\oint \mathbf{F} \times \widehat{\mathbf{n}} dA \equiv \oint \nabla \times \mathbf{F} \, dV$. (See Appendix G.5). In this way, we see that several of the integrals vanish: $\oint \widehat{\mathbf{n}} dA \equiv \mathbf{0}$ and $\oint \mathbf{r} \times \widehat{\mathbf{n}} dA \equiv \mathbf{0}$. We are left with

$$\mathbf{B} = g \left(\rho_L V_{in} + \rho_a V_{out}\right) \widehat{\mathbf{k}} \tag{4.16}$$

and

$$\mathbf{R}_B \times \mathbf{B} = g \left(\rho_L \oint_{in} \mathbf{r} dV + \rho_a \oint_{out} \mathbf{r} dV\right) \times \widehat{\mathbf{k}} \,. \tag{4.17}$$

Equation (4.16) contains Archimedes Principle, but he stated the relation in ancient and elegant Greek. A fair translation is "The buoyant force on a partially or fully submerged body equals the weight of the displaced fluid." In Eq. (4.16), both the water and the atmosphere generate buoyancy on the body.

Defining the position of the 'center of pressure' by

$$\mathbf{R}_C \equiv \frac{\rho_L \oint_{in} \mathbf{r} dV + \rho_a \oint_{out} \mathbf{r} dV}{\rho_L V_{in} + \rho_a V_{out}} \tag{4.18}$$

we can write

$$\mathbf{R}_B \times \widehat{\mathbf{k}} = \mathbf{R}_C \times \widehat{\mathbf{k}} \,. \tag{4.19}$$

There follows the connection,

$$\mathbf{R}_B = \mathbf{R}_C + Z\widehat{\mathbf{k}}, \tag{4.20}$$

where Z is arbitrary. As Z changes, this relation defines a line of buoyancy. If a completely submerged body is rotated, a new line of buoyancy is defined. The intersection of these lines, where $Z = 0$, defines the center of buoyancy, which matches the center of pressure: $\mathbf{R}_B = \mathbf{R}_C$, given by Eq. (4.18).

In the case of water and air, because the density of the water is about 830 times that of air, we can usually drop the smaller air term. In that case, the center of pressure is given simply by $\mathbf{R}_C = (1/V) \oint_{in} \mathbf{r} \, dV$, i.e. the geometric center of that part of the body under the liquid.

For a body only partially submerged, the center of buoyancy is not fixed as the body is tilted, but depends on the shape of the volume below the level of the liquid. (See Fig. 4.2.)

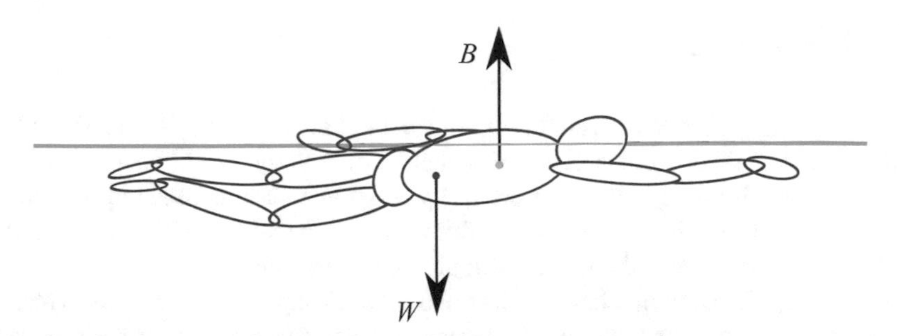

Fig. 4.2 The buoyant force on a swimmer is shown acting at the center of buoyancy and the weight of the swimmer is shown acting at the center of mass. The resulting torque can be opposed by torques produced by movement of the swimmer's hands and feet

A low density box can be floating upright and be stable, even with its center of buoyancy above its center of gravity. A denser box may tilt when placed in water, and float stably at an angle.

Stability can also be expressed in energy terms, because 'conservative forces' are expressible as changes of an appropriate energy with displacement.[3] A floating object will be stable if, in tilting it, its center of gravity rises, since this requires work and the input of energy.

Fish have evolved to take advantage of buoyancy, minimizing their effort to live in water. A fish with neutral buoyancy will have an average density matching the water. Fish, water mammals, and other marine animals can dynamically change their buoyancy by controlling the air in interior spaces. Neutral buoyancy, however, is not a stable condition within a fluid. If the body displaces upward, even slightly, the pressure acting on that body will decrease. Because the body must have some elastic behavior, its volume under reduced pressure will be greater. This will cause the buoyant force to be larger, so the body continues to rise. Similarly, if a body in neutral buoyancy displaces downward, it will continue to move downward. Marine bodies require active regulation to maintain a given depth in water, such as control of the volume of air sacks or slight movements of fins.

[3] In contrast, 'dissipative forces' produce heat. Because heat production causes a loss of mechanical energy, dissipative forces are also called "non-conservative forces." Formally, conservative forces are those which do no net work when acting on a body carrying it on any closed path: $\oint \mathbf{F} \cdot d\mathbf{r} = 0$. From Stokes' theorem, this also means $\nabla \times \mathbf{F} = 0$. In turn, this means that a conservative force field can always be expressed as the divergence of a scalar field: $\mathbf{F} = -\nabla V$. This V is called the potential energy function.

4.3 Fluid Dynamics

When physical principles are applied to understand the motion of fluids, the subject of the study is fluid dynamics. Examples abound in biology. Blood flow is an evident one. But there are many other important applications: How animals move through water, how birds and bees fly, and how fluids in cells circulate. While tackling these applications, we will use the underlying physical principles.

We will apply Newton's laws to describe biologically important fluid dynamics. These laws, together with the relationships between the physical properties of the fluid (known as the equations of state) contain all the classical predictions, including conservation laws and limitations on the behavior of fluids. Many of the predictions simply follow from simple circumstances. However, we should also appreciate that because Newton's laws for fluids are non-linear in the fluid velocity, they can have solutions involving chaotic states. (See Sect. 13.2 for a description of chaos.)

During fluid motion, one layer of fluid can slip over another. Such slipping introduces a short-range force of one layer onto the other in a way analogous to the material shearing force we have studied, except that instead of a static displacement of one layer relative to another in the solid, there is a continuous displacement of one layer relative to its neighboring layer.

4.3.1 *Viscosity*

Pressure forces of one fluid element on another are defined to be perpendicular to the element surface. But fluid elements can also slip past one another, creating a friction force tangent to the element surface, called a 'viscous force'. Figure 4.3 shows a fluid being 'sheared' by viscous forces.

The size of the viscous force f_v is directly proportional to the area of the surfaces of fluid slipping (if you double the area of contact A, the force doubles). The viscous force on a layer of fluid of thickness dx should be inversely proportional to dx (if you double the thickness, each half layer needs the original force to keep it slipping, but now there are two such layers to shear). Observation shows that the viscous force also grows linearly with the difference in the speed between the top and bottom layers as they are sheared by relative motion dv (as long as the difference in speed is not too large). Thus, we have

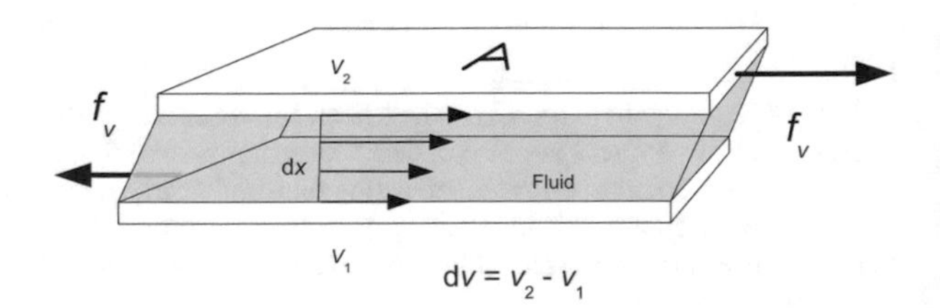

Fig. 4.3 Shearing a fluid with two parallel plates moving at different speeds

Table 4.1 Viscosity of some common fluids (in centipoise)

	Temp.	Viscosity
Hydrogen	0 °C	8.42×10^{-3}
Air	15 °C	17.9×10^{-3}
Methanol	25 °C	0.55
Ethanol	25 °C	1.074
Water	25 °C	0.89
Blood	37 °C	3–4
Glycerol	25 °C	938

$$f_v = \eta A \frac{dv}{dx} . \tag{4.21}$$

The proportionality factor η is called the 'dynamic viscosity' (also called the shearing viscosity)[4] The relation defines the shearing viscosity of a fluid. A fluid with a larger viscosity requires a larger force to continuously shear the fluid. The unit for viscosity in the cgs system is called a 'poise' which is 1 gm cm^{-1}sec^{-1} (Table 4.1).

Quite generally, viscous fluids have infinitesimal relative-motion with interacting fixed boundaries infinitesimally far away. This is the so-called 'no-slip' boundary condition. At a molecular level, this is understandable, since some fluid molecules will be trapped and stagnated for a short time by surface roughness, even if the roughness is only at a molecular scale.[5] Moreover, fluid molecules are often attracted to solid surfaces. These trapped fluid molecules, in turn, act on the next layer of fluid with a viscous force. Even air rushing past a wing has no motion right on the wing surface.[6] This is an important constraint in finding how real fluid moves with bounding surfaces present, such as a bird wing or the skin of a whale.

In practice, keeping the fluid between two flat plates is awkward for measuring viscosity. A more practical device for liquids is a 'viscometer'. In one such device, the fluid fills a gap between two concentric vertical cylinders. The inner cylinder is rotated, but not fast enough to cause turbulence. The torque needed to sustain this rotation determines the fluid viscosity.

We can apply the viscosity defining relation (4.21) to each thin layer of fluid of thickness dr from the inner cylinder radius, say a, to the outer cylinder radius, say b. The velocity of each cylindrical layer of fluid at each radius will be $v = \omega r$,

---

[4]A second kind called the 'bulk viscosity' applies when internal frictional forces act during the compression of a fluid. This will be noted in our discussion of the Navier–Stokes equation Sect. 4.3.5. We encountered this kind of viscosity in the behavior of viscoelastic materials.

[5]Exceptions for water solutions include hydrophobic surfaces. Liquid Helium-4 is another exceptional fluid, In fact, ${}^4_2He$ below 2.17 K is a quantum superfluid, without any viscosity. There is insufficient thermal energy to excite the atoms. It can escape an open glass jar by running up the inside by capillary action against gravity, with no friction, and then spilling to the outside.

[6]Dolphins can ripple their skin to cause a more intimate flow of water as they swim, reducing the retarding effect of wake turbulence. The dimples on golf balls have the same 'porpoise'.

where ω is its angular velocity. If the angular velocity were the same for all layers, the fluid would be rotating as a rigid body, with no viscous shear occurring. Thus, in the radial change of speed from one layer to the next, $dv/dr = r\,d\omega/dr + \omega$, only the first term contributes to the fluid shear. Each cylindrical layer of fluid will have a shearing force over its surface given by

$$f_v[r] = \eta(2\pi\, r\, h)\, r\, \frac{d\omega}{dr}\,, \tag{4.22}$$

where h is the height of the liquid between the cylinders.

For each layer of radius r with thickness dr, there will be a torque

$$(r + dr)\, f_r[r + dr]$$

on the outside of this layer which must match the torque

$$r\, f_r[r]$$

on the inner side, if the layer is not to have angular acceleration. This balancing of torques makes

$$\frac{d}{dr}(r f_r) = \frac{d}{dr}\left(2\pi\, \eta\, h\, r^3\, \frac{d\omega}{dr}\right) = 0\,. \tag{4.23}$$

This condition means $r^3\, d\omega/dr$ must be constant. At the inner radius a, we will put $\omega = \Omega$ while at the outer radius b, $\omega = 0$. These give

$$\omega = \left(\frac{1}{1/a^2 - 1/b^2}\right)\left(\frac{1}{r^2} - \frac{1}{b^2}\right)\Omega \tag{4.24}$$

for the angular velocity, and

$$\tau = 4\pi\, \eta\, h\, \frac{1}{1/a^2 - 1/b^2}\, \Omega \tag{4.25}$$

for the torque on the inner cylinder, so the viscosity can be found from

$$\eta = \frac{(1/a^2 - 1/b^2)}{4\pi\, h\, \Omega}\, \tau\,. \tag{4.26}$$

Alternative viscometers can take advantage of forces on bounding surfaces due to viscous drag, such as a sphere falling in the fluid, the rise time for a bubble, or the pressure drop in the fluid passing through a tube.

For most fluids, there is a strong dependence of viscosity on temperature. The viscosity of gases tends to increase with increasing temperature (proportional to square root of the absolute temperature), while ordinary liquids decrease viscosity with increased temperature (often exponentially). The viscosity of air at standard temperature and pressure (273.15 K and 1.033×10^6 dyne per square centimeter) is 1.8×10^{-2} centipoise, increasing by about 0.3% per Kelvin increase in temperature. Water has a viscosity of 1.0 centipoise at 20 °C and one atmosphere, decreasing by about 2% per degree as the temperature rises (Table 4.2).

A model by Arrhenius using molecular kinetics predicts that a simple ('Newtonian') fluid will have a Boltzmann factor $\exp(-E/RT)$ determining the dependence of the fluid viscosity on temperature, where E is an activation energy per mole and R is the gas constant. Using $1/T = 1/(T_o + T_C) \approx (1 - T_C/T_o)/T_o$, this behavior can be approximated by

$$\eta = \eta_0 \exp(-aT_C) \tag{4.27}$$

in a limited temperature range. T_C is the temperature in Celsius. For water in the range of 10–70 °C, a reasonable fit is $\eta = (37/25)\exp(-T_C/52)$ centipoise.

The so-called non-Newtonian fluids have a viscosity that depends on the shearing rate. Blood flow through capillaries is dramatically non-Newtonian, principally because the red-blood cells must squeeze through. (See Fig. 4.4.)

Table 4.2 Viscosity of water

Temp. (°C)	Viscosity (cpoise)
0	1.79
20	1.002
40	0.653
60	0.467
80	0.355
100	0.282

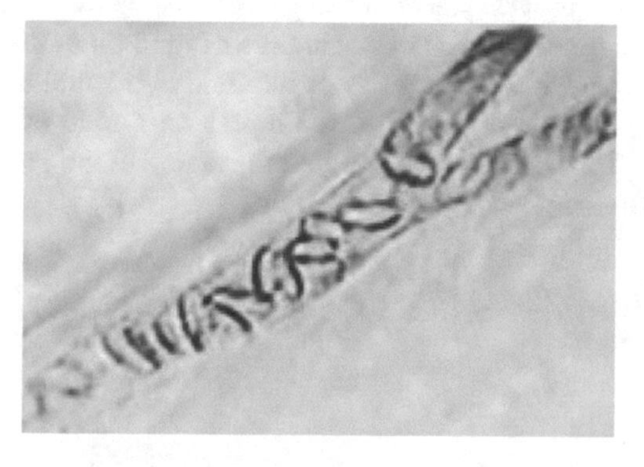

Fig. 4.4 Red blood cells squeezing through capillaries: the corpuscles are about 8 μm across, while the vessels are 7–15 μm in diameter. Turbulence around the cells enhances the dispersal of dissolved oxygen in the blood after the O_2 diffuses out of the blood cells, whose shape also minimizes the time for O_2 to diffuse from the hemoglobin within

The viscosity of blood strongly depends on the relative volume of red-blood cells to total blood volume, measured by 'hematocrit' in a centrifuge. In men, hematocrit is normally about $(45.5 \pm 4.8)\%$, in women about $(40.2 \pm 4.1)\%$, making the blood viscosity about five times that of water. Anemia can cause a decrease in hematocrit, and various vascular and bone-marrow diseases can increase the hematocrit above 60%, directly affecting blood flow because of the viscosity increase by as much as a factor of two. Hypothermia of the blood also increases its viscosity, so that intentional hypothermic operations must take into account viscosity changes in determining oxygenation of tissue.

A simple technique to determine the viscosity of a sample of blood is to measure the rate of fall, v, of a small ball (radius a and mass m) dropped in the blood. (See Fig. 3.6.) The ball stops accelerating almost immediately, so $6\pi \eta a v = mg$, which can be solved for η. This Stokes expression fails if the flow becomes turbulent behind the ball.

4.3.2 Fluid Flow in Tubes

Consider a fluid flowing through a tube, such as one of your arteries. The motion of a viscous fluid at the boundary walls usually becomes vanishingly small. The flow near the center of the tube will be higher than near the walls. This means each cylinder of fluid from one radius to the next within the tube rubs against the adjacent cylinder with a viscous force. The slipping layers generate heat, and energy is required to sustain the flow through the tube. This energy is supplied by an agent, such as your heart, which generates a force to push on the fluid at one end of the tube, with a lesser force from the distal fluid acting back on the other end. The difference between these two pressure forces on the opposite ends of the tube give a net force. As the fluid moves, power is being expended by the net force given by the pressure difference times the area of a thin tube of fluid times distance moved over time, or $p\, dA\, v$, where v is the velocity of the fluid in the thin tube. (See Fig. 4.5.) Under steady conditions,[7] we can balance forces on a fluid cylinder of length L and thickness dr. The viscous drag on the thin tube will be $f(r+dr) - f(r) = \eta L\, 2\pi(r+dr)v(r+dr) - \eta L\, 2\pi r v(r)$, giving

$$\Delta p \cdot 2\pi r dr = \eta(2\pi L)\left[-r\frac{dv}{dr}\right]_{r+dr} - \eta(2\pi L)\left[-r\frac{dv}{dr}\right]_{r}, \tag{4.28}$$

or

[7]Heart pumping, of course, is not steady. A more general treatment is handled by the time-dependent Navier–Stokes equation, as in, for example, W.E. Langlois and M.O. Deville, **Slow Viscous Flow**, Springer, Switzerland (2014). But as long as the reaction times in the system are short, a steady flow will approximate the behavior in longer segments of time.

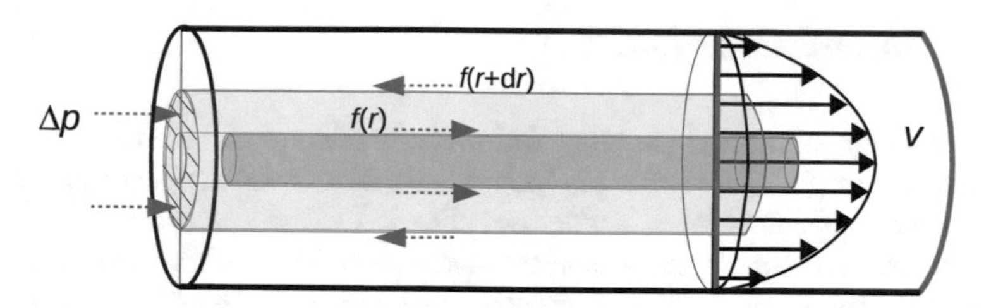

Fig. 4.5 Poiseuille parabolic velocity flow: the f are frictional forces due to viscous drag on the outside and inside surface of the thin tube of fluid

$$\frac{d}{rdr}\left[r\frac{dv}{dr}\right] = -\frac{\Delta p}{\eta L} . \tag{4.29}$$

The solution of this relation for the speed in a tube of radius a is

$$v = \frac{\Delta p}{4\eta L}\left(a^2 - r^2\right) . \tag{4.30}$$

The velocity has a parabolic profile across the tube.

We can find the total amount of fluid flowing (volume per unit time) by adding up the cylindrical flows at each radius:

$$\frac{\Delta \mathcal{V}}{\Delta t} = \int v\, dA = \int v\, 2\pi\, r dr ,$$

$$\frac{\Delta \mathcal{V}}{\Delta t} = \frac{\Delta p}{8\eta L}\pi a^4 . \tag{4.31}$$

This is Poiseuille's law for flow in a pipe. There is a very strong dependence on the radius of the tube. Ramifications on the flow of blood in a constricted artery follow.

Note that Poiseuille's law is a direct analog of Ohm's law in electricity. The fluid current $i = \Delta \mathcal{V}/\Delta t$ plays the role of the electric current $i = \Delta Q/\Delta t$. The fluid pressure difference Δp plays the role of the electric potential, ΔV. This analog is even closer: We will see when we discuss Bernoulli's law that the fluid pressure difference Δp is a fluid energy gained or lost per unit volume due to the elastic energy stored in the fluid by pressure. In electricity, the difference in potential energy gained or lost per unit charge is ΔV. The electric resistance R is the analog of the fluid flow resistance. Reading Eq. (4.31) as $i = \Delta p/R$, we see that the resistance in the pipe is

$$\mathcal{R} = \frac{8\eta L}{\pi r^4} . \tag{4.32}$$

Note the strong dependence of an artery's resistance on its radius.

4.3.3 Conservation of Mass

Conservation of volume in blood flow follows from the more fundamental conservation of mass-energy. In the realm of biochemistry, the conversion between mass and energy is not noticeable. Mass is effectively fixed in any closed system. As water is practically incompressible, mass conservation makes blood volume conserved.

The 'equation of continuity' expresses local conservation of mass within any fluid volume. It states that any change in the mass of a fluid volume must come from a flow of mass into or out of the volume. The rate of flow of mass through any surface is the density of the fluid times the volume which passes that surface in a given time. For a surface oriented in the xy plane, this will be $\rho v_z\, dt\, dx\, dy/\, dt$. For arbitrary orientation,

$$\frac{d}{dt}\int_V \rho dV = -\oint_S \rho\, \mathbf{v} \cdot \widehat{\mathbf{n}}\, dA = -\oint_S \rho\, \mathbf{v} \cdot d\mathbf{S}\,, \tag{4.33}$$

which is the integral form of the mass conservation law. Here, $d\mathbf{S}$ is an infinitesimal 'surface vector' $\widehat{\mathbf{n}}\, dA$, i.e. a vector with size equal to a small area dA on the bounding surface and pointing outward perpendicular to that surface, that direction denoted by the unit vector $\widehat{\mathbf{n}}$.

Applying Gauss' Theorem (See Appendix G.5) to the surface integral, and using the fact that the volume is arbitrarily chosen within the fluid will give

$$\frac{\partial \rho}{\partial t} + \nabla \cdot (\rho\, \mathbf{v}) = 0\,, \tag{4.34}$$

which is the 'local' form of the mass conservation law.

If the mass density, ρ, is taken as constant, the fluid is assumed effectively incompressible, and then, according to Eq. (4.34), the velocity field must satisfy

$$\nabla \cdot \mathbf{v} = 0. \tag{4.35}$$

This divergence-free condition on the velocity is equivalent to the statement that the volume of any mass of fluid does not change as the fluid moves. As we know, incompressible fluids do not exist, because there are no absolutely rigid bodies. However, many fluids, such as water, have a relatively large bulk modulus, making the density of water increase by less than 0.3% over increased pressure from atmospheric to 1000 times atmospheric going from the surface of the ocean to its greatest depths. The density of the salt water next to a deep-sea fish is not much different from fish in the ocean surf. For considerations of fluids consisting mostly of water, such as blood, assuming incompressibility introduces little error. An exception would occur if the fluid contained gas bubbles, which do get smaller under increased pressure.

4.3.4 Applications to Blood Flow in Arteries

Conservation of mass for flow in an unbroken artery means that whatever blood went into the artery minus what came out must be still inside. For a steady flow, we would have, from point 0 to point 1 along the flow

$$\int_0 \mathbf{v} \cdot d\mathbf{S} = \int_1 \mathbf{v} \cdot d\mathbf{S} . \tag{4.36}$$

If we let v_i be the average fluid velocity within the cross-sectional area A_i at the location i along the artery, then

$$\langle v \rangle_0 A_0 = \langle v \rangle_1 A_1 . \tag{4.37}$$

In the case the artery divides into N smaller arteries,

$$\langle v \rangle_0 A_0 = \sum_{k=1}^{N} \langle v \rangle_k A_k . \tag{4.38}$$

This relation is very useful in determining the speed of blood in the branches after the location of arterial bifurcation.

Even though blood flow is not uniform, under ordinary circumstances, the flow is sufficiently smooth that Poiseuille's law still applies approximately. The dramatic part of Poiseuille's Law is the very strong dependence of flow on the radius of the artery. If a constriction, such as that produced by arteriolosclerosis or plaque buildup, diminishes the diameter of an artery by half, the same pressure produces only an eighth of the original flow!

The nature of blood flow determined by physical laws leads to useful observations about our circulatory system. For example, using fluid volume conservation, we can estimate the number of capillaries in the body, a number difficult to get by direct counting. The net flow in the aorta should match the net flow through all the capillaries:

$$\langle v \rangle_a A_a = N_c \langle v \rangle_c A_c \tag{4.39}$$

Figure 4.6 shows how the cross-sectional area of the vessels carrying blood varies with distance from the heart, and also shows the average velocity of the blood at these same distances. Note that variations in blood pressure will induce small changes in the arterial cross-sectional areas along with the velocities. These variations are not shown in the figure.

The forcing term from a pumping healthy heart produces a pressure above atmospheric varying from systole of about 120 mmHg to diastole of 70 mmHg with a cardiac output of about 90 cc/cycle, with about 1.2 cycles per second. From this we can calculate by pressure times volume flow that the adult heart performs at

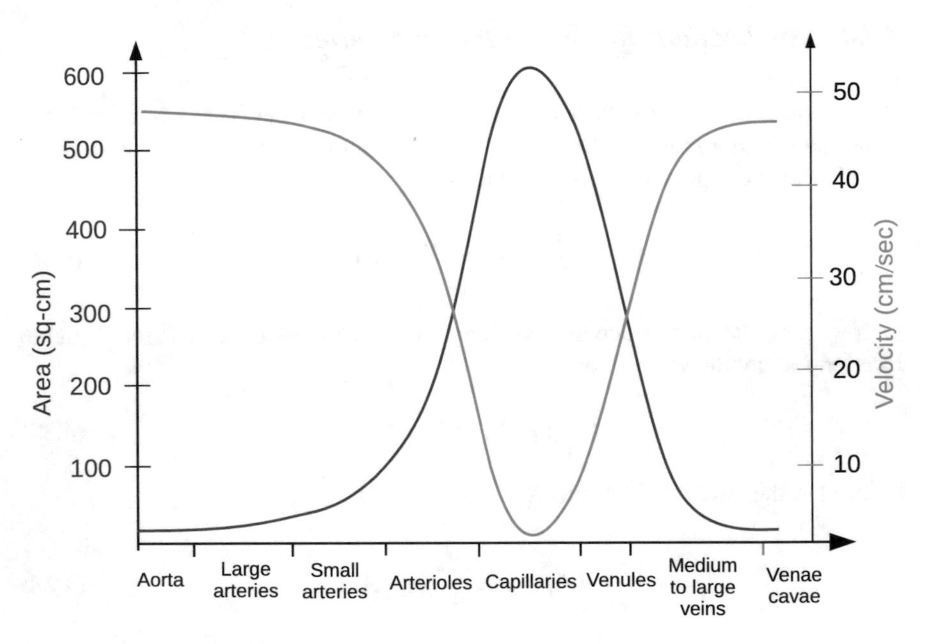

Fig. 4.6 Blood vessel mean areas and average blood velocities

about 1.3 W. The greatest load on the heart comes from the arterioles, over which the pressure has the largest drop (about 50 mmHg). (See Fig. 4.7.) Measuring the pressure in the aorta is a useful monitoring tool during medical operations and also serves in diagnosing heat problems. The pressure variations and valve sounds produce pressure waves which propagate down the blood in the aorta at a faster speed (100 cm/s) than the blood speed (30–50 cm/s), and experiences reflections from the arterioles. The effect of the reflected wave can be seen as a 'dicrotic notch' (downward pulse) in the pressure wave. (For more details on arterial pulse pressure waves, see Sect. 4.3.10.)

An optimal human circulatory system requires that the walls of the capillaries be of minimal thickness in order to transfer molecules to and from body cells and in order not to have to maintain underutilized tissue. This means that the pressure should be minimized in capillaries, to minimize rupture of the thin walls. This is achieved by having the arterioles, with thicker walls, sustain the arterial pressure (Fig. 4.8). By limiting the number and radii of such arterioles, the pressure drop across them is large. Blood which returns to the heart through the vena cava has a pressure just a little over atmospheric (by 2 or 3 mmHg).

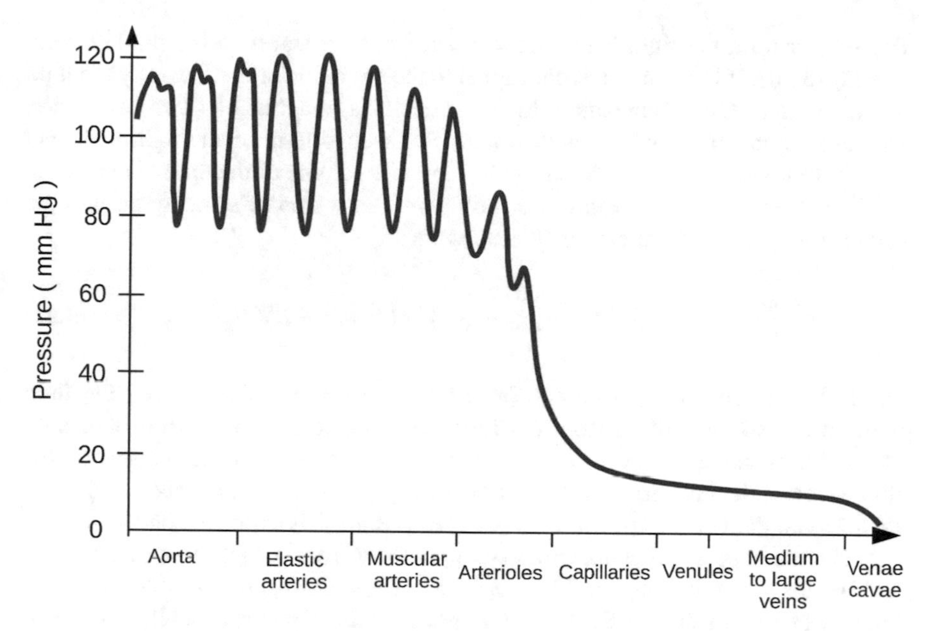

Fig. 4.7 Blood pressure in the vessels of the body

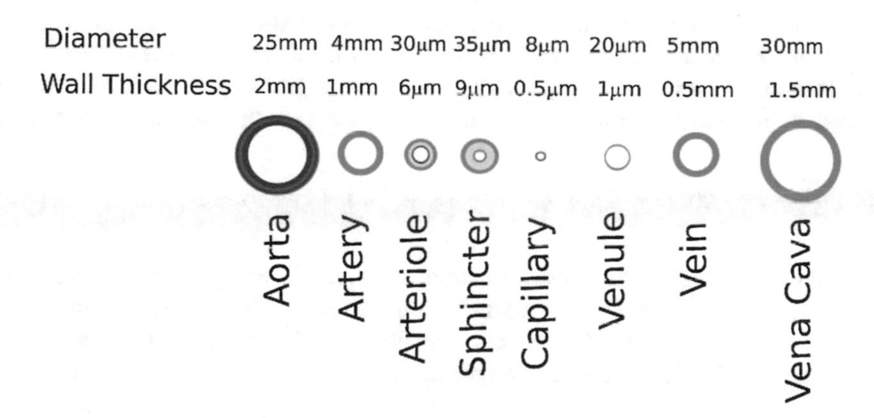

Fig. 4.8 Vessel diameters and wall thickness

4.3.5 The Fluid Equation of Motion

The Navier–Stokes equation is Newton's 2nd Law of motion applied to each small element of mass of a fluid. For a given small mass element in a fluid with local density ρ, The Navier–Stokes' equation reads

$$\rho \frac{d\mathbf{v}}{dt} = -\nabla \cdot \mathbf{S} + \mathbf{f}_{vol} . \tag{4.40}$$

The first term on the right is the net stress acting on surfaces of the fluid element (see Eq. (3.16). If the stress is proportional to the strain in the fluid, we say the fluid is 'Newtonian'. (Non-Newtonian fluids include lava, magma, blood in capillaries, and suspensions of cornstarch with water.) The second term on the right is the net external force per unit volume acting on the whole volume of the fluid element.

If the shearing forces within the fluid come only from viscous drag, and the volume force comes from gravity, there results

$$\rho \frac{d\mathbf{v}}{dt} = -\nabla p + \eta \nabla^2 \mathbf{v} + \left(\zeta + \frac{1}{3}\eta\right) \nabla \left(\nabla \cdot \mathbf{v}\right) - \rho \nabla \Phi_g \,. \tag{4.41}$$

The first term on the right comes from the pressure difference across the fluid element; the second comes from the force per volume due to shearing viscosity; and the third comes from a viscous force due to changes in the volume of the fluid element; the last term is the force of gravity per unit volume, where Φ_g is the gravitational field energy per unit mass.[8] The constant ζ is called the bulk viscosity.

For an incompressible fluid, we can drop the third term, because $\nabla \cdot \mathbf{v}$ vanishes. As we have described, incompressibility is not a bad approximation for water solutions under most circumstances. But it's not so good for air. There is a significant loss of sound energy in air into heat that comes from the volume variation term.

The velocity of a collection of fluid elements is a 'field' of vectors $\mathbf{v}(x, y, z, t)$. The total differential of the velocity in the left-hand side of Eq. (4.40) can be rewritten as two terms: a change in the velocity at a given point (x, y, z) plus the change in the velocity from one point in the motion of the fluid element to the next:

$$\rho \frac{\partial \mathbf{v}}{\partial t} + \rho \mathbf{v} \cdot \nabla \mathbf{v} = -\nabla p + \eta \nabla^2 v + \left(\xi + \frac{1}{3}\eta\right) \nabla \left(\nabla \cdot v\right) - \rho \nabla \Phi_g \,. \tag{4.42}$$

The first term on the left is called the 'transient inertial force per unit volume'. For steady flow, it vanishes. The second on the left is called the 'convective inertial force per unit volume'. For a fluid whose compressibility can be neglected in the behavior of its flow, the Navier–Stokes equation becomes

$$\frac{\partial \mathbf{v}}{\partial t} + \mathbf{v} \cdot \nabla \mathbf{v} = -\nabla (p/\rho) + (\eta/\rho) \nabla^2 \mathbf{v} - \nabla \Phi_g \,. \tag{4.43}$$

The ratio η/ρ is referred to as the 'kinematic viscosity'. (When ρ is taken as the density of water, then η/ρ is called the 'specific viscosity'.)

To find the motion of a fluid, we will also need to know how the pressure depends on density and how the internal energy of each fluid element depends on its specific entropy. These relations, called the 'constitutive equations', are equivalent to the

[8]Near a planet, $\Phi_g = -GM/r$, where $G = 6.67 \times 10^{-11}$ newton-m^2/kg^2, and r is the distance from the center of the planet. Near the Earth's surface, $-\nabla \Phi_g = -g\widehat{\mathbf{k}}$, with $\widehat{\mathbf{k}}$ being a unit vector pointing up.

equations of state for systems in equilibrium. We also need to know how the energy is transported across its boundaries, and how material is carried across boundaries.

Some marvelous properties of fluids in motion are predicted by the Navier–Stokes Eq. (4.43). The creation of vortices and turbulence comes about mathematically by the effect of the innocent-looking second term on the left. You will notice that all the other terms are either first power in the speed v or free of the speed. The second term, having $\mathbf{v} \cdot \nabla \mathbf{v}$, is second power in the speed. This makes the differential equation non-linear in the unknown velocity. In such systems, one can interpret the non-linear term as the system acting back on itself.

Non-linear differential equations are often notoriously difficult to solve. But the solutions of such systems are also often fascinating. In fact, some people can spend hours watching a babbling brook. Babbling brooks are one class of solutions of the Navier–Stokes equation. Hurricanes are another. People watch them also, but usually with the motive to learn how to avoid them.[9]

4.3.6 Chaotic Evolution of Fluids

Non-linear systems can show chaos. The idea behind chaotic systems was first described by Poincaré in the early 1890s in his work on the celestial three-body problem.[10] The idea was revived in the early 1960s by Edward Lorenz[11] in his attempts to predict the weather using none other than the Navier–Stokes equation, with the atmosphere as the fluid. Lorenz found that his computed solutions could differ widely when the initial conditions were changed only infinitesimally. He allowed his observation to be called the "butterfly effect". The flapping of the wings of a butterfly in Brazil could make the difference in whether a tornado occurs in Texas months later. In general, 'chaotic systems' have the property that two solutions which are arbitrarily close to one another at one time may diverge from each other exponentially at later times.[12]

In principle, the solutions of Navier–Stokes equation are completely determined by the initial conditions. If we had exact knowledge of initial states and exact solutions, we could make exact predictions. But if we do not know the exact initial

---

[9] I was once on a hiking trip near Boulder, Colorado, with a group of theoretical physicists. One half-jokingly said, after looking over at a distant landscape, "That is almost as beautiful as a Wightman function." A friend in the group responded that if he expressed further audacities in the face of the gods, his 'Wightman' may no longer function.

[10] Henri Poincaré, Les mèthods nouvelles de la mècanique cèleste, [Paris: Gauthier-Villars] Vol I (1892), Vol II (1893), Vol III (1899).

[11] Edward N. Lorenz, *Deterministic non-periodic flow*, J Atmos Sci **20**, 130–141 (1963).

[12] Mathematically, this is expressed as follows. Suppose a dynamical system depends on a parameter r. Express the solutions of the system as $y(t, r)$. Let δr be a small change in the parameter r. If the solutions diverges in time as $[y(t, r + \delta r) - y(t, r)] \propto \exp(\lambda t)$, the system is exhibiting chaotic behavior. The constant λ is referred to as the Lyapunov exponent.

temperatures, pressures, etc. across the fluid, our solutions grow in uncertainty. This makes long-time predictions difficult, both for meteorologists in predicting the weather, and for fluid biomechanics in predicting how a bumble bee flies.[13]

These days, we resort to computers to follow the development of the solutions. When chaos applies, the initial conditions would have to be known and the solutions would have to be tracked with extreme precision to be certain about how the system might evolve far into the future. As our computers have a finite memory capacity, we cannot store all the initial conditions nor follow solutions with infinite precision. With finite precision, we are forced to truncate answers at each iterative step of a calculation. So we lose more information about the solutions the more we iterate, even with perfectly known initial conditions.[14] We will return to the topic of chaos in Sect. (13.2.2).

Do not despair! Chaos is organized.[15] And, under the right conditions, some solutions of the Navier–Stokes equation are relatively simple. Also, we know of general conditions which the Navier–Stokes equations must satisfy, such as local energy, linear momentum, angular momentum, and mass conservation, which confines the kinds of solutions and their future behavior, chaotic or not.

4.3.7 *Scaling Law for Fluid Flows*

Notice that the Navier–Stokes equation (4.43) for an incompressible fluid has one parameter which depends on the intrinsic fluid properties, η/ρ, the specific viscosity, and an implicit constitutive equation giving how the pressure varies throughout the fluid.

Now consider the problem of finding the fluid motion around a stationary object of given shape in an incompressible fluid with speed v_f far from the object. The scale of size of this object is fixed by a single length, say L. For a given pressure distribution, the solutions for the fluid motion can only depend on three physical quantities, η/ρ, v_f, and L. These three numbers determine a unitless quantity, the Reynolds number:

[13]In Sect. 13.2, we will see that models of population growth and other physical models also can show chaotic behavior.

[14]Of course, we are assuming that Nature is smooth across space and time when we use a differential equation to describe her. If she were discrete (!), we might have difference equations instead of differential equations, and these difference equations might involve only rational numbers. Rational numbers which fit into available memory can be handled by computers without truncation loss, and, for such discrete equations, computers might be able to track solutions without errors, limited only by memory capacity. There is still a fundamental limitation: Tracking the whole universe would require a computer at least as big as the universe, which is a contradiction, as this computer must be in our universe!

[15]Not to be mixed up with 'organized confusion'.

$$\mathscr{R}_e = \rho v_f L/\eta \,, \tag{4.44}$$

a quantity to which we referred earlier when describing the drag on a body in a moving fluid (see Sect. 3.3.5). We may think of the Reynolds number as the ratio of a characteristic convective inertial momentum to the impulse on the body due to the viscous force.

If we measure all distances relative to L, all times relative to L/v_f, define a unitless pressure by $p' = p/(\rho v_f^2)$, and a unitless gravitational potential by $\Phi'_g = \Phi_g/v_f^2$, then the Navier–Stokes equation (4.43) will become

$$\frac{\partial \mathbf{v}'}{\partial t'} + \mathbf{v}' \cdot \nabla' \mathbf{v}' = -\nabla' p' + \frac{1}{\mathscr{R}_e} \nabla'^2 \mathbf{v}' - \nabla' \Phi'_g \,. \tag{4.45}$$

This is marvelous, because we can study the flow, say in a wind tunnel, for one set of values of L and v_f, and then re-scale to find a family of solutions for differing values of L and v_f, as long as we keep the same Reynolds number and boundary conditions. This can be done for model airplanes and model birds, with re-scaling to give solutions for real airplanes and real birds.

Another interesting dimensionless quantity for a fluid is the 'Rayleigh number' which, for given boundary conditions, determines if the fluid will have convective instability due to the buoyancy of heated fluid. It is given by

$$\mathscr{R}_a = \frac{\rho g \alpha \Delta T V}{\eta \kappa_D} \,, \tag{4.46}$$

where V is the volume of the fluid element being buoyed up, ρ its density, g the local gravitational acceleration, α the thermal expansivity of the fluid, ΔT the temperature decrease from bottom to top of the fluid element, η is the fluid viscosity, and κ_D the thermal diffusivity. For example, if you sit in a room of initially still air, a convective plume of heated air will rise above your head. As the Rayleigh number increases, the plume can divide into smaller convective zones, show flapping zones and puffing behavior, and finally chaotic turbulence.

4.3.8 Streamline Versus Turbulent Flow

Streamline flow is characterized by being steady, with adjacent fluid element following almost parallel paths. Such flow is also referred to as 'laminar flow', i.e. flow in layers.

Fluid flow is 'turbulent' if it is unsteady and irregular. A threshold size of the Reynolds number is observed to characterize fluid instability at the onset of a transition of fluid flow from streamline to turbulent. Figure 4.9 shows how a fluid such as water might flow past a ball (fluid enters from left). The top drawing depicts streamline flow ($\mathscr{R}_e < 1$), for which Stokes' law applies; the center drawing depicts

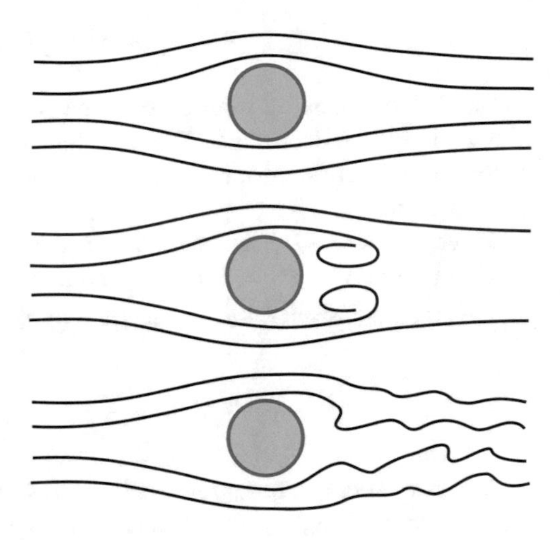

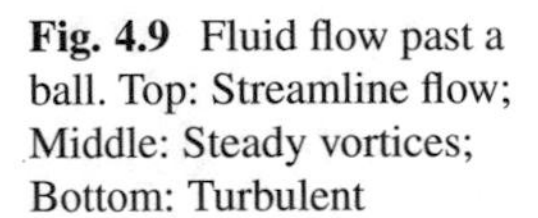

Fig. 4.9 Fluid flow past a ball. Top: Streamline flow; Middle: Steady vortices; Bottom: Turbulent

vortex flow ($1 < \mathscr{R}_e < 100$). The drag on the ball is less than for streamline flow. When $\mathscr{R}_e$ reaches about 150, the vortices oscillate back and forth. When $\mathscr{R}_e$ exceed about 2000, the flow past the ball becomes chaotic, varying strongly with time.

Streamline flow of water-like fluids in a pipe, such as blood, also becomes unstable when $\mathscr{R}_e$ exceeds about 2100.

Turbulent Flow of Blood in an Artery

Evidently, from Poiseuille's law, if an artery becomes constricted, say by arteriosclerosis, and the tissue being supplied needs a certain amount of blood flow, then the speed of blood in the constricted region must be increased. This increase may have the effect of causing the flow to undergo turbulence, since the Reynolds number is proportional to the blood speed. In turn, turbulence in an artery will cause a sizable increase in the back pressure, requiring the heart to work harder. The presence of turbulence in an artery can be heard with a stethoscope by the sound generated as fluid vortices bounce against the arterial wall.

An increase in the Reynolds number also occurs if the blood becomes 'thinner', i.e. less viscous. This can happen in cases of anemia, for which turbulent flow is more likely.

There is good evidence that the aorta is near optimal in diameter. (See Sect. 13.3.3.) If it were smaller, significant turbulence would occur during systole. If it were larger, the body would needlessly maintain extra tissue. Even so, there will be a small transverse component of the flow in the aorta of normal circulation during the pumping of the heart. However, there is insufficient time for this motion to generate turbulence before diastole.

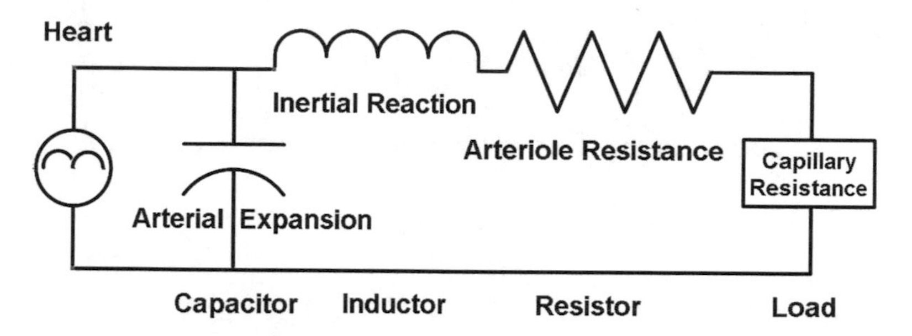

Fig. 4.10 Electric circuit representation of blood circulatory system

Turbulent Flow in Capillaries

As we pointed out before, the flow of blood through our capillaries is non-Newtonian for a good reason. The capillaries have diameters which barely let red corpuscles through (Fig. 4.4). The flow around the corpuscles becomes turbulent. This has the advantage of mixing plasma nutrients and oxygen, so they reach the capillary walls far faster than diffusion alone, and similarly for cell waste products and carbon dioxide passing into the plasma, since the gradient of concentration of these materials is greater across the capillary walls if they are strongly mixed in the blood.

4.3.9 Circuit Analogy for Blood Flow

A dynamical model of the arterial system with a pumping heart can be constructed using an electrical analog. This simple model for blood flow in the circulatory system has all the elements of an electric circuit, including resistance, capacitance, inductance, and, of course, a power source. (See Fig. 4.10.) The direct analog of electric potential (measured by voltage) is the fluid pressure (measured by newtons per square meter). Both represent a specific energy stored: In the electric case, the potential gives the energy stored in charges per unit charge; in the fluid case, pressure is the energy stored in the fluid volumes per unit volume. The analog of electric current (measured in amperes) is fluid volume flow, measured by the volume of fluid passing through a given area per unit time.

The passive elements in an electric circuit have the following analogs in fluid flow:

Flow Resistance Fluid viscous drag in arteries and veins act as resistors in series with the current flow. This drag is highest in arterioles and the capillary system. From Poiseuille's law,

$$\Delta p|_{resist} = \left(\frac{8\eta L}{\pi a^4}\right)\frac{dV}{dt}\,, \tag{4.47}$$

and using the electric analog, $V_R = iR$, the fluid resistance in an artery is

$$\mathcal{R}_a = \frac{8\eta L}{\pi a^4}\,, \tag{4.48}$$

which is Eq. (4.32).

Artery Capacitance Since the arterial walls have elasticity, when the blood pressure increases, the walls expand. This elastic expansion stores energy, just as a capacitor stores energy when the electrical pressure across the capacitor increases. The energy stored by the elastic expansion of an artery acts as a capacitor in parallel with each resistive and inductive element. If we let $\int dV$ be the net increase in volume by the elastic expansion of a segment of an artery whose volume was V_0, then, with the definition of distensibility (Eq. (3.24)), we will have

$$\Delta p|_{cap} = \frac{1}{DV_0}\int dV\,. \tag{4.49}$$

With the electric analog $V_C = Q/C$, the arterial capacitance measures the stored extra volume per unit extra pressure. Thus, the arterial capacitance is given by

$$\mathcal{C}_a = D\,V_0 = D(\pi r^2 L)\,. \tag{4.50}$$

The expression makes sense, in that a more distensible artery, and one with a greater volume, will have a greater capacitance for storing fluid and fluid potential energy. Note: For arteries, the expansion by pressure changes the radius, so that, for small radial changes, the distensibility is determined by $D = (2/r)\Delta r/\Delta p$. Thus, the capacitance of a segment of an artery is the increase in the volume of that segment per unit increase in pressure: $\mathcal{C}_a = (2\pi r\Delta r L)/\Delta p$.

Fluid Inductance In an electrical circuit, a coil of wire acts as an inductor. Electric inductors store energy by converting charge motion into a magnetic field. Faraday's law predicts that a back electrical pressure of size $L_e di/dt$ is created when the current i changes. For a fluid, the inertia of each mass element δm of fluid produces the Newtonian δma term, i.e. $\delta m dv/dt$, when the fluid is forced to accelerate. Thus, the inertia of fluid mass acts like an inductor in series with the current. For blood flow, attempts to change the velocity of the flow through pressure changes will generate an inertial back pressure. For an artery of length L and area A, we have $-\Delta pA = \delta m dv/dt$ which gives

$$\Delta p|_{induct} = -\frac{\rho L}{A}\frac{d^2V}{dt^2}\,. \tag{4.51}$$

We see that the arterial inductance due to inertia is given by

$$\mathcal{L}_a = \frac{\rho L}{A} . \tag{4.52}$$

The source of power for blood flow is an oscillating but rectified generator of pressure, namely, the heart. Instead of a single set of passive elements: resistor, capacitor, and inductor, the arterial system is better represented by a series of such elements on small segments of the arteries, with resistors and inductors in series, and capacitors in parallel. The heart can be represented in the electric analog by a wave generator with diodes representing heart valves. The model leads to a set of linear differential equations for the flow of blood in the arteries and veins.

4.3.10 Pulse Pressure Waves in Arteries

As arterial flow is non-uniform due to varying heart pressure, arteries undergo elastic expansion and contraction. As a result, elastic waves along the walls of arteries create a variation in the blood pressure. These pulse-pressure-waves travel significantly faster than the blood itself, with a speed

$$v_{PPW} = \sqrt{\frac{K\,\Delta r}{2\rho\, r}} \tag{4.53}$$

derived by Moens and by Korteweg in 1877 and 1888.[16] Here, K is the elastic modulus of the artery wall, Δr its thickness, r its radius, and ρ is the blood density. Note: For a uniform artery, the elastic modulus K is proportional to the artery length, increases strongly at higher blood pressures ('stiffening') and varies significantly from artery to artery.[17] Also, note that because of the $1/\sqrt{r}$ factor, the pulse-pressure-wave speed is larger for smaller vessels (about 1 m/s in the pulmonary artery; about 15 m/s in the small arteries). This pulse wave travels down the aorta, partially reflecting back from the arterial branches. The result is that the blood pressure pulse sharpens in amplitude some ten centimeters from the heart.

[16] A. Isebree Moens, *On the speed of propagation of the pulse*, (Ph.D. thesis) (in Dutch), [Leiden, The Netherlands S.C. Van Doesburgh] (1877); D.J. Korteweg, *On the propagation of sound in elastic tubes*, Annalen der Physik **241:12**, 525–542 (1878).

[17] The principal elastic components of an artery wall are collagen, elastin, and smooth muscles. For the choroid artery, these produce values of K in range 10^7 to 10^6, 10^5, and 10^4 pascal, respectively.

4.3.11 Living in Fluids with Low Reynolds Number

The flagella of bacteria have a low mass, and the flagella move in viscous water. This is an example of a situation where the conditions allow for the inertia term in the Navier–Stokes equation to be neglected.

As artfully described by E.M. Purcell,[18] microscopic marine life take advantage of the viscous properties of water. To move through water, an animal must be able to generate a net forward force on itself over a cycle of its locomotive organs. However, if the animal is microscopic and living in water, it cannot use its own inertia to carry it forward with each stroke of locomotive parts, as you or a fish might do. Viscous forces on the micro-organism dominate over inertia. As soon as the propelling force vanishes, the organism almost immediately stops.

To estimate the relative sizes of protozoan inertia versus viscous forces acting on them, take a spherical animal of radius $a = 1$ micron, density about that of water, and moving at a typical speed of 30 μ/s. Then the viscous force acting will be given by Stokes' law, $f = 6\pi\eta a v$. Acting alone, this viscous force will just about stop the animal in a time given approximately by $(4/9)\rho a^2/\eta = 50\,\mu$s. The stopping distance is only about 14 times the diameter of the hydrogen atom! Clearly, viscous forces dominate over inertia for these critters.

As we have noted, the unitless Reynolds number $\mathscr{R}_e$ for fluid flow is often used to give a threshold for the onset of turbulence under given boundaries for the fluid. But the Reynolds number, as a ratio of fluid inertial effects to viscous effects, also characterizes when viscous forces dominate over convective fluid inertial forces for bodies moving through a fluid. For a fish swimming in water, $\mathcal{R}_e$ is in the hundreds. For an ambulating protozoa, $\mathscr{R}_e$ can be as small as one ten thousandth. For a very small $\mathscr{R}_e$, the Navier–Stokes equation describing the fluid flow past a body in the fluid will be well approximated by dropping the fluid transient and convective inertial terms. This leaves

$$-\nabla p + \eta \nabla^2 \mathbf{v} \approx 0\,. \tag{4.54}$$

Under these conditions, if a paddle attached to a body is pushed one way, and then the other, the fluid will move for the reverse paddle motion just opposite to that of the forward motion, and the body just wiggles in place. A stiff paddle in a reciprocating motion is ineffectual in moving a micro-organism.

Protozoan move through water not with stiff paddles, or fins, but rather by flexible flagella, which can flail in the water, or even 'corkscrew' through the water, using a rotating shaft on a molecular motor.

---

[18]E.M. Purcell, *Life at Low Reynolds Number*, Physics and Our World: A Symposium in Honor of Victor F. Weisskopf, [American Journal of Physics Publishing] (1976).

4.3.12 Vorticity

The very names divergence and curl used in mathematics originated from fluid flow relations. Divergence of fluid in a small volume will exist if $\nabla \cdot \mathbf{v}$ is not zero within that volume. If a fluid has a curl within a small volume, then there is a 'vortex' in the fluid creating circular motion within that volume. With Stokes' integral theorem, $\oint \nabla \times \mathbf{v} \cdot d\mathbf{S} = \oint \mathbf{v} \cdot d\mathbf{r}$, we know that vorticity exists in a fluid wherever there is a net circular motion around a point in the fluid. The vector field $\boldsymbol{\Omega} = \nabla \times \mathbf{v}$ defines the 'vorticity' of the fluid near the point where the curl is calculated. Fluid motion without vorticity is called 'irrotational'.

We met the curl of material displacements before when we considered displacements which represent rotations rather than strain. (See Chap. 3, footnote 32.) Dividing the relation $\delta\boldsymbol{\theta} = (1/2)\nabla \times \xi$ by an interval of time, we find the local angular rotation of the fluid is given by

$$\boldsymbol{\omega} = \frac{1}{2}\nabla \times \mathbf{v} \,. \tag{4.55}$$

From the identity

$$(\mathbf{v} \cdot \nabla)\,\mathbf{v} = \frac{1}{2}\nabla v^2 - \mathbf{v} \times (\nabla \times \mathbf{v})\,, \tag{4.56}$$

the curl of the terms in Navier–Stokes equation gives, for an incompressible fluid,

$$\rho\frac{\partial\,\boldsymbol{\Omega}}{\partial t} - \rho\nabla \times (\mathbf{v} \times \boldsymbol{\Omega}) = -\eta\nabla^2\boldsymbol{\Omega}\,. \tag{4.57}$$

Equation (4.57), together with the incompressibility condition $\nabla \cdot \mathbf{v} = 0$ and boundary conditions, mathematically determine the velocity field of the fluid. We see that once the initial conditions are fixed, the later velocities of the fluid elements do not depend directly on the fluid pressure or the presence of gravity.

Moreover, if the fluid had no viscosity, and the initial vorticity vanished everywhere in a region, then $\partial\,\boldsymbol{\Omega}/\partial t$ would vanish, so it must vanish at any time later. A non-viscous fluid remains irrotational if it starts that way. Vortices cannot be created in such a fluid through fluid element interactions. In contrast, bumble bees can create useful vortices in air because the air is viscous.

For a fluid under irrotational conditions, the velocity can always be expressed as the divergence of some scalar function ϕ. The solutions for incompressible and irrotational fluids can then be found with standard techniques for solving Laplace's equation $\nabla^2\phi = 0$. In fact, since static electric potentials also satisfy Laplace's equation, if you know the potential around positive and negative charges, you also know how fluid will flow in analogous geometries. Mathematically, a point positive charge becomes a 'source' of fluid. A point negative charge becomes a 'sink' of fluid. The electric field lines become fluid flow lines.

Another simplifying case is a fluid with a large enough viscosity that the right-hand side of Eq. (4.57) dominates the second term on the left. Then the vorticity itself satisfies a diffusion equation. The larger the viscosity, the greater the diffusion of a vortex. Expanding smoke rings that are made by blowing smoky air through a circular hole are a good example.

4.3.13 Energy Transport in Fluids

To see how energy is transported through fluids, we first recognize that the energy can be stored in an element of the fluid in both kinetic and potential energy forms. The potential energy includes energy stored because the fluid has been compressed ('pressure energy'), energy stored by a force which pulled the fluid element away from the Earth (gravitational energy), and energy stored by having an external electrical force displace the fluid by acting on its net charge (electrical energy). We will let $u(x, y, z, t)$ be the potential energy stored per unit mass for a fluid element located at (x, y, z) at time t.

Each element of the fluid is in contact with adjacent elements or a boundary, and may be in interaction with external fields. This means the energy stored in that fluid element can change. We will keep our attention on a fixed mass of fluid. We will follow the small element of fluid with mass δm and internal energy per unit mass of u. As it moves, its energy, $\left((1/2)\rho v^2 + \rho u\right) \delta V$, changes over time by several mechanisms: Mechanical energy can be transported into the fluid element through the work done by external forces. Those forces which act on the surface of the fluid element ('contact forces') include the effects of pressure and viscous shear. The work done per unit time by a force is the power delivered, force times displacement per unit time, which is force times velocity. For the work done by pressure p, added up over the whole surface of the small fluid element, the power delivered by pressure is $\mathcal{P} = \oint p\, \mathbf{v} \cdot d\mathbf{A}$, where $\mathbf{v}dt = d\mathbf{r}$ is the displacement of the surface due to the pressure acting in the time dt. Gauss' theorem, $\oint \mathbf{F} \cdot d\mathbf{A} = \oint \nabla \cdot F dV$, let us rewrite the expression in terms of a volume integral, so that $\mathcal{P} = \oint \nabla \cdot (p\, \mathbf{v})\, dV$. Similarly, we can find the power delivered due to viscous shear.

To find the work done per unit time by an external force over the whole body, such as gravity, we must integrate over the whole volume of the fluid element, finding the work done per unit time on each piece within. When the external forces are conservative (as are the forces of gravity and electricity), each can be expressed in terms of the gradient of a potential. For gravity acting on the fluid element with volume δV, this reads: $\delta \mathbf{F_g} = -\rho \nabla \Phi_g\, \delta V$, while for electric forces, $\delta \mathbf{F_q} = -\rho_q \nabla \Phi_q \delta V$. The power delivered to a fluid element by such body forces is then

$$\mathcal{P}_\Phi = -\oint \rho \nabla \Phi \cdot (d\mathbf{r}/dt)\, dV$$

$$
\begin{aligned}
&= -\oint \nabla \cdot (\rho \mathbf{v} \Phi)\, dV + \oint \nabla \cdot (\rho \mathbf{v}) \Phi dV \\
&= -\oint \nabla \cdot (\rho \mathbf{v} \Phi)\, dV - \oint \frac{\partial \rho}{\partial t} \Phi dV \\
&= -\frac{d}{dt} \oint \rho \Phi \, dV + \oint \rho \frac{\partial \Phi}{\partial t} dV \,.
\end{aligned}
\tag{4.58}
$$

In many biophysical applications, the external gravitational and electrical fields will be constant in time: $\partial\Phi/\partial t = 0$. Also, within the body of the fluid, there is little or no charge in volumes with more than a few molecules, even with ions present, because opposite charges will cluster and tend to electrically neutralize each other.

Heat energy transported across the fluid element surface because of thermal gradients must also be included in the power loss. The topic of heat conduction will be treated in Chap. 9. From Eq. (9.8), the heat flux density[19] is given by $\mathbf{J}_Q = -\kappa_H \nabla T$.

Expressing each of the power loss terms as a flow of energy through the fluid surface, one can show that the rate at which a fluid element gains or loses energy can be expressed as

$$
\frac{d}{dt}\left[\left(\frac{1}{2}\rho v^2 + \rho u + \rho \Phi_g + \rho_q \Phi_q\right)\delta V\right] = -\nabla \cdot \left[\mathbf{v}p - \mathbf{v} \cdot \boldsymbol{\sigma}' - \kappa_H \nabla T\right]\delta V, \tag{4.59}
$$

where

$$
\boldsymbol{\sigma}' = \eta \left(\{\nabla \mathbf{v}\} - \frac{2}{3} \nabla \cdot \mathbf{v} \boldsymbol{\delta} \right) + \zeta \nabla \cdot \mathbf{v}\, \boldsymbol{\delta} \tag{4.60}
$$

is the viscosity strain tensor, $\{\nabla \mathbf{v}\}$ is the tensor with components $(\partial_i v_j + \partial_j v_i)$, and $\boldsymbol{\delta}$ is the unit tensor.

4.3.14 Bernoulli's Principle

A useful special case of the energy conservation law occurs under the following conditions: Negligible compressibility, negligible fluid viscosity; negligible temperature variation; and no unbalanced charges in the fluid.

In addition, steady conditions will be assumed. This means that the fluid motion looks the same later as now. Mathematically, at each point in the fluid, $\partial f/\partial t$ vanishes, for any of the fluid macroscopic variables.

[19] A flux is a rate of flow, and flux density is flux per unit area.

Steady conditions mean there will be no explicit time dependence of the energy density. With the above assumptions, Eq. (4.59) can be written

$$\mathbf{v} \cdot \nabla \left[\frac{1}{2}v^2 + u + p/\rho + \Phi_g \right] = 0 \,. \tag{4.61}$$

This can be read to say that the energy per unit mass does not change along a streamline. This is Bernoulli's Principle, more often expressed as

$$\frac{1}{2}v_1^2 + u_1 + p_1/\rho + \Phi_{g1} = \frac{1}{2}v_2^2 + \rho u_2 + p_2/\rho + \Phi_{g2} \,, \tag{4.62}$$

applying to two different points along one streamline. Near the Earth, $\Phi_g = gz$, where z is the height above the Earth's surface.

Bernoulli's Principle can be used to explain the behavior of a Venturi tube, namely a pipe with one end in a liquid fluid and the other in a gas with movement of the gas across the open end of the tube. In these circumstances, fluid is 'sucked' up the tube, and may be turned to a mist of tiny droplets at the open end. (You can see this by blowing across a transparent straw immersed in a drink.)

Perhaps the most famous application of Bernoulli's Principle is to give a simple elucidation of bird flight. Following along nearby streamlines, one passing under a wing, another passing over the wings, the faster speed of the air above the wing due to the wing curvature means the pressure is lower. The differences in pressure forces from the bottom of the wing minus those on top give a net upward force (lift) on the wing. More quantitative expressions for the lift can be found by solving the curl of the Navier–Stokes equation (which eliminates the pressure) together with the continuity equation and with wing boundary conditions to find the velocity field. Then return to the full Navier–Stokes equation to find the fluid pressure across the wing. Integrating the pressure times area over the wing surface will give the lift.

For blood flow, as the pressure from the contraction of a heart chamber causes each small mass unit of blood to accelerate, the pressure energy is turned, in part, to blood kinetic energy.

4.4 Measuring Blood Pressure and Flow

There are a number of devices which can be used to directly monitor blood pressure and which have a quick response time. Commonly, they are attached to a needle which has been inserted into an artery or vein.

Capacitive: By allowing the blood to exert a force on a conductive elastic membrane of area A near a second conducting plate, the distance d between the two conductors will change as the blood pressure changes. This change causes the capacitance, $C = \epsilon_o A/d$, to vary, which can be measured in an electronic circuit.

Inductive: If blood pressure acts on an elastic membrane attached to a magnet near a coil of wire, then variations in the blood pressure will induce a current in the coil, and that current can be measured. Note that this will give dp/dt, i.e. the variation of pressure with time.

Resistive: Blood pressure acting on a lever can cause a wire to elastically stretch. This causes the resistance of the wire to increase, which can be detected in a Wheatstone bridge circuit.

Blood flow can be monitored with small devices placed around an artery or vein:

US Doppler flow-meter: Ultrasonic waves generated by a small piezoelectric crystal can be used to measure blood flow speed v by detecting the Doppler shift of the sound back-scattered off blood corpuscles. The sound frequency is shifted by $\Delta f = 2v\cos\theta/v_{sound}$, where θ is the angle between the artery (or vein) and the sound-wave direction.

Hall-effect flow-meter: The Hall effect is the production of an electric field E transverse across a conductor and perpendicular to an external magnetic field B made perpendicular to the conduction direction. The electric field $E = vB$ is generated by the Lorentz force acting to shift charges to one side as they move with speed v through the magnetic field. As blood plasma contains ions, the Hall effect will occur when a magnetic field is placed across a blood vessel.

Problems

4.1 Suppose you hold your breath and dive down 5 m into the ocean. What will be the pressure in your lungs?

4.2 If a nitrogen bubble of diameter 0.3 μm were to form in a diver's bloodstream when at a depth of 200 m under the ocean, what would the diameter of that bubble be at the surface, assuming no more nitrogen diffuses into or out of that bubble?

4.3 An IV infuser bag is placed on a stand 1.5 m above the heart of a patient in bed. When the IV is operating, what is the infuser fluid pressure in the patient's vein where the cannula has been inserted? Compare this pressure with the patient's systolic pressure of 120 mmHg.

4.4 A certain fish weighs 530 g in air and has a body density of 1.05 gm/cm^3. How large a swim bladder does that fish need to be near buoyancy at a depth of 25 m in the ocean (with a density of 1.03 gm/cm^3)?

4.5 Estimate the change of density of water from the surface to one kilometer deep in the ocean. By what percent would this increase in density increase a fish's buoyancy compared to the buoyancy near the surface of the ocean?

4.6 At a depth of 10 m in the ocean, what will be the buoyant force on a person who weighs 40 kg and has a body volume of 40 l?

4.7 Suppose you need to quickly determine the viscosity of a jar of blood plasma. You decide to use the drag on a falling bead. You measure that the bead has a diameter of 2.075 cm and a mass of 4.80 gm. You find that it takes 10.0 s for the bead to fall 2.00 cm through the plasma. Calculate the plasma's viscosity.

4.8 If the pressure difference from one end of an artery to the other end were 10 N/m^2, the artery is 0.5 cm in diameter, with a length of 5 cm, what would be the flow rate (in cubic centimeters per second) in the artery?

4.9 How much power would the heart expend to push 100 cubic centimeters of blood per second through an artery of radius 0.5 cm and length 10 cm with a pressure difference from on end to the other of 30 mmHg?

4.10 A certain artery (A) of diameter 0.30 cm bifurcates into two arteries (B and C) of diameters 0.24 cm and 0.18 cm respectively. The speed of the blood in A along its axis was found to be 1.23 cm/s and the speed in B to be 0.86 cm/s. What speed would be measured in C along its axis? (Include the effects of blood viscosity.)

4.11 Estimate the number of capillaries in your body. Start with a velocity flow in the aorta of 110 cm/s and an inner diameter of 20 mm. Use a capillary diameter of 6 μm and a blood speed in the capillary of about 0.4 mm/s.

4.12 Suppose an artery of length 5 cm and diameter 2.6 cm expands in volume under an increase in pressure of 50 mmHg by 10%. Defining the capacitance of a given length of artery as the extra volume it can contain per unit pressure change, what is the capacitance per unit length of the given artery? The inductance per unit length of an artery is the ratio of the extra pressure due to a change in flow per unit change in flow with time. From Newton's law of inertia, this will be the density of the blood per cross-sectional area of the artery. Find a vibrational frequency of the artery of length 10 cm, using the formula for the natural vibrational frequency of a circuit with a capacitor and an inductor in series.

4.13 The speed of an eagle with spread wings and drifting horizontally was measured 34.0 m/s. The shape of the wings made the air on top move at 35.6 m/s. Estimate the pressure difference from bottom to top of the wing. If the bird weighs 19.2 N, and a third of the wing surface area has this effective pressure difference acting, what is the area of the bird's wings?

4.14 Blood speed in humans is generally less than 50 cm/s. Compare the kinetic energy per unit volume of blood moving with this speed to the energy stored per volume due to the pressure difference between your head and foot, when standing.

4.15 Some bacteria move by twirling their tail (flagellum) (up to 1000 revolutions per second). This propels them forward, at speeds up to 10 μm/s. Assume the bacterium is in the shape of a ball 1.0 micrometer in diameter, and use Stokes' law and the viscosity of water to estimate the power (joules per second) that this bacterium must expend to move at top speed. If there were a colony of one billion such bacteria, how much heat (in calories) would be generated per second just for their movement?

4.16 If the compliance of a lung is found to be 90 ml/mmHg, with a capacity of 1.2 l, how much pressure is needed to inflate the lung?

4.17 A non-viscous fluid of density 1.6 gm/cm^3 flows in a horizontal tube of varying diameter. The pressure drops by 0.010 mmHg from one end to the other. If the velocity starts at 7.0 cm/s, what is the velocity at the other end? By what factor does the diameter change?

4.18 Consider the design of a bionic heart. List those factors which might affect cardiac output, peak pressure, minimal turbulent flow, minimal erythrocyte rupture, heat dissipation, material stress, material failure, and possible power sources.

Chapter 5
Acoustics in Biology and Medicine

Everything I say is sound.

—wcp

Summary Sounds originate from vibrational waves traveling in a material. Animals take advantage of sounds to get and send notice of opportunities and impending dangers. Within the science of acoustics, we have applied mechanical principles to study how life forms detect and generate sounds, how these sounds carry information and energy, and how life utilizes sound, including for communication. Covered are acoustical analysis, biogeneration, biodetection, music and acoustical dosimetry.

5.1 The Nature of Sound Waves

The term sound was first used to refer to the physiological sensation accompanying hearing. Recognizing the origin of this sensation, we now use the term to mean any coherent vibration propagating in materials. Coherent here refers to a group-like behavior of numerous elements in a system. In contrast, incoherent microscopic vibrations occur in all ordinary materials above absolute zero in temperature. The statistical fluctuations in these incoherent motions produce thermal noise, an important consideration both in the design of transistors and in the design of your ear. Acoustics is the study of sound.

Consider a small segment of material of fixed mass δm at the position x in the material, with adjacent segments in front and back along the x-axis exerting opposing pressure-forces on the selected segment. Differences between these pressure-forces account for the acceleration of the segment mass. Suppose, in response to these variations in the pressure forces, the left surface of the segment is displaced an amount $\xi(x)$, while the right surface is displaced $\xi(x + \delta x)$. These displacements may depend both on the position x where the displacement originated in the material and on the time t when the displacement is observed. To describe the

W. C. Parke, *Biophysics*, https://doi.org/10.1007/978-3-030-44146-3_5

resulting wave,[1] we apply Newton's second Law to the mass δm between the two nearby surface segments:

$$(p(x) - p(x + \delta x))A = \delta m \frac{d^2\xi}{dt^2} \tag{5.1}$$

with A being the transverse area of the segments on which the pressures act. Dividing by the area A and the initial thickness between the surface segments, δx, then Eq. (5.1) reads

$$-\frac{\partial p}{\partial x} = \rho \frac{\partial^2 \xi}{\partial t^2} \, . \tag{5.2}$$

where ρ is the density of the material at the position x.

The pressure variations in the material arises from changes in the material density. For small relative displacements, the pressure will be the ambient pressure p_0 when there is no wave plus the pressure variation because of density variation as the wave passes. As a Taylor series, we have

$$p = p_o + \left[\frac{dp}{d\rho}\right]_o (\rho - \rho_o) + \frac{1}{2}\left[\frac{d^2p}{d\rho^2}\right]_o (\rho - \rho_o)^2 + \cdots \, . \tag{5.3}$$

Following Beyer,[2] this is often written as

$$p = p_o + \mathscr{A}(\rho - \rho_o)/\rho_o + \frac{1}{2}\mathscr{B}(\rho - \rho_o)^2/\rho_o^2 + \cdots \, . \tag{5.4}$$

The dimensionless ratio $\mathscr{B}/\mathscr{A}$ characterizes the degree of non-linearity the material shows in response to the passing wave. In Table 5.1, values of $\mathscr{B}/\mathscr{A}$ are given for some common materials.

Table 5.1 Beyer's parameters $\mathscr{B}/\mathscr{A}$ for common materials

Material	$\mathscr{B}/\mathscr{A}$	Temp. (°C)
Air	0.4	20
Distilled water	5.0	20
Salt water	5.4	20
Ethanol	10.5	20
Hemoglobin (50%)	7.6	30
Liver	6.5	30
Fat	9.9	30

[1]This will be a 'longitudinal wave, i.e. a wave whose displacement occurs along the same line as the wave moves. A wave with displacements perpendicular to the direction of movement of the wave is called a 'transverse wave'.

[2]R.T. Beyer, *Parameters of Non-linearity in Fluids*, J Acoust Soc Am **32**, 719–721 (1960).

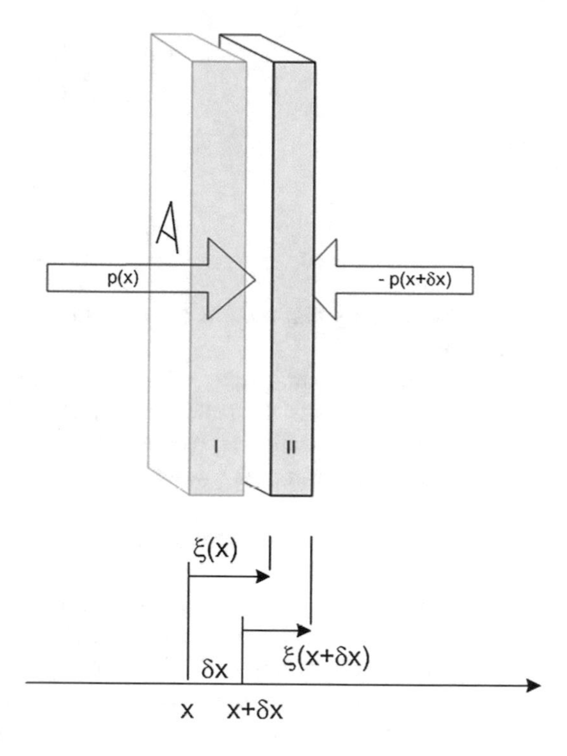

Fig. 5.1 Pressure waves: A thin rectangular volume of material moves from I to II under the influence of unbalanced pressures

The density of a mass segment as the wave passes changes because the volume of that segment changes (Fig. 5.1 shows such a volume change). Thus

$$\rho = \frac{\delta m}{\delta V} \simeq \frac{\delta m}{A(\xi(x+\delta x)+\delta x-\xi(x))} \simeq \frac{\delta m}{A\delta x}\left(1-\frac{\partial \xi}{\partial x}\right) = \rho_o - \rho_o \frac{\partial \xi}{\partial x} \,. \tag{5.5}$$

Combining the linear term in Eqs. (5.4) and (5.5) gives

$$p = p_o - \rho_o \left[\frac{dp}{d\rho}\right]_o \frac{\partial \xi}{\partial x} \,. \tag{5.6}$$

This equation is a disguised form of Hooke's law for springs, giving a linear relation between force and the resulting displacement. The 'stiffness' of the 'spring' is now measured by how much the pressure increases as one squeezes the material. This is reminiscent of the definition of the bulk modulus B (Eq. (3.22)). However, for longitudinal sound waves, the compression is not isotropic, but rather is along the direction of the wave movement, while for transverse sound waves, the material is being sheared.

Using Eq. (5.6) in Eq. (5.2) produces the famous wave equation, in this case a differential equation for the displacements within the material as sound passes:

$$\frac{\partial^2 \xi}{\partial x^2} = \frac{1}{v^2}\frac{\partial^2 \xi}{\partial t^2} , \tag{5.7}$$

where

$$v = \sqrt{\left[\frac{dp}{d\rho}\right]_o} . \tag{5.8}$$

We will see shortly that v is the speed of the longitudinal wave as the wave moves across space.

For an elastic band, piano wire, and tendons in the body, longitudinal and transverse waves satisfy the same wave equation. For stretches of material, the wave speed squared is the stretching tension divided by the mass per unit length.

The general solution of the wave equation (Eq. (5.7)) is

$$\xi(x,t) = G(x - vt) + H(x + vt) , \tag{5.9}$$

with the functions G and H arbitrary. The first term represents a wave with shape given by G traveling toward increasing x. The second term is a wave with shape H traveling toward decreasing x. Both of these waves move with the speed v. To see the direction, imagine 'riding' the wave $G(x - vt)$ as a surfer. As you travel, you try to keep your position (height) on the wave fixed, i.e. you keep the value of G constant. That means $x - vt$ is constant, or your position at any time is $x = vt +$ constant. This is the equation of forward linear motion at speed v. The speed of sound in a material is given by Eq. (5.8), provided the assumption of linearity of the pressure change to density change holds (i.e. $dp/d\rho$ remains constant). If not, the speed will also depend on the amplitude of the wave. The effect can be expressed as a series, giving[3]

$$\begin{aligned} v' &= \sqrt{(dp/d\rho)_o}\left(1 + \tfrac{1}{2}(\mathscr{B}/\mathscr{A})(\rho - \rho_o)/\rho_o + \cdots\right) \\ &\approx v + \tfrac{1}{2}(\mathscr{B}/\mathscr{A})\, p/(\rho_o v) . \end{aligned} \tag{5.10}$$

Isaac Newton presumed that as sound passes through air, its temperature would remain constant. However, it turns out that there is not sufficient time for the heat generated by compression to be conducted into the rarefied regions. Laplace realized that the process is closer to adiabatic.

The atmospheric air approximates an ideal gas, for which an adiabatic process follows the relation

[3] We should point out here that the presence of bubbles in the fluid has a dramatic effect on the speed of sound in the mixture, and on sound dispersion, since the gas in the bubbles can much more easily change in volume with pressure changes compared to fluids.

$$pV^{\gamma} = \text{constant} \tag{5.11}$$

where the so-called polytropic exponent,[4] $\gamma = c_p/c_v$, is the ratio of the specific heat at constant pressure to the specific heat at constant volume. In any material, the adiabatic approximation during sound transmission is a good assumption.[5]

To estimate the speed of sound in air, we use $v = \sqrt{dp/d\rho}$ applied to an ideal gas acting adiabatically to get

$$v_G = \sqrt{\frac{\gamma p_o}{\rho_o}} \tag{5.12}$$

$$= \sqrt{\frac{\gamma RT}{M}} \tag{5.13}$$

$$= 331.45\frac{m}{s} + 0.607\frac{m}{s}\left(\frac{T_c}{°\mathrm{C}}\right) \tag{5.14}$$

where M is the average mass per mole of gas, i.e. the molecular weight, and $T_c = T - 273.15°$ K. The third expression applies to air at atmospheric pressure for temperatures near 0 °C.

If you breathe helium gas instead of air, the speed of sound in your throat is almost three times that of air (being that He is about 1/7.2 the average molecular weight of air, and its γ is 5/3 while nitrogen is 7/5). This makes you sound like Mickey Mouse when you speak after inhaling from a helium balloon (but be careful, because helium balloons contain no oxygen).

The speeds of longitudinal (compressional) sound waves and transverse (shearing) sound waves in a solid turn out to be

$$v_C = \sqrt{(B + 4n_s/3)/\rho}\ , \tag{5.15}$$

$$v_S = \sqrt{n_s/\rho}\ , \tag{5.16}$$

where B is the material's bulk modulus, n_s is its shear modulus, and ρ its density. (Seismologists call these two kinds of waves 'primary' and 'secondary', because, after an earthquake, the compressional waves arrive at detectors before the transverse shearing waves.)

Liquids do not easily support a shearing wave. When only the compressional wave survives, it will have a speed given by

$$v_L = \sqrt{B/\rho}\ . \tag{5.17}$$

---

[4]For an ideal gas, $\gamma = 1 + 2/f$, where f, the 'degrees of freedom', is three for a monatomic gas, one for each translational degree of freedom, and is five for a diatomic gas, three translational, two rotational degrees of freedom.

[5]In thermodynamics, an adiabatic process requires that the local entropy S remain constant, so that the speed of sound is written $v = \sqrt{(\partial p/\partial \rho)_s}$.

Table 5.2 Speed of sound in various materials, and their density

Material	Density (g/cm^3)	Speed (m/s)
Air	1.39×10^{-3}	331.45
Helium	1.78×10^{-4}	972.
Ethanol	0.79	1207.
Fat	0.95	1450.
Water	1.00	1496.
Muscle	1.07	1580.
Bone of skull	1.91	4080.
Steel	7.86	5940.

The speeds in air and helium are for 0 °C, 1 atm. The speeds are for compressional waves

The measured density and speed of compressional sound waves in various materials of interest here are shown in Table 5.2.

The linear nature of the wave equation (i.e. ξ appears to the first power in every term) means that if we know two independent solutions, their sum will also be a solution. This is the great 'Principle of Linear Superposition' property of waves.[6] Suppose two children are playing with a long rope, held taut between them. Each wiggles the rope at their end with relatively low amplitude and then stops. The two wiggles move toward each other. They cross each other in the center, continuing to move to the opposite end. This is way the waves $G(x - vt)$ and $H(x + vt)$ behave. Even though G and H add in the center, they separate and move apart toward the opposite ends of the rope. Except for possible damping losses we have not included, they will have the same shape they originally had when they started and before they interfered in the center.

The Principle of Linear Superposition also implies that 'wave interference' occurs when two waves of the same type come together. The resulting wave at any point and any time is simply the sum of the two displacements at that point and time. The sum may be less than either or even zero if the two waves have displacements opposite to each other. This we call 'destructive interference'. If the sum is greater than either wave displacement, we say this shows constructive interference. Places where waves completely cancel during the whole time of observation are called 'nodes'. Those places where they maximally add over the whole time of observation are called 'antinodes'. An orchestral hall should be built to minimize the likelihood of a node of audible sound of important frequency being created where the ears of a member of the audience may be located.

If $\xi(x, t)$ can be expressed as a product of a function only of time and one only of space, $\xi(x, t) = \tau(t)\sigma(x)$, then ξ is said to be a 'standing wave' . A familiar example is the wave on a guitar string. It appears not to be traveling, making the

[6]'Large' amplitude material waves can show non-linear behavior.

string move transverse to its length. Even so, this wave can be expressed as a sum of two equal sinusoidal waves traveling in opposite directions.

As can be seen from Eq. (5.8), the speed of sound in a material will be greater for 'stiffer' materials and smaller if the mass density is greater. The first statement comes about simply because a stiffer material has stronger inter-material forces, allowing a disturbance to transmit its influence quicker than for a weaker force, while the second statement follows from the inertial property of mass: It is harder to get a more massive body to accelerate. (It is harder to change the motion of a heavier football player!) As a sound wave passes through a material, the material is sequentially displaced from equilibrium. Each local region can be observed to oscillate with time. As each region is coupled to the next, energy is transported through the material. For a sound wave, material only locally moves by wiggling, while the wave energy can be carried a long distance through the material.

An initial oscillation at one frequency will force the same frequency of motion into adjacent regions, although adjacent regions will not be synchronized in displacement, since there must be a time lag in the transferred motion. This idea works for all waves, including light. (Although the vitreous humor of your eyes changes the wavelength of light, it does not change its color, since color is associated with frequency.) As a wave passes from one material to another, or through a non-uniform material, the wavelength of the wave can change, since the wavelength will depend on the wave speed.

A special solution to the wave equation is a sinusoidal wave,

$$\xi(x, t) = A \sin\left((2\pi(x/\lambda \mp t/T) + \phi\right), \tag{5.18}$$

with a single 'wavelength' λ, where λ is defined to be the distance from the region with one value of displacement to the next region with the same value, and a single 'period' T, where T is defined to be the time for the material at x to oscillate through a complete cycle of motion.

A sinusoidal wave has a characteristic maximum displacement from equilibrium called the 'amplitude' of the wave A. The argument of the sine function is an angle called the 'phase' of the wave. The fixed angle ϕ is the initial phase (at x & $t = 0$). The sign between the x term and the t term in the phase depends on whether the wave is taken traveling toward increasing or decreasing x. A fixed displacement point on the wave has constant phase, and moves according to $x = \pm(\lambda/T)t +$ constant. Thus, $v = \lambda/T$ is called the 'phase velocity' of the wave.

These sinusoidal solutions are also called 'simple waves', 'monochromatic waves', or, for audible sounds, pure notes. Because of the connection of such sounds with music, a single frequency wave is referred to as a 'harmonic', and the motion of material with sinusoidal variation as harmonic oscillation. Sinusoidal solutions turn out to be very useful to study because of the principle of linear superposition. We will be able to add lots of them with different frequencies to form more complicated waves.

Suppose we follow a point that has a fixed amplitude on a sinusoidal wave as the wave carries the point forward. We will see that point move with the speed v

of the wave. This speed can be found by dividing how far the wave moves by the time taken to move this distance, which is its wavelength λ divided by the wave period T, so $v = \lambda/T$. A surfer on such a wave can measure both the wavelength (distance to the next bump) and period (time the water takes to move up and down at each location), and so can easily get the speed of the wave he or she is riding. The frequency f of a simple wave is the number of cycles per unit time, i.e. the inverse of its period (which is time per cycle), making $f = 1/T$.[7] This means we can calculate the speed of any simple wave by

$$v = f\lambda\,. \tag{5.19}$$

Because this relation comes from the definitions of the speed of a wave, its characteristic wavelength, and its period, it is universally true for all types of waves, including light.

In materials, coherent vibrations are meaningful only if the wavelengths are longer than twice the separation of the atoms or molecules. For liquid water, molecules are separated by about 3 Å. The speed of sound in water is about 1500 m/s, so sound frequencies in water above about 2.5×10^{12} Hz are not possible. Molecules in air at STP, standard temperature and pressure, [8] are about ten times further apart than in liquid water, and the speed of sound about four and a half times slower. This makes the maximum frequency in air about 6×10^{10} Hz. As cells and tissue contain a high percentage of water, the upper limit for sound frequency in soft tissue is about 2×10^{12} Hz.

Corresponding to frequency as the number of cycles per unit time, the 'spatial repeatance' κ is defined as the number of cycles per unit length, which is the inverse of the wavelength (length per cycle): $\kappa = 1/\lambda$.[9]

This gives an alternate way of expressing Eq. (5.19), namely

$$f = v\kappa\,. \tag{5.20}$$

Sometimes, the 'angular frequency' $\omega = 2\pi f$ and angular wave number $k = 2\pi\,\kappa$ are introduced so that the sinusoidal wave can be more simply expressed as

$$\xi(x,t) = A\sin(kx \mp \omega t + \phi).$$

[7]The physical unit for frequency is inverse seconds, but cycles per second is commonly used to remind the listener what is being talked about. The SI unit for one cycle per second is a Hertz, after the first sentient being to detect radio waves, almost 20 years after Maxwell predicted them. A move away from unit clarity still Hertz!

[8]Standard temperature and pressure (STP) are taken (since 1982) as 273.15 K and 1.00×10^5 N/m^2. Normal temperature and pressure (NTP), are $T = 293.15$ K and $p = 1.01325 \times 10^5$ N/m^2.

[9]The spatial repeatance, $\kappa = 1/\lambda$, is commonly referred as the wave-number, even though this terminology ambiguates with the angular wave number $k = 2\pi/\lambda$.

In materials, the speed of the sound may depend on its wavelength. If so, different wavelength waves can be made to disperse in that material, i.e. each would separate into a different direction. This behavior is more familiar for light when it passes through a prism, making red light separate from blue. We will see that the dispersion of waves is determined by how the wave speed depends on wavelength (a 'dispersion relation').

The wave equation (5.7) applies to waves with displacement allowed in just one direction, such as pressure waves and transverse vibrations of long thin materials. Sounds in solids, such as tissue and bones, can have simultaneous vibrations in three directions. Assuming an isotropic and homogeneous material, a straight forward generalization of our wave equation to three dimensions becomes

$$\frac{\partial^2 \xi}{\partial x^2} + \frac{\partial^2 \xi}{\partial y^2} + \frac{\partial^2 \xi}{\partial z^2} = \frac{1}{v^2} \frac{\partial^2 \xi}{\partial t^2} . \tag{5.21}$$

A shorthand version reads[10]

$$\nabla^2 \xi = \frac{1}{v^2} \frac{\partial^2 \xi}{\partial t^2} . \tag{5.22}$$

Waves on two-dimensional material surfaces and waves in three-dimensional space both show a property predicted by the wave equation called 'diffraction', which is the 'bending' of the wave into the shadow region of a barrier. If you set up two sponges on your bath water, separated by a gap, then water waves you generate on one side of the sponges will pass through the gap and then spread out on the other side. This spreading of the wave is diffraction. The wave just past the gap moves the water on the adjacent sides of the forward traveling wave, making the water wave 'leak' into what would be a shadow region. As a wave diffracts, adjacent waves can interfere, making a 'diffraction pattern'. The surface waves on your bath water will show this diffraction pattern behavior if you wiggle the water on one side of three floating sponges, with two gaps between them. The sound you hear through a doorway without being able to see the source comes from reflections and from diffraction through the opening of the doorway. From the properties of the wave equation, one can show that the angle of the spread of a wave past a boundary edge, the 'diffraction angle' is inversely proportional to the wavelength of the wave.

You may ask, "If waves passing through a gap show diffraction, and light is a wave, why do you not see diffraction of sunlight passing through a window?" The answer is: Diffraction does occur, but the effect is small, because the wavelength of visible light is much smaller than the window opening, and because there are many wavelengths of light in sunlight, making each color of light diffract into a different

---

[10]The operation ∇^2 sums the second derivatives in each of the directions (x, y, z), and is called the 'Laplacian operator'. It may be considered a vector dot product of the vector operator ∇ with itself. The shorthand is extra nice, as it no longer needs to refer to a particular Cartesian set of axes (x, y, z). To express the Laplacian in more general coordinates, see Appendix G.6.

angle. A laser source (near one frequency) passing through a thin slit (with a gap of a tenth of a millimeter, or so) will generate an evident diffraction pattern on a far wall.

Note that from Eq. (5.6) the pressure of a wave moving in one direction satisfies

$$p = p_o \pm \rho_o v \frac{\partial \xi}{\partial t} \; . \tag{5.23}$$

We will use this relation when describing the energy carried by a wave.

5.2 Fourier Decomposition of Sound

Sound is often generated as a complicated wave, i.e. one that contains many more than just one note. Your speech contains a variety of simultaneous waves at various frequencies and amplitudes. Short pulses of sound contain a wide distribution of frequencies. If all frequencies in a given range are present with nearly equal amplitudes but random phases, we call such a wave 'white noise', in analogy to white light, which contains all visible colors with about equal intensity. In contrast, the sounds from which we get information and pleasure have temporal structure. Musicians may even poetically call such sounds colorful.

In the analysis of sound, we can express properties of the sound according to what frequencies are present, and with what amplitudes. Mathematically, the frequency-amplitude analysis of a sound is a Fourier decomposition. A general discussion of the Fourier decomposition of waves is given in Appendix H. The idea is this: No matter how complicated a wave may be, that wave can always be decomposed into a summation of simple harmonically-related sinusoidal waves with various amplitudes. As the wave equation for sound is linear in the sound amplitude, a linear combination of solutions will again be a solution. So, if we can find simple sinusoidal waves which differ from each other only in frequency (with a corresponding wavelength), then we have found the general solution to the wave equation.

There are many practical examples of Fourier series. The decomposition of a pure sound (or, in music, a pure note), will have just one non-zero term in the series. In contrast, a repeating but short duration clicking sound will have numerous high frequency terms in the harmonic analysis. The decomposition of a 'square wave' shows a dominant low frequency wave, but also many high frequency waves, which are needed to make up the sharp rising and falling part of the wave.

In a Fourier series for a wave with period T, the amplitude of the harmonic component cosine wave of frequency $2\pi j/T$ is usually called a_j, while the amplitude of the component sine wave is called b_j. Figure 5.2 shows a graph of a square wave and the result of adding the first five harmonics present within the wave. If the line $t = 0$ in the graph is centered on a peak of the wave, then all

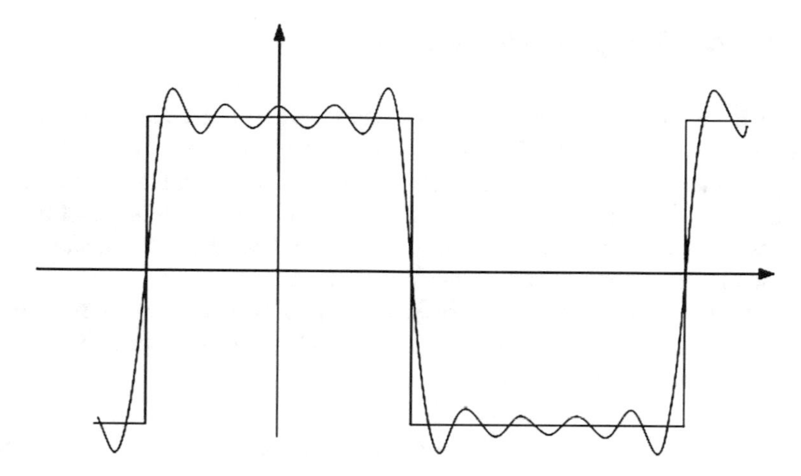

Fig. 5.2 Square-wave Fourier series: A square wave and a few of its harmonics. The curve shows the sum of the first five harmonics in the Fourier series for the wave

the b_j terms vanish. If the horizontal axis is centered on the wave, a_o vanishes. This leaves only the a_j terms for $j = 1, 2, \ldots$ non-zero. With Appendix Eq. (H.2) applied to such a square wave, the even a_j vanish and the odd are found to be given by $a_{2j+1} = (-1)^j (4/\pi)(1/(2j + 1))A$, where A is the amplitude of the square wave.

Both Fourier series and Fourier transforms are used extensively in biological and medical analysis, such as voice pattern recognition, heart murmur diagnostics, ultrasonic imaging, medical holography, X-ray crystallography and computed tomographic imaging (CT scan). Square waves are common in digital electronic circuits.

Early carbon microphones were not good at responding to high frequencies. This made the transmitted electrical wave representing the sound of a voice over a telephone 'cut off' at the vocal high frequency end. The effect is the same as leaving off the high-frequency components in a Fourier series. One can hear the consequence of this infidelity in early recording of the opera singer Enrico Caruso. The same reduction of high frequency components occurs in electrical circuits which have a relatively high inductance along the line or high capacitance across the input of the signal.

As we will see, the human inner ear mechanically performs a frequency decomposition of sound. In effect, the physiology of the Cochlea lets our brain generate a Fourier transform of the incoming sound.

As we have noted, the spectrum of sound frequencies possible is limited at the high frequency end only by the discrete nature of matter. For solids and liquids, wavelengths must be longer than twice the separation between the vibrating entities (atoms or molecules) as there is nothing to wiggle in between the atoms! For gases, the length of the mean-free path between collisions determines the shortest wavelength.

5.3 Phase, Group, and Signal Speed

You may have noticed in a bundle of waves spreading in a still pool of water from a stone's impact that the group of waves moves with a slower speed than the individual 'bumps' within the group. This is possible because as the bundle moves forward, new bumps grow at the rear of the bundle while the advanced bumps in front die out. As we will see, we can account for this behavior from the fact that waves in the bundle with different wavelengths have varying speeds.

First, in the spirit of Fourier, let us express any bundle of waves moving in one direction as a superposition of simple (sinusoidal) waves. The wave displacement for the bundle will then have the form

$$\xi(x,t) = \int dk A(\kappa) \exp i(\kappa x - \omega(\kappa)t) . \tag{5.24}$$

We will take a localized bundle of waves[11] with an amplitude $A(\kappa)$ dominated by wave numbers centered around some value κ_0. Expanding $\omega(\kappa)$ about these wave numbers gives

$$\xi(x,t) \sim \int_{near\ \kappa_0} dk \left[A(\kappa) e^{(i(\Delta\kappa(x - d\omega/d\kappa\ t)))} \right] e^{(i(\kappa_0 x - \omega_o t))} \tag{5.25}$$

i.e., the bundle has sinusoidal waves $\exp(i(\kappa_0 x - \omega_0 t))$ modulated by an 'envelope' with a more slowly varying phase, $\left[A(\kappa) \exp i\Delta\kappa((x - d\omega/d\kappa\ t))\right]$. From this expression, one can see that the wave with the faster-changing phase moves at the speed

$$v_p = \omega_o/\kappa_o , \tag{5.26}$$

called the 'phase velocity', while the envelope of the wave moves at a speed of

$$v_g \equiv d\omega/d\kappa , \tag{5.27}$$

called the 'group velocity' of the bundle of waves.

For a bundle of waves, there is a third speed one can define, which is the speed that the front of the wave packet is first detected by a distant receiver. For any wave packet, this speed is limited, according to Einstein's relativity theory, to be not greater than the maximum speed c, taken to be the speed of light. Wave pulses cannot carry signals faster than c. The speed of the leading edge of a wave pulse is called the 'signal velocity', v_s. The signal velocity may differ from both the phase velocity and the group velocity. In fact, either one of the latter may be greater than c.

---

[11] A bundle of waves is also called a 'wave packet'. If the wave packet is strongly localized, it is called a wave pulse.

The restriction $v_s \leq c$ applies even in the case of the so-called 'quantum information transport' and more fanciful 'quantum teleportation.' Within the formalism of standard quantum theory, there are instantaneous non-local elements. However, as soon as a measurement is made, information cannot be transferred faster than the speed of light. This must be so, because quantum theory satisfies Einstein's relativity postulates.

The relationship $\omega = \omega(\kappa)$ between the component frequencies in a wave and their wave numbers is called a 'dispersion relation', because the relation determines how the component waves of individual wavelengths separate from each other in speed and, during refraction, separate in direction. Since $\omega = v_p\kappa$, if the speed of a sound wave, $v_p = \sqrt{dp/d\rho}$ does not depend on wavelength of the sound, then $v_g = d\omega/d\kappa = v_p$, i.e. the phase velocity and the group velocity of the wave match. Quantum electron waves in relativity theory satisfy $\omega^2 = c^2\kappa^2 + (mc/\hbar)^2$, so $v_g = c^2/v_p$.

5.4 Energy Carried by Sound

Because sound causes material compression and rarefaction as well as oscillatory motion of material mass, sound energy has both potential and kinetic forms. This energy may be transported through the sequence of interactions of each layer of material with the adjacent layer as the sound propagates. We receive a small amount of energy each time we hear a sound. For any fluid, the energy per unit volume is given by $u = \delta E/\delta V = 1/2\rho v^2 + \rho\epsilon$, where ϵ is the internal energy of the fluid element per unit mass. As sound passes through the fluid, we will assume only small variations in the density, $\delta\rho$, and internal energy, $\delta\epsilon$, will occur. If the fluid was initially at rest, the speed is due solely to the vibrational motion. Expanding the energy of the fluid elements per unit volume in terms of the density variations will give

$$u = \frac{1}{2}\rho v^2 + \delta\rho\frac{\partial(\rho\,\epsilon)}{\partial\rho} + \frac{1}{2}(\delta\rho)^2\frac{\partial^2(\rho\,\epsilon)}{\partial\rho^2}\,, \tag{5.28}$$

in which terms of third order and higher in the density variation $\delta\rho$ are dropped.

Usually, as the sound moves through the fluid, there is little time for heat energy to be transported from the instantaneously pressurized regions to the rarefied regions of the sound before the wave moves half a wavelength. (The heat diffusion constant is usually much smaller than the wavelength of the wave times its speed.) Zero heat transfer defines an adiabatic process, and means that the entropy of each fluid mass element is unchanged. Thermodynamically, the internal energy of the fluid per unit mass increases only by heat input or work input, expressed as $d\epsilon = Tds + (p/\rho^2)d\rho$, where ds is the change in the entropy per unit mass in the fluid element, and Tds is the heat input. For our adiabatic processes, $ds = 0$. This makes $d\epsilon = (p/\rho^2)d\rho$. We have, at constant entropy per unit mass, $\partial(\rho\epsilon)/\partial\rho = \epsilon + p/\rho$, and

$\partial^2(\rho\epsilon)/\partial^2\rho = \partial(\epsilon + p/\rho)/\partial\rho = (\partial(\epsilon + p/\rho)/\partial p)(\partial p/\partial\rho) = v^2/\rho$. The energy per unit volume becomes

$$u = \frac{1}{2}\rho v^2 + \delta\rho(\epsilon + p/\rho) + \frac{1}{2}v^2(\delta\rho)^2/\rho\,. \tag{5.29}$$

For a repeating wave or a transient one, the second term on the right-hand side of the equation, averaged over the repeat or duration length, vanishes, since the density will return to its initial value. The two quadratic terms survive an average, being always positive. For a traveling wave, relation (5.23) gives $|\delta\rho| = \rho\,|\partial\xi/\partial t|\,/v$, making the two quadratic terms equal, so

$$\langle u\rangle = \rho\left\langle(\partial\xi/\partial t)^2\right\rangle.$$

For a single sinusoidal wave of angular frequency ω and amplitude A, the average energy density reduces to

$$\langle u\rangle = \frac{1}{2}\rho\,\omega^2 A^2\,, \tag{5.30}$$

since the average of the square of a sine wave is half its amplitude. For an arbitrary wave decomposed into a Fourier series, the average energy density will be a sum of such quadratic terms:

$$\langle u\rangle = \frac{1}{2}\rho\sum_i \omega_i^2 A_i^2\,, \tag{5.31}$$

because all the cross terms in the product $\sum_j\sum_k$ with $j \neq k$ coming from the Fourier series squared averaged over a period become zero, while those with $j = k$ have a squared sine wave which averages to $1/2$. Note that if the amplitude of a sound wave doubles, the energy it carries quadruples.

5.5 Intensity of Sound

The intensity I of a sound wave is the rate of energy transferred onto a unit area per unit time. This must be the net work per unit time per unit area done by the pressure. A given volume of material will have pressure acting on both of its sides (of area A), leading to an excess force of $(p - p_o)A$ compressing the volume in a time δt. We can write the rate of work done as the excess pressure times the speed of the moving layer of gas, so that

$$I = (p - p_o)\frac{\partial\xi}{\partial t} = -\rho_o v^2\frac{\partial\xi}{\partial x}\frac{\partial\xi}{\partial t}\,. \tag{5.32}$$

For waves moving in one direction, we can use Eq. (5.23) for sine waves to write

$$I = \frac{(p - p_o)^2}{\rho_o v} . \tag{5.33}$$

The two derivatives in Eq. (5.32) give a sine or cosine function squared factor, whose average is a half. This means that the average intensity delivered by a sound wave of a one-frequency wave is

$$I = \frac{1}{2}\rho_o \omega^2 A^2 v = v\, u . \tag{5.34}$$

which is the speed of the sound times the energy density in the wave. This relation also tells us that the wave intensity is proportional to the wave frequency squared and to the wave amplitude squared. For a given amplitude wave, doubling the frequency means four times more power is transmitted per unit area. The behavior of the intensity on the amplitude of a wave is typical of many types of waves, including those in electromagnetism.[12]

The relation between the intensity of a beam of sound and its energy density, $I = v\, u$, as given in Eq. (5.34), applies to waves and particles alike. If the wave is allowed to scatter around within a box, over some time, then the direction of the wave may become 'randomized'. Now, the chance that a wave arrives on a given small interior surface area is an average over solid angles about that area of the cosine between the surface normal and each possible beam direction, because as the beam which reaches that area tilts, less will hit the area in proportion to the cosine of the angle between the beam and the normal. Since $(1/4\pi)\int_0^{\pi/2} \cos\theta d\Omega = 1/4$, the intensity measured on any interior wall will now be $I = (1/4)v\, u$.

5.6 Sound Level

The impressive range of sound intensities for human hearing also makes intensity an inconvenient scale in describing our sensation of loudness. A logarithmic scale of intensity more closely matches that sense.

Physical 'loudness' is defined by

$$\beta = 10 \log_{10}\left(\frac{I}{I_{th}}\right) = 20 \log_{10}\left(\frac{p}{p_{th}}\right) , \tag{5.35}$$

where β is also called the 'sound level'.

[12] In quantum theory, the intensity of the waves are in proportion to the probability of finding a particle, each carrying a fixed amount of energy for each frequency. These intensities are also proportional to the wave amplitudes squared.

The loudness scale is often given in 'decibels'. Because β is an exponent, loudness has no physical units. Rather, the name decibel is a reminder of how the sound level is defined. The originally appended name for the loudness intervals of $\beta/10$ were called 'bels', named after Alexander Graham Bell for his work on the audiometer, a hearing-test instrument. By including the factor of ten, the loudness β has a convenient range from about 0 dB to about 120 dB, which is between barely audible and a level which can produce damage to the ears.

The threshold sound intensity measured and averaged over undamaged and healthy young ears comes to about $I_{th} = 10^{-12}\,\mathrm{W/m^2}$ for frequencies of 1000 Hz in air at 25 °C and 1 atm. This is the intensity adopted as the threshold of hearing. An average size mosquito makes this sound intensity at 3 m from the ear, with wing vibrations at about 600 Hz plus harmonics.[13] Now, to find to what pressure the threshold intensity corresponds, use Eq. (5.33) to form the time averages

$$\langle I\rangle = \frac{1}{2\rho_0 v}\langle(\delta p)^2\rangle$$

which gives a root-mean-square (RMS) threshold pressure for human hearing as

$$p_{th} \equiv \sqrt{\langle(\delta p)^2\rangle} = \sqrt{2\rho_0 v\langle I\rangle} = \sqrt{\rho_0 v I_{th}}\,, \tag{5.36}$$

where we have used the fact that for a sinusoidal wave, the maximum value of the intensity is twice its time average. The density of air at 25 °C and 1 atm is $\rho_0 = 1.184\,\mathrm{kg/m^3}$ and the speed of sound at this temperature and density is $v = 346\,\mathrm{m/s}$. This makes

$$p_{th} = 20.24 \times 10^{-6}\,\mathrm{N/m^2} \approx 20\,\mu\mathrm{Pa}\,.$$

A comparison of sound wave amplitude, pressure, intensity, and loudness is given in Table 5.3. Note: If your living room has a background sound level of 45 dB, you

Table 5.3 Conversion of sound amplitude, pressure, intensity, and loudness

Amplitude (cm)	Pressure (dyne/cm^2)	Intensity (W/m^2)	Loudness (dB)	Example
0.07	200	1	120	Jet exhaust
0.007	20	10^{-2}	100	Siren
7×10^{-4}	2	10^{-4}	80	Loud radio
7×10^{-5}	0.2	10^{-6}	60	Loud speech
7×10^{-6}	0.02	10^{-8}	40	Soft speech
7×10^{-7}	0.002	10^{-10}	20	Country home
7×10^{-8}	0.0002	10^{-12}	0	Quiet

[13] Male and female mosquitoes find each other by adjusting these harmonics. Humans can do this too, but not as well.

cannot comfortably hear the typical low sounds from a hi-fi audio system playing an orchestral piece with a dynamic range of 90 dB.

Assuming our 'small amplitude' linearization assumptions have some validity for intense sounds, the loudest sinusoidal sound wave that can be created in air would have a loudness of

$$\beta_{max} = 20\log_{10}\left[\sqrt{\langle(\delta p)^2/2\rangle}/p_0\right]$$

$$= 20\log_{10}\left[1.01325 \times 10^5\, N/m^2/(\sqrt{2} \times 20 \times 10^{-6}\,\mathrm{N/m^2})\right] = 191\,\mathrm{dB},$$

at which level the pressure variations go to zero absolute pressure ($p_o = 1.01325 \times 10^5\,\mathrm{N/m^2}$) on the low side. The denominator factor in the log is the peak threshold-of-hearing pressure, $\sqrt{2}\times(20\times10^{-6}\mathrm{N/m^2})$, which is $\sqrt{2}$ times the root-mean-square (RMS) threshold pressure. Such a wave generates a sound intensity of $1.2823 \times 10^7\,\mathrm{W/m^2}$, more than ten million times the intensity which can damage human ears, and more than ten thousand times the power delivered by sunlight on a square meter of the Earth's surface. Another limit occurs. Being a longitudinal wave, the amplitude of the wave cannot be greater than half the wavelength. Again with our 'linearization' assumptions, this happens when

$$I = (1/2)\pi^2\rho_o v^3 = 2.42 \times 10^8\,\mathrm{W/m^2},$$

which is 204 dB.

However, our linearization for the effects of pressure and density variations in the dynamics of sound will eventually fail as the sound intensity increases. We enter the world of non-linear behavior, wherein shock waves and other phenomena can reach more extreme levels of sound.[14] In liquids and solids, negative pressures can exist and be far larger than one atmosphere, because layers of liquids and solids can hold themselves together with adhesive forces.[15]

5.7 Acoustical Impedance

The transfer of sound energy from one material to another is important in hearing, sound insulation, ultrasonics, and other medical applications. The degree to which a material resists any kind of flow is called its impedance. This term applies to sound transmission just as well as electric current flow. In the case of sound, three

[14] An acoustical shock wave is a sound wave with strong variations in the supporting material speed over times much shorter than the period of the wave. A dynamical description of a shock wave involves forces non-linear in the speed variations.

[15] We return to this topic in the discussion of liquid cavitation by ultrasound, Sect. 5.24.2.

factors determine the impedance of a material: The inertial properties of the material as measured by its density, the degree of coupling between adjacent layers which determines the speed of sound in the material, and the dissipative processes within the material affecting the conversion of sound energy to heat or other forms, even before a sound wave can develop. We measure impedance by a ratio of the pressure needed to cause motion to the current that pressure causes.

The definition of impedance for sound transmission is perfectly analogous to impedance used to describe the general resistance to the flow of current in an electrical circuit. Within the subject of electricity, impedance is given by the relation $Z = V/i$, where V is the electric potential across a circuit (i.e. the difference in energy per unit charge), and i is the resultant current due to that difference in potential. The intrinsic electrical impedance, z, is defined to be the electric field needed to produce a current divided by the resulting current density: $z = E/J$. The quantity z gives an intrinsic measure of impedance since it does not depend on the size of the resisting material being used.

In analogy to the electric case, instead of electric energy per unit charge, in acoustics we use acoustical energy per unit volume, that energy determined by the mechanical pressure. Instead of electrical current density measured by the number of charges moving through an area per unit area and unit time, for fluids we use the volume of fluid moving through an area per unit area per unit time. If the response of a material to an external pressure is a smaller motion in comparison to a second material, then the impedance of the first material is larger.

These notions lead to a useful quantity for sound waves in a material, the specific acoustical impedance,

$$z \equiv -\frac{\delta p}{\partial \xi / \partial t} . \tag{5.37}$$

where the displacement speed, $\partial \xi / \partial t$ is due to the excess pressure $\delta p \equiv p - p_o$ acting above the ambient pressure p_o.

With Eqs. (5.23) and (5.18), the acoustical impedance for a sinusoidal wave can be expressed as the product of the material density times the speed of sound in the medium:

$$z_c = \rho v , \tag{5.38}$$

a value defined to be the 'characteristic acoustical impedance'. For compression waves, the impedance can also be expressed in terms of the bulk modulus (the inverse of the compressibility) and shear modulus of the medium using Eq. (5.15):

$$z_{cC} = \sqrt{(B + 4n_s/3)\rho} . \tag{5.39}$$

For transverse shearing waves,

$$z_{cS} = \sqrt{n_s \rho} . \tag{5.40}$$

The acoustical impedance in the mks system of units is often given the unit rayl, named after the physicist Lord Rayleigh. A rayl carries the fundamental mks units of kilogram per second per meter-squared. The acoustical impedance of air at room temperature (23 °C) is about 420 rayl; salt water is about 1.54×10^6 rayl and steel is 47×10^6 rayl. Some values of acoustical impedance for biological tissue (in cgs units) are shown in Table 5.7.

The fraction of the sound energy transmitted into a second material depends on the relative impedance of the materials. A wave which traverses a smooth boundary between two materials will be partially reflected and partially transmitted (and refracted if the incident ray is not normal to the boundary). Across the boundary, the wave amplitude and the pressure on either side of the boundary much match. For normal incidence, this makes the intensity of the transmitted wave, I_2, a fraction of the intensity of the incident wave I_1, according to

$$\frac{I_2}{I_1} = \frac{4z_1z_2}{(z_1 + z_2)^2} . \tag{5.41}$$

Proof At the boundary, let initial and reflected waves have amplitude A and B, and transmitted wave have amplitude C. Although the wavelength of the wave may change traversing the boundary, the frequency cannot, as the material is being forced to move at the incident frequency. The wave amplitude must have a single value at the boundary between the two materials, so that $A + B = C$. Also, a balancing of pressures (using (5.23)) makes $z_1(A - B) = z_2C$. These two relations give $A = (1/2)(1 + z_2/z_1)C$ and $B = (1/2)(1 - z_2/z_1)C$, so that $B/A = (1 - z_2/z_1)/(1 + z_2/z_1)$ and $C/A = 2/(1 + z_2/z_1)$. The wave intensities can be found from (5.34), i.e. $I_i = (1/2)z_i\omega^2A_i^2$, so that

$$I_2/I_1 = (z_2/z_1)C^2/A^2 = 4(z_2/z_1)/(1 + z_2/z_1)^2 \quad q.e.d.$$

As can be seen from Eq. (5.41), the percent of sound energy transmitted as a function of the impedance ratio peaks when $z_1 = z_2$, for which all the energy of the wave passes into the second material, with no reflected wave.

Consider the electrical analog: An amplifier with a output impedance Z_A transfers power to speakers with Z_S. If the amplifier power supply has a voltage V, the current through speakers is $i = V/(Z_A+Z_S)$, so the power supplied to speakers is $P = i^2Z_S = Z_SV^2/(Z_A + Z_S)^2$. For fixed V and Z_A, the power transferred is maximum when the slope of P as a function of Z_S vanishes. You can check that this occurs when $Z_S = Z_A$.

In acoustics as well as in electronic circuit design, adjusting the transmitter or receiver impedances to maximize power transfer is called 'impedance matching'. An example of poor impedance matching occurs when you to try to push air back and forth with the tip of your finger. A better match is to push air back and forth with a hand fan.

The fraction of energy reflected is $(z_1-z_2)^2/(z_1+z_2)^2$. As the impedance of the second material grows above the first, less and less of the initial wave is transmitted.

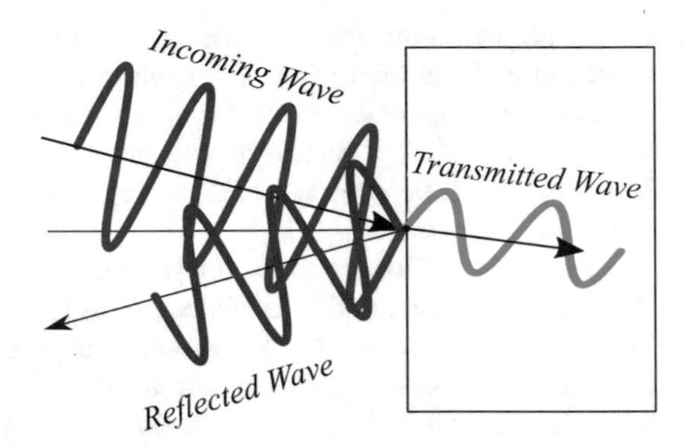

Fig. 5.3 Reflection and transmission of sound across a boundary from a lower to a higher acoustical impedance

Table 5.4 Impedance parameters for air and salt water

Air density (15 °C)	ρ_a	1.225×10^{-3} g/cm^3
Air speed of sound	v_a	3.43×10^4 cm/s
Air impedance	$z_a = \rho_a v_a$	41 g/(cm^2 s)
Salt water density	ρ_w	1.025 g/cm^3
Salt water speed of sound	v_w	1.5×10^5 cm/s
Salt water impedance	$z_w = \rho_w v_w$	1.54×10^5 g/(cm^2 s)

Instead, a larger and larger fraction of the transmitter energy is reflected back into the first material.

Consider the transmission of sound from air into the human body. We will approximate the impedance properties of human tissue by those of salt water. Figure 5.3 shows how the sound waves (along rays) reflect and refract. Taking numbers from Table 5.4, we have an intensity of transmission of only 0.1%. Thus, most of the sound energy in the air does not enter the body through the skin, but rather gets reflected. In order to effectively use an ultrasonic probe, the impedance of the probe end and the tissue should be more closely matched. This can be done by using a grease-gel between the probe and the skin.

5.8 Sound Propagation in Non-uniform Matter

As a coherent vibration within materials, sound waves are propagated through layers of matter as each layer interacts. Huygens' principle applies: Each location on a wave front acts as a source of new waves.[16] If the properties of the supporting matter change as the wave moves, the newly created Huygens wavelets may sum to a reflected and refracted wave. This occurs particularly at the boundary between two materials with differing sound velocities. The refraction of sound follows Snell's law in the form

[16]See Appendix B for more on Huygens' Principle.

$$(1/v_1)\sin\theta_1 = (1/v_2)\sin\theta_2\,, \tag{5.42}$$

where the θ_i are the angles between the boundary normal at the intersection of a sound ray (a line perpendicular to a wave front).

Snell's law is a property of any wave (acoustical or field wave, such as light) passing through regions of differing wave speed. Physical continuity requires the wave fronts to be continuous, and causality requires that the frequency of the wave remains fixed through the region. These two conditions are sufficient to derive the bending of rays[17] and Snell's law. Acoustical lenses can be constructed to converge or diverge sound waves.

5.9 Dispersion of Sound

The dispersion of a wave refers to the separation of waves of different frequencies. Refraction of sound waves in materials, including biological tissue, with that sound containing a number of Fourier components, can also show a separation in the direction of refraction for each of the component waves. Dispersion, or separation, of waves of different wavelengths may occur if the wavelength of the wave in a material depends non-linearly on the wave frequency. This is a property of waves for which the group velocity differs from the phase velocity. For a given frequency, the phase velocity, $v_p = f\lambda$, will be proportional to the wavelength, but the group velocity is $v_g = df/d\kappa$, where κ is the wavenumber, the inverse of the wavelength. So if the frequency is not proportional to the wavenumber, wave dispersion can occur. Table 5.5 shows the measured change in speed of sound per unit frequency in bovine heart tissue and in turkey breast tissue.

Since the absorption and re-radiation of sound in a material, such as in the interior of a cell, in tissue, or in bodily fluids, depends strongly on the natural vibrational frequencies of membranes, filaments, capsules, and bubbles, so too the dispersion of sound is strongly affected by those natural vibrational frequencies (resonances).

Table 5.5 Sound dispersion in biological tissue

Tissue	dv/df (m/s/Hz)	f
Bovine heart	0.63 ± 0.24	at 1.5 MHz
	0.27 ± 0.05	at 4.5 MHz
Turkey breast	1.3 ± 0.28	at 1.75 MHz
	0.73 ± 0.01	at 3.9 MHz

[17]Rays are lines perpendicular to the wave fronts pointing in the direction of propagation.

5.10 Non-linear Sound Waves

The linear wave equation (5.22) applies when the pressure amplitude is small compared to the ambient pressure, the variation of the material density is small compared to the ambient density, and the speed of the fluid as it moves in response to a sound wave is small compared to the speed of sound in the material.

Phenomena such as surface water waves breaking at the beach shore, shock waves produced by whips and supersonic aircraft, bumble bee flight, and other fluid turbulence produced by sound can be explained by the non-linear effects coming from the $(\mathbf{v} \cdot \nabla)\mathbf{v}$ term in the Navier-Stokes equation (4.42).

If density variations are not small, such as in fluids containing bubbles, the non-linear effects again become important. As sound passes, the pressure variation changes the volume of bubbles, and thus the density of the medium. Non-linear effects cause harmonics in the wave as it propagates. A wave in bubble water initially sinusoidal can evolve into a saw-tooth shaped wave as the wave progresses.

5.11 Absorption of Sound

The source of sound attenuation may be baffling.
—wcp

Besides impedance mismatch causing sound reflection, sound also tends to reflect better from hard surfaces than from soft ones, because during the reflection from a soft material, some sound energy may be lost by conversion into heat. For example, a piece of cloth will absorb energy when the fibers within the cloth flex and heat while responding to the incoming sound. The sound 'absorptivity' of a surface is defined as the fractional power dissipation when sound encounters that surface.

Even within one material, sound scattering and viscous loss causes sound energy to be dissipated. The loss of sound intensity as a sound wave propagates is referred to as 'sound attenuation' of the wave. For a traveling wave, scattering and viscous energy loss means that the further the wave goes, the more it loses kinetic and elastic potential energies. Since the loss over a short distance is proportional to the distance traveled and the energy present, the intensity through a homogeneous material must satisfy[18]

[18]This follows from $dI \propto I\,dx$, or $\int dI/I = -\alpha \int dx$, so $\ln I = -\alpha x + C$ or $I = I_0 \exp(-\alpha x)$. The relationship is quite 'universal'. The only important assumptions are that the absorbing material is homogeneous along the direction x and that the loss of intensity is proportional to the intensity. We will use this same behavior when considering the attenuation of light traveling through a lossy material (as in Eq. (7.14) for microwaves and in the Beer-Lambert Law Eq. (7.19)). There are exceptions, such as the intensity loss with distance for a beam of charged particles sent through tissue, because the loss depends non-linearly on the beam energy flux.

Table 5.6 Extinction length of sound in water and air

Frequency	Water	Air
1 kHz	1000 km	500 m
10 kHz	10 km	5 m
100 kHz	100 m	5 cm
1 MHz	1 m	0.5 mm
10 MHz	1 cm	5 m

$$I(x) = I_o \exp(-\alpha_L x), \tag{5.43}$$

at least for short distances. Here, α_L is called the 'intensity attenuation coefficient', and I_o is the intensity at $x = 0$. The attenuation coefficient determines the relative sound energy loss per unit length, and generally is strongly dependent on the sound frequency, particularly when the material has vibrational resonances.

The loudness of the sound is reduced over distance according to

$$\Delta\beta = -10 \log_{10}(e)\, \alpha_L x = -4.343\, \alpha_L x \equiv -ax\,.$$

The constant $a = 4.343\, \alpha_L$ is called the sound 'absorption coefficient'. most often given as decibels per meter. The inverse of the sound attenuation α_L is call the 'extinction length', giving the distance over which the sound intensity falls to $1/e$ of its initial value. Table 5.6 shows the extinction length for sound in water and air. The extinction length for sound in water is about 2000 times that of air. (Whales can hear one another across a good fraction of an ocean.)

In a gas and many liquids and solids, the energy loss per unit distance increases as the wave frequency squared. When liquids contain particles and bubbles whose sizes are comparable to the wavelength of the sound, sound scattering will be an important cause of attenuation. For sound with a frequency in the range of 5–10 kHz, seawater has an absorption coefficient $a \approx 1.5 \times 10^{-8} f^2$ dB s^2/km, while fresh water is only about $1/75$ of this value.

The absorption coefficient of sound in air depends on the temperature and humidity and increases as the square of the frequency for high frequencies. Table 5.7 gives values for the impedance and absorption coefficient for some materials of biological interest. Note that bone has a much larger absorption coefficient than soft tissue, and a stronger frequency dependence. The reason is due largely to viscous loss in friction between the bone matrix and softer components and to sound scattering by the heterogeneous material in the bone.

As an example of the frequency dependence of the absorption of sound with distance, if a 1 kHz sound travels 500 m in air before dropping 5 dB, then sound at 1 MHz will only travel $10^{-6} \times 500$ m $= 0.5$ mm before dropping 5 dB. Sound attenuation occurs in tissue by the following mechanisms:

Table 5.7 Impedance and absorption

	Speed (cm/s)	Impedance (g/(cm^2 s))	Absorption coeff. (dB/cm at 1 MHz)	Frequency dependence
Air	0.343×10^5	0.00041×10^5	1.2	f^2
Water	1.48	1.48×10^5	0.0022	f^2
Fat	1.45	1.38×10^5	0.63	f
Brain	1.54	1.58×10^5	0.85	f
Blood	1.57	1.61×10^5	0.18	$f^{1.3}$
Muscle	1.58	1.70×10^5	1.30 $\parallel$ (3.3 $\perp$)	f
Skin	1.73	1.99×10^5	0.44	$f^{1.55}$
Bone	4.08	7.80×10^5	>8.	$f^{1.7}$
Teeth	5.50	15.95×10^5		

- Viscous heat production as sound passes through materials. Low frequencies (below 20 Hz) are only weakly absorbed, and so have the smallest α. High frequencies (above 100 kHz) are strongly absorbed, even in fresh water. The presence of ionic salts increases absorptivity. In tissue, vibration of internal structures adds to the loss of sound energy. The resulting heat may cause coagulation and depolymerization, and other chemical changes in a cell.
- Molecular vibrations and rotations, which can be strongly excited by ultrasonic waves in fluids.
- Heat conduction from higher pressure regions to lower pressure regions. (Usually a small effect because the speed of sound is much greater than the speed of the heat wave.)
- Absorption by gas bubbles can cause considerable sound scattering if their size is comparable to the sound wavelength. Heat at the surface of the bubble will cause a loss of sound energy. Similarly, absorption by droplets in a gas occurs, such as in a fog.
- Inelastic vibration and scattering by inhomogeneous materials, including sols, gels, and network structures in cells, tissues, and bone.
- Inelastic vibration of boundary layers between tissue will absorb sound at the interface. Some energy from the sound is converted to other forms, usually to heat.
- Creation of 'cavitations' (partial vacuum bubbles) in fluids. The cavitation bubbles vibrate, and cause rapid fluid motion with strong eddies, severing protoplasm microtubules, converting gels to sols, and breaking cell membranes. (See Sect. 5.24.2.) Analogously, the creation of condensates in a saturated vapor as sound passes adds to sound absorptivity.
- Tissue rupture and material fracturing.
- Stimulation of chemical reactions.

5.12 Resonance

An important phenomenon in nature is the transfer of energy from one oscillatory system to another near a natural vibratory frequency of the receiving system. At this frequency, the energy transfer is greatest.

The phenomenon is called 'resonance'. To imagine resonance behavior, think of pushing a swing with a rhythmic shove. If your intent is to make the swing gain amplitude, your rhythm should match the swing's natural frequency. Over many cycles of the swing, when you push with the natural frequency, it takes only a little effort on each cycle of the swing to cause a build up in the energy stored in the swinging motion.

Alexander Graham Bell liked to demonstrate sound resonance by singing into a piano, thereby causing only those strings to respond which matched the frequencies in his singing voice. Singing certain notes in the bathroom can cause the air in the room to resonate as standing waves are set up between the rigid walls.

Here are some other important examples:

- Your radio, television, and cell phone use resonance circuits to tune to stations or channels (frequencies), while speaker cones are made to NOT resonate, in order to have a 'flat' response to various sound signal frequencies.
- Light arriving at your retina is resonantly absorbed by rhodopsin and other pigments in the rod and cone cells within the retina.
- Light energy is utilized by plants through resonant excitation of chlorophylls and accessory pigment molecules.
- In a process called 'auscultation', a doctor listens for resonant echoes when he thumps on your chest to cause an internal organ or cavity to vibrate. Organs such as the lung, the heart, soft and hard regions of the digestive tract all have particular resonant frequencies which can be heard on the skin as particular and identifiable sounds.
- The colors of your clothes result from light scattering by organic dyes which resonantly absorb the complementary colors.
- Biochemists use resonance in analyzing organic compounds by performing an infrared spectroscopy of the material. Light in the infrared band can resonantly excite the vibrational modes of atoms with particular bonds in a molecule. Detecting the frequencies of these excitations forms a signature of a particular kind of molecule.

Resonance can occur in the forced vibration of a material. For a small segment of mass m of the material experiencing a force from an external wave, Newton's second law lead to

$$\ddot{\xi} + 2\pi\beta\dot{\xi} + 4\pi^2 f_0^2 \xi = (F_o/m)\cos(2\pi f t) . \tag{5.44}$$

Here, the first term on the left-hand side of the equation is the acceleration of the mass (whose position measured from equilibrium is called ξ). The second term on the left is caused by a 'damping force', which we take proportional to the speed of the mass. The factor β is called the 'damping coefficient'. The third term comes from an elastic restoring force on the mass due to interactions by adjacent material. The number f_o is the natural vibrational frequency of the mass. The right-hand side is the forcing term (per unit mass), and is taken as a cosine wave. The number f is the frequency of the forcing term.

The solution to Eq. (5.44) has a piece which remains after an initial transient dies out, expressed by

$$\xi(t) = \frac{F_o/m}{4\pi^2\left[(f^2 - f_0^2)^2 + \beta^2 f^2\right]^{1/2}} \cos(2\pi f t - \phi) \tag{5.45}$$

where $\tan\phi = \beta f/(f_0^2 - f^2)$.

One can see that the amplitude of the motion, after the initial transient, will peak at $f_M = \sqrt{f_0^2 - \beta^2/2}$. As shown below, this is near the frequency of resonance when $\beta << f_0$. The strength and width of the resonance amplitude depends on the damping (dissipation) in the system, i.e. the mechanisms for loss of energy.

The steady energy transferred per cycle from the forcing term to the oscillator will be

$$\langle\mathcal{E}\rangle = \frac{F_0^2}{4\pi m} \frac{\beta f^2}{\left[(f^2 - f_0^2)^2 + \beta^2 f^2\right]} . \tag{5.46}$$

The energy that is lost from the forcing wave by dissipation leads to attenuation of waves as they move across an absorbing medium. The maximum energy transfer occurs at the peak of the function $\langle\mathcal{E}\rangle$, where $d\langle\mathcal{E}\rangle/df = 0$. This occurs when the frequency of the forcing term is $f_R = f_0$, i.e. the natural vibrational frequency of the oscillator.[19]

Figure 5.4 shows how the energy transfer to a harmonic oscillator becomes maximum when the forcing frequency f is at the resonant frequency f_R. The energy scale has been normalized to one at resonance and we have taken a damping coefficient of $\beta = (1/5) f_0$.

Besides the effect of tissue on acoustical waves, the effect of material on electromagnetic waves passing through uniform biological tissue can also be modeled by a set of damped harmonic oscillators made up of bound charges responding to the passing electric field. In this model, the imaginary part of the index of refraction which determines light absorption has the same functional dependence on f as $\langle\mathcal{E}\rangle$ does in Eq. (5.46).

[19]Note that the forcing frequency does not have to exactly match the natural frequency for there to be significant enhancement of absorption by the oscillator. This will be an important point when we discuss infrared spectroscopy and other resonant absorptions. For an electromagnetic wave forcing bound charges to wiggle, the dragging force may be due to re-radiation back to light rather than frictional drag producing heat.

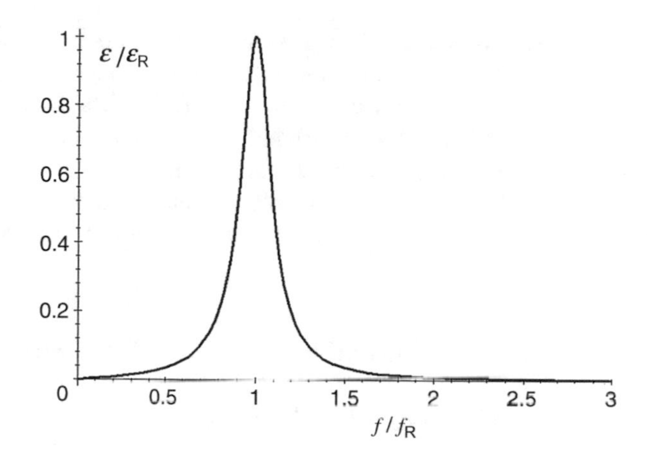

Fig. 5.4 Energy transfer to a harmonic oscillator as a function of the frequency of the driving external force

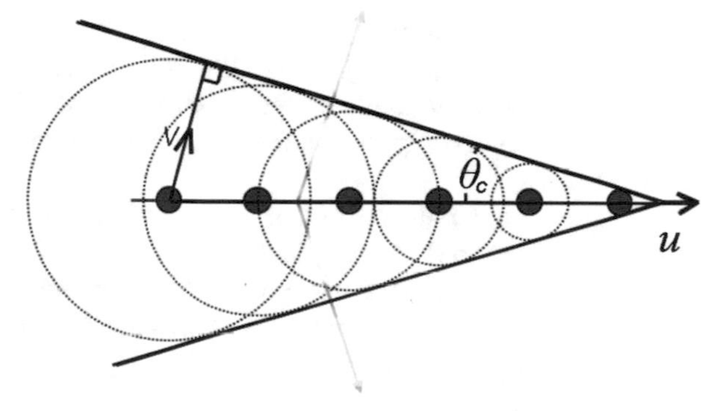

Fig. 5.5 Shock wave formation behind a supersonic object: The central ball is shown in time sequence, moving faster than the waves it created. The waves are shown as circles, which expand away from their creation point. These waves add coherently on a cone whose axis is the ball's trajectory. Note $\sin\theta_c = v/u$

5.13 Supersonic and Subsonic Motion

Supersonic motion occurs when an object exceeds the speed of sound, and subsonic refers to motion less than the speed of sound. The ratio of the speed of the object and that of sound is called the Mach number. The end of a whip which makes a cracking sound moves faster than the speed of sound in air. The Concord airplane moved faster than the speed of sound. Sources of sound from that aircraft created a cone shaped shock wave in the air, carrying sufficient energy to rattle windows on homes below, and creating a "sonic boom".

One can see by the following argument that the shock wave emanating from the supersonic motion of a localized object forms a cone with half vertex angle whose sine is the inverse of the Mach number (Fig. 5.5). Suppose a body periodically emits a sound while traveling supersonically at a speed u greater than the sound speed v. A given sound pulse will expand spherically into the medium, which we take at rest. Before the next pulse is emitted, the body has penetrated through the wave of the first pulse. The wave fronts for a series of past pulses will form a set of intersecting spheres, each displaced from the prior one by a distance uT, where T is the time between pulses, and each one has a growing radius of size vt, with t the time since the pulse was emitted. The greatest wave amplitude will occur on a cone, with axis along the body's motion, and vertex at the position of the body. The half vertex angle of the cone will be $\theta_c = \arcsin(v/u)$. The figure shows wave fronts from sound emanating from the moving source, adding coherently on the cone's surface. This is where the sound shock wave exists.[20]

If the object were a supersonic airplane (or a meteor entering our atmosphere), the lower part of the cone produces a 'sonic boom' across the ground.

5.14 Non-biological Sound Wave Production

5.14.1 Mechanically Generated Sounds

Vibratory mechanical motion in a medium produces sound. Sound-producing mechanical devices transform a fraction of their kinetic energy into sound energy. A jet engine produces sound by the whir of a turbine fan and the turbulent kinetic motion of hot air exiting the engine, producing sound power of the order of a few kilowatts.

Whistles and sirens operate by interrupting the flow of air, generating air turbulence. The rapid changes in air pressure near the instrument radiates sound waves. Sirens can produce extremely intense sound waves, sufficient to cause a cotton shirt to catch fire.[21] Sirens have frequencies from 0.1 to 100 MHz, and sound intensities up to 30 kW/cm^2 (205 dB).

The table below lists some of the highest power sounds produced in air. Their sound levels have been calculated at 1 km from the source.

Event	Sound power in Watts	Sound level at $r = 1$ km
Tunguska 1908 comet explosion	1.26×10^{27} W	320 dB*
Krakatoa 1883 eruption	5.00×10^{16} W	216 dB*
USSR Tsar Bomba	1.26×10^{13} W	180 dB
Saturn V launch	7.93×10^{11} W	168 dB
1 ton TNT explosion	3.16×10^{10} W	154 dB
Bell-Chrysler siren	5.00×10^{6} W	116 dB
Thunder clap	1.00×10^{6} W	109 dB

[20]Electromagnetic shock waves can also be produced. Charges moving through ordinary material faster than light moves in that material will radiate. This 'Cherenkov' radiation is observed as a blue glow in the water used to shield an operating nuclear reactor.

[21]This experiment is best done without a person wearing the shirt.

The 'starred' sound levels were not measured, but rather calculated from power. The maximum sine wave sound level at 191 dB occurs when a pressure variation matches the atmospheric pressure, so that the lowest pressure in such a wave is zero. Higher top pressures have finite zero-pressure regions with little air, caused by layers of air slamming together and then scattering apart. The maximum amplitude condition for a longitudinal wave gave an upper limit of 204 dB. (See Sect. 5.6.) Both of these limits are indicative only, since the waves become non-linear. (If the sound produces a ten per cent variation in the atmospheric pressure, it will be at a sound level of 171 dB.)

Much higher pressure variations can be sustained in water than in air. Whales can produce sound levels up to 188 dB.

Life on Earth suffers the vicissitudes of astronomical and geological turmoil. Earthquakes are an example which produce high-energy sound waves within the Earth that can destroy lives and environments.

Earthquakes are generated by the movement and slippage of large masses of material under our feet. Tectonic plate movement can build energy through increasing stress of solids in the Earth. When the stressed material finally breaks, the slippage is usually rather jerky, not smooth. There results large and energetic sound waves propagating outward from the slipping region. The sound reflects, refracts and gets partially absorbed through various solids and liquids in the Earth. The sound wave has both longitudinal and transverse components, each moving at different speeds, and when they reach the Earth's surface, surface waves propagate at still another speed.

The energy released by an earthquake can be rather well determined by sound-wave-detecting devices called seismometers. Because this energy measured in joules extends over many decades in powers of ten, a logarithmic scale is used to characterize the magnitude of the quake. Such a scale was invented in 1935 by Charles Richter.[22] Barely-detected quakes have a Richter magnitude of about 1.0, and quakes which produce total destruction of even sturdy buildings having a magnitude above 9.0.

To calculate the energy released when the Richter magnitude has the value $\mathscr{M}$, use

$$\mathscr{E} = 10^{1.448\mathscr{M}+4.7993}\ \text{J}\,. \tag{5.47}$$

A magnitude 9 earthquake releases 6.78×10^{17} J. If the quake lasts 10 s, the power generated would be about 6.78×10^{16} W, which is comparable to the sound power that was released to the air by the Krakatoa 1883 volcano eruption.

[22]C.F. Richter, *An instrumental earthquake magnitude scale*, Bulletin of the Seismological Society of America **25:1–2**, 1–32 (1935).

5.14.2 Electromechanical Devices

- *Electromagnetic Speakers*: An electromagnetic speaker produces sound by the vibration of a stiff cone of thin fibrous material pushing and pulling the air. A thin axial cylinder is glued to the cone near its vertex. The cone and cylinder are suspended at the large circumference of the cone by a very flexible material which allows the cone to move freely (to some limit) in the direction of the axis of the cone. Around the cylinder is a thin coil of copper wire, attached to the frame of the speaker by very flexible copper wires. Within the hollow cylinder is a fixed permanent magnet, with a north-south axis along the cone axis. The magnet is not centered in the coil, but displaced outward a small distance. When a current is passed through the coil, the magnetic field that current produces either attracts or opposes the permanent field, causing the cylinder to push the cone in or out. The cone acts as an impedance matching device, i.e. the cylinder and cone match the impedance of the surrounding air far more than the cylinder by itself.
- *Electrostatic devices*: If two metal plates are put in parallel with a gap between, they can act as a capacitor. If one plate is made thin and held elastically, it can be made to vibrate by an alternating current fed into the capacitor. The surface of the vibrating plate pushes and pulls air, producing sound.
- *Magnetostriction devices*: Magnetostriction occurs when there is a change in the length of a ferromagnetic material due to an external magnetic field. An unmagnetized rod will expansion if put into magnetic field. Thus, an alternating field will cause such a rod to vibrate at twice the frequency of the field. If the rod is already magnetized, the alternating field can generate vibrations in the rod with the same frequency as the field. If the field's frequency matches the natural frequency of vibration of the rod, resonance will occur, making the rod's vibration a maximum. Longitudinal sound waves are generated in the rod, which are emitted from the ends of the rod as sound waves in the adjacent material. Magnetostrictive devices have outputs up to 50 W/cm^2.

  A magnetostrictive generator is usually constructed around a inductor-capacitor circuit stimulated by an electronic oscillator. The coil of the inductor contains a rod of ferromagnetic material such as pure nickel, or nickel-copper alloys (invar and monel). The oscillator frequency $f_R = 1/(2\pi\sqrt{LC})$ can be adjusted by changing the inductance L of the coil, or the capacitance C. The natural frequencies of a rod vibrating longitudinally are given by $f_n = (n/2l)\sqrt{Y/\rho}$, where Y is the Young's modulus of the rod, ρ is its density, l is its length, and n is an integer giving the harmonic index for that frequency.
- *Piezoelectric devices*: Piezoelectric generation of ultrasound is possible by placing an alternating electric field in the same direction as the piezoelectric axis of the crystal. Resonance, and thus the maximum amplitude of vibration, occurs when the electric field frequency matches the mechanical vibrational frequency of the crystal along the axis of the field. A low frequency limit of about 10 kHz comes from the difficulty in exciting thick slabs of crystal. An upper frequency limit of about 10 MHz comes from having to make crystals too thin to sustain

vibrations. Because of the precision of the mechanical vibrational frequency, piezoelectric crystals are commonly found in clocks and computers, where they are placed in electrical timing circuits. Quartz, tourmaline, and Rochelle salt are commonly used.

5.15 Biological Generation of Sounds

5.15.1 Vocalization

Many species of animals living in air and water use sound as a method to communicate. Some use sound echoes to navigate, locate food, sense predators, and find other objects.

Birds communicate with alarm calls, flock-gathering hails, territorial defense cries, identifying signals, and courtship songs (Fig. 5.6). Instead of a larynx, birds have a syrinx. The syrinx does not have vocal cords, but rather muscle control of a series of cartilage and tissue rings surrounding the air passage above and below the juncture of the trachea with the two bronchi. Having control of the muscles within the two bronchi, birds can make two different sounds at once.

Whales can easily communicate across many kilometers, and under the right conditions, several thousand kilometers, i.e. across oceans, but then with perhaps an hour to wait for an answer! They have a sound frequency range of $f = 10$–31 kHz, with a loudness level up to 188 dB above a threshold level of 0 dB. Note that the threshold sound level in water is taken to have a pressure of 1 μPa (1 μPa = 10^{-6} N/m^2), while, as we have seen, the threshold level in air is set at 20 μPa based on human hearing.

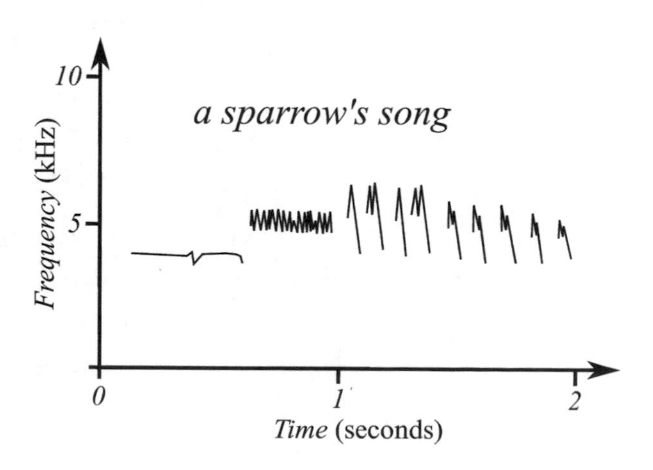

Fig. 5.6 Spectrogram of a sparrow song

5.15.2 Vibration of Elastic Biomembranes

Purposeful biosounds are produced by vibration of specialized elastic biomembranes and organs, as well as noise-generating turbulence. For loud sounds, these vibrations must be strongly coupled to the sound medium (air for land animals, water for aquatic animals). Coupling can be achieved by impedance matching the vibrating surface to the medium of sound. Our vocal chords are set in motion by air we push out of our lungs, but we use the larynx for better matching of the chords elastic impedance to the air impedance in the throat, lungs, mouth, and nasal passages. (See Fig. 5.7.)

As we have described, birds produce sounds by oscillations of membranes in their syrinx, but some also communicate by the sound of their fluttering wings.

Some insects scrap stiff body parts together to make sound, a technique called 'stridulation'. To match the impedance of air, the surfaces of the vibrating and radiating parts are flat, with wide areas. Adult male cicada produce the loudest sound among insects by flexing and snapping ribbed membranes ('tymbals') on their first abdominal segment (Fig. 5.8). The African cicada brevisana sounds has been measured with a sound level of 106.7 decibels at a distance of 50 cm.

Crickets make sounds by rubbing their wings together. Grasshoppers use stridulation to attract other grasshoppers, using sound far up in the ultrasonic range.

The buzzing sound of a mosquito also comes from stridulation, in this case the scrapping of a wing organ against a body organ. They hear with hair cells clustered in an organ on their antennae.

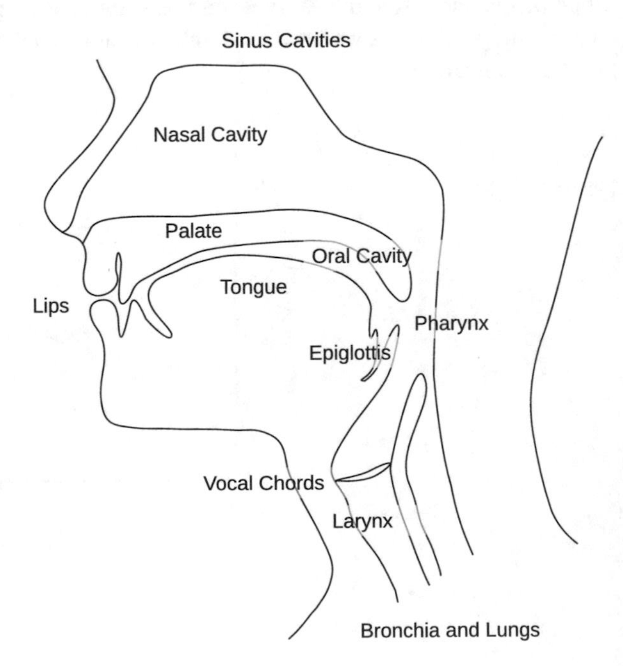

Fig. 5.7 Human vocal apparatus

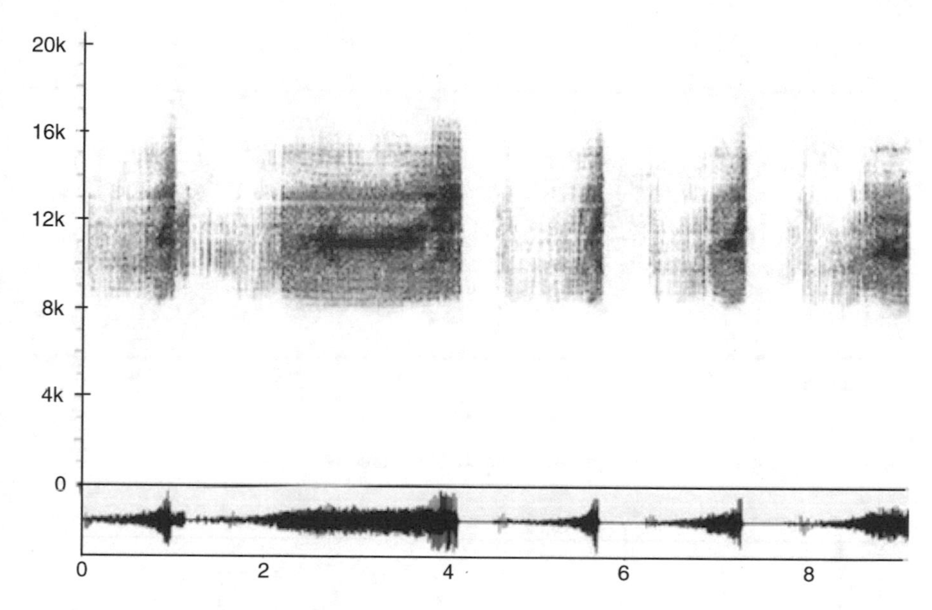

Fig. 5.8 Spectrogram (and oscillogram) of a European cicada. Vertical axis: Frequency, Horizontal axis: Time (s) (From www.cicadasong.eu)

Bats use the sound they generate for echolocation, which is the active use of sound for navigation and ranging (SONAR). Most bats produce sound with their larynx, but some click their tongue. The sound is emitted from the mouth, or, in the case of horseshoe bats, from specially shaped nostrils. The sounds range in frequency from 20 kHz to 200 kHz (a range above human hearing) (Fig. 5.9). This makes sound of wavelength in air in the range of 1.7 cm to 1.7 mm. Thus bats can resolve, with sound, objects in the millimeter size range. Their calls vary in frequency and pulse length, and in intensity (near mouth) from 50 dB to as high as 120 dB. The bat brain can detect Doppler shifts as small as 0.1 Hz, giving them information about the speed of their insect prey. Before emitting a call, bats protect their own ears as we do by contracting the muscles of the middle ear which decouple the three middle ear bones from the inner ear window. This action occurs about 6 ms before the bat emits a sound pulse of about 1 ms, and then 2 ms after, the muscle relaxes to allow the returning echoes to be transmitted to the inner ear. This means that bats can sense echoes from objects no less than 34 cm away. As the bat reaches it prey, the pulse rate increases to just under 200 a second. With a flight speed up to 2682 cm/s, the bat would reach its prey in just 13 ms after getting the last echo. To distinguish a call from those of other bats, the call pitch and its harmonics varies from bat to bat. Note that the bat ears also play a strong role in collecting sound from a particular direction.

Dolphins and some whales also use echolocation (Fig. 5.10). Because sound enters and leaves the prey's body due to close impedance matching, a dolphin can 'see' if a fish's heart is beating and other physiological properties related to the prey's health.

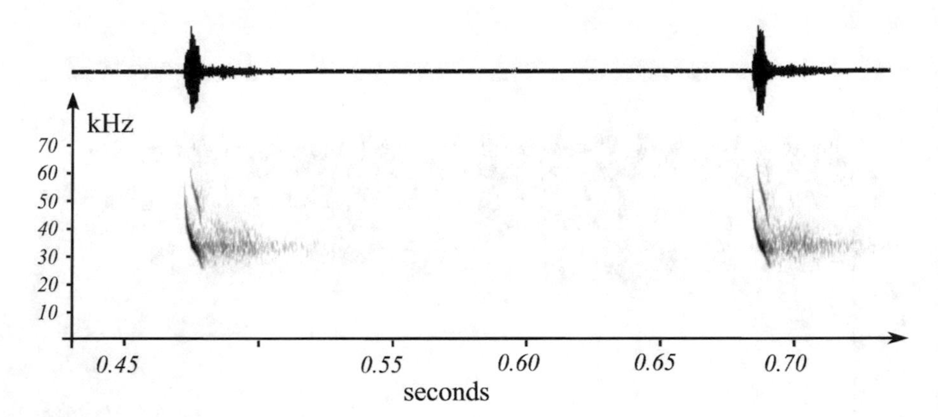

Fig. 5.9 Oscillogram and spectrogram of a brown long-eared bat

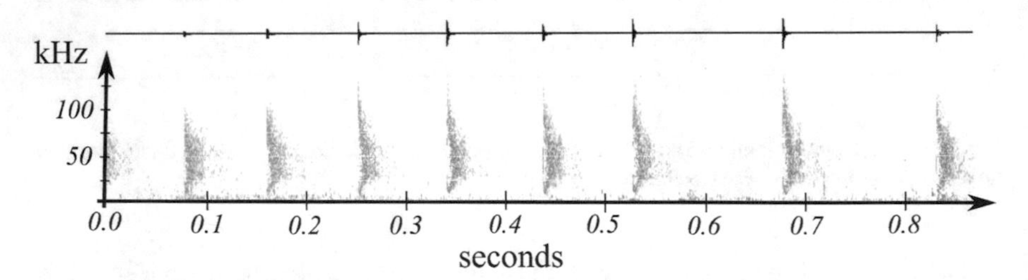

Fig. 5.10 Oscillogram and spectrogram of the clicks from a bottle-nose dolphin

A large variety of fish can communicate with popping and grunting sounds, with the purpose of keeping schools together, sending distress signals, or attracting the opposite sex.

5.16 Physical Sound Wave Detection

Any phenomena in nature capable of generating sound has an inverse: The same mechanisms can be used to detect sound. This reflexive property of such phenomena has its basis in the time reversal symmetry of the laws of nature at the level of atoms and molecules: Electrodynamics applied to atoms and electrons has the property that for every dynamical solution, changing the time variable t to $-t$ gives another valid solution. At a macroscopic level, the principle still holds, except it may happen that the time-reversed solution is highly improbable. Evidence for this possibility comes from the production of entropy in the forward process. If the forward process has little or no entropy production, then the process which has all the dynamical components with an odd power of time reversed (such as speed) is also a macroscopic possibility.

A speaker constructed with a stiff paper cone glued to a coil of wire put in a magnetic field has no intrinsic need to produce entropy (in this case, heat released divided by the background temperature), and so can also act as a microphone. As such, the air vibrations wiggle the cone. This wiggles the coil in the magnetic field, causing charges to be pushed through the wire of the coil. If the wires are attached to another speaker, that second speaker will respond and produce a sound. (More commonly, the first coil is attached to an amplifier to raise the sound level of the second speaker.)

5.16.1 Microphones

A device which converts one form of energy into another form is called a 'transducer'. A microphone is any device which can convert acoustical intensity into a proportional electric current. A faithful conversion will preserve the information content in the sound.

The first microphone made by Alexander Graham Bell's assistant used a conductive wire whose depth in an acid bath varied by the vibrations of an attached sound diaphragm. Current in the wire then changed according to the sound.

Early telephones used Bell's electromagnetic speaker and Thomas Edison's idea for a microphone: a metal diaphragm gently pressed against grains of carbon, with another stiff conductive surface on the other side of the carbon. When the carbon grains are squeezed, their overall electrical resistance drops. Sound impinging on the diaphragm produces a changing pressure, which varies the current between the two conducting surfaces.

'Voice-coil' microphones, as we have described, take advantage of the induction of a current by moving a conducting coil (usually copper wire) next to a permanent magnet. The coil is attached to a cone in order to better match the impedance of air. This is the reverse technology of a voice-coil speaker.

A ribbon microphone also uses Faraday induction, but in a conductive flexible ribbon placed in the field of a magnet. In this case, the surface of the ribbon itself provides impedance matching to the air. The frequency response in the range of human hearing can be adjusted by how flex lines are imprinted on the ribbon. Simple corrugation on a thin aluminum sheet is a popular design.

A condenser microphone uses a small parallel-plate capacitor, one side being thin and flexible enough to respond to sound by slight displacements. As the capacitance depends on the plate separation distance, a sound will cause variations in the capacitance, which can be detected by a suitable electronic circuit.

An electret microphone is a special kind of condenser microphone. It uses a dielectric material, called an electret (such as polytetrafluoroethylene), on which a 'permanent' electric charge has been induced. (The charge decays away in 2–10 years). Sound moves a conducting diaphragm a short distance in front of the electret material. The charges on the electret push and pull charges on and off the diaphragm, causing a small current. Because these microphones can be constructed with very small sizes, high sensitivity, omnidirectionality, and very little external power, they have come to be used in hearing aids and body-sound detectors.

Just as for radiators of sound, if the detection of a wide range of sound frequencies is desired, there should be no mechanical resonances in the frequency range of the detector. Also, the detector material, such as the diaphragm, must be in reasonable impedance-matching to the air (or water if microphone is embedded in tissue). These considerations limit the frequency range of the sound a given microphone is designed to detect, as well as the sensitivity of the microphone.

Piezoelectric crystals are good radiators and good detectors of single ultrasonic frequencies. A crystal, such as quartz, is first ground into a rectangular shape and to have a natural mechanical vibrational frequency at the desired frequency to be detected. Electric plates are then put on opposite sides of the crystal. A passing sound wave will squeeze and expand the crystal, causing an electric field to be generated in proportion. This field, in turn, causes charges to be pushed from or pulled onto the plates, making a small current. These devices are used in ultrasonic probes, both for generation and for detection.

Magnetostrictive devices use the change in shape of a magnetic material, such as cobalt, in response to a change in the external magnetic field to produce a sound. Such devices are capable of producing intense ultrasonics waves. They are used in underwater sonar and in surgical tools which employ high-intensity ultrasound for cutting tissue and stopping bleeding.

Sometimes the presence of sounds, including the human voice, can change the operation of electronic instruments. This occurs when variations in the sound pressure changes the distance between conductors in the circuit, and varying the capacitance between some of those conductors changes the operation of the circuit. For example, the frequency of an internal oscillator may be fixed by an 'LC' circuit, for which $f \approx 1/(2\pi\sqrt{LC})$, with L an inductance and C a capacitance. This undesirable phenomenon is call 'microphonics'.

Microphones with true noise cancellation require an auxiliary microphone immersed in the noisy region. The signal from the noisy pickup is compared to that of the primary microphone to isolate the sound which does not contain the noise. If one knows the frequency distribution of the noise, then frequency filtering (either analog or digital) can be employed to reduce the noise without using a second microphone. Of course, the source signal of the noise might also be reduced or eliminated. Wind shields on microphones keep moving air from whistling through microphone guards. ('Hums' are usually produced by poor grounding of amplifiers which use 60cps AC power sources.) Screeching comes from positive feedback between sound-producing speakers and nearby microphones.

Since sound is a material wave, destructive interference of that wave is possible, not just at one nodal point, but over a region. A matrix of active devices can be constructed which detects a wave over a surface and then generates a wave at each detected frequency and amplitude that is 180° out of phase with that incoming wave, effectively canceling most Fourier components. The matrix of detectors would have to be denser on the active surface than the shortest wavelength squared. This is the idea behind noise cancellation devices. To allow a signal to pass and noise to be attenuated, an independent set of microphones would be needed near the signal sources, spatially separated from the sources of noise. Headphones can pass a signal source to the ears from interior speakers, while an exterior set of microphones on the headphones detects the ambient noise to be used to generate the noise canceling waves generated inside the headphones. Buildings in earthquake zones which have active movable counterweights can be used to mollify passing earthquake waves.

Multiple pickup microphones (with at least four pickups on the vertices of a tetrahedral) can be used to electronically suppress the signals of one or more sources which are not at the same location. Such an arrangement can also be employed to determine the sound source locations from just the audio signals of the four microphones.

5.16.2 Radiation Pressure Devices

It is possible to directly detect sound pressure by recording that pressure on an elastic surface. If the surface responds to the pressure, then the displacement of the surface will be proportional to the sound pressure of an impinging wave.

The levitation of a small ball of fluid in a vertical standing wave of sound demonstrates that the sound pressure may be sufficient to overcome gravity. Such levitation devices are useful for the manufacture of small semiconductors levitated while cooling. The levitated objects must be significantly smaller than the wavelength of the sound. Ultrasonics is typically used, with sound levels above 150 dB.

The force on a small particle, such as a biological cell, smaller than the wavelength of the sound, is proportional to the negative gradient of the difference between sound energy in particle and the sound energy which would be in the same fluid volume without the particle. In turn, the sound intensity is proportional to the square of the sound pressure.

Acoustical radiation pressure has been used to move individual biological cells to desired positions, serving as acoustical tweezers.

5.16.3 Heat Detectors

Most commonly, sound energy is dissipated into heat (quasi-random motion in atoms and molecules). A 'bolometer' (or calorimeter) is a device which measures the heat produced by particle beams or radiation. To measure the heat production, a very sensitive thermometer can be employed, such as those constructed from a semiconductor. The detector material is weakly coupled to a heat reservoir at ambient temperature or colder. The absorbed radiation raises the temperature of the semiconductor, which changes the semiconductor conduction characteristics.

5.16.4 Light Scattering Detectors

Light is reflected and/or refracted at the interface between two materials with differing indices of refraction. If that interface is acoustically vibrating, then the analysis of the light which has interacted at the interface will give information about the sound.

Acoustical waves traveling over the surface of solids and liquids can be seen and measured by scattering light off that surface.

Some effects of sound waves in fluids containing biological cells can be analyzed by scattering light off the surfaces of those cells, which act as viscoelastic liquid drops in response to sound.

5.16.5 Detectors Using Electron-Phonon Interaction

The distortion of a semiconductor by a sound wave can produce changes in the electric potential energy experienced by electrons in the conduction band of the semiconductor. As the sound wave propagates through the semiconductor, it can even carry bunches of electrons with it. The transfer of sound energy into kinetic energy of electrons is sometimes put into the category of electron-phonon interactions (with a phonon being a quantum of sound energy in the solid).

5.17 Biodetectors of Sound

Over billions of years living on a planet with water and air, life systems have evolved a variety of method to detect sound, letting them communicate with each other, sense their environment, and search for food. The resulting detectors, or bioreceptors, are often quite sophisticated, particularly when there is an advanced brain to analyze the sound.

Animals with sound detection include:

- *Marine crustaceans*: External array of sensory hair cells and an internal statocyst are able to hear sound frequencies 30–500 Hz.
- *Fish*: Fishes have an inner ear and a mechanosensory lateral line system. The bodies of fish are nearly the same density as the surrounding water, so sounds easily enter. Hair cells in their inner ear move in response to sound, and send signals to the fish brain. (The inner ear structure in fish, including semicircular canals to sense orientation, is a precursor to that of land animals.) If the fish has an air bladder near the inner ear, the sound vibrations of that bladder will transmit to the ear, enhancing hearing for the resonance frequencies of the bladder. Fish also have a set of hair cells near the skin surface on a lateral line system called

Table 5.8 Range of hearing for selected animals

Animal	Frequency (Hz)	
	Low	High
Humans	20	20,000
Cats	100	40,000
Dogs	40	46,000
Horses	31	40,000
Elephants	16	12,000
Cattle	16	40,000
Bats	1000	150,000
Rodents	1000	100,000
Whales and dolphins	70	150,000
Seals and sea lions	200	55,000
Grasshoppers and locusts	100	50,000

a neuromast. These hair cells can detect movement of the surrounding water, including sounds.

- *Insects*: Insects detect sound with surface hair cells and sometimes with a set of tympanal organs. The tympanal organ has stretched membranes next to an air sac and connected to sensory cells. Some insects (such as moths) have evolved sensitivity to bat ultrasound, and evasion strategies. The moth ultrasound detection comes from mechanoreceptors, with a tympanic membrane over a chamber on the wall of the abdomen, and are most sensitive to the lower frequencies of ultrasonics (20–30 kHz).
- *Frogs*: Frogs have a middle and an inner ear and often an external ear drum, and sometimes with enhancements by lung or mouth cavities.
- *Reptiles, birds and other land dwelling animals*: Reptiles, birds (evolved from reptiles) and other land animals living in air have sophisticated hearing organs, with tympanic membranes, and an ossicular organ to transmit the sound to a cochlea which generates nerve pulses in accord with the intensity and frequency of the sound.

The frequency range of hearing for selected animals are given in Table 5.8.

5.18 Medical Instruments for Detecting Sound

5.18.1 Stethoscope

The stethoscope hanging on the neck of a physician has come to be a doctor's symbol. It is a simple device to listen to the sounds in the body. (As we have noted, listening to such sounds is called 'auscultation'). A stethoscope requires no electronics or power. The decisive piece of a stethoscope (aside from the attached physician) is an air chamber ('bell') with a thin flexible wall to put into contact with the skin of a patient. The chamber is designed to approximately match the sound impedance of the epidermis to that of the air in the chamber, permitting the skin, which ordinarily reflects most of the internal sounds, to send some sound energy into the chamber. In turn, the chamber connects to two air tubes (usually made from flexible rubber), carrying the sound to the physician's ear canals. Sounds from the heart, lung, and gastrointestinal canal are evident. By thumping on the body, a pulsed sound with a wide range of frequencies is generated by body organs and cavities near the thumping location. This pulsed wave can reflect off and cause resonant vibration of internal hard and soft interfaces. An experienced and good physician can recognize the character of such sounds, such as from trapped gas, enlarged organs, or possible tumors.

Because the passive stethoscope has no sound amplification, the physician can hear only heart sounds in the range of about 50–600 Hz. (See Fig. 5.11.) The principal 'heart sounds' are due to blood-flow turbulence when a heart valve closes. 'Murmurs' occur because of turbulent flow generated by back flow (regurgitation), and constriction (stenosis). Snaps, clicks and tissue rubs may also be detected. Other weaker sounds occur, but require an active stethoscope with amplified sound. The amplified heart sounds will have frequencies from 1 to 10,000 Hz. Wall movements make the lowest frequency sounds, 1–5 Hz. Diastolic sounds (third and fourth heart sounds) are in the range of 5–50 Hz. First and second heart sounds are in the range of 10–100 Hz. Valvular snaps and diastolic rumbles are in the range of 10–150 Hz. Systolic murmurs and diastolic basal murmurs are heard in the range of 150–1000 Hz. Our ears will not hear frequencies below about 20–30 Hz, but a hand or fingers placed on the chest can feel them. Using the hand in this way is called 'palpation'. Palpation may also sense 'thrills', sounds detected during palpation due to turbulent flow through a narrowed orifice into a receptive chamber. Thrills typically occur in the frequency range of 60–100 Hz, peaking in the range 75–85 Hz, although palpation is more sensitive to the lower frequencies.

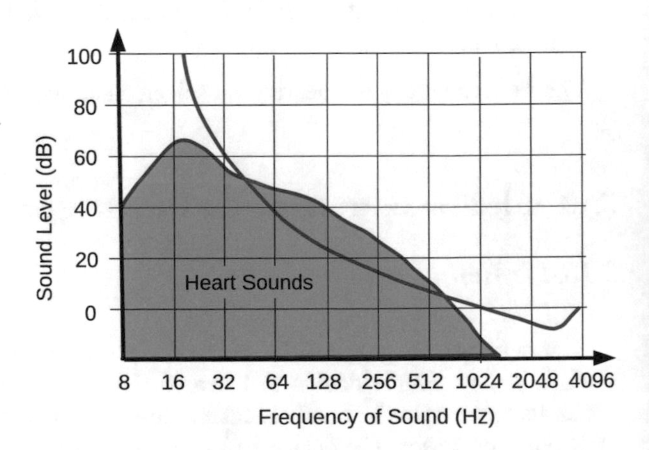

Fig. 5.11 Heart sound level at the chest over various frequencies. The top curve is the threshold of hearing level. Note that an unamplified stethoscope will only let humans hear the heart between about 50–600 Hz

Fetal heart sounds are detectable around the tenth week of gestation with an ultrasonic fetoscope. A regular stethoscope or 'Pinard' horn can detect fetal heart sounds after about 18 weeks. The normal fetal heart rate is 120–160/min.

Breathing sounds include clicking, bubbling, and rattling (collectively called rales, due to opening of air spaces), and rhonchi, stidor and wheezing (due to restrictive air passages). Lack of expected sounds is also a symptom.

5.18.2 Medical Microphones

Microphones are used in medical procedures to augment hearing, monitor heart sounds, hear low-level auscultation effects, and to listen for echoes from ultrasonic devices. When used externally to listen to internal sounds, the microphone's impedance must come close to that of the skin tissue. This can be done with a horn (a conical shape with the large end on the skin surface and the small end next to the microphone) or, if a piezoelectric microphone is used, by a gel on the skin.

Hearing aides are active devices (requiring a source of power) which transmit amplified sound to the inner ear. They can be made to fit the external ear, or made to be implanted in order to vibrate the cranium near the pinna (with a Faraday induction rechargeable battery), or to connect directly by electrical stimulation to the nerves in the cochlea. As electret microphones (see Sect. 5.16.1) can be miniaturized, they are used for implantation devices. Filtering of tissue noise in the mastoid due to vocalization, chewing, etc. helps in voice recognition.

5.18.3 Phonocardiogram

A phonocardiogram (PCG) is a recording of the amplified sounds made by the heart (Fig. 5.12). The dominant sounds come from the closure of heat valves, and is often characterized by the sound of "Lubb-bidub". In the cycle of heart activity, the first sound, S1, (the "Lubb") is produced by the closure of the atrioventricular valves (first mitral, then tricuspid) at the beginning of systole (contraction), and the second sound, S2, (the "bidub") is the sound of the closure of the aortic valve and then the pulmonary valve at the end of systole. Both S1 and S2 are similar in their spectral frequencies below 150 Hz, but are distinguishable above. A third low frequency sound, S3, may be present, particularly in children and in patients with congestive heart failure. S3 occurs in early diastole, due to increased atrial pressure causing increased flow rates. A low frequency sound (S4) just after atrial contraction may indicate an overly stiff left ventricle. Abnormal sounds such as murmurs, will also show.

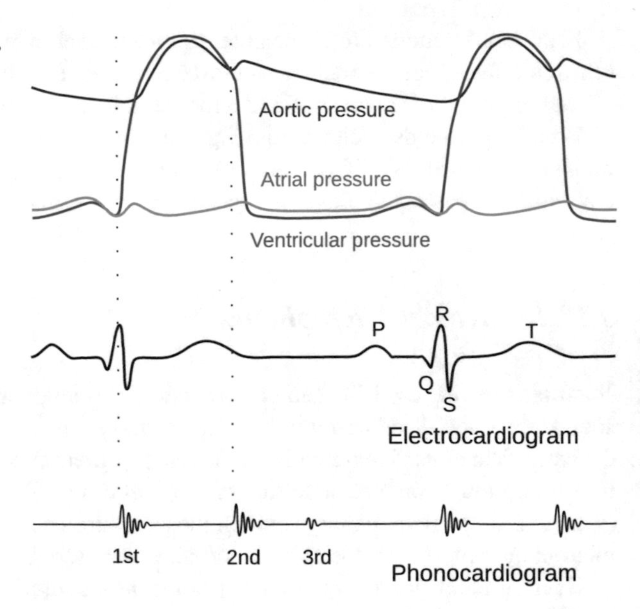

Fig. 5.12 Phonocardiograph, electrocardiogram, and heart pressures

5.18.4 Ballistocardiogram

When the heart ejects blood into the aorta with each heartbeat, the momentum of that blood must come from the impulse produced by heart muscles. The heart must, in turn, move in the opposite direction, imparting motion to the body, and causing the body to wiggle with the heartbeat. If a person lies on a bed with a suspended mattress held by springs oriented along the anterior-posterior axis of the person, then the motion of the bed will reflect the pumping of the heart. A record of this motion is a ballistocardiogram (BCG). (Respiratory components can be filtered out with a band-pass of 0.3–50 Hz.) Studying the wave forms in the recording can give information about the strength of the heart and therefore the possible presence of aortic valvular disease and of coronary artery disease.

5.19 Human Hearing

Before I speak, I have an important thing to say.
—Groucho Marx

5.19.1 General Properties of Human Hearing

The sensitivity of the human ear is quite remarkable. (Data on human hearing can be found in Appendix J.) A good ear can hear sound whose intensity is only 10^{-12} W/m^2 and a pressure variation of only 2×10^{-5} Pa. Over the eardrum (of area about 0.65 cm^2), this sound delivers 1.6×10^{-15} J/s. With this power, it would take 2.4 trillion years to make your morning toast (which takes 1000 W for 2 min). At a frequency of 1000 Hz in air, the amplitude of this barely audible sound is

$$A = \frac{1}{2\pi f}\sqrt{\frac{2I}{\rho_o v}} \tag{5.48}$$

$$= \frac{1}{2\pi(1000\,\text{Hz})}\sqrt{\frac{2 \times 10^{-12}\,\text{W/m}^2}{1.225(\text{kg/m}^3) \times 340(\text{m/s})}} \tag{5.49}$$

$$\approx 0.24 \times 10^{-10}\,\text{m}\,. \tag{5.50}$$

This vibrational amplitude is smaller than the diameter of hydrogen atoms! If our ear were a little more sensitive, thermal noise would be present in the auditory signals sent to our brain. Thermal noise creates a root-mean-square pressure of 0.0002 μbar, just about the value of the threshold pressure p_{th} (ref. *Acoustics*, p. 3–67 in AIP HB). Remarkably, owls have thresholds of hearing at 3000 Hz nine decibels lower than the human best ear. While in winter flight, the owl brain is able to pick out a signal of a mouse moving under the snow below, even in the presence of the thermal noise.

At the upper end of sound intensity, the ear attempts to protect itself from loud sounds by stiffening the middle ear muscles attached to the auditory ossicles which transfer sound from the eardrum to the inner ear. But we can tolerate intensities which carry a trillion times more energy than the weakest sounds we can hear, up to about 1 W/m^2. Sounds louder than this damage or destroy hearing. Mechanical sirens have been constructed which produce 10^4 W/m^2 at a frequency of 1000 Hz, which, as noted in Sect. 5.14.1, will cause cotton to catch fire.

5.19.2 Loudness

The human has more difficulty hearing very low frequencies (below about 30 Hz) and very high frequencies (above about 17 kHz), even when the sound level is fixed to a value easily heard at 200 Hz. Fletcher and Munson[23] measured the sensitivity

[23] H. Fletcher, and W.A. Munson, *Loudness, its definition, measurement and calculation*, J. Acoust Soc Am **5**, 82–108 (1933); see also D.W. Robinson, & R.S. Dadson, *A re-determination of the equal-loudness relations for pure tones*, Br J Appl Phys **7**, 166–181 (1956).

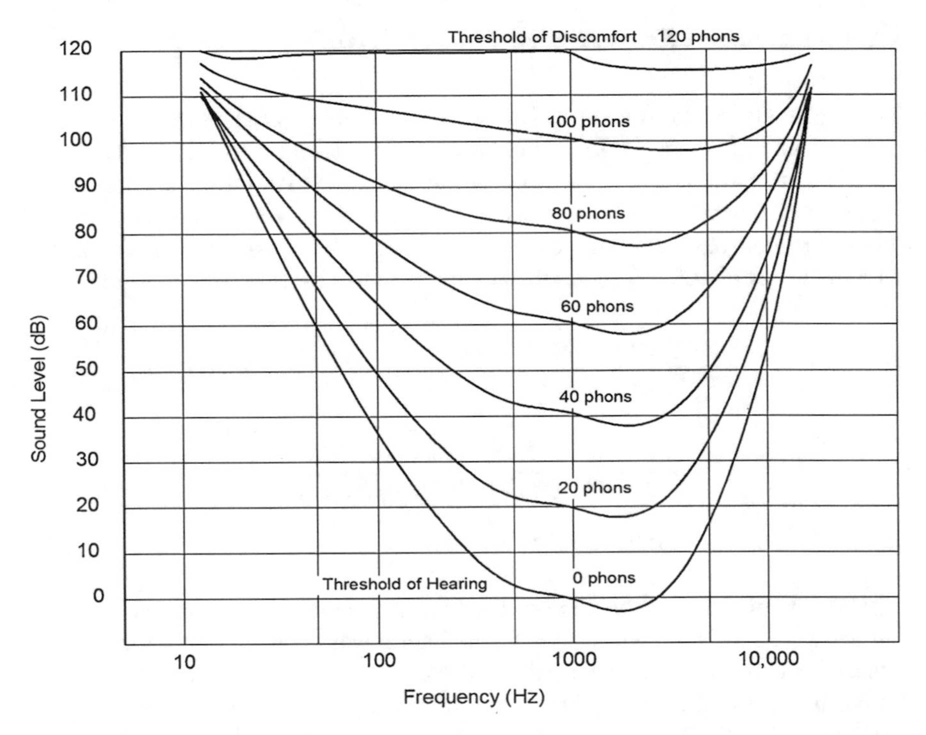

Fig. 5.13 Fletcher-Munson curves

of the adult healthy human ear as it depends on the frequency of sound. The contours shown in the graph give equal loudness levels, the lowest audible being the threshold-of-hearing curve, the highest the threshold-of-discomfort curve. Above the threshold-of-discomfort is the threshold-of-pain, which is just above 120 dB at all sound frequencies (Fig. 5.13).

The subjective sense of equal loudness sound levels are found by determining when the average healthy person thinks two sounds of different frequencies sound just as loud. The sound levels for subjective equal-loudness follow the curves plotted in the figure. This psychologically defined level of loudness is measured in 'phons', which are made equal to the physical loudness at 1000 Hz. Below about 30 Hz, and above about 17,000 Hz, sound intensities must be close to the threshold of discomfort to be heard. The slight increase in sensitivity one can see in the Fletcher-Munson curves at about 3000 Hz can be attributed to resonance in the ear canal which acts like an open-ended organ pipe (about 2.7 cm long and 0.7 cm in diameter), although sound interaction with the ear drum and pinna lowers and widens the resonance.

Humans can detect changes in sound level if the change is bigger than about one decibel. When the sound level increases by 10 dB, most humans perceive that the sound level has doubled. A 'sone' unit has been introduced to correspond to this doubling sensation, with one sone defined to be 40 phons, and 2^n sones to be $(40 + 10n)$ phons, with n integer.

5.19.3 *Pitch*

The musical term for frequency is 'tone'. The psychological sensation to a dominant frequency is 'pitch'. Two sounds may be judged to have the same pitch even though different mixtures of frequencies may be present. A person's sensation of frequency depends on the sound level, and there is a loss of sensitivity to frequency change at both the high and low frequencies. All tones which are perceived to have a single pitch are given one value for the pitch in a unit called a 'mel', defined to match the units of frequency when the sound level is perceived to be 40 phons. At middle sound levels, humans can detect if the frequency has changed by 5% of the interval between equal-temperament notes,[24] i.e., if $\Delta f/f > 0.05 \times (\sqrt[12]{2} - 1) = 3 \times 10^{-3} = 0.3\%$.[25] Because the sensation of frequency change is logarithmic, a "MIDI pitch" has been defined as the number

$$m \equiv 69 + 12 \log_2(f/440\,\text{Hz}) .$$

(The value 69 is arbitrarily assigned to the frequency 440 Hz, the note A above middle C, while other notes on the equal-tempered scale are labeled sequentially by the integers.)

Our sensation of pitch vs. frequency and pitch vs. sound level are shown in the graphs (Figs. 5.14 and 5.15). Our ability to sense changes of frequency diminishes as the frequency increases.

As the sound level lowers, we sense a lower pitch than the actual frequency, and as sound level increases, not only is our sensitivity to the loud sound dulled, but we hear an increased pitch, when no change in frequency has occurred. These false sensations are 'aural illusions'. The pitch illusion is explainable, at least in part, by the changes in the tension of the muscles of the middle ear. Other more involved illusions occur, such as the hearing of the fundamental note when a set of its harmonics is present, known as the 'residue effect'. and the reduced sensitivity to harmonics when a strong fundamental is present, known as the 'masking effect'. Small radios take advantage of the residue effect, in that small

---

[24]The equal tempered scale takes notes with frequencies in each octave as $f_n = (n + 1)^{1/N} f_0$, where N is the number of notes. An octave is the interval of frequency between one note and double the frequency of that note.

[25]Another way of describing this sensitivity is to give the number of different pitches that a good ear can distinguish. This number is about 2300, with a logarithmic separation between adjacent pitches. Taking the range of audible frequency to be from $2^4 = 16$ to $2^{14} = 16,384$, each distinguishable frequency in this range can be specified by $f_n = 2^{4+n/230}$ Hz, where $n = 0, 1, \cdots, 2300$. In this case, $\Delta f/f = 2^{1/230} - 1 = 0.3\%$.

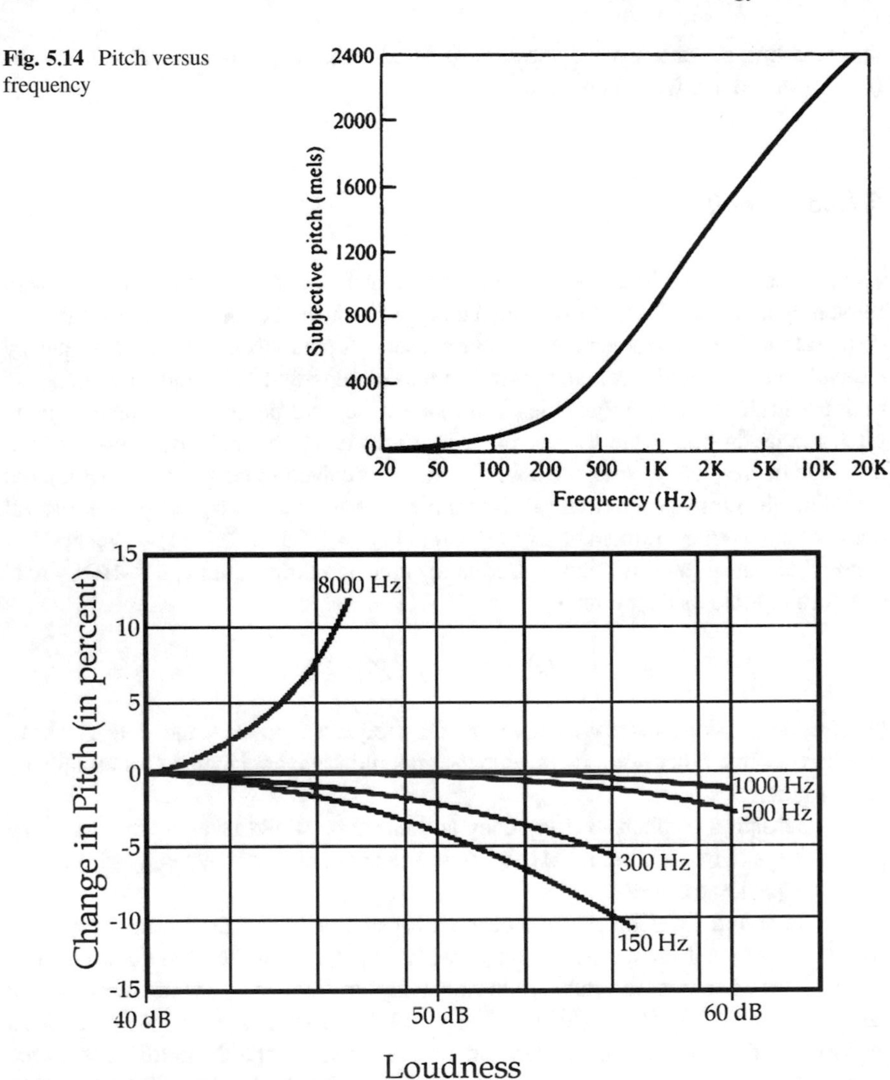

Fig. 5.14 Pitch versus frequency

Fig. 5.15 Pitch versus loudness

speakers cannot faithfully reproduce base notes. However, we humans think we hear them anyway when the harmonics are present. The residue and masking effects are psychologically, rather than physiologically, based.

5.19.4 Binaural Directional Sensation

Our sense of the direction from which sound originates relies on how information from one ear and the other determines a difference in (1) pulse time arrival; (2) the beginning of a phase of the wave (for low frequencies); (3) intensity of a sound.

5.19.5 The Human Ear

Important parts of the human ear include (Fig. 5.16):

Auricle:	('Pinna') The outer ear aids in collecting sound
Ear Canal:	The open passage from the outer ear to the ear drum
Ear Drum:	('Tympanic membrane') is set in motion by an external sound
Malleus:	('Hammer') is the middle ear bone in contact with the ear drum
Incus:	('Anvil') is the middle ear bone between the malleus and the stapes
Stapes:	('Stirrup') is the bone which contacts the inner ear cochlea window; the three middle ear bones match the impedance of the ear drum to that of the oval window and liquid behind
Oval Window:	A membrane which transfers sound energy from the middle to the inner ear
Round Window:	A membrane which allows the liquid in the inner ear ducts to move in response to sound
Vestibule:	A chamber which holds balance organs (utricle and saccule); midway between the cochlea and the semi-circular canals
Semi-circular canals:	Three perpendicular fluid-filled semi-circular organelles in the inner ear; the fluid moves in one or more canals when the head accelerates, causing hair cells to stimulate nerve cells
Cochlea:	An organelle in the inner ear which discriminates various sound frequencies and their intensities, and sends this information to the brain
Basilar Membrane:	Stiff membrane separating two liquid-filled ducts within cochlea
Tectorial membrane:	The vibratory membrane in the cochlea
Organ of Corti:	Auditory cells under the tectorial membrane with hairs to sense vibration
Scala Vestibule:	A liquid-filled duct for sound transmission in the cochlea from the oval window to the heliotrema, the end of the cochlear spiral

Scala Tympani: A second duct to carry sound vibration back from the heliotrema to the elastic round window

Cochlear Nerve: The auditory nerve from the cochlea to the brain

Eustachian tubes: The pressure-equalizing tubes from the throat to each middle ear

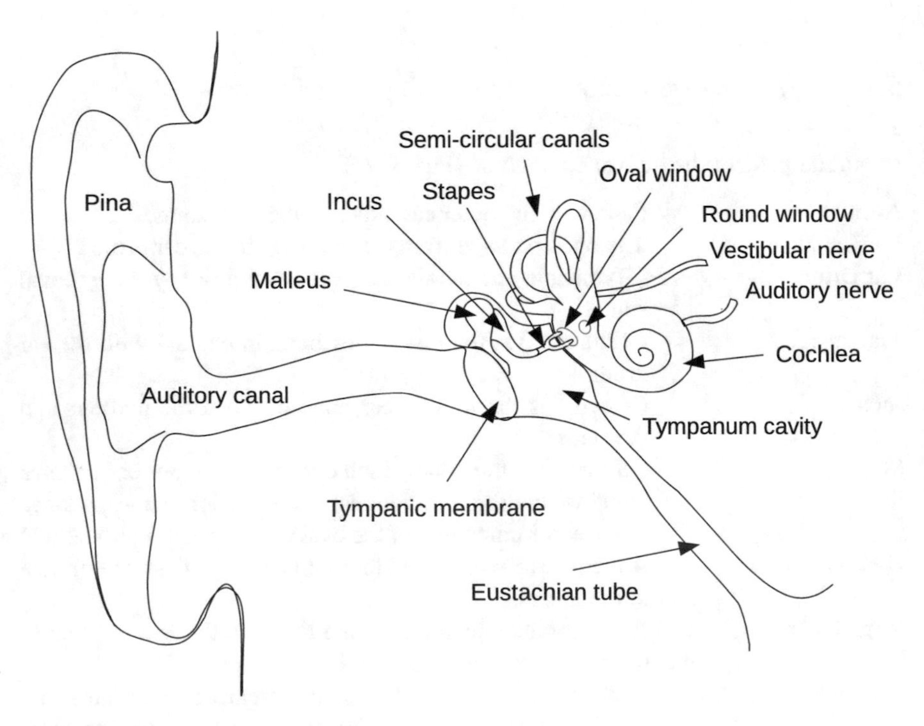

Fig. 5.16 The human ear

5.19.6 Music

Birds, whales, and most humans like music. Humans may even get 'chills down their spine' and 'goosebumps' listening to certain familiar and anticipated combinations of notes. Why this is so is a subject of genetics, hormonal reactions, and psychology, particularly memory responses. Songs are stored in the cerebellum as subprograms, and so might be replayed even when other cerebral functions have had damage or degeneration.

Musical tones or notes (sounds with a single dominant frequency) in the Western 'Equal-Tempered Scale' are taken from sets of standard frequencies. These sets are divided into 'octaves' having eleven notes each. The lowest note of one octave has a ratio to the lowest in the next octave of 2 to 1. In the equal temperament scale, the twelve notes within one octave and the beginning note of the next octave have frequencies in equal interval ratios, so that the notes within an octave, named C, $C^\sharp = D^\flat$, D, $D^\sharp = E^\flat$, E, F, $F^\sharp = G^\flat$, G, $G^\sharp = A^\flat$, A, $A^\sharp = B^\flat$, and B, are in ratio $\sqrt[12]{2}$. In this way, the corresponding notes in each successive octave are a factor of two above the previous one, making up harmonics of each other. The fifth audible octave is taken with the note A having frequency 440 Hz. The next C note up in frequency is called 'middle C'. The name octave, with a Latin root meaning eight, comes from the fact that tonal music is composed with the notes C, D, E, F, G, A, B, with the next C included in the count.

The lowest frequency of audible sound produced by a musical instrument is called its 'fundamental frequency'. Higher frequency sounds from this instrument are called 'overtones'. If the overtones are integer multiples of the fundamental, they are the harmonics, or 'partials' of the sound. Those harmonics which are 2^n times the fundamental are said to be from the n-th 'octave' above the fundamental. Simultaneous tones which are in ratio of 2^n are in the same 'musical key', but n octaves away.

Humans can become enthralled with rhythmic sounds if they have sufficient symmetry, but also if they have sufficient complexity to be evocative or challenging to the mind. Curiously, simultaneous notes which do not have simple ratios of frequencies are disturbing, perhaps because their 'beat frequency' (frequency difference) does not match another musical tone. Instruments which play sounds which are dominated by a mixture of integer-multiples of one frequency, together with a mixture of lower-level anharmonic frequencies, have attractive sounds. The 'richness' of a sound refers to the number of distinct component frequencies. The term 'quality of sound' is sometimes used to portray the presence of a set of tones having frequencies which are in approximate whole number ratio. Such sounds seem to be pleasing to us, much more than listening to a series of pure tones. In contrast, if the frequency ratios are irrational, the sounds are dissonant. The vibrational modes of drum heads and bells produce component sounds whose frequencies tend not to be in simple ratio.

A 'musical note' is a sound of short duration (seconds or less) with a definite 'sound envelope, determining how the amplitude of the note rises, is sustained, and then falls.

The spectral distribution of frequencies in a musical note determines its musical 'timbre'. which is also called its 'tonal quality' and tonal color. Physically, if two sounds have the same loudness, but they have a different frequency spectrum, they will differ in timbre. Musical notes with different sound envelopes but the same loudness and spectrum also are said to differ in timbre.

The inner ear and brain performs a kind of Fourier analysis on the incoming sound. The pitch of two sounds can be distinguished if the stimulation of the basilar membrane is separated by more than 50 μm (which is about six hair cells apart). Fourier decomposition can be performed by instruments, generating a 'spectrogram', i.e. a visual representation of the frequencies present in the sound as they vary in time. Spectrograms of speech and song are referred to as 'voice prints'.

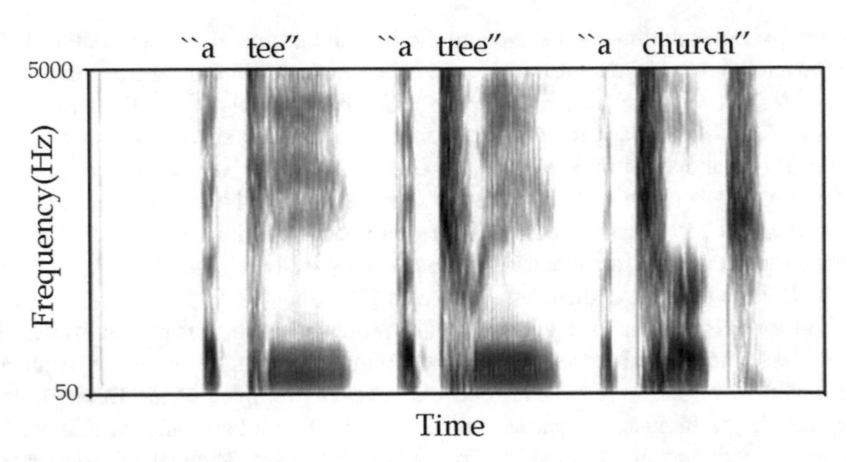

Fig. 5.17 An example of a human voice print

A voice print displays the various frequencies present in a sound versus time. The intensity of each frequency is shown as density in the print. Figure 5.17 shows a voice print of a person saying “a tee”, “a tree”, and “a church”. Vertical dark patches in the spectrogram are due to the dominant resonances in the sound-generating apparatus, and are called ‘formants’. For the human vocal system, as many as nine formants may be present from 50 to 8000 Hz. Even fricative hissing sounds will show formants because, as noise with a wide spectrum of frequencies present, they will stimulate all the resonances in the nasal cavities and membranes.

Figure 5.18 shows the voice print of a humpback whale. What he is saying has not yet been deciphered.

The following terms are used in reference to musical scales. The fact that there are so many such terms shows that the subject is mature, and that music is important to many humans: *Aeolian Mode*, *Chromatic Scale*, *Diatonic Scale*, *Dorian Mode*, *Diatonic Genus*, *Harmonic Minor Scale*, *Hexachord*, *Ionian Mode*, *Lydian Mode*, *Major Scale*, *Mean Tone Scale*, *Melodic Minor Scale*, *Myxolydian Mode*, *Pentatonic Scale*, *Phrygian Mode*, *Pythagorean Scale*, *Scientific Scale*, *Tempered Scale*, *Tonic Scale*. Ready for a tonic?

Here are some other useful musical terms and phenomena:

- *Subharmonic:* A frequency $f = f_0/2^n$ below a given frequency f_0, where n is a positive integer.
- *Frequency below the fundamental:* The fundamental frequency in a wave is the lowest frequency in its Fourier decomposition. If that wave interacts with a media having non-linear properties, subharmonics below the fundamental can be generated.
- *Vibrato:* Regular variation in amplitude and frequency imposed on an auditory sound, with a period much longer than the dominant frequency of the sound. Some humans feel vibrato adds ‘warmth’ to a sustained note. It certainly adds more information, such as hints as to the emotional state of a human musical

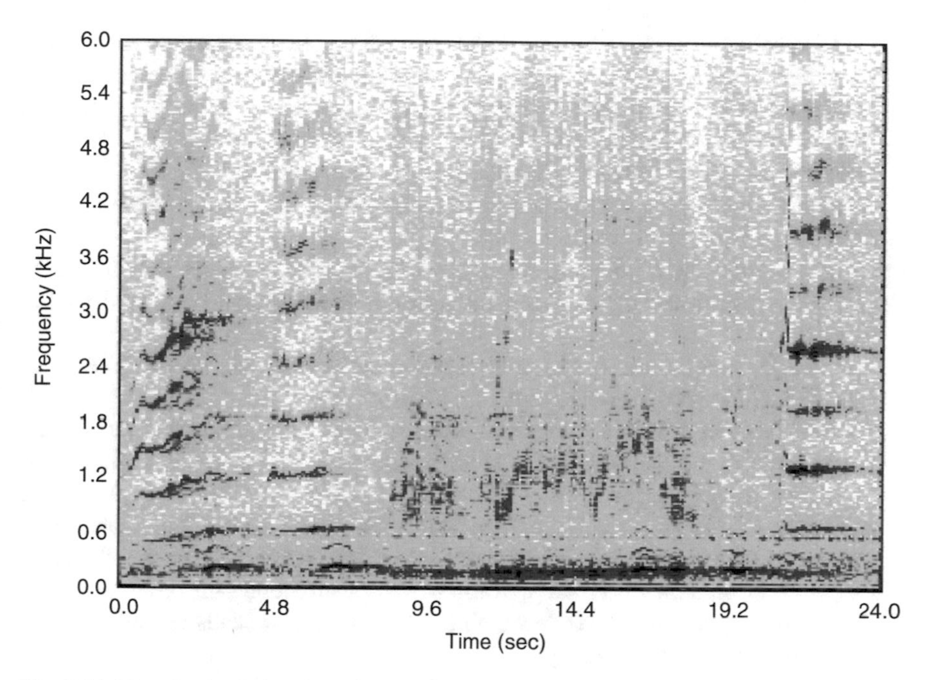

Fig. 5.18 Humpback whale voice print

instrument player. Involuntary quavering of the voice is indicative of tremors, brought on by heighten psychological responses or by neurological pathologies.

- *Beats:* Suppose two sinusoidal waves with the same amplitudes but slightly different frequencies, f_1 and f_2, are made to interfere. The result will be a wave which has a wavering amplitude of frequency $|f_2 - f_1|$. For example, with two audible sounds of frequency 440 and 442 Hz, one can hear a very low variation in intensity with a frequency of 2 Hz. One says that the two waves are 'beating' against each other, with a result that you hear 'beats'. The effect demonstrates the principle of superposition of waves.
- *Modulation:* Varying the amplitude (AM) or frequency (FM) of a given wave, called the 'carrier'
- *Masking:* The perceptional loss in hearing the fundamental of a sound when that sound has strong harmonics
- *Aural Illusion:* The human brain's re-interpretation of sounds which may not be present; evident when a small speaker plays music with little or no fundamental. The brain's auditory interpreter tends to add the low note in the sounds from a small speaker when it expects to hear one.
- *Phoneme:* Short distinguishable audible sound used to make up speech
- *White Noise:* Sound with the presence of a full spectrum of frequencies with intensities and phases which vary quasi-randomly over the spectrum and over time. Quasi-randomly distributed tones occurring at unpredictable times produce what is now called 'white noise', the term white borrowed from a similar

character of white light. White noise has a masking effect, not only blocking the sensation of other sounds, but diverting attention from other senses, including mild pain. White noise can be produced by converting to sound the amplified thermal fluctuations of electrons in electronic devices.

- *Speech Power:* The human vocal sound energy output per unit time
- *Audiometry:* A device for evaluating hearing through ears and possibly through the mastoid part of the temporal bone
- *Otoscope* A device for viewing the eardrum through the ear canal. A light source is used to illuminate the eardrum through the same optical path used by the viewer.
- *Acceptable noise levels:* A sound level range between 0 and 100 dB. The noise level in a quiet room is about 40 dB.
- *Anechoic room:* A room with sound absorbing walls, ceilings and suspended floors; used to test hearing and low level sounds. Humans can hear their own heartbeat in such a room. The effects are often psychologically disturbing. Anechoic rooms are usually constructed with cones coated with absorbing surfaces covering the walls, floor, and ceiling, all facing toward the chamber. People enter on a metal screen platform to keep their weight off the floor cones. Sound entering a cone will reflect a number of times as it heads toward the cone vertex, each time losing energy.
- *Stereo hearing:* Having two good ears gives us the ability to judge angular direction to a source of sound, i.e., a stereo-sense, also called binaural sound detection. Below about 3000 Hz, this deduction uses the phase difference between the waves arriving at the two ears, while much above 3000 Hz, information about the times of arrival of a pulse of sound to each ear comes into play.
- *Infrasonics:* Sounds below the frequency heard by humans, i.e. below about 20 Hz
- *Ultrasonics:* Sound above the frequency heard by humans, i.e. above about 20 kHz
- *Sound level change sensitivity:* The ability of humans to detect a small change in sound level. Humans can detect a change in sound level if the change is above one decibel.
- *Sound frequency change sensitivity:* The ability of humans to detect small changes in frequency. At 1000 cps, most humans with healthy ears can detect a change of $\Delta f = 0.3$ cps.
- *Sound direction sensitivity:* The ability to sense from where a sound came. This is also known as stereo sense, or binaural sense. Sensitivity is: 2° if sound come from near the front of the head, 10–15° if the sound comes from the side. The detection is via delay time, relative phase, and intensity variation from ear-to-ear.
- *Time resolution:* The smallest time detectable between two sharp pulses of sound, or the smallest detectable gap in a sound (about 1 ms for humans)
- *Sound Diffraction:* Bending of sound wave around boundaries (greater for wavelengths comparable or larger than boundary dimensions)

5.20 Reverberation

'Reverberation' is the collection of residual vibrations within a bound system due to wave pulse reflections which continue for some time even after a localized source of power has stopped. Enclosures, such as bathrooms, orchestra halls, and voids in the gut, reverberate from sound reflected back from the enclosing boundaries. The amplitude of the wall vibrations due to impinging sound waves depends on the stiffness of the wall, its inertia, and internal damping. In response, the vibrating surfaces produce a scattered wave, with a fraction of the sound energy lost by being conducted into the wall or converted to heat.

The time for the reverberation level to drop by 60 dB is called the 'reverberation time'. This figure for drop is used because the crescendo of a typical orchestra is about 100 dB, while the background hall noise is about 40 dB. At 500 Hz, the Boston Symphony Hall has a reverberation time of 1.8 s; the Vienna Musikvereinsaal, 2.1 s; the Royal Albert Hall, 2.8 s.

In 1895, Wallace Clement Sabine, the 'father' of modern architectural acoustics, showed[26] that the reverberation time for an enclosed room is given by

$$T_R = (0.161\,\text{s/m})\frac{V}{\Sigma\alpha_i A_i}\,, \tag{5.51}$$

where V is the volume of the room, α_i is the sound absorptivity of a surface in the room and A_i is its area. The constant 0.161 comes from $60/(10\log_{10}\{e\})(4/v)$ with v being the speed of sound.

The argument goes like this: Suppose a room is sufficiently large that standing waves do not develop in the room before the sound from a localized source diminishes by absorptive loss. Assume a pulse of sound with energy E_o is generated and then left to bounce around in a room. If the walls have sound absorptivity α, at each encounter with a wall a fraction $1-\alpha$ of the sound energy E is returned to the pulse. After N such encounters, the energy has dropped to $E_o(1-\alpha)^N$. The time between pulse hits will be L/v, where L is the average distance between the walls and v is the speed of the sound pulse, so in a time t there will be $N = vt/L$ such encounters. After a time t, the pulse has sound energy $E = E_o(1-\alpha)^{vt/L} = E_o\exp\{(vt/L)\ln_e(1-\alpha)\}$. One can show that the geometric average distance L between the walls of a box of volume V and total wall surface A is $4V/A$. For $\alpha << 1$, $\ln_e(1-\alpha) \approx -\alpha$, giving $E = E_o\exp\{-(v\alpha A/(4V))t\}$. With a variety of absorbing surfaces, we define an average sound absorptivity by $\bar{\alpha} = \sum \alpha_i A_i/A$.

The decrease in the sound level in a time t is then

$$\Delta\beta = 10\log_{10}(e^{-\frac{v\bar{\alpha}A}{4V}t}) = 10\left(\frac{-v\,\bar{\alpha}\,A}{4V}t\right)\log_{10}(e)$$

[26]**The Collected Papers on Acoustics** by Wallace Clement Sabine, [Harvard University Press] (1922).

Table 5.9 Sample absorptivity of various surfaces at 256 Hz

Material	Absorptivity
Window glass	0.025
Wood	0.060
Plaster walls	0.035
Carpet	0.10
Human being	0.45 (with area 1 m^2)

so

$$\Delta\beta = -1.086 \left(\frac{v\bar{\alpha}A}{V} \right) t\,.$$

The reverberation time, T_R (time required for the sound level to drop by 60 dB) is then

$$T_R = \frac{55.2}{v} \left(\frac{V}{\bar{\alpha}A} \right).$$

A few values of the absorptivity of various materials is given in Table 5.9.

5.21 The Doppler Shift

The Doppler shift is a change in the measured frequency of a wave due to relative motion between the wave source and an observer. For example, if a source of waves moves toward a detector, the frequency measured by the detector for the emitted waves is greater compared to that measured when there is no relative motion. The shift is named after Christian Doppler, who predicted the effect in 1842 for the light from moving stars. You know of this effect if you have heard a sound from a moving car or train change to a lower frequency as the car or train passed you.

Suppose a source of a sound with a single frequency heads toward your ear at a speed of $u < v$. In the time T that each wave is emitted from the source, the source has moved a distance uT toward the previous wave it has emitted, so that the wavelength between waves in that direction will be $\lambda' = \lambda - uT$. (See Fig. 5.19.) Replacing T by λ/v gives

$$\lambda' = \lambda(1 - u/v)\,.$$

In terms of frequency, this reads

$$f' = f/(1 - u/v)\,.$$

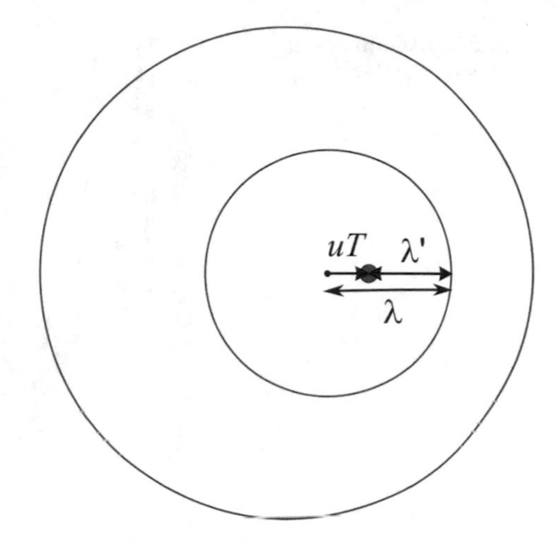

Fig. 5.19 Doppler effect: As a sound source moves at speed v while emitting sound with period T, the distance between the peaks of the sound wave, λ', differs from λ

You will hear a higher tone than if the same source were at rest. Similarly, if the source is moving away, you will hear a lower tone.

The Doppler shift is used in ultrasonic detection of movement within our bodies, such as the motion of heart valves and blood. We will come back to the Doppler effect in the application of ultrasonic waves to biology and medical technology.

5.22 Medical Applications of Ultrasonics

Ultrasonics has developed into a major tool in medical imaging. Although ultrasonic sound, as a wave, has weak resolution for small parts within tissue,[27] the intensities can be so low that damage to tissue is not expected. If intensities are high, potential dangers come principally from heating and cavitation. These properties contrast with those of X-ray imaging. X-ray images can be much finer in resolution, and can penetrate bone, but as an ionizing radiation, potential cell damage and the generation of cancer is ever present.

5.22.1 Acoustical Imaging

Imaging of internal organs and fetuses has the great advantage over X-ray imaging in that the sound waves are non-ionizing and therefore can have about zero probability

[27] The resolution of an image is better for smaller wavelengths but sound extinction lengths tend to drop by the inverse square of the wavelength, so that very high frequency ultrasound does not travel far in tissue.

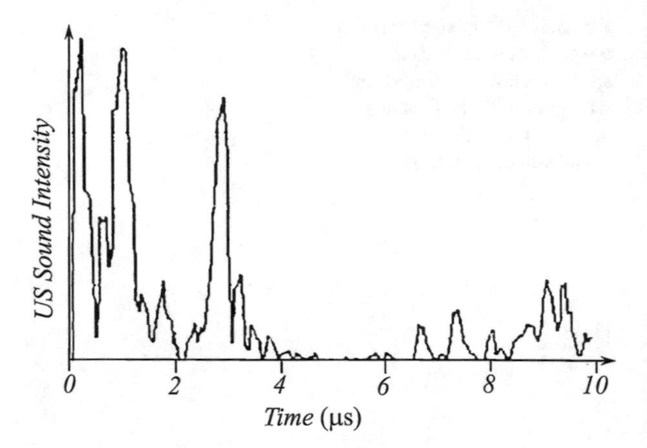

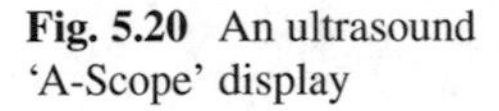
Fig. 5.20 An ultrasound 'A-Scope' display

of causing cancer or cellular damage. There are a variety of techniques in applying ultrasonics (US) in order to image surfaces within the body. One technique uses a US probe placed on the skin, with grease between the US transducer and the skin to match impedance. Pulses are sent into the body, and the echo of those pulses are received (often on the same probe).

In the 'A-scope' mode, a display (on a screen) is generated of the amplitude of the US pulse and echoes as a function of time. Major tissue boundaries (across which the acoustical impedance changes significantly) echo the pulse signal back to the probe. A display in time of the received echo is then equivalent to showing depth from the skin of the tissue boundaries (Fig. 5.20).

A so-called 'B-scope', is a display of the echo from two dimensional boundaries, i.e. both time and direction of the echoes are analyzed to get a 'x-y' display. These displays can show the position and shape of organs. Included in the analysis is the effect of sound attenuation as the pulse signal passes through tissue. Attenuation is caused by sound scattering and absorption.

From a series of US images, a computer-generated 3-D image can be constructed and displayed, including the ability to rotate the perspective view. One such view of a fetal face is shown in Fig. 5.21.

The 'acoustical resolution' of ordinary ultrasonic images is limited. The ideas of Rayleigh apply to the size resolvable in the transverse direction (see Sect. 7.5),

$$R_T \approx 1.2\,\lambda \frac{z}{L}\,, \tag{5.52}$$

where λ is the wavelength of the sound, z is the depth to the object location being imaged, and L is the ultrasonic-transducer transverse dimension. The resolution in the direction of the wave travel (called the axial resolution) is approximately

$$R_a \approx \frac{v_S \tau}{2}\,, \tag{5.53}$$

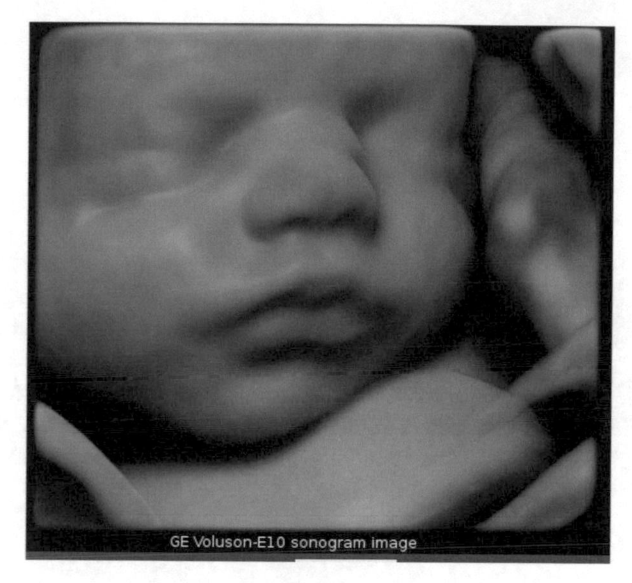

Fig. 5.21 3-D sonogram of a fetal face (Courtesy of the GE Corp.)

where τ is the width in time of the ultrasonic pulse. Pulse widths can be no smaller than $1/f = \lambda/v_S$, making $R_a \gtrsim 0.5\,\lambda$.

Non-linear effects in the propagation of ultrasound can be used to surpass the Rayleigh limit. The idea comes from the fact that sound waves in a material produce temperature and pressure variations. These cause the material viscosity and density to change, resulting in non-linear terms in the wave equation. Even if small compared to the linear terms, such non-linear terms will produce higher-frequency harmonics in the scattered wave, which would otherwise not be present. Being of higher frequency, the scattered sound wave will have components of smaller wavelength than the passing wave. With smaller wavelengths, smaller structural elements can be resolved in the ultrasonic image. An alternative but more invasive technique is to scatter ultrasound from very small bubbles injected into the tissue of interest. The injection material is referred to as a microbubble-contrast agent. The vibration of those bubbles also has a non-linear behavior, and will make shorter wavelength harmonics. The echoed sound can be easily filtered for the first harmonic.

Contrast-Enhanced Ultrasonics

Microbubbles in tissue can enhance ultrasonic image contrast, due to the fact that they can resonate at the ultrasonic frequencies used in diagnostics, sending sound waves back to the detector. Such imaging becomes comparable in resolution to those produced by magnetic resonance or X-ray computer tomography, but, with care, considerably safer (with reduced power levels and prior screening for any reaction against the agents used). The bubbles are injected as intravenous agents, and then dissolve after a few minutes. During resonance, they emit harmonic signals which can be selectively detected.

The imaging of focal liver lesions is an important application of 'contrast-enhanced ultrasonics' (CEUS).

Acoustical Holography

Appendix C describes the principles underlying holography.

An acoustical hologram of organs in the body can be generated by immersion of the body in water. A US emitter sends sound waves toward the body, and also toward an array of two-dimensional piezoelectric transducers. The interference pattern of the direct versus the reflected waves is recorded. That record can be used to reconstruct the reflecting tissue and organs in the body, as a three-dimensional image. This is best done on a computer and screen.

5.22.2 *Doppler Rheology*

As we have described, the Doppler effect is the shift in the frequency of sound because of the relative motion between the source and the observer.

When ultrasound projected into our body reaches a moving material, such as a heart valve or the blood in an artery, that material responds by vibrating at the ultrasound frequency f shifted by the Doppler factor. In addition, because the material is in motion, the sound wave it emits in response to the ultrasound will be detected at a different frequency f' than from the material emitting the pulse. The frequency detected, f', will be related to the US probe frequency by: $f' = 2f/(1 + \mathbf{u} \cdot \widehat{\mathbf{r}}/v)$, where u is the speed of the echoing material, $\widehat{\mathbf{r}}$ is the direction from the source of the ultrasound to the location of the echoing material, and v is the speed of sound in the body tissue.

The fastest moving internal material in the body is blood moving through the aorta, which travels at a maximum speed of about a meter per second. The maximum Doppler shift for a 1 MHz US in-coming wave is therefore $\Delta f \leq 2\, f\, u/v = 2 \cdot 1\,\text{MHz}(1\,\text{m/s}\; 1500\,\text{m/s}) = 1.2\,\text{kHz}$. (The echo from flowing blood will be audible) (Fig. 5.22).

5.22.3 *Physiotherapy*

Some medical applications of ultrasonics in therapy include:

- Ultrasonic Diathermy (Producing tissue heating with ultrasonics)
- Ultrasonic breakup of calculi (such as kidney stones and gallstones)

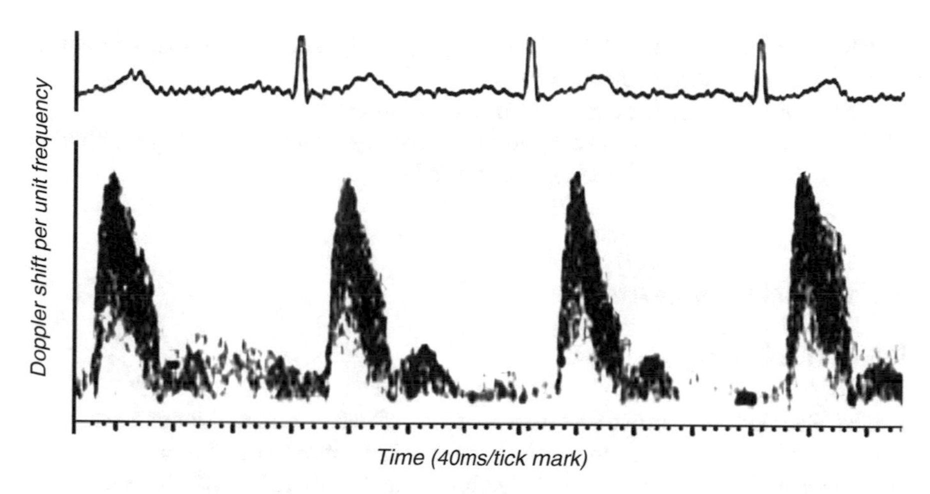

Fig. 5.22 Doppler rheology of aortic blood flow. The top trace is the corresponding EKG

- Ultrasonic breakup of cataracts (phacoemulsification)
- Ultrasonic-assisted lipectomy (i.e. the removal of fatty tissue)
- Ultrasonic measurements for elastography (to discern certain healthy from unhealthy tissue)
- Ultrasonic Gel-Sol Transformation for Drug Delivery (e.g. chemotherapy to brain cancer cells) 1–10 MHz, 0–20 W/cm^2
- Ultrasonic Cauterization: Such as in the treatment of Meniere's disease and prostrate tumors with high intensity focused ultrasound
- Sonic stimulation of bone growth

5.22.4 Ultrasound Diagnostics

Here are a few examples of ultrasonics for medical diagnostics:

- US Imaging: Using ultrasound to generate images showing the interior to the body without invading the skin. For soft tissue, this method is preferred over X-ray imaging, as US is non-ionizing.
- US Rheology and Ultrasonography: The Doppler shift of ultrasound reflecting off moving blood can be used to determine blood flow volume and speed.
- US Echocardiogram: Ultrasound reflecting from the moving valves and blood in the heart can be used to diagnose heart-value prolapse and regurgitation.
- Arterial sonography and venosonography: Imaging to determine sufficiency of blood flow and health of arteries and veins.
- Focused assessment with sonography for trauma (FAST): Used to find possible hemoperitoneum or pericardial tamponade after trauma, and to find possible gallstones or cholecystitis.

- Ultrasound Gastroenterology: Search for gas bubbles, fat deposits, and appendicitis along gastric track and organs.
- Ultrasound Neonatology: Search for fetal abnormalities.
- Sonographic Urology: Investigation of urinary functions, including fluid in bladder, pelvic organs, prostate, and testicles.

5.22.5 Other Applications

Dental Ultrasonics

Dental cleaning now often involves a stream of water which contains ultrasonic waves with frequencies in the 40 kHz range. The vibration in the water tends to shake loosely held material from the teeth, such as tartar and bacteria. This descaling process is generally more effective than mechanical scraping.

Ultrasonic Cleanings

High intensity US waves in a cleaning solution are effective in vibrating the cleaning solution near the surfaces of medical instruments. The vibration will emulsify soft material on the surface of harder materials, particularly if intensities are high enough to cause fluid bubble-cavitation.

Ultrasonic Study of Chemical Kinetics

Chemical reaction rates are quite sensitive to variations in temperature and pressure. As we have described, sound waves produce pressure variations and also temperature variations, since heat does not have much time to equilibrate during the ultrasonic wave vibration period. At high intensities, ultrasonic cavitation can create chemical reactions under extreme conditions. This application is called 'sonochemistry'.

Given that ultrasonic frequencies in the 10 GHz range are easily produced in liquids, the study of relaxation kinetics at the picosecond range is possible. (In contrast, laser pulse studies can reveal chemical kinetics at the femtosecond range.) The formation of mesoscopic structures in solutions occurs in the picosecond range.

5.23 Acoustical Dosimetry

Elderly humans are likely to have significant hearing loss for frequencies above 10,000 Hz, and individuals who listen to loud sound at rock concerts (which often have amplified sound above 120 dB) and even movie theaters will have diminished response to the corresponding frequencies of high-level sound. The loss mechanism is breakage of the auditory hairs (stereocilia) and loss of the hair cells in the inner ear (cochlea), usually first at the high-frequency end. However, even infrasound above 120 dB can damage hair cells. These hair cells do not appear to regenerate in mammals (although they do seem to get repaired in birds and fish).

Hand-held acoustical dosimeters measure sound decibel levels over a range of frequencies. A-weighting emphasizes the middle range of frequencies from about 200 Hz to 2 kHz. B-weighting includes more low tones down to about 50 Hz. The C-weighting accounts for all frequencies from 30 Hz to 10 Hz about equally. A D-weighting is like B, but is made more sensitive in the frequencies 2–8 kHz where aircraft jet engines have a peaked emission. If no weighting is used, the meter is on a Z (for zero weighting) setting.[28]

For persistent sound, such as in a workplace, standards are set by both national and international committees. For the United States, the Occupational Safety and Health Administration (OSHA) sets limits on exposure to noise to be 90 dB over an 8-h period, using an 'A' filter on the sound level.[29] For each increase in loudness by 5 dB, the maximum exposure time is halved. Taking these limits into account, OSHA defines a 'noise-exposure dose' to an individual according to

$$D_{NE} \equiv \frac{100}{8\,\text{h}} \int_0^T 2^{(L(t)-90)/5} dt \,, \tag{5.54}$$

where T is the duration of the noise (in hours) and $L(t)$ is the background sound loudness β with 'A' filtering, measured near the ears, as a function of time t. This noise-exposure dose is reported as a percent of the maximum noise permitted in an 8-hour period. As an alternative to noise-exposure dose, OSHA also defines a 'time-weighted-average loudness' by

[28]The 'equal loudness curve' at 40 phons can be electronically imitated by sending a 40 dB sound signal of fixed level and varying frequency from a good microphone through a wide bandpass filter centered at 2.5 kHz, with ~20 dB attenuation at 100 Hz, and ~10 dB attenuation at 20 kHz. In one implementation, the signal is multiplied by the two factors in

$$\left(\frac{1}{(f^2+20.6^2)(1/12200^2+1/f^2)}\right)\left(\frac{1}{\sqrt{1+107.2^2/f^2}\sqrt{1+737.9^2/f^2}}\right),$$

divided by the same factors with $f = 1000$, where f is measured in Hz. The factors were selected because they can be generated with resister and capacitor filters in an analog device. The corresponding 'C' filter, used to imitate the equal loudness curve near 100 dB, has attenuation of ~3 dB at 31.5 Hz and at 8 kHz, making the 'C' filter flatter in response than the 'A' filter at the low frequency end, multiplying the signal by just the first factor above normalized at 1000 Hz.

[29]When loudness has been modified with an 'A' filter, the loudness number is sometimes given the specification 'dBA'.

$$L_{TWA} \equiv \left\{(5/\log_{10} 2)\log_{10}(D_{NE}/100) + 90\right\} \text{dB}. \tag{5.55}$$

One can see that if the loudness is 90 dB over an 8 h period, $L_{TWA} = 90$ dB. In addition to the average noise limit of 90 dB over 8 h, OSHA sets a limit of 115 dB over a maximum of 15 min, and a limit of 140 dB for an 'impact' sound, such as that made by a mechanical hammer.

5.24 Ultrasonic Dosimetry

5.24.1 Absorption of Ultrasonic Waves in Materials

Sound decreases in intensity as it propagates through materials. The 'attenuation' is due to scattering of the sound and to various forms of dissipation and conversion of the sound energy to other forms of energy, including heat energy transfer. The rate of sound-intensity attenuation for waves moving in one direction is proportional to the distance the sound has traveled. 'Damping' is a generic term for the loss of wave amplitude due to vibrational dissipation.

Mechanisms for sound absorption in tissue include:

- *Ultrasonic thermal effects*: Local vibrations, particularly if gas bubbles, cavitation, or membrane vibrations occur, will generate heat as these materials are forced to wiggle and experience viscous dragging. In addition, local focusing effects due to reflection from concave bone surfaces can cause local hot spots at the focal points of the surfaces.
- *Ultrasonic sound scattering*: 'Sound scattering' is the phenomena of sound reradiation from objects forced to vibrate by ultrasonic waves.
- *Ultrasonic sound dispersion*: Sound dispersion refers to the frequency dependence of the sound speed in the material causing sound waves of different wavelengths to separate on refracting from layers of material. The strength of this dispersion comes from resonances in the vibration of tissue and fluids constituents.
- *Ultrasonic stimulation of material phase change*: Phase changes in materials can be caused by ultrasonic waves. For example, ultrasonics can be used to cause a transition between a gel and a sol in a biological cell, a process called liquefaction of the gel. The production of gas bubbles from gas dissolved in liquids can also occur. Ultrasonic wave absorption can vaporize liquids, and force liquid surfaces to produce a mist. The created bubbles greatly enhance sound dispersion.
- *Ultrasonic depolymerization*: Intense ultrasound can cause the breakup of chain molecules whose length is comparable or larger than the ultrasonic wavelength.
- *Ultrasonic coagulation effects*: Molecular vibration by ultrasonics may give the molecules sufficient energy to overcome anticoagulation tendencies.
- *Ultrasonic rupture of tissue*: The vibrational motion of membranes under intense ultrasonics may be sufficient to cause rupture of those membranes, including even

biological cell walls. The effect is enhanced if the membranes are in resonance with the sound.

5.24.2 *Effects of High-Level Ultrasonics*

Very high levels of ultrasonic waves can be disruptive and destructive to tissue. We have already given a general list of the mechanisms for sound attenuation in Sect. 5.11. We will now consider the most important ones for ultrasonics in medical applications.

Heat Production

If there is insufficient time for the heat generated in tissue from a pulse of sound to diffuse away, then the rise in temperature ΔT can be found from the definition of the specific heat of the tissue, c_h, namely

$$\Delta Q = c_h\, \delta M\, \Delta T\ , \tag{5.56}$$

where δM is the mass of a segment of the tissue. We will use δx to represent the small length through which the sound has passed. Now the intensity of the sound will drop as it passes through, according to $I = I_o \exp(-\alpha_L x)$ with α_L being the sound attenuation coefficient, the relative drop in the sound intensity per unit distance. (See Eq. (5.43).) We will assume that the energy loss in the sound passing through a segment of tissue of area A and thickness δx will go into heat ΔQ. This means, after a time Δt,

$$\Delta Q = -(dI/dx)\delta x\, A\, \Delta t\ . \tag{5.57}$$

Writing the mass of the segment of tissue, δM, as density ρ times its volume $\delta V = A\delta x$, we will have

$$\frac{\Delta Q}{\Delta t} = \alpha_L I \delta V = c_h\, \rho\, \delta V \frac{\Delta T}{\Delta t} \tag{5.58}$$

As a result,

$$\Delta T = \frac{2\,\alpha_L I_{us}}{\rho\, C_h} \Delta t \tag{5.59}$$

where I_{us} is the time-averaged ultrasonic (US) sound intensity. Given that some heat will dissipate, the expression for ΔT should be an upper limit. To get an estimate in a practical case, we will take the specific heat of wet tissue and its density close to that of water: $c_h \approx 1$ cal/gm-deg $= 4.18$ /J gm-deg and $\rho \approx 1.0$ gm/cm^3. For pulse-echo ultrasonic imaging, typically, I_{us} may reach as high as 500 mW/cm^2 but with a pulse duration of only 2 μs. At a frequency of 5 MHz, $\alpha_L \approx 0.25$/cm, we find that $\Delta T \leq 0.12°$ C. For a continuous wave with intensity of 1 W/cm^2, $\Delta T \leq 0.12$ °C in 1 s.

Commercial non-Doppler ultrasound devices operating with pulsed waves have averaged intensities ranging from 0.001 to more than 0.20 W/cm^2. Pulsed Doppler waves used for imaging can reach time-averaged intensities as high as 1.9 W/cm^2 next to the probe. Thus, the probe transducer for these intensities should not be held in one place on the body more than a few seconds.

Cavitation

Strong negative gauge pressure (i.e. absolute pressure minus ambient pressure) in short times can produce bubbles in liquids and viscoelastic solids. These bubbles will contain the liquid vapor. As the pressure increases, the bubbles tend to drop to zero volume (or collapse in times shorter than the period of the sound wave). Bubble formation is strongly dependent on the existence of 'nucleation' sites, where the phase change from liquid to gas occurs with less negative pressure than in the bulk of the liquid. Small particles and surfaces with micro-indentations can act as nucleation locations. The presence of various salt ions also enhances cavitation. When a cavitation bubble forms and when it collapses, it may disrupt a cell structure. Cavitation bubbles oscillate in size and position in response to the ultrasonic wave. These oscillating micro-bubbles can rupture membranes, including cell walls. They also cause stronger absorption of the wave intensity, through conversion of mechanical work into scattered sound energy and heat. More details on bubble vibrations can be found in Sect. D.3 of Appendix D.

If there were no dissolved gases or nucleation sites, the theoretical value of negative pressure to cause water cavitation is in the range of 100,000 atmospheres. The presence of nucleation sites (within the water or on a bounding surface) drops the required negative pressure to about 50 atmospheres. For water saturated with oxygen, only negative pressures >0.1 atm are needed to cause gas bubble cavitation.

Mammalian nervous system tissue is damaged by cavitation when exposed to sound of intensities greater than 1000 W/cm^2 ($\delta p_{rms} = 39$ atm)[30] for at least 1 μs. Mice suffer lung damage due to cavitation if the sound has pressure variations above 8 atm at 1 MHz.

[30] The connection between intensity I and RMS pressure $\sqrt{\langle(\delta p)^2\rangle}$ is given in Eq. (5.33), namely $I = \langle(\delta p)^2\rangle/(\rho v)$, where ρ is the material density and v the speed of sound in the material.

Ultrasonic Chemical Action

High intensity ultrasonic waves in solutions can effect the rate of chemical reactions in several direct ways: (1) By raising the temperature, (2) by increasing the pressure, (3) by causing a more complete mixture of reactants and (4) by generating vibrations in long molecular chains and membranes, causing them to be closer to reactants. Regarding the third effect: Mixtures of immiscible fluids and fluids with insoluble solids will likely have macroscopic regions within made of one kind of material. Intense ultrasonic waves will tend to break up these regions into a number of smaller such regions, allowing reactants to have greater surface areas on which to react. Moreover, the products of the reactions will not accumulate as much if they are dispersed once formed.

Fluid Streaming

Ultrasonic streaming: When high-intensity ultrasonic waves pass through a fluid, non-linear effects may induce macroscopic motion of the fluid. Ultrasonic streaming has been used to noninvasively carry medicines to the blood stream, to hasten chemical reactions through its stirring effect, and to study the composition of heterogeneous fluid mixtures such as contaminated blood.

5.24.3 Ultrasonic Dosimetry Standards

In 1972, The Food and Drug Administration (FDA) (also The World Federation for Ultrasound in Medicine and Biology (WFUMB)) gave the following guidelines:

- US heating should be restricted to $\Delta T < 1.5\,°\mathrm{C}$
- Peripheral vascular intensities: $730\,\mathrm{mW/cm^2}$
- Ophthalmic US intensities: $I < 17\,\mathrm{mW/cm^2}$
- Fetal US intensities: $I < 94\,\mathrm{mW/cm^2}$
- Cardiac US intensities: $I < 430\,\mathrm{mW/cm^2}$

The American Institute of Ultrasound in Medicine (AIUM) has proposed guidelines for limits below which ultrasound has been demonstrated to not show damaging effects. These guidelines include: A diagnostic exposure should cause less than one degree rise in temperature above normal. Exposure intensity should be less than $1\,\mathrm{W/cm^2}$ for focused ultrasound beams.

In 1992, the FDA mandated that ultrasonic instruments used for diagnostics with intensities that can reach $720\,\mathrm{mW/cm^2}$ have on-screen display of Thermal and Mechanical Acoustic Output Indices, defined below.

Diagnostic ultrasound systems have outputs ranging from $10\,\mathrm{mW/cm^2}$ for imaging to as high as $750\,\mathrm{mW/cm^2}$ for pulsed Doppler ultrasound. The intensity values limits are given for spatial peak and temporal averaged intensities (I_{SPTA}). Tissue exposure limits are also imposed on heating over time and on cavitation.

For temperature rise, a 'thermal index' ('TI') is calculated for the kind of tissue exposed. It is the ratio of the sonic energy delivered to a small volume of tissue divided by the sonic energy that would raise the temperature of that tissue by 1 °C. The recommended limit is given by:

$$TI \equiv \frac{E}{E(1^{\circ}\mathrm{C})} \le \begin{cases} 6 - (5/3)\log_{10}(t_e/1\,\mathrm{min}) & \text{adult,} \\ 5 - (5/3)\log_{10}(t_e/1\,\mathrm{min}) & \text{fetus}\,. \end{cases} \tag{5.60}$$

where t_e is the time of exposure, in minutes. Maximum fetal exposures are limited to $t_e \le 4\,\mathrm{min}$.

Cavitation risk is estimated by calculating a 'Mechanical Index' ('MI'), defined to be the maximum peak rarefaction pressure in the tissue, in MPa, divided by the square-root of the ultrasonic frequency, in MHz:

$$MI \equiv \frac{|p_r - p_o|}{1\,\mathrm{MPa}} \left(\frac{1\,\mathrm{MHz}}{f}\right)^{1/2} \le \begin{matrix} 1.9 & \text{no contrast microbubbles,} \\ 0.3 & \text{with contrast microbubbles.} \end{matrix} \tag{5.61}$$

The maximum rarefaction value of $|p_r - p_0|$ at the face of a US imaging transducer can reach 2 MPa (20 times atmospheric pressure p_o) over times smaller than 0.5 μs, five times the recommended limit within tissue at $f = 4\,\mathrm{MHz}$. Acoustical mismatch between the probe and tissue will reduce the pressure within the tissue. In these calculations, the pressure at some depth in the tissue is often found from the pressure at entry into the body by using an attenuation coefficient of 0.3 dB/cm/MHz. The factor $f^{-1/2}$ comes from fitting the observation of how inertial cavitation depends on the ultrasonic frequency. Cavitation is more likely at the lower end of the ultrasonic frequency spectrum used in medical imaging.

Problems

5.1 In your own words, describe the physical reason sound travels much faster in the human body than in air.

5.2 Consider muscle tissue of mass density 1.06 g/cm^3 and a measured speed of sound of 1440 m/s. From these figures, estimate the compression modulus of the tissue.

5.3 From an estimate of the mass density and Young's modulus for bone (along its natural axis), estimate the speed of sound in your tibia.

5.4 Consider the soft fatty tissue in our guts. From its mass density (0.9 g/cm^3) and the measured speed of transverse sound waves (1430 m/s) in this tissue, estimate the shear modulus of soft fatty tissue. (Assume the shear modulus is much larger than the bulk modulus, due to the difficulty in compressing water in tissue.)

5.5 Knowing that air has an effective molecular weight 7.24 times that of helium, and that the ratio of specific heats for helium is 1.67 while that of air is the 1.4, how many octaves above your regular speech will your voice sound when you breathe helium instead of air before speaking?

5.6 Is it always possible to decompose a given voice pattern (giving wave amplitude over a segment of time) into Fourier components? If so, what might the various Fourier components tell us? If we are given the Fourier components, can we always reconstruct the voice pattern?

5.7 Suppose a wave is found to have a dispersion relation $\omega = c\sqrt{k^2 + k_o^2}$, where k_o and c are constants. Show that the phase velocity for this wave is related to the group velocity by $v_g = c^2/v_p$.

5.8 Find the intensity of sound in air if the sound level is 65 dB.

5.9 If you hear a pure note of frequency 512 Hz with a level of 65 dB at your ear, what is the amplitude of the air vibration?

5.10 A 60 dB sound is sustained in a room with a size of 5 m by 6 m by 3 m. How much sound energy is contained in that room at any one time?

5.11 Suppose an ultrasonic wave is produced in the air and arrives at your skin perpendicular to the surface. Estimate the percent of the energy of that wave reflected back into the air.

5.12 If an ultrasonic wave goes from the air to tissue (with a speed four times that of air), what is the maximum angle of incidence?

5.13 If ultrasonic wave of frequency 2 MHz travels through muscle tissue, how far will the wave travel before being reduced to half its initial intensity.

5.14 If the skin of an abdomen has a sound impedance of 1.4×10^5 g/(s cm^2) while that of the surrounding air is 4 g/s cm^2, what percent of sound intensity arriving at the skin from the air enters the abdomen?

5.15 If a $4 \times 3\,\text{m}^2$ wall of a room has a sound impedance of 3×10^5 g/(s cm^2) (the figure for air is 4 g/(s cm^2)), how much sound power enters the walls, with an incident sound at 50 dB?

5.16 If some energy of an ultrasonic wave goes from the air to tissue (with a speed four times that of air), what is the maximum angle of incidence?

5.17 A acoustical converging lens typically has two concave sides, in contrast to an optical converging lens. Why are they so different in shape compared to optical lenses?

5.18 If the absorptivity of muscle tissue is 1.2 dB/cm/MHz, how far will an ultrasonic wave of frequency 3.4 MHz travel in that tissue before its intensity is reduced to one fourth of its original intensity

5.19 If all the sound energy of a 1 MHz ultrasonic continuous emitter with an intensity of 1 W/cm^2 is deposited into an isolated cubic centimeter of tissue, with specific heat of 1.2 calories per g-degree and sound velocity of 1.54 m/s, how much would the tissue rise in temperature after 1 min?

5.20 If a room of volume 800 m^3 were filled with noise at 40 dB level, and all the energy of this sound goes into heating the air each quarter second, how much heat would be produced each hour?

5.21 From the size of an African elephant's external ear, estimate the frequency that the elephant might be able to efficiently detect if the whole external ear vibrates with half a wavelength across its diameter.

5.22 Sonic booms are a hazard to animals. Why are sonic booms called shock waves? Do they shock animals?

5.23 What power (in watts) do you generate in sound if you yell to produce an average sound of 90 dB 1 m from your mouth emitted into a solid angle of 7.3 steradians?

5.24 What are the dominant sound frequencies of: (a) the human male voice; (b) the human female voice; (c) an orchestra playing Beethoven?

5.25 List all these physical characteristics which determine how you sound when you speak, that allows others to identify you. Now list the important factors which make how your voice sounds to yourself different than how others hear you.

5.26 Why do you sound like Mickey Mouse when you speak after inhaling helium gas? (Caution: You can easily faint from lack of oxygen if you breathe helium for more than a few tens of seconds. Without excess carbon dioxide, your brain does not get the normal signal it uses to alert the body to gag.)

5.27 What is the amplitude of air movement with a sound of 115 dB and a frequency of 256 Hz?

5.28 What is the wavelength in air of the highest frequency of sound you can hear?

5.29 If sound of frequency 1000 Hz is attenuated by 0.005 dB/m in 50% humidity air, by what factor does the intensity of this should drop at a distance of 20 m, assuming the sound is 60 dB measured 1 m from the source and spreads out spherically?

5.30 How much energy arrives at your eardrum per second with an area of 0.2 cm^2 when a sound of level 85 dB enters the ear?

5.31 If the sound of your voice is attenuated by 0.03 dB per 100 m of damp air, by what factor does the intensity of your voice drop at a distance of 10 m, assuming the energy of sound is sent out from your mouth with 60 dB 1 cm away, and is spread in all directions over a hemisphere (and not reflected from any surface)?

5.32 Describe how you might design and operate an instrument which can measure the threshold of hearing of an individual, minimizing the psychology of individuals who want to exaggerate or minimize their hearing abilities.

5.33 What is the largest difference in arrival time to each ear of a pulse of sound in air if the ears are 16 cm apart.

5.34 Most humans enjoy music. Speculate on how this characteristic might have come about in evolution.

5.35 Damage to the Broca area of the brain can cause aphasia, the inability to formulate speech. However, the ability to sing a song persists. Investigate how this is possible.

5.36 A room of volume 7820 m^3 has bounding acoustical surfaces described by

Material	Area	Sound absorption coefficient
Window glass	20 m^2	0.025
Wood	175 m^2	0.060
Plaster	1025 m^2	0.035
People (100)	100 m^2	0.450
		(data at 256 Hz)

Estimate the reverberation time in this room.

5.37 Ultrasonic waves can be used to produce images of organs in our bodies. If sound waves of frequency 1.2 MHz are used, estimate the smallest size arteriole which can be resolved. Why not use sound of frequency 120 MHz?

5.38 In your own words, explain why sound imaging by doctors and by dolphins uses ultrasonics rather than sound of lower or higher frequency.

5.39 Suppose a person in a room speaks with an average sound levels of 68 dB when you are 1 m from that person. If the noise in the room is 60 dB, estimate how far away from that person you can move (in meters) and still decipher their speech?

5.40 Suppose a pulse of ultrasonic sound with frequency near 1.6 MHz is emitted from a probe into your body and reflects off blood moving away at 0.56 cm/s, returning to the probe and detected. If the pulse intensity has a Gaussian shape over time, with a maximum intensity (in tissue) of 3.2×10^{-6} W/m^2 and a width (at half maximum) of 1.7 ms, what will the Doppler shift be of the returned signal?

5.41 If the largest relative Doppler shift in frequency from an ultrasonic probe monitoring the blood flow in an artery were found to be 1/4000, how fast would the blood be moving?

5.42 How much time elapses between the moment when an ultrasonic wave enters the body, is reflected by the mitral valve 4.5 cm away, and its echo returns?

5.43 What is the wavelength of a 2 MHz US wave in tissue if the wave has a speed of 1.4×10^5 cm/s? How small an object is resolvable by such a wave?

5.44 What is the smallest wavelength of ultrasound used currently in medical diagnosis? Why not smaller?

5.45 Ultrasonic probes can measure blood flow in the arteries of retina, and fetal heart pumping rate. Describe how you might construct and operate such probes.

5.46 The sound wave from a 1 MHz US probe is reflected 180° from a mitral valve moving away from the source at 3.5 cm/s. What will be the shift in frequency detected at the probe?

5.47 If the intensity of the received sound at a 1 MHz probe must be greater than 1% of the probe's emitted sound, what is the greatest depth that sound reflections can be detectable? Take the attenuation to be 0.5 dB/cm at 1 MHz.

5.48 If all the sound energy of an 1 MHz US continuous emitter with an intensity of 1 W/cm^2 is deposited into an isolated cubic centimeter of tissue, with a density of 1.1 g/cc, specific heat of 1.2 calories per gm-degree and sound velocity of 1.54 m/s, how much would the tissue rise in temperature after 1 min, assuming no heat escapes?

Chapter 6
Electric and Magnetic Fields in Life

Let us be neither drawn aside from the subject in pursuit of analytic subtleties, nor be carried beyond the truth by a favorite hypothesis.

—James Clerk Maxwell

Summary Humans wondered about sparks, lightning strokes and lodestones early in our history, but we did not have a logical and full description of electric and magnetic effects until James Clerk Maxwell's work in the 1860s. The properties of electromagnetism is of central importance in life systems on Earth. The very structure of atoms and molecules comes about through electromagnetic interactions of electrons and protons. Our nervous system uses electric pulses to send signals. Our cellular flagellae and cilia motors work through electric pumps. Electrodynamics is a necessary subject within any serious study of life.

6.1 Importance of Electromagnetism to Life

Electromagnetism is the term used to label one of the four principal types of interactions acting on clumps of matter. Just like gravity, electric forces diminish as the square of the separation distance between a pair of interacting charges.[1] Both gravitational and electromagnetic forces extend over a long range of space, and are measurable for bodies separated by laboratory-size distances.[2] But, for small charged chunks of material in our environment, the electric force acting on one from another is far greater than the gravitational force.[3]

[1] $F_G = Gm_1m_2/r^2$; $F_E = kq_1q_2/r^2$.

[2] The nuclear force, which hold protons and neutrons together, and the weak force, which accounts for radioactivity, diminish exponentially fast with distance, and have not been measured over distances much larger than a few nuclear diameters.

[3] Gravity from a nearby mass becomes important to life only when one or both of the chunks is of planetary size!

W. C. Parke, *Biophysics*, https://doi.org/10.1007/978-3-030-44146-3_6

Electromagnetic interactions are distinguished from gravitational ones by having both repulsive and attractive forces, while the gravitational force between localized regions of mass and energy has always been found to be attractive.[4] If the electric force of a distant body on a local one is measured, and then an identical second body is put with the first, the combined body will feel exactly twice the force from the distant one. Thus, the electric force on a body is proportional to a property carried by the body which we call its charge. Magnetism is associated with electricity, in that two charges moving relative to each other will each experience a force, in addition to the electric force, called the magnetic force from the other charge.[5] Magnetic forces between charges are weaker than electric forces by a factor of the relative speed between the charges divided by the speed of light. This makes electric forces rather than magnetic forces a dominating factor in biochemistry and biomolecular structures. Electromagnetism and quantum principles account for the existence of molecules, including those involved in life. If we add in Einstein's gravitation theory, we can account for the origin of the elements themselves. Our Universe was built with the ingredients to create life, even before there was matter.

Lightning and electric sparks no doubt fascinated our prehistoric forebears. Electricity is also evident in life: Life at the nanoscale cannot be understood without knowing about ions, charge transfer, polarization, and intermolecular electric forces. At the macroscopic level, we see bio-electricity at work in the electric eel and the ability of sharks to detect weak electric fields in the surrounding water.

The magnetic effects of lodestone and magnetized iron inspired many a child, even to become a scientist, because non-gravitational forces acting with strength across a visible gap are not commonly seen. Mystery evokes curiosity. But magnetism in life is subtle. Some organisms, such as homing pigeons, can detect the magnetic field of the Earth. Humans do not seem to have this built-in ability. Even so, the existence of a magnetic field around the Earth has profoundly affected our evolution through deflection of equatorial cosmic rays, shielding life from high levels of ionizing radiation.

6.2 Ideas Behind Electricity and Magnetism

A serious study of galvanic effects in biological systems began just over 200 years ago. After the discoveries of Oersted, Faraday, Ampere, and Gauss, linking the

[4]Einstein's gravity is always attractive.

[5]The conservation of a local charge together with Einstein's principle of relativity are sufficient to account for the behavior of all electromagnetic effects. As Dirac showed in a 1931 paper, if one adds quantum theory, then the discreteness of charge and the difficulty in finding an isolated magnetic monopole can also be explained.

behavior of electric and magnetic phenomena, James Clerk Maxwell, in 1865, succeeded in formulating a logically complete and unified description of electricity, magnetism, and light. The idea behind Maxwell's Theory of Electromagnetism is that certain kinds of particles have a long-range interaction which depends on an intrinsic and additively conserved local property that they carry called their charge.

Maxwell's electromagnetism theory was allied with Dirac's relativistic quantum theory in 1948. The resulting description, called quantum electrodynamics, has been tested more extensively than any other devised by humans, with no deviations as yet found in comparison to observation.[6] In principle, quantum electrodynamics is capable of predicting all of chemistry, starting with nuclei and electrons. This includes biochemistry and, therefore, the mechanisms of life. Of course, in practice, we use higher-level models and descriptions in biochemistry which already assume atoms and molecules as their basic structures.

New relationships can evolve in the behavior of large-scale systems in chemistry and biology because there is a natural ordering that occurs in these open systems. We will discuss this tendency in Sect. 9.25 and Chap. 15. The basis of biochemistry is essentially the quantum electrodynamics of a large number of interacting atoms and molecules acting near statistically likely states, or states that are quasi-stable, building on dynamical pathways 'discovered' by nature during evolution.

6.3 Maxwell's Theory of Electromagnetism

Maxwell's theory is most often applied in two-steps. First, one determines the location and speed $\mathbf{v}$ of the "source charges", described by their density in space, ρ, and their motion, described by their current density, $\mathbf{J} = \rho\mathbf{v}$. We say that these charges create a disturbance in the surrounding space, named the electromagnetic field. This field is expressed in terms of the electric field, $\mathbf{E}$, and the magnetic field, $\mathbf{B}$, produced by the charges. Second, the electromagnetic field acts on other charges q' moving with velocities $\mathbf{v}'$. All dynamics of how charges behave in electromagnetic fields are predicted by Maxwell's equations[7]:

$$\nabla \cdot \mathbf{E} = 4\pi\rho \tag{6.1}$$

[6] The most accurate tests of quantum electrodynamics measure atomic transition frequencies and the electron magnetic moment, and show agreement to at least one part in 10^{11}. In contrast, because gravity around here is relatively weak, most tests of general relativity are not so precise. An exception is the period of a neutron star in a binary system measured by the Hulse-Taylor for the pulsar PSR 1913+16, giving an agreement with Einstein's theory including gravitational waves to at least one part in 10^{14}!

[7] Here, we have used the cgs system of units, and abbreviated $c \equiv 1/\sqrt{\epsilon_0\mu_0}$, where ϵ_0 is the electric permittivity and μ_0 the magnetic permeability of a vacuum. The first Maxwell equation is the differential form of Gauss' law. The fourth contains Faraday's law but adds the effect of a time-changing electric field.

$$\nabla \times \mathbf{E} + \frac{1}{c}\frac{\partial \mathbf{B}}{\partial t} = 0 \tag{6.2}$$

$$\nabla \cdot \mathbf{B} = 0 \tag{6.3}$$

$$\nabla \times \mathbf{B} - \frac{1}{c}\frac{\partial \mathbf{E}}{\partial t} = 4\pi \frac{1}{c}\mathbf{J} \tag{6.4}$$

together with the equations of motion for the particle with charge q'. In the classical case, these are

$$\frac{d\mathbf{p}'}{dt} = q'(\mathbf{E} + (\mathbf{v}'/c) \times \mathbf{B}) \,. \tag{6.5}$$

The right-hand side express the Lorentz force law and $\mathbf{p}'$ is the momentum of the particle with charge q'.[8]

One can move a couple of "test charges" through the region of interest with two different velocities, and determine by their observed motion what the strength of **E** and **B** must have been. Because of how the electric and magnetic fields are defined, and the linear dependence of the electric and magnetic forces on the size of the test charge, **E** and **B** do not depend on the size of the test charge, but rather the electric and magnetic fields determine a property we assign to the space around the source of these fields. Moreover, because Maxwell's equations have only first powers of the electric and magnetic field, solutions for the fields outside the regions where there are charges and currents will satisfy the principle of linear superposition: The sum of two solutions will again be a solution. This property leads to great simplifications in the solutions to the equations. For example, you can always express the general solution as a sum of solutions found for individual point charges and currents.

It can be shown with the Special Theory of Relativity that if an interaction is transmitted by a field at the fastest allowed speed, c, then the force produced by interaction must diminish as the square of the distance between the interacting particles. Maxwell's theory is consistent with the Special Theory, so that the solution of the first of Maxwell's equations for a point charge also gives an inverse-square law for how the electric field drops with the distance from that charge. Quantum field theory can be used to show that the quantum of the electromagnetic field, the photon, carries a unit quantum of spin, and that in consequence[9] electromagnetic interactions are both attractive and repulsive.

[8]In quantum theory, one uses a relativistic wave equation for the 'primed' particle, with the electromagnetic potential field entering with the particle's energy and momentum operators. From the wave function, average positions, velocities, energies, etc. can be determined.

[9]A. Zee, **Quantum Field Theory in a Nutshell**, 2nd ed [Princeton University Press] (2007).

6.4 Electric Fields in Biology

Because of the essential importance of electric interactions in biological systems, strong electric fields in the vicinity of life systems can have overwhelming consequences. So far, none of the extensive electromagnetic and quantum theory predictions disagree with experiments. With such agreement over wide realms, we expect that all observed properties of biochemical systems, including molecular structure and biochemical reactions, as well as the nature of the electrical pulses used by advanced biological organisms to transfer information between subsystems, follow current theory.

Most biological materials are close to neutral in charge. This fact comes about because in the temperature range within which water is liquid, oppositely charged particles can stay bound together. Electric interactions dominate the force between such charges, and it typically takes several electron volts of energy to pull a negative charge from a positive one when they are initially separated by atomic distances. The energy available by thermal collisions is about $1/25$ eV. Life systems use electric forces present in molecular and chemical interactions to do purposeful work by pulling charges around.

Our everyday experience with electricity comes from the transfer of a very small fraction of the charge from one body to another. The flux of one tenth of a trillionth of a percent of your electrons traveling as a spark from your finger to a doorknob can produce a painful searing of a nerve ending. (If the gap was half a centimeter, the electric potential from finger to knob would have been about 15,000 V.) At the microscopic level in biology, electric fields arise when charges are separated, such as within molecules (giving them an "electric polarization"), by cell walls, through biochemical reactions of molecules and ions, between ions in solutions, and by ion pumps in cell walls. Some molecules, like water, have built-in electric polarization. The strongly polarized water molecules make water a good solvent for other polarized molecules and ionic salts. Other molecules become polarized by being stressed in external electric fields ("induced electric polarization"). Charge separation and transfer is often used in life systems to drive biochemical reactions. Even so, electric forces in biosystems tend to be limited in range, because within the protoplasm of our cells, excess charge on a molecule will attract ions of the opposite charge, shielding the molecule from other long-range electric forces. However, significant charge separation can be maintained across cell membranes. Even so, the change in the electric potential across a resting nerve cell is only about 0.07 V.

The strength of an electric field is not normally reported in newtons per coulomb, but rather in volts per meter, in deference to the importance of energy. Energy is exchanged whenever the action of an electric field causes a charge to moves over some distance. If the field pushes the charge over some distance, the charge gains energy taken from the agent producing the electric field. The energy gained by a unit test charge is call the 'electric potential', ΔV. Inversely, the electric field is the negative gradient of the electric potential: $\mathbf{E} = -\nabla V$. For example, an active pump carrying a sodium ion across a cell membrane (with a typical thickness of 10 nm) against an electric field of 0.07 V per 10 nm expends 0.07 eV of energy. (The hydrolysis of ATP to ADP, which supplies energy to the pump, makes about 30.5 kJ/mole = 7.29 kcal/mole $\approx$ 0.326 eV per ATP molecule.)

6.5 Measuring Electric and Magnetic Fields

6.5.1 Instruments

There are a wide variety of instruments for detecting electric and magnetic fields. They all take advantage of the fact that these fields are defined by their effect on test charges. Specialized instruments can be used to measure biologically generated electric and magnetic fields.

Electroscope

When charges are separated, we characterize the surrounding space with an electric field. If a test charge is placed in the region, it will experience an electric force. Suppose a set of negative test charges are taken as electrons on the end of a probe connected to a conducting wire. If the probe is put in an electric field produced by a negative charge, the electrons will move from the probe into the wire. The field from a positive charge will pull probe electrons into the probe from wire.

Now imagine connecting the wire to a pair of conductors suspended in the Earth's gravity and hinged together at the top. When charged, the two conductors will push each other apart. The angle of separation can be used to calibrate the strength of the field. This is an 'electroscope'.

Voltmeter

An ordinary 'voltmeter' allows a small amount of charge to flow from one end of an electric probe (a conducting pointed material) touching one point on an object to a second point that allows charges to move toward or away from the first point. The resulting current (if small enough) is proportional to the potential difference between the two locations. The small current can be measured by using the current's magnetic field to deflect a magnetized needle. Of course, our intent is not to significantly change the system we are trying to measure. This limits the amount of current we can use. These days, very small currents can be amplified to much larger currents which in turn deflect needles or are calibrated digitally and this information is stored.

Electrometrics

Electrometers are devices used to measure electric potentials, electric fields, and/or electric currents.

A valve electrometer uses a vacuum tube with a grid and a plate. Currents as low as one femtoamp (10^{-15} A) can be measured.

Solid-state electrometers that used field-effect transistors (FETs) have an input impedance of about $10^{14}\,\Omega$, again making the required trickle current during measurement extremely low, and give fractions of a picoCoulomb sensitivity. These kinds of sensitivities are useful for detecting individual nerve actions.

Magnetometrics

The first detection of a magnetic field came from the observation of pieces of lodestone acting on each other. Soon after, when a piece of lodestone was allowed to rotate in a horizontal plane, the first compass was constructed in China at about 300 BC.

These days, magnetic field detectors, referred to as 'Gaussmeters', often use a moving coil of wire or a solid-state Hall effect device. If a coil of wire is rotated in a static magnetic field, an alternating electric field is induced in the wire, and will generate a current which can be measured. The Hall effect occurs when a conductor carrying a current is placed in a magnetic field. The field pushes the moving charges to the side (from the Lorentz force $q\mathbf{v} \times \mathbf{B}$). The potential from side to side of the conductor will be of magnitude $V = LvB$, where L is the width of the conductor. Thus, B will be proportional to the induced voltage.

Hall effect magnetometers can be made small enough to be incorporated onto the circuit boards of mobile smart-phones, thus giving them compass direction capabilities.

Alternative magnetometric devices include ones that use the fact that external magnetic fields can affect the energy states of orbiting and spinning charged particles in bound states of atoms and molecules, since such moving and spinning charges will feel and be affected by magnetic forces. The effects on energy states due to typical magnetic fields are small, but easily detectable using resonance measurements with an external electromagnetic wave stimulator. These devices can give magnetic field measurements accurate to 1 part per million. A common type of magnetometer uses proton-spin-precession. Water or other hydrogen bearing material is placed within a coil of electric wire capable of producing a uniform 5–10 mT magnetic field. When the coil is energized, protons within tend to 'align' their spins in the direction of the field. When the coil current is switched off, the protons spins tend to flip to the direction of the external field, releasing photons as a radio wave which can be detected. The frequency of this wave depends on the strength of the external magnetic field. Commercial proton-precession magnetometers are sensitive to about a tenth of a nanotesla.

6.5.2 *Electrophysiology*

Intracellular Electrophysiology

To investigate electrical activity within individual biological cells, a microprobe can be constructed as a glass pipette containing an ionic fluid or a chlorided silver wire, and tapered to a size smaller than the cell. The probe is inserted into the live cell. The circuit is completed by a second probe or silver plate in contact with an ionic extracellular solution. The strength of the electric field and variations in the electric potential across the cell wall are measurable. For nerve cells, these variations can be as large as 100 mV.

Electroencephalography

An electroencephalograph (EEG) is a recording of the electrical activity on the skin on skull. That activity reflects the synchronous electric field of a large number of neurons firing near the surface of the brain, i.e. the cortex. Such activity can show the presence of epileptic seizures, as well as states of sleep or thinking. Skin voltages are in the range of 10–100 μV, and have characteristic frequencies from a few cycles per second to greater than 100 Hz. The activity of individual cells requires alternate methods, such as neural probes inserted or electrodes implanted into the brain.

Electrocardiography

Electrocardiology is the study of the electric activity of the heart. An electrocardiogram is a device for measuring and recording that activity. Our heart pulse is regulated by nerve impulses. The electrical activity of the heart can be monitored by small electrically conducting pads placed on the chest.

An electrocardiogram (ECG or, from the German, EKG) records the electrical activity of the heart by measuring the cutaneous electric potential across the thorax. When the muscle cells of the heart electrically depolarize, they contract. During normal heart beating, the heart muscle cells of the atrium of the heart are stimulated to contract in unison by the electrical activity of the sinoatrial node. The depolarization synchronously spreads quickly to the right atrium and then to the ventricles. The voltage changes involved cause a change in the electric field on the skin of the thorax surrounding the heart. These changes are detectable by sensitive voltmeters, and produce typically voltages of 0.1–15 mV. Atrial depolarization can be seen as a rise and fall in voltage over a time of about an eighth of a second, producing what is called the 'P' wave. The ventricular depolarization produces the 'QRS' wave. The 'Q' and 'S' parts are voltages opposite to the atrial initial voltage, with 'S' usually stronger than' Q', while the 'R' part is a strong positive voltage spike (much larger than that of the 'P'-wave). Following the 'QRS' wave, a small 'T' wave can be seen, which is cause by the ventricular re-polarization

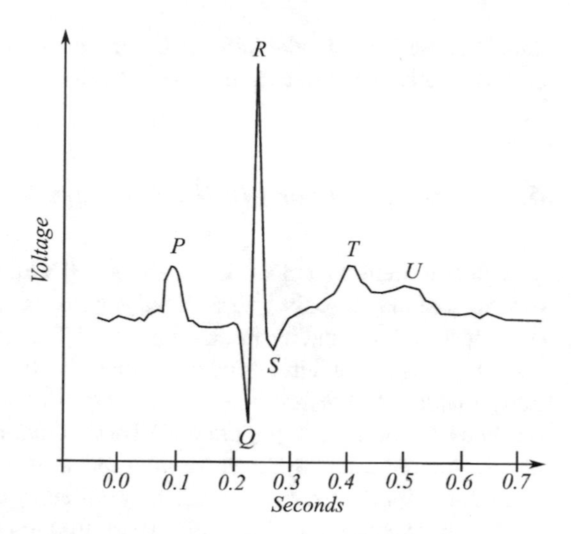

Fig. 6.1 A typical electrocardiograph

and an even smaller (or absent) 'U' wave, probably caused by repolarization of the interventricular septum. (See Fig. 6.1. Also, compare with Fig. 5.12.)

In practice, voltage differences are measured across several positions on the chest relative to the hands and feet. Subtracting a pair of voltages can eliminate the voltage across the skin because of currents induced by fields of external AC current sources, such as fluorescent bulbs and power chords in the room.

Direct measure of heart muscle electric activity (with invasive intracardiac electrodes) is also used to monitor activity and search for atypical behavior.

Defibrillators

Heart defibrillators employ a pulse of current through the thoracic tissues in order to re-establish rhythmic and coherent contraction of heart muscles. Typically, the pulse lasts about 100 ms, with a sinusoidal cycle of 10 ms, and several hundred volts to several thousand volts at the skin. The energy delivered can be 40–400 J. To get this much energy as an electric pulse from a portable device, a battery is used to charge a large high-voltage capacitor. The procedure is referred to as 'electroshock' treatment, although this phrase should be avoided as it resonates with electroshock therapy used and abused for brain illnesses.

Electric Fields and Bone Growth

Bone growth is faster where the bone material is negatively charged. The effect may be associated with the creation of material stress, which is known to promote bone strengthening. There may be some utility in stimulating bone growth by external electric fields. Vibrational waves also seem to help, perhaps again for the additional small stresses, and for facilitating blood flow. However, since the effects is weak, caution is advised until careful science evidence is collected.

6.5.3 Electric Fields in the Atmosphere and Lightning

The electric field on a clear day is about 100 V/m downward. At 10 km up in clear air, the field may be only 5 V/m. Now, you ask, why do we not feel this electric field of almost 200 V standing in an open field? The answer is that with salty water in our bodies, we are a fairly good conductor. Electrons come from the ground into our body, making the electric field near us perpendicular to our skin, and the surface of our body becomes an equipotential. There remains no field parallel to our skin to electrocute us! Sticking up above the ground does, however, make us an attractive target for a lightning strike. The lower portions of nearby thunderclouds are usually negatively charged, pushing a very small fraction of our conducting electrons into and away from the ground, leaving us positive. (See Fig. 6.2.) Of course, we are not feeling so positive when we get struck by lightning! Lying down on the ground or getting into a car (which acts as a 'Faraday cage') is a better strategy than getting under a tree, which tends to get more strikes than open ground.

The charge separation in thunderclouds is caused by the upward motion (sometimes 60 mph) of warm air, with adjacent downward flow of air, ice, and water droplets, which lose electrons through friction with the air. For one thundercloud which reaches 10 km high, the total charge may be as much as 40 coulombs separated by about 6 km. The electric field inside a thundercloud has been measured to be up to 200,000 V/m. The dielectric strength of dry air is 3 million V/m. Even though water vapor within the interior of the cloud reduces the required field for a discharge, it does not reduce it enough. The field near water droplets and ice crystals might be strong enough. In addition, an exotic mechanism may be at play. Cosmic rays generate high energy seed electrons. These can then cause an avalanche

Fig. 6.2 A man standing under a thundercloud has 'bubbles' of constant electric potential surfaces surrounding him. The electric field is perpendicular to these surfaces and is strong where the surfaces are close together

runaway effect, because the effective 'Bethe drag'[10] on relativistic electrons is less than on slower ones. Measured X-ray and gamma-ray emissions support this idea.

A lightning strike can carry 100,000 A and release billions of joules of thermal energy in a few microseconds.[11] The energy comes from a portion of the work done by the hot air rising into the cooler upper troposphere, pulling charges apart all the while. The rising moist hot air releases heat by cooling, but more significantly, the moist air releases additional heat from the condensation into water droplets and ice. In this way, the warmed air can push its way up to the edge of the stratosphere.

Lightning strikes to the Earth all over the globe give the Earth a negative charge relative to the atmosphere. There are also discharges upward from the tops of stratospheric thunderheads toward outer space into the ionosphere. Lightning strikes between a cloud and the ground are typically 5 km long, first along a channel called a plasma tube from cloud to ground, then by one or more return discharges, followed by two or three down strokes. The rapid heating of the air (to about 40,000 °C in a tube about 4 cm in diameter) produces a sound shock wave, which we hear first as a crackling sound of the leader discharges, and then as thunder. The sound wave persists over seconds as it refracts and reflects off clouds and thermal air gradients.

Most of the lightning energy is release as heat and light, with only a few percent into sound. Next to the strike, the sound level can be higher than 120 dB, causing temporary deafness or even rupturing of the eardrum. Of course, more serious is the passage of current through the body.

Lightning rods are pointed conductors which are attached to structures to either slowly dissipate built up charge (via "Saint Elmo's fire"), or, if necessary, quickly discharge, when a lightning strike releases charge to the Earth along a good conductor, such as a large diameter copper wire. (A point on a conductor is used because the electric field at a point is larger than on a blunt shape. A larger electric field will more likely ionize the air and allow charge to leak, thereby avoiding some lightning strikes.)

6.5.4 Typical Static Electric and Magnetic Fields

Here are the sizes of the electric and magnetic fields we might encounter:

- Electric fields in homes emanating from wiring: A few volts per meter.
- Electric fields near power lines: A few volts per meter (220 VAC lines) to several thousand (high voltage lines: up to 765 kV).
- Electric field of an electric eel shock: Up to 430 V/m, releasing about 2 J in 2 ms (not expected to be lethal to a human).

[10] As described in Sect. 8.13.3.

[11] One ton of TNT releases 4.184 billion J.

- Electric field of a heart defibrillator: 300–500 V/m, producing a 700 mA in pulse lasting 30 ms, delivering about 100 J to the tissue.
- Equatorial magnetic field of Earth at surface: About half a gauss.
- Magnetic fields near power lines: A few tens of milligauss.
- Magnetic field near appliances with motors: Several hundred milligauss.
- Magnetic field next to strong 'refrigerator magnet' or speaker magnet: About 2000–10,000 gauss (1 T).
- Magnetic field of an MRI scanner: About 5000–30,000 gauss.

6.5.5 Biogenerated Strong Electric Fields

Some life forms have developed the ability to create large enough electric fields and currents to stun their prey. Eels can produce up to 600 V across their bodies using a series of specialized muscle cells. Other 'weak electric fish' make electric pulses or electric field waves around themselves so that any other object, or another fish, can be detected with special skin organs by the disturbance of the electric field they produce. Sharks also have this capability.

6.5.6 Biodetection of Electric Fields

Sharks have sensitive bioelectric sensors giving them 'electroreception', used for location relative to other nearby sharks or fish, and for detecting nearby objects. The shark generates a surrounding external electric field from its head to its tail using an electric organ within its body near its tail. The field can be individually tailored for shark-to-shark identification. The bioelectric sensors are found on the skin of the head. Objects nearby with a different electric impedance than sea water will affect the field. Significant changes within a body length are detectable by the head sensors.

Electroreception has also been found in some fish and in bees.

Sadly, we humans do not seem to have any biosensors for electric fields until the field becomes strong enough to do damage.

6.5.7 Biodetection of Magnetic Fields

There is evidence that pigeons, turtles, worms, and some bees have magnetic field sensors to orient themselves in the Earth's magnetic field. The magnetic particles, nanosize magnetite ($FeO{\cdot}Fe_2O_3$) crystals, in the nerve tissue of the nasal cavity of homing pigeons, respond to external fields by rotating.

Humans seem to be rather insensitive even to very large magnetic fields (thousands of times greater than the Earth's field). Unless, of course, you have a steel needle buried in you, or surgically implanted ferric metal.

6.5.8 *Static Field Dosimetry*

Being critical to the functioning of life systems, there should be no surprise that life systems can be easily disturbed, harmed, and even killed by strong fixed electric fields. The direct effects of strong fixed magnetic fields can be far milder.[12]

A general rule: Neither voltage nor current kills; but their product over time does. After all, an electric voltage gives the electric pressure, which does not determine whether that pressure can harm. Some people are scared by hearing how high a voltage is. A very high voltage, with very little power, is harmless. It is voltage times current that determines the power delivered. One can pull 20 extra electrons from some material, squeeze them tightly onto a conductor to a million volts of electric pressure and store them there. Even if all those electrons with a million volts are released toward a person, that person will probably feel nothing. If those electrons were directed toward a single cell, the cell is likely to survive, perhaps sustaining damage to a few molecules on its surface. If, instead, 20 Coulomb of charge (1.25×10^{20} electrons) is stored at a million volts, then directed toward a person as a lightning bolt, not only is that person likely to die, but the energy released is 20 million J, enough to completely vaporize an adult. An analogy with water pressure and water flow is useful: A hypodermic needle can eject water under high pressure, but with little volume of water flowing, and harm no one. In contrast, an open fire hose, with the same water pressure, can knock a person to kingdom come.

The student volunteers who hold onto the ball of a van der Graaff electrostatic generator and have their hair stand on end (due to electric repulsion) with a voltage generated of several hundred thousand volts relative to the ground are not likely to be damaged, physically. This follows from the facts that there is only a small total charge and that the electrons, under high pressure, move to the surface of the body, discharging from there, rather than forming a current through the interior during discharge.

Because skin and underlying tissue have a relatively high resistance (hundreds of thousands of ohms from one hand to the other), no damaging current will flow when the electric pressure is a few volts. Touching the terminals of a 9 V battery to the tongue is mildly painful. However, with a 100 V sustained by a power company and put across your two hands, the current becomes on the order of a milliamp, depositing perhaps a quarter of a millijoule per centimeter per second. This is

[12]Dosimetry of electromagnetic waves is also an important topic, but will be postponed until the character of such waves is elucidated in Chap. 7.

sufficient to cause pain in nerve endings, to generate muscle contraction, heart arrhythmia, and to damage cells. Moreover, uncontrollable muscle contraction may make voluntary release of a voltage line impossible.

The resistance of surface skin is greater than the subcutaneous tissue. Wet salty skin can reduce the hand-to-hand resistance by a factor of two or three compared to dry skin, increasing the damage to the tissue conducting the current. Standing in salty water, particularly water that has a conduction path to the ground, and then touching the exposed black end of a 110 VAC power line, is a life ending event.

With smaller voltages, the measurement of resistance across the palms can be used to gauge the emotional reaction of the individual, such as during questioning, since second-by-second, the sweat glands on the palms, as well as other locations, open and close in response to emotional stress. Typically, resistances from one palm to the other range from several megaohms without sweat to 100,000 ohms or less after sweating. Measuring this resistance is one of the parameters used in a 'lie detector'.

A secondary effect of spark discharge in strong electric fields is the production of ozone and nitrogen oxides. Continuous arcing in confined spaces can cause a health hazard through these accumulated gases and free-radicals produced in the air, which can damage exposed membranes, such as bacteria cell walls and lungs in humans.

Biological cells are diamagnetic materials. Unlike paramagnetic or ferromagnetic materials, a diamagnetic material tends to exclude external magnetic fields, producing a repulsive force in a field which has a spatial gradient. Using this property, a live frog has been suspended in a magnetic field of about 15 T, which is about 300,000 times stronger than the Earth's magnetic field. (A tesla is ten thousand gauss.) The fact that the frog physically survives with no apparent after effects demonstrates that life systems are largely unharmed by static magnetic fields, even those far greater than any that occur naturally in the vicinity of planets, and far greater than those which are typical in laboratories. Magnetic resonance imaging devices use fields in the range of half a tesla to three tesla between the poles of a magnet. The danger is not the field, but rather in being hit by stray iron and nickel, which will be pulled violently toward one pole of such a magnet, or by movement of any foreign ferromagnetic material buried in the body.

We have found fields near some distant neutron stars reaching 10^{11} T. Life as we know it could not survive in such a field. In a field of 10^{11} T, atoms have long needle shapes aligned with the field. This completely changes chemistry, and molecules behave as one dimensional objects. It is likely that life needs three-dimensional structures. Certainly brains do.

6.6 Spin Resonance Devices

Relativistic quantum theory predicts that particles can carry an intrinsic quantized spin **S**, with magnitudes given by $\sqrt{s(s+1)}\hbar$, where s is zero or a positive half integer: $0, 1/2, 1, 3/2, \cdots$, and $\hbar$ is Planck's constant divided by 2π. If the particle is bound with an orbital motion around some center, its angular momentum **L** is also quantized, with magnitude $\sqrt{\ell(\ell+1}\hbar$, but now only the values $0, 1, 2, \cdots$ are allowed for ℓ.

We use the index s to label the particles. Relativistic quantum theory also requires that the integer spin particles are 'bosons' (with any number in one quantum state) and the half-odd integer spins are 'fermions' (with only one such particle permitted in each quantum state). Electrons and protons carry an intrinsic spin with $s = 1/2$. Because they also carry a charge, their spin gives them an intrinsic magnetic moment μ.[13]

Having a magnetic moment, the spins of electrons and protons will tend to anti-align or align with an external magnetic field, just as toy magnets do. But quantum theory makes the projection of the electron or proton spin along the external field have only two values when $s = 1/2$, given by $m_s\hbar = \pm(1/2)\hbar$. (See Fig. 6.3.) The number m_s is called the 'magnetic quantum number', taking the value $\pm 1/2$ for an electron or proton. (For toy magnets, the projection of magnetic field along a measurement axis is also quantized, but the number of possible values is astronomically big.)

In an external magnetic field **B**, there will be an energy stored by a magnetic dipole $\boldsymbol{\mu}$ given by

$$\mathscr{E}_B = -\boldsymbol{\mu} \cdot \mathbf{B}$$

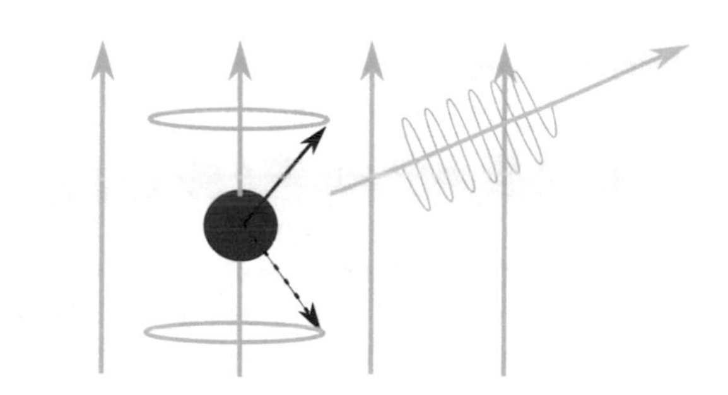

Fig. 6.3 Spin-flip in a magnetic field. The emitted 'wavelet' represents a photon released by the central particle as its magnetic moment (dark 'arrow') flips from anti-aligned to aligned with the vertical magnetic field

[13] Quantum electrodynamics predicts the electron magnetic moment along the direction of its spin to be (T. Aoyama et al., Phys Rev D **91 3**, id.033006 (2015))

$$\begin{aligned}\mu_e &= -\frac{|e|\hbar}{2m_e}\left(1 + \frac{1}{2}\left(\frac{\alpha}{\pi}\right) + \left[\frac{197}{144} + \frac{1}{12}\pi^2 - \frac{1}{2}\pi^2 ln(2) + \frac{3}{4}\zeta(3)\right]\left(\frac{\alpha}{\pi}\right)^2 + \cdots\right) \\ &= -928.4764619 \times 10^{-26}\text{J/T}.\end{aligned} \tag{6.6}$$

where $\alpha = e^2/(\hbar c) = 1/137.035999139$, the fine structure constant. Experimentally, the measured value of the electron magnetic moment is $-928.4764620 \times 10^{-26}$ J/T. The agreement of theory with the accepted experimental value is one part in a billion. As we have noted in Sect. 6.2, quantum electrodynamics is the most extensively and precisely-tested theory humankind has yet produced.

in addition to any other energy the system may hold. The magnetic dipole moment $\boldsymbol{\mu}$ comes from the motion of orbiting charge (just as the current in a coil of wire creates a magnetic field), and from the intrinsic spin of the particles. Classically,

$$\boldsymbol{\mu} = q/(2mc)\mathbf{L}\,.$$

In quantum theory, this becomes

$$\boldsymbol{\mu} = \mu_M\,\left(g_L\mathbf{L} + g_S\mathbf{S}\right)\,.$$

The number g_L is called the 'orbital gyromagnetic ratio' and g_S the 'spin gyromagnetic ratio'. For the electron, the quantity

$$\mu_M = \mu_B \equiv |e|\hbar/(2m_e c)$$

is called the Bohr magneton, while for protons,

$$\mu_M = \mu_N \equiv |e|\hbar/(2m_p c)$$

is the called the nuclear magneton. For the electron,[14]

$$g_S = g_e \equiv -2.002319304\,,$$

while for the proton,

$$g_S = g_p \equiv 5.58569469\,.$$

Now, let's specialize to spin flip. If the electron or proton is forced away from its low-energy state, the particle can give up the energy it stored by flipping back to the opposite alignment. The energy released typically is radiated as a photon. (See Fig. 6.3.) Conservation of energy gives

$$hf = \Delta\mathcal{E}_B = 2\mu_M g_S(1/2)B = \mu_M g_S B$$

so the frequency of the emitted light will be

$$f = \mu_M g_S B/h\,.$$

This is a resonant frequency, since an external electromagnetic field with this frequency will likely cause the electrons to flip to the higher energy state (aligned

[14]A minus sign is incorporated into g_S to take care of the sign of the charge on the electron, i.e. the electron spin is pointed opposite to its magnetic dipole direction. We note that quantum theory forbids a spin-1/2 particle from having any higher moments than a dipole.

with B), but external waves far from this frequency will not be effective in causing a transition between the two energy states.

6.6.1 Electron Spin Resonance

For electron spin resonance (ESR) devices (also called electron paramagnetic resonance devices (EPR)), the frequency of the stimulating wave is taken in the microwave region (3–400 GHz) for convenience in the ease of their generation and detection. Radio wave detection generally requires bigger antennae (comparable to $\lambda/4$), while much higher frequencies need prohibitively larger magnetics. With $f = 9.3882$ GHz, the B field is 0.3350 T (3500 gauss). For a 1 T field, the electron-spin-flip resonance occur at $f = 28.025$ GHz.

In application to biological tissue, ESR works when there are significant numbers of unpaired electrons in the tissue. A pair of electrons in one atomic or molecular orbital typically have oppositely directed spins, causing their magnetic fields to effectively cancel. Free radicals tend to have an unpaired electron. The concentration of free radicals in normal cells is too low for ESR to work effectively, but that concentration can be enhanced by UV irradiation. Alternatively, a 'spin-probe' can be used, i.e. a molecule built with an unpaired electron and with a uniquely-identifiable ESR resonance behavior. A relatively stable organic radical called nitroxide was synthesized in 1965. Nitroxide can be bound to proteins and lipids. ESR with such molecules incorporated into cells then gives information about the mobility of polypeptide chains and about lipids in biomembranes.

The atomic and molecular environment of the unpaired electron may affect its ESR frequency, since this environment can generate its own magnetic field which adds to the external one. As the other 'environmental electrons' are paired, the dominant environmental magnetic field comes from the nuclei of atoms that have a magnetic moment. Hydrogen is plentiful in organic material, and its proton has a relatively large magnetic moment. The nucleus in carbon-12 has no magnetic moment, so it will not affect the ESR resonance frequency. In Table 6.1, magnetic moments for the principal atoms in organic materials are given in units of the nuclear magneton $\mu_N = |e|\hbar/(2m_p c) = 0.15155\, e\,\mathrm{fm}$. One can surmise that the effect of hydrogen atoms bound within organic molecules will dominate the observed electron spin resonances for these molecules.

The interaction of the electrons with nearby nuclei due to a magnetic dipole-magnetic-dipole interaction causes what is called 'hyperfine splitting' of the resonance frequency observed for the unpaired electrons. The number of frequencies into which the resonant frequency is split depends on the spin of the nucleus. The strength of the resonances is determined by how far the interacting nucleus is from the unpaired electron. A study of these splitting and their strength can reveal the geometry of the molecule holding the unpaired electron(s) (Fig. 6.4).

Table 6.1 Nuclear spins and magnetic moments (in units of the nuclear magneton) for atoms common to organic molecules (abundance by fraction of number of atoms in human)

Isotope	Spin	μ	Abundance
^{1}H	1/2	2.79278	0.62
^{12}C	0	0	0.12
^{14}N	1	0.40375	0.011
^{16}O	0	0	0.24
^{23}Na	3/2	2.217	0.00037
^{24}Mg	0	0	0.000070
^{31}P	1/2	1.1317	0.0022
^{32}S	0	0	0.00038
^{35}Cl	3/2	0.82181	0.00024
^{39}K	3/2	0.391	0.00033
^{40}Ca	0	0	0.0022

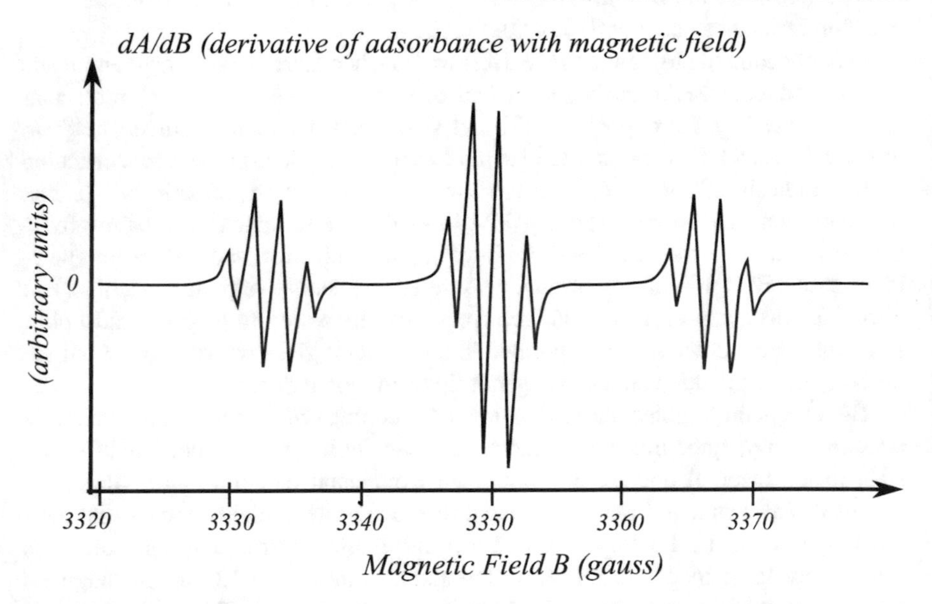

Fig. 6.4 Electron paramagnetic resonance (EPR) simulated spectrum of methoxymethyl radical, $H_3COCH_2^*$. Commonly, the derivative of the absorption with respect to the magnetic field is measured. The center of the resonance occurs where this derivative vanishes while going negative. Here, the electron spin-flip resonance is split into three groups of four. Each distinct resonance ('energy splitting') occurs because the local magnetic field at the location of the unpaired electron in OCH_2 depends not only on the exterior magnetic field, but also the magnetic field of nearby hydrogen protons

In Pulsed ESR, microwave pulses are used to measure the evolution of the coupled state between the unpaired electrons and the nuclei. The half-life for an interacting pair of dipoles to relax back to their initial configuration is determined by the nearby atomic environment which dissipates the stored magnetic energy.

6.6.2 Nuclear Magnetic Resonance and Imaging

Just as in the case of Electron Spin Resonance (ESR), isolated protons will flip their spins by resonantly absorbing a photon at frequency

$$f_p = eg_pB/(2\pi m_pc) = (42.5781\,\text{MHz/T})\,B.$$

For B of 1 T, $f_p = 42.5781$ MHz, which is in the radio frequency band of the electromagnetic spectrum. Typical measurements use a sweeping frequency in the range 300–1000 MHz, with a fixed magnetic field with a strength about 1 T.

As water is plentiful in soft biological tissue, protons in water molecules will dominate the 'nuclear magnetic resonance' (NMR) of soft tissue.

To generate a 'Magnetic Resonance Image' (MRI), the tissue is placed in a strong magnetic field with a gradient, i.e. a field which increases in a particular direction. A broad-band radio frequency pulse is passed through the tissue. The absorption of this pulse is monitored over each segment of tissue perpendicular to the direction of the gradient of the magnetic field and at various angles in this plane. After scanning all segments, the absorption data is computer analyzed to find what the densities of absorbing protons are for particular magnetic environments.

MRI is relatively safe for live tissue,[15] not least because there is no ionizing radiation involved. There are quite large magnetic fields, so any ferromagnetic materials (such as might be in implants or pacemakers) are prohibited. The radio waves used to stimulate the protons must also be limited to prevent overheating. Specific Absorption Rates have been established for this purpose, determined by the temperature rise in a given mass of tissue. (See the Table at the end of Sect. 6.7.2.)

6.7 Biological Effects of Low Frequency EM Fields

Electromagnetic waves in the radio and microwave range are considered low frequency (0–300 GHz). The photons of these waves have insufficient energy per photon to ionize atoms or to break molecular bonds. Biological tissue can be affected because the electron orbits are weakly distorted by the passing wave. This distortion produces a weak electric dipole moment, and the atoms and molecules respond by movement in synchrony with the wave frequency, but with some phase delay. If the atoms or molecules are only weakly held by intramolecular forces, their resulting vibrational motion can be dissipated into heat, causing some absorption of energy from the passing wave. Ions and quasi-free electrons will also easily respond to passing electromagnetic waves.

[15]The word 'nuclear' has been removed from the title of this procedure by doctors and equipment manufacturers due to its negative connotation in some patients. In fact, the nuclei involved undergo no nuclear transitions, so there is no nuclear radiation generated.

The dosimetry for non-ionizing (low frequency) electromagnetic radiation will be described in Sect. 7.31.

6.7.1 Strong Static Electric Fields

Electric fields, such as those under a thundercloud, can reach 1000 kV/m, sufficient to cause lethal current discharge within tissue, with damage largely due to heating. Electric fields above around 100 mV/m across tissue has measurable consequences. The field particularly affects the behavior of cell membranes since cells develop charge separation across their cell wall as an energy source to power membrane motors, pumps, and nerve pulses.

For humans, however, because strong electric fields will cause charges on the skin to accumulate, little static electric field penetrates the skin, being attenuated by a factor of a trillion by the epidermis.[16] In effect, the skin acts as a Faraday cage.

Strong static electric fields are detectable by humans by the effect those fields have in aligning hairs on the skin. The 'pin-prick' sensation caused by tissue heating from charges accumulated on the body discharging when touching a grounded piece of metal is secondary evidence that a strong external electric field was present, sometimes as large as 3000 kV/m. However, there are so few charges producing this field that their discharge generates little total energy. Large fields exist near operating cathode ray tubes ('CRT'), up to 20 kV/m at a distance of 30 cm from the tube.[17]

6.7.2 Strong Static Magnetic Fields

The magnetic fields used in MRI medical imaging, measured in 1–8 T range, are relatively large compared to typical environmental magnetic field strengths. (The field of the Earth is more than 20,000 times smaller.) The Food and Drug Administration sets limits on not only the strength of the magnetic field to which patients are subjected, but also the rate of change of these fields in both time and across space. Gradients over space are necessary for imaging. They are produced by radio waves at about 64 MHz with powers at the source below 25 kW. Time variations occur when the magnetic field is switch on and off. These variations in the magnetic field produce local electric fields, also varying with time.

Important issues include:

---

[16]Polk C, Postow E., **Handbook of biological effects of electromagnetic fields, 2nd ed.** Boca Raton: [CRC Press] (1996).

[17]The greater danger here is not the strength of the field, but the X-rays produced by electrons of kinetic energy of 25 keV hitting the phosphor and glass in a color CRT.

- Ferromagnetic materials, such as implants or accidental inclusions, could be rotated and, in a spatial gradient magnetic field, could be dragged through tissue.
- Ferrous and non-ferrous metals in the body can be heated by eddy currents when the magnetic field changes in time and from radio waves for field gradient induction.
- Rapid changes in the strength of the magnetic field can induce electric fields capable of disrupting nerve action. Near the limits of rates, patients can feel tingles, vertigo, metallic tastes, cardiac stress, and even pain from this effect.

The FDA and industry standards give the following limits:	
Magnetic field strength, ordinary	< 2 T
Field strength, under med. supervision	< 4 T
SAR (Specific Absorption Rate), ord.	$< 0.5\,°$C or < 2W/kg
SAR, med. supervision	$< 1.0\,°$C or < 4 W/kg
SAR, head	$\leq 38\,°$C or 3.2 W/kg
SAR, torso, any 10 g	$\leq 39\,°$C or 10 W/kg
SAR, extremities, any 10 g	$\leq 40\,°$C or 10 W/kg
Field variation dB/dt	< 2 T/s
Field gradient dB/dx	< 45 mT/m

Problems

6.1 Argue against, and then for, the proposition that there may be as yet undiscovered forces affecting how living cells operate.

6.2 Compare the energy available in the hydrolysis of ATP to thermal energy in a living cell.

6.3 What is the strength of the magnetic field next to the north pole of a proton due to its spin? (Take this distance to be 0.5 fm from the center of the proton.)

6.4 How strong is the magnetic field of a proton (due to its spin) in the vicinity of the electron in a hydrogen atom when the electron is at a distance of 10^{-10} m?

6.5 If a proton in a magnetic field is made to flip its spin with a radio pulse ($f = 10^5$Hz), what is the strength of this field?

6.6 If an MRI machine uses a magnetic field of 5 T, estimate the frequency of EM waves need to resonantly flip a proton spin inside a patient in that field.

6.7 Estimate the energy of solar protons after they are captured in circular motion with radius 10 km by the magnetic field of the Earth where the field strength is $B = 0.2$ gauss.

6.8 In recording brain waves on the skull, at least three electrode contacts are made. Why would two contacts not work very well?

6.9 When an electroencephalograph is used to monitor brain activity, how is an alpha wave created?

6.10 An eel can stun prey with just a 600 V shock, but that same animal can suffer a 100,000 V spark with little effect. Why is this?

6.11 If a person with a steel needle buried under her skin enters a MRI machine with magnetic field gradient of 0.04 T/m, what force might be exerted on that needle when it gains an induced magnetic moment of 50 J/T and it is aligned with the field?

6.12 Estimate the electric potential across an artery due to the Hall effect produced an MRI field of 2 T, with a blood flow of 1 cm/s and an artery diameter of 1 mm.

Chapter 7
Light in Biology and Medicine

"Researchers in this field have thrown great darkness on the subject, and, it is probable that if they continue, we shall soon know nothing about it."

— Mark Twain

Summary Humans were unaware of radio waves until 1885, when Heinrich Hertz detected them after Maxwell's prediction in 1864. We now realize how important all electromagnetic waves are in the Universe and in their actions on life forms, including the effects of infrared 'heat' waves, microwaves, visible light, ultraviolet light, X-rays, and gamma rays. Today, our technological society utilizes all such waves, including for biological research and medical applications.

7.1 Electromagnetic Waves

James Clerk Maxwell, using the logic of his theory of electromagnetism, predicted the existence of self-sustaining electromagnetic waves, with all possible frequencies. In empty space, the curl of each side of Eq. (6.2), combined with Eq. (6.1) gives

$$\frac{1}{c^2}\frac{\partial^2 \mathbf{E}}{\partial t^2} - \nabla^2 \mathbf{E} = 0 \,. \tag{7.1}$$

Similarly, Eqs. (6.4) and (6.3) give

$$\frac{1}{c^2}\frac{\partial^2 \mathbf{B}}{\partial t^2} - \nabla^2 \mathbf{B} = 0 \,. \tag{7.2}$$

These are wave equations, just like the ones we encountered for sound waves.

W. C. Parke, *Biophysics*, https://doi.org/10.1007/978-3-030-44146-3_7

In addition, Maxwell's equations lead to the following expression for the work done by an electric field (magnetic fields can do no work) in moving a set of charges in a small volume, per unit time per unit volume:

$$\mathbf{J} \cdot \mathbf{E} = -\frac{\partial}{\partial t}\left(\frac{1}{8\pi}\left(\mathbf{E}^2 + \mathbf{B}^2\right)\right) - c\,\nabla \cdot \left(\frac{1}{4\pi}\mathbf{E} \times \mathbf{B}\right) . \tag{7.3}$$

The first term on the right is the loss of field energy per unit time per unit volume, and the second term is the flux of field energy flowing into the volume per unit volume. That flux of energy through the closed boundary of the volume turns out to be the speed of the wave, c, times the momentum flow through the boundary. In this way, Maxwell knew that the general expression for the energy density, u, of an electromagnetic field is

$$u = \frac{1}{8\pi}\left(\mathbf{E}^2 + \mathbf{B}^2\right) \tag{7.4}$$

and that an electromagnetic field can carry a momentum whose flux density,[1] is given by the 'Poynting vector.'

$$\mathcal{P} = \frac{1}{4\pi}\mathbf{E} \times \mathbf{B} . \tag{7.5}$$

This momentum flux density points in the direction of $\mathbf{E} \times \mathbf{B}$, showing the direction that an electromagnetic wave travels. For all electromagnetic waves, Maxwell's equation (6.2) makes $\mathbf{B}$ perpendicular to $\mathbf{E}$. (See Fig. 7.1.)

It follows that the intensity of the wave arriving at a surface, i.e. the energy arriving per unit time per unit area, is given by

$$I = \frac{c}{4\pi}|\mathbf{E}|^2 = cu , \tag{7.6}$$

and the momentum arriving at a surface of area A per unit time will be

$$dp/dt = IA/c . \tag{7.7}$$

If this momentum is transferred to the surface, then this expression also gives the force of the wave on the surface.

By calculating the speed of electromagnetic waves using the measured values of the electric permittivity and magnetic permeability of air (which are close to their vacuum values), Maxwell in 1864 found 3.1074×10^8 m/s, a number close to the

[1]The momentum flux density is defined as the momentum arriving at a given area per unit time per unit area. The momentum flux density multiplied by c is the energy flux density.

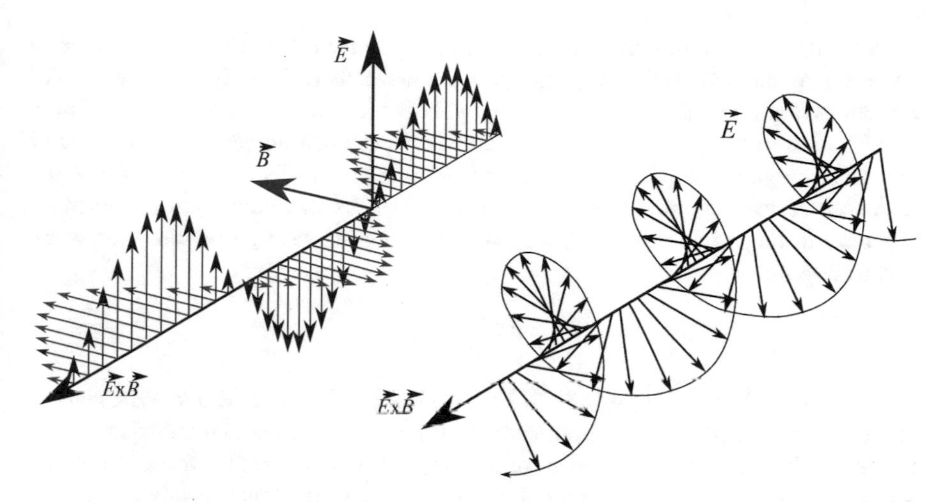

Fig. 7.1 Depiction of two plane electromagnetic waves. The first is linearly polarized. The second is left-circularly polarized. (If you point your left-hand thumb in the direction of propagation, **E** × **B**, your curled fingers show the direction of rotation of the electric field.) For clarity, the magnetic field vectors for the circularly polarized wave are not shown. Unpolarized light passing through sugar solutions will have more right-circularly polarized light absorbed than left-circularly polarized light

best measured value for the speed of light at the time, done by Foucault in 1862,[2] Maxwell deduced[3] that light is an example of an electromagnetic wave. Almost 20 years later, Hertz showed that invisible Maxwellian waves with wavelengths much longer than light could be generated and detected across a room. These are now called radio waves. Maxwell's theory predicts that all measurable lengths can be wavelengths of electromagnetic waves. As far as we know, the range of possible wavelengths observed for electromagnetic waves covers the entire breath of observed lengths in the universe. These lengths extend from below 10^{-15}–10^{26} m.

Electromagnetic waves are generated whenever an electric charge undergoes acceleration. The electric and magnetic fields of an electromagnetic wave are always perpendicular to the direction of propagation of the wave and perpendicular to each other. Such waves whose oscillation is perpendicular to the direction of propagation are said to be 'transverse'.[4] One defines the 'polarization' of the electromagnetic field as the angle in the plane perpendicular to **E** × **B** between the electric field

[2]The modern value of $c \equiv 2.99792458 \times 10^8$ m/s is set exact so that length can defined by c times the time for light to traverse the given length.

[3]This was one of the few moments in the history of humankind that the existence of an important phenomenon in nature was first predicted through thinking about how certain observations could fit together logically. Other examples of such predictions were given in Sect. 1.2.

[4]Note: The strength of the fields oscillate in time at each point in space and from one point in space to the next with field strength transverse to the line drawn in the direction of propagation. There is no motion of any substance in free space.

vector and a selected direction in the same plane. Since the electric field must lie in a plane perpendicular to the propagation of the wave, there are only two independent directions of polarization of the wave (any polarization can be represented as a linear combination of two vectors perpendicular to the propagation vector). The speed of all electromagnetic waves in a vacuum is $c = 2.99792458 \times 10^8$ m/s. Since the distance between maxima along a given wave (i.e. its wavelength) is traversed by the wave in its period $T = 1/f$, we have the universal relation relating frequency and wavelength:

$$\lambda/T = f\lambda = c.$$

As Maxwell's wave equations (7.1) and (7.2) are linear in **E** and **B**, a general solution can be expressed as a Fourier series (or Fourier transform if there is no finite period for the wave). A single term (of one frequency) will have the form of a 'plane wave', i.e. the surfaces of constant phase for the wave are planes. Again because of linearity, we can us the convenience of the complex form[5] for these waves, since both the real and the imaginary parts will be solutions. The expressions

$$\mathbf{E} = A\widehat{\boldsymbol{\varepsilon}} \exp(i\mathbf{k}\cdot\mathbf{r} - i\omega t) \tag{7.8}$$

and

$$\mathbf{B} = A\widehat{\mathbf{k}} \times \widehat{\boldsymbol{\varepsilon}} \exp(i\mathbf{k}\cdot\mathbf{r} - i\omega t) \tag{7.9}$$

will be a plane wave solution of Maxwell's equations (some distance away from the source charges), provided $\mathbf{k}^2 = \omega^2/c^2$. Here, as usual, ω is angular frequency of the wave, related to the ordinary frequency by $f = \omega/(2\pi)$. To get physical solutions, one can take either the real or the imaginary part of the complex solution.

Besides the rather limited frequencies of light we humans detect with our eyes, nature produces both higher and lower frequencies (longer and shorted wavelengths) with biologic consequences. The common names for various regions in this spectrum are (with increasing frequency): radio, microwave, infrared, visible, ultraviolet, X-ray, and gamma ray. (See Fig. 7.2.) Approximate wavelength ranges, as well as frequencies and photon energies, are given in Table 7.1. The ranges for these names are rather arbitrary. The ranges for a given name overlap, with the choice depending on the usage. Man-made radio waves were first detected by a spark across a gap of a fraction of a millimeter in a circle of wire tuned to a corresponding transmitter.

Humans initially produced and detected X-rays by scattering electrons off metal, after the electrons were accelerated in a vacuum by an anode placed at thousands of volts relative to a cathode. Gamma rays were first observed in the decay of

[5]Here, 'complex' refers to the mathematical combination of a real number and an imaginary number: $a + ib$, or in polar form, $\sqrt{a^2 + b^2} \exp(i \arctan(b/a))$.

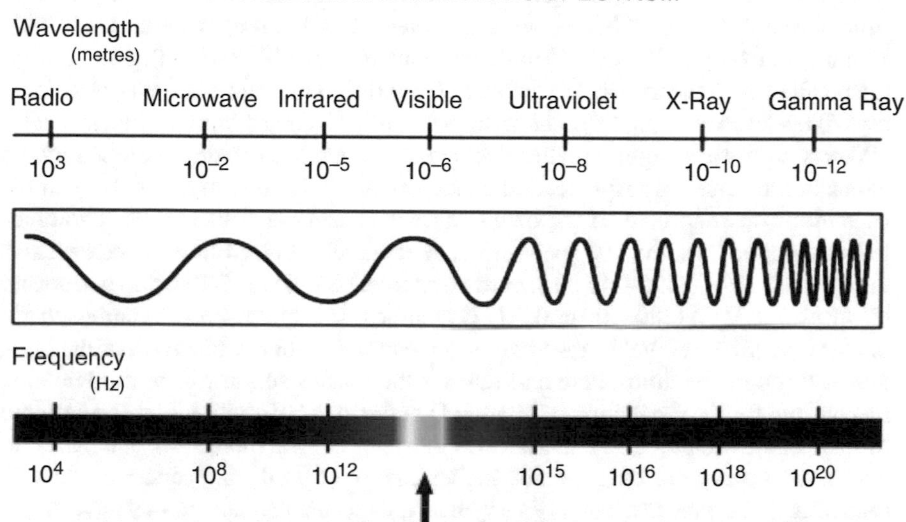

Fig. 7.2 Electromagnetic spectrum

Table 7.1 Our divisions of the electromagnetic spectrum

	Wavelength (m)	Frequency (Hz)	Photon energy (eV)
Radio	>0.1	$<3 \times 10^9$	$<10^{-5}$
Microwave	$0.1-10^{-4}$	$3 \times 10^9-3 \times 10^{12}$	$10^{-5}-10^{-2}$
Infrared	$10^{-4}-7 \times 10^{-7}$	$3 \times 10^{12}-4.3 \times 10^{14}$	$10^{-2}-2$
Visible	$7 \times 10^{-7}-4 \times 10^{-7}$	$4.3 \times 10^{14}-7.5 \times 10^{14}$	2–3
Ultraviolet	$4 \times 10^{-7}-10^{-9}$	$7.5 \times 10^{14}-3 \times 10^{17}$	$3-10^3$
X-ray	$10^{-9}-10^{-11}$	$3 \times 10^{17}-3 \times 10^{19}$	10^3-10^5
Gamma ray	$<10^{-11}$	$>3 \times 10^{19}$	$>10^5$

radioactive nuclei. We can make gamma rays by accelerating charged particles to high energy and then scattering them from nuclei.

The range of electromagnetic waves with frequencies below visible have been further subdivided, and include long and short wave radio, long and short wavelength microwaves, and long and short wavelength infrared. Because of the quantum nature of light, all of these forms of light are non-ionizing, i.e., in the range of intensities encountered in our environment, this light will have an infinitesimally small chance of causing ionization of molecules.

Infrared light is particularly important to biological systems, because it is strongly absorbed and emitted by molecules. Infrared resonant absorption and emission spectra can be used to identify molecules. Infrared light is a necessary component in the thermal dynamics of biophysical systems. Infrared's alternate name is heat radiation.

Above the visible light frequencies, there are 'near ultraviolet' (400–300 nm), 'mid ultraviolet' (300–200 nm), 'far ultraviolet' (200–121 nm) 'vacuum UV' (200–10 nm), 'extreme UV' (121–10 nm), and soft X-rays (10–0.01 nm), hard X-rays (also called low-energy gamma rays) (0.01–0.0001 nm), and high-energy gamma rays (0.0001 nm and smaller). The term 'vacuum UV' comes from the fact that solar UV-rays with wavelengths smaller than about 200 nm are strongly absorbed by the ozone in our upper atmosphere, and so are important in solar exposure only in the vacuum of outer space or if the ozone layer is missing or depleted. For photobiological applications, the UV spectrum is alternatively divided into the ranges: soft ultraviolet (UV-A) (400–315 nm), hard ultraviolet (UV-B) (315–280 nm) and severe ultraviolet (UV-C) (280–100 nm). UVC is almost totally absorbed by atmospheric ozone; over 90% of UVB is absorbed by atmospheric ozone, while most of the UVA passes through the atmosphere and reaches the Earth's surface. Near ultraviolet is responsible for the production of vitamin D in our skin.[6] Far ultraviolet and extreme ultraviolet electromagnetic radiation will ionize biological molecules, and therefore exposure to these will have a much higher chance of producing cancer of the skin than near UV. Never-the-less, UVA erythemal exposure can lead to melanomas and skin aging. If not treated, some melanomas can be invasive into the body.

Sunscreen creams are often rated by an SPF, Sunscreen Protective Factor. The number represents the inverse of the fraction of UVB reaching the skin through an expected layer of lotion. So, an SPF of 50 means 1/50th of incoming UVB would penetrate the expected layer of lotion on the skin. This is 2/100, i.e. only 2% of UVB reaches the skin. The SPF does not apply to UVA protection. Rather, a broad-spectrum sun blocker (UVA and UVB), such as ones with zinc oxide, does stop some UVA, with ratings designated by UVA-PF (UVA protection factor) or UV-PA showing' protecting grade' or by an indication of at least 1/3 for the blockage of UVA in ratio to UVB.

Since the beginnings of quantum theory at the turn of the twentieth century, we have recognized that light can be seen to behave both as a wave and as a particle, depending on the method of detection. Max Planck and Albert Einstein found that light is emitted and absorbed only by a 'quantum' of energy proportional to the frequency of the light, according to

$$E = h f \,, \tag{7.10}$$

where h is called Planck's constant. The quantum of light is now called the 'photon'. The interaction of matter with light may always be described by the absorption and emission of photons. A large number of photons at nearly one energy and with

[6]People who came from Africa to inhabit the northern latitudes of the Earth survived better if they had lighter skin (less melanin), to produce needed vitamin D.

'phase coherence' will act similarly to a classical electromagnetic wave.[7] We knew since Maxwell's work in 1865 that light also carries momentum. (See Eq. (7.7).) With Einstein's work, we know that this momentum is imparted by photons, each with momentum

$$p = E/c = hf/c = h/\lambda \,. \tag{7.11}$$

Note that the energy carried by an electromagnetic wave is always proportional to the electric (or magnetic) field squared (refer to Eq. (7.4)). This energy can be measured by the intensity of the wave, which is the energy arriving at a unit area per unit time. In quantum theory, light intensity is proportional to the flux of photons detected or emitted. Sunlight at ground level has an intensity of about 0.1 W/cm^2, dangerous to the human eye if the light enters directly. Lasers, which we will discuss later, have been made with light intensities of petawatts per square centimeter[8] in very short pulses (and a total energy of 680 J). With a duration of a picosecond, a 10^{12} W/cm^2 laser would deposit about 0.6 keV within the area of an atom, enough to knock all the electrons off low-Z atoms.[9]

7.2 Atmospheric Absorption of EM Waves

The Earth has sufficient gravity, a magnetic shield against solar wind, and the right range of temperatures to hold a relatively dense humid nitrogen-rich atmosphere for billions of years. (Mars is a counter example.) Our atmosphere serves as a partially protective shield for life on the surface of the Earth. X-rays from the Sun and outer space are largely blocked. Hard ultraviolet rays are attenuated, ever since plant life produced oxygen in the atmosphere. In the stratosphere, ozone is produced by hard ultraviolet light acting on oxygen molecules (O_2). Ozone has a strong absorption of hard UV. Oxygen and ozone also have molecular absorption bands in the infrared and microwave ranges. Water vapor (H_2O) and carbon dioxide (CO_2) have a number of infrared resonant absorption bands and strong microwave absorption bands.

Because of the absorption characteristics of the atmospheric gases, there is a 'window' in the electromagnetic absorption spectrum of the atmosphere in the range of visible light. (The same 'window' exists between ultraviolet light and infrared light in many homogeneous solids, such as glass.) Another 'window' exists for radio waves. (See Fig. 7.3.) Except for the sharp nuclear resonant frequencies

[7] A set of waves near one frequency will be coherent if the phase angle of each wave closely matches those of all the other waves. The closer the phase angles are to each other, the more uncertain will be the number of photons detected in the wave.

[8] For example, the Lawrence Livermore Petawatt Laser.

[9] The atomic number is Z, measuring the number of protons in the nucleus of an atom, and the number of electrons in the neutral atom.

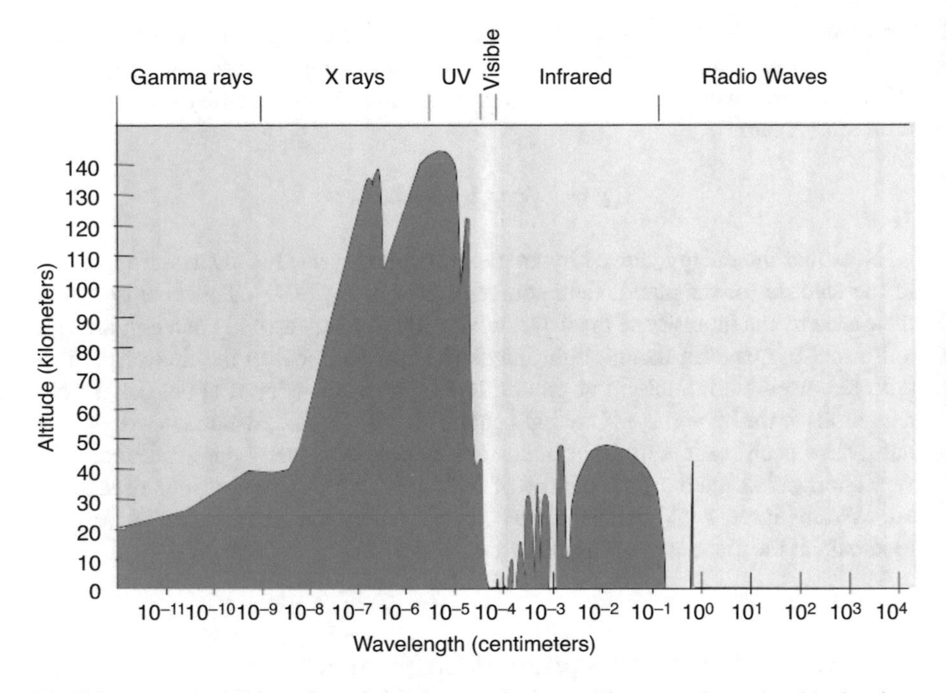

Fig. 7.3 Atmospheric absorption of electromagnetic waves. The curve shows the altitude where the radiation attenuation has become 50% due to molecular absorption, with strong resonant absorption due to water vapor, carbon dioxide, ozone, and methane. The effect of clouds and dust has not been included

in atmospheric atomic nuclei, gamma rays have a good chance of reaching the ground.[10] Even so, their absorption in biological tissue is low.

Cosmic rays and solar wind containing mostly energetic protons and electrons are deflected by the Earth's magnetic field, except for a smaller flux of very high energy protons and electrons.[11] These can reach the upper atmosphere. There, they are likely to be inelastically scattered by atmospheric atoms, producing secondary showers, including pions, neutrons, and muons. The pions do not go far before being taken up by nuclei or by decay. The radiation hazard of energetic neutrons comes from (1) their kick of hydrogen protons in tissue, (2) their ability to transform nuclei of atoms into radioactive nuclei, and (3) the fact that they themselves decay (with

[10]For example, Nitrogen-14 has a resonant absorption for gammas at an energy 9.17 MeV, but the width of this resonance is only 1.28 eV, and the effective radius from the nucleus into which the gammas must enter is 0.00874 Å, compared to the atomic radius of 6.5 Å.

[11]Protons from a solar flare reach the Earth with a flux density as high as 10^5/s/cm^2, and with proton energies in the 100 MeV range. Total solar irradiance at the Earth actually increases by about 0.3% during solar sunspot minimum compared to maximum, since the Sun's magnetic surface activity is reduced, decreasing proton capture near the Sun. Galactic protons can carry even more energy, but with flux densities below 1/s/cm^2.

half-life of 10.3 min), producing energetic protons and electrons.[12] The biological effectiveness of neutrons can be 10 times that of gamma rays of the same energy. Some muons reach the ground before interacting or decaying, with a flux at the surface of the Earth of about one muon per second per square centimeter. They cause a radiation dose to humans of about 0.31 mSv/yr, compared to a total background radiation dose of 2.4 mSv/yr.

As visible light does penetrate the atmosphere and can be used to form images which resolve details on objects in sizes above about 10,000 Å, larger life forms have evolved eyes to take advantages of long-range vision. Since the wavelengths of radio waves are much larger than the sizes of animals that gravity allows, radio eyes did not evolve, remaining unknown to life on Earth until Maxwell.

7.3 Polarization of Light

Maxwell's theory predicted that electromagnetic waves are generated by accelerating charges, such as electrons forced to move up and down in a wire. That wire becomes an antenna.[13] The electromagnetic wave itself is an oscillation in the strength of the electric and magnetic fields measured some distance from the originating charges. The wave moves out from the source at the speed of light. In the radiation zone (far from the source), the electric and magnetic fields are always perpendicular to each other and to the direction the wave is traveling.

These properties can be seen in the plane-wave solutions (i.e. single frequency solutions) given in Eqs. (7.8) and (7.9), and depicted in Fig. 7.1. Electromagnetic waves can be 'linearly polarized' by filtering the electric field oscillation to be along one direction. The unit vector $\hat{\boldsymbol{\varepsilon}}$ in Eqs. (7.8) and (7.9) is the polarization direction of that wave. Light containing waves in a coherent mixture of polarizations are said to be 'elliptically' polarized. Light reflecting from shiny surfaces with 90° between refracted and reflected waves will be linearly polarized with an axis parallel to the reflecting surface. Reflection at other angles will make elliptically polarized light. If light passes through sugar solutions, it becomes, in part, 'left-circularly polarized', which means that the direction of polarization in this light rotates across space, instead of being fixed in direction across space. Light can be left or right-circularly polarized. Maxwell's equations and Einstein's photon relation $E = hf$ can be used

[12] Secondary neutrons cause carbon-14 to build in the atmosphere. Carbon-14 gets incorporated into plants. When the plants die, their carbon-14 slowly decreases by radioactive decay (with half-life of 5730 years), making carbon-14 an important component in determining the age when the plant died.

[13] The design of an optimal antenna is still an art, because the requirements for directional sensitivity, efficiency, frequency range, size limitation, and effective segment dimensions as well as nearby reflector and passive repeater design, are all variables. Curiously, the optimal design for an antenna which most efficiently receives a wide band of frequencies is a wire in the form of a fractal. Fractal antennas are used in cell phones.

to show that a circularly polarized beam of light of frequency f carries a spin angular momentum density of $u/(2\pi f)$, where u is the light energy density.

In fact, individual photons detected or emitted always have their spins aligned or anti-aligned with the direction the light traveled. The ‘helicity’ of a particle is the projection of its spin vector along the direction of its momentum. Relativity and quantum theory requires that photons have helicity of $\pm 1\hbar$.

If all the photons had their helicities aligned in a beam of light, the light was fully right-circularly polarized. If they were anti-aligned, the light was fully left-circularly polarized. A coherent and equal combination of the two kinds of photons meant that the light beam was linearly polarized. Light from the Sun and other thermal bodies is produced with photons with a variety of frequencies and photon helicities, and little coherence. Light detected from a laser may show billions and billions of photons concentrated into a small set of quantum states, making the light highly coherent.

Human eyes are not sensitive to the direction of polarization. However, if you wear polaroid glasses, you will exclude one direction of polarization in favor of polarization perpendicular to the excluded one.[14]

The scattering of sunlight by air molecules also produces partially polarized light. You can see this by rotating your polaroid glasses in front of one eye while looking at the blue sky in various directions.[15] Some insects can detect the direction of polarization of light from the sky, and use this information to orient themselves.

7.4 Linear Superposition of Waves

One of the universal properties of waves of a single kind is their ability to combine when they pass through one location at the same time. We see this as interference, sometimes constructive, sometimes destructive. This adding of waves works for both material (acoustical) and field waves. Waves on the surface of water make a good visual example. We can see waves generated by stones dropped into a quiet lake. If the stones are separated, the waves may spread from each source, combining in the middle region. If, acting alone, the pressure in the wave from the first stone would push the water up at some mid location, while reduced pressure in the wave from the second stone would let the water down, the combined pressures will tend to cancel, making destructive interference. If the two waves push the water in the

[14]These glasses work by stretching long chain molecules having polar side groups parallel to one another onto glass, as a thin layer. Visible light passing through the glass can easily vibrate the molecules perpendicular to the chain by acting on the polar side groups, causing a loss of energy to heat. The same light cannot so easily wiggle the chain molecules along the axis of the chains, as the molecules are a lot stiffer in this direction. This allows waves with polarization parallel to the molecules to pass through the glass.

[15]Some ‘3-D’ glasses used in viewing 3-D films will not work seeing the polarization of the sky: They let ‘circularly polarized’ light through, either left-circularly polarized for one eye, or right-circularly polarized light for the other eye.

same direction, they will cause constructive interference. After the two waves pass each other, they appear in the water as they did before arriving at the interference region.

As we observed, the standing wave in a guitar string can be thought of as two traveling waves of equal amplitude and frequency, but traveling in opposite directions. The two waves interfere to make nodes where complete destructive interference occurs.

According to Maxwell's Theory of Electromagnetism, light waves and all other kinds of electromagnetic waves obey the linear superposition principle in regions where the electric permittivity and magnetic permeability are independent of the strength of the electric or magnetic fields, such as in a vacuum.

7.5 Diffraction of Light Waves

Diffraction is the bending of waves around corners and other barriers. We ordinarily do not observe the diffraction of light waves through windows because the size of the diffraction angle depends on the wavelength of the wave divided by the size of the window. But with tiny openings of a few microns in width, diffraction of visible light is evident. Biologists who look through microscopes at small objects also see the effect as diffraction patterns around tiny structures. An optical spectrometer takes advantage of diffraction to separate light into component frequencies and measure the corresponding light wavelengths.

A 'diffraction grating' is a surface used to spread in angle the various frequencies of a wave. It does this by diffracting the wave with a grating, a surface with parallel groves or dark lines separated by distances comparable to the wavelength of wave being used. It is straightforward to show that light of wavelength λ which arrives perpendicular to a diffraction grating with spacing between groves of d will be diffracted to angles

$$\theta_m = \arcsin\left(m\frac{\lambda}{d}\right) \tag{7.12}$$

where m is an integer.

Diffraction gratings have largely replaced prisms in optical spectrometers as they physically take less space and the separation of frequencies is wider, more uniform and more easily controlled by the grating construction.

For two point-like sources emitting light at wavelength λ observed with an instrument with circular aperture of diameter d, the Rayleigh criterion,

$$\sin\Theta_R > 1.220\frac{\lambda}{d}\,, \tag{7.13}$$

gives the limiting angular separation Θ_R, called the Rayleigh limit between the objects which will still allow one to see two peaks in the light diffraction pattern, and therefore be able to distinguish two objects.[16]

7.6 Electromagnetic Waves in Materials

The behavior of electromagnetic waves in interaction with biological material can be followed with solutions of the wave equation, a result of applying Maxwell's electromagnetic relations in regions with varying electric permittivity and magnetic permeability but removed from the source charges and currents.

Another property of the wave solutions of Maxwell's equations is that waves scattered from a set of molecules can coherently add if those molecules are in an ordered array. Scattering from dust or incoherently distributed molecules leads to Rayleigh scattering, in which the intensity of scattering is proportional to the frequency to the fourth power and the amplitude of the incoming wave squared. However, for ordered molecules, such as on the smooth surface of a solid (see Fig. 7.4), the waves from molecules which are within an area with a diameter about the size of the wavelength of the incoming light will add coherently (constructive interference) some distance from the surface, making the amplitude proportional to

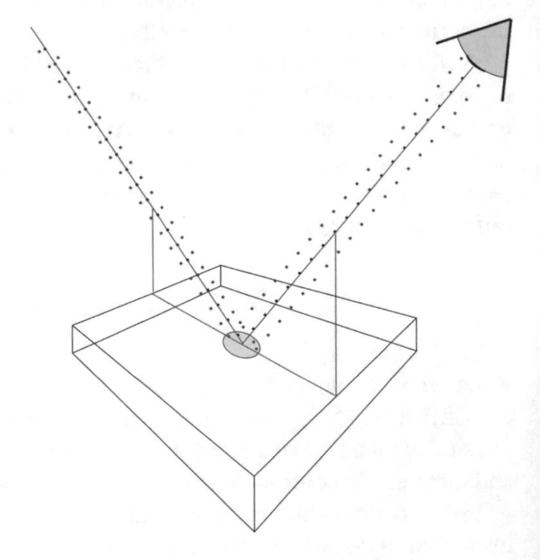

Fig. 7.4 Fresnel zone (greatly magnified and drawn as a circle): Rays of light seen reflected from a smooth solid surface will add coherently from those atoms within about half a wave length of the light around the location of reflection

[16]To define the diffractive limit, Rayleigh calculated the diffraction pattern produced by a pair of distant objects emitting light of one color, with that light focused by a convergent lens toward a focal point. In the focal plane, he determined where the first minimum in the diffraction pattern occurs. The angle formed between rays of light from the lens to the central maximum intensity of the diffraction pattern and those rays to first dark ring minimum defined the resolution angle.

the size of this area called the 'Fresnel zone', i.e. the amplitude is proportional to the wavelength squared, so that the intensity becomes proportional to wavelength to the fourth power times the strength of radiation from a single molecule. But for each molecule, the radiation intensity is proportional to frequency to the fourth power. The two factors, wavelength to the fourth and frequency to the fourth, just cancel, so that the color of the light reflected from the surface is not changed. A glass surface reflects visible light with no discernible change in color.

The scattering of sunlight by molecules in our atmosphere is another good example of Rayleigh scattering. Blue gets scattered by molecules in the atmosphere far more effectively than red, making our sky blue. See Fig. 7.5. For visible white light scattered from very small particles, we can estimate the intensity ratio for blue versus red light scattering using

$$I_{blue}/I_{red} = (f_{blue}/f_{red})^4 = (\lambda_{red}/\lambda_{blue})^4 = (7000/4000)^4 = 9.4.$$

The Sun is whiter than most people think! At sunrise and sunset, the sunlight has more atmosphere to pass through to get to your eyes, so there is even more blue scattered away, leaving more red.

The clear sky is blue because molecules in air will scatter blue light better than red light. The light from the Sun that scatters in the atmosphere will reach your eye even when you are not looking in the direction of the Sun (which is a bad idea anyway). Since blue light is scattered more effectively than red, you will see blue.

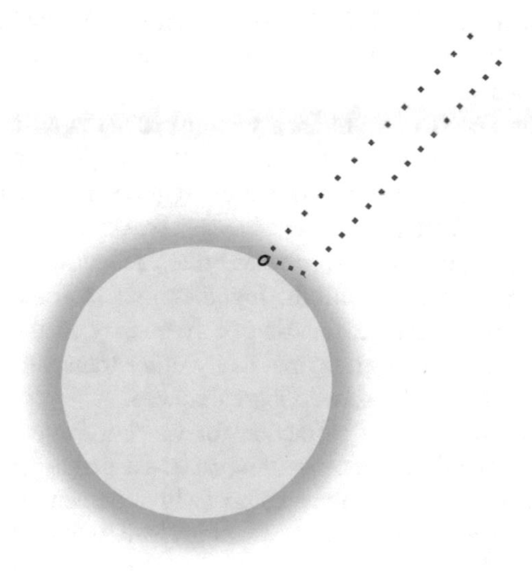

Fig. 7.5 Blue sky: Blue light from the Sun is more likely to be scattered by atmospheric molecules than red light, so when an observer (the 'O' in drawing) looks away from the sun toward a clear sky, it will look blue. (The thickness of the atmosphere is exaggerated in this figure. It should be about a fortieth the radius of the Earth, making a thin blanket.) The atmosphere of Saturn's moon Titan looks orange rather than blue because of the presence of hydrocarbons

The color of oceans and lakes combines the colors reflected from the sky with the color of material in the water at its surface.

Scattering from large spheres (Mie scattering) is, for the same reason, also frequency independent. A good example is light scattered from large water droplets, such as in clouds. Small particles in the air, such as fine smoke, in contrast, will have a blue shade. This is a case in which the particle diameter is comparable to the wavelength of the light being scattered.

7.7 Dispersion of Waves in Media

Light, even when it seems to pass straight through a transparent material, such as the protoplasm of a cell, is nonetheless affected by the charges in the material. Charges jiggle in the material in response to the passing wave. If their jiggling has a natural vibrational frequency matching that of the passing light frequency, resonance will occur. The oscillating charges will radiate light, causing some light scattering. In effect, a fraction of the light passing by is absorbed and then re-radiated by those charges. Since there is a time delay between absorption and re-radiation, the light wavelets generated by the material charges will be out of phase with the light passing by, with the phase of the wavelets shifted back from the originating light. Adding the scattered light to the light passing by unscattered gives the resultant wave. The sequential phase shifts causes the resultant wave to be delayed, and the light takes a greater time to reach the other side of the material than it would have taken across a vacuum. This is why the speed of light in glass, or in the vitreous humor of your eye,[17] is less that in a vacuum.

The ratio of the speed of light in a vacuum to the speed in the materials is called the 'index of refraction, n' of the material: $n = c/v$. Because the resonances of charges in the material occur at different frequencies, the index of refraction may have a complicated frequency dependence. Water (and therefore many dilute organic solutions) has ultraviolet resonances due to electron motions and infrared resonances due to molecular vibrations. Together, these resonances make the index of refraction for water rise as a function of frequency in the visible part of the spectrum. The same is true of glass, and many other transparent crystals. So, blue light in water and glass will be slowed more than red.

The behavior of the index of refraction for varying frequencies, $n = n(f)$, is called a 'dispersion relation', because this variation makes glass prisms disperse white light into its component rainbow colors (with blue bent to a greater angle than red, being that blue is a higher frequency, and is close to the UV resonances).

[17] Some light-headed people may be slowed by vicious humor, but that's a different story.

7.8 Methods for the Production of Electromagnetic Waves

As indicated above, an accelerated charge will generate an electromagnetic wave. This charge may be a proton inside the nucleus of an atom, an electron in an atom, an accelerating electron in a piece of metal, ions in a plasma, a charged particle in a galactic magnetic fields, or a myriad of other possibilities. An electromagnetic wave can also be made by flipping particles with intrinsic magnetic fields. This behavior follows the same prediction of radiation by changing the orientation of circulating charges, such as electrons in atomic and molecular orbits. However, one can show that the intrinsic magnetic field of electrons and protons cannot come from internal charge rotating, which always produces integer spin quantum numbers, rather than the half-integer characterizing electrons and protons.[18]

For each frequency range, typical production mechanisms and techniques are summarized in Table 7.2.

7.9 Generation of Radio Waves

Radio antennae, TV transmitters, and cell phone towers all generate radio waves by electronic circuits. They all contain resonant oscillators to generate a 'carrier wave' using a varying current of the appropriate frequency. This wave is modulated with an analog or digital variation in either the frequency or amplitude of the carrier wave. An analog modulation is considered a continuous variation of an original carrier, while a digital modulation discontinuously changes the carrier.

The modulation of the carrier wave contains the information being transmitted, called the 'signal'. Transmission and reception is by antennae. To increase the range of transmission, the signal is often amplified in a separate part of the electronic circuitry. If the antenna is a straight wire, it is made of length about a fourth the wavelength being transmitted or received, so that an antinode will exist where electric energy is supplied to the circuit. Digitized signals have an advantage over analog signals in that any intervening or intrinsic noise can more easily be separated from the signal, and digitized information is more easily stored in packets with addresses to individual receivers.

Electronic circuits work with currents having oscillation frequencies up to a few tens of gigahertz, but do not work well with microwaves, because at microwave frequencies (with wavelengths in the centimeter range), the size of the electronic components and connections are comparable to the microwave wavelengths (in the

[18]Dirac's relativistic quantum theory predicts that electrons with half-integer spin can exist, and will have intrinsic magnetic moments. The Dirac theory exactly predicts the value of the electron's magnetic moment, indicating that electrons, if they have internal structure, that structure is not yet detectable even when probed to distances down to 10^{-18} m.

Table 7.2 Typical EM Radiation Sources

Radio	Electric currents oscillating in a wire
	Large-scale plasma accelerations
	Large-scale cyclotron radiation
Microwave	Klystron and magnetron tubes
	Electron paramagnetic resonance
	Hydrogen paramagnetic resonance ($\lambda = 21.106$ cm)
	Big-bang relic radiation (black-body at 2.73 K)
Infrared	Hot bodies with $T \approx 150$–1000 K
	Some molecular vibrational de-excitation
	Some molecular rotational de-excitation
	GaAs crystal semiconductor transitions
	Stimulated emission (IR lasers)
Visible	Hot bodies with $T \approx 1000$–10,000 K
	Fluorescence and phosphorescence
	Some delocalized-electron molecular transitions
	Electron transitions in some semiconductors (LEDs)
	Stimulated emission: visible-light lasers
Ultraviolet	Hot bodies with $T \approx 10{,}000$–100,000 K
	Electronic transitions in atoms (e.g. Hg lamps)
X-rays	Atomic transition to K-shell in heavy atoms
	Bremsstrahlung from electrons scattering off heavy nuclei
	Radioactivity ('Low energy' nuclear transitions)
	Radiation from particle colliding beams
	Compton scattering
Gamma rays	Radioactive nuclei ('High energy' nuclear transitions)
	Electron-positron annihilation ($f > 1.236 \times 10^{20}$ Hz)
	Nuclear scattering
	Energetic cosmic rays

centimeter range), so that the electrons lose a significant fraction of their oscillation energy to radiation from the circuit.

A receiving antenna for a passing electromagnetic wave will transfer some power to drive current in the antenna. Curiously, at resonance (natural vibrational frequency of the electrons matches that of the passing wave), the power transferred can be much greater than the wave intensity times the area of the antenna. In Maxwell's theory, this effect is explained by noting that as the electrons in the antenna wire oscillate, they will also radiate. That reaction radiation will be out of phase with the passing stimulating wave, canceling some of its intensity. So, surrounding a receiving antenna will be a cylinder of reduced intensity for the passing wave, whose cross-sectional area can be much larger than the antenna wire area itself. This means your cell phone antenna can be relatively small in area.[19]

7.10 Radio Waves and Biological Systems

Until Maxwell's predictions, life forms on Earth were oblivious to radio waves. With wavelengths much larger than typical organisms, radio waves from outer space did not have much effect, nor did organisms on Earth evolve to use radio waves to communicate.

These days, humans blast radio waves to space, making this planet an obvious location of intelligent life, if any other such intelligent life forms were to monitor our transmissions. Our planet radiates billions of watts of man-made radio waves into space, containing our signals.[20] An additional one billion watts in radio waves radiate from currents in aurora near the Earth's poles.[21] We expect advanced civilizations to use radio to transmit and receive information, although, once they start using global wires and fibers, their intelligent radio radiation with high power would diminish. Moreover, unless their transmitters focused directly toward Earth, radio noise from stars would obscure the signal, even over the distance to nearby extrasolar planets.

It is also possible to safely transfer large amounts of power via radio and microwaves. Solar power collected in space or on the Moon can be sent to Earth over microwave beams, and collected by elevated wire grid antennae covering several acres of land. These grids would let sunlight through, so cows could eat grass under them.

Small remote sensing devices, such as chemical, electrical, thermal, and video probes, encapsulated in pills or implantable devices, are routinely used. They transmit information and receive instruction via radio waves.

Radio waves have a relatively low absorptivity in biological tissue. However, those individuals with implanted pacemakers have to be cautious around radio transmitters. The stray signals can induce currents in any implanted metal, including the wires of the pacemaker. These currents, even though small, may be enough to interfere with the operation of the pacemaker, or stimulate the heart into arrhythmia through the wires. Pacemakers themselves have an average current drain of only about 20 microamps.

---

[19]For cell phones, the length of the antenna can also be reduced by making the antenna from a substance with a large electric permittivity, decreasing the resonance length.

[20]TV signals in the 100 MHz range at one time dominated our radio radiation, spreading in a bubble with a radius in light-years equal to today's date minus 1950.

[21]Donald Gurnett, *The Earth as a Radio Source: Terrestrial Kilometric Radiation*, J Geophys Res **79**, 28 (1974).

7.11 Radio Wave Absorption in Tissue

Generally, biological tissue requires extraordinarily intense radio waves before absorption effects are deleterious. This follows from the fact that such tissue has no appreciable resonances at radio frequencies, the wavelengths of radio waves is large compared to the tissue extent, and no harmful currents are ordinarily induced. An exception occurs when metal is present, such as fillings in teeth. Individuals with such fillings who live near a commercial radio transmission antenna might be able to hear transmissions by the vibrations from their filling in their teeth caused by the passing radio waves stimulating currents. The interaction of these currents produce forces which vibrate the fillings not at the carrier-wave frequency, but rather at the much lower AM signal frequency due to a non-linear response to the induced voltage. The resulting sound vibration spreads through the upper jaw into the inner ear.

Radio waves are absorbed in tissue largely by stimulating ions to vibrate and by causing polar molecules to rotate. The dominant effect of this absorption in tissue is heating, and not chemical transformations, and thus induced cancer in cells is not likely. The tissue absorption is calibrated using the 'Specific Absorption Rate' (SAR) (α_R), i.e. the rate at which radio-wave energy is absorbed per unit mass. Below $f = 10\,\text{MHz}$,

$$SAR \equiv \alpha_R \approx 2\,(f/10\,\text{MHz})^2\ \text{mW/kg}$$

for human tissue.[22]

Radio Wave Dosimetry

To minimize harm, there is general agreement that the resulting SAR should be less than 1.6 W/kg for radio waves and less than 0.4 W/kg for microwaves. The U.S. recommended safe limit for the incoming radio wave intensity is $10\,\text{mW/cm}^2$. More specific recommendations are imposed on Magnetic Resonance Imaging (MRI) scanners.

Damaging effects of radio waves produced by power transmission lines are much less than microwaves of the same intensity, since the absorption in tissue is significantly less. Moreover, as the frequency of the alternating current in a transmission line is 60/s, the radiation from transmission wires is far less than wires carrying current with a higher frequency of oscillation, since the radiative power from a given segment of wire is proportional to the fourth power of the frequency. The (2016) standard limit for cell phone radiation is 1.6 W/kg of tissue. A typical

[22] *A Practical Guide to the Determination of Human Exposure to Radio-frequency Fields*, NCRP Report **119** (2015).

value for a cell phone power within the brain tissue is less than 0.3 W/kg, but only a fraction of this power is absorbed. (Cell phones use frequency bands within 800–900 MHz and 1.8–1.9 GHz. The adult human brain has a mass of about 1.3–1.4 kg.)

7.12 Generation of Microwaves

Microwaves are usually generated in a resonant microwave cavity stimulated by electrons. (A microwave cavity is an enclosure with conductive walls which can support a standing electromagnetic wave with microwave frequency.) Within a 'klystron', a beam of electrons in a vacuum is accelerated and decelerated by a cavity containing a weak microwave, generated by a stimulating electronic oscillator or from positive feedback from the output. The electrons travel to a second cavity, but in between, the difference in speed of the various electrons causes a bunching of electron density which grows to a maximum in the length between the cavities. These bunched electrons induce a strong microwave resonance in the second cavity, which is used as a source of emission. Klystrons can have good control over the frequency of the generated microwaves, but can only create a maximum of around a factor of ten amplification from the original weak microwaves in the first cavity.

'Magnetrons' have an electron emitter ('cathode') constructed as a filament made from a conductive wire, heated with an electric current. The cathode is placed at the central axis of a cylindrical microwave cavity, with a conductive wall some distance away acting as the anode. Electrons from the cathode must pass through a strong magnetic field which is oriented axially. The magnetic field forces the electrons to spiral outward rather than just accelerate radially. At a certain radius, the metal anode has a series of axial cavities. The spiraling electrons passing by the holes of these cavities get jiggled by the induced charges on the walls of the anode. This jiggling motion has a period close to the resonant period of microwaves within the cavity, so some kinetic energy from the electrons passes into microwave energy. Magnetrons can have energy efficiencies in the 70% range (while klystrons are typically in the 30% range) and generate powerful microwaves, but the spread and variation of the microwave frequency is not as sharp as in a klystron.

Microwaves are ubiquitous in our technological world. They are used in communications (1–300 GHz), microwave ovens (2.45 GHz), cell phones (824–894 GHz and 1850–1990 GHz), global positioning, electron paramagnetic resonance imaging, microwave diathermy, and power transmission via microwave antennae.

Microwaves are preferred over radio for communications for several reasons:

- Microwaves have a shorter wavelength than radio, and so, with parabolic reflecting surfaces, they can be more directional between a transmitting antenna and a receiver.
- Microwaves can carry more information than radio waves, being at a higher frequency.

In common with radio waves,

- Microwaves can penetrate long distances through non-metallic materials, such as air.
- Microwaves are non-ionizing radiation, and, at low intensities, will do little or no harm to life forms.

7.13 Microwave Absorption

7.13.1 Absorption in Water

Biological tissue can strongly absorb microwaves principally because 10–75% of this tissue is water (particularly muscle which has the highest fraction). When microwaves pass through liquid water, the polar nature of the water molecule makes the whole molecule rotate, tending to align and then anti-align with the passing wave polarization direction. But the molecule is not free to rotate when in close proximity to other water molecules. Hydrogen bonding tends to make liquid water form quasi-bound nanostructures. The binding is not tight enough to hold under thermal motion, so the structures are in dynamic fluctuation. When a microwave forces a water molecule to twist, that molecules interacts with adjacent ones and the added rotational energy will be partly converted to thermal energy. As seen in Fig. 7.6, for water near 0 °C, the absorption is strongest when the microwave has a

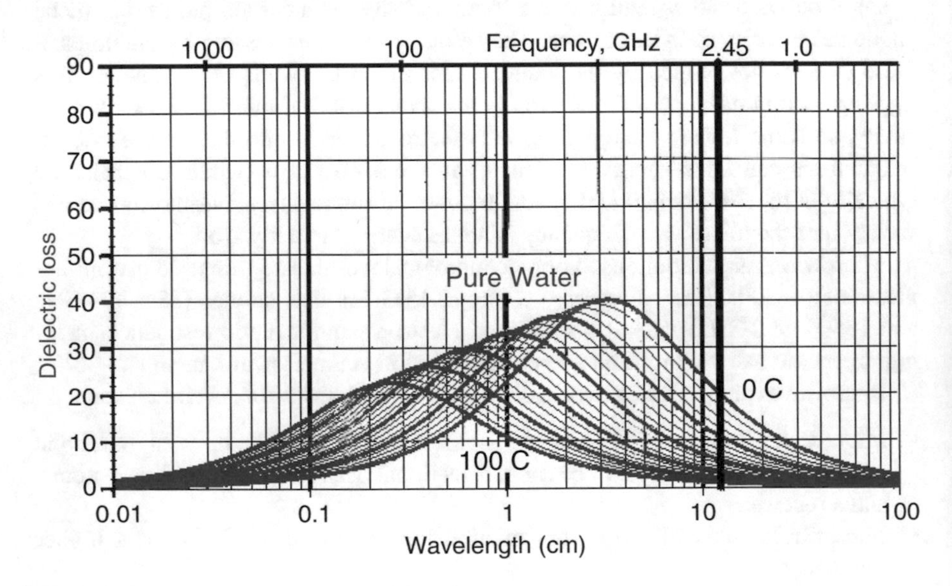

Fig. 7.6 Microwave absorption in liquid water

frequency of about 10 GHz, and has maximum absorption at even a higher frequency as the water heats.

A lower frequency of 2.45 GHz is used in microwave ovens so that the wave can penetrate the food, rather than just heating the outer layer. The part of the wave that is reflected and transmitted is reflected back and forth a number of times in the oven cavity, so that a higher fraction of the microwave energy is converted into heat in the food. Moreover, a metal deflector fan is used to shift the nodes and antinodes of the microwaves in the oven to heat the food more uniformly.

7.13.2 Absorption in Materials and Tissue

Because of absorption in materials and tissue, microwaves drop in intensity the further they travel through uniform tissue, according to

$$I = I_0 \exp\left(-2x/\delta_p\right) , \tag{7.14}$$

where δ_p is called the 'penetration depth', giving the distance of travel in the tissue which causes a drop of intensity by a factor $e^{-2} \simeq 0.135$. (Half this distance is called the extinction length, introduced in our description of sound attenuation in materials, Eq. (5.43).)

Table 7.3 shows the penetration depth of microwaves for various materials and organic substances that might be put in a microwave oven operating at 2.45 GHz.

Microwaves of moderate intensity are used in 'medical diathermy', i.e. to increase blood flow and metabolism by warming, which can promote healing of tissue.

7.14 Microwave Spectroscopy

Some properties of molecules can be investigated by observing their absorption frequencies in the microwave region of the electromagnetic spectrum (6–24 GHz),

Table 7.3 Penetration depth for common materials (at 25 °C) in a microwave oven ($f = 2.45$ GHz)

Material	Penetration in cm
Aluminum	0.00016
Potato	0.8–0.9
Meat	0.9–1.2
Water	1.4–5.7 at 95 °C
Paper	20–60
Glass	35
Quartz	1600

because these absorptions are typically due to excitations of rotations of the molecule. To allow free rotation, a gas cell is usually employed. The molecules in the cell are excited by a short pulse of microwaves, and absorption measured by receiving the microwave decay signals from the molecules. Thus, the molecular moments of inertia can be determined, from which the inter-atomic spacing in the molecule may be revealed, particularly when several atomic isotopes are employed in separate measurements.

7.15 Radiation from Hot Bodies

Infrared light is produced by heating a body, using the thermal motion of the atoms and molecules within to generate light. After all, atoms and molecules contain charges. Jiggling them jiggles the charges, and such accelerated motion produces radiation. Thermal radiation, also called blackbody radiation,[23] follows the theory given by Max Planck[24] in 1901. His explanation required that light be emitted in finite energy units (quanta, later called photons), with an energy proportional to the light frequency given by Planck's relation $E = hf$. Planck's constant, h, has the value 6.626×10^{-34} Js $= 4.13567 \times 10^{-15}$ eV/s.[25] The small size of h compared to laboratory units suggests why quantum effects were not noticed in experiments involving billions and billions of quanta typical of macroscopic interactions. But without the quantization of light emission, a hot body would quickly release its thermal energy by radiation at high frequencies. Quick release is impeded by quantization of emission because thermal collisions do not have enough energy to make photons of arbitrarily high energy. The Planck result for the intensity dI of thermal radiation intensity emitted per unit frequency interval df for a blackbody at temperature T is[26]

$$\frac{dI}{df} = \frac{2\pi f^2}{c^2} \frac{hf}{\exp\left(hf/(k_B T)\right) - 1} . \tag{7.15}$$

A blackbody, a perfect absorber of radiation, will emit this thermal radiation as a perfect radiator (Fig. 7.7). For a less-than-perfect radiator, an 'emissivity' factor ε is included. All humans (irrespective of their skin color) have a skin emissivity of about 0.97 at infrared frequencies.

[23]The phrase 'blackbody radiation' is used to emphasize that this radiation would be emitted even if the body had no material color.

[24]Max Planck, *On the law of the energy distribution in the Normal Spectrum*, Annalen der Physik **4**, 553–563 (1901).

[25]As yet, there is no fundamental explanation for the size of the unitless constant $e^2/(\hbar c)$, i.e. we do not yet know how or why Planck's constant is connected with electric charge.

[26]A derivation of this result is given in Appendix F.

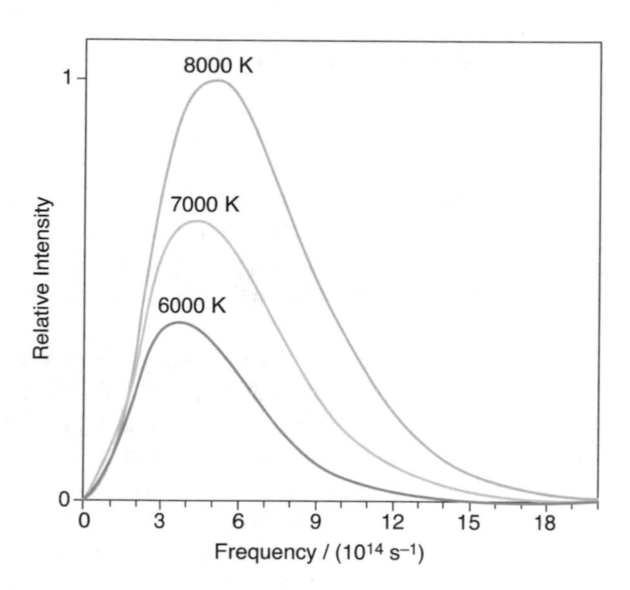

Fig. 7.7 Planck's Distribution. The vertical axis is the intensity of light per unit frequency, divided by the peak intensity at 8000 K

Wien's Law,

$$\lambda_{max} = \frac{0.00289776855\,\mathrm{m}\cdot\mathrm{K}}{T};$$

can be derived from Planck's distribution. It gives the wavelength for the greatest light intensity per unit wavelength. Wien's relation shows that the wavelength at the peak intensity increases inversely with the temperature.

Light intensity from a blackbody at all frequencies follows by integration of the Planck expression, giving Stefan's Law:

$$I = \varepsilon \int_0^\infty \frac{2f^2}{c^2} \frac{hf}{\exp(hf/(k_B T)) - 1} df \tag{7.16}$$

$$= \varepsilon \frac{2\pi h}{c^2} \left(\frac{k_B T}{h}\right)^4 \int_0^\infty \frac{x^3}{\exp(x) - 1} dx \tag{7.17}$$

$$= \varepsilon \sigma T^4 \tag{7.18}$$

where σ, 'Stefan's constant', has the numerical value[27] 5.6704×10^{-8} W per square meter per Kelvin^{-4}. The constant ε is the surface emissivity of the material emitting light through thermal motion of its molecules.

[27]Fortuitously for those who often use Stefan's law, the digits of Stefan's constant in the *mks* system of units follow the sequence 5678, with the 8 being the negative power of ten.

7.16 Physical Infrared Detectors

Just as semiconductors can be used to generate IR light, they also can be used to detect various wavelengths of IR radiation. The basis for this technology is the fact that valence electrons in a semiconductor are bound to atoms, but are not very tightly bound, so that relatively small inputs of energy will release electrons to become conducting. This energy requirement per electron is typically in the IR photon energy range (but also in the visible light range, and with germanium, up to UV light).

A bolometer is a device for measuring the temperature change due to incoming electromagnetic radiation. Bolometers for detection and measurement of the intensity of infrared radiation are a good example. This is often done by using a blackened thin sheet of metal which is insulated from the environment except for a thermal conductor. The metal can act as a thermistor, i.e. change resistance by temperature changes, or the metal can be thermally coupled to a separate thermistor. By comparing the resistance of the exposed metal-thermistor to that left at a fixed colder temperature, the heating effect of the radiation can be measured. In 1880, the inventor Samuel Pierpont Langley constructed a bolometer which could detect a cow a quarter of a mile away.

7.17 Infrared Instruments

7.17.1 IR Absorption Spectroscopy

Absorption spectroscopy has become an indispensable tool for the analysis of complex molecules. Table 7.4 shows the range over which absorption IR spectra are obtained.

For chemical solutions, absorption is measured by passing light through a given concentration of the substance to see how much of the light gets absorbed, and then compensating for the absorptivity of the solute and for the light that is scattered and not absorbed. The schematic design of an infrared spectrometer (also called an IR spectrophotometer) is shown in Fig. 7.8. The reference cell contains the solution without the absorbing solutes, so that the solution spectra can be subtracted from the sample spectra. The diffracting grating spreads the spectra into its component frequencies, one of which is picked by the slit.

Table 7.4 IR spectroscopy range

	Wavelength	Frequency	Frequency/c	Photon energy
Far IR	<1 mm	3.00×10^{11} Hz	$10\,cm^{-1}$	0.00124 eV
Near IR	>700 nm	4.28×10^{14} Hz	$14286\,cm^{-1}$	1.77 eV

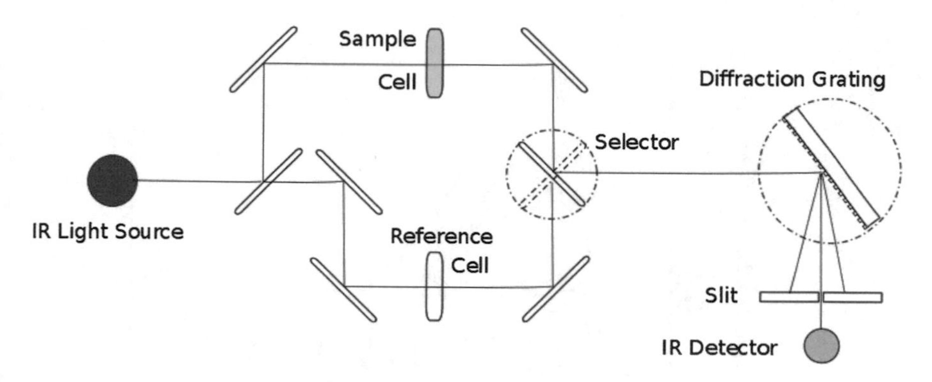

Fig. 7.8 Spectrophotometer design

The intensity I, initially at I_0, will drop with distance in proportion to the intensity of light at that location according to $I = I_o \exp(-\mu x)$. The absorption coefficient μ will depend on the molar concentration $[C]$ of the absorbing material, leading to the Beer–Lambert Law

$$I = I_0\, 10^{-\epsilon_a [C] x}\ , \tag{7.19}$$

where the constant ϵ_a is called the 'molar extinction coefficient' or the 'molar absorptivity'. $T \equiv I/I_0$ is called the transmittance and the combination $A_o = \epsilon_a [C] x = \log_{10}(I_0/I)$ is called the 'absorbance'.[28]

From the above, the molar absorptivity is calculated by

$$\epsilon_a = \frac{1}{[C]L} \log_{10}\left(\frac{I_o}{I}\right), \tag{7.20}$$

where L is the length of the light path through the sample cell. Usually, $[C]$ is measured in moles per liter and L in centimeters, so ϵ_a has units of inverse M-cm. In the present context, the symbol M as a unit represents moles per liter (molarity).

Absorption of a photon passing through an organic solution is strongest at or near the excitable resonances of the atoms and molecules present, where the photon energy $hf = \Delta E$ matches the difference between two excitation energies in the molecule. Differences of excitational frequencies for stretching, bending, and twisting of molecules are likely to be in the range of infrared light. As the force needed to stretch a bond is typically greater than that needed to bend it, and that

[28] If the concentration of the absorber molecules becomes high enough, there will be an overlap of the effective cross-sectional areas of the absorbing molecules from the perspective of the incoming beam of light. In this case, the absorption no longer increases linearly with concentration, but rather flattens out.

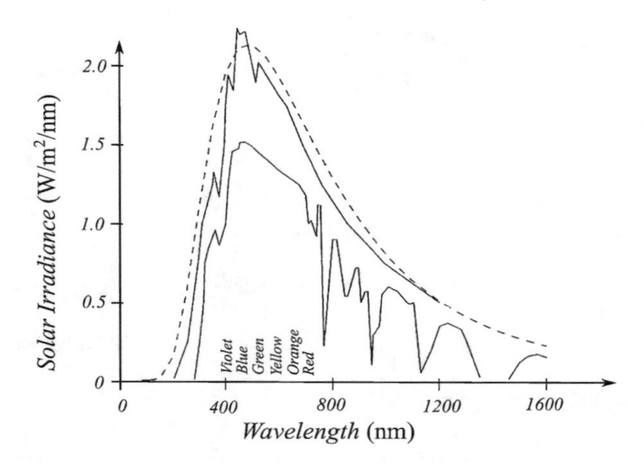

Fig. 7.9 Solar irradiance near and at the Earth. Smooth dashed curve shows radiation intensity for a blackbody at 5800 K. Upper ragged curve shows solar irradiance above the Earth's atmosphere; lower ragged curve at sea level. Hard UV absorption below 300 nm is due largely to ozone O_3. The dips in the IR region above 700 nm are mostly caused by photon absorption in the water molecule H_2O. Carbon dioxide CO_2 has significant absorption bands in the IR region above 1300–11,000 nm. Not shown is IR absorption by methane that occurs near 3500 and 8000 nm and by nitrous oxide N_2O at 5000 and 8000 nm

force is typically greater than that need to twist a bond, the absorption frequencies for the production of each of these types of vibrational distortions correspondingly is from high to low IR.

The position of the excitation energy levels is determined by what are called the 'normal modes' of vibration. The normal modes of a molecule are the steady-state single-frequency solutions to the dynamics of the molecule involving periodic motion of the coupled atomic centers. Any other motion can be represented by a combination of normal modes. The infrared photon absorption will be stronger for vibrational modes which have higher electric polarization.

Once a particular substance has been measured in the infrared region, it has, in effect, been "fingerprinted", since the nature of all the chemical bonds in a given chemical is unique (see Figs. 7.9 and 7.10).

Note that in Figs. 7.11 and 7.12 for the IR transmission of liquid ethanol and propanone, there are sharp dips in the transmission at certain frequencies. These are due to photo-excitation of various vibrational modes of the molecule, which are oscillation of the angles and separation in the bonds C–C, O–H, C–H, C–O, and C=O. (See Fig. 7.10.) The energies of these vibrational modes depend on the position of the excited pair in the molecule and on nearby solute molecules, if present. This means that the combined absorption will sometimes be spread over a region of wavelengths, causing a trough instead of a sharp dip in the transmission across wavelengths. Notice that the stretching modes of vibration occur above wave-numbers of about 1600/cm, while the deformational modes are below.

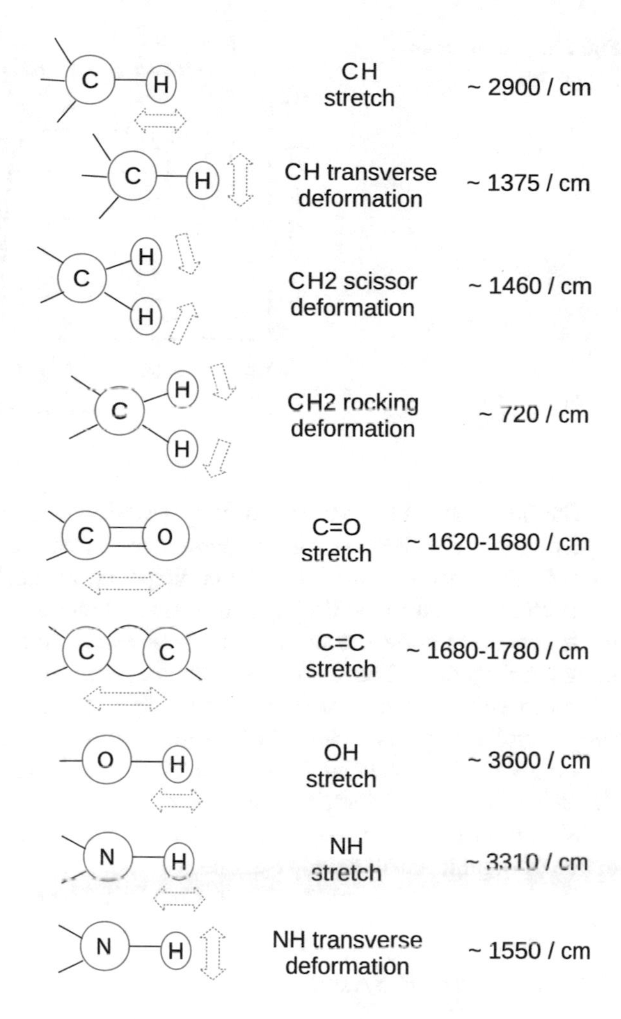

Fig. 7.10 Some vibrational normal modes of some organic bonds, together with their typical frequencies

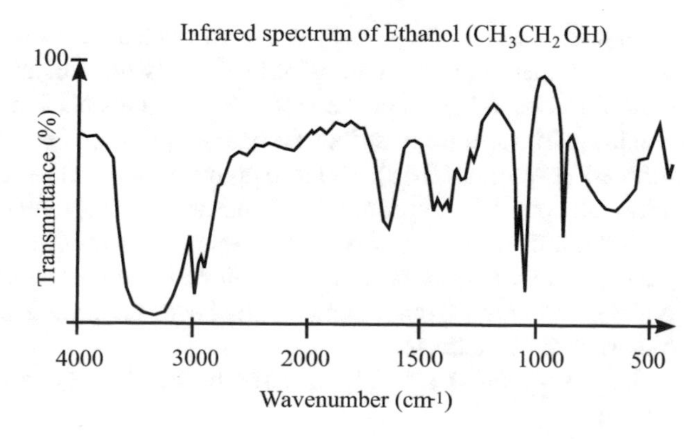

Fig. 7.11 IR spectrum of ethanol

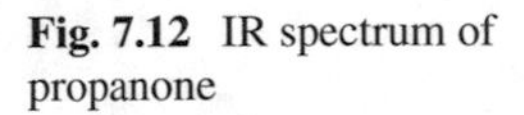

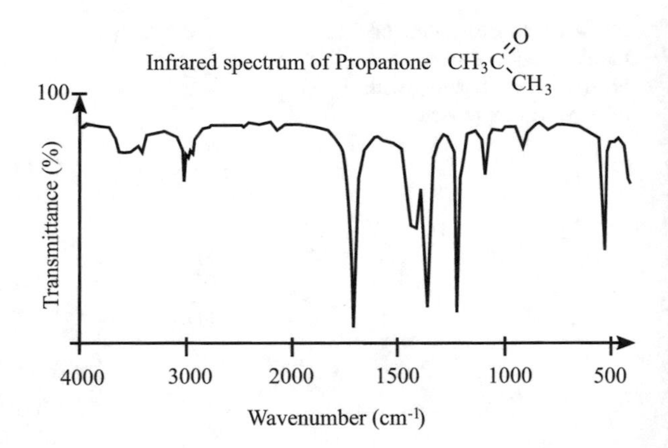

Fig. 7.12 IR spectrum of propanone

The frequency of a vibration, for low excitational energies, is proportional to the square root of the effective spring constant and inversely proportional to the square root of the reduced mass of the two participating atoms. We expect the bonds that are stretched to have a 'stiffer' spring constant than the effective spring constant for those bonds that allow the atoms to move sideways, so the stretching modes will be higher in frequency. The higher wave-number regions are therefore 'diagnostic' for the presence of various atomic bondings, while the lower wave-numbers are in the 'fingerprint' region characterizing how the molecule's chain of atoms is configured.

We give in Figs. 7.9 and 7.13 examples of visible light transmission spectra for the atmosphere and absorption spectra for chlorophyll a and b.

We will elaborate on the subject of IR Absorption Spectroscopy for the identification of organic substances in Sect. 11.8.1.

7.17.2 Near IR Scans

In tissue, water strongly absorbs infrared radiation at the longer wavelength end of the IR spectrum, and hemoglobin strongly absorbs at shorter wavelengths (see Figs. 7.14 and 7.15). In between, the higher transmission of IR is termed the 'NIR window'. The spectral distribution of absorption in the NIR region is called a Near Infrared Spectrum (NIRS). Melanin pigmentation in the skin absorbs strongest in the ultraviolet part of the spectrum, with absorption dropping down exponentially going toward the IR region (approximately as $\lambda^{-1.23}$) (See Fig. 7.18). Such information as the oxygenation of blood can be monitored from outside the body. Looking at the NIRS of radiation from regions of the head, the presence of a stroke, or a nearby tumor, might be detected.

IR absorption is strongly affected by the IR resonances in water, blood, and melanin.

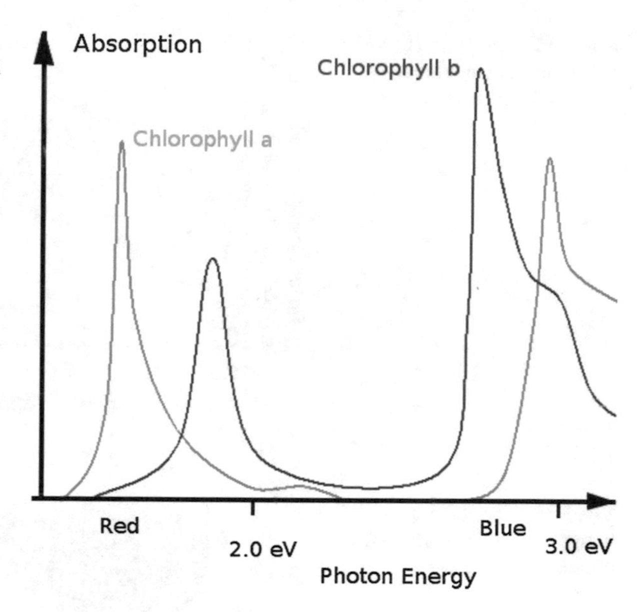

Fig. 7.13 Chlorophyll *a* and *b* absorption spectra. Adapted from Daniele Pugliesi's wiki image

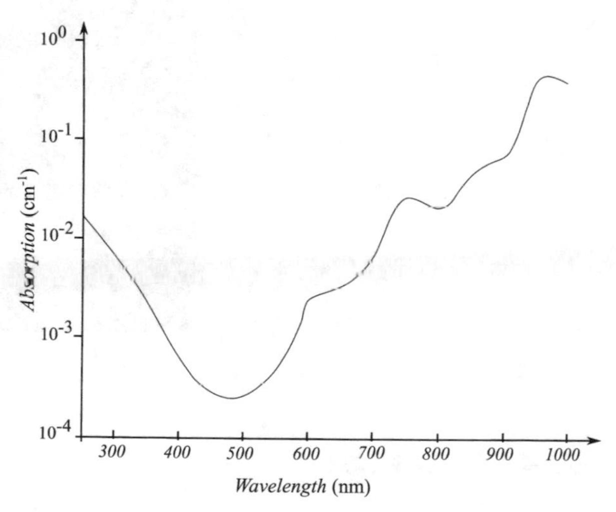

Fig. 7.14 Absorption spectrum of water

7.17.3 IR Imaging

Thermography and thermal imaging is a useful way to detect abnormal blood circulation near the skin's surface. Fevers, inflammations, loss of blood circulation, and even subcutaneous tumors can show in infrared images of patients. Figure 7.16 shows a hot spot within a breast visible by the IR radiation from the skin.

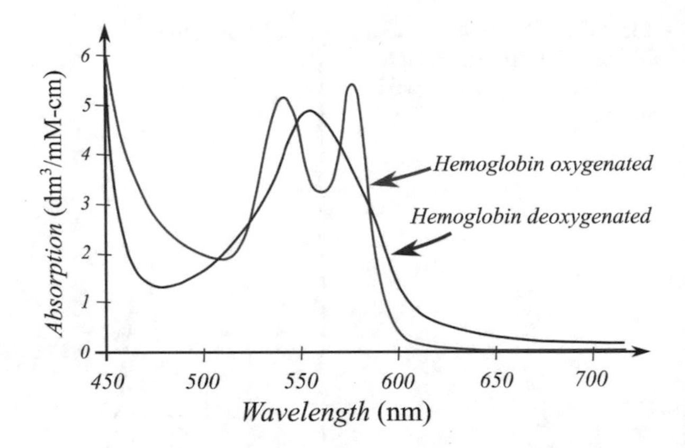

Fig. 7.15 Absorption spectrum of hemoglobin

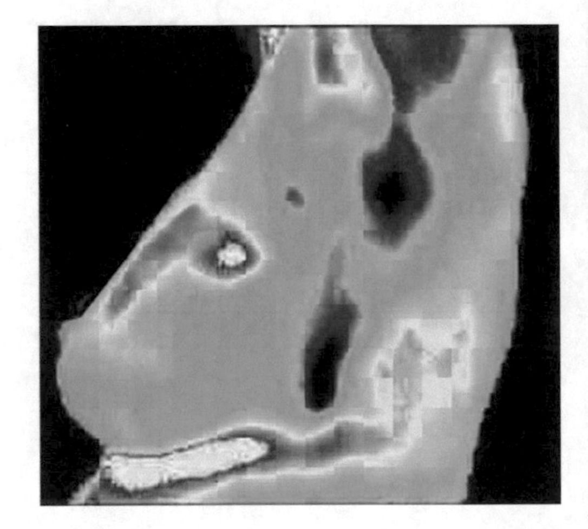

Fig. 7.16 Thermographic image showing a hot spot in a breast

7.17.4 Night Vision

Using an IR image intensifier on a pair of binoculars, a night scene may be seen as bright as day, but with hotter bodies brighter than colder ones. This ability follows from the fact that all bodies above absolute zero in temperature radiate electromagnetic waves. This so-called heat radiation (also called thermal radiation and black-body radiation) follows Planck's distribution, with a total emission proportional to the temperature to the fourth power (Stefan's law), and a peak wavelength inversely proportion to temperature to the first power (Wein's law). (See Appendix F for further details on the Planck's distribution.)

7.17.5 IR Motion Detection

By detecting the shift in frequency of emitted infrared radiation, one can detect the motion of the reflecting objects. This is possible because the returned waves are Doppler shifted in frequency by $f' = f/(1 - v \cos\theta / v_s)$, where f is the frequency of the emitted wave, v is the speed of the object, v_s is the speed of light in the medium in which the light is propagating, and θ is the angle between the direction of motion of the object and the direction from the object to the detector.

7.17.6 Infrared Communications

Infrared light can carry information, as any wave can. But infrared light is easily generated and detected by solid-state devices. The ubiquity of remote control devices shows how useful this form of communication is. Being limited in air to 'line-of-sight' is an advantage when the range of the device is intended to be relatively short. IR light absorption in the air is strongly dependent on the IR resonances in carbon dioxide, water, ozone, methane, nitrous oxide, and other molecules in the air, and on any dust or aerosols present.

7.17.7 IR Solid-State (LED) Emitters

A light-emitting diode (LED) uses an electric field across an n-p junction[29] in a solid-state diode to cause some electrons in the n-type semiconductor to gain enough energy to surmount the n-p junction potential difference. When those electrons drop in energy on entering the p-type semiconductor, the energy they lose can go into light. If the n-p juncture is close to the surface of the p-type material of the semiconductor, and the conductor on the surface is transparent, striated, or only partially covers the crystal surface, then the light can escape. The process is called 'electroluminescence', and typically produces infrared or visible light, depending on the crystal material and its doping. The first LEDs were made from gallium-arsenide-phosphide semiconductors, and emitted in the infrared (about 900 nm).

[29] If the semiconductor silicon or germanium, which have four valence electrons in each atom, is doped with arsenic, phosphorous, or antimony (all with five valence electrons) it becomes an n-type semiconductor, i.e. electrons available for conduction is higher than for pure silicon or germanium. If silicon or germanium is doped with boron, aluminum, gallium, or indium which have only 3 valence electrons in each atom, the material becomes a 'p-type' semiconductor, i.e. it has fewer available electrons for conduction than the pure silicon or germanium.

7.18 Infrared Biodetectors

People have infrared sensitivity on their skin. Next time you are on the beach at night with some exposed skin, stand far enough away from a bonfire so that you are not warmed by the temperature of the air. Close your eyes. As you rotate your body, your skin will tell you the direction of the fire. Thermoreceptors (both cold and hot) in your skin send information to the brain about the temperature of the skin. Temperature sensitivity of nerves occurs when active channels for the inward flow of ions such as Na^+ and/or the outward flow of K^+ ions are constructed to be strongly temperature dependent.

Pit vipers have evolved organs to detect infrared radiation, useful for catching mice in the dark. These organs can detect radiation in the wavelength range from 5 to 30 μm, with acuity sufficient to see various glowing body parts of the mouse. Directional focusing is performed by head position relative to two pit organs, one on each side of the head, and by the shape of pit as well as the interior sensitive membrane and cavities. Pythons, boas, and vampire bats also have evolved directional IR detectors.

7.19 Production of Visible Light

To make visible light, one need only to employ charges made to oscillate with frequencies in the range 10^{14}–10^{15} cycles/s, or use Doppler shift of non-visible light into the visible spectrum. There are several techniques for visible light generation in non-biological systems.

7.19.1 Thermal and Atomic Generation of Light

Visible light is evidently made by sufficiently hot bodies (above about 1000 K) including our Sun and tungsten filaments in old light bulbs. Some atomic and molecular transitions of electrons from excited states to lower energy states will make visible-light photons. If the excited states are above some intermediate states, then de-excitation may occur step wise: From the excited state to an intermediate state, and then from the intermediate state to the lowest state ('ground state'). (See Fig. 7.19.) The material is then call 'fluorescent'. Fluorescent bulbs have mercury as a gas inside a partially evacuated glass tube. The mercury is excited by an electric current. But its radiation is largely in the UV part of the spectrum. On the inside surface of the tube is a set of 'phosphors', chemicals such as zinc sulfide which have intermediate states between those which are excited by UV. Transitions from these intermediate states make visible colors.

If the intermediate state is 'metastable', i.e. it has a relatively low probability of making a transition to a lower state, then the material is called 'phosphorescent'. Metastable decay times for various materials range from a few thousandth of a second to hours.

7.19.2 Solid-State Generation of Light

In solid materials, many of the electrons are tightly bound to atoms and molecules. The least tightly held are referred to as valence electrons, as they can take part in the chemistry of bonding. Weakly held electrons may jump between atoms, or even become delocalized across the material, making that material a good electrical conductor. The possible electron quantum states for the valence electrons of all the atoms are numerous. If a large number of quantum states have nearby energies, they are said to be in an "energy band". The valence electrons are in the 'valence band'. With the Pauli exclusion principle at work, the valence electrons fill the possible quantum states with only one per state. At room temperatures, the lowest of these quantum states will be just about completely filled. Quantum states with energy far above $k_B T$ will likely be unoccupied. The energy at which the occupancy probability takes the value of one half is called the 'Fermi energy level',

If the electrons near the Fermi level are free to roam throughout the material, they are called conduction electrons, and occupy the 'conduction band'. These are the electrons which carry current in the metal. They also cause polished metals to appear shinny (by reflecting light), and to be good conductors of heat (which is carried quickly by the 'gas' of electrons in the metal).

In insulators and semiconductors, there is an energy gap between the valence band and the conduction band. (Good conductors have no such gap.) In semiconductors, the gap is less than about 4 eV, so that a few volts placed across the semiconductor can pull electrons from the valence band into the conduction band. For silicon, the gap is 1.2 eV, close to the energy of photons at the solar intensity peak, 1.4 eV. Thus, silicon is useful for generating electric energy from sunlight as a photoelectric solar cell.[30]

Similarly, as described in Sect. 7.17.7, light-emitting diodes (LEDs) use semiconductors to generate visible light.

---

[30]Photoelectric solar cells with frequency filters above optimized semiconducting surfaces can reach energy conversion efficiencies of 35% (Zhao et al., *A solar photovoltaic system with ideal efficiency close to the theoretical limit*, Opt Express **20(1)** A28–A38 (2012)). Plant photosynthesis can be up to 2% efficient.

7.19.3 Bioemitters and Bioluminescence

Bioluminescence has evolved over forty different times across the varieties of living organisms, including certain species of fungi, bacteria, marine animals, worms, and insects. All use a variety of light-emitting proteins called luciferin together with an enzyme luciferase and an energy source. At one time, coal miners used dried fish skins as a source of safe illumination.

Lightning bugs emit light from their lower abdomens within special cells containing luciferin. With the aid of the catalyst luciferase and the presence of a cofactor of magnesium ions, luciferin uses ATP to make luciferyl adenylate and adenosine pyrophosphate which, in turn, combines with oxygen to form oxyluciferin, carbon dioxide and adenosine monophosphate with the emission of light of wavelength of 510–670 nm (yellow-green to red). The reaction sequence can be 80% efficient in converting ATP energy to light.

More generally, the color of bioluminescent emission depends on the binding of the luciferase with luciferin and any associated fluorescent proteins. Placing the gene for producing luciferase in other organisms lets researchers use light to track ATP usage.

7.20 Light Detectors

The listing below show some methods of detecting the full range of electromagnetic waves of various frequencies, starting with the lowest:

Radio	Antennae (metal extended to lengths near wavelength/4)
Microwave	Bolometers (that detect heat production); microwave “dishes”; microwave cavities
Infrared	Thermistors; IR-sensitive semiconductors; heat-sensitive molecules (Note: In semiconductors, the energy gap between the valence band and the conduction band tends to be in the IR and the low frequency end of the visible spectrum)
Visible	Chemical transformations (pigment molecules, omititia, photochemical emulsions); photocells; kinescopic tubes; photomultipliers; photovoltaic cells; photoconductive cells
Ultraviolet	Atomic excitation; crystal excitation; fluorescence; phosphorescence
X-rays	phosphor screens; scintillation crystals; ionization chambers; Geiger counters; photomultiplier tubes
Gamma-rays	Geiger counters; ionization chambers; scintillation materials; photo-nuclear reactions; nuclear absorption

Wouldn’t it be nice if we could ‘see’ over the whole range of the electromagnetic spectrum? Likely, no one kind of biosensor would work. With our technology, external sensors can do the job.

7.21 Light Absorption Characterization

7.21.1 Light Extinction in Materials

Because of light scattering and absorption in materials, the intensity of a light beam as it travels a short distance dx through a uniform material will drop in proportion to dx and in proportion to the intensity $I(x)$ within that segment of material: $dI \propto Idx$ (See Fig. 7.17). It follows that the light intensity after a distance x through homogeneous material will be

$$I = I_o \exp(-\mu x) , \tag{7.21}$$

the same relation we saw for the attenuation of sound (Eq. (5.43)), microwaves (Eq. (7.14)), and IR light (Eq. (7.19)). Strong light scattering will occur near resonances in the induced motion of charges within the material. In most ordinary materials, these resonances occur in the IR (molecular), UV (atomic), X-ray (heavy element K and L shell atomic transitions), and gamma-ray (nuclear) regions of the spectrum, with 'windows' of transmission in the radio and visible band. Just as for sound, the number μ is called the 'absorption coefficient' and $1/\mu$ is the 'extinction length', i.e. the distance over which $1/e$ of the light intensity has been lost. For visible light through ocean water, we have a hard time seeing from the surface through ocean depths more than about 100 m, even with clear water.

7.21.2 Polarization and Rotary Power

The rotary power of a transparent or translucent material is the degree to which the material causes scattered light to become circularly polarized, usually due to left or right-handedness (also called 'left- or right-handed chirality')[31] of some organic molecule in solution. Such material is called 'optically active'.

Rotary power is measured by the angle that linearly-polarized light is rotated by a solution of the material. As this angle depends on the distance the light has traveled

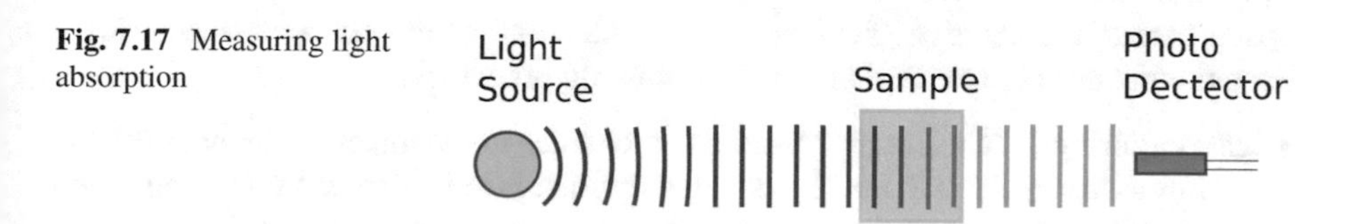

Fig. 7.17 Measuring light absorption

[31] A molecule is chiral if its mirror image differs from the original molecule. Two distinct chiralities for a molecule and its mirror image are always chemically possible, but often only one form is found in life systems on Earth, largely because DNA happened to be right-handed.

in the material, and on the concentration of a solute, the 'specific rotary power' is the rotation angle α divided by the optical length and the solute molar concentration: $[\alpha] = \alpha/(L\,[c])$.

The reason for the polarization is the following. Unpolarized light may be thought of as a combination of the two possible linearly-polarized light components or a combination of the two possible circularly polarized light components. An electromagnetic wave passing along a helical molecule with some electric dipoles aligned along the helix will lose energy in distorting those dipoles for that component of the wave whose electric field has the same sense of rotation through space as the molecule. That component of the electromagnetic wave with the opposite sense of rotation will lose less energy. For example, unpolarized light scattered by a sugar solution will become polarized, as can be seen by passing the scattered light through a quarter-wave plate, and then viewing with a linearly-polarizing sheet. Sugars have right-handed chirality, as does DNA. Right-handed chiral molecules are also called dextrorotary, while left-handed ones, such as most cellular proteins, are called levorotary.

7.22 Light Absorption by Biological Matter

The only way to make electromagnetic waves is through the acceleration of charges or by forcing changes in the direction of intrinsic magnetic fields of particles. The only way to detect them is through their absorption by charges accelerated by the wave or by the wave rotating particles with intrinsic magnetic fields.[32] For each enumerated technique we have for generating a certain frequency range of electromagnetic waves, there is a parallel technique for absorbing that wave. But the technology for generation can widely differ from that for absorption.

The molecules in cells which are identified or selected as light absorbers are called 'chromophores'. Chromophores which are indigenous to a particular tissue are called endogenic. Those injected or ingested are called exogenic. Figure 7.18 shows the absorption coefficient for important materials found in tissue.

Light is scattered and absorbed by biological materials through a number of important mechanisms, all of which involve the acceleration of charges or intrinsic spin flips associated with the incoming electric field oscillations and, to a much lesser extent, the magnetic field oscillations. Generally, light changes on entering a material because of one or more of the following processes:

Light scattering: The energy absorbed by charges in the material can be given off immediately as scattered light, often divided into incoherent scattering (such as Rayleigh scattering) and coherent scattering, (such as Mie scattering).

[32]If we find that nature has intrinsic magnetic charge, then these kinds of charges also are able to emit and absorb electromagnetic waves. So far, such 'magnetic monopoles' have not been found. Electrons, protons, and neutrons all have magnetic dipole fields.

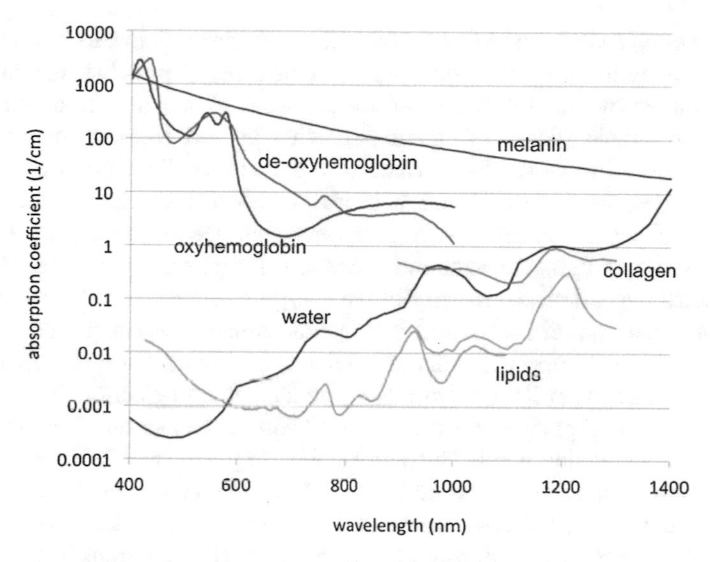

Fig. 7.18 Endogen photo-absorption

Delayed re-radiation: Re-radiated after an internal transition (including Raman scattering, "fluorescence" and "phosphorescence")
Photo-conversion: Light energy can be stored in an atomic or molecular excitation and then converted into some other form of energy, such as a chemical bond or heat.

There are many distinct light absorption mechanisms[33]:

1. *Free or quasi-free electron absorption*: Electrons in a vacuum and electrons in a material, in response to electromagnetic waves whose photon energies are much greater than any electron binding energy, move as if no material force is acting on them. They act as free particles. Then, energy and momentum conservation applies to the photon and electron alone. This makes re-radiation of light a necessary consequence (Compton scattering). For incoming X-rays or gamma rays, even inner-shell atomic electrons can be effectively free (since their binding energy can be small compared to the incoming photon energy).
2. *Excitation of delocalized electrons*: Some electrons in molecules may not be localized around only one or two atoms, but roam around many atoms with relatively low binding energy. Since light which excites such electron states is often in the visible range, molecules with delocalized electrons are used in dyes. Humans can make their world more colorful with such dyes. Flowers attract insects with colors, even in the ultraviolet range, invisible to us.

[33] As we will see, and despite a good book with the subtitle, "Fifteen causes of color," there are more than fifteen physically distinct ways to produce color!

3. *Metal band transitions*: Many of the valence electrons in metals are delocalized over the entire substance, occupying a dense set of possible quantum states determined by the closely packed metal atoms. Metals at a finite temperature have an ongoing dance of electrons moving up and down in energy near the Fermi level. Exposing the surface of the metal to light gives extra energy to electrons near the conduction band. The electrons quickly release excess energy, causing a scattered wave. If the metal surface is smooth compared to the wavelength of light, the scattered waves will form a reflected wave. The colors reflected depend on the density of the energy levels near the Fermi level. The valence electrons of copper and gold include atomic d-waves (i.e. electron wave functions which carry two units of angular momentum). These d-wave electrons form a 'd-band' in the quantum energies available to electrons in copper and gold. Incoming photons need about 2 eV to excite electrons from the d-band into the conduction band. This corresponds to photons in the blue part of the spectrum. Some energy carried by these 'blue photons' is not returned as blue light, but rather gets transferred to other electrons. This makes copper and gold have an orange-yellowish color. Silver metal also has electrons in a d-band, but they require ultraviolet photons to release them to the conduction band.
4. *Semiconductor band transitions*: Semiconductors have a very small population of electrons in the conduction band, with most of the electrons tied to individual atoms. Those that are least tied to atoms fill the valence band close to the conduction band, and can be excited into the conduction band by incoming light in the infrared and visible range of frequencies. Doping the silicon or germanium with phosphorus or arsenic adds excess electrons, while doping with boron causes unfilled electron energy levels (called "holes"). The dopants create a new "donor" or "acceptor" band of quantum states between the conduction and valence bands. The position of the bands can be selected to match the photon energies which the material will absorb.
5. *Transitions between unfilled atomic inner shell states in transition metals*: Transitions of electrons in the unfilled "d" and "f" atomic orbitals taking part in ligand bonding can resonantly absorb visible light, giving color to such metal ions and metal complexes.
6. *Photoinduced charge transfer*: Light-sensitive chemicals change chemical state by exposure to light. In the process, photons absorbed from the light give sufficient energy to cause electrons to jump from one atom to another. Silver bromide in photographic film becomes silver and bromine by exposure to light. Chlorophyll *a* and *b* absorb light energy, causing a series of charge transfer reactions.
7. *Color centers*: Impurities and dislocations in some otherwise clear crystals can cause them to exhibit color, as electrons localized near those impurities (or the absence of electrons, i.e. 'holes') may have less binding energy than the electrons taking part in the covalent or ionic binding of the crystal. Exciting one of these electrons or holes can often be done with visible light, rather than ultraviolet.

8. *Excitation of valence electrons*: Electrons at the surface of atoms, the valence electrons, take part in chemical bonds, and are typically excited by ultraviolet radiation, but sometimes by visible light. If the latter, then the material will show color.
9. *Ionization*: An incoming photon may have sufficient energy to release a bound electron in a material. Atomic ionization processes typically requires UV or X-ray rays. Apart from the valence electrons, the other electrons in atoms are held tightly, and require tens to many thousands of electron-volts of energy to release them. Photons with those energies have frequencies ($f = E/h$) in the ultraviolet to X-ray part of the spectrum. Ions created by radiation inside cells easily make free radicals. These can alter the chemistry of the cell, almost always in unhealthy ways.

   Light can also release electrons from the surface of metals. This is the photoelectric effect. Although the conduction-band electrons are not bound to individual metal atoms, it does take energy to pull one of those electrons from the metal. Part of this energy comes from the work done in pulling the electron away from the opposite charge the electron creates on the surface of the metal by its repulsion of nearby conduction electrons. Einstein, in 1905, used energy conservation and the photonic properties of light to write $hf = (1/2)mv^2 + W$, where v is the speed of the ejected electron, and W is the work needed to get the electron away from the metal. Einstein could explain all the properties observed in the process, while classical theory had failed to do so.
10. *Absorption through atomic or molecular electron-orbital distortions*: Electrons in bound orbitals can be perturbed by the action of an oscillating electric field (of an electromagnetic wave). This may distort the orbitals rather than cause quantum transitions to other bound states. This distortion occurs in optical materials as light passes through. The oscillating electrons generate new radiation of the same frequency as the wave passing by. As we have described, this accounts for the index of refraction of transparent materials.
11. *Absorption by molecular excitations*: When the frequency of the incoming light is close to the natural resonant frequency for charge centers in the molecule, the probability for absorption is enhanced and the photon energy can be used to excite a particular atomic or molecular state. Because this light energy is stored in the molecular system for a time longer than typical scattering times (e.g. 10^{-6} s vs. 10^{-9} s), the index of refraction is significantly larger at this frequency and the "absorption coefficient" (a ratio of outgoing vs. incoming intensity along the direct beam) also increases. There are several general types of molecular excitations:

    (a) *Molecular rotational excitations*: Small molecules with a large moment of inertia and electric dipole moment can be induced to rotate by exposure to light of an appropriate frequency. This frequency can be calculated as follows: The energy stored in a rotating system is given by $E = L^2/(2I) = l(l+1)\hbar^2/(2I)$ where I is the moment of inertia of the molecule about its axis of rotation, $\hbar$ is Planck's constant over 2π, L is the angular momentum of system, and l is the angular momentum quantum number ($l = 0, 1, 2, \ldots$). For a molecule with a line axis of symmetry, such as carbon dioxide, generating spin around the symmetry axis will only be seen in exposure to the far ultraviolet, which has sufficient energy to distort the molecule to bulge radially. Otherwise, there would be no way to detect the spin. Rotation of the molecule perpendicular to the axis of symmetry is far easier, and occurs by exposure to infrared light.

In exciting the molecule to rotate, the photon, which always carries one unit of angular momentum, $1\hbar$, changes l for the molecule to $l \pm 1$. Thus, the resonant absorption frequencies will be given by $f = (l+1)h/(4\pi^2 I)$ and will show a linear increase in their separation in the absorption spectrum. Taking, for example, carbon monoxide, we have an I of about 2×10^{-40} g/cm^2 giving a light frequency (for $l = 0$ to 1 transition) of $f = 8 \times 10^{12}$ Hz and a wavelength of 1.2×10^{-3} cm. Rotational "lines" or bands in the absorption spectrum of small molecules typically lie in the far infrared. So, if one sees a set of IR transmission dips whose separation increases linearly with absorption frequency, one should suspect IR rotational excitations are the cause for this part of the IR absorption spectra.

(b) *Molecular vibrational excitations*: Small and large molecules can be excited into vibrational modes by light. Since the natural vibrational frequencies depend very sensitively on the strength of the various bonds which are stretching during vibration, the absorption spectra of a substance is an important tool in identifying the molecular structure of the material. To estimate the frequency of light required to excite vibrational modes, we can use the quantum mechanical result that the energy stored in a vibrational system with a small amplitude of vibration is given by $E = (1/2 + n)hf_o$, where n is a quantum number ($n = 0, 1, 2, \ldots$) and f_o is the natural vibrational frequency of the system. Note that the energies are evenly spaced, so the absorption spectrum will show evenly spaced absorption dips.

If the bond length is changing during the vibration, vibrational frequencies are given by $(1/(2\pi))\sqrt{|\partial^2 V/\partial x^2)|_{x_0}/(2\mu)}$, where μ is the reduced mass of the two objects forming the bond and $V(x)$ is the potential energy stored in the bond as a function of bond separation. The second derivative is evaluated at the average bond separation, x_0, when the system is in its lowest vibrational quantum state. (Quantum fluctuations do not allow a fixed separation.)

An isolated water molecule has three vibrational modes: a symmetric stretching mode with wave-number ($k = f/c$, in units of cm^{-1}) 3657; an antisymmetric deformation mode with wave-number 1595; and an antisymmetric stretching mode with wave-number 3756.

Figure 7.10 shows some vibrational modes found in organic molecules, along with their photon absorption wave-numbers. (These wave-numbers vary when nearby molecules affect the bond being distorted.)

(c) *Molecular orbital excitations*: A valence electron in a molecular orbital (i.e. a quantum state of an electron not confined to a single atom) can absorb a photon if there is sufficient energy carried by that photon to excite the electron to a higher energy molecular orbital. Transition to this new electron state may cause the molecule to change its configurational state.

12. *Atomic excitations*: Electrons in inner-shell atomic orbitals can be excited to unoccupied higher energy states by light. The energies necessary can be estimated by the Coulomb binding of electrons to nuclei, which is of the order of $Z_{eff}e^2/r$, where r is the distance to the positively charged nucleus and Z_{eff} is the effective charge number for the core of the atom (Z, the nuclear charge number, minus the average number of electrons inside the orbit of the excited one). Using $e^2/\hbar c = 1/137.0$, we find the wavelengths of the light needed is of the order $\lambda = 10^{-5}$ cm. We expect, then, that ultraviolet light frequencies will be necessary to excite these levels.
13. *Fluorescence and Phosphorescence*: A 'phosphor' is a substance which emits light by atomic or molecular transitions some time after being excited. Phosphors can emit light by being exposed to a source of particle or light radiation. If the time delay in emission is perceptibly long (greater than about a microsecond), then the material is 'phosphorescent'. By selection of the phosphor, the length of light persistence on the old cathode-ray tubes used to generate images could be adjusted to reduce flickering, since the electron beam took about 2 ms to return to the same spot on the screen. X-ray phosphorescent screens are used to monitor X-ray sources and view images produces by X-rays. Some phosphors can "glow in the dark" seconds and even minutes after being exposed to a radiation source.

If there is no perceptible time delay between energy absorption and light emission, but not all the energy is re-emitted, then the material is 'fluorescent' The coating inside a fluorescent tube converts ultraviolet light generated by a current passing through a vaporized gas of mercury (together with an inert noble gases) into visible light. Commercial tubes use non-toxic halophosphate-based phosphors to coat the inside of the tube, but the tubes still contain small amounts of toxic mercury.

Phosphors stimulated by light must have at least two special unoccupied excited states in the atoms or molecules. One for electrons to go when first excited by the incoming radiation. A second intermediate energy state is needed into which an excited electron can 'fall'. Transitions from the first to the second state must have a high probability relative to transition from the first back to the initial ground state. In turn, if the intermediate state has a high probability to 'decay' into the ground state, the material will be fluorescent. Typical such decay times are less than a microsecond range. If the intermediate state has a low probability of decay, then the electron will take some time before falling back to the ground state, and the material will be phosphorescent, with decay time in the milliseconds to hundreds of seconds.

Excited electron states in atoms or molecular will be long-lived if electromagnetic effects cannot easily force the electron to make a transition. Such states are said to be 'metastable'. For example, suppose both the excited state and the ground state are quantum s-waves. Then light is not easily generated by a transition from this excited state to the ground state because photons must transfer a unit of angular momentum, and the two states have no angular momentum to give. The transition probability may still be finite, however, because such quantum states can have a small admixture of p or d-waves through interaction with neighboring electrons.

Let the energy of the ground state be E_0, the second state be E_2, and the third state be E_1 (as in Fig. 7.19). Let the detected photons have energy hf_i each and each emitted photon have energy hf_e. Then energy conservation demands: $hf_i = E_2 - E_0$ and $hf_e = E_1 - E_0$. Since $E_2 > E_1$, the frequency of the emitted light must be lower than that of the incident light. In fluorescent tubes, UV light from excited mercury atoms is converted to visible light.

14. *Rayleigh, Mie, Raman, and Brillouin Scattering*: Light may interact with the electrons in molecules by distorting their orbitals, without changing their quantum state. The resulting vibrational motion of the orbits causes light to be re-emitted, without change of frequency (making the photons scatter elastically). This is the basis of the scattering of light by small particles or individual molecules as described by Rayleigh. One important property of Rayleigh scattering from small particles (whose size is smaller than the wavelength of the light) is the strong dependence of the intensity of the light on the frequency of light being scattered, namely the intensity goes as the frequency to the fourth power.[34]

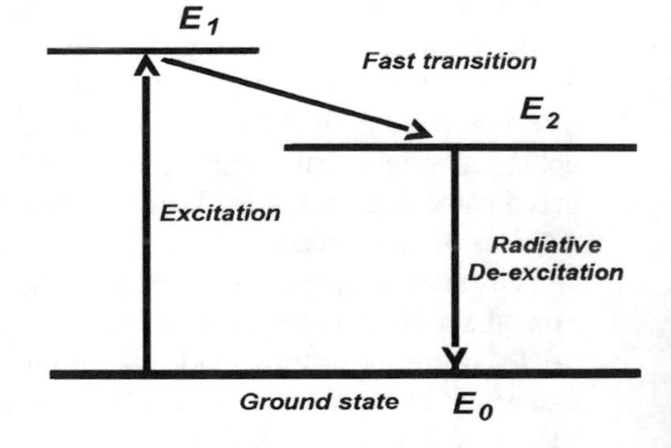

Fig. 7.19 Energy levels for fluorescence: If the state with energy E_2 is quasi-stable, then the system acts as a phosphor (causing 'phosphorescence'). If a light wave passes with frequency $f = (E_2 - E_0)/h$ while the E_2 state is populated, then stimulated emission is likely

[34]This dependence follows from Maxwell's equations applied to a single oscillating point-like charge.

Gustav Mie studied elastic light scattering from spherical particles whose size is larger than the wavelength of the light.[35] In this case, there is little or no frequency dependence of the scattering intensity. Scattering of sunlight from water droplets in clouds is a good example, as is scattering of light by large smoke particles.

Raman considered what could be learned about materials from the inelastic scattering of light. The loss or gain of energy comes from the available mechanisms for energy transfer in the material, which in turn depends on molecular structure.

Brillouin scattering of light arises from the charge density variations in a material, such as would occur if an acoustical wave were present, or temperature variations caused changes in density.

All of these kinds of scattering are used by biochemists to study organic molecules.

15. *Electron spin flip*: A photon (which carries a unit of angular momentum) can cause a valence electron spin to flip, i.e. the electron's magnetic moment vector is changed from being aligned with an external magnetic field to being anti-aligned with that field. The Pauli exclusion principle impedes this flip if there is a second electron already occupying the first electron's molecular orbital, but with opposite spin direction. If a flip occurs, there may be an accompanying change in the molecular bonding. When an electron in a 1 T magnetic field is forced to flip spin by absorbing a photon, that light will have a microwave frequency.
16. *Nuclear spin flip*: Like electrons, protons have an intrinsic spin, as do some nuclei. Associated with their spin **S** is a magnetic dipole field of strength $\boldsymbol{\mu}$. For a proton, $\boldsymbol{\mu} = (|e|/(mc))\mathbf{S}$. If the proton or other nucleus is put in a magnetic field, either from local fields of other nuclei and electrons, or from an external magnetic field, the 'north' direction of the nucleus will tend to align with the magnetic field. Like electrons, protons can only have two measurable directions of their spin: Either aligned or anti-aligned with the direction of a magnetic field (so the spin projection is either $+(1/2)\hbar$ or $-(1/2)\hbar$). However, some nuclei have greater spins, allowing a range of projections from $s\hbar$ to $-s\hbar$. When a dipole is twisted away from the direction of an external magnetic field **B**, it requires an energy $\Delta(\boldsymbol{\mu} \cdot \mathbf{B})$. For the proton, the change in the energy will be $2\,\mu B$. If the change is caused by a photon absorption, then $hf = 2\,\mu B$ so the electromagnetic radiation being absorbed will have a frequency of $f = 2\,\mu B/h = (eB/(2\pi mc))$. With magnetic field strengths in the tesla range, a proton spin-flip will radiate in the radio region of the electromagnetic spectrum.
17. *Dispersion, Diffraction, and Interference*: As light passes through transparent materials, the variation of the index of refraction cause colors to separate,

[35]In the first decade of the 1900s, Mie studied the solution to Maxwell's equations when a light wave interacts with a spherical dielectric ball.

generating dispersion and rainbow effects. The separation of colors by a prism demonstrates dispersion.

Light passing opaque objects whose size is comparable to the wavelength of the light will produce diffraction patterns. The location of the diffraction patterns will depend on the color of the light, and so color fringes will be generated from white light by such structures. The iridescence of pearls is due to diffraction from a fine-scale structures near the surface of the pearl. Interference of light reflecting from closely spaced layers in a material can also show colors. Oil slicks on water and the iridescence of bird feathers and insect wings are further examples.

Far up in the electromagnetic spectrum, and therefore not affecting visible light color, gamma rays will also be absorbed by materials in ways similar to atomic and molecular absorption. Such absorption tends to be dominated by:

1. *Nuclear rotations*: The absorbed gamma ray may cause the nucleus to change its rotational state.
2. *Nuclear vibrations*: The absorbed gamma ray may cause the nucleus to change its vibrational state. With heavy nuclei, the protons may undergo collective vibrational motion relative to the neutrons.
3. *Nuclear excitations*: An incoming gamma ray into a nucleus can cause individual protons to be excited into unoccupied proton quantum states at higher energy than their state before the disturbance. These resonant absorptions can be relatively sharp in photon frequencies, due to the lifetime of the excited states. For example, iron-57 (^{57}Fe) has an excited state at 14.4125 keV above its ground state, and this level has an energy width of only 4.66 neV wide, corresponding to a lifetime of 141 ns.
4. *Nuclear transformations*: A gamma ray absorbed by a nucleus can cause some nuclei to undergo a nuclear transformation. Nuclear fragments can be knocked out (such as a proton or an alpha particle), or the gamma ray can induce a weak interaction, changing the nuclear atomic number, and cause the emission of an electron or positron together with an antineutrino or neutrino.

The nucleus of most atoms can be excited into rotational, vibrational, and excitational modes by gamma ray energy photons. In the process, isolated nuclei are kicked by the gamma ray. Rudolf Mössbauer discovered that if the nuclei are anchored in a crystal, some gamma absorption will be effectively recoilless because the whole crystal can take up the gamma ray's momentum. The sharpness in energy of the nuclear states lets us change the gamma absorption by a small motion of the crystal, because the Doppler effect changes how the nuclei absorb the gammas. Since the nuclear state energies are affected by the molecular electrons in s-states (no angular-momentum states), the position of the gamma absorption frequencies are correspondingly affected. Such measurements for organic molecules give information about those molecules' configurations. An application to biochemical research will be given in Sect. 8.7.2.

7.23 Development of Lasers

In 1917, Albert Einstein[36] showed that stimulated emission of light was possible. Einstein deduced that excited electronic states can be stimulated to release their energy by light passing by. Any passing light which has about the frequency of the light that would be released by an electron transition will stimulate that release.

However, no serious thought emerged on how to apply the idea until the 1950s. In 1953, searching for a good source for intense microwaves near one frequency, Charles Townes, at Columbia University, acting on an idea of Joseph Weber, at the University of Maryland, developed the first MASER (Microwave Amplification by the Stimulated Emission of Radiation) by 'pumping' excited ammonia molecules into a microwave cavity with a resonant frequency of 24.0 GHz.[37] Townes recognized that a LASER (Light Amplification by the Stimulated Emission of Radiation) could be built, and helped develop the technology. Theordore H. Mariman at Hughes Research Lab in Malibu California built the first operational Laser in 1960.

By populating electronic metastable states in a material placed between two parallel mirrors, light waves spontaneously released by a few of the excited atoms will stimulate others, amplifying the wave. If the wave happens to be heading perpendicular to the parallel mirrors, it will reflect back and forth, amplifying on each pass, continuing in this manner as long as there is energy to maintain the 'population inversion' of the metastable states, i.e. more than half should be excited at any one time. (See Fig. 7.20.)

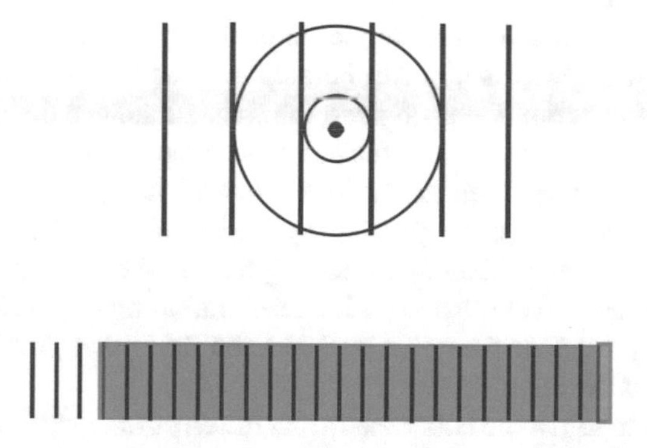

Fig. 7.20 Laser operation: The atom depicted as a dot, initially in a metastable excited state, is stimulated to emit light by a passing wave at just the emitting wavelength. The device shown has waves reflected back and forth between two parallel mirrors, with the mirror on the left only partially reflecting, letting some waves pass through

[36] A. Einstein, *The Quantum Theory of Radiation* Phys Zeits **18**, 121–128 (1917).

[37] J. Gordon, H. Zeiger, C. Townes, *The Maser—A New Type of Microwave Amplifier, Frequency Standard, and Spectrometer*, Phys Rev **99:4**, 1264–1274 (1955).

A laser produces high intensity light waves with close 'phase coherence'[38] among the waves of nearly one frequency. In contrast, ordinary light sources, including incandescent and fluorescent, produce light with little or no phase coherence among the waves produced by the radiating atoms or molecules. Laser devices rely on the existence of metastable states in a material which can emit light when that material releases energy by changing to a lower energy state. This release is stimulated by passing light through the material.

Generally, metastable states are energy states of a system which have a higher energy than a lower energy state of that system, but the system does not 'fall' quickly into the lower energy because the available forces which might carry the system to the lower energy are weak, or there is a barrier that the system has to overcome in order to reach the lower energy.

For a laser, the metastable states are usually electron excited states which have 'weak coupling' to the ground state of the system. Because of the nature of electromagnetic interactions, if an excited electron state can be reached quickly by an electron absorbing a photon from the ground state, that same excited state will be 'strongly coupled' to the ground state, and therefore is not suitable as the metastable state of a laser. However, there may be an intermediate state the system may fall into which is only weakly coupled to the ground state. Electron metastable states may be excited by electrons 'falling' from higher-energy excited states or directly by non-radiative electromagnetic interactions, such as collisions with other atoms. Figure 7.19 can be used to picture the three levels used for laser operation. In this case, electrons are pumped into state with energy shown as E_1. The level with label E_2 is the metastable state.

By populating metastable states in the atoms or molecules of a selected material, a light wave near one frequency can be amplified as the wave reflects back and forth through the material. On each pass, the light stimulates the metastable atoms to emit, amplifying the wave. To maintain the amplification, more atoms must be in the metastable state than in the lower ground state, producing a population inversion. Otherwise, the wave passing by will have a greater chance of being reduced in intensity, rather than amplified. Optical flashers, atomic collisions after acceleration, and electron collisions after acceleration by an electric field have all been successfully used as 'laser pumps' to keep more than 50% of the working atoms in a metastable state.

Laser beams are said to be coherent, in that the electromagnetic waves have close to a single phase. As such, the average intensity of the beam is in proportion to the square of the sum of the component waves from the emitting atoms. If there are N such waves coming from N atoms, each of amplitude A, the intensity will be proportional to N^2A^2. By contrast, the sum of waves with no phase relationship

[38]Phase coherence means that the phases of the waves produced at different locations have a definite relationship to each other.

will make an average intensity proportional to NA^2. The ability to make highly intense laser beams relies on the extra factor of N in the generation of the light.[39]

Because optical lasers use parallel reflecting surfaces, the light produced have parallel wave fronts over the extent of the surfaces. (This light is released by having one of the reflecting surfaces only partially reflecting, such as is a half-silvered mirror.) Being parallel, the laser beam comes out of the laser as a well-defined columnar beam. Such a wave can be easily focused by a converging lens to a region about the size of the wavelength of the light. The focused beams can have intensities exceeding 10^{24} W/cm^2.[40] For the safety of the human retina, laser pointers are not to have beam power greater than 5 mW, which is about 80 mW/cm^2 over the beam area. The intensity on the retina can be much larger than that in the laser beam, as the lens of the eye can focus the beam.

7.24 Laser Applications

Because of the ability to sharply focus laser beams and because the beam can be made intense, lasers can be used for such purposes as laser-induced photochemistry; laser scanning microscopy; laser ablation molecular spectroscopy; laser surgical cauterizing and cutting; laser trabeculoplasty (treatment of glaucoma); laser ablating of tooth decay, tattoos, sunspots, moles, and melanomas; laser lithotripsy (breaking up hardened substances); laser angioplasty, laser neurological surgery; laser attachment of retinae; laser microsurgery; laser subcutaneous imaging; medical holograms; and minor soft tissue imaging using the near infrared (NIR) 'window'.

7.24.1 Optical Tweezers

It is now possible to use focused laser beams to trap, move, and manipulate cells, molecules, and even individual atoms.

The idea is this: Cells, molecules, and atoms all have internal charges. In an electromagnetic field with a strong gradient, these particles, even though neutral overall, will have a net force on them. We can see this from the Lorentz force law and Maxwell's equations. The mechanical force on a neutral particle to first order in the field gradients will be (in the cgs system of units)

---

[39]Consider the sum: $\left\langle\left(\sum_i A\cos\left(\phi_i(t)\right)\right)^2\right\rangle$, averaging over time. If the phases ϕ have no correlation, only the square cosine terms will survive the average, so the result becomes $(1/2)NA^2$. However, if the phases are all the same, the 'direct' and the 'cross terms' of one cosine with another add together, making $(1/2)(NA)^2 = (1/2)N^2A^2$.

[40]Sunlight at ground level has an intensity of about 50 mW/cm^2.

$$\mathbf{F} = (\boldsymbol{\delta} \cdot \nabla)\mathbf{E} + \frac{1}{c}\frac{d\boldsymbol{\delta}}{dt} \times \mathbf{B} + (\boldsymbol{\mu} \cdot \nabla)\mathbf{B} \, ,$$

where $\boldsymbol{\delta}$ is the electric dipole moment of the particle, and $\boldsymbol{\mu}$ its magnetic dipole moment. Now suppose that the electric dipole moment is induced by the electric field, so that $\boldsymbol{\delta} = \alpha\mathbf{E}$. he number α is the electric polarizability of the particle. For simplicity here, take the particle's magnetic moment in negligible. The mechanical force becomes

$$\mathbf{F} = \alpha(\mathbf{E} \cdot \nabla)\mathbf{E} + \frac{\alpha}{c}\frac{d\mathbf{E}}{dt} \times \mathbf{B} \, ,$$

$$= \alpha(\mathbf{E} \cdot \nabla)\mathbf{E} + \frac{\alpha}{c}\frac{d}{dt}(\mathbf{E} \times \mathbf{B}) - \frac{\alpha}{c}\mathbf{E} \times \frac{d}{dt}\mathbf{B} \, .$$

If we assume the particle has little or no speed, then

$$\frac{d}{dt}\mathbf{B} \approx \frac{\partial}{\partial t}\mathbf{B} = -c\nabla \times \mathbf{E}$$

from one of the four Maxwell's equations. The force can now be written as

$$\mathbf{F} = \alpha(\mathbf{E} \cdot \nabla)\mathbf{E} + \frac{\alpha}{c}\frac{d}{dt}(\mathbf{E} \times \mathbf{B}) + \alpha\mathbf{E} \times (\nabla \times \mathbf{E}) \, ,$$

$$= \alpha(\mathbf{E} \cdot \nabla)\mathbf{E} + \frac{\alpha}{c}\frac{d}{dt}(\mathbf{E} \times \mathbf{B}) + \frac{1}{2}\alpha\nabla(\mathbf{E}^2) - \alpha(\mathbf{E} \cdot \nabla)\mathbf{E},$$

so

$$\mathbf{F} = \frac{1}{2}\alpha\nabla(\mathbf{E}^2) + \frac{\alpha}{c}\frac{d}{dt}(\mathbf{E} \times \mathbf{B}) \, . \tag{7.22}$$

For an electromagnetic wave impinging on the particle, the first term is proportional to the gradient of the electromagnetic field intensity, and points in the direction of the greatest increase in the field intensity. The second is proportional to the rate at which the field momentum flux charges per unit time (see Eq. (7.5)). Both terms would wiggle the particle at twice the laser frequency, but for a macroscopic particle, inertia makes this wiggling imperceptible. The second term vanishes when averaged over time. The first does not. Light can also scatter from the particle, creating a force pointing away from the origin of the light.[41]

In 1970, Arthur Ashkin, working at Bell Labs, observed the confining behavior of a light beam on micron-sized particles. He knew that a strong gradient in the

[41]The radiative reaction force which occurs in light scattering was not included in the Lorentz force law used here.

field might confine a particle. According to Eq. (7.22), there will be forces on a particle in an electric field from weaker field regions toward the direction of the strongest electric field. By sharply focusing a laser beam, the electric field intensity will have a peak at the focal point. As long as the rate of momentum transfer from the beam to the particle is less than the gradient force, the particle will be trapped at the laser focal point. Later, he and his colleagues showed how to use laser beams as optical tweezers, i.e. using the beam to trap and move micro and nano-sized particles. Ashkin received a Nobel Prize in Physics (2018) for his work. He shared the prize with two other physicists working on pulsed lasers, with wave packets highly confined in time. (This idea and a few of its applications will be described in the next section.)

As an example of a biophysical application, optical tweezers are now used to make measurements of the mechanical force needed to stretch, bend, and twist macromolecules. We now can see the jerking force of attachment and detachment of cell cargo-transporting kinesin proteins as they walk across microtubule filaments.

7.24.2 *Laser Pulse Amplification*

Half of the 2018 Nobel Prize in Physics went to Donna Strickland and Gérard Mourou, who developed, in the 1980s, a practical method of generating high-intensity, ultrashort laser pulses, known as 'chirped pulse amplification' (CPA).

Very high intensity laser pulses (with a power of more than 100 TW), using laser amplifiers, can be constructed, but the apparatus is large and expensive. The main difficulty is that laser amplification to intensities above 700 gigawatts per square centimeter reach the non-linear region of the laser amplifier, causing pulse compression and destruction of the amplifier material. Strickland and Mourou realized that if the laser pulse is first spread out over time according to the frequency components of the pulse, the resulting wave pulse would contain the same total energy but be much less intense. Dispersion devices can spread pulses by a factor of up to 100,000. After spreading, laser amplifiers act on the pulse, and then an inverse dispersion is performed to resharpen the pulse. All this with a 'tabletop' apparatus which can produce a terawatt light pulse with a femtosecond width.

Because a CPA enhanced laser beam can be so precisely focused with high intensities, it has enabled Lasik eye surgery, a procedure for correcting vision by reshaping the cornea of the eye. CPA technology produces clearer and sharper remote spectroscopy of atmospheric gases than previously available, and lets us more easily study the non-linear optics of materials. The prospect for laser-blasted detonation of fusion pellets as an energy source is brighter.

7.25 Optical Holography

A hologram is an interference pattern produced on a surface that is formed as a combination of a wave scattered from an object and a reference wave. The resulting interference pattern is capable of reproducing a three-dimensional image of the original object if a ‘single-frequency’ light wave, such as a spread-out laser beam, is passed through (or reflected from) the hologram (Fig. 7.21).

Optical holograms can be produced on photographic film by reflecting laser light from an object onto the sensitized film while part of the laser light is sent directly to the film. The result is an interference pattern recorded on the exposed and developed film, with the darkest regions occurring where the two light waves constructively interfered. No lenses need be involved. Note that for interference to work, the light must be close to one wavelength (which lasers are) and the film must not move or vibrate more than about a quarter of the wavelength of the light (and thus, no appreciable sound is allowed in the optical holography room!).

Fig. 7.21 A single hologram of a mouse, viewed from two angles. Photo by Geog-Johann Lay, 2008

It is also possible to generate a similar hologram by a computer calculation ('digital hologram') of what the interference pattern should be and then printing that pattern on a two-dimensional surface. At the simplest level, scratching a surface with arcs of a compass whose center is placed at various points of an object will make the equivalent of a Huygens-constructed interference pattern. Illuminating those scratches with a point-like source filtered for one color reveals a holographic image. This construction is essentially what the computer algorithm does.

The fact that a three-dimensional (3-D) image can be recorded on a two-dimensional surface can be thought of in following way: Imagine light passing through a window. One can look through the window and see a three-dimensional world beyond. Now imaging recording all the information that passes through the window in a very short time. Then that information should be capable of reproducing the three-dimension view one would have seen in that short instance of time. Indeed, in viewing a hologram, one does not look at the surface where the interference pattern exists, but rather one looks 'through' the film at the image beyond (i.e. the eyes must focus beyond the film of the hologram). Moving one's head to a new position while looking through the hologram will allow viewing the object from a different angle than before. Also, if part of the window becomes obscured, one can still see objects through the unobscured part of the window. Holograms have the same property. This means they have significant redundancy, and store delocalized information. (See Appendix C for a description of how wave interference creates a hologram.)

Digital Holographic Microscopy

If monochromatic light scattered from regions on and within a three-dimensional object is used to create a recorded interference pattern on a two-dimensional digital sensor array, the system is called a digital holographic microscope. Unlike traditional photographs recording just the intensity of light on a surface, information stored holographically contains the intensity and phase information sufficient to reconstruct a three-dimensional representation of the object, and, due to its speed of recording, time sequences can be stored (sometimes referred to as a 4-D imaging).

A 'digital in-line holographic microscope' (DIHM) is constructed using a simple LED laser, with its light passing through a pinhole (about a micron in diameter) into a medium with objects or microbes of interest (about a millimeter or more from the hole), and then to a CCD attached to a recorder. Image reconstruction is performed by a computer (rather than secondary illumination of a hologram). Apart from the computer, the device itself can be made relatively cheaply and compactly. DIHMs have successfully and non-invasively recorded, in situ and in real time, live microscopic marine organisms, with both lateral and depth resolution far better than microscopes with just lenses.

7.26 Optics

Geometric optics is a study of the properties of light in interaction with matter under conditions in which the wavelength of the light can be assumed much smaller than the dimensions of macroscopic structures with which the light is interacting. It then becomes possible to follow the behavior of light by how light rays[42] are affected by substances. When one assumes small wavelengths (in the sense above), Maxwell's equations become the "eikonal equation".[43]

There are several useful and general properties of light propagation in the geometric optical limit, all derivable from the eikonal equation. (In the following statements, angles are measured between a light ray and the normal to an involved surface.)

Law of Reflection When light reflects from a smooth surface,[44] the incident ray, the normal to the surface, and the reflected ray are all in one plane. In addition, the incident angle and the reflected angle are the same.

When light enters at an angle into a transparent medium, we say the light is refracted (bent) if the direction of light rays changes on entering.

Law of Refraction As a ray crosses the boundary between two materials, the incident ray, the normal, and the refracted ray are in the same plane. In addition, the sine of the incident angle and the sine of the refracted angle are in the same ratio as the average speed of light in the first material in which the incident light originated to the second material into which the light entered. This statement is 'Snell's Law', and can be expressed as $n_i \sin\theta_i = n_r \sin\theta_r$, where n is called the index of refraction, defined to be the ratio of the speed of light in a vacuum to that in the material.

With the law of reflection and refraction, image formation by mirrors and lenses follows. A mirror is a cylindrically-symmetric opaque device with a reflective surface that causes rays arriving parallel to the axis of symmetry to the surface to be reflected and then converged toward or diverged from a focal point. A lens is cylindrically-symmetric transparent device that refracts rays and that causes rays arriving parallel to the axis of symmetry to either converge or diverge toward or from a focal point. The distance from the mirror or lens to the focal point is called

---

[42]Light rays are directed lines constructed perpendicular to light wave fronts and point the way the wave fronts are moving.

[43]Eikonal comes from the Greek word $\epsilon i\kappa\acute{\omega}\nu$, meaning image. The eikonal equation from which all of geometric optics follows, is derived from Maxwell's equations, and can be expressed as $(\nabla S)^2 = n^2$. Given the behavior of the index of refraction $n(x, y, z)$ (i.e. the speed of light in a vacuum divided by its speed in the material at (x, y, z)), one can solve for the 'eikonal' S. Then the equation $S(x, y, z) = constant$ will define a surface in space that is everywhere perpendicular to optical rays. A light ray, specified by $\mathbf{r}(x, y, z)$, will be a solution of $n\, d\mathbf{r}/ds = \nabla S$, where ds is an infinitesimal distance along the ray. For more details, see Sect. A.1.

[44]In this context, smooth means that any bumps on the surface are much smaller than the wavelength of the light.

the focal length f, taken negative if the affected rays are sent away from the focal point. We will call lenses and mirrors optical devices. Note that a plane mirror or a flat piece of glass will have an infinite focal length.

Plane mirrors and spherically-surfaced mirrors are relatively easy to make, using silver deposited on glass. (This is best done on the front of the glass, but for protection of the coating, non-technical mirrors have the reflective surface on the back side of the glass, with the disadvantage that reflection off the glass itself produces faint secondary images.) A mirror with a negative f has a convex surface, and is called diverging.

A piece of bulk glass spherically ground on opposite sides, with the center of the spheres along one axis, becomes a lens. Either side may be convex, planar or concave. Whether a lens is converging or diverging depends on the curvature of both sides of the lens, and on the index of refraction of the glass relative to the index of refraction of the substance surrounding the glass.[45]

The inverse of a lens' focal length f is called its 'optical power', and is given a unit name 'diopter' when the focal length is given in meters. Note that the word power here is easily misconstrued, as optical power refers neither to an energy rate nor to the lens' magnification. If a lens is diverging, f is negative, and so is the optical power.

Eyeglass lenses are usually constructed to be concave on the inside and convex on the outside, to accommodate the eyelash movements. Both diverging and converging lenses can be constructed this way. In the case of the diverging eyeglass lenses, the glass is thinner nearer the optical axis. An eye which is astigmatic (i.e. its focal length along the horizontal axis differs from that along the vertical axis) requires an eyeglass lens with a different curvature in each of the two axis directions. Eyeglass prescriptions usually give three numbers, the 'spherical power', the 'cylindrical power', and the axis of astigmatism, where the spherical power is the optical power needed along the axis ('principle meridian') with the least optical power; the cylindrical power is the added optical power needed along the perpendicular meridian; and the angular orientation of the principal meridian, measured in degrees starting at zero for a horizontal meridian, 90° for a vertical meridian measured from zero counterclockwise looking at the eye.

With the laws of optics above, it is straightforward to construction light rays from an object, intersect a mirror or lens, and then leave it. Rays of light from each point on the object, after reflection or refraction by the optical device, converge toward a point, producing a real image point, or divergence from a point, making a virtual image point. It is relatively easy to show geometrically that for a thin optical device,

[45]The focal length of a thin lens obeys the 'lens maker's equation', $1/f = (n/n_0 - 1)(1/R_1 - 1/R_2)$, where n is the index of refraction of the glass, n_0 the index of refraction of the surrounding material (e.g. air), R_1 the radius of curvature of the lens on the ray entrance side, and R_2 the radius of curvature of the lens on the ray exit side. If the center of curvature is further along the optical axis in the direction a ray is traveling after intersecting the lens, then the radius of curvature is positive; if the center of curvature is behind from the direction of travel of the ray at refraction, then the radius of curvature is negative.

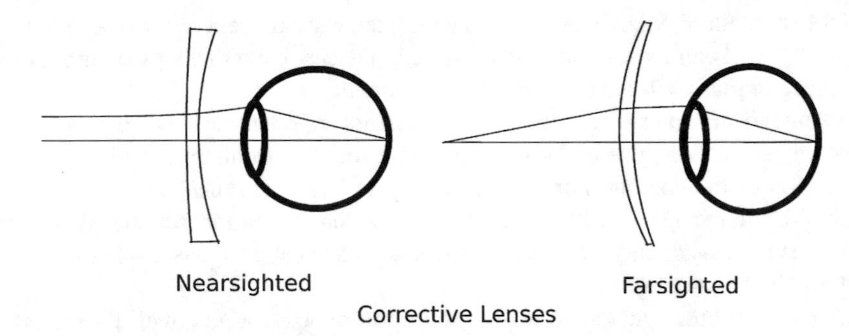

Fig. 7.22 Corrective lenses for the human eye

such as a thin lens, the inverse of the distance p from the device to the object along the axis of the device, (i.e. the 'object distance', taken positive when light goes from the object toward the lens) added to the inverse of the distance q from the lens to the image along the axis of the lens (i.e. the 'image distance', taken positive when light goes from the lens to the image) is the inverse of the focal length f of the lens. This is the so-called 'thin lens equation', often expressed as:

$$1/p + 1/q = 1/f \ .$$

The magnification of a thin lens is given by $M = i/o = q/p$, where i is the image height and o is the object height.

By placing a diverging lens of the appropriate focal length in front of an eye which is nearsighted (myopic), a distance image can be made to focus on the retina. Similarly, a converging lens of the right focal length can correct the image location in a farsighted eye (hyperoptic) so that nearby objects are in focus. (See Fig. 7.22.)

7.27 Vision and the Eye

7.27.1 The Eyes Have It

Humans get a good fraction of the information about their environment through visible light entering their eyes.[46] As shown in Fig. 7.23, light rays (dashed lines) from a distant point on a visible object enter the pupil, become refracted by the cornea (about 40 diopter equivalent lens), refracted again by the eye lens (from 20 to 30 diopter equivalent lens) to the retina. The aqueous humor has an index of refraction of 1.337, the cornea about 1.377, and the vitreous humor 1.336. The eye has an equivalent lens with a focal length of about 17 mm when relaxed.

[46]The information transfer rate of a given sensory input can be measured by the number of bits/second sent to the brain by the input devices for that sensation.

Fig. 7.23 Human eye

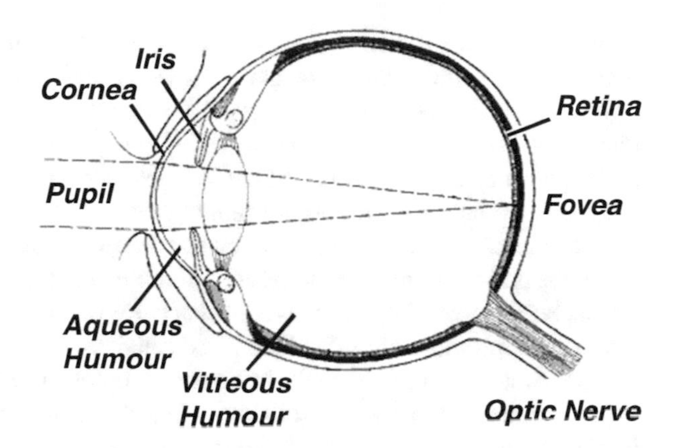

Images are 'upside down' on the retina, but of course the brain circuitry interprets the image as 'right side up'. The shape of the eye lens is controlled by radial ('zonule') muscles and circumferential ('ciliary') muscles. The lens varies in index of refraction from typically 1.406 in the central layers to 1.386 at the peripheral layers.

Nearsightedness can be caused by an increase in the thickness of the refractive elements or by an increase in the axial length of the eye itself. This causes rays of light from a distant object point to converge in front of the retina. In the case of farsightedness, rays from a nearby object point would converge beyond the retina. The healthy human eye without glasses has a horizontal field of view approximately 180°. Acuity comes from light reaching the fovea and not the surrounding region of the retina. The acuity drops to half for light that reaches 5° off from the center of the fovea.

Diffraction of light waves fundamentally limits the ability of any optical instrument to distinguish two nearby object points. Instead, if the two objects points are close together, but far from the viewer, they will appear as only one fuzzy object point. This is the Rayleigh limit discussed in Sect. 7.5. Object points separated by much less than one minute of arc cannot be resolved by the human eye, but rather become indistinguishable. The pupil of our eyes have a variable diameter, from $0.8\,\text{cm} < d < 1.6\,\text{cm}$, controlled by the light intensity reaching the retina. A typical wavelength for visible light is 500 nm. (Inside the eye, the wavelength drops in size to about 3/4's of its value in air.) These give a Rayleigh diffraction limit angle of $\Theta_R > 0.3$ arc minutes.

7.27.2 *Visual Acuity*

The ability to see sharply is called 'visual acuity', and involves not only the eye optics, but also the cellular structure in the retina, the dynamics of muscle control of the eye, nerve transmission to the brain, and brain analysis. The internationally adopted definition of acuity, α_V, is the inverse of the angular gap size in arc minutes between discernible black objects on a white background (a 'chart'), with chart lighting of 480 lux (32 W/m^2). If the objects are separated by a distance d and are a distance L from the eye, then $\alpha_V = (360 \times 60)/(2\pi) \times (d/L)$. Thus 20/20 vision corresponds to a visual acuity of $\alpha_V = (360 \times 60)/(2\pi) \times 0.07\,\text{in}/(20 \times 12\,\text{in})$, which, as we indicated above, is 1 min of arc.

This visual acuity limit depends on the density of photoreceptors in the fovea of the retina and the number of ganglion cells connected. The fovea where sharp vision occurs holds most of the about six million color photoreceptors (called 'cone' cells) while the approximately 120 million 'rod cells' are distributed over the whole retina and are more sensitive to light, but have no color discrimination. There are three types of cones, with greatest sensitivity to red, green or blue colors. This gives the human the ability to distinguish about 10 million colors. The peak sensitivities are at wavelengths 564 nm (red), 534 nm (green), and 420 nm (blue).

The indication 20/20 for the acuity of a human eye means that eye must be no greater than 20 ft from an object to resolve details that an eye with good vision can resolve at 20 ft. A person with 20/40 vision can resolve alphabetic letters at 20 ft that a person with good vision can resolve at 40 ft. To make a test of vision, letters are displayed on a Snellen chart, and a set of lenses are used to see what lens power is required to correct fuzzy vision due to near or farsightedness.

A person with 20/20 vision can only distinguish the separation of contours in letters at 20 feet if they are separated by more than 7 hundredths of an inch. Such resolvable contours make an angle of 1 arc minute at 20 ft. Good human vision is no better than three times the diffraction limit.

7.27.3 The Eagle Eye

Eagle eyes each have two foveae, one pair for binocular vision and the other to spot objects to the side of the head. They also can see into the soft ultraviolet part of the electromagnetic spectrum. The typical eagle eye pupil is larger than a human eye pupil, so it has a better optical resolution, with a corresponding greater density of fovea cones. The combination means the eagle has an acuity four times as good the best human eye, putting the eagle eye resolution close to the diffractive limit.

7.27.4 More Colorful

Rainbow colors are called 'primary', as each is near a single frequency. Mixtures of two rainbow colors produce 'secondary' colors, such as brown. An equal intensity mixture of all three of red, green, and blue light makes white light. A combination of just red and green light will appear yellow. Such combinations of light are called additive mixtures. (If you look closely at a TV screen, you can see the phosphor dots emit red, green and blue. Yellow is constructed.) On the other hand, a surface which appears colored by illumination has pigments that absorb certain ranges of light at various frequencies, leaving others to be scattered. Thus, mixtures of paints involves absorption of light, and their combination are called subtractive mixtures. The subtractive mixture of red and green is brown.

7.27.5 Photo-Reception Celebrates Our Vision

Within the retina's photoreceptors, light stimulates large proteins molecules (opsins) to change shape. This activates a number transducin molecules, which in turn activates the enzyme cGMP phosphodiesterase. That enzyme can catalyze the reduction of many cGMP by cleavage. The effect of the original photon is amplified in the production of modified transducin molecules, and in the multiple catalysis (about 1000) of cGPM break-up by each enzyme molecule. The reduced level of cGMP causes sodium ions channels to close. The photoreceptor membrane becomes 'hyperpolarized' as the charge on the inside becomes more negative (−40–65 mV). This change in potential causes the membrane channels for the flow of calcium ions into the cell to close. The decrease in calcium in the cell slows the release of the neurotransmitter glutamate. Depending on the cells across the postsynaptic gaps, each of those cells will either be stimulated or inhibited. The results are pulses sent through the optic nerves to be analyzed by the cerebral cortex.

7.28 Visual Illusions, Distortions, and Hallucinations

Who are you going to believe, me or your lying eyes?
— Groucho Marx

Human perceptions of the world are known to be influenced by actions of the mind. Where our attention focuses and how we interpret visual input depends both on our genetics and on our expectations and experiences. Moreover, the brain can give us the impression of seeing in dreams and even when awake, especially under sensory deprivation and other stressful conditions.

Our memories of what we saw is not the same as a digital recording of a scene. In fact, the information that is sent to the brain about an object codes its lines, orientation, motion, color, and so forth, rather than pixels. This information is compared to past visual memories, in sophisticated pattern-recognition algorithms. There is even a special area dedicated to the recognition of faces. When the visual information is analyzed, recognizable parts that are considered important are imprinted onto subsets of neurons, together with connections to past associated memories, such as sounds, feels, smells, past pleasures and pain, etc. Stimulation of these subsets of neurons evokes some memory of the original visions.

When visual information is partial, ambiguous or fuzzy, then the brain makes inferences. For example, in print with missing parts to characters, the brain can subconsciously fill in the missing parts while the person reads on. If an image contains more than one interpretation, the subconscious brain doing the visual analysis makes an assumption of which interpretation to adopt. A drawing of the edges of a transparent cube has two interpretations about what faces of the cube are in front. Only one is 'seen' at a time.

Cases in which the appearance of an object differs from the actual object are called optical illusions. Visual illusions, such as mirages, are generated by unexpected external physical conditions. Some illusions are physiological, caused by a breakdown of the mechanisms for vision. Some are psychological, caused by the brain's faulty interpretation and interpolation, or by a purposeful misinterpretation. Sometimes, a brain may cause a sensation of vision without any corresponding input from the eyes. Dreams and hallucinations are examples.

A visual hallucination occurs if the brain generates the perception of an external visual stimulus when none exists. Such an event can be generated by brain sensory deprivation and by pathological conditions in the brain, such as blood imbalance, drugs, lesions, and impending seizures.

Bright light can cause rods and cones to be depleted of photochemical agents, causing a subsequent darkening in locations where the light was bright. Quick recovery from this fatigue is expected if the light has not damaged the retina.

If the cause of an optical illusion is due to the brain's analysis, then the effect is called a cognitive illusion. The cube drawing is an example of a cognitive ambiguous illusion. If associated lines and shapes are misinterpreted, then a cognitive distortion illusion has occurred. If the brain juxtaposes or reinterprets objects in a scene into a physically impossible configuration, this is a cognitive paradox illusion. The brain can also add color information where there was none, producing cognitive chromatic illusion. (This was discovered by Edwin Land, the inventor of the Polaroid instant camera.)

Most cognitive illusions can be understood as the brain trying to make sense of what is seen, so that subsequent analysis or action is possible. In Fig. 7.24, the Escher staircase drawing puzzles the brain when incongruity is perceived. The 'old woman/young woman' shows that the brain makes just one interpretation of a visual input, although it can flip between two or more.

In some cases of cognitive illusions, there may be physiological hyperfunction. For example, in grapheme color synesthia, black and white letters on paper are seen tinged with different colors for different letters. One plausible explanation is 'cross-talk' between networks of the brain. This idea could also explain why those with synesthia are disproportionately intelligent.

Fig. 7.24 Famous illusional drawings: Left: M.C. Escher's staircase; Right: An 1888 German postcard of a young woman/old woman

7.29 Optical Instruments

What we can see with our 'naked' eyes, compared to what can be seen with 'dressed-up' eyes, is infinitesimal. Visible light from objects as small as 4000 hydrogen atoms lined up can be seen, given the right instrument. There is a world of active life at the 'mesoscopic scale' (about 400 nm–4 μm) about which human kind could only speculate for millions of years.

Optical instruments are designed to give us the ability to

- See a magnified view of nearby small objects, or far-away objects that are small in angular size, with as large a field of view as is practical;
- See objects with better resolution than ordinary vision affords;
- See very dim objects or objects with low contrast between parts;
- Separate and better distinguish colors of objects;
- Analyze the light from objects to investigate their nature.

7.29.1 The Lens

A drop of water placed in a circular orifice acts as a lens, and can produce a magnified image of an object placed nearby. (The fact that such a device was not constructed eons ago is likely due to cultural strictures in a world with meager communication of discoveries.) Glass balls were used in the Middle Ages to read small print, and eyeglasses soon followed.

There are several important properties which characterize lens systems: (1) Field of view; (2) Magnification; (3) Quality of images (i.e. degree of distortion, resolution, and aberrations). The field of view is the largest angle formed by the light from any point on an object to the lens system which can be seen in the image. The linear magnification is the ratio of the image size to the object size. The angular magnification is the angle formed by the size of the whole image perpendicular to the optical axis divided by the distance from the last lens to the image. The quality of an image is determined by its resolution (number of distinguishable points across the image compared to the number across the object), by a measure of how well an image matches an object by a single overall scaling factor, and how well colors are brought to a single focus.

7.29.2 A Simple Microscope

With a 'magnifying glass', i.e. a single lens microscope, magnification is limited because as the lens' angular magnification ($\theta \equiv \arctan(o/p)$) increases, the distance from the object being viewed and the lens decreases (the image is virtual and, for comfortable viewing, the lens is positioned so that the image distance is large, which means $p \approx f$, so $\theta \approx \arctan(o/f)$). As the lens has a finite width, say $2w$, we must have $p > w$. This limits the angular magnification to $\arctan(o/w)$. The first microscope, made by van Leeuwenhoek in about 1670, had only one small glass-ball lens. The smallest ball had a magnification of 275×. The field of view was strongly limited, due to spherical aberration of the image. Still, van Leeuwenhoek must have been enchanted by what he saw for the first time in the small world. Modern visible-light microscopes have many lenses, and can magnify up to 1000× with resolution down to about 200 nm.

A good microscope is constructed to produce a magnified image with (1) low distortion, (2) high resolution, (3) an adequate field of view, and (4) sufficient light and contrast to differentiate small structures. Having two or more lenses gives the instrument the name 'compound microscope'. Refinement and expense comes in making lenses which focus both blue and red the same way (reducing chromatic aberration) and in making high magnification with little distortion.

Dyes and fluorescent tags (markers) supplement the microscope methods in the biology lab for making the invisible visible. Fluorescent organic molecules can be used as tags in order to follow the location and timing of particular intercellular reactions.

From the Rayleigh limit Eq. (7.13), one can show that the smallest distance that can be resolved by a microscope is given by $d = 0.61\lambda/(n \tan\theta_i)$, where n is the index of refraction of the object material and θ_i is the angle formed by the image to the objective lens. The depth of focus (i.e. the depth into the object material that will still be in focus) is given by $\delta z = \lambda/(4n \sin^2(\theta_i/2))$.

7.29.3 Phase-Contrast Microscope

As many biological cells are mostly transparent, their structures are difficult to see in an ordinary microscope. In 1934, in order to increase the contrast of images of living cells without using dyes, the Dutchman Frits Zernike invented the 'phase contrast microscope'. He knew that the indices of refraction in the cell and its transparent structures are slightly greater than that of water, so that the speed of the light is slightly less while passing through these structures. This causes the phase of light waves from the illuminator to be retarded after traveling through the cell, and thus slightly shifted back in phase relative to the illumination outside the cell. Moreover, the illuminating light refracted by passing through the cell is dispersed to greater angles than the background light passing through the media surrounding the cell, so that light dispersed by the cell can be separated from the background light.

Suppose a plane wave from an illuminator in the microscope passes through a mostly transparent specimen. After passing, the wave will have a change of phase that depends on the variations in the index of refraction in the object. In complex number notation, the new wave will be proportional to $A(x, y) = \exp(i\phi(x, y))$, where (x, y) are locations in the plane just past the specimen. Note that the intensity of this light, proportional to A^*A, will have no phase variation. The fact that the specimen's index of refraction differs only slightly from the background water means that ϕ is small, so that $\exp((i\phi) \approx 1 + i\phi$.

To get the phases of the wave to cause an intensity variation, Zernike used a quarter-wave plate[47] to shift the phase of the separated background waves by $-90° = -\pi/4$ radians. Recombining the phase-shifted background waves (whose amplitude of 1 changed to $exp(-i\pi/4)$) with the specimen-dispersed waves in effect changes $A \approx 1 + i\phi$ into $A' = -i + i\phi$. This makes the intensity of the light depend on phase variations: $A'^*A' \approx 1 - 2\phi$, and the image now has light and dark regions depending on the index of refraction variations. The phase term can be enhanced relative to the background term by having some absorption in the quarter-wave plate, so that the amplitude of the background light is changed by transmission with a factor $a \exp(-i\pi/2)$, where the absorption coefficient a^2 is less than one. On recombination with the separated specimen light, the new light intensity becomes proportional to $a^2 - 2a\phi$. The ratio of the phase-contrast term to the direct term is $-2\phi/a$, so the phase term is enhanced by a factor $1/a$ compared to a non-absorbing case when $a = 1$. Of course, this also means the overall intensity is reduced by a^2, so there is a playoff of enhanced phase contrast with reduced overall intensity. Color fringes should also be expected when white light rather than one color is used.

[47] A quarter-wave plate is constructed by depositing a thin dielectric film onto a glass plate. The dielectric causes the phase of the light of the selected wavelength to shift back by 90°. Adding a little metal to the film causes an additional absorption.

7.29.4 Ophthalmoscope

The ophthalmoscope is a simple illumination device that permits a physician to see the fundus of the eye, i.e. the surface opposite the lens. Ordinarily, looking at the eye of a subject shows a dark pupil. The ophthalmoscope sends extra light along the ophthalmologists viewing direction into the eye of a patient. The light is refracted by the patient's eye cornea and lens onto the fundus. Some is reflected, and goes back through the eye lens, which directs much of the light rays back to where they originated. However, suppose the source of illumination comes from a reflecting plate placed at a 45° angle to the line of sight, and that reflecting plate has a small hole at the line of sight. Then the ophthalmologist can see reflected light coming back, thereby allowing examination of the fundus, including the retina, fovea, optic disc, and small blood vessels. A direct ophthalmoscope uses no lens; typically just a light source and mirror.

The indirect ophthalmoscope has lenses in the path of light from the retina. Those lenses can enlarge the field of view and provide for magnification (up to $15\times$) of the image of the fundus. With a converging (objective) lens, at a position from the patient's eye a little greater than the objective lens' focal length f_o, and assuming the patient's eye is relaxed ('emmetropic', which occurs when focusing on a distant object), then the physician's field of view will have an enlargement factor of D/f_o, where D is the diameter of the cone of returned rays of light at the objective lens' position.

After light reaches the patient's fundus, then, for an emmetropic eye[48], the image of the retina will be a large distance and in front of the patient's eye. We can estimate the magnification of the ophthalmoscope as follows: Suppose the objective lens of the ophthalmoscope has a focal length of f_o and placed at a nearby distance from the patient's eye. Applying the thin-lens equation twice, once for the patient's eye, with focal length f_e, and once for the objective lens in the ophthalmoscope, the retina will appear to the physician magnified by $M \approx f_o/f_e$.

7.29.5 Otoscope

The otoscope is designed to view the eardrum and affords a diffuse view of the middle ear. A flexible tube fits into the entrance of the ear canal, and conducts light into the ear, and also is wide enough to allow viewing through the tube. The image of the eardrum and shadows of the malleus in the middle ear is made visible to the physician at a comfortable distance from the physician's eye.

[48]Emmetropic vision requires no correction.

7.29.6 Fiber Optics

Optical light pipes include any transparent cylindrical rods used to conduct light. Plastic and glass rods serve this purpose. Any light sent into a glass fiber at an angle below the critical angle of the material of the rod will internally reflect and propagate down the rod.

Optical light pipes clustered together as filaments create an optical fiber bundle. Glass filaments engineered to have low loss of transmitted light are widely used conductors of information over multiple kilometer distances. Such glass conduits of light also have a variety of applications for medical remote viewing.

The intensity of light in one fiber is most often given in terms of the dB level, defined analogously to that for sound, by $\beta = 10 \log_{10} (I/I_0)$, where I_0 is a base intensity. Typically, these intensities are in the micro- to milliwatts over an effective area (such as 40 $(\mu m)^2$) of the glass core of an individual fiber, with light generated and detected by semiconductors made from silicon and germanium. When I_0 is taken as 1 mW for a given area fiber, then the decibel level β is appended with the name 'dBm'.

Information can be coded digitally on the light beam by pulsing the light on and off. The light wave is then called the 'carrier' wave. Digital encoding has a distinct advantage over analog encoding (which historically used amplitude modulation or frequency modulation of the carrier) in that low-level noise and other unwanted variations of amplitude do not spoil the decoding of digital signals. The frequencies of light used in fiber optic systems range from ultraviolet to the near infrared, 400 nm to 1600 nm. Wavelengths of 850 nm, 1300 nm, and 1550 nm are often used in fiber optical systems. Losses along glass fibers can be as low as 0.2 dB per kilometer, making it possible to lay hundreds of kilometers of fiber without repeaters.

The wavelengths used correspond to frequencies in the 10^{14}–10^{15} Hz range. In principal, encoded pulses as rapid as 10^{15} Hz (i.e. a petaHz) can be used, but loss in the glass and the use of electronic digital circuitry makes upper practical limits in the multiple terahertz range. This also gives an upper limit on the digital bit rate. (Divide by (8 plus 2 redundancy check bits) to get an estimate of the bytes per second rate.)

There are important uses of fiber optics in medicine. Here are a few:

- Transporting an image through a small-diameter conduit, from one local location, such as inside a patient, to another local location, such as a viewing screen for a physician. At one end, light is focused by a small lens onto the face of one end of a collection of fibers, which we will call the 'business' end. If the fiber is to make images in an initially dark region (e.g. inside a body cavity), then illumination comes by sending light from the viewing end to the business end. The fibers can be held fixed in position by epoxy and then cut flat and polished. The image is divided into a pixelated set according to the number of individual fibers, with a density sufficient to discern structures in the image. (Typically, about 50,000 individual fibers are used to produce an image with 50,000 pixels, all within a flexible rod as thin as 1 mm in diameter.) This image is, in effect, passively transported to the other end of the fiber tube, which has the individual fibers arranged in the exact same pattern as the business end. The tube itself is usually flexible, and therefore able to round gentle corners. The curvature can be controlled by a thin metal wire with a curved end placed within the housing of the endoscope, so the physician can steer the end. Additional manipulator wires might be placed in the same housing. Note that no electronics or power is needed at the business end.

- *Information sharing*: The availability of medical information makes it possible to make medical diagnoses based on the full spectrum of study of diseases, both rare and common, both past and newly emerging. Consultations can occur across widely separated locations. Also, medical histories can be stored in 'cloud servers' and retrieved from across the country when needed in an emergency.
- *Distal examination and treatment*: The transmission of high-resolution images for patients some distance from a physician is possible.
- *Laser-light therapy*: Laser light can be redirected by passing the light through a flexible optical fiber. In therapy, this allows laser light to be used for

  - Removing Sun spots, birthmarks, tattoos, moles, warts, and unwanted hair;
  - Surgical cutting and cauterizing;
  - Killing cancer cells on the skin or in the body via endoscopy.

  Depending on the application, intense laser light can be generated with a carbon-dioxide gas laser, argon gas laser, or Neodymium-doped yttrium aluminum garnet ($Y_3Al_5O_{12}$) solid-state laser, with a frequency selected to give the greatest light absorptivity in a given tissue.

7.29.7 Endoscopes

An endoscope is any device built to view interior regions of the body. Fiber-optic endoscopic imaging produces visual images using fiber conduits inserted into cavities and organs. Note that glass fibers have a distinct advantage over other endoscopic techniques because they are biologically inert, their light intensities and frequencies are harmless to tissue, and they can produce highly localized images.

A capsule endoscope is an untethered enclosure holding a micro-CCD camera, a radio transmitter, and a battery, together with optics, often including plastic light pipes.

7.29.8 Image Intensifiers

When light levels are too low for effective viewing, an image intensifier may be employed. The same kind of device may be used to make low-level infrared,

X-ray and gamma-ray images visible. Video transmissions were first performed by kinescopic tubes, which projected a light image on a specially coated plate in a vacuum which would become locally charged when light hit it. An electron beam was swept in raster fashion across the plate to release and read the charge. The collected electrons produced an electric signal which was in proportion to the charge at the location the beam was directed. An amplified signal used to control the current sent to the phosphor of a Cathode Ray Tube (CRT), with the electron beam of the CRT swept across the phosphor mirroring the beam in the kinescopic tube, made up an indirect image intensifier.

Direct image intensifiers cause incoming photons absorbed on a matrix of pixel elements to emit electrons. These electrons are then accelerated in an electric field, often along microchannels, so that when they collide with a phosphor, they emit visible light with an intensity far higher than the original light. Cascading the sequence of electron accelerations creates additional amplification of the light.

With intensifiers having light amplification, lower doses of X-rays for dental and organ imaging is possible. In cameras and video systems, the initial conversion from photons to electrons is made by Charge-Coupled Devices (CCDs) rather than phosphors, and amplification is done electronically.

CCDs and electronic flat screens have largely replaced kinescopic tubes and CRTs. This technology has dramatically lowered X-ray doses needed for diagnosis.

7.29.9 Fluorescent Tags

Biomolecular dynamics can be studied by taking advantage of 'Förster Resonance Energy Transfer' (FRET). (Also called fluorescence resonance energy transfer.) Donor chromophores send energy to nearby acceptor chromophores, which then radiate. The sites for the energy transfer are called 'fluorophores', and may be inserted into the molecules of interest by the experimentalist. Because the interaction between chromophores is predominantly dipole-dipole, energy transfer requires separation distances within a few nanometers. Small changes in distances of this size become visible by the amount of light emission.

In single molecule FRET (smFRET), two fluorescent dye molecules are attached to a biological molecule, such as each side of the DNA backbone. The interaction between the two dyes can tell us about how far apart they are, say, during replication. This distance information can be used to measure the shape changes of a single molecule during its functional activity.

Applied in vivo, FRET has been used to detect the location and interactions of genes and cellular structures, and to study metabolic pathways.

7.30 Solar Radiation and Biology

The Sun supplies about 10^{24} kcal of energy over the surface of the Earth in 1 year. Above the atmosphere, this solar power amounts to about 1370 W per square meter. (Your toaster uses about 1000 W of power while in operation.) With the Sun overhead on a clear day, plant leaves would receive about 1000 W per square meter. Including the solar ray angle variations and day/night cycles, the average plant on Earth gets only about 200 W/m^2. Using photosynthesis, a green plant[49] can convert about 1–2% of the arriving solar energy into stored chemical energy in organic molecules, such as glucose.[50]

This could explain why you are not green. Assuming you could expose a quarter of your green skin to sunlight, you would produce less than about 0.03×400 W/m^2 $\times$ 0.5 m^2 $=$ 6 W of power. But your metabolism will require an average of 160 W. A person is supported in his/her energy needs by at least 27 m^2 of productive leaf surfaces. That's about 17,000 healthy tree leaves. A medium size maple tree can have that many, but lots of the leaves are in partial shadows, so we should look for a large maple tree to sustain just one of us.

Currently, photovoltaic cells can convert about 10% of the incoming solar energy into chemical energy in the disassociation of water into hydrogen and oxygen. Your metabolic energy requirement could be fulfilled by a solar panel aimed at the Sun and covering about 8 m^2.

Plants fix about 150 trillion kilograms of carbon each year; i.e. about 10^{18} kcal of equivalent glucose energy per year. Largely to support industry and food production and distribution, man is using about 4×10^{17} kcal per year in fossil fuel, the result of plants geologically converted over hundreds of millions of years.

The chemical energy used by organisms is stored in carbohydrates, lipids, waxes, and proteins. The energy stored in these molecules can be later converted into the energy in a phosphate bond of adenosine triphosphate (ATP), which is the ready fuel for the activity of cells.

The analysis of energy exchanges in organisms follows the principles of non-equilibrium thermodynamics. (See Sect. 9.23.)

---

[49]Chlorophyll (*a* and *b*) absorb in the red and blue part of the spectrum, while green light tends to be reflected. See Fig. 7.13.

[50]The pigment-protein complex which absorbs photons through electron excitation acts as a highly efficient antenna. However, It is possible that quantum coherence plays a role in this transfer of energy, as described by Engel et al., *Evidence for wavelike energy transfer through quantum coherence in photosynthetic systems*, Nature **446**, 782–786 (2007).

7.31 Dosimetry for Non-ionizing Electromagnetic Waves

7.31.1 General Considerations of Dangers of Non-ionizing Radiation

People living in populated areas of modern cities are bathed in man-made radio waves from radio transmitters, and microwaves from repeater towers and cell phone devices. We also receive such waves from natural sources, including solar radio frequency radiation from flares and eruptions, and radio waves from the Earth's atmospheric electric discharges and aurora. Above frequencies of about 10 MHz to about 10 GHz, man-made electromagnetic waves now dominate natural sources in our environment.

There is also a pervasive background microwave radiation coming from outer space as the relic radiation left over from the big bang. Because the Universe cooled by expansion, that radiation is now the same as that from a blackbody at the temperature of 2.725 °C. From Stefan's law (Eq. (7.18), the intensity of this radiation is

$$I = \sigma T^4 = 5.67 \times 10^{-8} (2.725)^4 \,\mathrm{W/m^2} \simeq 3\,\mu\mathrm{W/m^2}\,. \tag{7.23}$$

But the microwave intensity near our cell phones is a million times greater, in the range of 3 W/m^2.

Intensities of background radio and microwaves are usually low enough that no health effects are evident.[51] There are exceptional instances. A person accidentally encountered dangerous levels when standing in front of a powerful microwave antenna inadvertently turned on by a co-worker.[52]

Table 7.5 shows the Federal Communication Commission recommended upper limits for non-ionizing electromagnetic wave power densities in human tissue. Surface power density, a term used by engineers, is simply another name for

Table 7.5 FCC recommended upper limits to radiation power density impinging on humans

EM-wave	Limit
Radio	$\sim 10^4$ W/cm^2
Microwave	10 mW/cm^2
Infrared	1 W/cm^2
Visible	0.1 W/cm^2
Soft ultraviolet	0.001 W/cm^2

[51] After all, we humans have lived on Earth for millions of years unaware of the existence of radio waves. Then Maxwell predicted from pure deductive logic that radio waves must exist from his equations for the behavior of electric and magnetic fields.

[52] Joseph J. Carr, **Microwave & Wireless Communications Technology** p. 9, [Newnes (Elsevier), Boston] (1997).

intensity, namely. the energy arriving at or leaving from a surface perpendicular to the direction of propagation of the wave per unit time per unit area.

The theoretical description of radio and microwave radiation absorption in materials begins with Maxwell's equations. Absorption can be treated by taking the electric permittivity ϵ and magnetic permeability μ as complex numbers. As magnetic effects are much weaker than electric effects in biological tissue, the magnetic permeability can be taken real and the same value as in a vacuum. However, because electric polarizabilities of bound charges and induced electric currents in tissue can be significant, the real part of the electric permittivity can be large (causing a slowing down of the wave), and the imaginary part can also be large, causing heat production. In fact, by applying Maxwell's equations, the imaginary part, ϵ_b, comes from the motion of bound charges on molecules and membranes. Their coupling to adjacent molecules dissipates their energy, generating heat. We write

$$\epsilon = \varepsilon_r \epsilon_0 - i\epsilon_b , \tag{7.24}$$

where ε_r is called the relative permittivity[53] and ϵ_0 is the vacuum electric permittivity.[54]

The electric field in the tissue induces a current density $\mathbf{J}$ that will have contributions from real currents and bound currents, expressed as

$$\mathbf{J} = (\sigma + 2\pi f \epsilon_b)\mathbf{E}. \tag{7.25}$$

In this relation, a form of Ohm's Law, σ is the conductivity of free charges (electrons and ions) within the tissue, f the frequency of the passing electromagnetic wave, and $\mathbf{E}$ the electric field of the wave.

As yet, there is no indisputable evidence that low frequency radiation to the public has caused cancer. Static magnetic fields can be detected by some animals, such as pigeons, who use the Earth's field to help their navigation. The Earth's magnetic field is around 20–70 μT, and is biologically detectable by the effect on microferrous materials embedded in nervous tissue. Very strong static magnetic fields, tens of thousands times stronger than the Earth's field, still do no damage to human tissue. Strong (several tesla) low frequency magnetic fields can cause effects to nerve tissue from the induced electric currents in the tissue. These effects can also be caused by movement of the body tissue through a static magnetic field. The current near the sinoatrial node of the heart induced by a 5 T field is about 100 mA/m^2, which is below the cardiac excitation threshold.[55] Oscillating magnetic

[53] Relative permittivity is the same as the older term, 'dielectric constant'.

[54] It follows, again from Maxwell's equations, that the tissue's index of refraction squared is given by $n^2 \simeq \varepsilon_r$.

[55] Kinouchi et al., *Theoretical analysis of magnetic field interactions with aortic blood flow*, Bioelectromagnetics **17:1**, 21–32 (1996).

fields as low as 5 mT at 20 Hz can induce faint flickering in vision, called magnetic phosphenes. Evidence suggests the cause is induced electric current in the retina.

7.31.2 Special Considerations for Microwaves

As can be seen in Table 7.5, limitations on microwave intensities are more severe than those of the adjacent bands of frequencies. This follows from two facts: (1) Microwaves are easily absorbed by human wet tissue and (2) Microwaves penetrate millimeters to centimeters into the body when incident on the skin.

Since microwave photons are much too low in energy to directly cause chemical bond breakage, the predominant mechanism for energy loss from the wave is the stimulation of oscillations of polar molecules, vibration of polarized membranes and ionic motion. The kinetic energy of these oscillations and motions dissipates into heat. It is the consequent excessive rise in temperature that can cause damage, particularly to organs with low blood circulation, such as cornea of the eye. A small rise, less than a degree Celsius, is easily tolerable, and may even be helpful as a diathermy treatment (refer to Table 7.6 for tissue thermal properties).

The penetration depth δ_p, (introduced in Sect. 7.13.2), for microwaves arriving at the skin, strongly depends on the water content of the skin and underlying tissue, its salt content, and on the frequency f of the wave. An analysis from Maxwell's Theory gives

$$\delta_p = \sqrt{\frac{1}{\pi \mu \sigma f}} , \tag{7.26}$$

where μ is the magnetic permeability and σ the electric conductivity of the tissue. The magnetic permeability differs little from the vacuum value of $\mu_0 = 4\pi \times 10^{-7}$ H/m. However, as seen in Table 7.7, the conductivity is tissue dependent, and the electric permittivity can be far greater than its vacuum value: $\epsilon_0 = 8.854 \times 10^{-12}$ farad/m. (A larger ϵ makes the speed of the microwave smaller than its vacuum value $c = 2.9979 \times 10^8$ m/s, and correspondingly a shorter wavelength. The frequency f of the wave is unchanged, as the material charges are being forced to wiggle at the incoming wave frequency.)

Table 7.6 Some thermal properties of tissues

Name	Density kg/m^3	Specific heat J/kg · K	Thermal conductivity W/m · K
Skin	1100	3663	0.20–0.30
Fat	950	2348	0.20–0.21
Muscle	1065	3421	0.50

Table 7.7 Some electrical properties of human tissue

Frequency (MHz)	Permittivity (V/m)	Conductivity (S/m)
Skin		
900	41.4 ϵ_0	0.87
1800	38.9 ϵ_0	1.18
2450	38.0 ϵ_0	1.46
Muscle		
433	53 ϵ_0	1.4
915	51 ϵ_0	1.6
2450	49 ϵ_0	2.2
5800	43 ϵ_0	4.7
Fat		
433	5.6 ϵ_0	0.08
915	5.6 ϵ_0	0.10
2450	5.5 ϵ_0	0.16
5800	5.1 ϵ_0	0.26

Table 7.8 FCC microwave limits (f in MHz)

Frequency (MHz)	E-field (V/m)	B-field (A/m)	Power density (mW/cm^2)
0.3–1.34	614	1.63	100
1.34–30	$824/f$	$2.19/f$	$180/f^2$
30–300	27.5	0.073	0.2
300–1500	–	–	$f/1500$
1500–100,000	–	–	1.0

In Table 7.8 are recommended Maximum Permissible Exposure (MPE) within human tissue, expressed in terms of the 'Power Density', which is just the intensity of the microwave (energy arriving at a unit area per unit time).

The wave energy absorption rate is calibrated by the 'Specific Absorption Rate' (SAR), already introduced in Sect. 7.11 for radio waves. The SAR is calculated by

$$SAR \equiv \alpha_R \equiv \frac{1}{V}\int \frac{\sigma E^2}{\rho} dV\ . \tag{7.27}$$

In this relation, σ is the conductivity, E^2 the mean square of electric field, ρ the density, and V a small volume, all within the tissue. The SAR is used to predict the temperature increase of the tissue, knowing the specific heat, blood flow, heat conductivity, and tissue density. Sample thermal properties of tissue are shown in Table 7.6. Clearly, blood circulation is a major factor. The FCC requires that cell phones have a SAR level below 1.6 mW per gram.

The U.S. Occupational Safety and Health Administration set a limit on energy delivered by microwaves to less than 1 mW/h/cm^2. (This is an average intensity of 10 mW/cm^2 over 6 min.)

The U.S. Food and Drug Administration limits the emission of microwave ovens to no more than 5 mW/cm^2 beyond 5 cm from the oven surface.

Because the eyes have low blood circulation, they are the first tissue in the body to be damaged by thermal heating due to microwaves. Cataracts occur for sustained intensities above about 100 mW/cm^2.

Microwaves of frequency 5.8 GHz with intensity of 30 mW/cm^2, will cause an increase of facial skin temperature by 0.48 °C. With the same intensity, the corneal surface of the eye heats by 0.7 °C.

7.31.3 Cell Phone Radiation

Cell phones use frequencies in the 800 MHz and the 1900 MHz bands. At these frequencies, the penetration depth of microwaves into biological tissue is less than 2 cm. Cell phones typically emit radio wave power of less than one watt, with intensities typically less than 0.5 W/cm^2 near the phone. Due to wave reflection and absorption in the skin and hypodermis surrounding the skull, brain exposure is usually less than 1 mW/cm^2.

Some attention has been given to the effects of cell phone radiation (with frequencies in the GHz region) and the radiation of power lines (with frequency of 60 Hz), since these radiations, although relatively weak, are pervasive and long-term.

Bluetooth devices transmit at frequencies about 2.4 and 2.5 GHz, but with far less radiation energy than cell phones.

Problems

7.1 Laser light has been made with an intensity of 10^{12} W/cm^2. What is the energy density in this light?

7.2 Two identical polarizing sheets are placed perpendicular to a light beam, but one is rotated 23° relative to the first. If unpolarized light arrives at the first sheet, what will be the percent loss in intensity of light after it passes through both sheets?

7.3 Most microwave ovens operate at a frequency of 2450 MHz. What is the wavelength of this wave? How much energy (in electron-volts) does each such microwave photon carry? Will such a photon cause atomic ionization?

7.4 The application of Stefan's law to our skin includes radiation in the radio end as well as the higher frequency end of the electromagnetic spectrum. Use Planck's expression for the intensity of light emission per unit frequency interval to numerically estimate the percent of radiation that the skin emits just in the radio part of the spectrum (up to 4 GHz) compared to the total radiation over all frequencies. Assume a surface temperature of 37 °C.

7.5 A set of evenly spaced IR spectral lines are found for an unknown compound, spaced apart by 1430 cm^{-1} (inverse wavelength). If these are due to vibration of the hydrogen atoms bound to carbon, estimate the effective spring constant between the hydrogen and the carbon.

7.6 Predict the light absorption frequency which would cause molecular nitrogen, N_2, to start rotating in its first rotational quantum state. (Use the known separation of the nitrogen atoms and the mass of atomic nitrogen.).

7.7 Suppose light of wavelength 820 nm were observed to be Raman scattered by a protein molecule, with a change in color of the light to wavelength of 910 nm. How much energy did the protein molecule take?

7.8 Coherent visible light can be generated with LASERs. Why is the production of coherent x-rays so much more difficult?

7.9 Lasers can be used for skin resurfacing and in brain surgery. Identify why lasers are better than other devices for such purposes. What frequencies would be appropriate and why? What intensities are used?

7.10 As an exercise in the mathematics of holography, find the two-dimensional Fourier transform of the following 'Gaussian':

$$f(x, y) = A \exp(-(x^2 + y^2)/a^2).$$

7.11 Estimate the smallest separation of two objects that can barely be resolved by an optical microscope using light of frequency 10^{15}/s.

7.12 The cross section of one chlorophyll molecule is about 10^{-21} m^2 at resonance. Find the number of chlorophyll molecules per square centimeter needed to capture 30% of the incoming light energy with frequency in the resonance range.

7.13 The chlorophyll molecules in plants have a red resonant absorption at photon energies near 1.8 eV with a resonant width of 0.2 eV, and another in the blue at 2.8 eV with a width of 0.3 eV. The overall photon absorptivity in the red resonance region is about half that in the blue region. About 30% of the incoming photons in the resonance regions are absorbed. How much power (in watts) is captured by chlorophyll molecules found in a plant leaf cell. If the photosynthesis process has chemical efficiency of 82%, how much leaf surface is needed to produce 1000 W.

7.14 Compare the absorption of a photon of energy 1.8 eV by a chlorophyll molecule of mass 28,000 Da to the energy due to thermal motion. If the molecule were free to move by the kick of an absorbed photon, how fast would it move?

7.15 Estimate the intensity of radiation arriving at our skin from the microwaves left over from the big bang.

7.16 From a power input of 700 W, estimate the energy density of microwaves within a typical microwave oven when it is operating in continuous high mode. If the microwaves were to leak from a hole of area $2\,\mathrm{cm}^2$, what would be the intensity of the microwaves at that hole? Use Bethe's hole formula (H.A. Bethe, *'Theory of Diffraction by Small Holes'*, Phys Rev **66**, 163–182 (1944)) for the transmitted intensity ratio through a sub-wavelength hole of radius r, namely

$$\frac{I}{I_0} = \frac{64}{27\pi^2}\left(\frac{2\pi r}{\lambda}\right)^4 . \tag{7.28}$$

What intensity would reach your face 10 cm from the hole? How does this intensity compare with the recommended upper exposure limits to humans?

7.17 An operating cell phone may use one watt of power to generate transmission from its antenna. The antenna is generally no more than 50% efficient, and when placed next to your head, the antenna's efficiency drops to about 25%. About 40% of the radiation enters the head. If the brain tissue absorbs some of this radiation, estimate the intensity of the radiation entering the brain, and the power deposited per kilogram. Compare with recommended upper limits for humans.

7.18 If visible light of wavelength 500 nm is reflected from a polished quartz eyeglass surface, estimate the number of silica molecules SiO_2 contributing to the coherently scattered waves. Take the density of SiO_2 to be $2.2\,\mathrm{gm/cm}^3$.

7.19 A camper exposed to the environment experiences a sudden cold front. She decides to enclose herself inside a sleeping bag and a thermally insulating sheet. She weighs 56 kg, and has a resting BMR of 90 W. If no heat can pass through her covering, how long before her temperature rises to a dangerous 41 °C from 37 °C? (In your estimate, you can use an average body tissue specific heat of 3.5 kJ/kg/°C.)

7.20 By accident, an astronaut becomes separated from her spacecraft, carrying only her LED flashlight. She remembers from her biophysics course that light carries momentum. She is 10 m from her spacecraft, and not moving relative to it. Her mass is 40 kg. Her flashlight, of mass 0.25 kg, produces a directed beam with a brightness of 4500 lumens (50 W) over a cross-sectional area of $75\,\mathrm{cm}^2$. What force will the flashlight exert on her when she points it away from the spacecraft, and how long before she gets back to it. (Of course, she could get back quicker if she threw the flashlight, but then if she were a little off in direction, she would have no recourse.)

Chapter 8
Ionizing Radiation and Life

All the good experimental physicists I have known have had an intense curiosity that no Keep Out sign could mute. Physicists do, of course, show a healthy respect for High Voltage, Radiation, and Liquid Hydrogen signs.

—Luis Alvarez

Summary Ionizing radiation includes all electromagnetic waves from ultraviolet light to gamma rays, for the simple reason that the localized interaction of light with matter is by photons with energy proportional to frequency. With frequencies below ultraviolet light in the spectrum, the photons have insufficient energy to break typical organic molecular bonds. Subatomic particles moving with high kinetic energy can also be ionizing, and therefore potentially damaging to organic tissue. These realizations make radiation dosimetry an important topic. On the positive side, mild radiation facilitates genetic mutation, necessary for evolution, and strong radiation can be used in therapy to purposely kill aberrant cells.

8.1 Detection of Ionizing Radiation

There are a wide variety of techniques for detecting ionizing radiation, corresponding to the various ways particles lose energy while traveling through materials, including solids, liquids, and gases.

Each of the following schemes have found wide use in detecting ionizing radiation:

- *Photographic materials*: Henri Becquerel discovered radioactivity by the effect that uranium salts had on photographic film. The basic idea is this: Some compounds form structures that are changed by exposure to light. A silver bromide crystal is such a compound sensitive both to visible light and to light of higher frequencies. An incoming photon converts a $AgBr$ pair into native silver and bromine. The silver atoms, with a large quadrupole moment from its d-shell electrons, can migrate through the crystal with little activation energy (about

W. C. Parke, *Biophysics*, https://doi.org/10.1007/978-3-030-44146-3_8

0.3 eV), and accumulate on the surface of the crystal (where they are more stable than in the bulk crystal). In the photographic process, a silver bromide emulsion is prepared with micrometer or smaller grains of $AgBr$ salt crystals in a gelatin that is spread in a dark room on a surface such as glass. After exposure to radiation, the film is placed in a 'developer' solution to release the metallic silver from the exposed crystals. After development, a 'fixer' solution (sodium thiosulfate) dissolves the unexposed silver bromide, and the film is then washed. Now the film will have black silver where light has reached the film, and be clear where it has not. On 8 Nov 1895, Wilhelm Röentgen noticed that barium platinocyanide gave off a faint flickering light near a Crookes tube.[1] He named the suspected unknown emissions from the Crookes tube 'X-rays' ('X' for unknown). Experimenting with the discovery, he soon found that X-rays would penetrate human tissue and bone, exposing a photographic plate behind. (See Fig. 8.1.)

- *UV detection*: UV can be converted to visible light using a fluorescence material such as a rare-Earth phosphor; solid-state photodiodes (e.g. AlN, diamond, Ga_2O_3) are capable of changing resistance in the presence of UV light; dyes exist which change color on exposure to UV.
- *Thermoluminescent devices*: LiF absorbs radiation by changing structure; heating the exposed crystal releases light (first describe in 1663 by Robert Boyle).
- *Fluoroscopic screens*: A phosphorescent or fluorescent screen is a flat surface covered with a material which emits visible light when UV waves, X-rays, or gamma rays impinge on the material. Images are produced by having the variations of the radiation field proportional to variations of radiation absorption

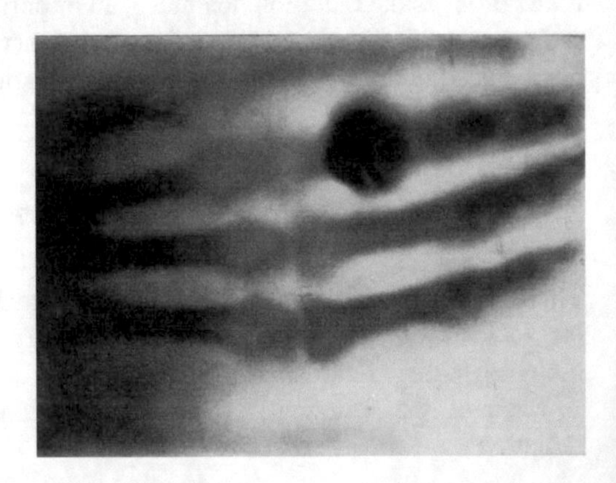

Fig. 8.1 First 'medical' X-ray image: Röntgen's wife's left hand, with wedding ring, in December 1895

[1]A Crookes tube is a partially evacuated glass enclosure containing an anode and a cathode at opposite sides of the tube. A high voltage (typically 10 kV) is connected across the electrodes. Gases inside the tube between the cathode to the anode light up in various colors. In 1897, J.J. Thomson showed that the ionization of the gas in the tube is caused by an as yet undiscovered particle with negative charge and a mass about eighteen hundred times less than a hydrogen atom. These particles we now call electrons. Röentgen's X-rays are produced by the collision of these electrons with the metal anode.

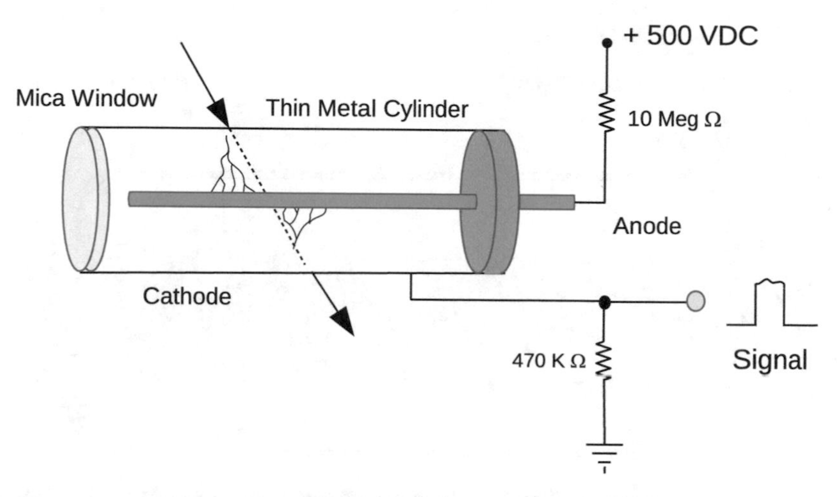

Fig. 8.2 A representation of a Geiger-Müller tube: a passing ionizing particle causes an avalanche of electrons due to a strong inward radial electric field on the interior gas

within some object. If a fluorescent material is used, then the image will persist for a time dependent on the material's fluorescent half-life.

- *Simple dosimeters*: Ionizing radiation can be detected by the use of 'film badge dosimeters', by passive electroscopes, or by other electrostatic discharge devices, such as a sealed chamber with charged polyethylene balls.
- *Ionization counters*: Geiger and Müller constructed a device (depicted in Fig. 8.2) which can count individual events that occur by the effects of ionizing radiation after entering a partially evacuated tube. The tube operates by letting the ionizing radiation produce a charged track in a sparse gas within the tube. An electric field from a thin conducting material on the outside cylinder of the tube to a central wire anode causes the ions and electrons in the track to accelerate in opposite directions. They collide with other atoms, causing a cascading effect, and a measurable current pulse between the anode and cathode. (Too large a maintained electric field will cause the ion track to become a continuously ionized trail.)
- *Scintillation detectors*: A scintillation detector responds to ionizing radiation by emitting visible or near visible light. For examples, NaI doped with Thallium will emit visible light in response to gamma rays. Zinc sulfide is a scintillant which responds to X-rays. Polystyrene will fluoresce under UV light. Tetraphenyl butadine gives light by interacting with alpha particles. Crystals of Cesium Iodide scintillates when protons or alpha particles penetrate. Lithium Iodide will emit light on exposure to a neutron beam. The emitted light can be measured by a photosensitive semiconductor or a photomultiplier for high sensitivity.
- *Photomultiplier tubes*: These are vacuum tubes used to detect photons. Within, a series of positively-charged plates are configured to amplify the effect of a single photon on the first plate. The primary low-energy photon, (or secondary photon made in a scintillant by a high-energy photon) is used to release an electron,

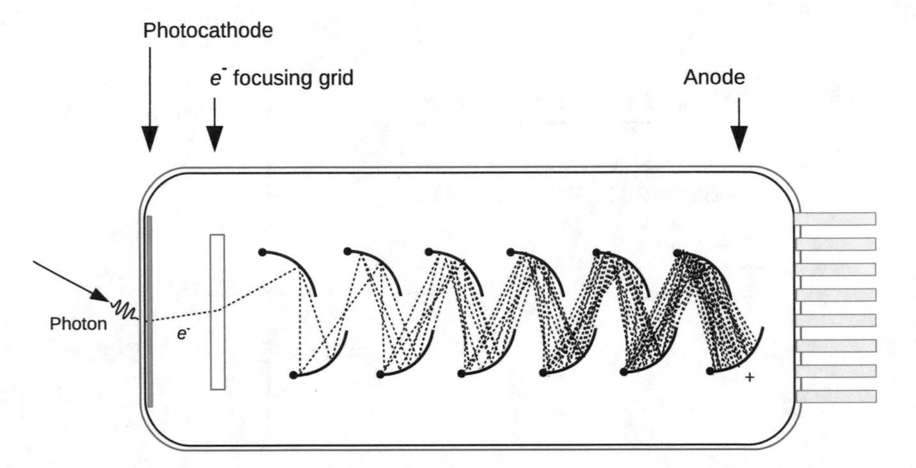

Fig. 8.3 Workings of a Photomultiplier Tube: a photon enters into an evacuated glass chamber, hitting a photosensitive surface which emits an electron. The electron is focused toward the first metal 'dynode' which carries a positive charge. The accelerated electron hits the dynode with enough energy to release several more electrons. The process continues as a cascade until a significant current is produced at the last plate, the anode. The various metal parts are connected to exterior metal pins shown on right. The pins to successive dynodes from left to right are put at increasing electric potential

by the photoelectric effect. That electron is accelerated by the electric field between the first plate ('dynode') and the second, so that by the time the electron reaches the second dynode, it has enough energy to knock away a number of other electrons. These, in turn, are accelerated to a third dynode, and so forth, producing a cascade effect. The result is a measurable current flowing from the last dynode, generated by a single photon. (See Fig. 8.3.)

- *Crystal color centers*: A crystal exposed to a fast-traveling particle or electromagnetic radiation may develop color centers, location where a negative ion has been pushed aside, leaving a trapped electron, or a positive ion is missing, leaving a positive 'hole'. These defects may absorb visible light at various frequencies, making what was a clear crystal colored.
- *Particle track detectors*: Bubble chambers, cloud chambers (Wilson, 1911), spark chambers. A pure liquid with no surfaces on which bubbles can grow can be heated just a little beyond its boiling point, in a condition called superheated.[2] An ionizing particle passing through the liquid can cause bubbles to form along its track. The track can then be photographed. Inversely, a vapor can be cooled to a temperature just below the condensation point. A passing charged particle can cause and leave a track of small droplets. Tracks of a passing ionizing particle

[2]Even water heated on a stove in an unscratched glass pot can be superheated, so that if a spoon is dipped in, it may cause an explosive burst of steam bubbles.

are also formed in a gas put in a strong electric field. The tracks are generated by a cascade of electrons following the ionization path of the particle, i.e. a spark.

- *Charge-discharge quartz-fiber dosimeters*: In these devices, a gold-plated thin quartz fiber is used as one electrode in an electroscope. Charging causes the fiber to bend away from a fixed electrode electrically connected to the gold on the fiber. Ionizing radiation discharges the device.
- *Neutron detection*: Neutrons can be detected using the absorption of neutrons in light nuclei with the emission of a detectable alpha particle, such as in the reactions $^{10}B(n,\alpha)^{7}Li$ and $^{6}Li(n,\alpha)^{3}H$. The gas BF_3 in a tube will absorb slow neutrons. Fast neutrons can be slowed in graphite and then detected. (Graphite is a good moderator of neutrons because ^{12}C nuclei can easily scatter neutrons, but will not easily absorb them.)
- *Plastic tract-etch detectors/recorders*: An energetic charged particle passing through a clear plastic can leave a track of defects. Etching the plastic can make these tracks visible. The tracks through the helmets of astronauts have been used to gauge radiation exposure.[3]
- *Cherenkov detectors*: A charged particle can move through a material faster than the speed of light in that material. In doing so, an electromagnetic shock wave is created, generating light.

Electromagnetic radiation with frequencies above about 10^{15} Hz will contain photons capable of breaking the bonds in many organic molecules. Just as in rain, for which the chances that two drops to hit your nose at once is quite small, the chance that an individual molecule absorbs the energy of two photons at once is also small compared to the chance for a single hit, except for extraordinary intensities of light. The discrete nature of light absorption means that a molecule is damaged not by the buildup of the light wave energy, but rather by individual hits of photons. A single photon with enough energy will suffice.

8.2 Characterizing Ionizing Radiation

Radioactive substances occur naturally, but also are manufactured. Background radiation, which also includes secondary cosmic rays, affects biological systems and even plays a role in the rate of biological mutation and therefore evolution. Man-made ionizing radiation is used in medical diagnostics and therapy, as well as in power devices and other technologies.

[3]See Comstock et al., *Cosmic-ray tracks in plastics: The Apollo helmet dosimetry experiment*, **Science**, Apr 9 1971. They find that 8–60 cells in the cerebellum per million would be killed by cosmic radiation after spending a year in space near the Earth.

Measures of Radioactivity

Radiation from radioactive materials can be characterized in a variety of ways, indicated below.

- *Activity* (A): The activity of a radioactive substance is defined to be the number of radioactive disintegrations per unit time. One curie (1 Ci) is the activity of 1 gm of radium, or 3.7×10^{10} disintegrations per second. A becquerel (Bq) is one disintegration per second. Since the disintegrations from nuclei to nuclei within a substance are uncorrelated, $dN \propto Ndt$, or

$$N = N_o \exp(-t/\tau)\,.$$

  Activity at any time is then $A = (N_o/\tau)\exp(-t/\tau)$. The number τ is called the 'lifetime' of the radioactive substance. In a time τ, $1 - 1/e \sim 63.2\%$ of the substance has been transformed into daughter products. Alternatively, $N = N_o\,(1/2)^{t/t_{1/2}}$, where $t_{1/2}$ is called the 'half-life' of the substance, since in each time $t_{1/2}$, half the nuclei initially present have disintegrated. The half-life is related to the lifetime by $t_{1/2} = \ln_e(2)\,\tau = 0.693147\,\tau$.

  If a radioactive material is ingested, its lifetime in the body will be determined not only by its physical lifetime, but also by its metabolic lifetime, i.e. the time for $1 - 1/e$ of the substance to be expelled from the body after it is digested and bound to tissue. The two rates add, so that $N = N_o\,(1/2)^{t(1/t_p+1/t_m)}$, where t_p is the physical half-life, and t_m is the metabolic half-life for the material as if it were not radioactive.
- *Exposure* (X): Exposure measures the ionization charge produced by a radioactive substance per unit mass. One 'röentgen' (1 R) was originally one electrostatic unit per cc of dry air at STP. This is equivalent to 2.58×10^{-4} coulombs per kilogram.
- *Absorbed Dose* (D): The Absorbed Dose (also called the Physical Dose) is defined by the energy deposited by radiation per unit mass. One 'rad' is defined as the deposition of 100 ergs per gram. One 'gray' (Gy) is one joule deposited per kilogram (100 rads).
- *Equivalent Dose* ($H = QD$): The Equivalent Dose is the physical dose multiplied by a "Quality Factor" Q, and is measured in 'rem', standing for 'Radiation-Equivalent-Man'. The factor Q is selected so that QD will have the same physical radiation damaging effect as the damage caused by X-rays with the same initial energy, whether the radiation consist of electrons, protons, alpha particles, or other particles. For example, Q is about 20 for 1 MeV alpha particles, indicating that 1 MeV alpha particles leaving a physical dose of 1 Gy will produce the same physical damaging effect as X-rays at 1 MeV whose intensity and duration leaves a physical dose of 20 Gy, reflecting the fact that α particles are far more effective in leaving a trail of ions than X-rays at the same energy and intensity. (See Fig. 8.4).
- *Relative Biological Effectiveness*: The RBE of a given radiation replaces the older Quality Factor Q. RBE is defined by the ratio of absorbed dose of the

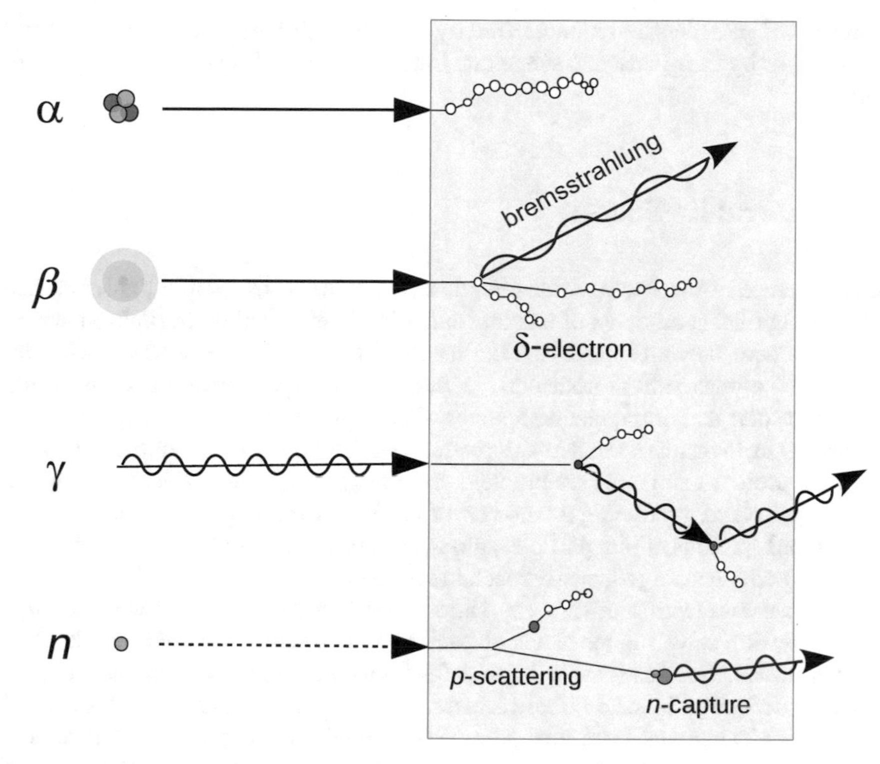

Fig. 8.4 Ionizing radiation impinging on matter, showing energetic alpha ($^4He^{++}$), beta (electron), gamma, and neutron particles entering material leaving a trail of destruction. A delta electron is simply a secondary scattered electron

given radiation divided by the absorbed dose of photons which cause the same biological damage at the same energy per particle. Protons and charged pions have a RBE of about 2. Alpha particles have a RBE of about 20.

- '*Dose Equivalent*': The product of the absorbed dose D (in Gy = joule/kg) and the RBE is the Dose Equivalent, DE, measured in a unit called the 'sievert' (Sv). The older (but still widely used) unit, 'rem', is taken to be 10 milli-sievert (mSv).

Figure 8.4 shows the typical ionization tracks left by energetic alpha, beta, gamma, and neutron particles passing through materials, including tissue. Because the alpha particle has two positive charges (protons), it leaves a trail of ionization denser than that produced by the other three types of radiation with the same initial energy. The beta particle (a fast-moving electron) can kick out other electrons (these 'secondaries' are called 'delta electrons') and also produces gamma rays as 'Bremsstrahlung'). Gamma rays have a much greater penetration depth, scattering off nuclei, which then move through the material leaving an ionization trail. Neutrons scatter from protons (hydrogen nuclei), which then produce an ionization trail, or are captured by nuclei, emitting gamma rays.

Beams of particles are characterized by their intensity (energy per unit area per unit time) or by their particle fluence rate (number of particles per unit area per unit time).

8.3 Natural Radiation

At the surface of the Earth, solar radiation is dominated by visible light, but also contains significant amounts of infrared and ultraviolet radiation. In addition, above the atmosphere, the Sun exposes the Earth with X-rays and a solar wind made mostly of a plasma of energetic protons and electrons. Much higher-energy particles come from extrasolar and extragalactic sources. Our atmosphere and our magnetic field protects us from much of the damaging effects of these ionizing radiations. The very energetic protons can penetrate the field, but then get inelastically scattered by the upper atmosphere, producing a 'shower' of particles with less energy, including X-rays, muons, protons, alpha particles, pions, electrons, positrons, and neutrons. The neutrons, being uncharged, easily reach the ground.

The dose rate from a cosmic-ray shower varies with height and with latitude. Airline crews receive higher doses of background radiation per year than nuclear power workers. The dose rate is highest for flights near the polar regions, because magnetic deflection is least effective there. As a way of comparison, the sea level annual dose to humans from cosmic rays is about 0.30 mSv per year. The annual dose, including all natural sources, averages to about 3.4 mSv per year. Thus, cosmic sources at sea level are only about 1/10th the natural background. An abdominal CT Scan gives a dose of 6–20 mSv. A single flight from London, England, to Abilene, Texas, can make a cosmic-ray dose of over 0.03 mSv.

The natural background includes radon inhalation ($\lesssim$2.3 mSv/yr), terrestrial radiation ($\lesssim$0.5 mSv/yr), and ingested radiation, such as Potassium-40, Thorium-232, and elements from the uranium series (0.29 mSv/yr). Radon-222 and, to a lesser extent, radon-220, are the largest contributors to the natural background radiation. They come as a decay product of Uranium-238, Radium-226, and Thorium-232 radioactivity within the Earth. Being an inert gas, radon seeps to the surface.[4] A square kilometer of soil down 40 centimeters contains about one gram of Radium-226, a decay product of uranium. The Radon-222 gas released has a half-life of only 3.8 days, but it is continually replenished, and is denser than air. Radon accumulating in basements with poor ventilation is an increased hazard to people living there. If inhaled into the lungs, it will emit alpha particles and leave other radioactive nuclides such as polonium, also an alpha emitter, lodged in lung tissue. After smoking, radon is the second leading cause of lung cancer (causing about 20,000 deaths per year in the United States).

[4]Even though uranium atoms are heavier than iron, uranium oxide complexes are less dense, concentrating in the lower mantle of the Earth and carried to near the surface by magma flow.

8.4 Ultraviolet Radiation

Ultraviolet (UV) radiation consists of electromagnetic waves above the visible-light frequencies of the normal human perception but below X-rays. Some animals can see UV, including some birds, bees, butterflies, and even arctic reindeer. 'Soft ultraviolet' wavelengths are usually taken from the end of the violet colors, 400 nanometers, to about 200 nanometers. A frequency in the region 1.5×10^{15} Hz to 3×10^{16} Hz corresponds to 'hard ultraviolet' radiation, ranging in wavelength from 200 nanometers to about 10 nanometers, where X-ray wavelengths begin. The UV spectrum is also divided into UV-A: 400–315 nm; UV-B: 315–280 nm; and UV-C: 280–100 nm. (See Table 8.1.) UV-A penetrates skin a few cell layers deep, and triggers the body's defense mechanism to produce UV-absorbing melanin. UV-B does not penetrate as far as UV-A (hardly reaching the dermis through the epidermis), but is more likely to cause cell damage, including cancer. UV-C is ordinarily stopped by the ozone layer in the upper atmosphere.

Suntan is a reaction of the skin to UV radiation, causing skin cells to generate the brown pigment melanin. Melanin will block UV radiation, converting the photon energy to heat. Skin cells also have repair mechanisms in response to UV free-radical formation and DNA damage. However, some damage is irreparable, and the cell may die, or worse, become cancerous.

The danger of ultraviolet light to living cells is demonstrated by its use as a germicide. The light for this purpose commonly comes from a discharge tube made from quartz (which will pass some UV light; ordinary glass does not) and filled with argon and mercury vapor through which an electric current is passed.

Individuals who have had an eye lens removed (because of cataracts, for example) are able to see some UV light. Claude Monet at age 82 had an operation to remove the lens of one eye. Thereafter, he could see a blue ting to what appeared to him before as white flowers.

Table 8.1 Names and ranges for ionizing electromagnetic waves

	λ (nm)	f (Hz)	E	λ (nm)	f (Hz)	E
Near UV	400	7.5×10^{14}	3.10 eV	300	1.0×10^{15}	4.13 eV
UV-A	400	7.5×10^{14}	3.10 eV	315	9.5×10^{14}	3.94 eV
UV-B	315	9.5×10^{14}	3.94 eV	280	1.1×10^{15}	4.43 eV
Mid UV	300	1.0×10^{15}	4.13 eV	200	1.5×10^{15}	6.20 eV
UV-C	280	1.1×10^{15}	4.43 eV	100	3.0×10^{15}	12.4 eV
Far UV	200	1.5×10^{15}	6.20 eV	121	2.5×10^{15}	10.2 eV
Vacuum UV	200	1.5×10^{15}	6.20 eV	10	3.0×10^{16}	124 eV
Extreme UV	121	2.5×10^{15}	10.2 eV	10	3.0×10^{16}	124 eV
UV-X	100	3.0×10^{15}	12.4 eV	88	3.4×10^{15}	14.1 eV
Soft X-rays	10	3.0×10^{16}	124 eV	0.01	3.0×10^{19}	124 keV
Hard X-rays	0.01	3.0×10^{19}	124 keV	0.0001	3.0×10^{21}	12.4 MeV
γ rays	0.0001	3.0×10^{21}	12.4 MeV	$<10^{-4}$	$>3 \times 10^{21}$	>12.4 MeV

8.4.1 Production of Ultraviolet Light

Ultraviolet light (UV) is typically made by atomic transitions in atoms such as mercury in fluorescent tubes or by heating gases to temperatures of above 4000 °C, such as in carbon arcs. If the inner surface of a fluorescent bulb is coated with a phosphor, the UV can be converted to visible light. Alternatively, if the tube is made of quartz glass, the UV radiation will pass out of the tube.

Very hot bodies, such as our white-hot Sun, produce ultraviolet light, and blue stars make proportionately even more. Carbon arcs are sometimes used by welders to melt metals. They also were once used in movie projectors to make intense white light. They consist of two metal-coated carbon rods which are touched together, letting a large current pass through. The current vaporizes the carbon which becomes hot enough to emit a significant fraction of its thermal radiation in the visible part of the electromagnetic spectrum, but also UV, causing nearby tissue to be sunburned.

A 'black light' is a light source which makes ultraviolet light in the range of UV-A. This so-called 'black light' is produced by passing current through mercury vapor in an inert gas, with a surrounding tube made of quartz. The quartz tube allows UV-A to pass through, unlike ordinary glass. The electrons in the current created between an anode and cathode collide and excite the gas atoms. Excited mercury atoms will emit light at wavelengths of 365.4 nm in the UV-A spectrum, and of 184.45 nm and 253.7 nm in the UV-C spectrum,

UV-C is sometimes called 'germicidal UV', since bacteria are killed by exposure to UV-C. Air exposed to UV-C will generate ozone. This will be evident by the distinct smell accompanying ozone near an air purifier equipped with a Ultraviolet Germicidal Irradiation (UVGI) device. UVGI devices are also used to help purify water, although bacteria can survive if embedded in small particles in the water.

Light-emitting diodes are available which generate UV light at a variety of frequencies in the UV-A to UV-C ranges. They have application in finding fluorescent materials, such as tagged biomolecules, and in the quick curing (polymerizing) of dental resin-based composites.

8.4.2 Ultraviolet Light in Biology

Although humans cannot normally see UV, a wide variety of animals, including many birds, reptiles, and insects, can see and use ultraviolet light. Flowers have developed UV-reflecting pigmentations to attract insects.

Biologists can use phosphate and urine fluorescence (rat urine fluoresces yellow, human glows blue-green) to find animal trails. Fluorescent materials can be used as tracers and labels in studies of organic processes, even down to the molecular level.

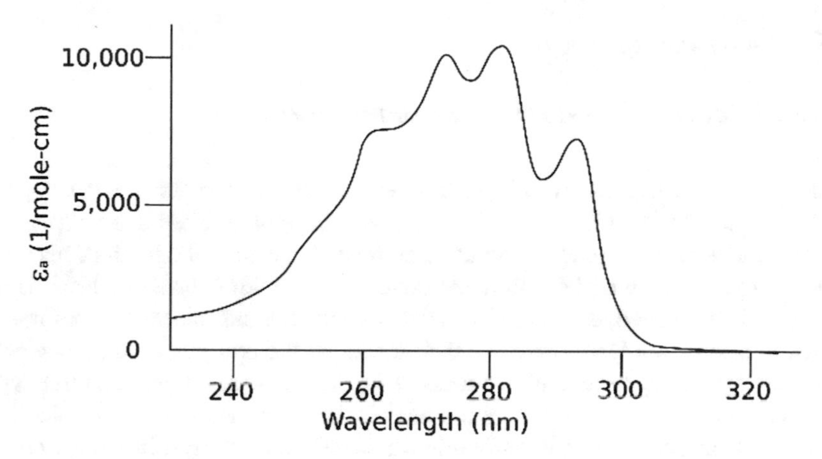

Fig. 8.5 UVB molar absorptivity of 7-dehydrocholesterol

As UV is ionizing, biological systems have evolved strategies to protect cells from deleterious chemical changes, including pigmentation and repair enzymes. UV-B can cause thymine dimers to form in DNA, disrupting faithful reproduction. It is likely that early prokaryotes could not have evolved into surface life on Earth until plants had produced enough oxygen in the atmosphere to generate an ozone blanket, absorbing over 96% of the UV-B and UV-C from the sunlight reaching the surface. Vitamin D_3 is produced in our skin via absorption of UVB by 7-dehydrocholesterol (Fig. 8.5).

Some applications of UV light in medicine and biology, arranged by wavelength, are given below, together with animal vision ranges:

240–280 nm	Disinfection (DNA absorption has a peak at 260 nm)
250–300 nm	Forensic analysis, drug detection
270–300 nm	Protein analysis, DNA sequencing
280–400 nm	Cell Imaging (Rayleigh limit $UV < visible$)
300–365 nm	Polymer curing (such as in dentistry)
300–320 nm	Light therapy in medicine
350–370 nm	Bug zappers (flies are attracted to light about 365 nm)

340–660 nm	Fish eye vision
360–700 nm	Reptile eye vision
300–650 nm	Insect eye vision
300–700 nm	Bird eye vision
400–700 nm	Human eye vision
5–30 μm	Reptile pit organ

8.5 X-ray Radiation

Natural Sources of X-rays and Gamma-Rays

X-rays were discovered by Wilhelm Röentgen in 1895 when he developed a photographic plate that had never been exposed to light, but had been placed near a high-voltage vacuum tube. A month later, he had made a radiograph of his wife's hand on a photographic plate, showing bones and a wedding band. In 1903, Ernest Rutherford collimated the radiation from radioactive substances by putting the source in a hole in a lead brick. He then separated the emissions using an electric field, and later a magnetic field. He was able to divide the radiation into three types: α, which he later showed to be helium nuclei, β, which he showed to be electrons, or positrons, and an uncharged radiation which he named γ rays. In 1912, Max von Laue, William Bragg, and his son Lawrence Bragg, deduced that X-ray radiations were electromagnetic waves by exhibiting their diffraction effects. Rutherford and Edward Andrade confirmed, in 1914, that gamma rays were electromagnetic waves. They showed that radium gamma rays form diffraction patterns in small-angle scattering from rock salt (NaCl) crystal surfaces.

X-rays and gamma rays are copiously produced by stars and violent events in the Universe, but our atmosphere largely shields life on the surface of the Earth from these astronomic sources. Radioactive elements in the Earth do expose life forms, but the level is tolerable, and perhaps even useful to evolution by producing mutations.

In the early 1950s, before good radiation standards, children who watched TV close to the high-voltage tube for energizing the Cathode Ray Tube (CRT) display screen were exposed to X-rays from electrons scattering within the tubes. In the late 1940s, a shoe store might have a machine that used X-rays and a fluoroscope for patrons to see how their shoes fit. Kids, instead, would look to see their foot bones wiggling.[5] Those same kids might have been wearing a radium-dial watch, with radium mixed in a phosphor painted on the numbers and hands, to make them glow. Workers who did the painting employed a fine brush, which they might shape by licking the tip. These workers had a higher-cancer rate than similar non-radium workers.

X-ray wavelengths are of atomic sizes, while gamma rays have wavelengths comparable to nuclear diameters, or less. With the rough rule of thumb that wiggles of naturally bound charges within regions of length L will likely generate/absorb electromagnetic waves with wavelength of the order of L, we should expect electrons in atoms to be good X-ray absorbers, and charges in nuclei to be good gamma ray absorbers. But given the small size of the nuclei compared to their separations in ordinary material, gamma rays can penetrate lead several centimeters thick.

[5] I was one of them.

Jiggling the K-shell electrons in atoms with relatively high Z tends to produce X-rays. Conversely, tissue which contains relatively high Z atoms will be better absorbers of X-rays compared to tissue which does not.

Gamma rays were found by Paul Villard in 1905 coming from radium. We now know that many radioactive materials emit gamma rays.

8.6 Man-made X-rays

8.6.1 X-rays Via Characteristic Emission

Among the many ways atoms radiate electromagnetic waves, electron transitions from a higher atomic shell (labeled by 'n') to the lowest-energy shell (the 'K' shell, for which $n = 1$) are a source of X-rays when the atom has sufficiently high Z (the atomic number and also the number of protons in its nucleus). This transition will occur spontaneously if the K shell is missing an electron. (That electron may have been kicked out by another passing energetic electron.) Among the possible atomic transitions from higher-energy shells to the K shell, the most likely is from the next higher shell, called the L shell, with $n = 2$. The emitted photon carries a unit of angular momentum. This means that a $2s$ to $1s$ transition requires the electron to flip its spin, a magnetic transition. Such transitions are far less likely than the much stronger electric dipole $2p$ to $1s$ transition. The released photon produces what is called the 'K_α' peak in the X-ray emission spectrum. The next most likely is a transition from the M shell to the K shell ($n = 3$ to $n = 1$), making the 'K_β' peak. The Table 8.2 shows the K_α and K_β energies and wavelengths for a selection of anode materials. (The conversion from energy to wavelength comes from $E\lambda = (hf)\lambda = hc = 1.23980\,\text{keV-nm}$.)

The energy released into a photon by atomic electron transitions must, by energy conservation, be in the amount $hf = E_n - E_1$. The energy of atomic electron orbitals are known experimentally, and, with sufficient patience, the energies E_n can

Table 8.2 Characteristic X-rays

X-ray characteristic emission					
Element	Z	$hf\ K_\alpha$ (keV)	$\lambda\ K_\alpha$ (nm)	$hf\ K_\beta$ (keV)	$\lambda\ K_\beta$ (nm)
Cu	29	8.048	0.15405	8.9053	0.139220
Ga	31	9.252	0.13401	10.2642	0.120790
Mo	42	17.50	0.07093	19.608	0.063228
Rh	45	20.21	0.61327	22.724	0.054559
In	49	24.21	0.05121	27.276	0.045454
Ag	47	22.16	0.05594	24.942	0.049707
W	74	59.32	0.02090	67.244	0.018437

be calculated in quantum theory, including the energy differences due to electron orbital angular momentum and spin motion. For now, we will estimate the energy from the Bohr model. The K-shell electrons have no angular momentum, and their spins are anti-aligned. Each will feel the full force of the nucleus except for the shielding effect of its mate. As for the effect of the atomic electrons spread uniformly 'above' the K-shell electron, Gauss' law tells us they will have a net zero force. Two factors weaken this Gauss' law argument: First, some outer electrons, namely those with no angular momentum (called 's-shell' electrons) do not stay outside the K-shell orbit; Second: They may not be uniformly distributed. The second point we will neglect, expecting the electron cloud will have a degree of spherical symmetry. The first point can be estimated by finding how often a higher shell s-wave electron gets closer to the nucleus than a K-shell electron. In 1930, Slater calculated such probabilities using simplified Gaussian electron wave functions. He reported the result by giving the effective nuclear charge, $Z_{eff} = Z - s$, that acts on the various electrons in a specific orbital. For the 1s electrons, $s = 0.3$. For $n > 1$ s or p electrons, $s = N_2 + 0.85\,N_1 + 0.35\,N_0$, where N_2 is the number of electrons in lower principal shells $n - 2$ and smaller, N_1 is the number of electrons in lower principal shells $n - 1$ and smaller, and N_0 is the number of other electrons in the same shell n. With one K-shell electron missing, the Bohr model energy for the remaining $1s$ electron gives

$$E_1 \approx -Z^2\alpha^2(m_ec^2)/2 \approx -13.6\,\text{eV}\,Z^2 \tag{8.1}$$

where $\alpha = e^2/(\hbar c) = 1/137.036$ is the fine structure constant (in cgs units), and $m_ec^2 = 511.0\,\text{keV}$. The L shell will have $2n^2 = 8$ electrons, six in p orbitals and two in s orbitals. To make a photon (spin one), transitions from $s \to s$ are suppressed, so the transition is likely from $2p$ to $1s$. With one $1s$ electron missing, the $2p$ electron ($n = 2, l = 1$) will feel the effect of the nuclear Coulomb charge with $Z_{eff} = Z - 0.85 - 0.35 * 7 = Z - 3.3$, so

$$E_2 \approx -(Z - 3.3)^2\alpha^2(m_ec^2)/(2(2)^2) \approx -13.6\,\text{eV}\,(Z - 3.3)^2/4 \tag{8.2}$$

Our estimate for the K_α X-ray energy will be

$$hf \approx 13.6\,\text{eV}\,Z^2\,(1 - (1 - 3.3/Z)^2/4)\,.$$

Taking $hf = 124\,\text{eV}$, we see that soft X-rays will be produced for Z as small as three (lithium). A tungsten target ($Z = 74$) is predicted by this simple model to make the emitted K_α have an energy of about 57.5 keV, compared to the experimentally measured value of 59.3 keV.

8.6.2 X-rays Via Bremsstrahlung

X-ray machines are often constructed with a vacuum tube device for producing the X-rays. (See Fig. 8.6.) A hot cathode emitter produces a large current (one to a thousand milli-amperes) drawn to a water-cooled tungsten or molybdenum anode which is kept at 20 to 150 thousand volts relative to the cathode for precise periods of time (typically fractions of a second). The electrons from the cathode therefore gain 20–150 thousand electron volts of kinetic energy by the time they reach the anode. When they collide with the tungsten, some electrons are bent around the tungsten nucleus by Coulomb attraction. This produces high acceleration of the electron. The resultant radiation from the electron is called 'Bremsstrahlung', ('breaking radiation') and can be in the X-ray range of frequencies when the electrons have keV energies (Fig. 8.7). Some X-rays produced near the surface of the tungsten will be emitted and pass out of the tube. However, about 99% of the electron energy is lost to heat in the anode, so cooling of the anode, which may have to dissipate several kilowatts of energy, is critical. Many X-ray tubes in computer axial tomographic imaging devices and angiography use a rotating anode with a focused electron beam hitting one small spot (about 1 square millimeter) near the edge of the anode in order to make a 'point' source of X-rays. The heated spot on the anode can cool after it rotates out of the beam during one revolution. These tubes might require as much as 100 kW of power and heat dissipation.

8.6.3 X-rays Via Synchrotron Radiation

A synchrotron is a device to accelerate and contain moving charges by forcing them to travel in a vacuum on a cyclic path. Magnetic fields are used to deflect the motion

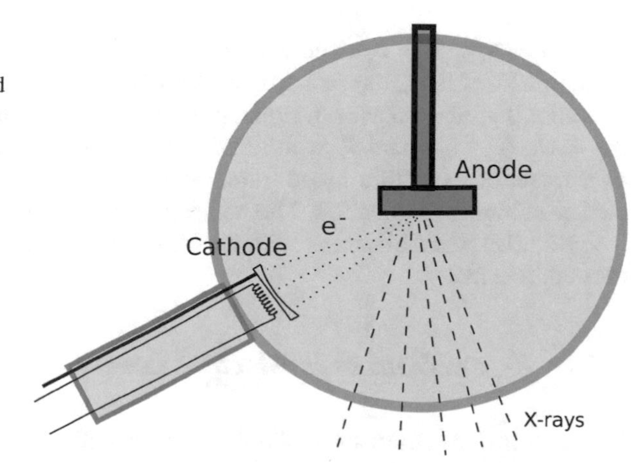

Fig. 8.6 X-ray tube elements: Evacuated tube, heating filament, cathode, and anode (possibly water-cooled). The anode is held at tens of thousands of volts relative to the cathode, causing electrons to accelerate from the cathode and collide with the atomic electrons and nuclei in the anode, producing X-rays

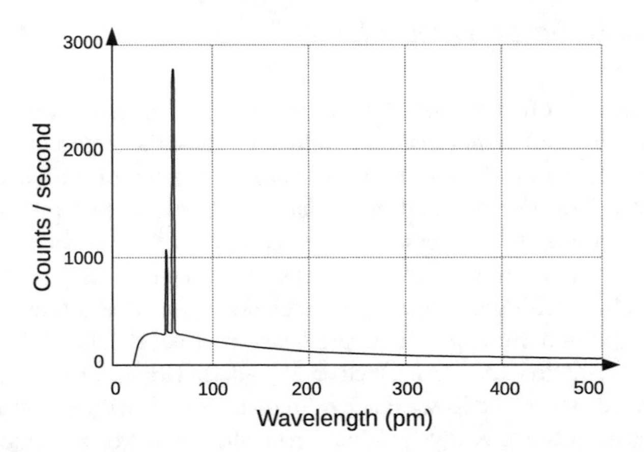

Fig. 8.7 Emissions from an X-ray tube with a rhodium anode operating at 60 kV: The horizontal axis, wavelength, is in units of picometers (10^{-12} m). The smooth part of the emission is from Bremsstrahlung, while the spikes are characteristic K-shell emissions from rhodium which occur at 61.33 and 61.76 pm, corresponding to $2p_{3/2}$ to $1s$ and $2p_{1/2}$ to $1s$ atomic transitions. Note: If all the energy of a 60 keV electron is released to an X-ray photon after hitting the tube anode, that photon has a wavelength of 20.7 pm, the threshold in the figure

of the charges into a curve, and electric fields, often in the form of a radio wave in a cavity, are used to accelerate the charges.

Synchrotron radiation occurs whenever a charge is forced to follow a curve. On that curve, the charge necessarily has an acceleration toward the center of the curve. We know from Maxwell that all accelerating charges radiate. The power radiated by each charge moving at the velocity **v** and with acceleration **a** in a curve is given by

$$\mathcal{P} = \frac{2e^2\gamma^4}{3c}\dot{\boldsymbol{\beta}}^2 \tag{8.3}$$

where $\boldsymbol{\beta} = \mathbf{v}/c$, $\dot{\boldsymbol{\beta}} = \mathbf{a}/c$ and $\gamma = 1/\sqrt{1-\boldsymbol{\beta}^2}$. When the speed of the charges is close to that of light, the radiation is strongly confined to a narrow forward beam of width $2/\gamma$, and exists as a signal pulse in a detector in a very short time, about $\Delta t = R/(c\gamma^3)$, where R is the radius of the curve. The radiation has a spread over frequencies, with a broad hump at about $1/\Delta t = c\gamma^3/R$ and then a sharp decline at higher frequencies. The hump frequency is in the low X-ray band (about 3×10^{16} Hz) when electrons with a kinetic energy above 230 MeV are bent in a curve of one meter.

8.6.4 X-rays from Free Electron Lasers

X-rays from characteristic emission, Bremsstrahlung, and synchrotron radiation are largely incoherent, i.e. each charge emits with little or no correlation to other emitters. But if we can make the emitters emit together, the light will be coherent, and also we will be able to produce a far more intense light. (If one emitter produces radiation with wave amplitude E_1, the incoherent intensity from N emitters will be proportional to NE_1^2, while the coherent intensity will be proportional to $N^2E_1^2$.) Lasers have coherent light because the emitters are stimulated to emit just as waves pass by. (See Sect. 7.23.) If we try to make an X-ray laser in the same fashion as a visible-light laser, we find that the X-rays are too strongly absorbed in the base material to get a lasting pulse. However, by producing bunched charges in a vacuum, accelerating them, and then back scattering light from a visible-light laser, the back-scattered light can be made to be Doppler-shifted into the X-ray bands. Because the incoming light was coherent, the back scattered light will also have coherence. These source of X-rays are called 'free electron lasers', although the X-ray radiation is not directly amplified.

Because the X-rays from free electron lasers are coherent and can be made with high intensities, the imaging of proteins and other molecules need no longer be restricted to those materials which can be ordered at the molecular level, such as in a crystal. Images of individual nanoscale particles in dynamical setting are now possible. But there is a caveat: With the needed intensities, and because individual X-ray photons transfer keV energies, each 'picture' you might capture effectively destroys the subject! However, you can learn a lot about what structures WERE there and how they might have been associating.

An estimation of the intensities you might require to illuminate a macromolecule follows: we will probably want a resolution of 1 nm. For a molecule of area $A = 1000\,\text{nm}^2$, that means we will want at least 1000 X-ray photons to be scattered or absorbed. The chance that a particle is scattered by a scattering center is measured by its 'cross section', σ. The cross section is the effective area around a scattering center perpendicular to the incoming beam such that if the scattered particle enters that area, it will be scattered into a given range of angles or it will be absorbed. Since K and L shell electrons resonate with X-ray photons for the high Z atoms, their atoms will be good scatterers as well as absorbers of X-rays. The probability that an X-ray photon gets scattered or absorbed will be about σ/A. The probability that N photons get scattered will be about $1 - (1 - \sigma/A)^N \approx 1 - \exp(-N\sigma/A)$. This number should get close to one by selecting N sufficiently large. This means we will want an intensity bigger than about $I \approx Nhf/\sigma\Delta t$ for an X-ray pulse of duration Δt. For carbon atoms, the cross section is of the order of half a kilobarn in the X-ray region, i.e. about $0.5 \times 10^{-7}\text{nm}^2$. This makes $N \approx A/\sigma > 10^{10}$ for our molecule with carbon as the principal scatterer/absorber.

8.6.5 Penetration of X-rays Through the Body

Because the incremental energy loss of an X-ray beam must be proportional to the incremental distance of penetration, the intensity of the radiation behaves with penetration distance as

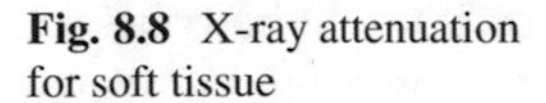

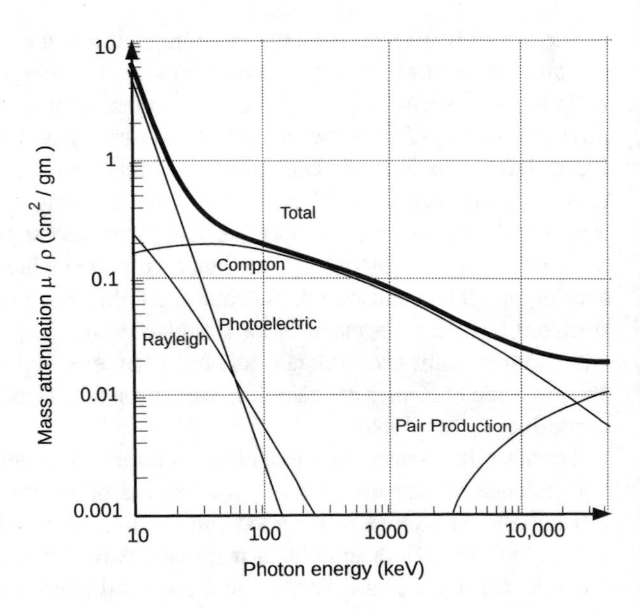

Fig. 8.8 X-ray attenuation for soft tissue

$$dI/dx = -\mu dx.$$

The quantity μ is called the attenuation coefficient.

For a homogeneous material, the penetration depth, defined by μ^{-1}, will not depend on x, so we will have

$$I = I_o \exp(-\mu x).$$

Figure 8.8 shows the mass attenuation (μ/ρ, where ρ is the material density) typical of soft tissue. The contribution from various ways in which scattering removes X-rays from the passing beam are shown. These include Rayleigh scattering, the photoelectric effect, Compton scattering, and e^+e^- pair production.

Alternative to the penetration depth μ^{-1} is the so-called Half-Value-Layer (HVL), which is the distance that reduces the X-ray intensity by a half. Putting $\exp(-\mu x) = (1/2)^{(x/HVL)}$ gives $HVL = 0.693\ \mu^{-1}$.

8.6.6 Research and Diagnostic Applications of X-rays

As we have indicated, X-ray beams have a variety of medical applications, including the study of macromolecular structure and the imaging of body organs.

X-rays produce diffraction patterns when scattered from a regular array of atoms and molecules because X-rays have wavelengths comparable to the distance between atoms and because X-rays are scattered by atomic electrons (DNA diffraction pattern is shown in Fig. 8.9). As the X-ray photons have energies higher than the binding energies of most electrons in atoms, the scattering can be approximately described by Thompson scattering, i.e. the elastic scattering of a photon off a free electron. Classical electrodynamics describes the same process as the effect of the light's electric field in jiggling charges at the frequency f of the incoming light. Those jiggling charges are forced to radiate at the same frequency f, causing outgoing wavelets from each atom.

X-rays as emitted from an X-ray tube have a wide spread of frequencies. Even so, Max von Laue (in 1912) showed that distances between layers of atoms in a crystal could be measured by the pattern of dots created on a photographic plate exposed to X-rays after a beam of X-rays was scattered by the crystal. William Lawrence Bragg then successfully reasoned that the diffraction pattern can be thought of as the interference of X-ray waves produced by their reflection from various layers of atoms in a regular array. If only one set of parallel planes of atoms dominate the X-ray scattering, then, as one can infer from Fig. 8.10, the scattered waves will have constructive interference at the angles given by arcsin $(n\lambda/(2d))$, where n is a positive integer, λ is the X-ray wavelength, and d is the separation distance between the plane of atoms.

In 1937, Dorothy Crowfoot Hodgkin used X-ray crystallography to reveal the structure of cholesterol. She went on to unveil penicillin (1946), vitamin B12 (1956), and insulin (1969). In such studies of large biomolecules, the challenge is to order or crystalized the substance in a configuration that matches its *in vivo* form, and then to decipher its X-ray diffraction pattern. The helical structure of DNA was revealed in 1952 by a diffraction pattern of a humidified 'B-form' of DNA stretched into fibers, crystalized and X-rayed by Rosalind Franklin's student Raymond Gosling. (See Fig. 8.9.) The structure was deciphered by Watson, Crick, and Wilkins in early 1953.

We will describe how X-ray diffraction patterns can unveil macromolecular structure. To display the essence of the argument, several simplifications can be

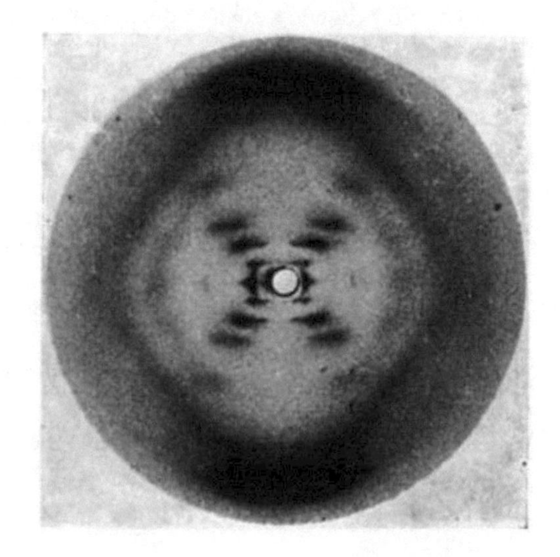

Fig. 8.9 The 1952 Franklin-Gosling X-Ray Diffraction Pattern of Hydrated, Stretched, and then Crystallized DNA (Imagefromwikipedia.org)

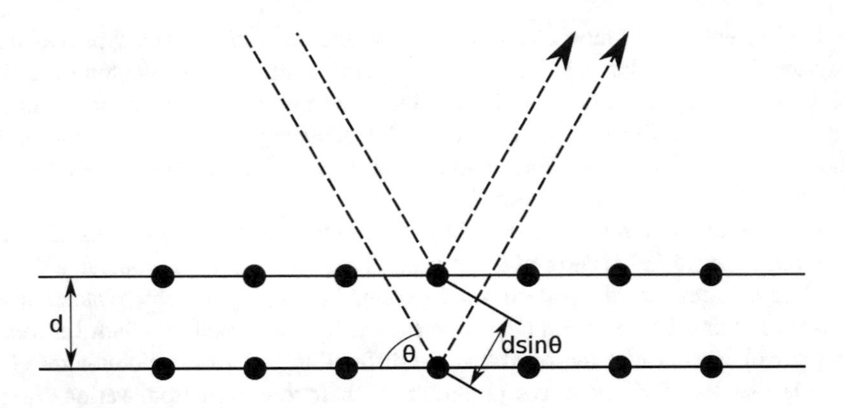

Fig. 8.10 Bragg planes in a crystal reflecting X-rays

made. First (1), explicit reference to the vector nature of the X-ray electric field will be hidden. Second (2), the incoming X-ray wave which passes through the sample will be taken as a plane wave with a single frequency. For convenience, we will use complex notion, so that the beam plane wave can be represented by $\phi_0 \exp(i\mathbf{k} \cdot \mathbf{r}_0 - i\omega t)$, where the X-ray beam is aligned along the **k**-direction with a wavelength $\lambda = 2\pi/k$ and a frequency of $f = \omega/(2\pi)$. Third (3), we will assume that the incoming X-ray is scattered elastically from essentially free electrons. The second assumption means all the waves in the sample have the same time dependence, namely, in complex notation, a factor $\exp(-i\omega t)$. At one location in space, we need only deal with the spatial factor that makes the wave. The third assumption will break down if there are high Z atoms in the molecular structure. The K and L shell electrons are strongly bound in high Z atoms, with binding energies comparable to the X-ray photon energies.

Now, because the atomic electrons in the sample elastically scatter the passing X-ray, each will produce a scattered wavelet radiating radially outward and adding to the electromagnetic field of the X-rays in the region. The strength of these wavelets will be in proportion to the strength of the passing X-ray wave from the beam. The amplitude of the net radiation arriving at a location **r** on a distant screen not in the path of the initial X-ray beam will be the sum of all the created wavelet amplitudes at that location on the screen:

$$\begin{aligned}\phi(\mathbf{r}) &= S\phi_0 \int \exp(i\mathbf{k} \cdot \mathbf{r}_0) \cdot \exp(i\mathbf{k}_0 \cdot (\mathbf{r} - \mathbf{r}_0))\, f(\mathbf{r}_0) d^3x_0 \\ &= S\phi_0 \exp(i\mathbf{k}_0 \cdot \mathbf{r}) \int e^{i(\mathbf{k}-\mathbf{k}_0)\cdot\mathbf{r}_0} f(\mathbf{r}_0) d^3x_0\,.\end{aligned} \tag{8.4}$$

Here, the factor S represents the probability amplitude that an electron will scatter the passing X-ray. The factor $f(\mathbf{r}_0)\, d^3x_0$ is the number of electrons within the volume d^3x_0 centered at the location $\mathbf{r}_0$ in the sample.

We now recognize that with $\mathbf{q} \equiv \mathbf{k}_0 - \mathbf{k}$ the integral in the second line of Eq. (8.4) is proportional to

$$\tilde{f}(\mathbf{q}) = (2\pi)^{-3/2} \int e^{-i\mathbf{q}\cdot\mathbf{r}_0} f(\mathbf{r}_0) d^3x_0 \,, \tag{8.5}$$

which is the three-dimensional Fourier transform of the electron number density. (See Appendix H for a discussion of the Fourier transform.)

Using the fact that a Fourier transform of a Fourier transform restores the original function (apart from a sign flip of its argument), we might hope to be able to use recorded diffraction patterns to recover the positions of the atoms. However, the recorded X-ray diffraction pattern has an X-ray intensity in proportion to the magnitude square of the Fourier transform of the electron number density in the sample material being X-rayed. The phase of the Fourier components in such a pattern will be lost.

There are a number of strategies for getting back phase information. These include (1) Replace certain atoms by others with different electron densities and that chemically do not significantly change the molecular structure and then make new images; (2) Replace certain molecular groups with similar ones and make new images. In each case, the difference between the original image and the new ones depends on changes in the phases, so that analysis now gives some phase information.

Each method is often accompanied by generating a set of reasonable molecular models, calculating what X-ray diffraction pattern they would make, and comparing the calculation with measured intensities on the measured diffraction. Also, each method is helped by collecting a larger amount of data from the X-ray scattering at different orientations of the sample. Except for simple crystals, the task can be daunting.

Sharp diffraction patterns encrypting the position of X-ray scattering centers require nearly monochromatic X-rays with some spatial phase coherence. In the case of X-ray tubes, near monochromaticity is achieved by a filter which only allows nearly one frequency (usually a strong emission in the spectrum) to pass but absorbs the others. For example, copper has characteristic K-shell emissions at 1.54 and 1.39 Å, while nickel has an absorption edge at 1.49 Å, so that nickel can be use to absorb the 1.39 Å X-rays from copper, but pass its 1.54 Å waves. Passing the X-rays through a small hole will produce X-rays with some phase coherence.

The advantage of using crystallized forms of a sample, or at least some repeating order of molecular groups, is that the superposition of low intensity X-rays of scattering from many 'unit cells' can build up the intensity in the resulting diffraction pattern.

One strategy for finding the tertiary structure of protein nanostructures which make up larger organelles, such as a bacterium flagellar motor, is to first break down the organelles into the component proteins, and then isolate, crystallize, and x-ray each crystal of protein units.

As noted in Sect. 8.6.4, with synchrotron X-ray generation and free-electron lasers, high intensity X-ray beams with near one frequency are available and can be used to study micro and nanoscale structures, with the realization that the intensities required may destroy the object being studied, so that a copy of that object has to be used each time a new image is needed.

8.6.7 Small-Angle X-ray Scattering

There are circumstances when we wish to get structural information at the nanoscale for biologically interesting molecules and systems of molecules in environments as close to their natural state as possible. By scattering X-rays of wavelengths near 0.15 nm with glancing angles from biological nanoscale objects distributed in solution, some structural information about the shape of these objects can be extracted.

Detection of small angle scattering is used to capture only those X-rays which were scattered by the surface of the objects, as the interest is getting at their size and shape. However, unlike X-ray crystallographic studies, now the detected X-rays come from a sum of scattered waves made by a large number of rather arbitrarily oriented identical objects in solution. (The contribution due to scattering from water can be subtracted.) By trying various configurations of the molecules in a theoretical calculation of the resulting X-ray scattering intensity as a function of the scattering angle and then comparing with experiment, the possible configurations which nature picks can be selected. In this way, the tertiary and quaternary protein structures in situ can be unveiled.[6]

8.6.8 X-ray Spectroscopy

X-ray Spectroscopy X-ray beams or electron beams, with energies in the keV range, and sent through materials, are preferentially taken up by electrons in the inner (K, L shells) of heavy atoms, kicking those electrons to a higher energy level or completely out of the atom. When another electron falls back into the 'hole' left by the kicked-out electron, radiation is emitted that is characteristic of that atom. Studying this radiation gives information about not only the presence of the atom, but about the environment around the atom affecting the electron orbital energies. For example, the stereochemistry of anti-cancer drugs can be studied by seeing how the drug makes attachments to cancer-active molecules.[7]

The emitted X-ray photon energies can be detected by various techniques, including Bragg diffraction from a crystal (giving the photon wavelengths) and

[6]See, for example, Jessica Lamb, Lisa Kwok, Xiangyun Qui, Kurt Andresen, Hye Yoon Park, Lois Pollack, *Reconstructing three-dimensional shape envelopes from time-resolved small-angle X-ray scattering data*, J Appl Crystallography **41**, 1046–1052 (2009).

[7]See Czapla-Masztafiak et al., *X-Ray Spectroscopy on Biological Systems*, IntechOpen (2017).

by semiconductors (giving photon energies). The result is called 'X-ray emission spectroscopy'.

With precise measurement of the emitted X-rays which came from transitions from a higher to a lower atomic level, the technique is called 'Resonant X-ray Emission Spectroscopy' (RXES).

If just the absorption by a material of the X-rays of particular wavelengths is measured, the technique is called 'X-ray absorption spectroscopy'.

Auger Emission Spectroscopy When a photon or electron beam excites or ionizes the K or L shell of a heavy atom, the electron that falls into the created hole, while falling, may interact with an outer electron, causing it to be ejected from the atom. This is called the Auger effect, discovered by Lise Meitner.[8] Because the lower-energy ejected electron has a short mean-free path in a solid material, one usually detects these electrons as they are ejected from the surface of the material placed in a vacuum.

8.6.9 Mammography

Mammography uses low-energy X-rays ($\approx$30 kV peak) to view the interior of the human breast, looking for potentially cancerous masses. The X-rays used in mammograms deposit up to 5 mGy per exposure, with an average of 4 mGy. Natural background is 2–3 mGy per year.

8.6.10 CAT Scans

A Computer Assisted Tomographic scan (CAT scan), also called a Computed Tomography (CT) scan, applied to humans, is a computerized axial tomographic view of body organs, tissue and bones. The idea is that the absorption and scattering of X-rays depends on the electron density around atoms and molecules and also on the presence of K-shell electrons in heavy elements. A selected heavy element is sometimes administered beforehand to increase the scan contrast in certain organs. By sending X-rays from one direction and confined to a plane which extends through the body, the absorption pattern in that plane can be measured by detecting the intensity of the transmitted X-rays. By rotating the direction of the X-rays (but keeping their plane), a sequence of such intensity patterns is recorded. From these, a two-dimensional cross-sectional density pattern can be constructed by a computer analysis. Combining the density patterns for a sequence of adjacent planes, a three-

[8]Lise Meitner, along with Otto Hahn and Otto Robert Frisch, discovered nuclear fission in 1939.

dimensional image can also be generated. This computer-generated 3-D image can be viewed from arbitrary angles on a two-dimensional display screen.

In a CT scan, differences in electron density less than 1% can be discerned. However, radiation doses may be 100–1000 times higher than conventional chest X-rays, and so involves an increased cancer risk. Exposure of a fetus to 10 mGy increases its risk of cancer by the age of 20 from 0.03 to 0.04%.

8.7 Gamma Rays

Gamma rays are made naturally by radioactive materials and by cosmic rays hitting the atoms in the upper atmosphere. They are made by humans in particle accelerators by synchrotron radiation, by the scattering of high-energy nuclear particles, by nuclear reactors, and by particle-antiparticle annihilation. In the following, we will focus on the application of gamma rays in biological research and medicine.

8.7.1 Positron-Emission Studies

Positron-emission studies takes advantage of the fact that two oppositely directed gamma rays are emitted by positron annihilation, so that the location of the emission region can be pinpointed. For example, in proton beam therapy against cancer tumors, a small number of radioactive nuclei can be produced. Their beta decay can produce positrons, which then annihilate with local electrons, producing two gamma particles moving in opposite direction (without much interaction with the surrounding tissue). Detecting these gammas lets one map where the energy deposit of the protons is concentrated.

Positron-emission tomography will be described in more detail in the Sect. 8.9.

8.7.2 Mössbauer Techniques

As an example of the use of gamma rays in biochemical research, we will describe how the Mössbauer effect can be employed in the investigation of organic molecular structure.

Because of their short wavelength (of nuclear dimensions), gamma rays are not emitted by electron transitions in molecules and atoms. But gamma rays can be emitted by the acceleration of charges in the nuclei of atoms. It is therefore remarkable that this kind of emission can be used to study the behavior of valence electrons in molecular bonds. However, even the valence electrons can briefly visit the interior of nuclei, those small heavy dots at the center of atoms, a few ten thousandth the diameter of the atom. The electrons which penetrate nuclei are the so-called s electrons, which have no angular momentum relative to the central nucleus. Due to those visits, the s valence electrons slightly change the energy levels of nucleons in the nucleus. Now suppose that nucleus is in an excited state, ready to decay, and give off a gamma ray. Then the energy (and therefore the associated frequency) of that emitted gamma ray will be affected by the s valence electrons, which, in turn, influence the chemistry of the molecule where that nucleus resides.

Gamma rays emitted from a nucleus of an atom have sufficient momentum to kick the nucleus backward. This causes the gamma ray to suffer a Doppler shift down in frequency. Rudolf Mössbauer discovered (in 1958) that the kick of the gamma ray from the nucleus of an atom bound in a solid might not cause the individual atom to recoil, but rather the recoil momentum is taken up by many atoms at once. This is a pure quantum effect, in that the bound atom must either remain at rest relative to its neighbors, or take up a fixed integer number of vibrational quanta ('phonons'). The phenomenon was important in Mössbauer's research in that the emitted photon, little diminished in energy, could be subsequently absorbed by an unexcited nucleus of the same kind as the emitter, but some distance away, while, because of the sharpness of the nuclear levels, the Doppler-shifted gamma ray is little absorbed.

Gamma rays from excited states in nuclei can have very sharply defined frequencies, typically a few parts in 10^{12}. For resonant absorption of those gamma rays, the incoming gamma must be within a very small frequency range. Given the Doppler shift on emission, gases do not show subsequent gamma absorption after emission. But, as Mössbauer demonstrated, solids can.

A good case is ^{57}Fe, which can be excited by a 14.37 keV gamma ray emitted from an excited state of ^{57}Fe created by the beta decay of ^{57}Co (Fig. 8.11).

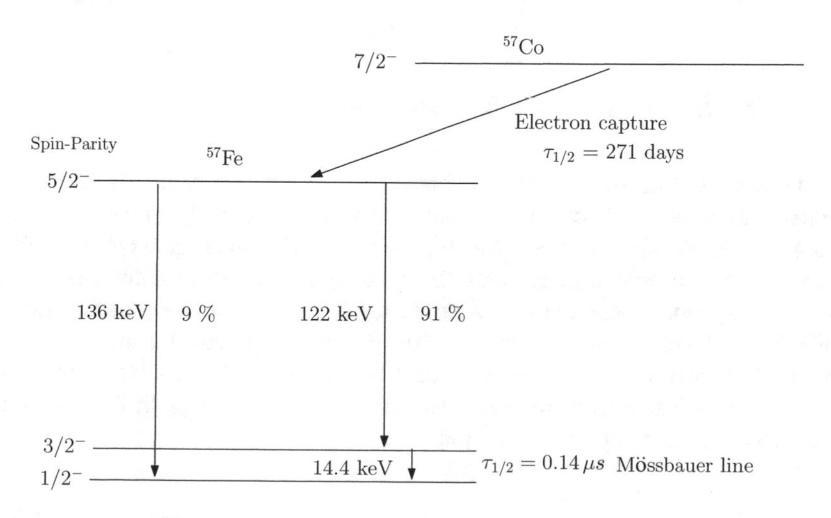

Fig. 8.11 Decay of ^{57}Co. Shown are the relevant nuclear energy levels in cobalt-57 and iron-57

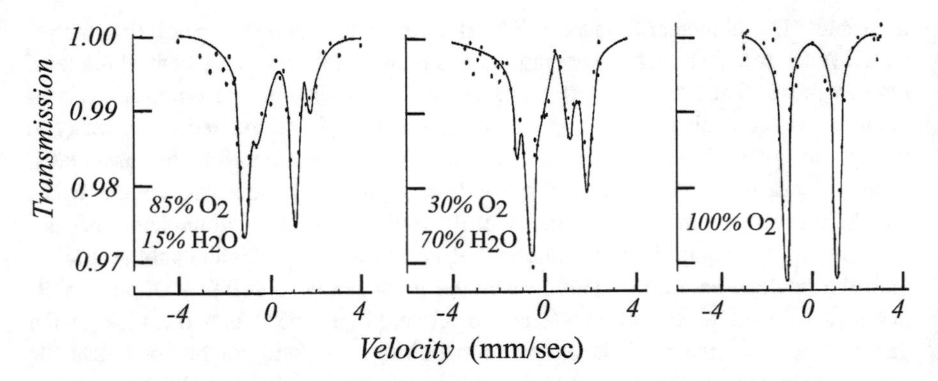

Fig. 8.12 Mössbauer γ-ray absorption by hemoglobin iron with different concentrations of O_2 ligand. Curves are a fit from theory (see U. Gonser and R.W. Grant, Biophys J **5** 823 (1965))

The absorption into this energy level of the iron nucleus shows a very sharp resonance behavior. The Mössbauer effect has been extensively applied to the study molecular structure through the effect that valence s-shell electrons have on nuclei. For example, the structure of some organic molecules which contain iron, such as hemoglobin, can be investigated this way. For hemoglobin, the valence s-electrons cause a slight change in the nuclear energy states in the iron atom. This change can be detected by giving a small speed to the source of gamma rays and therefore a slight shift in the energy of those gammas, because a Doppler shift in frequency also shifts the emitted photon energy.[9] Monitoring the absorption as a function of the Doppler speed determines the new absorption energies. Now, any change in the structure of the hemoglobin will, in principle, be detectable by watching changes in the energy of selected nuclear states of the iron! (See Fig. 8.12.)

8.7.3 Nuclear Resonance Spectroscopy

The development of strong tunable sources of X-rays and gamma rays via synchrotron radiation (see Sect. 8.6.3) means that gamma ray nuclear resonant absorption, as in the Mössbauer effect studies, can be used to investigate electron orbital states in organic molecules without the need of a radioactive source of gamma rays. With synchrotron radiation, a spectrum of frequencies of gamma rays are available, and with much higher intensities than is practical from radioactive sources. Typically, nuclear resonance spectroscopy uses time-resolved gamma-ray detection rather than energy-resolved detection, taking advantage of the short pulse width of synchrotron radiation (less than 100 ps).

[9]Amazingly, the source speed necessary can be produced by the small oscillation of a speaker coil.

8.8 The Electron Microscope

An electron microscope is a device capable of forming images at the nanoscale using the fact that electron waves can have a much shorter wavelength than visible light, and therefore allowing the microscope to have a much higher resolution, and the fact that electron beams in a vacuum can be focused with electric and magnetic fields (Fig. 8.13). The wavelength of an electron wave is given by the DeBroglie relation $\lambda = h/p$, where h is Planck's constant (6.626×10^{-34} Joule-s) and $p = mv$ is the electron's momentum. In both the optical and the electron-beam case, the resolution of an image formed by single-wavelength waves was given by Rayleigh as the smallest angle between two points on the object which can be resolved: $\Theta_R \approx 1.22\,\lambda/d$, with d the diameter of the focusing device. If f is the focal length of the focusing device, then the smallest size one can see will be of length $\Delta\ell \approx 1.22(f/d)\,\lambda$. (There is further discussion of the Rayleigh limit in Sect. 7.5.)

There are disadvantages. First, specimens should be dead, dried, and perhaps dressed with a gold coat,[10] before being placed in a vacuum. Second, electrons do not penetrate far in materials unless they carry energies which will break up molecules. Third, the electron microscope tends to be big and expensive in order to hold all the devices needed to make and sustain a good vacuum, and to generate, focus, and detect streams of electrons in that vacuum (Fig. 8.13).

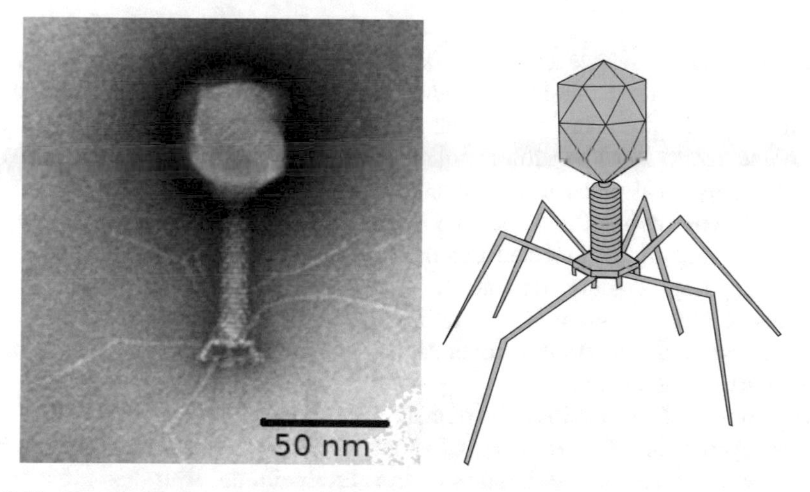

Fig. 8.13 Electron microscope image of a T4-Phage virus and model
Adapted from M. Wurtz, *Bacteriophage Structure*, Electron Microsc Rev **5 (2)** 283–309 (1992)

[10]The gold scatters electrons effectively, and keeps the specimen from building up a repulsive negative charge.

Electron microscopes can resolve objects as small as one ångströms (0.0001 μ), while a (UV) light microscope is limited to about 0.2 μ. As an example, to get a resolution of 10 Å, one would need electrons with an energy of the order of 2 keV.

If the specimen is cut very thin, an electron wave image might be generated by having the electron beam pass through the object. This technique is used in a 'transmission electron microscope'. Alternatively, the initial electron beam can scatter back from the object. The beam is often highly focused and then scanned across the specimen. This forms what is called a 'scanning electron microscope'.

In an electron microscope, the electrons are generated either by thermionic emission (boiling electrons off a hot metal, such as LaB_6-coated tungsten), or by field emission from the sharp tip of a cold cathode. However, field emission requires the chamber to have an ultra-high vacuum (less than 10^{-8} torr). Images can be created by having the electrons from the specimen hit a fluorescent screen, or through the analysis of signals from one or more solid-state detectors, such as a CCD, placed under a scintillator which converts the relatively high electron energies into light. (The electrons in the beam have energies in the keV range, which is too large to impinge directly on a silicon device, which only requires a few eV to release an electron, and would be damaged by keV electrons.)

8.9 PET Scans and Gamma Emission Tomography

When information is needed about body functions, such as specific metabolic processes, a 'Positron Emission Tomography' technique is one option when simpler options are ruled out. The idea is that positrons (anti-electrons) can be generated by the nuclear decay of specific isotopes. Almost immediately, those positrons are annihilated by local electrons, producing two gamma rays with energy of about 0.511 MeV sent in directions opposite to each other. Such gamma rays easily pass through tissue, and are detected external to the body by a set of detectors on a surrounding ring. Tracing the line of emission of one of these gamma rays (in coincidence with the second) will define a thin cylinder in the body where the emitting tissue must lie. Analysis over angles and along the tissue produces an image of the positions of the emitters.

The source of the positron in clinical applications is often ^{18}F within fluorodeoxyglycose, in effective doses of about 14 mSv. This compound accumulates in cells with a high glucose uptake, such as the brain, the liver, and most cancer cells.

Brain neuroimaging is effected by using ^{15}O, but with a half-life of only 2 min, the scan must be done adjacent to a cyclotron production facility.

Isotopes which are positron emitters and that are used in medicine are shown in Table 8.3. Carbon-11, Nitrogen-13, and Oxygen-15 are used in studying brain physiology and pathology, in particular for localizing epileptic focus, and in dementia, psychiatry and neuropharmacology studies. They also have a significant role in cardiology. Note that the very short half-life of ^{15}O means that a medical

Table 8.3 Some positron emitters used in medicine

	Product	Half-life	Max E_β	Application
^{11}C	^{11}B	20.4 m	0.96 MeV	Trace serotonin
^{13}N	^{13}C	10.0 m	1.19 MeV	Brain physiology
^{15}O	^{15}N	2.05 m	1.72 MeV	Brain physiology
^{18}F	^{18}O	109.6 m	0.635 MeV	Detect cancer
^{38}K	^{38}Ag	7, 6 m	2.68 MeV	Heart studies
^{52}Fe	^{52}Mn	8.275 h	0.8 MeV	Iron usage
^{73}Se	^{73}As	7.16 h	1.32 MeV	Sulfur tracing
^{124}I	^{124}Te	4.176 d	2.135 MeV	Thyroid uptake

cyclotron for the production of ^{15}O via $^{14}N + d$ must be nearby. PET scans with ^{18}F typically uses fluorodeoxyglucose (FDG) to monitor glucose uptake in cells. As cancer cells are rapid growers, FDG-PET is widely used in clinical oncology.

Scanners similar to PET that can detect single gammas are also used to generate images after ingestion of compounds labeled with a particular radioisotope. Technetium-99m is such an isotope, made from the decay of ^{99}Mo. The m indicates that ^{99m}Tc is metastable, decaying via gamma emission of energy 143 keV, with a half-life of only 6.0072 h. Compounds formed with ^{99m}Tc are used to image the skeleton and heart muscle in particular, but also for brain, thyroid, lungs (perfusion and ventilation), liver, spleen, kidney (structure and filtration rate), gall bladder, bone marrow, salivary and lacrimal glands, heart blood pool, infection and numerous specialized medical studies. Technetium-99m is the most common radioisotope for diagnosis, accounting for over 80% of such scans.

8.10 Radiative Bactericide

Ionizing radiation is effective in killing bacteria. Ultraviolet radiation is often used to reduce bacterial growth on public surfaces, in operation rooms, and in meat processing. Radiation in the wavelength range of 250–290 nm at intensities of 30 mW per square centimeter are typical.

To reach bacteria deeper than the surface, electron beam, X-ray and gamma radiation are used. Sterilization of milk and meat is done with a dosage of 25–70 kGy. Food that is sealed and then radiated have shelf lives much greater than otherwise.

Radioisotopes are also used in commercial sterilization procedures. As a gamma ray emitter, a ^{60}Co source serves this purpose. Strong gamma irradiation is effective against bacteria in surgical gloves, bandages, and in transplant materials such as bone and skin.

8.11 Radiotherapy

Radiotherapy is the use of implanted or ingested radioactive substances in treatment of disease. (In distinct contrast, radiation therapy uses external beams of ionizing radiation focused on diseased tissue or organs.) When the radiation source is placed near or within an intended target, the treatment is called 'brachytherapy'.

As an example of radiotherapy, consider cancer treatment of the thyroid with radio-iodine. Iodine-131 is a beta-particle emitter, and therefore damaging to cells, particularly those dividing. High ingested doses have been used to treat thyroid cancer. ${}^{131}_{53}I$ has a physical half-life of $\tau_{1/2} = 8.0198\,\text{d}$, decaying via beta and gamma emission according to

$$ {}^{131}_{53}I \rightarrow \beta(0.3338\,\text{MeV}) + {}^{131}_{54}Xe^{*}(7/2+, 0.636989\,\text{MeV}) \;\; 7.27\% $$

and

$$ {}^{131}_{53}I \rightarrow \beta(0.6063\text{MeV}) + {}^{131}_{54}Xe^{*}(5/2+, 0.364489\,\text{MeV}) \;\; 89.9\%. $$

The excited xeon subsequently decays into

$$ {}^{131}_{54}Xe(3/2+, 0\,\text{MeV}) $$

by the release of a γ. The biological half-life of ingested iodine is patient dependent, but 66–80 days is typical if a functioning thyroid is present. Thus, the physical half-life of 8 days means that physical decay predominates the loss of the isotope. During the first few weeks after ingestion, caution must be taken in the handling of urine and excretion.

Other examples of radiotherapy are shown in Table 8.4.

Table 8.4 Radioisotopes used in radiotherapy

	Product	Emits	Half-life	Max E	Application
${}^{125}I$	${}^{125}Te$	γ	59.4 d	35 keV	Ocular implant[a]
${}^{131}I$	${}^{131}Xe$	β, γ	8.02 d	606 keV	Cancer of thyroid
${}^{192}Ir$	${}^{192}Pt$	β, γ	73.8 d	1.46 MeV	Brain, breast implant
${}^{90}Y$	${}^{90}Zr$	β, γ	64.1 h	2.28 MeV	Liver via μ-spheres[b]
${}^{177}Lu$	${}^{177}Hf$	β, γ	6.73 d	497 keV	Imaging, targeted therapy[c]
${}^{153}Sm$	${}^{153}Eu$	β, γ	46.3 h	807 keV	Metastatic bone disease
${}^{32}P$	${}^{32}S$	β	14.3 d	1.709 MeV	Blood disorders
${}^{223}Ra$	${}^{219}Rn$	α	11.4 d	5.78 MeV	Targeted alpha therapy

[a]Diagnostic imaging
[b]Delivery of radioisotopes via impregnated microspheres called radioembolization [c]Radioisotopes attached to specific chemicals target specific cells

8.12 Rate of Loss of an Ingested Radioactive Isotope

Because a radioisotope can be expelled from the body, its effective lifetime in the body, τ_{eff}, may differ from its physical lifetime, τ_{phys}. The rate of loss must be the sum of the rate of expelling biologically plus the disintegration rate. Thus

$$\frac{1}{\tau_{\text{eff}}} = \frac{1}{\tau_{\text{phys}}} + \frac{1}{\tau_{\text{bio}}}, \tag{8.6}$$

where τ_{bio} is the biological lifetime, i.e. the time that $1 - 1/e$ of the substance would be eliminated if the isotope itself were stable. The same relation applies to the half-lives.

As an example, consider polonium ingestion, which is easily lethal. Polonium-210 is an alpha particle emitter, with a physical half-life of 138.38 days, and an approximate 50 day biological half-life. The emitted alpha particle has an energy of 5.30438 MeV. There is also likely a gamma ray emission with energy 803 keV but with low probability (less than 1%). After ingestion, 50% or more is eliminated in feces. 45% gets into the spleen, kidneys, and liver, muscles, and 10% into the bone marrow. Liver and kidney damage occur from alpha particle damage to cells. Victims often experience vomiting, diarrhea and hair loss, just as in radiation sickness. Polonium-210 in a dose of 5 Sv is lethal. This dose would come from the ingestion of just under a tenth of a microgram of Polonium.

It is the alpha particle emission that makes Polonium-210 highly radiotoxic if ingested or taken into the lungs. The 'radiotoxicity' (as opposed to chemical toxicity) of a substance refers to its potential capacity to cause damage to living tissue due to its radioactive emissions. To establish the radiation dose due to exposure to Po-210, one must differentiate between external exposure and internal exposure. Since Po-210 is mainly an alpha emitter, and these alphas do not penetrate the skin, the main damage is from internal exposure either by ingestion or by inhalation. Once in the blood stream, polonium disperses through the entire body giving rise to a whole-body radiation dose which kills or damages cells, tissues, and organs. To determine the biological hazard, the so-called effective dose coefficients should be used. These relate the biological hazard (measured in sievert, Sv) to the activity of at intake (in becquerel, Bq) of the radionuclide. The effective dose coefficients for intake of Po-210 (ICRP 72) are as follows: The effective dose coefficient for ingestion is 1.2×10^{-6} Sv/Bq; The effective dose coefficient for inhalation is 4.3×10^{-6} Sv/Bq. To obtain the dose from intake of Po-210 from ingestion of this material, the activity of intake (in Bq) is multiplied by the effective dose coefficient for ingestion. For a mass of Po-210 of 0.1 μg (corresponding to an activity of 1.7×10^{7} Bq), the dose is given by: 1.7×10^{7} Bq$\times 1.2 \times 10^{-6}$ Sv/Bq= 20 Sv/0.1 μg or 200 Sv/μg. This is a very high dose. To put this into perspective, a radiation dose of 5 Sv received over a short period will cause death in 50% of cases within 30 days (Lethal Dose, LD 50/30).

Mr. Alexander Litvinenko, a former Russian dissident died on the 23rd of November 2006 from purposeful polonium poisoning. In 1956 Irène Joliot-Curie, whose parents first isolated polonium, died from leukemia, attributed to chronic poisoning following the bursting of a capsule containing polonium 10 years earlier. Polonium-210 is produced by irradiating natural Bismuth-209 with neutrons in a nuclear reactor. The result is Bismuth-210, which beta decays into Polonium-210 with a half-live of 5 days. Polonium-210 is also present in cigarettes, being a daughter isotope of the decay of uranium in the soil. Polonium-210 may account for a large fraction of all smoking related lung cancers. When Polonium-210 is combined with Beryllium, it creates neutrons though the reaction $^{9}Be + \alpha \rightarrow {}^{12}C + n$. The alpha particles from micrograms of Polonium reduce static charge buildup on insulators. Polonium is even used in spacecraft for the heat it produces.

8.13 Mechanisms for Energy Loss by Light and Fast Moving Particles

8.13.1 Radiation Tracks and Trails

The Apollo astronauts reported seeing a flash of light every few minutes. By studying the tracks left in their plastic helmets, the flashes correlated with cosmic ray particles, and were likely produced by Cherenkov radiation within the eye.

The mechanism for the production of cosmic ray tracks can be understood by the electric interaction of energetic charged particles as they pass through substances, and by the action of gamma rays in producing such energetic charged particles.

As ordinary matter is made up of charged particles, an ionizing stream of particles passing through that matter will have a likelihood of inelastically colliding with stationary charges. This collision can deflect the incoming particle and kick one of the material's particles into motion, as well as produce secondary radiation and even create matter-antimatter pairs, also moving away from the scattering event. With sufficient incoming energy, a cascade effect occurs, with each of the secondaries engaged in their own sequence of scattering events. Side branches of a track are called 'spurs', often produced by secondary scattered electrons (called 'deltas', named to follow the sequence α, β, γ introduced by Rutherford for radioactive particles).

If the material is a solid, a series of tracks may be formed with atoms no longer occupying their original positions. The tracks, being of a different character than the surrounding unaffected material, can often be etched out to make them more visible. In a supercooled gas, the charges in the tracks may induce condensation, making a vapor trail. In a superheated liquid, the track maybe visible by induced gas bubbles.

In water, an ionizing beam of particles will make a trail of hydroxyl radicals (OH).

The distance of penetration of a moving charged particle until its energy is depleted is called its 'range', provided its forward motion is not significantly deflected by multiple scattering. Angular deviations which cause the penetration to vary from one particle to the next is called 'straggling', measured by the variation in the range. The ability of a material to slow a moving charge, measured in the charge's energy loss per unit distance per unit material density, is called the material's 'stopping power' for the given radiation: $(1/\rho)dE/dx$.

8.13.2 Energy Loss in Materials from Ionizing Beams

The mechanisms by which an ionizing beam interacts with a material can be categorized in the following terms:

- *Excitation of bound electrons by light*: The interaction of a passing ionizing particle with an atom may transfer sufficient energy to excite an electron or a nuclear particle to a higher energy bound state.
- *Ionization of bound electrons by light*: The interaction of a passing ionizing particle with an atom may transfer sufficient energy to knock out an electron out of the atom.
- *Photoelectric effect (ejection of non-localized electrons by light)*: A photon may have sufficient energy to eject an electron held by a material in a non-localized state, such as the conduction electrons in metal.[11]
- *Thomson scattering (of light)*: Electromagnetic waves can be elastically scattered from the charges in a material when a group of those charges respond collectively. Electrons can act together as a larger mass, coherently re-radiating any absorbed energy. This process is predicted by classical electrodynamics for such a collection of charges reacting to a passing electromagnetic wave. Thomson scattering involves no energy loss.
- *Compton scattering (of photons by electrons)*: If an individual energetic photon hits an individual electron, with the photon energy significantly more than the binding energy that the electron might have, then the ejected electron will carry off kinetic energy, making the scattering process inelastic. The outgoing photon will be reduced in energy (and have a lower frequency and a longer wavelength).[12]
- *Møller scattering (of electrons by atomic electrons)*: A relativistic electron will be scattered while passing through a material through electron-electron interactions. The material electron is likely to be ejected from any bound state and will take up a fraction of the incoming electron's energy.

[11]Einstein, in 1905, was the first to properly describe this process, extending to absorption the new idea of Planck that light is emitted in bundles of energy called quanta.

[12]Compton, in 1923, used the photon idea and energy-momentum conservation to explain how that part of X-ray scattering which ejected atomic electrons also produced longer wavelength X-rays.

- *Mott scattering (of electrons by nuclei)*: A relativistic electron can be scattered by the nuclei of atoms. (The probability is highest when the electron wavelength is comparable to the nuclear size.) Mott scattering does not include the emission of a photon in the scattering process. Since nucleons are 1836 times more massive than electrons, even electrons with kinetic energy up to the order of GeV's will not kick the nuclei very hard. Rather, the electrons will be scattered away by the Coulomb force while hardly moving the nuclei of atoms.
- *Bhabha scattering (of positrons from electrons)*: Energetic positrons, in beams or from radioactive emissions, can scatter from electrons in a material before they are captured and annihilated by combining with electrons.
- *Bremsstrahlung (of electrons with nuclei)*: An electron, while going around a nearby nucleus, necessarily accelerates. In quantum theory, this implies that the electron has a finite probability of emitting a photon. This kind of emission is called Bremsstrahlung ('breaking radiation'), and is one of the ways X-rays are produced. The process is also a mechanism for energy deposit into a material by a charged particle beam.
- *Particle Pair Creation (by a photon near a nucleus)*: If a photon passing nearby to a nucleus has an energy greater than 1.022 MeV, it has a chance of emitting a pair of particles, the electron and the positron. These particles, in turn, contribute to the ionization produced by a gamma-ray beam.
- *Cherenkov radiation (by charged particles moving through materials)*: If a charged particle traveling through a material at speed V moves faster than the speed of light in that material, v, an electromagnetic field shock wave is created over a trailing cone with vertex at the particle's location and with an opening half angle $\theta_c = \arcsin(v/V)$. (This is analogous to an airplane creating a sound shock wave by traveling faster than sound in the air. See Sect. 5.13.) The blue glow from water surrounding a neutron-rich radioisotope is due to the Cherenkov radiation from the beta emissions. This blue glow can also be seen from the water pool used in some nuclear reactors to moderate neutrons.

8.13.3 Linear Energy Transfer

In 1930, Hans Bethe calculated the rate of energy loss of a fully-ionized nuclear particle beam passing through matter. For momenta of the beam particle in the range $0.05 < p/(mc) < 500$, the energy loss of the beam particle is dominated by collisions of the ions with electrons that act almost free relative to the incoming ion. Bethe's expression[13]

[13] H. Bethe, *Zur Theorie des Durchgangs schneller Korpuskularstrahlen durch Materie*, Annalen der Physik **397**, 325–400 (1930).

$$\frac{dE}{dx} = -\frac{4\pi k_C^2 z^2 e^4 n_e}{m_e c^2 \beta^2}\left[\ln \frac{2m_e c^2 \beta^2}{I(1-\beta^2)} - \beta^2\right] \tag{8.7}$$

gives the Linear Energy Transfer (LET) of the ion to the material through which the ion is passing.[14] In the formula, $k_C = 8.98755 \times 10^9\,\text{N}\cdot\text{m}^2/\,\text{coul}^2$ is the Coulomb constant, z is the atomic number of the ion, e is the charge on the electron, m_e is the electron rest mass, c is the speed of light, $\beta = v/c$ is the speed of the ion over the speed of light, I is the mean electron excitation energy in the material ($I \approx 16\,\text{eV}\cdot Z^{0.9}$, where Z is the material atomic number). The number of electrons per unit volume in the material is $n_e = \mathcal{N}_A Z\rho/A$, wherein $\mathcal{N}_A$ is Avogadro's number, Z is the atomic number, A is the atomic mass, and ρ is the mass density of the material. The overall constant factor

$$\frac{4\pi k^2 e^4}{m_e c^2}\mathcal{N}_A = 0.31102\,\text{MeV-cm}^2\ . \tag{8.8}$$

The Bethe's LET formula predicts that the beam-particle ions lose more energy at the end of their motion through tissue, when they are moving with energies comparable to the material ionization energies, than when they first enter the material. This dramatic rise in dE/dx at the end of the beam-particle's path is called the 'Bragg peak'. See Figs. 8.14 and 8.15.

8.14 Radiation Therapy

8.14.1 X-rays in Radiation Therapy

Radiation in the form of X-rays, gamma rays, particle beams, and that from radioisotopes can kill cancer cells. Such radiation also can kill normal cells, but usually at a lower rate. Some cancer cells, such as leukemic, are highly radiosensitive.

The most consequential effect of radiation on cells is the damage that radiation causes to the cell's DNA. (Appendix E shows how a simple mathematical model can predict the broken DNA fragment sizes caused by an X-ray or gamma ray beam.) This includes direct single and double strand breaks in the DNA sugar-phosphate backbone helices as well as indirect damage from free radicals that are formed,

[14]Often also included are additional small corrections analyzed separately through the years by Felix Block, Enrico Fermi, Lev Landau, R.M. Sternheimer, and W.H. Barkas. A high-energy correction is added within the bracket of Eq. 8.7 given by $-\delta/2 = -(1/2)\ln(\beta^2/(1-\beta^2)) - \zeta/2$. This term is due to limits on material polarization as the beam particle passes. A low-energy correction, $-C/Z$, is also added in the bracketed expression to account for atomic electron shells, where Z is the atomic number of the material atoms. The ζ and C are constants dependent on the material.

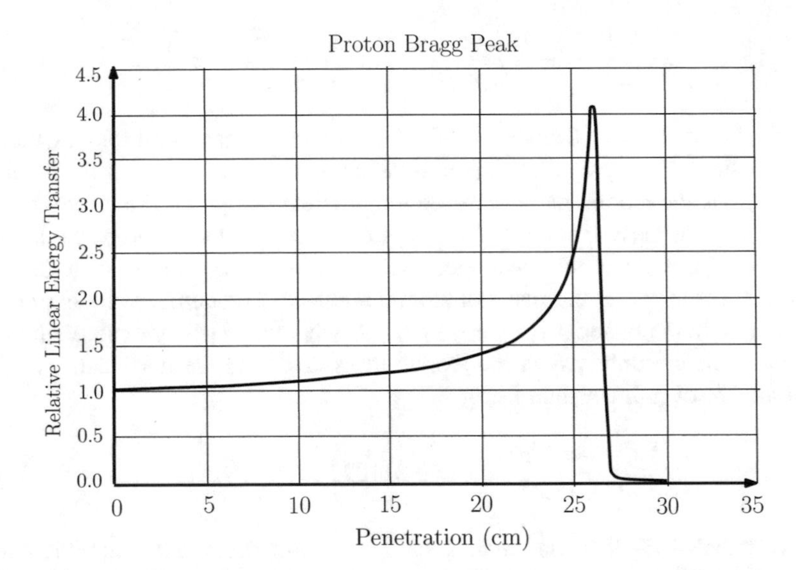

Fig. 8.14 Bragg curve for 205 MeV protons: range in high-density ($\rho = 0.97$ g/cm^3) polyethylene is 26.10 cm where the peak of the curve occurs. The Linear Energy Transfer (LET) at the entrance point is 0.4457 keV/μm in water

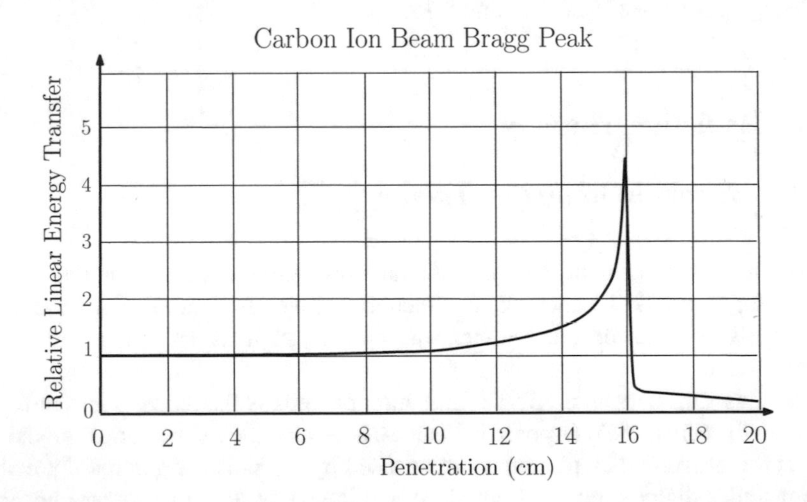

Fig. 8.15 Bragg curve for 292.7 MeV carbon ions: the range is 15.95 cm in high density polyethylene. Linear Energy Transfer (LET) on entrance is 24.33 keV/μm in water. Beyond the Bragg Peak at 16 cm you can see the tail produced by low-Z fragments

such the hydroxyl OH, in the nucleus of the cell. Before chemical reactions deplete OH radicals (with cellular lifetimes in the milliseconds), they can damage DNA.

Cancer cells reproduce faster than normal cells, and so have greater likelihood of having DNA vulnerable to damage by radiation, and they have diminished ability to repair breaks in DNA. Because cells can repair in minutes single-strand breaks of the DNA and some effects of radicals, the longer-term damage to cells depends non-linearly on the intensity of the radiation exposure. Radiation may also induce apoptosis (cell suicide).

If a cancer is concentrated in a tumor, then radiation can be focused at the tumor location from a set of angles, reducing damage to healthy tissue. For tumors on the skin, 'superficial X-ray' therapy can be employed, using low energy X-rays (30–60 keV) which do not penetrate far below the skin (less than 4 mm). Low-energy X-rays can be produced by X-ray tubes (using Bremsstrahlung in a metal anode). 'Orthovoltage X-rays' (also called deep X-rays), having photon energies 200–500 keV, will penetrate up to 6 cm below the skin. Orthovoltage X-rays can also be produced in X-ray tubes, but with special construction to isolate the high voltages used. X-rays produced by the inelastic scattering of electrons accelerated to millions of electron volts are referred to as 'megavoltage X-rays', and are used to reach tumors on internal organs.

Localized radiation dosages range from 20 to 80 Gy, often 'fractionalized' over a series of lower dose sessions.

8.14.2 Particle Beam Therapy

Cancer therapy using heavy-particle beams have a distinct advantage over X-ray beams in that the energy loss due to ionization by a charged heavy particle has a peak near the end of the particles motion (when it is slowed so much that it no longer moves forward), at a location in dE/dx as a function of x called the Bragg peak introduced in Sect. 8.13.3.

A common form of particle-beam therapy uses protons. Heavier particle accelerators, such as for ^{12}C nuclei, are more expensive, but have advantages over protons: They have a stronger Bragg peak and the beam is not scattered from its direct line as much. All such heavy-particle beams have the risk of producing new cancer in healthy tissue along the beam track.

8.14.3 Systemic Radiation Therapy

Systemic radiation therapy refers to targeted radioactive-substance ingestion or infusion.

Cancer of the thyroid can be treated by the injection or ingestion of a high dose of iodine-131, which is partly taken up by the thyroid. Iodine-131 is radioactive, with a physical half-life of 8.02 days, emitting an electron of maximum energy either 333.8 keV (7.27% of the decays) or 606.3 keV (89.8% of the decays) and then a gamma ray (of energy either 637.0 or 364.5 keV, resulting in xenon-131. It is the electron that does most of the damage to nearby tissue. Longtime exposure to low doses of iodine-131 can induce cancers. For patients with a fully functional thyroid gland, the biological half-life of I-131 is 5.5 days. If part of the thyroid is not functioning or excised, the biological half-life is longer than 5.5 days.

8.14.4 Brachytherapy

Brachytherapy (brachy is Greek for short distance) refers to the use of localized capsule containing a radiation source which is placed in or near a tumor, exposing that tumor to a high radiation dose, while reducing exposure of healthy tissue compared to beam therapy.

A number of other radioisotopes are used for cancer treatment (and diagnostics), including Bismuth-213, Holmium-166, and Lutetium-177.

8.15 Radiation Dosimetry

8.15.1 UV Light Limitations

Since ultraviolet light does not penetrate far below the cutaneous layer of skin, measures of UV radiation effects are often calibrated by the energy deposited per unit area of the skin rather than energy per unit mass used for Equivalent Dose.

By the way, do not let your dentist convince you that UV light on one sunny day is equivalent to the X-ray exposure he/she is about to give you. X-rays are penetrating; UV is stopped by the skin. You are likely to be unaware of the development of a deep tumor, but a melanoma is visible. The issue for getting an X-ray is whether, for a given intensity, the chance of a later cancer is more of a concern than the present treatment you may need. A soldier may decide to cross a field of bullets if his present danger is high, or he decides whether the risk is worth the hoped-for outcome. X-ray photons act as bullets.

One 'Standard Erythemal Dose of UV' (SED)[15] is a measure of the energy, in joules, deposited over each square meter of skin. One SED is 100 joules per square meter.

The 'UV Index' is an adopted value intended to indicate to the public the UV sunburn danger. The raw value is the integrated ground-level solar UV light intensity in each frequency interval (dI/df) times a skin sensitivity factor called the 'erythemal action spectrum'.[16] The raw UV index is then normalized to have a value of 10 for a clear sky mid-latitude zero-altitude midday solar exposure. Finally, the

---

[15]Erythea is a reddening of the skin due to inflammation. Typically, there are several hours after Sun exposure before reddening becomes evident.

[16]The erythemal action spectrum (EAS) is the ratio of the intensity per unit frequency of UV to just barely produce erythema, divided by the intensity per unit frequency at wavelength 290 nm.

index is corrected for the effect of expected clouds, altitude, latitude, air pollutants, surface albedo, and stratospheric ozone layer holes.

Because the radiation effects are dominated by single photon events, over a wide range of intensities, the damaging effects of UV radiation will be proportional to the UV index. For example, a value of 5 would produce the same sunburn in an hour as a value of 10 in half an hour. Exposure of light skin to solar UV with index of 10 will produce a sunburn in about 12 min. If the UVI is 5, sunburn should be expected after 24 min of exposure.

As UV is an ionizing radiation, damage to DNA can be expected. For UVB, damage is mainly caused by the formation of a thymine-thymine cyclobutane dimer mutation.

8.15.2 X-ray and Gamma Ray Damage

Although Wilhelm Röentgen and both Pierre and Marie Curie knew early in their investigations that their newly discovered radiations from radioactive materials could cause damage to the skin, the extent of the dangers from radiation-induced cancer came some years later. In 1927, Hermann Muller showed that X-rays can produce inheritable genetic damage. Even so, X-rays were being used often without regard to safety precautions until the radiation effects of the atomic bomb in 1945 raised the public consciousness.

A single energetic ionizing particle can cause a double-strand break (DSB) of a DNA molecule.[17] DNA damage may cause genetic instability or cell unviability. Biological-cell systems have remedies, as indicated here:

8.15.3 Reactions by Cells to Radiation Damage

1. DNA repair mechanisms;
2. Radical scavenging;
3. Radiation-induced apoptosis;
4. Signals to neighboring cells to enhance their defenses;
5. Immune destruction of aberrant cells.

The remedies are known to be more effective at low dose levels. Counteracting the defenses, cancers cells can mask their deviant behavior, and induce new blood supply through their uncontrolled growth. These facts lead to the proposition that humans will have imperceptible long-term damage by radiation if the dose is only about 100 mGy.

The EAS is rather flat between 250 and 300 nm, but then drops rapidly for longer wavelengths, being 0.1% at 330 nm as the EAS at 290 nm.

[17] See Appendix E.

8.15.4 Ionizing Radiation Exposure to Humans

Cosmic rays	0.7 mSv/yr
Cosmic rays at 2000 meters, equatorial	0.8 mSv/yr
Geological background	0.8 mSv/yr
Radioactive isotopes ingestion	0.29 mSv/yr
Cigarettes, one pack a day, by polonium-210	500 mSv/yr
Cancer over lifetime for the entire population	5% per Sv
Lethal dose delivered in seconds	5.0 Sv

A 'lethal dose equivalent' is defined as the dose equivalent that would cause death to 50% of the exposed population within 30 days.

With quantum theory, we know that energy is delivered in 'bundles' called 'quanta'. For light, the quanta are called 'photons'. Each photon carries an energy of size $E_\gamma = hf$, where f is the frequency of the radiation. For X-rays, photon energies are in the keV range. Thus, X-rays, like fast-moving particles, can easily break chemical bonds, which only have binding energies in the 3–11 eV range. The X-rays 'rain down', and the question of whether a bond is broken is just a matter of the probability of a hit.

The accumulation of body exposure to ionizing radiation, whether natural or human-made, and the period of time of those exposures, determines the level of damage the radiation might cause. At low levels, if there is sufficient time, repair mechanisms may be effective in reducing the damage. If the accumulated dose reaches about 500 rad= 5 Sv, about half the exposed population will die in 30 days.

Measuring the total body dose of ionizing radiation comes under the subject of dosimetry. The following devices can be used to measure the accumulated dose:

- Film badges hold photographic film between light-proof material. Ionizing radiation converts silver halide in the film emulsion into silver.
- A capacitive device can measure dose by the amount an internal capacitor is discharged due to the passage of ionizing radiation between the plates of the capacitor. The ionized air after a single ionizing particle passes allows some charge to flow for a short time. Such devices can be small enough to wear.
- Metal-Oxide-Semiconductor-Field-Effect-Transistors (MOSFET) devices measure the small current through a silicon dioxide crystal. Conduction is increased by the presence of crystal defects. Defects are produced by ionizing radiation.
- Scintillant detectors, as describe in Sect. 8.1.

Maximum dose standards have been formulated by the National Academy of Science's Biological Effects of Ionizing Radiation (NAS-BEIR) reports, by the International Commission on Radiological Protection (ICRP), the U.S. Occupational Safety and Health Administration, and through the Nuclear Regulatory Commission (NRC) Radiation Exposure Information and Reporting System (REIRS).

Table 8.5 Ionizing radiation dose

Federal Limits (except medical)	
Adult occupational worker/yr	50 mSv
Under 18 years old worker/yr	5 mSv
Fetus during gestation (9 months)	5 mSv
All man-made radiation on an average adult/yr	1 mSv
Medical radiation doses	
Hand X-ray	0.001 mSv
Dental X-ray (flat)	0.005 mSv
Dental X-ray (3-D)	0.05 mSv
Chest X-ray	0.65 mSv
Spinal X-ray	1.5 mSv
Radiotracer (Barium)	7.7 mSv
Radiotracer (Tc-99m)	15 mSv
CT abdominal scan	20 mSv
PET/CT scan	25 mSv
Therapy X-ray dose to eye	150 mSv
Therapy X-ray dose to a body organ	500 mSv
Other Ionizing Radiation Doses	
Airport X-ray scanner	0.0001 mSv
Airplane trip Paris-Tokyo	0.08 mSv
Ingested radioactive minerals/yr	0.35 mSv
Cosmic radiation background	0.3 mSv
Background radiation from rocks/yr	0.2–0.5 mSv
Inhalation of home radon gas/yr	0.1–2.3 mSv
Space Station 6 mo mission	160 mSv
Projected Mars mission (6 yrs)	1200 mSv
Lethal body dose (50% death in 30 days)	5000 mSv

The dosages in Table 8.5 are given in millisievert, which is a tenth of a rem. These limits should be compared to the dose from natural background radiation and other man-made sources, some of which are shown in the bottom part of the table.

The issue of cancer risk is not just one of radiation dose, since body cells initiate repair mechanisms when damaging radicals are formed or DNA and other important molecules are incapacitated. The repair mechanisms can take minutes or hours, so the duration of radiation is an important factor. Thus, comparison of a medical X-ray to exposures during air travel can be misleading. The medical X-ray takes seconds; air travel can take hours.

Evaluating the dose effect of medical X-rays also must include peripheral radiation exposure and secondary scattering from nearby objects. The importance of our thyroid and the gonads means dentists and radiologists should have strategies for protection against unnecessary exposure, including columnation of the X-ray beam and lead aprons. Hair follicles, gut lining, bone marrow, and equatorial eye-lens epithelial cells are particularly susceptible to damage because they have rapid cell growth, adding exposure risk to unraveled DNA.

There are treatment strategies if a person has been exposed to high levels of radiation, although there are no treatments which can stop or prevent all damaging effects. The following lists important elements of these strategies:

1. Find protected and un-radiated shelter, with an effective filter on the air intake to the shelter;
2. Wash whole body of any radioactive materials;
3. Drink pure water and eat only food that was protected from radiation, such as food in cans;
4. Ingest Prussian blue to help purge radioactive Cesium and Thallium (reducing their biologic half-lives: t_{Cs} from 110 to 30 days, t_{Tl} from 8 to 3 days);
5. Ingest stable iodine (as KI) to replace any radioactive iodine in thyroid;
6. Take diethylenetriamine pentaacetate (via IV) to help purge Plutonium, Americurium, and Curium;
7. Inject filgrastim to stimulate an immune system damaged by radiation.

8.16 Radiation Hazards of Space Travel

Astronauts face a number of dangers and debilitations by being in space without the radiation protection afforded by the Earth's atmosphere, without atmospheric incineration of passing meteors and debris, without the charged-particle shielding of the Earth's magnetic field, and without g-forces. The degeneration which may be caused by zero gravity can be easily compensated by regimented exercise and living within a rotating spacecraft. Psychological disorientation due to social isolation can be overcome.

For home-bodies, the Earth's field stops 99.9% of the harmful ionizing radiation from outer space. Our atmosphere equals about 1 m of thick metal.

As for space travelers without such shielding, the radiation hazards depend on the dose received and the duration. The dose itself depends on the intensity and energy of the radiation, and on the length of exposure time. The result of a high dose is:

Dose	~0.5 Sv	Nausea, vomiting
Dose	5–12 Sv	Nausea, vomiting, diarrhea, dehydration, electrolytic imbalance, loss of digestion ability, bleeding ulcers. Death in 30–60 days from sepsis (inflammation throughout body)
Dose	>20 Sv	Central nervous system breakdown, with symptoms include loss of coordination, confusion, coma, convulsions, shock. Also evident are the symptoms of degradation of the hair follicles, of blood-forming bone marrow cells and of the gastrointestinal tract, due to the fact that cells in these organs divide faster than other cells, and so their DNA is more vulnerable to radiation. No survivors expected.

Near the Moon, the galactic cosmic radiation (GCR) accounts for ~91.4% of the total absorbed dose in astronauts, with GCR protons responsible for ~42.8%, GCR alpha particles for ~18.5%, and GCR heavy ions for ~30.1%. The remaining

~8.6% of the dose at Lunar Reconnaissance Orbiter altitudes (~50 km) arises from secondary lunar species, primarily "albedo" protons (3.1%) and electrons (2.2%). Other lunar-nuclear-evaporation species contributing to the dose rate are positrons (1.5%), gammas (1.1%), and neutrons (0.7%).

The great solar storm of August 4, 1972, had it occurred during the Apollo lunar landings in April 1972 and in December 1972, would likely have caused acute radiation sickness in the astronauts. The galactic cosmic rays consist of protons p, helium nuclei ^4He (with a fluence of about 1/4th that of protons), carbon (with a fluence of about 1/6 that of He), iron (about 3/8 of carbon) with peak energy at about 1 GeV, dropping off exponentially with a power of about 10^{-2}. The dose would have been: Protons: 6.21 cGy, Helium: 3.02 cGy; Carbon: 0.83 cGy; Oxygen: 1.37 cGy; Magnesium: 0.66 cGy; Silicon: 0.69 cGy; Iron 1.56 cGy; with a total 14.34 cGy.

The solar particle radiation is dominated by protons, giving about 0.17 Sv per year at solar sunspot min, 40 Sv per year at solar sunspot max, to exposed astronauts in space near the Earth.

A Solar Particle Event (SPE) may occur during a solar magnetic storm. For example, a July 2000 event produced 1.7 Sv in the vicinity of the Earth, and an August 1972 event made 3.4 Sv. The SPE can last for several hours.

Thin shielding may be worse than none, as secondary radiation by proton scattering from the shield occurs. However, a thickness of 10 cm of Al shielding to the Aug 1972 SPE would have reduced the dose from 3.4 cSv to 0.40 Sv. There are a number of possible shields: Hull shields, deployable water shields, and deployable high-density polyethylene (HDPE) slabs. Active magnetic or electrostatic shielding is an untried alternative.

During the Apollo missions, which were 4 day missions to the Moon, the astronauts experienced flashes in eyes every few minutes. These were caused by cosmic rays traversing through the vitreous humor in front of retina.

Astronauts on a Moon mission for 6 months would incur 50–2000 mSv. (The Moon has no significant magnetic field, nor an atmosphere.)

A 3-year Mars mission would incur about 1200 mSv. (1000 mSv makes a 5.5% increased risk for fatal cancer.) Mars has no significant global magnetic field, and little atmosphere to protect against solar wind and solar flares.

The Curiosity and Odyssey Mars rovers detected an average of about 240 mSv per year, compared to a dose on Earth of about 3.5 mSv per year (from both cosmic and geologic sources), so Mars at its surface has about seventy times more radiation than the Earth. A human habitat on Mars would have to employ radiation shields (initially layers of dirt), until a good atmosphere can be manufactured, perhaps by robots and genetically engineered bacteria, that can take Mars' radiation.

In addition to solar proton flux, there are energetic heavy ions from the Galaxy. Galactic High Z Energetic ions (HZE) include carbon, oxygen, magnesium, silicon, and iron ions. Their mean energy is from 1 to 10 GeV/amu. Damage to tissue is comparable to solar flare protons from the sun. Galactic cosmic rays are a hazard due to dosage near and far from the Earth, and near and far from the Sun. In addition, damage occurs to electronics and solar panels. Strategies are being considered for minimizing the damage, both short and long-term. For further investigations, see the space physiology studies of Dr. Mary Anne Frey and her colleagues at NASA.

First colonization away from Earth will more likely be in near-Earth space stations rather than Mars. The detailed engineering for such a venture was worked out by Princeton Professor Gerald O'Neill and his students, in the late 1970s.[18] Using large rotating cylinders about 500 meters in diameter, radiation shielding would come from the cylinder walls and the encapsuled air.

If you wish to drop your lifetime radiation exposure by a fourth, put meter-thick purified lead walls, ceiling, floor and doors surrounding your bedroom. (Unpurified lead is no good, as it has radioactive substances mixed in.) Of course, then you would have to be careful to cover the lead, so you do not scrape or lick the walls. But that's not all. You would have to acquire special breathable air that has no radon gas. This could be produced by containing a volume of air for many days to allow radon decay, or by liquefying and then boiling above -183 °C but below -61.7 °C to release nitrogen and oxygen, but leaving liquid radon.

Problems

8.1 We know that each human on Earth has about 10^{14} neutrinos from the Sun passing through their bodies every second. However, a solar neutrino will pass through about 8×10^{15} meters of tissue with only a 50-50 chance of hitting anything. Estimate the number of neutrino hits in your body per day.

8.2 How much work must be done by an electric field to ionize nitrogen molecules in air?

8.3 What is the lowest frequency of light that can break an organic chemical bond formed by the release of 4 eV of energy?

8.4 If the physical half-life of iodine-131 is 8 days and the biological half-life of iodine for a certain person is 70 days, calculate the percent of ingested iodine-131 left in the person after 10 days.

8.5 Suppose a 100 microcurie radioactive source of half-life 8 days and which emits 0.2 MeV beta particles is placed within some tissue. How many disintegrations occur each second in this source, initially? What is the activity of this source after 10 days?

8.6 A spectrophotometer passes light of wavelength 280 nm through a cuvette of inside length 1.0 cm and holding a water solution with an unknown concentration of a protein with an absorptivity of 59,000/M-cm. If the absorbance is measured to be 0.0352, what was the molar concentration?

[18]Gerald K. O'Neill, **The High Frontier: Human Colonies in Space**, [New York: William Morrow & Company] (1977).

8.7 The UVB molar absorptivity of provitamin D3 (7-dehydrocholesterol) is shown in a figure of this chapter. If your epidermis had no other absorbing compounds except provitamin D3 at a concentration of 20 μm/L, what depth would UVB light of wavelength 280 nm reach before dropping in intensity by 90%?

8.8 Look up the typical binding energy of the carbon-carbon bond in an organic molecule (usually given in kilocalories per mole.). Convert from kilocalories per mole to electron-volt per molecule. Calculate the frequency of the photon which will have sufficient energy to break this chemical bond. Calculate the wavelength for this frequency, in nanometers. Look up where in the electromagnetic spectrum this wavelength occurs, i.e. what is the name of this type of radiation? Look up the full range, in nanometers, of this type of radiation. Are there named subcategories for this type of radiation? If so, what are their names? Electric arc welders can get burns on their face from this radiation. Why? Which frequencies in this radiation range are likely to cause the most severe burns?

8.9 What makes infrared light more useful to investigate organic molecules than ultraviolet light?

8.10 If the electron current in an X-ray tube is 50 mA, accelerated by a 12 kV potential, how much heat might be generated in the tungsten target? What is the highest frequency of X-ray produced?

8.11 Electrons, accelerated in a vacuum to 10 keV, hit a block of tungsten over an area of 0.2 square centimeters. What is the shortest wavelength of X-rays produced? If the electron current is 100 mA, estimate the maximum intensity the X-rays could have just outside the tungsten?

8.12 What is shortest wavelength x-ray which can be emitted by an x-ray tube that employs Bremsstrahlung production if the electrons are accelerated toward the target anode by a 20 kV electric potential?

8.13 If a dentist's X-ray machine uses 15 kV across its X-ray tube, what is the minimum wavelength of X-rays from that machine.

8.14 Suppose x-rays of a certain wavelength drop in intensity through bone with an absorption coefficient of 15/cm. How thick would the bone have to be to cause the x-ray intensity to diminish to 1/10th of it original value?

8.15 An electron in an atom of oxygen in the water of your tissue is hit by an x-ray photon of wavelength of 0.014 nanometers. If the electron was bound in the oxygen by 46 eV, how much kinetic energy will the electron gain after being ejected?

8.16 X-rays of wavelength 1.4 Å are reflected from a series of parallel Bragg planes in a crystal of DNA. If the reflected waves add coherently at an angle of incidence and reflection of 25°, what is the minimum separation between the planes?

8.17 If a Computed Tomography (CT) procedure delivered 850 mrem into a patient's abdomen of mass 6.4 kg and density 0.94 gm/cm^3, estimate the amount of energy that was deposited in the volume of a DNA molecule, taken as $\pi(10\text{Å})^2 \cdot 3.4\,\text{Å} \cdot 3 \times 10^9 \ = 3 \times 10^{-18}\text{m}^3$

8.18 Why are gamma rays less dangerous to biological tissue than alpha particles if both have an energy per particle of one MeV and the same intensity?

8.19 As part of a treatment of a brain tumor, a 67.5 MeV beam of protons is used to destroy cancerous tissue inside a volume of ten cubic millimeters. The Bragg peak occurs at a distance from 35 to 37 mm from the entrance point. When the protons enter the Bragg peak region, they have energies less than 1 MeV. Estimate the beam flux in order to deposit an accumulated dose of 100 rads in 1 h.

8.20 Suppose you wish to form an image of a cancer tumor in the gut with either ultrasonic imaging or an X-ray CT scan. Describe the relative merits and disadvantages of each.

8.21 If beta particles of energy 606 keV from 2.4 millicurie of radioisotope Iodine-131 (with half-life of 8.02 days and 89.9% emission of this β particle) initially produce 54 ions per beta particle in 20 grams of tissue, what is the exposure (in röentgen) after the first hour? If all the energy of the beta particles is deposited into the 20 grams of tissue over a time of 30 days, what dose (in millirads) was given?

Chapter 9
Bioenergetics

For those who want some proof that physicists are human, consider the idiocy of all the different units which they use for measuring energy.

—Richard Feynman

Summary The activity of life requires a flow of energy into and out of the life system. Because life systems maintain and add to their internal order, the energy source may allow an 'entropy' flow into the life system, and, in turn, the energy flow out must carry entropy into the environment. The concepts of temperature, heat, and entropy are defined in the subject of thermodynamics, at first developed in the 1800s through macroscopic observational principles, but now is supported by the physics and statistics of a large number of molecules in microscopic interaction. This chapter emphasizes the macroscopic behavior of thermal systems, including phase changes, osmosis, diffusion, active transport across cell membranes, and biochemical reactions. The next chapter gives the statistical arguments behind thermodynamics, with applications to cellular systems.

9.1 Biology and Energy Flow

Life, having replication capabilities, must be able to generate order from more disordered states.[1] Microscopic dynamics and statistical ideas are sufficient to understand how open subsystems naturally become more ordered, and to understand the limitations to the amount of order which can be expected. These ideas are also able to explain how energy spontaneously flows through open systems, which includes all life forms, if the 'sources' of energy are more ordered than the 'sinks'. Individual organisms become more ordered by building internal structures. This means the organism necessarily must release heat into its environment.

[1] As we will see, order is a measurable quantity.

W. C. Parke, *Biophysics*, https://doi.org/10.1007/978-3-030-44146-3_9

A deeply seated idea about nature discovered over 160 years ago[2] is that a quantity we call energy is conserved.[3] With quantum theory, the internal energy E of any system has the simple expression

$$E = \sum_i n_i \epsilon_i \,, \tag{9.1}$$

in which n_i is the number of entities in the energy state ϵ_i. The energy states are numbered to make $\epsilon_{i+1} > \epsilon_i$. We use the term 'energy level' for the values ϵ_i. The n_i are often referred to as the 'occupation numbers', as each tells how many entities occupy a possible energy level ϵ_i.[4] Complex life systems will have a large variety of possible energy levels available to a relatively large number of entities. With energy conservation, we know that the total internal energy E can only change if energy is transferred into or out of the system.

9.2 Thermodynamics Applied to Life

Thermodynamics is the study of macroscopic systems near equilibrium. Equilibrium in this context means 'thermal equilibrium', defined to be a state of a macrosystem in which all mechanisms that might allow internal energy transfers have effectively come to balance. Systems with a variety of entities, which we will label by α, a variety of interactions, and a number of possible energy states, will have a set of different times to reach equilibrium, call their "relaxation lifetimes" τ_α.[5] Some metastable states may have much longer lifetimes than typical observation times. Some may have cyclic states which never relax. If there is a wide separation between the shorter relaxation times and the longer ones, then thermodynamics can still be applied to systems that reach quasi-equilibrium states during the first relaxation time. Life systems come into this category, in that the natural decay time

[2]The energy conservation principle which, by necessity, includes transformations into heat, was stated in quantitative form by the physician Julius Robert von Mayer in 1842, inspired by how the oxygen supply of blood maintains body temperature.

[3]Using Nöther's Theorem, described in Sect. 3.2.3, energy conservation can be derived for any localizable system which has time translation symmetry, i.e. the same laws which apply now to the system also apply later.

[4]We take the summation over i to start at $i = 0$, corresponding to the lowest energy level, ϵ_o. We expect that the number of particles in a system $N = \sum n_i$ is finite, so $i \leq N$ and the range of i is finite.

[5]If a macrosystem reaches equilibrium through statistical processes among their microstates then the drop in number of microstates of variety α which are out of equilibrium is proportional to the number which are left and to the time. This makes their number approach zero exponentially: $n_{0\alpha} \exp(-t/\tau_\alpha)$, where $n_{0\alpha}$ is the initial number out of equilibrium and τ_α is called the relaxation time.

for molecules built into structures is long, while times for metabolic processes are short.

Systems made from a large number of individual entities which interact and exchange energy will likely evolve into an equilibrium state. To be alive, an organism cannot be in a settled equilibrium state. There are processes in us that keep us away from such a state. However, we can still apply most of the formalism of thermodynamics during the ordinary processes within our cells and between cells. This ability is reflected by the fact that thermodynamic variables, such as temperature and pressure can be meaningful in life subsystems, because their fluctuations may be much smaller than the averages used to define them.

9.3 Definition of Temperature

Observation shows and theory suggests that if two macroscopic systems are placed in 'thermal contact',[6] the net flow of energy between them will cease after some 'relaxation' time. We then say the two bodies are in thermal equilibrium.

The following definitions of temperature are all equivalent:

1. All bodies which, when put in thermal contact, show no transfer of macroscopic energy, have a common temperature. We find that if A_1 and A_2 have the same temperature, and body A_2 and A_3 have the same temperature, then A_1 and A_3 will have the same temperature. This idea is sometimes referred to as the 'zeroth law of thermodynamics'. Also, if body A is put into thermal contact with body B and energy flows from A to B, then we say body A is hotter. We find that if body A is hotter than B and B is hotter than C, then A is hotter than C. In this way, temperature is both transitive and reflexive, and we can categorize all bodies in equilibrium into sets according to their temperature.
2. Some thermometers take advantage of the fact that most materials change volume with changes of temperature. A common thermometer uses a selected liquid material, such as mercury, put into a long thin closed cylindrical tube, with a vacuum in the distal part of the tube. Changes in temperature can be seen by the expansion or contraction of the material within the tube. Difficulties with such thermometers arise if the temperatures needed to be measured are beyond the range for a single phase of the material, or the material, in some range of temperature, reduces in volume with increased temperature. For example, water increases in volume while dropping in temperature from 4 to 0 °C.
3. A more universal thermometer, but less convenient, uses an 'ideal' gas.[7] A certain amount of gas is sealed in a volume, whose pressure can be monitored. To measure temperature, the gas is put into thermal contact with a material whose

[6]Two bodies in thermal contact are allowed to exchange energy, but not by work of one on the other.

[7]Dilute helium approximates an ideal gas.

temperature is desired. When no more thermal energy is transferred, the gas and the material have reached the same temperature. Of course, the gas volume must be small enough to not significantly change the temperature of the material being observed. Using the ideal gas law in the form $T = pV/nR$, the material temperature is determined from the gas pressure. In fact, in the basement of some hospitals, there use to be a thermometer calibration unit, which measured the pressure of a dilute inert gas within a known volume while a more portable thermometer needing calibration is inserted in the gas.

4. For a wide range of temperatures, the internal energy E stored in an ideal monatomic gas made from N monatomic molecules is purely kinetic. According to the kinetic theory for such gases,

$$pV = (2/3)N\left\langle (1/2)mv^2 \right\rangle,$$

so $E = N\langle (1/2)mv^2\rangle = (3/2)Nk_BT$. Thus, for such a gas, temperature is a measure of the average kinetic energy of the molecules. This does NOT mean that the temperature of your tongue, or any other system, represents the average kinetic energy in that system! This is a very common misconception. The statement connecting temperature to average kinetic energy applies ONLY to systems which have no intermolecular interaction, such as an ideal gas. When you put a thermometer on your tongue, the thermal energy transferred to the thermometer comes from BOTH kinetic energy and potential energy in the molecules of your tongue.

5. There is another variation on the definition of temperature. From the theoretical side, it is the cleanest definition: It does not depend on any of the properties of any particular material. However, it is the least intuitive. Temperature is the rate at which the internal energy of a system changes per unit change in the disorder of that system while the volume and particle number of the system are held fixed:

$$T = \left.\frac{\partial E}{\partial S}\right|_{V,N} . \tag{9.2}$$

Because the volume is held fixed during the change of the internal energy, no work is done by or to the system. To interpret this definition, we'll have to wait until disorder is associated with a measurable quantity.

Related to this last form of the definition of temperature is the observation that the quantity $S_2 - S_1 = \int_1^2 \delta Q/T$, a sum of small quantities of heat into a given system divided by the temperature of that system, taken over a reversible path,[8]

[8] A thermodynamically 'reversible process' is a sequence of small changes in a thermodynamic system that keeps the system near equilibrium for the whole process *and* can be taken along the same sequence of thermodynamic states, but in reverse order, to carry the system back to its initial condition.

is, in fact, process independent, and only dependent on the endpoints. For this reason, the inverse temperature has been called the 'integrating factor of heat'.

For each of the definitions, a common scale for temperature is introduced. Looking from our current perspective, the most natural scale for temperature would have been to adopt energy units. However, to communicate with our fellow humans, we use the Celsius scale[9] for which, at first, the freezing point and boiling point of water were set at 0 and 100 °C measured at one atmosphere, but then absolute zero was set at −273.15 °C and the triple point of water, was set at 0.01 °C. The Kelvin unit is the same as the Celsius unit, but starts with 0° at absolute zero. Conversion to energy units is effected by multiplying with Boltzmann's constant

$$k_B \equiv 1.380649 \times 10^{-23}\ \text{J/deg.}$$

9.4 Definition of Heat

The energy transferred between two macrosystems due to temperature differences is heat. Through the definition of temperature, heat always goes from a hotter to a colder body. The fact that a temperature can be identified for each of the systems implies that in this definition, each system is at or near thermal equilibrium while heat is being transferred. Conventionally, heat which goes into a given system is taken as positive, while heat which goes out is taken as negative.

The internal energy of any macrosystem can always be expressed in terms of the energies of the microstates, as shown in Eq. (9.1). Small changes in the internal energy of the system can occur by energy transfer into or out of the system. From the expression for E, a small energy transfer can be divided into two distinct terms: One caused by small changes in the numbers of entities n_i (such as particles) in the energy states with energy ϵ_i, and the second by changes in the energies ϵ_i of the states of the system:

$$dE = \sum_i dn_i \epsilon_i + \sum_i n_i d\epsilon_i\ . \tag{9.3}$$

Since the n_i are integers, a calculus notation dn_i requires explanation. Most of the n_i for macrosystems near equilibrium are typically astronomically large, so dn_i can be integer and still be much smaller than n_i. i.e. infinitesimal by comparison. The microscopic distinction between heat and work is pictorially represented in Fig. 9.1.

The volume dependence of the total energy in a macrosystem comes into the internal energy expression Eq. (9.1) through the energies ϵ_i. For example, consider a box confining a particle bouncing between the walls of the box. In quantum

[9]The term Celsius has replaced Centigrade.

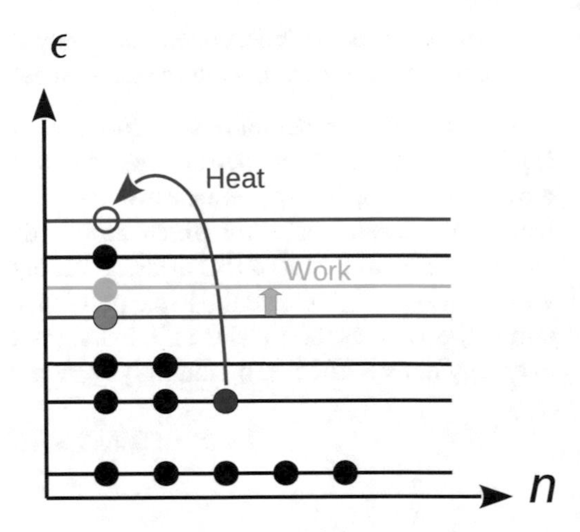

Fig. 9.1 A representation how heat and work input affects the microscopic internal energy states and their occupancy. An exchange of work requires a change in the energy levels, while heat exchange is due to the redistribution of the entities among the fixed energy levels

theory, the momentum of the particle is quantized because its wave must have nodes at the walls, making $i_x \lambda_{i_x}/2 = L$, the length across the box. Here, i_x is the number of nodes in the wave in the x axis direction. If the box is allowed to slowly increase in size, the quantum standing wave describing that particle could expand, while preserving the number of nodes in the wave. With a longer wavelength, the momentum of this wave ($p_{i_x} = h/\lambda_{i_x}$) would become smaller. The energy of the particle is

$$\epsilon_{\{i\}} = \frac{p^2}{2m} = \frac{h^2}{8mL^2}\left({i_x}^2 + {i_y}^2 + {i_z}^2\right) = \frac{h^2}{8mV^{2/3}}\left({i_x}^2 + {i_y}^2 + {i_z}^2\right) . \tag{9.4}$$

This shows, for a particle in a box, the dependence of the energy of a given state on the volume of the box. The particle energy diminishes as the volume increases.

A set of such particles exerts pressure on the walls of the box. The expansion of the volume of the box, dV, means the pressure $p = -\partial E/\partial V$ of the set of particles does work $\delta W = pdV$.[10] We conclude that the second term in Eq. (9.3) is the total work done by the system:

$$\delta W \equiv pdV = -\sum_i n_i d\epsilon_i . \tag{9.5}$$

All 'adiabatic processes' i.e. a change of a system involving no transfer of heat, do not change the occupation numbers n_i. Rather, such processes change the value of one or more energy levels ϵ_i, requiring work.

[10]The strikeout bar in the delta symbol is a reminder that this delta is not a differentiation. The work produced by a system is not a thermodynamic function. The work out depends on how the system evolved to produced that work. Thermodynamic functions depend only on the current state of a system, and not on the history used to reach a quasi-equilibrium condition.

By transferring energy in the form of heat, open systems can lose energy to another system without doing any work on the second system. This means we can identify the first term in Eq. (9.3) as the heat input, δQ:

$$\delta Q = \sum_i dn_i \epsilon_i \,. \tag{9.6}$$

We see that heat transfer causes a redistribution of the occupation numbers of the possible energy levels in the system, without changing the energies of the microstates. Later, we will associate such redistributions as a change in the disorder of the system.

9.5 Heat Transfer Mechanisms

We can identify two distinct methods for the transfer of energy by heat: conduction and radiation. Some authors also include convection; others add phase change. But convection is macroscopic transport of material carrying energy that does not necessarily come from heat transfer.

Phase change also may involve heat absorption or release, but the energy released or absorbed does not have to come from heat transfer. Material undergoing phase changes, such as water going from a solid to a liquid, or a liquid to a gas, may produce or absorb heat energy while spatially reconfiguring their atoms or molecules, without any chemical change. A phase transformation alters the microscopic energy states ϵ_i and may change the occupation numbers n_i of the atoms or molecules, but not necessarily so. If several phases are mixed together, and each involves heat energy transfer to change, then under ordinary circumstances the temperature of the mixed system is preserved until conversion to one phase is complete. A glass of ice water accepting heat demonstrates this behavior.[11]

The distinction between heat conduction and radiation is one of spatial range. Conduction occurs when short-range forces, usually electric, act between incoherently moving particles, causing a net energy transfer when one set of particles carries more energy than a second set. Heat transfer by radiation is caused by long-range interaction between two sets of particles, that interaction being the exchange of electromagnetic waves. The waves are generated by the incoherent jiggling of the particles in both sets, causing charges with the particles to accelerate and radiate.

[11] Under more extraordinary circumstances, substances can become 'supercooled' or 'superheated', i.e. not change phase even when cooled or heated past the ordinary phase-change temperature. This condition may occur when 'nucleation' sites are missing. For example, in the case of supercooled gases, there would be no locations where adhesion of molecules to a surface brings those molecules together over a sufficient time for weak bonding to form between them, and then quasi-stable clusters of molecules can form.

Heat conduction causes a change in the occupancy of the microscopic energy levels as energy is being transported across a material, but, by definition, heat energy transfer does not include coherent motion of material particles, which also carries energy. In ordinary materials, the layers of material may interact though local electric and magnetic fields, through incoherent collisions and by close-by encounters as particles thermally move and vibrate. Through such proximity effects, there will be a net tendency for energy to be transferred from the hotter layer to the colder one, 'the thermodynamic downhill direction'.

The flow of energy by heat conduction is a dissipative process, in that there is a generation of more disorder in the colder material than a loss in the hotter one.[12] If a process is carried out with each change during the process only infinitesimally far from an equilibrium state, then the sequence of states is called a 'quasi-static processes'. A quasi-static process may also have the property that if the system is carried back to some initial state through the same quasi-equilibrium states, no net entropy is produced in the overall process. In this case, the process is called 'reversible'. Otherwise, the total disorder increases, and the process is irreversible. A reversible process must be quasi-static, but a quasi-static process need not be reversible.

Strong and dense materials have close-by molecules and relatively strong molecular interactions, permitting quick and large energy transfers. Diamond is a classic example of such a good conductor of heat. It is a very good conductor of heat because the carbon atoms in diamond are in tight and close bonds. The heat conductivity of metals is dominated by valence electrons which are free to quickly carry thermal energy across the material, because the electrons are so much smaller in mass compared to atoms. These electrons act like a gas in the metal.

Because of the pseudo-random nature of the heat-transfer process at the microscopic level and because of the large number of particles involved, the rate of heat transfer $\dot{\delta Q}$ is proportional to the area A of material layers in contact, and inversely proportional to the thickness dx of a given layer conducting heat. After all, with all else the same, doubling the area of a layer will mean twice the heat will be transferred. Similarly, doubling the thickness of a layer will result in twice the time for the heat to transfer through both layers, at least for thin layers.

If the temperature throughout a layer of the material is constant, no heat will transfer. One should expect,[13] and for most materials one finds, that for small differences in temperature, the transfer rate is proportional to the difference in temperature. Clearly, the rate of heat transfer will also depend on the material 'carrying' the heat.

---

[12] A formal definition of disorder and its measure will be given in the chapter describing the statistical basis of bio-energetics, Chap. 10.

[13] A Taylor expansion of the heat-transfer rate through a layer of material as a function of temperature difference across that layer will have a first term ΔT, provided the first derivative does not vanish. If the first term does not vanish, the next term, proportional to ΔT^2, can be made negligible by taking the thickness of the layer, and therefore ΔT, sufficiently small.

The behavior of the material in conducting heat is incorporated into the number κ_H, called the heat conductivity, introduced as the proportionality constant in our statements above about the conduction rate. Written quantitatively, the statements reduce to:

$$\frac{\delta Q}{\delta t} = \kappa_H A \frac{dT}{dx} . \tag{9.7}$$

The heat conductivity of pure diamond is 2000 W/m-K. Cast iron is 55 W/m-K, human skin[14] about 0.1 W/m-K, and air at STP 0.025 W/m-K.

The three-dimensional form of Eq. (9.7) in an isotropic medium is[15]

$$\mathbf{J}_Q = -\kappa_H \nabla T , \tag{9.8}$$

in which J_Q is the flux of heat per unit area per unit time, $\delta Q/\delta A\, \delta t$, pointing in the direction of the greatest downward change in temperature. Eq. (9.8) is referred to as Fourier's law of heat flow.[16]

If there is a net heat flow into a system, the extra energy can cause an increase in the kinetic and vibrational energy of the molecules. This extra energy may also be sufficient to overcome any weak binding of the molecules. If so, that unbinding will dominate the energy absorption. The most likely state of the molecules has a balanced distribution of available energy.[17] Unbinding a macroscopic system of molecules without breaking the molecules themselves is a phase change. When water changes from ice to liquid, or liquid to gas, the water molecules held together largely by hydrogen bonds are being unbound. Water takes 80 cal/g to change from solid to liquid phase, and 540 cal/g to change from liquid to gas at 100 °C. As the temperature will not change during a thermodynamic phase change of a material, ice water will remain at 0 °C as long as the ice has not completely melted. When rain turns to sleet or snow, it releases heat, slowing the drop in temperature that caused the freezing.

If the residual energy deposited in a material by heat conduction does not do work such as producing a phase or chemical change, the temperature of the material increases. As we will see in the Chapter on statistical mechanics (Chap. 10), the ability of a material to absorb heat depends on the variety of ways that energy can be re-distributed among the available quantum states.

---

[14]Skin tissue is a dynamic organ, with mechanisms to control heat loss by controlling the circulation of blood, making the effective thickness of the skin between outside temperatures and body temperatures variable.

[15]In material such as a crystal or a mixture with oriented molecules, the flow of heat may not be in the direction of the thermal gradient. In these cases, Eq. (9.8) is replaced by $\mathbf{J}_Q = -\boldsymbol{\kappa}_H \cdot \nabla T$, where $\boldsymbol{\kappa}_H$ is a 3×3 positive-definite symmetric tensor.

[16]Joseph Fourier, **Théorie Analytique de Chalier** [Paris] (1822).

[17]This point will be made more exact when we expand on the subject of the statistical basis of bioenergetics in Chap. 10.

If the heat flow into a material only causes a redistribution of the molecules among the possible energy states, and does not change the energy states themselves, we define the ability of the material to take in heat while changing temperature one degree to be its heat capacity. The specific[18] heat is the heat capacity c_p per unit mass of material, held at constant pressure:

$$\Delta Q = c_p m \Delta T \, . \tag{9.9}$$

Water has a relatively high specific heat, as far as common substances go. Because we contain a lot of water (about 70% of our body mass), our tissue correspondingly has a relatively high specific heat. For example, we lose more heat by conduction from our skin for a given drop in temperature than dry wood under the same conditions. In our environment, air with high humidity becomes far more difficult to change in temperature from night to day than dry air. It is the moisture in the air which requires more heat to change the air temperature. In this respect, Arizona, with very dry air most of the year, has much larger variations in temperature from night to day than Washington, D.C., which often has high humidity, particularly in the summer.

If the material being heated is a layer within a thicker object, we can analyze the heat transport in the material by first imaging a division of the material into thin layers. Start with a thin layer of thickness δx, with adjacent layers left and right. Take the layer on the left hotter than the middle layer, and the middle hotter than the one on the right, making a flow of heat from left to right. Any energy left in the middle layer will then be accumulating at a rate

$$\frac{\delta Q_L}{dt} - \frac{\delta Q_R}{dt} = c_p \delta m \frac{dT}{dt} \, . \tag{9.10}$$

From Eq. (9.7), the rates in and out on the left can be written as shown below, and we use $\delta m = \rho A dx$ to write

$$-\kappa_H A \left(\left. \frac{\partial T}{\partial x} \right|_L - \left. \frac{\partial T}{\partial x} \right|_R \right) = c_p \rho A \delta x \frac{\partial T}{\partial t} \, , \tag{9.11}$$

where ρ is the density of the material. The minus sign on the left takes care of the fact that each of the two gradients of the temperature are negative if heat is flowing to the right. Now the parenthetical factor on left-hand side, divided by δx, is the second derivative. There results the rather famous 'heat equation':

[18]Formally, the modifier 'specific' in front of a measurable quantity means the measure for the given material divided by the measure for a standard material, usually water. The calorie was defined using water, so the value of the specific heat of a substance is the heat capacity divided by approximately one calorie per degree. Rather than dividing, most researchers and references report the specific heat in calories per gram degree.

$$\frac{\partial^2 T}{\partial x^2} = \frac{1}{D_H}\frac{\partial T}{\partial t}\,, \tag{9.12}$$

with $D_H \equiv \kappa_H/(\rho c_p)$ being the 'heat diffusion-constant' for the material. In three-dimensional form, it becomes

$$\nabla^2 T = \frac{1}{D_H}\frac{\partial T}{\partial t}\,. \tag{9.13}$$

The same equation describes how perfume spreads into a room. Instead of T, the density of the perfume appears. It also describes how a concentration of some chemical spreads into a homogeneous and isotropic solute within cells, or oxygen spreads across plasma from red-blood cells to capillary walls (endothelium). Heat spreads in materials in the same way as particles spread by diffusion. This should not be a surprise, as heat energy is transported through materials by pseudo-random interactions, just as particles are carried by diffusion, since the diffusive process occurs through pseudo-random collisions of the diffusing particles with the background particles.

Suppose a layer of tissue were made to have a steady rate of heat flow. Then $\partial T/\partial t$ would vanish. For the one-dimensional case, $d^2T/dx^2 = 0$, so $T = a + bx$, the temperature follows a linear profile with distance. In three dimensions, the steady temperature field satisfies Laplace's equation, $\nabla^2 T = 0$.

The Laplace equation is familiar to those who study the electric potential in electrostatic problems. For given boundary conditions, the same form of solutions apply. Except now, the potential is the temperature, and the electric field is replaced by the negative of the temperature gradient, which points in the direction of the heat flow. Positive charges act like the source of heat, and negative charges act like the sinks of heat.

Now consider an initially thin hot layer at the left of a much larger volume of colder material. It is straightforward to show that the Gaussian

$$T = \frac{A}{\sqrt{t}}\exp\left(-x^2/(4D_H t)\right) + T_B \tag{9.14}$$

is a solution to the heat equation. This function near the surface $x = 0$ and $t \sim 0$ is very sharply peaked. As time progresses, the peak flattens and lowers. Eventually, T approaches the background temperature T_B. (Without a boundary, the extra energy in the hot layer will spread and dissipate to large distance.) The same happens to perfume initially concentrated in a small volume within an open air space.[19]

The analysis of how heat diffuses from brain tissue, or from the heart muscles, or from a tumor, can be investigated with the heat equation.

---

[19] Curiously, if t is replaced by $\sqrt{-1}\,t$, the heat/diffusion equation becomes the quantum wave equation for a free particle. In fact, the probability distribution for a particle initially confined to a small space and left alone 'diffuses' out with time.

9.6 Life Under Extremes of Temperature

Life forms have found a wide variety of strategies to keep their essential systems from being irreversibly terminated from extremes of temperatures (hyper- and hypothermia). At present, life systems on Earth are dependent on liquid water where organic reactions can occur. Thus, the temperature range for the activity of organic life is just below 0 °C to just above 100 °C. At elevated temperatures above 100 °C, proteins first become 'denatured' (unraveling and losing their structural integrity), and then chemically disassociated. Some bacteria living near hydrothermal vents can survive temperatures up to 121 °C.

For temperatures much below 0 °C, cellular life forms can be destroyed by the growth of ice crystals which can break through cell walls. (Very rapid freezing reduces this risk.) For nanoscale life, such as viruses, water molecules can diffuse into the virus' capsid structure, but this water will not become crystalline ice, and so viruses are not threatened by very low temperatures. Although they become inactive at low temperatures, viruses are revived when returned to liquid in host cells.

Some life forms use a strategy of active control over the size of any ice crystals that may form, and become inanimate during below-freezing temperatures. If the temperature is not far below freezing, some plants seeds survive intemperate environments by first expelling water from cells (directed desiccation), binding most of the remaining water in cells, producing osmolytes, such as ethylene glycol, to protect protein structures, and secreting 'antifreeze' compounds ('cryoprotectants' into intercellular regions. Some animals use 'suspended animation', during which they become dormant, but with a minimum of metabolic activity ('metabolic depression'). Some warm-blooded animals with stored fat and insulation use hibernation. A low level of metabolic heat keeps the animal from freezing.

9.7 Temperature Regulation in Humans

Humans are warm-blooded animals. Our body attempts to maintain a temperature of about 37 °C so that our activity and metabolism can consistently function at a rate compatible with our life style and survival. The temperature regulation mechanisms are shown in the Fig. 9.2. Whether cold or warm blooded, life activity requires the release of heat. This follows from the fact that life orders molecules in cyclic processes. We will elaborate on this kind of metabolic and other microscopic activity in life systems later.

Warm-blooded animals tend not to be as efficient 'heat engines' as cold-blooded ones, sacrificing some efficiency in favor of power and activity. Humans release from about 35 kcal/m^2 of skin surface per hour (sleeping) to 600 kcal/m^2/h (running). For an adult, the power output in heat ranges from about 80 W to about 1400 W. Our bodies selectively control the flow of blood in the skin, the extremities, internal organs, and the more critical organs. Our temperature is also controlled by

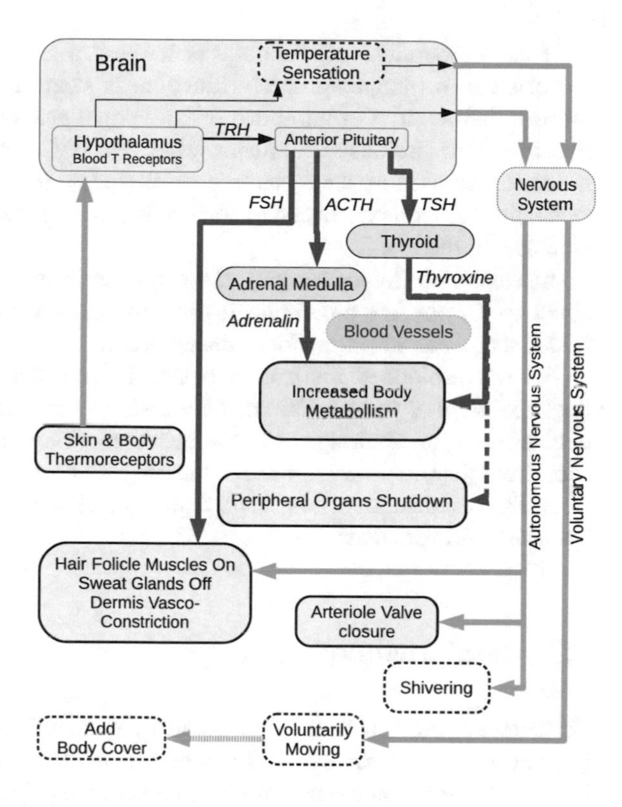

Fig. 9.2 Temperature regulation in the human: response to cold. *TRH* stands for Thyrotropin-Releasing Hormone; *TSH* is Thyroid Stimulating Hormone; *ACTH* is Adrenocorticotropic Hormone; *FHS* is Folicle Stimulation Hormone

evaporation through perspiration, by coverings, by hair erection, by shivering, by cell mitochondrial thermogenic control, by folding to reduce exposure, and by how we exhale. We have a set of mechanisms to release and regulate our heat output, all assuming an environment can be found with a lower temperature than the skin, i.e. lower than 28–37 °C.

Humans have a special problem: In any time interval, their brains can consume about 25% of the available energy stored in the blood. This energy is converted to the work of nerve functions, with a by-product of heat carried away by blood circulation. In the process, the heat must be released from the body.

Perspiration ('sweat') releases about 602 cal/g of water that vaporizes from the skin (539.4 cal/g plus 63 cal/g for the water to go from skin temperature to evaporation temperature). We have about two and a half million sweat glands. The rate of sweating can reach 30 g of water per minute, expelling as much as 18 kcal/min of heat from the body. (By comparison, the 'burning' of one mole of glucose, 180 g, into carbon dioxide and water produces 686 kcal.) However, if the relative humidity is 100%, sweating gives no cooling effect.[20]

[20]Here's where I ask my students: "Why do we not have an in-built humidity sensor to shut down the loss of water when there is such high humidity?"

If the temperature of our body is lowered much below 30 °C (hypothermia), muscles can no longer properly function. In deep hypothermia, the temperature drops to below 20 °C (but above 0 °C). People can survive without breathing air (but in an unconscious state) under water near 0 °C for up to 45 min. Hypothermia is sometimes used by surgeons to operate (for up to an hour) with a stopped heart and brain. Temporary induced hypothermia can help preserve nerve cells when there is trauma in the brain.

Above 42 °C (hyperthermia), the central nervous system breaks down, and at about 44 °C, our free protein begins to undergo denaturalization, i.e. open from its folded shape and lose function, causing death.

For warm-blooded animals, exposure to cold and wind adds to the chances for hypothermia, since wind not only displaces the warmed air next to the body, held by hair or clothing, but also increases skin evaporation, a cooling effect. Meteorologists, as a way to warn of the dangers, have devised various 'heat indices', 'feels like temperatures, and 'wind chill' indices to communicate the temperature a human would perceive.[21]

9.8 Basal Metabolic Rate

For humans, internal heat is generated by metabolic activity. Empirical formulae have been devised for the 'Basal Metabolism Rate' applied to the average adult human. The BMR is the rate of heat production dQ/dt in the body due to minimal metabolism for life functions. Perhaps the simplest reasonably accurate empirical formula is that of Mifflin et al. (1990)[22]:

$$\frac{dQ}{dt} \approx 5\left(\frac{2}{\text{kg}}M + \frac{5}{4\,\text{cm}}h + \frac{1}{\text{yr}}a + s\right)\frac{\text{kcal}}{\text{day}}\,, \tag{9.15}$$

where M, h, and a is the person's mass, height, and age, while s is $+1$ for males and -33 for females. We note that as M goes to zero, dQ/dt should also go to zero. But Eq. (9.15) was intended for human adults, not for babies or muskrats! One may

[21]The formula devised by the Joint Action Group on Temperature Indices and adopted by the U.S. National Weather Service in 2001 is: $T_{chill} = 35.74°\text{F} + 0.6215T - (35.75°\text{F} - 0.4275T)(V/(1\,\text{mph}))^{0.16}$, with T the actual temperature in Fahrenheit and V the wind speed in miles per hour. The formula does not apply if the wind speed is less than 3 mph or the temperature is greater than 50 °F. The formula assumes low humidity. Also, even if $V = 0$, there is chilling by an assumed walking motion of a bare face at about 3.1 mph. For hot environments, the National Weather Service uses a calculated heat index and a 'feels like' temperature.

[22]M.D. Mifflin, S.T. Jeor, L.A. Hill, B.J. Scott, S.A. Daugherty, and Y.O. Koh, *A new predictive equation for resting energy expenditure in healthy individuals*, Am J Clin Nutr **51:2**, 241–247 (1990).

think of Eq. (9.15) as a straight-line approximation, near human adult masses, to a full BMR function of mass.

The condition $\lim_{M\to 0} dQ/dt = 0$ is satisfied by a second even simpler but less accurate empirical formulae given by

$$dQ/dt \approx 91.6\,\text{kcal/day}\,(\text{M/kg})^{3/4}\,. \tag{9.16}$$

This second relation agrees with the scaling law presented in Sect. 13.4 and so applies over a wide range of masses. Note: When a process depends linearly on size, there is an 'isometric' relationship, i.e. a proportionality. If the process depends non-linearly on size, the relationship is called 'allometric', from the Greek meaning different (i.e. not proportional). Amazingly, Eq. (9.16) for the basal metabolism rate also works reasonably well for other animals from mice all the way to elephants. (For details, see Sect. 13.4.)

The resting metabolic rate for a male of 53 kg, a height of 188 cm, and an age of 25 years is typically 75 kcal/h, which is 1800 kcal/day or 87 W. This corresponds to the consumption of about 16 L of oxygen per hour from the burning of carbohydrates, fats, and proteins. Inversely, one can measure the oxygen consumption to find the metabolism rate as

$$\frac{dQ}{dt} = 4.69\frac{dV_{O_2}}{dt}\frac{\text{kcal}}{\text{L}}\,. \tag{9.17}$$

The BMR can be also measured directly by immersion of a patient into an insulated bath of water whose temperature is recorded.

During sleep, the metabolic rate drops to 65 kcal/h. Hard thinking adds about 22 kcal/h to the BMR.[23] During the act of running, the metabolic rate rises to as much as 700 kcal/h, but only about 25% of this produces mechanical work during exercise, i.e. you are hard-pressed to develop more than about 200 W of mechanical power, which is about a fourth of a horsepower (746 W).

One can determine the overall metabolic heat production by combining the oxygen consumption, carbon dioxide release, methane release, and nitrogen in urine through the Brouwer equation (1965). Using data for roosting chickens, he finds

$$Q = 3.865[O_2/L] + 1.200[CO_2/L] - 0.511[CH_4/L] - 1.431[N/g] \tag{9.18}$$

given in kilocalories.[24] Smart farmers know this.

---

[23] Your brain should be getting hotter about now!

[24] E. Brouwer, *Report of the Subcommittee on Constants and Factors*, Energy Metabolism. Proceeding of the 3rd Symposium, 441–443 [Academic Press] (1965).

9.9 An Introduction to Entropy

In this section, we will summarize the central ideas which led to the second law of thermodynamics and the concept of entropy. In Sect. 9.10, the logic behind the second law will be given. In the history of the subject, heat engines played a central role. The heat from burning wood supplied the energy for the first steam engines. Engines which required less wood to produce the same amount of work were desirable. The advantage of engines to drive industry with reduced manpower and replacing beasts of burden was evident. This impetus drove innovation and also new thinking about how to make heat engines more efficient.

In the abstract, a heat engine is any system, running in a cycle, which converts some fraction of the heat input into work output. If one imagines metabolic conversion of carbohydrates into usable energy as a 'burning' process, then we are heat engines.

In 1822, Sadi Carnot carefully thought through the behavior of a simple engine which runs by extracting heat Q_H from a hot reservoir at temperature T_H, produces work W, and expels heat Q_C into a cold reservoir at temperature T_C. Carnot used the following observation, now taken as a form of the second law of thermodynamics:

> Heat does not spontaneously flow from a colder body to a hotter one.

With this statement, Carnot proved that the most efficient heat engine is a reversible one, i.e. an engine which, if carried in its cycle in reverse order, will have the same values of Q_H, W, and Q_C, except reversed in direction.[25] The efficiency is the generic term for what you wish to get out divided by what you put in. For a heat engine, this is the work out divided by the heat in: $\mathcal{E} = W/Q_H$. Carnot showed that the efficiency of reversible heat engine is the maximum efficiency of any heat engine, and can be written in terms of the temperatures of the source and sink of heat with the expression $W/Q_H = (T_H - T_C)/T_H$.

Later, William Thomson (Lord Kelvin) and Rudolph Clausius, following Carnot's work, concluded, among other things, that the efficiency of any engine could not reach 100%. Their statements are now taken as another form of the second law of thermodynamics:

> No heat engine, running in a complete cycle, can convert all of its heat input into work.[26]

Because of energy conservation, $Q_H = W + Q_C$. Thus we can present the efficiency in the equivalent form $\mathcal{E} = 1 - Q_C/Q_H$. As we will see from Carnot's analysis, all reversible engines, independent of how the engine is constructed, have an efficiency given simply by $\mathcal{E}_{\mathcal{R}} = (T_H - T_C)/T_H = 1 - T_C/T_H$. Because this is the maximum efficiency, there follows the relation

[25] A heat engine run in reverse is a refrigerator.

[26] From our microscopic perspective, a 100% efficient engine would convert the chaotic motion in heat transfer into the coherent motion characteristic of work, with no other change occurring.

$$Q_H/T_H - Q_C/T_C \leq 0$$

for each cycle of any engine. From this, Clausius saw that if one takes heat going into the engine as positive and the heat coming out of the engine as negative, and includes in the thermodynamic cycle variations in temperature by using a sequence of Carnot engines running between the system and the environment, the relation above becomes:

$$\oint \frac{\delta Q}{T} \leq 0\,. \tag{9.19}$$

The equality will apply only for a reversible process.

Now suppose we divide the full cycle in the process into two: Start at some state of the system, called A and proceed to state B. Then go from state B back to A. In (9.19), we can move the second part of the cyclic integral to the right-hand-side, to get

$$\int_A^B \frac{\delta Q}{T} \leq -\int_B^A \frac{\delta Q}{T}\,. \tag{9.20}$$

Now if one selects a sequence of changes, called a path, from B to A to be a reversible process, then

$$\int_A^B \frac{\delta Q}{T} \leq \int_{A\ rev}^B \frac{\delta Q}{T}\,. \tag{9.21}$$

The equality holds if the process used for the integral on the left-hand side is also reversible. But even then, the relation is not trivial, because the thermodynamic 'path' that one takes going from A to B may be different.

In performing the integral, the temperature T is that of a series of reservoirs in thermal contact with the system, and the system must be allowed to reach the same temperature as the reservoir during each infinitesimal step of the sum. The reservoirs, by definition, must be capable of supplying and absorbing heat without significant changes in their own temperature. Even with heat transfer, the process can be reversible, because the state of any one reservoir will be restored by reversing the direction of the heat transfer at fixed temperature.

For the reversible process cases, we see that the integral in Eq. (9.21) does not depend on the path from A to B, just on the endpoints:

$$S_B - S_A = \int_{A\ rev}^B \frac{\delta Q}{T}\,. \tag{9.22}$$

The state A can be selected as a standard state, so the quantity S at any other thermodynamic state of the system can be defined by integrating $\delta Q/T$ from the standard state to the given one:

$$S = S_o + \int_{rev} \frac{\delta Q}{T} \,. \tag{9.23}$$

Clausius, recognizing he had discovered a new thermodynamic property of systems, gave the name ‘entropy’ to the quantity S.[27] From relation (9.21), we have a general form of the second law of thermodynamics:

$$S_B - S_A \geq \int_A^B \frac{\delta Q}{T} \,. \tag{9.24}$$

The entropy change of a system for any process which causes the system to pass from one state A to another B will never be less than the change of the entropy of that same system when it is carried from the state A to the state B by a thermodynamic process. An explosion may carry nitroglycerin from a cool liquid to a hot gas of nitrogen, carbon monoxide, water, and oxygen, but this process will not be thermodynamic. There is no well-defined temperature or pressure for the material in an explosion. However, we can find a more gentle process that takes nitroglycerin from the liquid to the product of the reaction. During the gentle (‘quasi-static’) process, the temperature and pressure of the components can all be measured. Our statement is: The entropy produced during the explosion is necessarily greater than the entropy change by gentle processes carrying the reactants to their products.

For a closed system, $\delta Q = 0$, so relation (9.24) gives

$$\Delta S \geq 0 \,. \tag{9.25}$$

The entropy of a closed system can never decrease.

This is a famous conclusion from the second law of thermodynamics, and perhaps the most abused one. Some wish to misinterpret and mislead by claiming that “life violates the second law of thermodynamics because the entropy of an organism decreases when forming structure such as proteins and memory,[28] and is therefore miraculous.” But life DOES NOT violate the second law. Life systems are NOT closed. One does not need life to see the entropy of subsystems drop spontaneously. A supercooled liquid can spontaneously turn to a mixture of crystals and remaining

---

[27] In 1865, Clausius wrote (in translation), “I suggest that S be given the Greek words ή τροπὴ, the translation to be called the *entropy* of the body. The word entropy was deliberately formed as similar as possible to the value of energy for the two magnitudes which are to be used by these words are so closely related to one another in their physical meanings, that a certain similarity in the designation seems to me to be expedient.” (from Clausius, R., *Über verschiedene für die Anwendung bequeme Formen der Hauptgleichungen der mechanischen Wärmetheorie*, Annalen der Physik **201**, 353–400 (1865)) Entropy translates from the Greek as ‘intrinsic direction’. It is likely Clausius used the letter ‘S’ out of respect for the work of Sadi Carnot.

[28] When we connect entropy to disorder, we will see that, indeed, forming structure from building blocks does reduce the entropy of the original materials.

liquid. The newly formed crystals release heat.[29] The crystal subsystem has less entropy than the originating molecules in the liquid. But because of the heat released, the mixture of crystals and liquid has more entropy than it had as a pure liquid. An egg becomes a chick by natural processes. The chick has far less entropy than the original material in the egg. But during development, the growing embryo releases heat into the environment, making the sum of the entropy released to the air and the negative entropy produced in the chick still positive. A chicken egg MUST expel more heat than it takes from the mother hen if the egg embryo is to develop![30]

9.10 The Laws of Thermodynamics

Those are my principles. If you don't like them, I have others
—Groucho Marx

The laws of thermodynamics apply to macroscopic systems, which are systems with a large number of entities, such small subsystems, sharing the available energy through quasi-random processes which are allowed to distribute that energy among the possible microstates of the system. The first law is simply energy conservation. The total internal energy of a system is designated E. Expressed microscopically,

$$E = \sum n_k \epsilon_k \tag{9.26}$$

where ϵ_k, an "energy level," is one of the energies (both kinetic and potential) available to the entities in the system. The energy levels are labeled by an index $k = 0, 1 2, \cdots$. The n_k, are the "occupation numbers," giving the number of entities which have the designated energy ϵ_k. For example, in a biological system, there will be lots of water molecules in their ground state, whose energy we might call $\epsilon_o^{\text{water}}$. For a tiny biological cell, the corresponding occupation number for water molecules in this state will be astronomical, perhaps $n_0^{\text{water}} \approx 10^{14}$. The system energy can change by changes in the occupation numbers (a "re-population"), by a shift in the energy levels themselves, or both. A shift in the energy levels can be caused by a change in the forces exerted on the system. A re-population of the levels can be caused by a chemical reaction, or by contact with another system at a different temperature. In the first case, the changing forces cause a change in the strains within the system, and work is exchanged with the environment. The second case defines heat exchange with the environment of the system. Formally,

[29] Sodium acetate acts this way and is used in pocket warmers.

[30] This is a surprise to many people, because they thought the hen was supplying heat! Instead, she keeps the embryo warm, but the embryo gives more heat than the hen does.

$$dE = \sum dn_k \epsilon_k + \sum n_k d\epsilon_k \tag{9.27}$$

or

$$dE = \delta\!\!\!{}^{-} Q - \delta\!\!\!{}^{-} W \ . \tag{9.28}$$

This is one form of the 'first law of thermodynamics': The total internal energy of a system can only change by the input of heat of the output of work. Of course, each of these changes can be negative, flipping the words input/output. The signs correspond to the convention that heat into a system is taken as positive, while work out is taken as positive.[31]

In the first law of thermodynamics, there is no implication that heat or work are properties of the system. Rather, they are energies transferred. In fact, there is NO such thing as "heat in a system" or "work in a system". For this reason, the infinitesimal energies of heat in and work out are not denoted with a differential of any function, such as df.[32] The usual implication of the symbol df is that there is a function we are differentiating with respect to some variable. But that is not what is meant by $\delta\!\!\!{}^{-} Q$, because Q is not a function of any of the measurable 'thermodynamic variables', such as pressure and temperature. In contrast, the energy in a system, E, *is* a property of the system. It can be expressed in terms of the system pressure, volume, temperature, etc. Thermodynamic variables such as E, S, N, p, V, and T, (energy, entropy, number of particles, pressure, volume, and temperature) do not depend on how the system arrived at these values. In contrast, the heat into a system and the work out of a system do depend on what process was used.

Systems with all the following characteristics are said to be thermodynamic systems:

- All subsystems of the system are in interaction;
- The time for energy to be transferred by work or heat between subsystems is not greater than observation times;
- Any measurable properties of subsystems vary smoothly across the system;
- Each of the measurable properties of subsystems must have fluctuations much small than their average values.

Explosive events are not easily characterized by thermodynamic quantities, although the materials before the explosion and the products after may be. Disconnected subsystems are not thermodynamic. But our cells and many of their subsystems are close enough to being thermodynamics systems that we can successfully apply the consequences of thermodynamics principles to them.

Thermodynamic variables which depend on size are called 'extensive'. The internal energy E is an example of an extensive variable. If a system is duplicated,

[31]This convention comes from the development of the subject of thermodynamics by the study of heat engines.

[32]We use a strike-through of the delta to emphasize that the symbol is not a differential.

the combined system has twice the internal energy of each half. Another example is the volume V of the system itself. 'Intensive' variables, such as the system pressure p or its temperature T, do not depend on how much of any subsystem is used to make the measurement.

The work out of a system when the system pressure expands the volume of the system can be found from

$$W = \int p dV . \tag{9.29}$$

The amount of this work will depend on how heat is allowed into the system during each step of the process. For this reason, we say that the integral depends on the "path" one takes. For a small change in volume,

$$\delta W = p dV . \tag{9.30}$$

As we will see, the second law of thermodynamics will also allow us to express the heat exchanged in terms of a path-dependent integral over thermodynamic variables.

Such expressions will no longer have meaning if the system goes non-thermodynamic, such as fluids in high chaos. One cannot use $\int p dV$ if the pressure cannot be measured or fluctuates wildly. However, we can devise thermodynamic paths for our integral by taking the system slowly through quasi-equilibrium states. In this way, we can find how much a thermodynamic variable has changed.

Through logical thinking about heat engines, which are systems which run in a cycle to take in heat and produce work, Sadi Carnot (1796–1832), laid the foundation for the second law of thermodynamics.

Figure 9.3 shows a pair of abstract heat engines. The essence of a heat engine: A source of energy, a 'sink' to discard more disordered energy, and a system which uses some of that energy to do work, and a scheme to carry the engine in a cycle, so that it can be used over and over. Later, we will be more specific about the term 'disordered'. Loosely, if all the particles in a system are moving in the same direction and with the same speed, i.e. with coherent motion, we say they are perfectly ordered. Such would be the case of a solid body you might push, or the motion of the piston in your car. If the particles are moving in so-called random directions with a distribution of speeds, they are disordered, and have incoherent motion. Thermal motion of molecules is an example of incoherent motion.

Heat engines abound in biology, including your muscles and the operation of flagellae in bacteria. Each takes in energy (such as through a chemical reaction which re-arranges energy occupations), does work, and expels heat.

Carnot introduced the idea of a reversible engine. A reversible engine can not only be made to run through each of its states in reverse order, taking heat from a cold reservoir by doing work, and pumping that heat into a hot reservoir, but when it does run in reverse mode, the *same* heats and work occur as when run in its forward cycle, except they each are reversed in sign. Carnot realized that this kind of engine has special idealized properties.

Using the idea of a reversible heat engine, and with pure logical thought alone, Carnot came to very far-reaching conclusions. These conclusions lead to the concept of measurable order and disorder in all thermodynamic systems.

The efficiency of a heat engine is defined as

$$\mathcal{E} = \frac{W}{Q_H} . \tag{9.31}$$

Suppose two engines are put in parallel, with the work of the first used to drive the second (as in Fig. 9.3), with the second being a reversible engine and run as a heat pump. The efficiencies of these two engines are:

$$\mathcal{E}_1 = \frac{W}{Q_H^{\ 1}}; \ \mathcal{E}_2 = \frac{W}{Q_H^{\ 2}} \tag{9.32}$$

so that the ratio of efficiencies is:

$$\frac{\mathcal{E}_1}{\mathcal{E}_2} = \frac{Q_H^{\ 2}}{Q_H^{\ 1}} \tag{9.33}$$

If two such engines are linked, with one driving the second, and run so both complete a cycle and return to the same state they initially had, then nothing in the combined system has changed in this full cycle except for the possibility that heat has flowed from one reservoir to the other. But we never observe heat flowing from a colder body to a hotter one spontaneously. This is one form of the second law of thermodynamics:

Heat does not spontaneously flow from a colder body to a hotter one.

As a result,

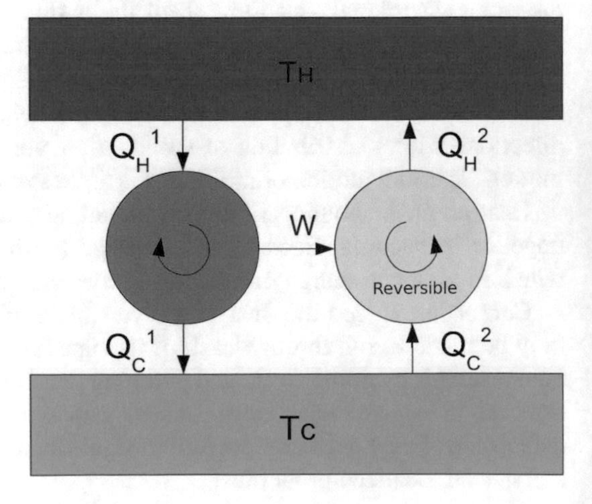

Fig. 9.3 Two heat engines in parallel

$$Q_H^1 \geq Q_H^2 \tag{9.34}$$

making

$$\frac{\mathcal{E}_1}{\mathcal{E}_2} \leq 1\,, \tag{9.35}$$

i.e. the reversible engine (number 2) is never less efficient than any other engine. If both are reversible, then reversing both engines inverts the relation, so that only the equality can apply:

$$\mathcal{E}_1^{rev} = \mathcal{E}_2^{rev} \tag{9.36}$$

If all reversible engines have the same efficiency (irrespective of how they are made), then the efficiencies can only depend on the temperatures of the reservoirs: $\mathcal{E}^{rev} = \mathcal{E}^{rev}(T_H, T_C)$, and not on what materials the engines are made from or how the engines are constructed.

Now imagine coupling two reversible engines in series, as in Fig. 9.4. Let Q be the heat transferred from the first to the second engine. Then energy conservation gives

$$Q_H = Q + W_1 \tag{9.37}$$

$$Q = Q_C + W_2 \tag{9.38}$$

so the efficiency of the two is

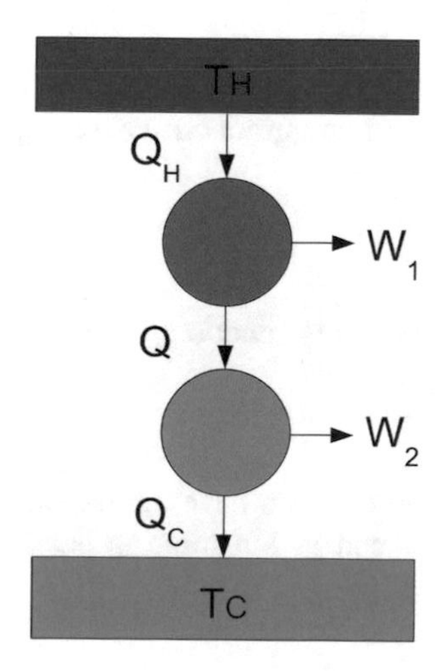

Fig. 9.4 Stacked Carnot engines

$$\mathcal{E} = \frac{W_1 + W_2}{Q_H} = \frac{Q_H - Q_C}{Q_H} = 1 - \frac{Q_C}{Q_H} \tag{9.39}$$

Let $\eta = 1 - \mathcal{E}$. We will then have

$$\eta_1 \eta_2 = \frac{Q_C}{Q} \frac{Q}{Q_H} = \frac{Q_C}{Q_H} = \eta \tag{9.40}$$

or, using the observation that the efficiencies of all reversible engines depend only on the hot and cold reservoir temperatures,

$$\eta\,(T_H, T)\,\eta\,(T, T_C) = \eta\,(T_H, T_C) \tag{9.41}$$

Let's see if we can solve the functional equation

$$\eta\,(x, y)\,\eta\,(y, z) = \eta\,(x, z)\ . \tag{9.42}$$

First, if we put

$$\eta = \exp\,(f)\,, \tag{9.43}$$

then the product of the η's becomes a sum of their exponents:

$$f(x, y) + f(y, z) = f(x, z)\ . \tag{9.44}$$

It follows that

$$\frac{\partial^2 f\,(x, y)}{\partial x \partial y} = 0\ . \tag{9.45}$$

Now integrate once over y to get

$$\frac{\partial f(x, y)}{\partial x} = g'(x) \tag{9.46}$$

with any function $g'(x)$ of x. Integrating the result over x gives

$$f\,(x, y) = g(x) + g_2(y)\,, \tag{9.47}$$

so now we have the exponent function f as the sum of two arbitrary function of x and y. Substituting back into (9.44) requires $g_2\,(z) = -g\,(z)$. Define $h\,(x) \equiv exp(-g(x))$, so

$$\eta\,(x, y) = \frac{h(y)}{h(x)}\ . \tag{9.48}$$

Now, because $0 \leq \mathcal{E} \leq 1$, we have $0 \leq \eta \leq 1$. Also, $T \leq T_H$. Thus, with h positive, $for\ all\ x > y,\ h(x) \geq h(y) > 0$, i.e. $h(x)$ is any non-zero smooth non-decreasing function of x.

To find $h(T)$, consider a reversible heat engine running with an ideal gas confined by a cylinder, a fixed wall, and a piston. By slowly changing the gas along either adiabatic or isothermal conditions, the process can be made reversible. One then finds $h(T)$ is proportional to the temperature T.

Here's the argument: For an ideal gas, the internal energy depends only on the temperature of the gas

$$E = nc_vT \tag{9.49}$$

(c_v is the heat capacity per mole) and

$$pV = nRT \tag{9.50}$$

In general, the 1st Law of Thermodynamics, i.e. energy conservation for a system exchanging heat and work with its environment, reads

$$dE = \delta Q - \delta W \tag{9.51}$$

But here E depends only on T. For any isothermal process of an idea gas, $dE = 0$, so we must have

$$W = Q\,. \tag{9.52}$$

The work done isothermally compressing the gas from state 1 to state 2 (this work is the area in Fig. 9.5 under the isothermal curve from state 1 to state 2), will be

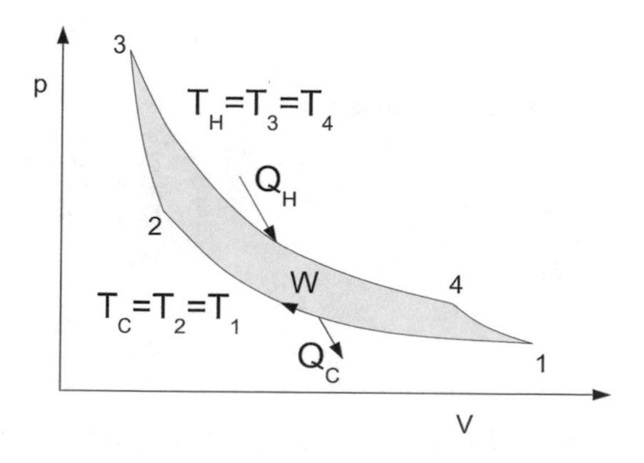

Fig. 9.5 Carnot pV cycle

$$W_{12} = \int pdV = nRT_C \ln\left(\frac{V_1}{V_2}\right) = Q_C \tag{9.53}$$

and the work gained from the isothermal process state 3 to state 4 is

$$W_{12} = \int pdV = nRT_H \ln\left(\frac{V_4}{V_3}\right) = Q_H \tag{9.54}$$

The two isothermals are connected by adiabats (i.e., no heat exchange). For these adiabatic processes,

$$dE = -pdV \tag{9.55}$$

or

$$c_v dT = -\frac{RT}{V} dV \,. \tag{9.56}$$

This means that during an adiabatic process,

$$TV^{\gamma-1} = constant \tag{9.57}$$

where

$$\gamma \equiv \frac{c_v + R}{c_v} \tag{9.58}$$

With the adiabats connecting the isothermals, we have

$$T_C V_2^{\gamma-1} = T_H V_3^{\gamma-1} \tag{9.59}$$

$$T_C V_1^{\gamma-1} = T_H V_4^{\gamma-1} \tag{9.60}$$

In ratio it follows that

$$\frac{V_2}{V_1} = \frac{V_3}{V_4} \tag{9.61}$$

Thus,

$$\frac{Q_C}{Q_H} = \frac{T_C}{T_H}, \tag{9.62}$$

showing that the efficiency of an ideal-gas Carnot engine is

$$\mathcal{E}^{rev} = 1 - \frac{T_C}{T_H} = \frac{T_H - T_C}{T_H} \,. \tag{9.63}$$

There results the famous Carnot efficiency theorem:

$$\mathcal{E} \leq 1 - \frac{T_C}{T_H} \tag{9.64}$$

with the equality holding for a reversible engine.

In addition, from Eq. (9.62), we have

$$\frac{Q_C}{T_C} = \frac{Q_H}{T_H} . \tag{9.65}$$

By using a series of reversible engines, there follows

$$\oint_{rev} \frac{\delta Q}{T} = 0 \, . \tag{9.66}$$

This was the basis for Clausius' introduction of entropy (as described in Sect. 9.9).

9.11 The Thermodynamic Entropy of a System

We have shown, with the arguments of Carnot and Clausius, that there must exist a thermodynamic property of systems called their entropy, given by Eq. (9.23):

$$S_B = S_A + \int_A^B \frac{\delta Q}{T} \bigg|_{\text{reversible path from } A \; to \; B} \, . \tag{9.67}$$

The infinitesimals δQ are the small amounts of heat which go in (with a positive value) and out (with a negative value) of the system for the selected process ('path'), and T is the temperature when the heat δQ is being exchanged with the environment of the system.

More generally, for any changes in the system made close to equilibrium, we have the 'entropic form' of the second law of thermodynamics (Eq. 9.24):

$$S_B - S_A \geq \int_A^B \frac{\delta Q}{T} \, , \tag{9.68}$$

with equality only for a reversible path from A to B.

We also see that since the entropy is a property of the system and not on its history, the entropy change of any system for any process that returns the system to its initial state (i.e., any cyclic process) will have a net zero change in entropy. For a cyclic process, the net change in the entropy of the system and its environment must not be negative. If the change is zero, this cyclic process is reversible.

If the system is closed, then $\delta Q = 0$, and any change in the system changing it from state A to state B either preserves its entropy or the entropy increases. Thus, following Classius, we have found that the one statement of the 2nd Law of Thermodynamics, "heat never flows from a cold body to a hotter one spontaneously," can be used to prove the existence of an extensive thermodynamic variable called the entropy and that the entropy of a closed system never decreases. The spontaneous progression of a closed system from any state to an equilibrium state cannot cause a decrease of its entropy. The equilibrium state must have a maximum entropy under the constraints there are on the closed system. If the system can change internally to a state of greater entropy, it will do so.[33]

For example, if a part of the system could expand doing work while taking in an equivalent amount of heat energy, it will happen, thereby increasing the total entropy of the system.

As we indicated, life systems may be considered as thermodynamic systems, even though there are some energy occupations of metastable molecules which are highly excited. Thermodynamic considerations are possible for cellular entities because water as a background solution is plentiful. Not a large portion of the energetic macromolecules change state in observational times. The pressure and temperature do not fluctuate significantly when measured over small regions of a cell. In fact, these two intensive variables are nearly constant from one region to the next. Metabolic processes, to good approximation, take place at fixed p and T.

Even though the entropy of a growing biological system decreases, *life systems do NOT violate the second law of thermodynamics.* They are OPEN systems. As such, they can drop in entropy, at the expense of their environment. The net change in the entropy of a life system and its environment does not decrease. Building a chicken from an egg, building proteins from amino acids, and making memory all involve a decrease in entropy of the life system being described. While these processes are occurring, there MUST be heat expelled to the outside. For example, if this heat Q is released into an environment at a fixed temperature T_E, the entropy change of the environment will be $\Delta S_E = Q/T$. If the life system has correspondingly dropped in entropy by ΔS_L, then the drop in entropy in the life system cannot be greater than Q/T.

With the new extensive quantity S characterizing a system, we can finally express the heat into a system in terms of a path integral over quasi-static changes of the system, since $\delta Q = TdS$. The changes in the energy of an open system is now expressible as[34]

[33] But the time for any given process is not given by thermodynamics. Some processes may take 'forever'.

[34] Added work terms would be included if the system is in an electric field acting on electric charges and dipoles, a magnetic field acting on magnetic dipoles, or has a bounding surface with surface tension. They all can be put into the form $f dX$, where f is a generalized force, and dX a

$$dE = TdS - pdV + \sum_k \mu_k dN_k \,, \tag{9.69}$$

where the system is allowed to exchange heat, work, and material with the environment. The N_k can be taken as the number of particles of species labeled by 'k' added to the system, which carries an energy μ_k per unit particle into the system. The intensive quantities μ_k are called the 'chemical potentials'. Differences of chemical potentials across space are associated with particle flow, just as differences in temperature across space are associated with heat flow.

The energy relation Eq. (9.69) is appropriate for chemical reactions among a subset of molecules in the system, wherein some of these molecules are converted into others. If the system is closed, then conservation of atoms ($N = \sum N_k$ fixed) will constrain the changes dN_k with $\sum dN_k = 0$. (To see how this is applied, see the discussion of biochemical reactions in Sect. 9.21.)

From our microscopic form for heat, Eq. (9.6), and now the connection $\delta Q = TdS$, we can see that entropy production involves the redistribution of the particles within the system. This observation will be made more explicit later by following Boltzmann's statistical interpretation of entropy.

Note that the expression (9.69) for dE tells us that E itself 'naturally' depends on the extensive variables S, V, and all the N_k. We say that the 'natural' variables[35] of the thermodynamic energy are entropy, volume, and particle numbers. Each of the coefficients of the differentials on the right-hand-side of the above expression for dE are intensive quantities, and must also depend only on S, V, and the N_k. The relation for dE can be used to *define* these measurable quantities:

$$T = \left(\frac{\partial E}{\partial S}\right)_{V,N} \tag{9.70}$$

$$p = -\left(\frac{\partial E}{\partial V}\right)_{S,N} \tag{9.71}$$

$$\mu_k = \left(\frac{\partial E}{\partial N_k}\right)_{S,V} \,. \tag{9.72}$$

These intensive quantities are 'slopes' of the extensive energy function E, plotted in the planes defined by holding all but one of its extensive variables (S, V, $\{N_k\}$) fixed.

For example, the temperature of a system can be determined by observing how its internal energy changes by varying its entropy, while holding its volume and constituents fixed. This is not a very convenient way to measure the temperature of a system, nor would you use the phrase to give others what the notion of temperature

---

generalized displacement, making $dE = \sum_i f_i dX_i$. The pair $\{f_i, X_i\}$ are called thermodynamic conjugate variables.

[35] Natural variables are also called by some authors 'characteristic variables'.

is all about. However, it is the most fundamental way to define temperature of any system, and it does not depend on the particular properties of a thermometer.

We now take advantage of the extensive nature of certain thermodynamic functions. The internal energy, the entropy, the volume, the total particle number, are all extensive quantities. In fact, if a system is reproduced λ times, then the energy of the combined system satisfies

$$E(\lambda S, \lambda V, \lambda N) = \lambda E(S, V, N) , \tag{9.73}$$

i.e. λ appears as an overall factor on the right, and to the first power. The 'degree of homogeneity' is this power, so the internal energy is 'homogeneous to degree one'. As we will see, a consequence is that we can express the internal energy in terms of its slopes. The argument goes as follows.

If a function is homogeneous to degree $\mathbb{h}$, then:

$$f(\{\lambda X_k\}) = \lambda^{\mathbb{h}} f(\{X_k\}) , \tag{9.74}$$

is called a homogeneous function of degree $\mathbb{h}$. By differentiation with respect to λ, and then setting $\lambda = 1$, we will have

$$\mathbb{h} f(\{X_k\}) = \sum_k \frac{\partial f}{\partial X_k} X_k . \tag{9.75}$$

This is called the Euler relation.[36] It applies to both intensive ($\mathbb{h} = 0$) and extensive ($\mathbb{h} = 1$) thermodynamics variables.

Applying the Euler relation to the internal energy, we have

$$E(S, V, N) = TS - pV + \sum_k \mu_k N_k . \tag{9.76}$$

We now see that if the functions $T(S, V, N)$, $p(S, V, N)$ and $\mu(S, V, N)$ are known, then the complete thermodynamic properties of the system are known.[37] These functions are called equations of state.

Given the equations of state, or given the function $E(S, V, N)$, all other measurable thermodynamic quantities can be expressed. For example, the heat capacity at constant volume becomes

[36]Actually, Euler discovered many important relations! Perhaps the most profound is $\exp(i\theta) = \cos\theta + i\sin\theta$. The latter contains arguably the most magical formula in all mathematics, $(2.71828\cdots)^{\sqrt{-1}(3.14159\cdots)} = -1.000\cdots$.

[37]These three functions determining the equations of state are not independent (see Sect. 9.11.2), so that only two of the three are needed.

$$C_v = \frac{\partial E/\partial S}{\partial^2 E/\partial S^2} \,. \tag{9.77}$$

Expressions equivalent to the fundamental equation of state $E = E(S, V, N)$ can be found by favoring a set of thermodynamic variables other than (S, V, N) in which to express the measurable quantities. For example, $S = S(E, V, N)$ serves just as well.

9.11.1 Equations of State for an Ideal Gas

As an example of equations of state, we will give them for an ideal gas:

$$k_B T = (2/3)u\,,$$

$$p = (2/3)u/v\,,$$

and

$$\mu = -u \ln\left[\left(\frac{4\pi m}{3h^2}\right) u \cdot v^{2/3}\right], \tag{9.78}$$

where $u \equiv E/N$, $v \equiv V/N$, m is the mass of the ideal gas particle, and h is Planck's constant.[38]

Later, we will find it useful to have the chemical potential expressed as a function of temperature, pressure, and particle numbers rather than S, V, and N. Substituting the first and second relations of Eq. (9.78) into the third, we have, for a mixture of ideal gases,

$$\mu_i(T, p, \{x_i\}) = k_B T \ln\left[\left(\frac{h^2}{2\pi m_i}\right)^{3/2} \frac{1}{(k_B T)^{5/2}} \cdot x_i p\right], \tag{9.79}$$

where

$$x_i \equiv N_i/N \tag{9.80}$$

is the 'partial fraction' of substance 'i' in a mixture of substances, so $0 < x_i \leq 1$. The pressure due to just one of the gases in the mixture is called the partial pressure

$$p_i \equiv x_i p\,, \tag{9.81}$$

satisfying $\sum_i p_i = p$, since $\sum_i x_i = 1$

[38]The constant multiplying $u \cdot v^{2/3}$ was first calculated in 1912 independently by Otto Sackur and Hugo Tetrode when they found the absolute entropy of an ideal gas, taking advantage of Arnold Sommerfeld's ideas on how to extend the Planck quantum hypothesis. Their prediction was confirmed in experiments with mercury vapor. A modern treatment is given in Chap. 10.

Our expression for the chemical potential of an ideal gas, Eq. (9.79), will be derived in Chap. 10, resulting in Eq. (10.158).

9.11.2 Gibbs–Duhem Relations

As we saw, the equations of state determine the state of the system uniquely because of the homogeneity of the internal energy E in the dependence on the extensive variables S, V, and the N_k, as shown in Eq. (9.73). But this same homogeneity implies that not all the equations of state are independent.

We can see this if we differentiate E in Eq. (9.76) and subtract dE from Eq. (9.102), to find

$$0 = SdT - Vdp + \sum_k N_k d\mu_k \,. \tag{9.82}$$

This is called a 'Gibbs–Duhem' relation. Because of this connection, T, p, and the μ_k cannot be varied arbitrarily. With r different components in the system (i.e. $1 \leq k \leq r$), there will be $r + 1$ independent equations of state among the $r + 2$ variables $T, p, \mu_1, \mu_2, \cdots, \mu_r$.

To see how the relationship works, let us again consider an ideal gas made from a mixture of different gases. Since the first law of thermodynamics can be written as $dE = TdS - pdV + \sum \mu_k dN_k$, the equations of state should determine T, p, and the μ_k as a function of S, V, and N. Kinetic theory applied to an ideal gas tells us that $E = Nc_vT$ and $p = nRT/V = NkT/V$, where c_v is the gas specific heat at constant volume. For a single component gas, we do not need to give a third equation of state for $\mu(T, p, N)$ as a function of three other thermodynamic variables because we can use the Gibbs–Duhem relation. If we take, for the moment, a constant-temperature process to do the integration, then Eq. (9.82) gives

$$\mu = \int \frac{V}{n} dp = RT \int \frac{dp}{p} \,.$$

For any temperature and pressure,

$$\mu(T, p, \{N_i\}) = RT \ln \frac{p}{p_o} + \mu^o(T, \{N_i\}) \,, \tag{9.83}$$

where $\mu^o(T\{N_i\})$ is a function of temperature but does not depend on pressure.[39]

[39] The chemical potential is often given in units of joules per mole. If we want the chemical potential in units of energy per molecule, we must replace R by k in the Eq. (9.83).

If we had a mixture of ideal gases, all acting independently, we would have for each:

$$\mu_i(T, p, \{N_i\}) = RT \ln \frac{p_i}{p_o} + \mu_i^o(T, \{N_i\}), \tag{9.84}$$

where $p_i = n_i RT/V = (n_i/n)p = x_i p$ is the partial pressure for the gas component labeled by the index 'i', and x_i is the 'mole fraction' of that component. As a result, we have the alternative expression:

$$\mu_i(T, p, \{x_i\}) = RT \ln x_i + RT \ln \frac{p}{p_o} + \overline{\mu}_i^o(T, \{x_i\}). \tag{9.85}$$

9.12 Entropy Maximum and Energy Minimum

Consider a system characterized by an internal energy $E(S, V)$ dependent only on its entropy and volume. If we solve for the value of the entropy S, we have the identity

$$E = E(S(E, V), V) \tag{9.86}$$

By differentiation, small deviations from the equilibrium volume will give

$$0 = \frac{\partial E}{\partial S}\frac{\delta S}{\delta V} + \frac{\partial E}{\partial V}, \tag{9.87}$$

which gives $\partial E/\partial V$, the slope of the energy function as the volume deviates from equilibrium. The curvature of the function can be found by differentiating with respect to V again and rearranging to get

$$\begin{aligned}\frac{\partial^2 E}{\partial V^2} &= -\frac{\partial^2 E}{\partial S^2}\frac{\delta S}{\delta V} - \frac{\partial E}{\partial S}\frac{\delta^2 S}{\delta V^2}\\ &= -\frac{\partial^2 E}{\partial S^2}\frac{\partial S}{\partial V} - T\frac{\partial^2 S}{\partial V^2}\end{aligned}$$

But at equilibrium, we know that S has reached a minimum as a function of volume, so $\frac{\delta S}{\delta V} = 0$, and $\frac{\delta^2 S}{\delta V^2} < 0$. Thus, since $T > 0$, we have $\frac{\partial^2 E}{\partial V^2} > 0$. (See Fig. 9.6.) This means the energy as a function volume is at the bottom of a parabola when near equilibrium. The energy of the system reaches equilibrium when it is at a minimum as a function of volume, holding entropy fixed.

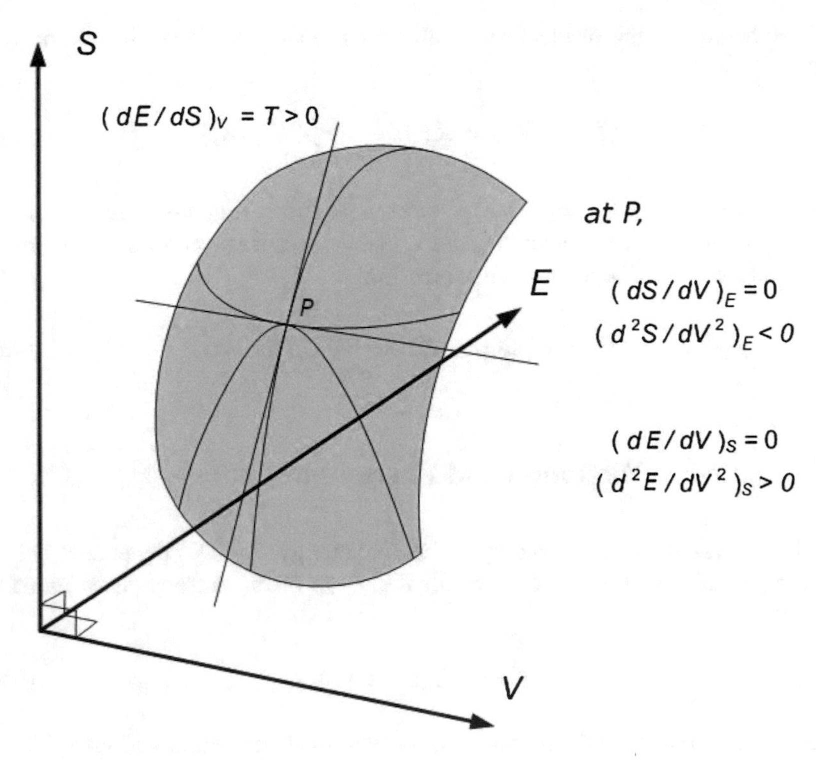

Fig. 9.6 Equilibrium surface for extensive variables

9.13 Free Energies

9.13.1 Enthalpy

It is useful to have a quantity which characterizes processes near equilibrium at constant pressure. The fluid in the cells of our body are at nearly constant (atmospheric) pressure. The largest variation comes from blood pressure adding to atmospheric, which reaches to about a 16% increase.

Consider a system variable called the enthalpy[40] defined by

$$H \equiv E + pV = TS + \sum_k \mu_k N_k , \tag{9.88}$$

so that for reversible processes

[40]It is unfortunate that H is the symbol commonly used for $E + pV$. That excludes H as a symbol for the Helmholtz free energy. Instead, $A \equiv E - TS$ is most often used for the Helmholtz free energy.

$$dH = TdS + Vdp + \sum_k \mu_k dN_k \,. \tag{9.89}$$

This relation tells us that if the enthalpy is expressed in terms of entropy, pressure, and the amount of each substance in the system, then the thermodynamics of the system is completely determined. For this reason, the set $\{S, p, N\}$ are the 'natural variables' for the enthalpy function.

The enthalpy change is used in chemistry to describe the heat energy released (or absorbed) during a reaction for the following reason. Often, chemical reactions in a laboratory test tube and in the cells of our body occur at constant pressure (usually atmospheric). Under constant pressure conditions, the change in the enthalpy ($\Delta H \equiv H_f - H_i$) after a reaction that releases an amount of heat Q will be

$$\begin{aligned} \Delta H &= \Delta E + p\Delta V \\ &= -Q - p\Delta V + p\Delta V \\ &= -Q \,. \end{aligned} \tag{9.90}$$

The minus sign on Q comes from the convention that heat *into* a system is positive. Thus, under isobaric conditions, the negative of the heat released will be the change in the enthalpy for the system.

As a system tends toward thermal equilibrium at constant pressure, the enthalpy will decrease toward a minimum. This can be seen from the first law in the form $\delta E = \delta\!\!\!{}^-Q - p\delta V + \sum \mu_k \delta N_k$ so, at constant pressure,

$$\delta H = \delta\!\!\!{}^-Q + \sum_k \mu_k \delta N_k \le TdS + \sum_k \mu_k dN_k \tag{9.91}$$

which gives $\delta H \le dH$.

Under constant pressure and fixed particle numbers, the enthalpy change ΔH of a system as the temperature changes can be found by measuring the system's molar specific heat (at constant pressure) c_p:

$$\Delta H|_p = \int dH|_p = \int T \left.\frac{\partial S}{\partial T}\right|_p dT = n \int c_p dT \,, \tag{9.92}$$

where n is the number of moles of substance in the system. Starting at a standard state (denoted by a superscript ($\ominus$)), the enthalpy at temperature T will be

$$H(T) = H^{\ominus} + n \int_{T_o}^{T} c_p dT \,. \tag{9.93}$$

9.13.2 Helmholtz Free Energy

The Helmholtz free energy[41] is defined by

$$A \equiv E - TS = -pV + \sum_k \mu_k dN_k \,, \tag{9.94}$$

so that for reversible processes

$$dA = -SdT - pdV + \sum_k \mu_k dN_k \,. \tag{9.95}$$

Now, for an isothermal process, the negative of the change in the Helmholtz free energy gives the work done by the system. As the system tends toward equilibrium at constant volume, its total energy tends toward a minimum, so that the Helmholtz free energy goes toward a minimum.

Gibbs Free Energy

The Gibbs free energy is

$$G \equiv E + pV - TS = \sum_k \mu_k N_k \,, \tag{9.96}$$

so for reversible processes,

$$dG = -Vdp - SdT + \sum_k \mu_k dN_k \,. \tag{9.97}$$

The 'natural variables' for G are $\{T, p, N\}$. The Gibbs free energy is a useful function for isothermal and isobaric processes, which is often the case for biological processes and chemical reactions in solution.

Note that the relation

$$G = \sum_k \mu_k N_k \tag{9.98}$$

shows that the Gibbs free energy is determined by the chemical potentials and the number (or concentration) of entities present. Since $G = H - TS$, the change in the Gibbs free energy for an isothermal process going toward equilibrium will be

[41] The phrase 'free energy' refers to energy available to do work in some specified closed cycle. If the cycle is not closed, ALL the energy is available, in principle.

$$\delta G = \delta H - T\delta S\,. \tag{9.99}$$

For an isobaric change of a materially closed system, $\delta H = \delta\!\!\!{}^{-} Q$. If the system is adiabatically closed, $\delta\!\!\!{}^{-} Q = 0$, so $\delta G = -T\delta S$. Since the entropy must not decrease, the Gibbs free energy must not increase. Spontaneous reactions in a closed system at fixed T and p must proceed in the direction that makes δG smaller.

From the 2nd Law, near equilibrium, $dG|_{T,p} \geq 0$. This means the Gibbs free energy has reached a minimum, and any small change in the particle numbers will cause G to increase.

9.14 Summary of the Classical Laws of Thermodynamics

Thermodynamics relies on the observation that most systems near but not at equilibrium, when left isolated, will tend toward equilibrium. In the process, local excesses in pressure may be relieved by changes in volume, and local excesses in temperature causes a redistribution of particles among the available energy states.[42] Macrosystems (systems with a large number of interacting particles), when near or at thermal equilibrium, satisfy:

- The "zeroth" law of thermodynamics is taken as the observation that if system A has the same temperature as B and B has the same temperature as C, then A will have the same temperature as C.
- The first law of thermodynamics, taken as the energy conservation statement, is often written as

$$dE = \delta\!\!\!{}^{-} Q - \delta\!\!\!{}^{-} W\,. \tag{9.100}$$

  The left side is the change in the internal energy of a system. On the right is the heat into the system minus the work done by the system. For systems with a mixture of chemical components, the work term can be generalized to include the energy transferred by the binding, unbinding, or recombining components.
- The second law may be stated as: There exists a property of a thermodynamic system, called its entropy, S, found by integrating the heat input to the system divided by its temperature through a reversible process from a standard state with a standard value of the entropy. For a CLOSED system, the entropy never decreases. In general,

[42] The changes of a system to relieve stresses within is often called 'Le Châtelier's' principle. Physically, the principle comes from the action of natural forces and of the tendency toward greater disorder predicted by probability theory.

$$S_B - S_A \geq \int_A^B \frac{\delta Q}{T} . \tag{9.101}$$

For a reversible process, the equality holds.[43]

- A third law of thermodynamics, first proposed by Walther Nernst in 1905,[44] states that no thermodynamic process can reduce entropy of a system to its zero-point value, nor reduce the temperature of a system to zero.

For thermodynamic processes, including those which might involve changes in the particle components, the first and second law lead to

$$dE = TdS - pdV + \sum_k \mu_k dN_k . \tag{9.102}$$

in which the entropy is calculated for a reversible process. The last term accounts for energy gain or loss due to the flow of material into or out of the system, or by internal changes in the amount of substances as might occur in a chemical reaction. The coefficient μ_k is called the chemical potential for the substance labeled by 'k' with a total particle number of N_k. The chemical potential determines how much energy per particle is added to the system when the corresponding particle number increases. Note that the internal energy can be changed only through changes in thermodynamic extensive variables: entropy, volume, and particle number. If we know $E(S, V, N)$, all other thermodynamic properties can be found.

The intensive/extensive pairs (T, S), (p, V), and (μ_k, N_k) in the energy differential are called thermodynamic conjugates. This pairing also appears in 'Maxwell's relations', to be presented shortly.

9.14.1 Entropy Maximum from the Second Law

A reformulation of the second law of thermodynamics can now be stated:

> If a system, closed to matter and energy flow, is in a state of non-equilibrium and allowed to relax, it will evolve toward equilibrium with its entropy reaching a maximum.

Mathematically, near equilibrium, any variation of S under any virtual changes of the extensive variables, apart from the total number of particles or total energy, will

[43] Government taxes should be based on excess production of entropy, rather than energy use. After all, if you transform energy with 100%e efficiency, you have not made any of it unavailable. But if you are wasteful, you irreversibly disorder the Universe more than necessary.

[44] W. Nernst, **The theoretical and experimental basis of the New Heat Theorem**, [translation: E.P. Dutton & Co., N.Y.] (1912).

have $\delta S = 0$ and $\delta^2 S < 0$, i.e. the slope of S must vanish at equilibrium and the curvature near equilibrium must be negative, so that S is at a maximum. (See Fig. 9.6.)

9.14.2 Energy Minimum from the Second Law

The entropy maximum formulation of the second law of thermodynamics can be reformulated in terms of an energy minimum principle:

> If a system, closed to matter and held to fixed entropy,[45] is put into a state of non-equilibrium and allowed to relax, it will evolve toward equilibrium with its internal energy reaching a minimum.

Near equilibrium, any variation of E under virtual changes in one or more of the unconstrained extensive variables, apart from the entropy and total particle number, will have $\delta E = 0$ and $\delta^2 E > 0$, i.e. the slope of E must vanish at equilibrium and the curvatures near equilibrium must be positive, so that E is at a minimum.[46] Similarly, if, near equilibrium, the curvature of the entropy under virtual changes in any extensive variable other than energy is negative, and the temperature is positive, then the curvatures of the energy function will be positive. (See Fig. 9.6.)

9.15 Utility of the Gibbs Free Energy

During chemical reactions, in the body and elsewhere, the entropy and the volume for any subsystems are usually not fixed. Fixed entropy means no heat exchange and no internal dissipation. Having perfect insulation is practically difficult. Also, having a fixed volume during a reaction is also difficult in practice for liquids and solids. Correspondingly, it is hard experimentally to keep the internal energy of a system fixed during a thermodynamic process. However, J. Willard Gibbs exploited the fact that the quantity

$$G \equiv E - TS + pV,$$

changes only if the temperature, pressure, or some particle numbers change, such as in a chemical reaction. This property follows from the differential of G:

---

[45] No heat is allowed to be transferred from the environment into or out of the system.

[46] Quantum systems with a finite number of possible energy states can have negative temperatures: The addition of energy causes a decrease of entropy as particles are forced into limited upper energy states.

$$dG = -SdT + Vdp + \sum_k \mu_k dN_k \tag{9.103}$$

showing that the Gibbs free energy is a function of temperature, pressure, and the particle number for each species which might undergo a change.

In the laboratory and in live organisms, the temperature and pressure of a variety of subsystems are nearly constant. Their variations are easily controlled. This makes the Gibbs free energy convenient to use in the description of chemical changes under natural conditions.

Now, since $dE = \delta Q - pdV + \sum \mu_k dN_k$, the second law of thermodynamics in the form $\delta Q \leq TdS$ leads to a statement evidently useful for chemical reactions:

> At constant temperature and pressure, quasi-static reactions proceed in the direction that lower the Gibbs free energy. At equilibrium, The Gibbs free energy is a minimum.

At constant temperature, changes in the Gibbs free energy,

$$G = H - TS,$$

come about by changes in the enthalpy H and/or by changes in the entropy S:

$$\Delta G = \Delta H - T\Delta S, \tag{9.104}$$

each change being calculated from their values after the reaction subtracted from the values before the reaction. When we study chemical reactions under constant temperature and pressure, we will see that at equilibrium, the final concentrations of the products of the reaction are related to the initial concentrations of the reactants through a Boltzmann factor $\exp(-\Delta G^o/(k_B T))$ that appears explicitly in the 'law of mass action', or implicitly as the 'equilibrium constant'.

In small regions of biological cells in which temperature and pressure have values which match the environment of the cell, any variation in the Gibbs free energy $dG = SdT - Vdp + \sum \mu dN \rightarrow \sum \mu dN$ comes about from changes in the amount of each substance in that region. These changes may occur due to phase changes, chemical reactions, or transport into or out of the region. Near equilibrium, small variations in G must vanish. For example, for a system with two varieties of two substances (with labels for substances A, B, and labels for varieties (such as phases or molecular configurations) 1, 2, then $dN_1^A = -dN_2^A$ and $dN_1^B = -dN_2^B$, since the total number of particles of each kind of substance is conserved. In this case, near equilibrium, $\delta G = (\mu_1^A - \mu_2^A)\delta N_1^A + (\mu_1^B - \mu_2^B)\delta N_1^B = 0$. As a result, at equilibrium, where $\delta G = 0$, the chemical potentials for each species must match across phases of that species:

$$\begin{aligned} \mu_1^A &= \mu_2^A\,, \\ \mu_1^B &= \mu_2^B\,. \end{aligned} \tag{9.105}$$

The chemical potential for a species of particles determines the 'pressure' to cause a change in the number of those particles, just as ordinary pressure causes changes in volume. More specifically, the chemical potential is the energy gained by a system per unit particle number (or moles) when those particles increase in the system, just as pressure is the energy gained in a system per unit volume loss due to pressure.

The balancing of the chemical potentials we will see also applies when a chemical reaction generates a transfer of material from one type to another. When the chemical potentials do not match across regions in a cell, material will flow until they do. If they do not match in a mixture of compounds, a chemical reaction may be available to equalize the chemical potentials. Whether the reaction occurs in times short compared to other activities is outside the purview of thermodynamics, but rather requires more detailed molecular dynamics, becoming the subject of chemical kinematics, a topic of Sect. 10.11.

After introducing Maxwell's thermodynamic relations, we will use the equality of chemical potentials in different phases to derive the vapor pressure of water and of a gas dissolved in a liquid, and the equality across a semi-permeable membrane to derive the osmotic pressure difference across the membrane. Such topics are clearly important for plants and animals.

9.16 Maxwell Relations

For quasi-static changes that cause a 'smooth' change in the energy of a system, we will have

$$\frac{\partial^2 E}{\partial S \partial V} = \frac{\partial^2 E}{\partial V \partial S}, \tag{9.106}$$

which can be written

$$\left.\frac{\partial p}{\partial S}\right|_V = -\left.\frac{\partial T}{\partial V}\right|_S . \tag{9.107}$$

This is the first of such 'slope' connections called 'Maxwell relations' based on the 'smoothness' of the thermodynamic functions in their variables, apart from regions of phase change. This one is useful in relating an isochoric process (fixed volume)[47] to an adiabatic one (no heat transfer).

Three more Maxwell relations come from the smoothness of the enthalpy, the Helmholtz free energy and the Gibbs free energy defined as follows. Consider the enthalpy,

[47] Isochoric is from the Greek: *isos choros*, 'equal space'.

$$H \equiv E + pV \; ;$$

the Helmholtz free energy

$$A \equiv E - TS \; ;$$

and the Gibbs free energy

$$G \equiv E + pV - TS = H - TS = \sum \mu_k N_k \; .$$

From the relation

$$dH = TdS + Vdp + \sum_k \mu_k dN_k \; . \tag{9.108}$$

we have already observed that the 'natural variables' in which to express the enthalpy are entropy, pressure, and particle numbers. It follows from the smoothness of $H(S, p, \{N_k\})$ that

$$\left.\frac{\partial T}{\partial p}\right|_S = \left.\frac{\partial V}{\partial S}\right|_p \; . \tag{9.109}$$

The Helmholtz free energy, with the differential:

$$dA = -SdT - pdV + \sum_k \mu_k dN_k \; , \tag{9.110}$$

is useful for describing changes of a system at constant temperature and volume. Being that the differential dA of the Helmholtz free energy is naturally expressed in terms of dT, dV, and the dN_k, the 'natural variables; of A are temperature, volume, and particle numbers. The smoothness of $A(T, V, \{N_k\})$ gives

$$\left.\frac{\partial S}{\partial V}\right|_T = \left.\frac{\partial p}{\partial T}\right|_V \; . \tag{9.111}$$

Finally, the Gibbs free energy reversible changes are given by

$$dG = -SdT + Vdp + \sum_k \mu_k dN_k \; . \tag{9.112}$$

showing that $G = G(T, p, \{N_k\})$. The smoothness of $G(T, p, \{N_k\})$ gives

$$\left.\frac{\partial S}{\partial p}\right|_T = -\left.\frac{\partial V}{\partial T}\right|_p \; . \tag{9.113}$$

In thermodynamics, we often wish to express some derivatives which are difficult to measure in terms of ones that are easier to measure. This is done by using relationships between the thermodynamic variables. For example, suppose we have a prediction for $\partial p/\partial T|_S$, i.e. the variation of the pressure in a system with temperature at constant entropy. Experimentally, holding a system at constant entropy may be difficult, but holding the system at constant temperature or constant pressure is likely to be easier. But as we will show,

$$\partial p/\partial T|_S \equiv -\frac{\partial S/\partial T|_p}{\partial S/\partial p|_T}\,, \tag{9.114}$$

so, making two separate measurements (one for the numerator and one for the denominator), we can calculate the left-hand side and compare with our prediction for this quantity.

Here is a general case. Suppose we know a relationship between the three variables x, y, z, of the form $z = f(x, y)$, where f is differentiable in both x and y. Then

$$dx = \left.\frac{\partial x}{\partial y}\right|_z dy + \left.\frac{\partial x}{\partial z}\right|_y dz\,, \tag{9.115}$$

$$dz = \left.\frac{\partial z}{\partial x}\right|_y dx + \left.\frac{\partial z}{\partial y}\right|_x dy\,. \tag{9.116}$$

Putting the dx from the second relation into the first will make an identity, expressed as

$$dx \equiv \left(\left.\frac{\partial x}{\partial y}\right|_z + \left.\frac{\partial x}{\partial z}\right|_y \left.\frac{\partial z}{\partial y}\right|_x \right) dy + \left.\frac{\partial x}{\partial z}\right|_y \left.\frac{\partial z}{\partial x}\right|_y dx\,. \tag{9.117}$$

As dx and dy are independent, we have

$$\left.\frac{\partial x}{\partial y}\right|_z = -\left.\frac{\partial x}{\partial z}\right|_y \left.\frac{\partial z}{\partial y}\right|_x\,, \tag{9.118}$$

and

$$\left.\frac{\partial x}{\partial z}\right|_y \left.\frac{\partial z}{\partial x}\right|_y = 1\,. \tag{9.119}$$

These are often written as

$$\left.\frac{\partial x}{\partial y}\right|_z = -\frac{\partial z/\partial y|_x}{\partial z/\partial x|_y}\,, \tag{9.120}$$

and

$$\left.\frac{\partial x}{\partial z}\right|_y = \frac{1}{\partial z/\partial x|_y} . \qquad (9.121)$$

The relation Eq. (9.114) comes from applying Eq. (9.120).

9.17 Pressure in Multistate and Multicomponent Systems

Given that life systems on Earth rely on water for their biological activity, the nature of evaporation and condensation of water from exposed surfaces is of considerable importance. Also, because biological cells hold interior concentrations of some salts and other compounds that differ from their exterior concentrations, osmotic pressure across cell walls affects how the cell operates. Biological cells have developed systems for both passive and active control of each of these pressures.

The air in your lungs is quickly saturated by water vapor and other gases from the inner surface of alveoli. The surface of the tiny sacks used to transfer oxygen to your blood, and carbon dioxide out, have a wet membrane lining next to the ventilating gases of the air. A variety of volatiles dissolved in that wet membrane can move across the liquid-gas boundary, changing the gas vapor pressure.

For another example, your skin evaporates water into the air, as does a lake. Just above your skin, and just above the lake, the vapor pressure of water approaches a saturation value..

When there are a multitude of species of materials in several phases, we can ask how many independent intensive thermodynamic variables, f, are needed to describe the thermodynamic equilibrium and near equilibrium states. Willard Gibbs derived the result: $f = c - p + 2$, where c is the number of component species, p the number of phases. The relation is called the 'Gibbs' phase rule'. For example, applied to water (one component species) in three phases (solid, liquid, and gas), $f = 1 - 3 + 2 = 0$, which means that if all three phases of water are in equilibrium, there is a unique state (called the 'critical point' with fixed values of the temperature and pressure: $T_c = 647°\mathrm{K}$, $p_c = 22.064\,\mathrm{MPa}$). If there are two phases, then $f = 1$, which allows equilibrium points on a line (phase boundary) in a pressure vs temperature (p-T) plot. When two phases exist near equilibrium, any heat added to the system goes into energy needed to change the phase (e.g. liquid water to vapor water), and not to changes of temperature. (Excursions into thermodynamically unstable regions are possible to observe, such as in superheated liquids and supercooled gases .) Water in a single phase ($f = 2$) can be near equilibrium for points on a surface in a p-V-T plot, as depicted in Fig. 9.7.

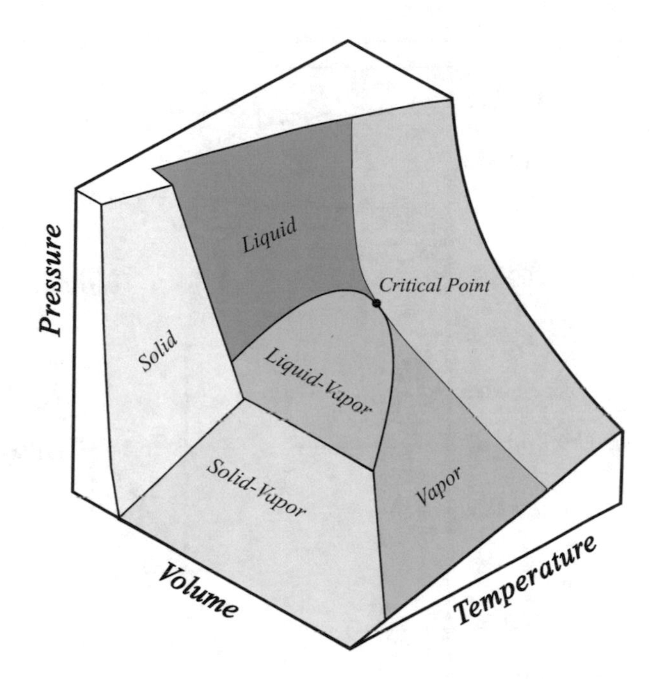

Fig. 9.7 Simplified water equilibrium PVT surface

9.18 Vapor Pressure

Life on Earth relies on liquid water for its existence. Plants and animals utilize gases dissolved in water solutions, particularly carbon dioxide and oxygen. The atmosphere of the Earth is intimately involved, as these gases are absorbed and are emitted by the oceans as vapor.

To see what is involved in vapor pressure[48] at a water solution-vapor interface, we will first analyze a system holding a single substance in two phases, such as a liquid and its gas, at a fixed temperature and pressure.

When the system reaches equilibrium, the chemical potential of the two phases must match across the phase boundaries: $\mu_1 = \mu_2$, where we have labeled the two phases by 1 and 2. This relation can be used to determine the pressure when both phases coexist. First, we determine the slope of a boundary line in a pressure vs temperature plot, such as the lines shown in Fig. 9.8, by following how the chemical potential changes going from one point on the phase boundary ∂b to an adjacent equilibrium point, i.e. using

$$d\mu_1|_{\partial b} = d\mu_2|_{\partial b} \,, \tag{9.122}$$

$$\frac{\partial \mu_1}{\partial p}dp + \frac{\partial \mu_1}{\partial T}dT = \frac{\partial \mu_2}{\partial p}dp + \frac{\partial \mu_2}{\partial T}dT \tag{9.123}$$

[48]The phrase ‘vapor tension’ for vapor pressure is now considered obsolete.

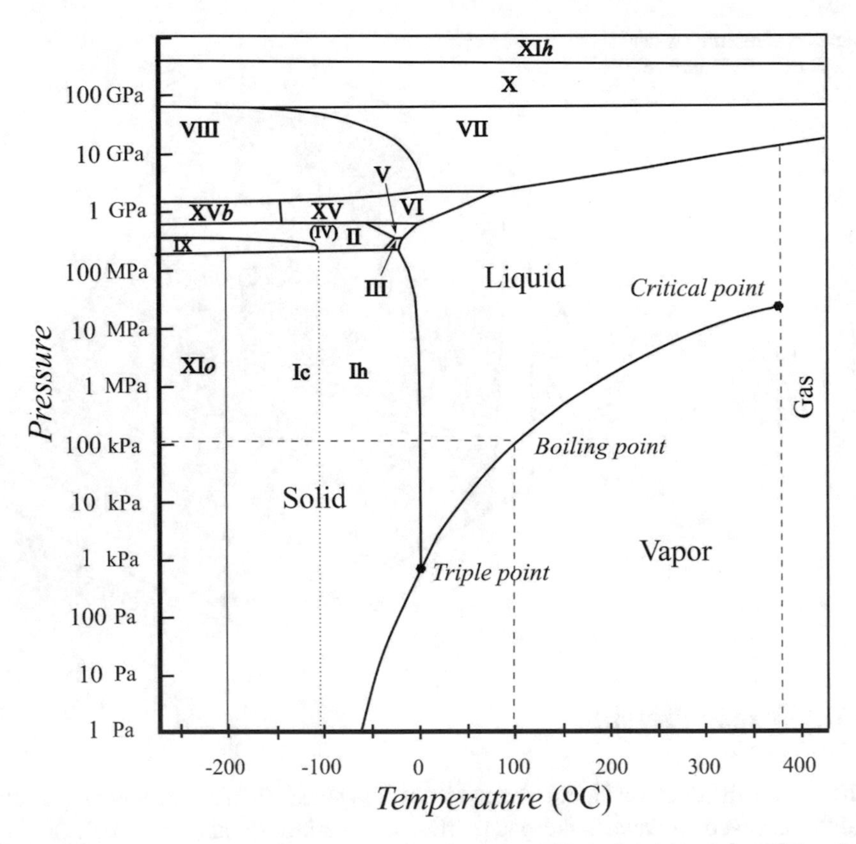

Fig. 9.8 Pressure versus temperature phase plot for water. This is the view of the PVT surface looking back along the volume axis. Kurt Vonnegut's ice-nine does not show

which gives

$$\frac{dp}{dT} = -\frac{\partial \mu_1/\partial T - \partial \mu_2/\partial T}{\partial \mu_1/\partial p - \partial \mu_2/\partial p} . \tag{9.124}$$

We will apply Maxwell relations to the partial derivatives, with the caution that these derivatives may become discontinuous if we attempt to cross the phase-change boundary in the p-T plot. In the resulting expression, the derivatives with respect to N_i are taken on the 'side' of the phase boundary in the i-th phase.

$$\frac{dp}{dT} = \frac{\partial S/\partial N_1 - \partial S/\partial N_2}{\partial V/\partial N_1 - \partial V/\partial N_2} . \tag{9.125}$$

Using

$$dN_2 = -dN_1 ,$$

$$\Delta S = (\partial S/\partial N_1)dN_1 + (\partial S/\partial N_2)dN_2,$$

and

$$\Delta V = (\partial V/\partial N_1)dN_1 + (\partial V/\partial N_2)dN_2,$$

we can write

$$\frac{dp}{dT} = \frac{\Delta S}{\Delta V}, \tag{9.126}$$

where ΔS and ΔV are the discontinuous changes in the entropy and volume because of the phase change. The entropy production per unit mass, Δs, occurs at constant pressure, so $\Delta s = \Delta h/T \equiv \lambda/T$, where Δh the enthalpy change per unit mass when a phase change occurs, and λ is the 'latent heat of vaporization per unit mass'.

For liquid water to vapor, λ_w is 539.3 cal/g = 2256.4 kJ/kg at 100 °C and 1 atm. The temperature dependence of λ_w above the freezing point is empirically well represented by Watson's formula[49] $\lambda_w = \lambda_o(1 - T/T_c)^{0.38}$, where $T_c = 647.3$ K is the critical temperature of water, and $\lambda_o = 3127.5$ kJ/kg. Condensation releases the same amount of heat that is absorbed by vaporization. Water going from liquid to solid releases 79.71 cal/g= 333.5 kJ/kg at 0 °C and 1 atm, the heat of fusion for water.

Two more simplifications are possible for liquid-gas and solid-gas mixtures. The volume of the gas for a given number of molecules that vaporize is often much greater than the volume they had as a liquid or solid, so, taking phase 1 as the gas, $\Delta V = V_1 - V_2 \approx V_g$. In addition, the gas is often well described by the ideal gas law, so $V_g = n_g RT/p$. We will use $\overline{\rho}_g$ as the mass per mole of the gas. The result for the slope of the phase boundary is

$$\frac{dp}{dT} = \frac{\overline{\rho}_g \lambda}{RT^2} p, \tag{9.127}$$

where $\overline{\rho}_g$ is the mass per mole in the gas. This relation was found by Clapeyron even before entropy was discovered, and then later more rigorously from thermodynamics by Clausius.

Over small regions around the temperature of the human body, λ is nearly constant. If it is taken so, then Eq. (9.127) can be easily integrated, giving the vapor pressure as a function of temperature as

$$p = p_o \exp\left(-\frac{\overline{\rho}_g \lambda}{RT}\right). \tag{9.128}$$

[49] K.M. Watson, *Thermodynamics of the Liquid States, Generalized Prediction of Properties*, Ind. Eng. Chem., **35**, pp.398–406 (1943).

Improvements to account for the temperature dependence of the latent heat can be represented by fits to semi-empirical formulae, such as the following for saturated water vapor over liquid water and ice for temperatures in the range $-100\,°\mathrm{C} < T < 100\,°\mathrm{C}$:

$$p = p_o \left(\frac{T_o}{T}\right)^{\alpha} \exp\left(\beta\left(1 - \frac{T_o}{T}\right)\right). \qquad (9.129)$$

Here, p_o is 6.0495 mPa, T_o is 273.15 K. The numbers α and β are 5.09 and 24.938 for vapor over liquid, and 0.555 and 23.051 for vapor over ice.[50]

When there is more than one volatile component in a system, each dissolved in one solvent, we can still use the balancing of chemical potentials, one for each component, to find their partial vapor pressures. Table 9.1 shows the approximate partial vapor pressures for the important gases in healthy human lung alveoli. Shown are absolute pressures, not gauge pressures. The sum of these partial pressures in the alveoli is the pressure in a lung with a relaxed diaphragm.

The concentration of each component in the solution is often characterized by its molecular fraction $x_i \equiv N_k/\sum N_i$, its mole fraction $n_k/\sum n_i = x_i$, or by its molar concentration $[C_k] = n_i/V = (n_k/n)(n/V) = x_k/\overline{V}$, where $\overline{V}$ is the volume per mole for all dissolved substances.

If the mole fraction of a substance in solution is close to one, as is the case of the solvent in dilute solutions, the other substances in the solution have little effect on the ability of dominant molecules to escape or enter the interfacing surfaces. Near equilibrium of the solution with the gas of the solvent, the partial pressure of the dominant substance in the saturated gas will be near its vapor pressure p_k^{pure} as a pure liquid times the mole fraction x_k of that substance in the solution:

$$p_k \approx x_k p_k^{pure}. \qquad (9.130)$$

This relationship (9.130) is called Raoult's law. It follows from a balancing of chemical potentials provided that no heat is released or drawn when the solutes are mixed with the solvent, and the solvent gas acts like an ideal gas. When Raoult's law applies, one calls the solution 'ideal'.

Table 9.1 The partial pressures of principle gases in the alveoli and in the atmosphere

Gas	p (alveoli)	p (atmosphere)
N_2	568 mmHg	593 mmHg
O_2	105 mmHg	159 mmHg
H_2O	40 mmHg	(3) mmHg
CO_2	47 mmHg	0.31 mmHg

[50]See Craig F. Bohren and Bruce A. Albrecht, **Atmospheric Thermodynamics**, Oxford University Press, New York (1998).

For substances with a low concentration ($x_k << 1$) in a solution, the ability of the molecules of that substance to escape from the solution as a gas depends on how 'sticky' the molecules are with the various materials at the solution-gas interface. Still, one should expect that the vapor pressure of the dissolved substance in low concentration in solution will be proportional to the substance's concentration in the solution:

$$p_k \approx x_k K_k^H . \tag{9.131}$$

This is Henry's law. The proportionality constant K_k^H is Henry's coefficient for the substance labeled by k, also called its 'volatility'. For dilute concentrations of the substance in solution, the inverse of Henry's coefficient is close to the 'solubility' of the gas. (Formally, the solubility of a material is the ratio of the maximum amount of material that can be dissolved in a given amount of solvent, both usually measured in moles.) When a gas is in equilibrium with a liquid, the solution will have become saturated with the gas. The solubility can range from almost zero (immiscible) to infinite (perfect miscibility).

Henry's coefficient is not easy to predict, since it depends on how the substance forms bonds with the solvent, or becomes ionic. For example, carbon dioxide will react with water to form hydronium and carbonate ions. These ions must reform CO_2 at the liquid-gas interface before CO_2 is released into the vapor state (Table 9.2).

Henry's coefficient varies with temperature. From Eq. (9.131), we have

$$\frac{d \ln K_s^H}{dT} = \frac{d \ln p_s}{dT} \tag{9.132}$$

to which we can apply the Clausius–Clapeyron equation (9.127) to find

$$\frac{d \ln K_s^H}{dT} = \frac{\overline{\rho}_s \lambda_s}{RT^2} . \tag{9.133}$$

Table 9.2 Henry's law coefficient for various gases in aqueous solution near 293.15 K

Gas	K^H (in atm)	$\overline{\rho}_s \lambda_s / R$ (in K)
N_2	9.1×10^4	1300
O_2	4.3×10^4	1700
Ar	4.0×10^4	1300
CO_2	0.16×10^4	2400
CO	5.8×10^4	1300

The second column determines the heat of dissolution λ_s. From R. Sander, *Compilation of Henry's law constants (v.4.0) for water as solvent*, Atmos Chem Phys, **14 (8)**, 4399–4981 (2015).)

Here, $\overline{\rho}_s$ is the mass per mole of the given solute, and λ_s is the energy required per mole to release the solute as a gas from the solution ('heat of dissolution'). Just as for pure liquid vaporization, λ_s is approximately constant over a range of temperatures, such as from 0 to 20 °C. As a consequence, Henry's coefficient for common atmospheric gases is predicted to rise in a range of temperatures between the freezing and boiling point for water. Correspondingly, the solubility of atmospheric gases will decrease with increasing temperature. As the ocean temperature rises, it can hold less oxygen and carbon dioxide, adding stress to fish and exacerbating the greenhouse effect.

As one application of Henry's law, we will apply the law to estimate the amount of nitrogen gas released from the blood when a diver ascends too quickly from the deep. The resulting bubbles can clog the capillaries, starving tissue of oxygen. The ascendant divers feel pain in their muscles as well as other symptoms. The syndrome is called 'the bends', since some divers suffering from this 'decompression sickness' cannot straighten their joints.

We will focus on a small volume V_b of blood in a capillary. Because water is practically incompressible, V_b will not vary with the pressure p on the blood. Let n_{N_2} be the number of moles of nitrogen gas that has been dissolved into the blood holding n_w moles of water. From Henry's law, the number of moles of nitrogen gas dissolved will be $n_{N_2} = x_{N_2} \cdot n \approx x_{N_2} \cdot n_w = p_{N_2} \cdot n_w / K_{N_2}$, where K_{N_2} is Henry's constant for nitrogen dissolved in blood. The partial pressure of the nitrogen in the air will be $p_{N_2} = fp$, where f is the mole fraction of nitrogen in the air. When the diver is at a depth D, the breathing air is at a pressure $p = p_o + \rho g D$. After ascending, the pressure in the breathing air is $p_o = 1$ atm, and the blood will hold less dissolved nitrogen. The amount that must have escaped is $n_{N_2}^{\text{esc}} = n_{N_2} - n_{N_2}^o = f(p_o + \rho_w g D) \cdot n_w / K_{N_2} - f_o p_o \rho_w g D \cdot n_w / K_{N_2}$. We will assume f remained nearly constant during the descent, and that K_{N_2} for blood is about that for water, since nitrogen gas does not react much with any of the substances in blood. If we apply the ideal gas law for the nitrogen as a gas, and neglect bubble surface tension, then the volume of nitrogen gas that came out of the blood into bubbles at atmospheric pressure will be

$$V_{N_2}^{\text{esc}} = \frac{n_{N_2}^{\text{esc}} RT}{p_o} = \frac{f \rho_w g D}{K_{N_2}} \frac{RT}{p_o} n_w = \frac{f \rho_w g D}{K_{N_2}} \frac{RT}{p_o} \frac{\rho_w}{\mathcal{M}_w} V_b \tag{9.134}$$

in which $\mathcal{M}_w$ is the mass per mole of water and V_b is the volume of the blood releasing the gas. We will take a deep-sea diving distance of 50 m. With the following values: $f = 0.78$, $\rho_w = 10^3$ kg/m^3, $g = 9.8$ m/s^2, $D = 50$ m, $R = 8.314$ J/mole, $T = 310$ K, $p_o = 1.01 \times 10^5$N/m^2, $\mathcal{M}_w = 18 \times 10^{-3}$ kg/mole, $K_{N_2} = 9.1 \times 10^9$ N/m^2, we have $V_{N_2}^{\text{esc}} / V_b = 0.06$, so about 6% of the volume of the blood is nitrogen gas bubbles. The difficulty of having tiny bubbles throughout the blood is magnified by the small space that the blood plasma has between the red-blood cells and capillary walls, since the red-blood cell diameter is comparable to the capillary diameters. Bubbles in the plasma will significantly add to the viscous drag on the blood flowing in those capillaries.

Divers have learned to ascend slowly, giving time for the blood to release its dissolved nitrogen to the breathing air, or, on ascent, before symptoms show, to quickly enter a hyperbaric chamber, which slowly (over hours) brings the air from the diving pressure down to the normal one atmosphere of pressure.

Astronauts breathe air at the same pressure as at the surface of the Earth. Pressures below about 20% sea-level will cause hypoxia in humans, even if they breathe 100% oxygen. The cabins of commercial passenger airplanes flying at 10 km (33,000 ft), where the outside pressure is just 26% of that at sea level, are pressurized to about 75% of sea-level pressure.

9.19 Osmosis and Osmotic Pressure

If a biological cell is immersed in a solution not matching its interior fluids, the concentration gradients across the cell wall will cause a flow of material into or out of the cell, for all substances which can pass through the cell-wall. This passage might be through pores and/or by diffusion through the wall. The concentrations of these substances (number of molecules per unit volume) on either side of the wall will tend to equalize, assuming no other processes are involved. However, a flow of material into a cell, acting alone, will expand the volume of the cell. The cell walls respond by pressing back on the interior fluid, creating added interior pressure. That added pressure pushes material out of cell, perhaps through the same pores. Equilibrium can be established when the flow in by a concentration gradient matches the flow out by a pressure gradient.

Imagine a cell with a solution in its interior, and with a cell wall permeable only to water,[51] i.e. the cell wall is a 'semi-permeable membrane'. Assume also that the cell wall has some elasticity and strength. If we put such a cell into a larger volume of water, water will diffuse into the cell because the substances dissolved in the interior water makes that water less concentrated than the water on the exterior. With flow in, the cell will swell in volume. The back pressure from the cell walls will act to push water molecules out of cell.[52] We should expect that the osmotic pressure will be proportional to the concentration gradient. Let's see if this follows from thermodynamical considerations.

We will use Gibbs' balancing of chemical potentials to find a relationship between the extra pressure across the cell wall, called the 'osmotic pressure', to the concentration gradient.

[51]For biological cells, with their bilayer-phospholipid walls, water can slowly diffuse through, but the walls also have aquapin pores that pass water molecules more rapidly than diffusion.

[52]The water molecules moving into the cell due to the differences in concentration of water act like an expanding gas, and will tend to cool the interior, but the molecules being pushed out by pressure generate heat by compression. The two effects mean the combined processes can take place at a fixed temperature.

The total Gibbs free energy for the cell and the volume of surrounding water (the 'bath' water!) will be a sum of the energies for each subsystem:

$$G = G_c + G_b = \mu_c N_c + \mu_b N_b \,. \tag{9.135}$$

The subscripts 'c' and 'b' stand for 'cell' and 'bath'.

Following Eq. (9.105), we balance the chemical potential for water in the cell with its value outside using

$$\mu_c^w(T, p_c, x_c^w) = \mu_b^w(T, p_b, 1)\,, \tag{9.136}$$

The fraction x_c^w is the number of water molecules inside a given volume of the cell divided by the total number of molecules, water and solutes, in that same volume. The corresponding fraction of solute molecules will be designated x_c^s. These fractions add to one: $x_c^w + x_c^s = 1$. The fraction x_c^s is connected to the molar concentration of solute by $[C_c^s] = (N_c^s/\mathcal{N}_A)/V = x_c^s(n/V)$. $\mathcal{N}_A$ is Avogadro's number. Outside the cell we assume there is only water, with no flux of solute out of the cell.

To see a few of the implications of the equilibrium conditions, we will use some simplifying assumptions. The chemical potential for the water in the cell will be expressed in terms of the pressure in the cell by

$$\mu_c^w(T, p_c, x_c^w) = \mu_o^w(T) + RT\ln\left(x_c^w \varphi_c p_c/p_o\right), \tag{9.137}$$

in units of joules per mole. This expression matches that for an ideal gas (Eq. (9.79)), except for the added factor, φ_c, called the 'fugacity coefficient,' which accounts for the deviation from non-ideal behavior. But even for materials in liquid phase rather than a gas, at ordinary temperatures and pressures, the fugacity coefficient is close to one. An 'ideal liquid' will have $\varphi = 1$, which we will assume. For the bath water, $x_b^w = 1$, so

$$\mu_b^w(T, p_b, 1) = \mu_o^w(T) + RT\ln\left(p_b/p_o\right). \tag{9.138}$$

Now assume that the extra pressure in the cell is not large compared to the pressure outside, and express $p_c = p_b + \Delta p$, with $\Delta p << p_b$. Expanding the chemical potential for the water in the cell in a power series around p_c, we have:

$$\mu_c^w(T, p_c, x_c) \approx \mu_c^w(T, p_b, x_c) + \left.\frac{\partial \mu_c^w}{\partial p_c}\right|_T \Delta p\,. \tag{9.139}$$

A Maxwell relation can be used to write the derivative

$$\left.\partial \mu_c^w/\partial p_c\right|_{n_c} \text{ as } \left.\partial V_c/\partial n_c^w\right|_p \equiv \overline{V}_c,$$

the molar volume density of the water.

With a balancing of chemical potentials for the water (Eq. (9.136)) we find

$$\overline{V}_c \Delta p + RT \ln x_c^w = 0 . \tag{9.140}$$

Specializing further, we apply $x_c^w = 1 - x_c^s$ for small concentrations of solute. This makes the approximation $\ln(1-x) = -x$ reasonable. Also, as water is nearly incompressible,

$$\overline{V}_c \equiv \partial V_c / \partial n_c = \partial(\mathcal{M}_c n_c / \rho_c) / \partial n_c = V_c / n_c .$$

With the definition of concentration of the solute as $C_c^s \equiv n_c^s / V_c = x_c^s / (V_c / n_c)$, we have, finally,

$$\Delta p = RT [C_c^s] . \tag{9.141}$$

$[C_c^s]$ is the molar concentration of the solute. We started with the concentration of the solute outside taken zero: $[C_b^s] = 0$. If we had put the cell in a solution of the same solute, but different concentration than the interior of the cell, then we would have gotten the more general expression

$$\Delta p = RT \left([C_c^s] - [C_b^s]\right) = RT \Delta[C_c] . \tag{9.142}$$

We have that the osmotic pressure is proportional to the concentration gradient of the solute. (This relationship was presented by van't Hoff in 1884 and so is named after him.) For non-ideal solutions, wherein interactions between solute particles come to play, the osmotic pressure may depend non-linearly on the concentration gradient of the solute.

Good examples of osmosis in biology abound. Osmotic pressure within red-blood cells varies as they transport gases. (Depiction in Fig. 9.9.) Water loss from

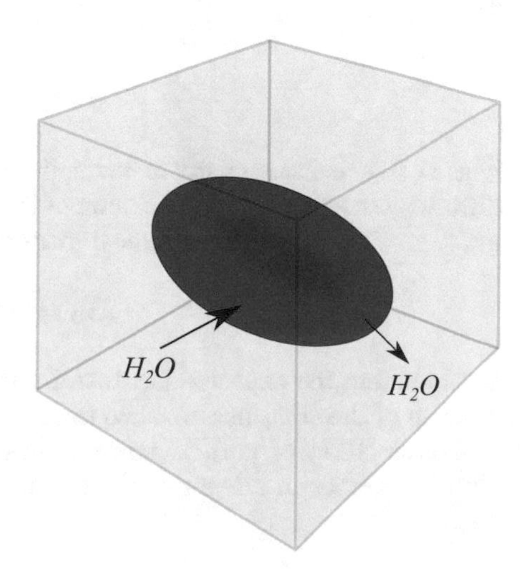

Fig. 9.9 Osmosis of water into a red-blood cell

plant leaf cells is controlled by stomate guard cells that expand or contract by water osmosis. Osmotic pressure in plant cells from ground water helps to keep plant stems stiff. Since our cells have a salinity greater than fresh water, when we bathe or swim in fresh water, there will be an intake of water through our skin into our cells. If we attempt to satiate our thirst with salt water, the effect in our gut is likely to be a loss of water, not a gain. Within protozoans, contractile vacuoles take up water by osmosis, and then discharge their contents through the cell wall, relieving the cell of built-up osmotic pressure.

9.20 Diffusion

Suppose there is a concentration gradient of small particles in a solution, such as macromolecules in water. These particles are in constant motion from thermal collisions. If there is a concentration of such particles, there is a statistical probability that thermal collisions will spread the particles from higher density regions to lower density regions. We measure the density of particles by their concentration C (in number per unit volume). The flux of those particles moving by diffusion is denoted J_D and is defined by the net number of particles which pass through a fixed small area per unit time per unit area. The particles will diffuse with some average speed we will denote by v_D. This is not individual speeds of particles, but the speed of the macromolecules drifting through the water averaged over times large compared to the time between collisions with the background water molecules. The flux of these macromolecules will then satisfy

$$\mathbf{J}_D = C\mathbf{v}_D \,. \tag{9.143}$$

Now we should expect that the diffusing particles should drift toward regions of lesser concentration, with a flux proportional to the gradient of that concentration. This is Fick's Law:

$$(J_D)_x = -D\frac{\partial C}{\partial x} \tag{9.144}$$

where $(J_D)_x$ is the net particle flux in the direction x. Fick's Law defines the particle diffusion constant D. It is an analog of the rate at which heat flows across a thermal gradient. (See Sect. 9.5.) In three dimensions, Fick's law reads

$$\mathbf{J_D} = -D\,\nabla C\,, \tag{9.145}$$

incorporating the expected fact that diffusion occurs in the direction of the largest gradient of the concentration, and toward lower concentration.

Conservation of particle number means that the net change per unit time in particle number in a layer of fluid of thickness dx is determined by the net flux of particles into the layer:

$$(J_x(x) - J_x(x+dx))\,A = \frac{\partial C}{\partial t} A dx\,. \tag{9.146}$$

The left-hand side divided by $-Adx$ defines the derivative of the flux. Now we apply the relation to the diffusion flux, Eq. (9.144), to find

$$D\frac{\partial^2 C}{\partial x^2} = \frac{\partial C}{\partial t}\,. \tag{9.147}$$

This is the diffusion equation. In three dimensions, it reads

$$\nabla^2 C = \frac{1}{D}\frac{\partial C}{\partial t}\,. \tag{9.148}$$

Let us consider a particular solution C of the diffusion equation. First, the diffusion equation in three dimensions is explicitly:

$$\frac{1}{D}\frac{\partial C}{\partial t} = \frac{\partial^2 C}{\partial x^2} + \frac{\partial^2 C}{\partial y^2} + \frac{\partial^2 C}{\partial z^2}\,. \tag{9.149}$$

A solution to this equation for the particle concentration C initially localized near the origin of a large volume is given by the Gaussian

$$C = N\left(\frac{1}{4\pi D t}\right)^{3/2} \exp\left(-\left(x^2+y^2+z^2\right)/(4Dt)\right)\,. \tag{9.150}$$

The spread of these molecules after a time t, can be measured by the root-mean-square width of the concentration peak, $r_{RMS} \equiv \sqrt{\langle r^2\rangle}$, which is given by $\sqrt{6Dt}$, so that the time of spread to a given r_{RMS} is

$$t = \frac{r_{RMS}^2}{6D}\,. \tag{9.151}$$

More generally, if diffusion occurs in a space of dimension d, such as within a cellular microtubule ($d = 1$) or on a organelle membrane ($d = 2$), then

$$t = \frac{r_{RMS}^2}{2d\,D}\,. \tag{9.152}$$

Einstein, in 1905, showed[53] how the rate of diffusion of particles in a fluid depends on the temperature of the fluid, its viscosity, and the size of the particle. Suppose an external force f, such as from gravity or an electric field, acts on the particles in the fluid, causing a concentration gradient in the x direction. Then particles in the fluid will tend to diffuse toward lower concentration, following Fick's Law, while the external force will drag them in the opposite direction. At equilibrium, according to Boltzmann, the distribution of particles is determined by the stored energy of each particle, i.e.

$$C = C_o \exp\left(-f\,x/\left(k_B T\right)\right) . \tag{9.153}$$

It follows that the concentration gradient at equilibrium is given by

$$\partial C/\partial x = -\left(f/\left(k_B T\right)\right) C \tag{9.154}$$

so that, from Fick's law,

$$J_D = -D\left(f/\left(k_B T\right)\right) C \tag{9.155}$$

The particles which feel an external force will, through collisions, reach a terminal drift speed v_f. Newton's law for each particle takes the form $m\ddot{\mathbf{r}}+(1/\mu)\dot{\mathbf{r}} = \mathbf{f}$, where μ is called the mobility of the particle in the surrounding medium and characterizes the drag on the particle as it moves. After the particle stops accelerating because of drag, it reaches the terminal speed is $v_f = \mu f$. The net flow of particles in the direction of the force will then be $J_f = v_f C$. Now, at equilibrium, the flow of particles due to diffusion toward lower concentrations will be balanced by the flow due to the external force: $J_D + J_f = 0$. With Fick's law, this balance of flow becomes

$$-D(f/(k_B T))C + v_f C = 0. \tag{9.156}$$

Einstein realized that the same relation applies to particles within the fluid diffusing through collisions and being impeded by viscous drag. By observing diffusion of small particles in a fluid, the diffusion constant becomes

$$D = \frac{v_f}{f}\, k_B T = \mu k_B T . \tag{9.157}$$

For a large spherical molecule of radius a diffusing in the fluid, f can be approximated by Stokes' law: $f = 6\pi\eta a v_f$, where η is the viscosity of the fluid. The Einstein relation becomes

[53] A. Einstein, *Über die von der molekularkinetischen Theorie der Wärme geforderte Bewegung von in ruhenden Flüssigkeiten suspendierten Teilchen*, Annalen der Physik **322:8**, 549–560 (1905).

$$D = \frac{1}{6\pi\eta a} k_B T \ , \tag{9.158}$$

referred to as the Stokes-Einstein equation, introduced in Sect. 2.2.2.

It is sensible that the rate of diffusion increases with temperature, and decreases with viscosity and with the size of the molecule. However, the diffusion 'constant' is not proportional to temperature, since the viscosity also depends on temperature. Corrections are needed to the Stokes-Einstein relation if diffusion near to fluid boundaries is important, such as within microtubules.

Combining the relation for the diffusion constant (Eq. (9.158)) and the expected root-mean-square of particle drift distance in terms of D (Eq. (9.152)), Einstein gave a way to measure Avogadro's number by observing the diffusional drift of particles in a fluid:

$$\mathcal{N}_A = \frac{RT}{3\pi\eta a} \frac{t}{\langle x^2 \rangle} \ . \tag{9.159}$$

In 1909, Perrin showed experimental verification of this relation, and as a consequence gave strong support for the molecular basis of matter.[54]

Large protein molecules in cells have slow diffusion rates: A 30 kDa protein takes about 10 s to traverse a 20 μm cell. For this reason, cells have evolved to incorporate protein transport 'motorized highways', with small motor proteins (kinesins) which carry larger vesicles along microtubules.

9.20.1 *Diffusion Across Alveolar Membranes*

Diffusion of oxygen and carbon dioxide across membranes in the lung is critical to our ability to breath. The following laws, derivable from statistical thermodynamics, apply.

- Fick's Law: The diffusion rate is proportional to partial pressure difference across the membrane, the area of the membrane, and inversely proportional to the thickness of the membrane. Lung alveoli: $\sim$100 square meters; thickness $< 10^{-3}$ mm.
- Graham's Law: Diffusion of a gas in a liquid is proportional to the solubility of the gas and inversely proportional to the square root of the mass of the gas molecule. CO_2 has 22 times the solubility of O_2, but is more massive, making the relative diffusion about 19 times greater.
- Henry's Law: The amount of gas which dissolves in a liquid is proportional to the partial pressure for that gas, as well as the solubility of the gas in the liquid. (The

[54] J. Perrin, *Brownian movement and molecular reality*, Ann Chimi Phys **18**, 5–114 (1909).

Table 9.3 Resting potentials

Cell type	Potential (mV)
Skeletal muscle	−95
Smooth muscle	−60
Astroglia	−80 to −90
Neuron	−60 to −70
Erythrocyte	−8.4
Photoreceptor	−40

solubility depends on how the gas molecules interact with those of the solute.) As we have noted, the law works well only for low concentrations of gas.

Diffusion Across a Charged Membrane

Generally, biological cells have walls with a difference in electric potential from the inside to the outside. This difference is maintained by active cell-wall pumps, largely to pump excess sodium ions out. Potassium ions are allowed into the cell through pumping action and their permeability. The resultant electric field affects the ability of other ions to cross the membrane wall. The resting (equilibrium) potentials (i.e. the difference of the outside electric potential to the inside electric potential) across common cell types is given in Table 9.3.

In the bulk fluid of the cell, the strong Coulomb force between positive and negative charges means that ions in solution will attract their opposite charge. This effect builds a polarized charge region nearby to ions, effectively screening the ions in solution from any appreciable long-range charge interactions. The local electric field 10 nm from an isolated ion will be ke/r^2 which is 14 million volts per meter. The Coulomb attraction dominates the local field in comparison to the relatively weak external fields in biological tissue, so the overall net charge in any region on the scale of tens of nanometers and up will be close to zero. This observation of little net charge over regions large compared to the size of the ions and molecules in solution is referred to as the '*Principle of Neutrality*'. The idea is that the average charge within volumes large compared to molecular sizes is near zero in regions where charges can easily move, such as in bulk fluid. In contrast, charges put on the surface of membranes may be held by the opposite charge on the other side of the membrane, and so significant charge separation may occur across such membranes. Even so, the number of ionic charges that can be built up on a cell membrane is a very small fraction of the bulk number of ions within the cell.[55] For this reason, the concentration of such ions is only slowly depleted after hundreds of thousands

[55]The same is true of the charge on you when you scuff your rubber-sole shoes on a rug on a cold dry day. That amount of that charge, coming from the displacement of electrons, is also a very small fraction of the number of electrons contained in your atoms.

of nerve impulses across a given nerve cell, principally involving the movement of sodium and potassium ions across the nerve cell wall. The potential built up across the wall by active pumps (using ATP) becomes a ready source of energy powering nerve pulses.

Now, suppose a biomolecular membrane has ionic charges separated across the membrane. Then an electric field E_x will exist across it. If the membrane has a thickness of Δx, then the separated ions will have created a potential difference across the membrane of $\Delta V = -E_x \Delta x$. Any positive ions within a diffusion channel in the membrane will be pulled in the direction of the electric field, and negative ions will be pulled in the opposite direction. If they can move, they will reach a terminal drift speed.[56] But the terminal drift speed is also influenced not only by electric fields, but also by thermal motion, which will generate diffusive drift of ions in all directions, including in the opposite direction to the field. The net effect is determined by the Boltzmann distribution of ion energies contained in Eq. (9.153).

Alternative to a Boltzmann argument (as in Sect. 9.20), consider the ions that can move through the membrane channels responding to the electric force $F_x = -q\partial V/\partial x$. While being pulled, collision of the ions with neighboring particles will create a drag on the ions. If the electric field were the only agent causing ion motion, the ion's average speed, after many collisions, could be taken in proportion to the dragging force f due to collisions with other molecules or the channel walls, acting on the ions in the channels. For low velocities, $f = \mu^{-1} v$, where μ is the ion mobility. When the ion reaches a terminal speed, the drag force must balance the electric force: $\mu^{-1} v = q E_x$. This is a disguised form of Ohm's law. In this case, the current density is qCv, where q is the ion charge, C the ion concentration, and v the average ion speed.

Now consider the effect of a concentration gradient on the motion of ions. Neglecting ion-ion interactions, which is a good approximation for dilute ion concentrations, a concentration gradient causes a diffusive flow in proportional to the negative gradient of the concentration, i.e. $v_D C = -D\, \partial C/\partial x$, where D is the diffusion constant. (This is Fick's law introduced in Eq. (9.144).) This flow includes diffusion in a direction opposite to that caused by an external electric field.[57] At equilibrium, there should be no net flow of the ions, so at any point x in the channel, $Cv = -Cv_D$, where

$$v_D = -\frac{D}{C}\frac{\partial C}{\partial x} = -v = -\mu\,(qF_x) \tag{9.160}$$

[56] This drift speed is averaged over many channels through which the ions drift. The motion through a single channel will be quite chaotic.

[57] Diffusion also causes particles in a solution to move upward when the mixture is standing in a gravitational field, even though the particles may be denser than the fluid. At equilibrium, a concentration gradient of those particles in the solution will exist.

$$-\frac{D}{C}\frac{\partial C}{\partial x} = (\mu q)\frac{\partial V}{\partial x}\,. \tag{9.161}$$

This differential equation for C is solved by

$$C = C_0 \exp\left(-q\Delta V/\left(\mu D\right)\right)\,. \tag{9.162}$$

We now use Einstein's relation, $D = \mu k_B T$, and solve for the potential difference across the membrane, giving the Nernst–Planck equation

$$V_2 - V_1 = \frac{k_B T}{q}\ln\frac{C_1}{C_2} \tag{9.163}$$

This relation is often expressed in the form:

$$V_2 - V_1 = \frac{RT}{\mathcal{N}_A\,|e|\,z}\ln\frac{C_1}{C_2} \tag{9.164}$$

$$= \left(\frac{RT}{F}\right)\frac{1}{z}\ln\frac{C_1}{C_2} \tag{9.165}$$

$$= \left(\frac{RT\ln 10}{F}\right)\frac{1}{z}\log\frac{C_1}{C_2} \tag{9.166}$$

$$= 61.54\,\frac{\text{millivolts}}{z}\log\frac{C_1}{C_2}\quad at\ 37\,^\circ\text{C} \tag{9.167}$$

$$= 57.97\,\frac{\text{millivolts}}{z}\log\frac{C_1}{C_2}\quad at\ 19\,^\circ\text{C} \tag{9.168}$$

in terms of the gas constant $R = 8.3145$ J/deg, Faraday's constant $F = 96{,}485$ coul (the charge of a mole of protons), and z the number of electronic charges on the ion (with z negative if a negative ion).

Biological cell membranes control the ions Na^+, Cl^-, K^+, and Ca^{++}. Their interior concentration are important to the functioning of the cell. For each of these ions, the voltages inside the cell due to any one of these ions are given by

$$V_{Na} = \frac{RT}{F}\ln\left(\frac{[\text{Na}^+]_O}{[\text{Na}^+]_I}\right)$$

$$V_K = \frac{RT}{F}\ln\left(\frac{[\text{K}^+]_O}{[\text{K}^+]_I}\right)$$

$$V_{Cl} = \frac{RT}{F}\ln\left(\frac{[\text{Cl}^-]_I}{[\text{Cl}^-]_O}\right)$$

$$V_{Ca} = \frac{RT}{F}\frac{1}{2}\ln\left(\frac{[\text{Ca}^{++}]_O}{[Ca^{++}]_I}\right) \tag{9.169}$$

In biological tissue, the ion channel-pumps typically allow only a few ions through at a time. One might ask how diffusive statistics applies. The answer is that these relations apply not to one, but to the average of a large number of channels across a given membrane.

The tables below show the result of applying the Nernst equation (9.169) to various ions and nerve cell membranes:

Mammalian resting nerve cell at 37 °C:

Ion	*Out*	*In*	*ratio*	ΔV
Na	145 mM	12 mM	145/12=12	$61.5\log_{10}(12) = +66.4$ mV
K	4 mM	155 mM	4/155=0.026	$61.5\log_{10}(0.026) = -97.5$ mV
Cl	123 mM	4.2 mM	123/4.2=29.3	$-61.5\log_{10}(29) = -89.9$ mV
Ca	1.5 mM	10^{-4} mM	$1.5/10^{-4}$=15000	$(61.5/2)\log_{10}(15000) = +128.4$ mV

Spinal motor neurons resting at 37 °C:

Ion	*Out*	*In*	*ratio*	ΔV
Na	150 mM	15 mM	150/15=10	$61.5\log_{10}(10) = +61.5$ mV
K	5.5 mM	150 mM	5.5/150=0.037	$61.5\log_{10}(0.037) = -88.3$ mV
Cl	125 mM	9.0 mM	125/9.0=13.9	$-61.5\log_{10}(13.9) = -70.3$ mV

Frog muscle cell at 20 °C:

Ion	*Out*	*In*	*ratio*	ΔV
Na	109 mM	10.4 mM	109/10.4=10.5	$58\log_{10}(10.5) = +59.2$ mV
K	2.25 mM	124.2 mM	2.25/124.2=0.0181	$58\log_{10}(0.0181) = -101.0$ mV
Cl	77.5 mM	1.5 mM	77.5/1.5=51.7	$-58\log_{10}(51.7) = -99.4$ mV
Ca	2.1 mM	10^{-4} mM	$2.1/10^{-4} = 2.1 \times 10^4$	$(58/2)\log_{10}(21000) = +125.3$ mV

Squid axon at 20 °C:

Ion	*Out*	*In*	*ratio*	ΔV
Na	440 mM	50 mM	440/50=8.8	$58\log_{10}(8.8) = +54.8$ mV
K	20 mM	400 mM	20/400=0.05	$58\log_{10}(0.05) = -75.5$ mV
Cl	560 mM	110 mM	560/110=5.10	$-58\log_{10}(5.10) = -41.0$ mV
Ca	10 mM	10^{-4} mM	$10/10^{-4} = 10^5$	$(58/2)\log_{10}(10^5) = +145.0$ mV

9.20.2 Active Transport Across Membranes

Now consider the non-equilibrium case in which ions are pumped across the membrane, resulting in a potential difference of $V_m = V_1 - V_0 = \Delta V$ across a thickness Δx.

The diffusive and forced flow for each ion α gives a resultant flow of

$$J_\alpha = -D_\alpha \left(\frac{\partial C_\alpha}{\partial x} \right) - \mu_\alpha q_\alpha \frac{\partial V}{\partial x} C_\alpha \ , \tag{9.170}$$

where we used the Einstein's relation $\mu = D/\left(k_B T\right)$.

Now, we will find the electric field within the membrane channels. With the Principle of Neutrality, there should be little electric field in the ionic solution inside cells, and little electric field in the ionic solution outside cells. However, there can be significant electric field across membranes because of the charge built up on one side of the membrane relative to the other. This charge separation creates a constant electric field within the membrane.

Across a charged membrane, we have an electric field given by $E_x = -\partial V/\partial x = \left(V_1 - V_0\right)/\Delta x = \Delta V/\Delta x$. From Eq. (9.170), the ion concentration then satisfies

$$\frac{\partial C_\alpha}{\partial x} + \left(\frac{q_\alpha}{k_B T} \frac{\Delta V}{\Delta x} \right) C_\alpha = -\frac{J_\alpha}{D_\alpha} \ . \tag{9.171}$$

For a steady flow, J_α is constant, so we can solve this differential equation for the concentration as a function of x :

$$C_\alpha\left(x\right) = A \exp\left(-\frac{q_\alpha}{k_B T} \frac{\Delta V}{\Delta x} x \right) - k_B T \frac{J_\alpha}{q_\alpha D_\alpha} \frac{\Delta x}{\Delta V} \tag{9.172}$$

If the concentration at $x = 0$ is called $C_{\alpha o}$, then

$$C_\alpha\left(x\right) = C_{\alpha o} \exp\left(-\frac{q_\alpha}{k_B T} \frac{\Delta V}{\Delta x} x \right) - k_B T \frac{J_\alpha}{q_\alpha D_\alpha} \frac{\Delta x}{\Delta V} \left(1 - \exp\left(-\frac{q_\alpha}{k_B T} \frac{\Delta V}{\Delta x} x \right) \right) \tag{9.173}$$

This means the concentration $C_{\alpha 1}$ at $x_1 = x_0 + \Delta x$ is

$$C_{\alpha 1} = C_{\alpha o} \exp\left(-\frac{q_\alpha}{k_B T} \Delta V \right) - k_B T \frac{J_\alpha}{q_\alpha D_\alpha} \frac{\Delta x}{\Delta V} \left(1 - \exp\left(-\frac{q_\alpha}{k_B T} \Delta V \right) \right) \tag{9.174}$$

The steady flow for ions of type α is therefore

$$J_\alpha = -\frac{q_\alpha D_\alpha}{k_B T} \frac{\Delta V}{\Delta x} \frac{C_{\alpha 1} - C_{\alpha o} \exp\left(-\frac{q_\alpha}{k_B T} \Delta V \right)}{1 - \exp\left(-\frac{q_\alpha}{k_B T} \Delta V \right)} \tag{9.175}$$

This flow still has both diffusive and forced contributions. Note that in this case, Ohm's law, which has a linear relationship between the electric current and the electric field, holds only for weak fields.

During steady conditions, the total charge carried by the ions across the membrane must add to zero, which makes

$$\sum_{\alpha} q_{\alpha} J_{\alpha} = 0\,. \tag{9.176}$$

We have

$$\sum_{\alpha} q_{\alpha}^2 D_{\alpha} \frac{C_{\alpha 1} - C_{\alpha o} \exp\left(-\frac{q_{\alpha}}{k_B T}\Delta V\right)}{1 - \exp\left(-\frac{q_{\alpha}}{k_B T}\Delta V\right)} = 0\,. \tag{9.177}$$

We can use this relation to find the potential across the membrane which must be maintained to keep the given concentrations of ions. The maintenance is achieved by active ion pumps.

If all the ions have the same magnitude of charge q, then the above becomes a linear equation for the exponent, and we can solve for the resulting voltage across the membrane, ΔV. Let the negative ions be labeled with β instead of α. The relation separates into two sets of terms

$$\sum_{\alpha} D_{\alpha} \frac{C_{\alpha 1} - C_{\alpha o} \exp\left(-\frac{q}{k_B T}\Delta V\right)}{1 - \exp\left(-\frac{q}{k_B T}\Delta V\right)} + \sum_{\beta} D_{\beta} \frac{C_{\beta 1} - C_{\beta o} \exp\left(+\frac{q}{k_B T}\Delta V\right)}{1 - \exp\left(+\frac{q}{k_B T}\Delta V\right)} = 0 \tag{9.178}$$

or

$$\sum_{\alpha} D_{\alpha} \left(C_{\alpha 1} - C_{\alpha o} \exp\left(- \frac{q}{k_B T} \Delta V \right)\right) + \sum_{\beta} D_{\beta} \left(C_{\beta o} - C_{\beta 1} \exp\left(- \frac{q}{k_B T} \Delta V \right)\right) = 0 \tag{9.179}$$

The terms for negative ions have the same form as those for positive ions, except for the reversal of the concentrations. This means we can write

$$\Delta V = \frac{k_B T}{q} \ln \frac{\sum_{\gamma} D_{\gamma} C_{\gamma i}}{\sum_{\gamma} D_{\gamma} C_{\gamma j}} \tag{9.180}$$

where $C_{\gamma i} = C_{\gamma 1}$ when γ labels a positive ion, and $C_{\gamma 0}$ when γ labels a negative ion, while $C_{\gamma j} = C_{\gamma o}$ when γ labels a positive ion, and $C_{\gamma j} = C_{\gamma 1}$ when γ labels a negative ion. This is the Goldman–Hodgkin–Katz voltage equation.[58]

[58] David E. Goldman, in his 1943 PhD thesis from Columbia University, derived this equation. Further development in the late 1940s came from the Alan Lloyd Hodgkin and Bernard Katz, who

Neglecting the relatively low contribution of calcium, the potential across a nerve cell, with active pumps maintaining a given set of concentrations, is

$$V_m = \frac{RT}{F} \ln \left(\frac{P_{Na}[\mathrm{Na}^+]_O + P_K[\mathrm{K}^+]_O + P_{Cl}[\mathrm{Cl}^-]_I}{P_{Na}[\mathrm{Na}^+]_I + P_K[\mathrm{K}^+]_I + P_{Cl}[\mathrm{Cl}^-]_O} \right) . \quad (9.181)$$

The $P_\alpha \equiv D_\alpha/\Delta x$ is the 'permeability' of the ion type α (with units length over time). The potential V_m represents the electric potential on the inside of a cell (compared to the outside), and the indices I and O refer to the inside and outside of the membrane. The effective diffusion constant D_α for ions moving through cell-wall channels is difficult to calculate, so the permeabilities (for both open and closed channels) are usually determined experimentally. Because a ratio appears in the Goldman Equation (9.181), the permeabilities can be converted into fractional permeabilities $f_i = P_i/\sum P_k$ or percents of the overall permeability by all ions able to move across the cell membrane.

Approximate permeabilities for the mammalian nerve cell and squid axon are given in the following:

Ion	*Permeability Mammal Nerve*	*Relative Permeability Squid Axon*
K^+	5×10^{-9}m/s	1.0
Na^+	5×10^{-11}m/s	0.03
Cl^-	1×10^{-10}m/s	0.1
Ca^{++}	1×10^{-12}m/s	0.001

Nerve action potentials are generated by active diffusion gates which can change their permeabilities according to the voltage across the cell membrane. See Sect. 14.2.2.

As an alternative to the GHK equation, conductances of the ion channels can be used instead of permeabilities. We model each set of ion pumps as a resistor and a battery in series. With this picture, energy conservation around a current loop including an assumed steady current through the channels for ions α is expressed by

$$i_\alpha = g_\alpha \left(V_m - V_\alpha^{\mathrm{equil}} \right) , \quad (9.182)$$

where g_α is the conductance of the channel and V_α^{equil} is the channel 'battery' potential, which is the energy per unit charge given to an ion when it moves through its channel, and V_m is the potential maintained across the membrane (from the inside

later were each awarded a Nobel prize for separate work in nerve cell action: Eccles, Hodgkin, and Huxley in 1963, Katz, Euler, and Axelrod in 1970. The Hodgkin-Huxley model is described in Sect. 14.2.2.

to the outside). The potentials V_α^{equil} can be found from the Nernst equation for the given ion. Under steady state conditions, $\sum i_\alpha = 0$, so

$$V_m = \sum_\alpha \left(\frac{g_\alpha}{\sum g_\beta} \right) V_\alpha^{\text{equil}} . \tag{9.183}$$

This is the sometimes referred to as the Millman equation,[59] or the chord-conductance equation.

Clearly, the conductance of an ion channel and the membrane permeability to an ion are related, but not so easily predicted, given the rather sophisticated construction of biological ion channels.

The conductances can be varied in time according to the mechanisms and chemical environment which determine the opening and closing of a set of channels, thereby modeling nerve impulses. Permeabilities are more easily determined experimentally than conductances, making the GHK equation preferable over the Millman equation.

To include calcium ions (with charge of $2q$,), the following quadratic system must be solved to find the membrane potential in terms of ion concentrations. With $x \equiv \exp(e\Delta V)/k_B T$,

$$0 = (1+x) \sum_\alpha D_\alpha (C_{\alpha 2} - C_{\alpha 1} x) + 4 \sum_\alpha D_\alpha \left(C_{\alpha 2} - C_{\alpha 1} x^2 \right) \tag{9.184}$$

i.e.

$$0 = c + bx + ax^2 \tag{9.185}$$

where

$$c = \left(\sum_\alpha D_\alpha C_{\alpha 2} + 4 \sum_\alpha D_\alpha C_{\alpha 2} \right) \tag{9.186}$$

$$b = -\left(\sum_\alpha D_\alpha C_{\alpha 1} - \sum_\alpha D_\alpha C_{\alpha 2} \right) \tag{9.187}$$

$$a = -\left(\sum_\alpha D_\alpha C_{\alpha 1} + 4 \sum_\alpha D_\alpha C_{\alpha 1} \right) . \tag{9.188}$$

The quadratic can be solved for $x = \exp(e\Delta V/k_B T)$ and then for $\Delta V = (RT/F) \ln x$.

---

[59] After Jacob Millman's equation, published in 1940, for the voltage in a network.

The calcium ion permeability for a typical cell is about 0.001 that of the potassium ion. The concentration of calcium inside a cell is about a million times less than potassium, while on the outside, their concentrations are comparable. Voltage-dependent calcium channel gates play an important role in muscle cells and in heart pacemaker cells.

9.21 Biochemical Reactions

Live systems take in energy and material, convert less ordered material into more ordered substances, and expel heat and waste material. This process on Earth occurs in life systems through organic chemistry, i.e. the chemistry of carbon compounds. All biochemical processes are consistent with the laws of thermodynamics and the known interactions of molecules.

A biochemical reaction is a process which chemically transforms organic molecular type. All chemical reactions are driven by a combination of natural molecular forces and the tendency toward greater disorder inherent in the statistical behavior of macroscopic systems. Moreover, during reactions of any kind, the symmetries of nature require that certain quantities determined before a reaction will be preserved during a reaction. The symmetries of space-time cause total energy, momentum, and angular momentum to be conserved. Internal symmetries in the interactions between particles require that certain additive charges are conserved during reactions, such as the electric charge, the leptonic charge, and the baryonic charge. Electrons carry leptonic charge, and atomic nuclei carry baryonic charge equal to the number of protons and neutrons present.

Reactions can transform the molecular components within a system, changing the number of particles in individual components or their arrangement. Chemical reactions involve only the outer electrons of atoms and molecules (the 'valence' electrons). Thus, a chemical reaction may rearrange the state and binding of atoms, but it preserves the total number of each kind of atom. Since the outer electrons are held to atoms with binding energies from zero to less than about 20 electron-volts (about 24 kcal/mole), energy transfers during a chemical reaction are also typically in this range.

The energy released in a chemical reaction is far less than the Einsteinian energy stored in the mass of the reactants, so the mass before a chemical reaction is very close to the mass after, usually to within less than ten parts in a billion. The energy transferred from a system during a chemical reaction necessarily changes the mass of that system, but chemical scales are not sufficiently sensitive to detect that change. The condition of approximate mass preservation will be insured by the stoichiometric balancing of numbers of each type of atom before and after a reaction.

As chemical reactions in the laboratory and in life systems on Earth take place predominantly at nearly constant temperature and pressure, the Gibbs form of the second law of thermodynamics is convenient to characterize such reactions and activity. Reactions at constant temperature and pressure will proceed spontaneously if the Gibbs free energy[60] can be lowered under the constraints on the system. Equilibrium will occur when the Gibbs free energy reaches a minimum. We say that if a reaction results in $\Delta G < 0$, the reaction was exergonic ('work producing'). If $\Delta G > 0$, the reaction was endergonic ('work absorbing'). Note that since the enthalpy is $H = E - TS$, changes of the Gibbs free energy during a reaction at constant temperature can be expressed as $\Delta G = \Delta H - T\Delta S$, so a reaction will proceed if $\Delta H - T\Delta S$ is less than zero. For a closed system at constant pressure, the first term, $\Delta H = \Delta E + p\Delta V = \Delta Q$, will be the negative of the heat released by the reaction ('exothermic') and ΔH will be positive if energy is absorbed by the reactants ('endothermic'). The second term, $T\Delta S$, is the temperature times the entropy increase. Both energy release and entropy increase will 'drive' a chemical reaction forward.

We should not expect the entropy production to depend much on temperature, as the production of order or disorder in a reaction among atoms and molecules comes from how atoms become re-arranged, collected, or dispersed.[61] Also, we should expect that the bulk of the enthalpy change comes from changes in binding interactions, which have little temperature dependence.

The enthalpy of a substance depend on its temperature, reflected by the heat capacity at constant pressure, expressed by

$$H(T) = H(T_o) + \int_{T_o}^{T} C_p(T')dT' \,. \tag{9.189}$$

This means that the temperature dependence of the reaction ΔH, i.e. the enthalpy of the products of a reaction minus the enthalpy of the reactants, can be found from

$$\left.\frac{\partial(\Delta H)}{\partial T}\right|_p = \sum_J \nu_J C_{pJ} \,, \tag{9.190}$$

(called 'Kirchhoff's law of thermochemistry') where the C_p are the heat capacities for the compounds taking part in the reaction, and ν_J are the stoichiometric coefficients for the reactants and products, taken negative for the reactants).

But since, for constant pressure, $\Delta H = \Delta E + p\Delta V$, the heat required or released comes from changes in volume and from individual or collective binding of particles. For biochemical changes within a cell not involving phase changes,

[60]The Gibbs free energy $G = U - TS + pV$, defined through the internal energy E of the system, will change (for processes near equilibrium) according to $dG = -SdT - Vdp + \sum \mu dN$, so that for constant temperature and pressure, $dG = \sum \mu dN$, i.e. only the behavior of the chemical potentials μ are needed to determine how the system's Gibb's energy changes.

[61]The definition of the entropy (and disorder) from the Boltzmann expression $S = k \ln W$ gives $\Delta S = k \ln (W_f/W_i)$ which shows that the entropy change depends on how particles are redistributed, with the count of the number of ways to distribute them after being W_f and before being W_i.

volume changes are small, so ΔH should be no more than weakly dependent on temperature. (Good estimates of ΔH can be made from 'binding enthalpies' alone.)

The 'law of mass action', to be discussed later, determines the equilibrium concentration of a product of a biochemical reaction through the value of the 'equilibrium constant', $K = \exp(-\Delta G/(k_B T)$, where $\Delta G/T = \Delta H/T - \Delta S$ is the Gibbs free energy of the products minus the Gibbs free energy of the reactants, divided by the temperature T. The strong temperature dependence of K comes largely through the explicit temperature T in the heat term $\Delta H/T = \delta Q/T$. This term will be negative for an exothermic reaction. However, even if the process were endothermic, the reaction could still be spontaneous if positive entropy (disorder) is produced and the temperature were high enough to suppress the contribution of the heat absorption term $\Delta H/(K_B T)$.

As an example of a simple reaction, consider the oxidation of methane:

$$CH_4(g) + 2\,O_2(g) \rightarrow CO_2(g) + 2\,H_2O(l)\ . \tag{9.191}$$

A parenthetical symbol $\{g, l, s\}$ indicates a gas, liquid, or solid.

The overall enthalpy change for a reaction is often calculated by adding the enthalpy changes to produce each molecular compound in the reaction, starting at a standard state for this compound. This adding of heats at constant pressure was suggested by Germain Henri Hess in 1840 and is now called Hess' law. It follows from the fact that the enthalpy is a state function, and so its change in value starting from a standard state is independent of the process used to pass from the initial to the final state. The 'standard conditions' for a compound is defined to be a pressure of 10^5 N/m^2 $\approx$ 1 atm and a temperature of $T = 298.15$ K= 25 °C, with the compound in its natural and most stable phase. If the compound is in solution, its standard condition is defined to have a concentration of one mole per liter.

From tables of the standard enthalpy change needed to produce a mole of each of the compounds involved from their elemental form, we record

$$\begin{aligned} CH_4 : \Delta H^{\ominus} &= \ -75 \text{ kJ} \\ O_2 \ \ : \Delta H^{\ominus} &= \quad\ \ 0 \text{ kJ} \\ CO_2 : \Delta H^{\ominus} &= -393 \text{ kJ} \\ H_2O : \Delta H^{\ominus} &= -285 \text{ kJ}\ . \end{aligned} \tag{9.192}$$

Following Hess' law, we find the enthalpy change for the whole reaction from

$$\Delta H = \sum_J \nu_j \Delta H_J\ . \tag{9.193}$$

Implicit is the assumption that the pressure before the reaction is the same as the pressure after. In the present case, $\{\nu_J\} = \{-1, -2, 1, 2\}$. The enthalpy change is calculated to be $\Delta H = -288$ kJ. The minus sign indicates that heat would be released (an exothermic reaction).

More generally, given the reaction

$$\sum_J \nu_J \{M_J\} = \{0\} \ , \tag{9.194}$$

where M_J is the molecule symbol, then each of the additive conservation laws can be expressed symbolically as

$$\sum_J \nu_J \{m_J\} = - \{q\} \ , \tag{9.195}$$

with $\{m_J\}$ as an ordered set of numbers giving the amount of the conserved quantity in the molecule, ion, or electron of component type J. The quantity q is the amount of the conserved quantity released into the environment. If the conserved quantity is expressed in terms of basic units, the size of those units is not made explicit, as each side of the equation can be divided by that basic unit. For quantities that are quantized by integers times a basic unit, one can use just the integers in the balance equation.

One conserved and quantized count is the number each particular atom in the molecule. The number of each kind of atom is conserved during a reaction because of baryon number conservation and the fact that atomic nuclei are not transformed by a chemical reaction. Because chemical energies are much less than the binding energies stored in atomic nuclei, balancing of atomic kind in a chemical reaction also ensures that mass will be practically unchanged. Energy as an additively conserved quantity can then be applied to the chemical energy stored in each type of molecule and to the energy released or absorbed in the reaction. Electric charge too must additively balance. (For chemical reactions, charge conservation and baryon number conservation will automatically account for lepton number conservation.)

Total charge conservation reads:

$$\sum_J \nu_J Q_J + \nu_e(-1) = 0 \, , \tag{9.196}$$

where Q_J is the number of charges on molecule of type J, including sign (if Q_J is non-zero, the molecule type would be called ionic), and ν_e is the number of electrons in the balanced reaction, with negative numbers for the electrons on the reactant side, and positive for those on the product side (in conformity with the convention on the stoichiometric numbers).

For example, in the oxidation of water,

$$2\,H_2 + O_2 \rightarrow 2\,H_2O \ , \tag{9.197}$$

we have $\nu_1 = -2$, $\nu_2 = -1$, and $\nu_3 = 2$. The conservation of atoms for each species in this reaction can be expressed as

$$-2 \cdot \{H_2\} - 1 \cdot \{O_2\} + 2 \cdot \{H_2O\} = \{0\} \tag{9.198}$$

or as

$$-2 \cdot \{2, 0\} - 1 \cdot \{0, 2\} + 2 \cdot \{2, 1\} = \{0, 0\} . \tag{9.199}$$

Energy conservation reads

$$-2 \cdot (-B_{H_2}) - 1 \cdot (-B_{O_2}) + 2 \cdot (-B_{H_2O}) = -q . \tag{9.200}$$

The B's are the binding energies of the molecules. The energy B_J (positive) is released when each molecule is disassociated into its atoms. The value q is the heat released into the environment.

If the heat energy released is positive, the reaction is exothermic; if negative, endothermic.

As another example, with an explicit electron transfer, is the following reaction:

$$OH + e^- \rightarrow OH^- \tag{9.201}$$

The reaction equation can be expressed as

$$-1 \cdot \{OH\} + 1 \cdot \left\{OH^-\right\} - 1 \cdot \left\{e^-\right\} = 0 . \tag{9.202}$$

Atomic conservation gives

$$-1 \cdot \{1, 1\} + 1 \cdot \{1, 1\} - \{0, 0\} = \{0, 0\} \tag{9.203}$$

while charge conservation gives

$$-1 \cdot 0 + 1 \cdot (-1) - (-1) = 0 . \tag{9.204}$$

Some quantities in nature are conserved as vectors or higher rank tensors instead of as scalars. A good example is angular momentum. With no external torques acting, angular momentum is conserved as a vector, in the form: $d/dt \sum_i \mathbf{J_i} = \mathbf{0}$. The conservation of angular momentum is an important consideration in photochemical reactions, since photons transfer spin angular momentum. Conservation of angular momentum limits the possible transitions allowed for molecular and atomic electrons excitations. For example, photo-transitions from an electron 's' state to another 's' state with no electron spin flip is not allowed, since, by definition, an 's' state has no orbital angular momentum.

In the classical theory, the conservation of a tensor can be expressed and observed as the conservation of each component, so that each can be considered an additively conserved quantity. However, in quantum theory, it often occurs that the order of observation of one component changes the expected value for the observation of another. This is a property of 'non-commuting observables'. It happens that the components of the angular momentum do not commute. One gets different results in averaging a sequence of measurement of L_x followed by L_y than the average measurements of L_y followed by L_x. But L^2 does commute with one of the components, so one can measure $L^2 = l(l + 1)\hbar^2$ followed by $L_z = m\hbar$ in either order without a subsequent change in the state of the system.

As stated above, the direction of a chemical change is determined not just by the energetics for the process, but also by the entropy change. During the reaction, $\delta G = G_{products} - G_{reactants} = \sum \mu_J \delta N_J$ is the energy which must be expended to change the component numbers, i.e. to make the reaction proceed. If $\delta G < 0$, the reaction will proceed spontaneously. We know that for constant temperature and pressure, a macrosystem continues to change if the Gibbs free energy can be lowered by spontaneous changes in concentration. But at constant temperature, $\delta G = \delta H - T\delta S$. Thus, if the enthalpy is lowered or the internal entropy is increased (or some combination of the two) by the reaction, with a resultant lowering of the Gibbs free energy, the reaction will continue. The enthalpy change of a system during a reaction at constant pressure is the negative of the heat energy that would be released if the reaction proceeded by itself. Biological systems have come to use this kind of reaction energy rather efficiently in a series of coupled reactions.

The entropy change term, $-T\delta S$, in the expression for δG comes from changes in the disorder within the system, due to the reaction. Reactions can proceed without any change in enthalpy simply by having the products more disordered that the reactants. A model example: Suppose you filled a box with lots of balls, linked in pairs by open hooks. If you shake the box for a while ('thermal motion'), you would find that most of the balls have unhooked. This is a more disordered state. The hooks themselves did not store any energy, so no energy was released by shaking the box.

Thermodynamics can give us equilibrium conditions for chemical reactions. But if we want to understand and utilize chemical reactions, we must also consider the kinetics of a reaction, and what determines its speed. The oxidation of hydrogen to produce water is a good example. If you wait long enough, a mixture of hydrogen and oxygen will produce water, even at room temperature. But the time scale needed for some reactions may be eons. (This may be a good thing: The diamond in a diamond ring will eventually turn to graphite!)

The *rate* of a reaction is determined by the nature of the microscopic interactions, the statistical probabilities that reactants and other participants encounter each other, and the natural lifetimes of the various components. At given temperatures and pressures, some reactions take microseconds to reach equilibrium; others take billions of years. Using balls again as an example, suppose the balls in a box were coupled together with open hooks, but now with springs pushing the balls apart. If the springs and hooks are strong enough, shaking the box may not unhook any pair of balls. However, if there were objects in the box to which the balls might attach (such as balls and objects with Velcro surfaces) that make the balls get a little closer together, the hooks might be more likely to unhook, and then the balls are propelled by the spring to fly off the object. These objects act as a catalysis to the releasing of the balls and the energy they carry.

Life systems on Earth have learned to use catalysts, inhibitors, enhancers, regulators, chaperones, etc., to control reaction rates. A catalyst is a substance which becomes chemically involved in one or more reactions, increasing one or more reaction rates. After a finite sequence of reactions, the catalyst is restored, ready to take part in a repeat performance. Biochemical enzymes are an example of a catalyst. There may also be temporary inhibitors that attach to a catalyst to slow or stop its catalytic action, or attach to a reactant to slow or stop a reaction; or enhancer proteins that facilitate catalysts; or parallel reactions that change the pH; or mechanisms that carry reactants together.

As a reaction proceeds near equilibrium, with fixed temperature and pressure,

$$\delta G = \sum_J \mu_J \delta N_J \,. \tag{9.205}$$

But the number of molecules of one type may change because of the reaction. As type J molecule is consumed by a given reaction, the amount of some other type K in the reaction must also change. From the definition of the stoichiometric coefficients, this observation can be expressed as: $\delta N_J/\nu_J = \delta N_K/\nu_K,\ J \neq K.$

To see this ratio condition in an example, consider again the oxidation of water

$$2\,H_2 + O_2 \rightarrow 2\,H_2O\,, \tag{9.206}$$

taking place in a gas with a large number of each molecule. If just one oxygen molecule reacts with two hydrogen molecules to make two molecules of water, then the amounts of each molecule change by: $\delta N_1 = -2$, $\delta N_2 = -1$, and $\delta N_3 = +2$. This means

$$\frac{\delta N_1}{-2} = \frac{\delta N_2}{-1} = \frac{\delta N_3}{+2} \,. \tag{9.207}$$

In the general case,

$$\frac{\delta N_1}{\nu_1} = \frac{\delta N_2}{\nu_2} = \cdots = \delta\xi \,, \tag{9.208}$$

where we have introduced a value for the common ratio, with the name $\delta\xi$. The value of $\delta\xi$ characterizes the progress of the reaction, and is the stoichiometric fraction of the number of any one type of molecule that has engaged in the reaction after a short time.

The 'rate of the reaction' is $\delta\xi/\delta t$, with values being the number of particles involved per unit time t. For example, if $\delta\xi/\delta t$ were N particles per second, then $N\nu_r$ of molecules of reactants of type r have reacted with $N\nu_p$ of type p product molecules each second. In the reaction rate, the detailed time-dependent dynamics of the molecular interactions come to play, including the interactions with other materials.

The change $\delta\xi$ must decrease toward equilibrium, and reach zero at equilibrium. We have

$$\delta G = \left(\sum_J \mu_J \nu_J\right) \delta\xi\,. \tag{9.209}$$

and the condition for equilibrium becomes

$$\sum_J \mu_J \nu_J = 0\,. \tag{9.210}$$

The simplest case of a reaction is $A \leftrightarrow B$. Then $\nu^A = -1$, $\nu^B = 1$, so that

$$\mu^A = \mu^B\,.$$

Systems with molecules which can exist as two different structures or as two different phases would be examples. When a mixture of two phases of a material reach equilibrium, the chemical potentials of each phase must be equal.[62]

As another example, the equilibrium condition across a cell membrane with a difference in ion concentration, producing a membrane potential V_I, is given by

$$\Delta G_{\text{conc}} + \Delta G_{\text{volt}} = 0 \tag{9.211}$$

or

$$-RT\ln\frac{[C_o]}{[C_i]} + zF\Delta V_{ion} = 0 \tag{9.212}$$

Here, z = number of charges on ion, and F = Faraday's constant $= |e|\,\mathcal{N}_A =$ 96,485 coul, the charge in one mole of singly-charged ions.

At the temperature of 37 °C,

$$\begin{aligned}\Delta V_{ion} &= 2.3026\frac{RT}{zF}\log_{10}\frac{[C_o]}{[C_i]}\\ &= \frac{61.5\text{ mV}}{z}\log_{10}\frac{[C_o]}{[C_i]}\end{aligned}$$

This is the Nernst–Planck equation discussed earlier (see Eq. (9.163)). For a nerve cell, potassium concentration is 140 mmol/dm^3 on the inside, 5 on the outside; sodium is 15 on the inside, 150 on the outside. These give $\Delta V_K = -89$ mV and

[62]A phase change in a material is a change in the ordering of molecules without changing how atoms are bound within the molecules. However, the description of the statistics for chemical reactions will also apply to phase changes which can be described by molecular reordering.

$\Delta V_{\rm Na} = 61.5\,\rm mV$. The Nernst–Planck equation balances the diffusion of ions across a membrane in one direction with the forced motion due to the electric field across the membrane carrying the ions in the opposite direction. As we have noted, for living cells, transport of ions across cell membranes involve very few ions compared to the number in the liquid near the membrane. This means that the concentrations of the ions do not significantly change near the nerve cell membrane due to ion transport, except after very long times of nerve cell firing.

A nerve membrane sustains a potential difference of about $\Delta V = -70\,\rm mV$ using ion pumps. (The interior of the cell has a lower potential than the outside.) Ninety percent of the energy used by the brain is for these sodium pumps. A resting brain uses about 20 W. An active brain can require 100 W.

The value of ΔV across a cell membrane comes from not only the concentrations of the various ions, but also the permeability of a variety of ion channels. (See the Goldman–Hodgkin–Katz equation (9.180).)

9.21.1 Equilibrium of Chemical Reactions

If a system has a number of particle species, the amount of any one component in one phase can change due to phase changes or by a chemical reactions. As such a change progresses, small changes in the number of a certain species of molecules M_i must be in proportion to the stoichiometric coefficients ν_i, (See Eq. (9.208))

$$dN_i = \nu_i d\xi \tag{9.213}$$

where $d\xi$ measures the progress of the reaction.

At constant temperature and pressure, the Gibbs free energy changes only by changes in the particle numbers:

$$dG|_{T,p} = \sum \mu_i dN_i\,. \tag{9.214}$$

For a reaction,

$$dG|_{T,p} = d\xi \sum \mu_i \nu_i \tag{9.215}$$

At equilibrium, the Gibbs free energy reaches a minimum, so at equilibrium, we have the condition

$$\sum \mu_i \nu_i = 0\,. \tag{9.216}$$

If the reactants and products are each in an ideal gaseous state, then the chemical potentials are expressible in terms of the partial pressures $p_i = N_i k_B T/V$ as

$$\mu_i = k_B T \ln \frac{p_i}{p_o} + \mu_i^o(T) \tag{9.217}$$

where $\mu_i^o(T)$ is the chemical potential of the i-th component at pressure p_{io}. The p_i is the partial pressure of the i-th component, defined by $p_i \equiv (N_i/N)p$. Relation (9.217) is a generalization of Eq. (9.83) for a mixture of ideal gases. We derived it using the Gibbs–Duhem relation.

For a mixture of ideal gases in a chemical reaction, or for substances in a dilute solution acting like an ideal gas, the equilibrium condition now reads

$$\sum \mu_i \nu_i = k_B T \sum \nu_i \ln \frac{p_i}{p_o} + \sum \nu_i \mu_i^o = 0 . \tag{9.218}$$

Exponentiating, we have

$$\prod_i \left(\frac{p_i}{p_o}\right)^{\nu_i} = \exp\left(-\sum \nu_i \mu_i^o/(k_B T)\right) = \exp\left(-\Delta G^o/(k_B T)\right) \equiv K(T) \tag{9.219}$$

The unitless number, $K(T)$, called the 'equilibrium constant' for the reaction, depends on temperature, but not on the partial pressures. Writing

$$p_i = N_i k_B T/V = \mathcal{N}_A n_i k_B T/V = n_i RT/V,$$

separating the stoichiometric coefficients for products (ν positive) and reactants (ν negative), and using molar concentration

$$[C_i] \equiv \frac{n_i}{V},$$

we find the familiar 'law of mass action' in chemistry[63]

$$\frac{\prod_{prod}[C_p]^{|\nu_p|}}{\prod_{react}[C_r]^{|\nu_r|}} \equiv K_c(T)/K^{\ominus} . \tag{9.220}$$

The power factors on the concentrations reflect the fact that for low concentrations, the chances for the molecules to arrive at a reaction site in the stoichiometric numbers needed for the reaction to occur is proportional to the concentrations to the power of those stoichiometric numbers. The constant $K^{\ominus}$ is selected to make K_c a unitless quantity, as is $K = \exp(-\Delta G^o/k_B T)$. This is done by using specific concentrations, $[C]/[C^{\ominus}]$, referring to a 'standard state.' The standard state for a

[63]For real gases, liquids, solids, and solutions, we will see in Chap. 10 that, for particles reacting independently, the pressures p_i in Eq. (9.219) should be replaced by the so-called activities a_i of the reactant or product.

gas has a pressure of $p^{\ominus} = 10^5\,\mathrm{N/m^2}$. For substances in solutions, the standard state concentration, $[C^{\ominus}]$, is one molar (one mole per liter).

In biochemical reactions, it is conventional to leave out certain concentrations in the law of mass action that remain essentially constant, such as the concentration of water in a cell, even when water takes part. We will expand on this topic later in this Section.

If the concentrations of reactants and products do not represent equilibrium values, then the Gibbs free energy will change by

$$\begin{aligned} dG &= \sum_i \mu_i dN_i \\ &= \sum_i \mu_i \nu_i d\xi \ . \end{aligned}$$

Using our expression for μ (Eq. (9.217)) applied to dilute solutions, we have

$$dG = \left(\sum_i \nu_i \mu_i^o + k_B T \ln \prod_i [C_I]^{\nu_i} \right) d\xi \ . \tag{9.221}$$

The rate of reaction, $d\xi/dt$, is not determined by statistics alone, but rather by the dynamics of the reaction, including the presence of interacting materials such as boundaries where short-term adhesion occurs, and catalysts in the bulk of the materials. Reactions which are energetically unfavorable might locally combine with ones that are favorable, producing an overall favorable set of reactions. Within a biological cell, the hydrolysis of ATP can couple with reactions which would be unlikely, making the combined reactions occur.

Rather than using N_i, the number of molecules of the kind i, we will introduce a number more commonly used in chemistry, the number of moles:

$$n_i \equiv N_i / \mathcal{N}_A \ ,$$

where $\mathcal{N}_A$ is Avogadro's number. Correspondingly, we define the molar change

$$d\overline{\xi} \equiv d\xi / \mathcal{N}_A$$

to be used instead of the particle change $d\xi$. Also, instead of the using the chemical potential as the change in the Gibbs free energy per change in particle number N_i: $\mu_i \equiv \partial G/\partial N_i$, we use the change in the mole number n_i: $\overline{\mu}_i = \partial G/\partial n_i$. Then the change in the Gibbs free energy per mole during a reaction will be

$$\frac{dG}{d\overline{\xi}} = \sum_i \nu_i \overline{\mu_i^o} + RT \ln \prod_i [C_I]^{\nu_i} ,$$

written as

$$\Delta g = \Delta g^o + RT \ln \left(\prod_i [C_i]^{\nu_i} \right) . \tag{9.222}$$

The standard way to assign $\Delta g^o = g^o_{products} - g^o_{reactants}$ is to measure the difference in the Gibbs free energy per mole of each of the molecules taking part in the reaction at a concentration of one molar and standard temperature and pressure minus the Gibbs free energy per mole of the atoms that make up the molecules, letting the Gibbs free energy of widely separated atoms be zero. Tables of the Gibbs free energy per mole can be found for most molecules of interest.[64] When we can neglect long-range interactions of the participating molecules in a chemical reaction, then we can calculate Δg^o by summing the tabulated Δg^o for each of the molecules taking part in the reaction.

Before equilibrium is reached, the concentration products in Eq. (9.222) we will call K_{ne}, while at equilibrium, we will call K. Evidently, when equilibrium is reached (i.e. the left-hand side vanishes), we recover the equilibrium condition on the concentrations, with an equilibrium constant

$$K = \prod_i [C_i]^{\nu_i} = \exp\left(-\Delta g^o / RT\right) . \tag{9.223}$$

Separating products from reactants, this becomes

$$K = \frac{\prod_p [C_p]^{|\nu_p|}}{\prod_r [C_r]^{|\nu_r|}} = \exp\left(-\Delta g^o / RT\right) , \tag{9.224}$$

which we saw earlier as the 'law of mass action' (Eq. (9.220)). So, specifying the equilibrium 'constant' is equivalent to giving the standard value for the change in the Gibbs free energy per mole.

A common problem in dealing with chemical reactions is to find the equilibrium concentrations after a reaction, given the initial concentrations and the equilibrium 'constant' K for a given temperature and pressure.

For example, consider the reaction of carbon monoxide with hydrogen to make methyl alcohol. This reaction has an equilibrium constant of $K = 14.5$ per square mole at 600 K. (A catalyst consisting of granules of copper, zinc oxide, and aluminum oxide is used to speed the reaction, but its presence does not change the final equilibrium concentrations.) If 1.0 molar concentrations of CO and of H_2 are mixed, find the equilibrium concentration of alcohol. The stoichiometric-balanced reaction is

[64] See, for example, Robert A. Alberty, *Calculating apparent equilibrium constants of enzyme catalyzed reactions at pH 7*, Biochemical Education **28**, 12–17 (2000).

$$CO + 2\,H_2 \rightleftharpoons CH_3OH \tag{9.225}$$

Initially,

$$K_{ne} = \frac{[CH_3OH]}{[CO][H_2]^2} = 0 \tag{9.226}$$

With a large positive equilibrium constant, the reaction causes K_{ne} to grow, eventually reaching K. Let $x = [CH_3OH]$. Since twice as many hydrogen molecules are consumed as carbon monoxide molecules, we have, at equilibrium,

$$K = \frac{x}{(c-x)(c-2x)^2}, \tag{9.227}$$

where, in this case, $K = 14.5/\mathrm{M}^2$, $c = 1.0\,\mathrm{M}$. The concentration x is a root of a cubic equation, which has the three roots:

$$x = 0.3941, 0.7011, 0.9048\,\mathrm{M}.$$

Only the first root satisfies the condition that $x < c/2$. Here, $x = c/2$ would mean that all the hydrogen is used up, which cannot happen for a finite K.

The restriction to one root, even though the equilibrium condition may be a high-dimensional polynomial in an unknown concentration x, is quite general. If the initial conditions are physically realizable, then equilibrium is reached as $K_{ne} = N(x)/D(x)$ progresses toward K.[65] The solution, x_s, for x occurs when $P(x) \equiv K\,D(x) - N(x)$ shifts toward zero. As each of these polynomials, $K\,D(x)$ and $N(x)$, are single-valued curves as a function of x, changing x sequentially and infinitesimally in the direction that makes $|P(x)|$ drop in value, will eventually reach one value of x_s, a root of $P(x) = 0$, or else the reaction cannot occur. Unphysical values also include roots with $N(x_s) \le 0$, $D(x_s) \le 0$, or $x_s < 0$.

Now consider how the energy released by a reaction and the entropy produced affects the direction a given reaction proceeds: Toward the products or toward the reactants. As seen in Eq. (9.224), if $\Delta g^o < 0$, the exponent will be greater than one, so the numerator on the left must be greater than the denominator. This means the reaction proceeded toward the products, which is called a spontaneous reaction. If Δg^o is greater than zero, the exponent will be less than one, so the reaction favors the reactants (non-spontaneous reaction).

Note that under isothermal conditions, $\Delta G = \Delta H - T\Delta S$. Also, that under isobaric conditions, $H_r = H_p + Q$, i.e. $\Delta H = H_p - H_r = -Q$, where Q is the heat released in the reaction. We conclude that the reaction goes forward ($\Delta G < 0$) if

[65] If a concentration appears with a fractional order, then K_{ne} should be replaced by a power of K_{ne} which makes both $N(x)$ and $D(x)$ polynomial.

$$\Delta S > -Q/T \, .$$

For a reaction to spontaneously precede forward, the entropy production by the reaction will not be less than the negative of the entropy generated in the environment due to the heat expelled by the reaction.

We see that the condition is a form of the second law of thermodynamics: Changes within a closed system cannot decrease the entropy of that system. If we include the region within and around a cell as the environment, then during a reaction the change in entropy of the cell plus the environment satisfies

$$\Delta S_{cell} + \Delta S_{envir} \geq 0 \, .$$

Since $\Delta S_{envir} = Q/T$, it follows that

$$\Delta S_{cell} \geq -\frac{Q}{T} \, .$$

Even for endothermic reactions, for which $Q < 0$, there may be more entropy produced than $-Q/T$, and the reaction proceeds forward anyway. For example, this might happen if there are more particles produced than were present initially. An endothermic reaction which does not progress forward at one temperature might do so at a higher temperature.

9.21.2 Suppressing Constant Concentrations

Within biological cells, the concentration of hydronium and other ions, such as Magnesium, is important to biological reactions. Mechanisms within the cell maintain the concentrations of these ions, so that their concentrations are effectively constant. Also, water, being the dominant material in the cell, changes little in concentration, even as hydrolysis and dehydration reactions occur. The standard biological cell conditions are defined by the interior of the cell having a concentration of H_2O of $[H_2O] = 993.33\,(\text{g/ml})/(18\,\text{g/mole}) = 55.19\,\text{M}$ @ $37\,°\text{C}$, a hydronium concentration of 10^{-7}moles/L (so the pH ('parts of hydrogen ion') is $= -\log_{10}[H_3O^+] = 7$, $[H_3O^+] = 10^{-7}$moles/L, which is 'neutral', neither acidic nor basic), and a concentration of Magnesium ions of $[\text{Mg}^{++}] = 1$ mM. As before, the statistics for biochemical reactions can be followed, but with the added constraint of a fixed value for the concentrations of water, hydronium ions, and magnesium ions. As these concentrations hardly change, the Gibbs free energy change during a biological reaction is hardly affected by them. When the concentrations of these are taken out of the left-hand side Eq. (9.224) and incorporated on the right-hand side into the Gibbs free energy, the new change in the Gibbs free energy per mole is given a prime: $\Delta g'$. Now the new 'reduced' equilibrium constant is related to $\Delta g'_o$ by

$$K'_{\text{equib}} = \frac{\prod'_p [C_F]^{|\nu_p|}}{\prod'_r [C_r]^{|\nu_r|}} = \exp\left(-\Delta g'_o/RT\right) . \tag{9.228}$$

In this way, for biological reactions, concentrations in the law of mass action for water, hydronium ion, or magnesium ion are not included when calculating the change in the Gibbs free energy per mole.

Biochemical reactions on Earth occur within water solutions or on boundaries of macromolecules in contact with water solutions. Active transport can carry reactants through membranes and along protein 'highways', and within capsules. In water solutions, isolated reactants and products are carried by thermal diffusion. In most cases, the reactant molecules are far less numerous than water, making the solutions dilute.

In those cases for which molecules involved in a reaction act independently, the application of statistical mechanics shows that the law of mass action is expressible in terms of what are called reaction 'activities', a_i, for each reactant and product, rather than concentrations. For dilute solutions with no long-range interactions between particles, the activities can be expressed in terms of concentrations with

$$a_i = \gamma_i [C_i]/[C_i^o] ,$$

where the γ_i are called the reaction activity coefficients (note they are unitless), and $[C_i^o]$ is a standard concentration for molecule kind labeled by 'i'. Activity coefficients are found in statistical mechanics from the so-called 'partition function' for the system, or modeled from experimental data. The arguments are presented in Chap. 10.

Even for dilute solutions, if reactant or product molecules become surrounded by other particles, such as by inhibitors or by ions attracted to participating charged macromolecules, then the activity coefficients for dilute solutions become less than one. As an example, we will discuss the Debye–Hückel model in Sect. 10.9.5.

9.21.3 Membrane Dynamics and Active Transport

Equilibrium ideas can be applied to a semi-permeable membrane, such as a cell wall. Suppose component s is allowed to come into equilibrium across this membrane. Let the system variables inside the cell be labeled by 'I', and those outside by O. If some dN_k of the molecules of component s move from the inside to the outside, then the Gibbs free energy for the combined system will change by

$$dG = (\mu_k^O - \mu_k^I) dN_k . \tag{9.229}$$

At equilibrium, $dG = 0$, and the chemical potentials must balance:

$$\mu_k^I = \mu_k^O \,. \tag{9.230}$$

Cellular membranes, made from a phospholipid protein bilayer, act as an electrical insulator. But water and small ions may be allowed to pass through the membrane through membrane channels. These are special pores in the membrane constructed with proteins. They include diffusion channels, leakage channels, voltage-gated channels, ligand-gated channels, mechanical-stress-gated channels, temperature-gated channels, ionic pumps, assisted exchangers, and facilitated transporters. Those channels that carry particles against a concentration gradient require the input of energy, such as by the work done by the membrane potential, or directly by the action of ATP at the location of the inside of the channel. Cells membranes can also release or engulf vacuoles of material in a process of exo- or endocytosis.

Suppose that in diffusing from the inside to the outside through passive channels, ions must pass through an electric field generated by a difference in concentration of ions across the membrane. If the electric potential across the membrane is $V_O - V_I$, then the average energy of positive ions on the outside will be less than the average energy of those positive ions on the inside by $e(V_O - V_I)$. Now to compare the energy states on the outside to those of the inside, we let $\epsilon_i^O = \epsilon_i^I - e(V_O - V_I)$. Then $Z^O = \exp(-(V_O - V_I)/k_B T))Z^I$. The table below show typical concentrations of sodium, potassium, and chloride ions interior and exterior to the cell, in units of millimoles per liter: The net potential across a cell membrane due to charge

Ion	Out	In
Na^+	155	15
K^+	5	150
Cl^-	110	10

separation creates an electric field in the membrane which the cell uses to do work associated with the membrane, including the operation of active channels in transporting molecules. The pumping of protons across the membrane powers the movement of flagellae.

Each Na^+/K^+ channel pumps three sodium ions out and two potassium ions in. Other channels open and close depending on the variety of passive and active mechanisms.

The membrane is semi-permeable to sodium and potassium ions, but the chloride ions are too large to pass through the membrane ion channels, as are the even larger protein anions in the cell. Rather, chloride and proteins have their own special transport channels and carriers.

Section 14.2.2 has a description of nerve membrane behavior as described by Hodgkin-Huxley model of a neuron action potential.

9.21.4 ATP Pumps

As an example of applying equilibrium considerations to an important biochemical set of reactions, let's look at the steady-state conditions of sodium and potassium ions across a neuron. Active sodium-potassium pumps, other active pumps, and passive switching-diffusion channels in the neuron membrane wall create an electric field across the membrane of about $-70\,\text{mV}$ under resting conditions. The wall has a thickness of about 3.5 nm, so the electric field across the membrane is about 2×10^7 N/C toward the inside. The approximate concentration of sodium and potassium ions for an inactive neuron membrane given in the above Table. Adenosine Triphosphate (ATP) supplies the energy for the active transport pumps. Only the energy stored in the phosphate group's binding to one ATP molecule is needed for each cycle of the transport mechanism.

For each of the sodium-potassium pumps, three sodium ions are forced across a neuron membrane through an especially dedicated protein channel in the wall of the cell membrane. At the same time, two potassium ions are allowed to pass into the cell through this same channel. The reactions can be represented by:

$$\text{ATP} \rightarrow \text{ADP} + \text{P}_i \tag{9.231}$$

$$3\,\text{Na}^+_{\text{in}} \rightarrow 3\,\text{Na}^+_{\text{out}} \tag{9.232}$$

$$2\,\text{K}^+_{\text{out}} \rightarrow 2\,\text{K}^+_{\text{in}}\,. \tag{9.233}$$

The shortened symbol P_i is used for the inorganic phosphate H_3PO_4. The first reaction, the hydrolysis of ATP, releases about 30.5 kJ/mole of Gibbs free energy under standard conditions, i.e. $\Delta g'^o = -30.5$ kJ/mole.[66] At a temperature of 310 K, each mole of ATP inside the cell which takes part in a reaction will release

$$\begin{aligned} \Delta g_{ATP} &= \Delta g'^o + RT \ln \frac{[\text{ADP}][\text{P}_i]}{[\text{ATP}]} \\ &= -30.5\,\text{kJ/mole} + 2.5786\,\text{kJ/mole} \ln \frac{(0.01)(0.001)}{0.01} \\ &= -48.3\,\text{kJ/mole}\,. \end{aligned} \tag{9.234}$$

From the Nernst–Planck equation, the pumping of each sodium ion out of the cell requires an expenditure of Gibbs free energy given by

[66] The participation of water, which stays at about 55.5 Molar, is implicit. The concentration of magnesium ions and hydronium ions are also important, but their concentrations are also implicit. Magnesium bound to ATP is just under 10 mM, while free ions have a concentration of about 0.5 mM. In the cellular environment, the concentrations are typically [ADP] $\approx$ 0.01 Molar, [P$_i$] $\approx$ 0.001 Molar, and [ATP] $\approx$ 0.01 Molar.

$$
\begin{aligned}
\Delta g'_{Na} &= -zF\Delta V + RT \ln \frac{[\mathrm{Na}^+_{out}]}{[\mathrm{Na}^+_{in}]} \\
&= -(+1)(96.5\,\mathrm{kJ/mole}\cdot\ \mathrm{volt})(-70\,\mathrm{mV}) \\
&\qquad +(2.5786\,\mathrm{kJ/mole} \ln \frac{0.155}{0.015} \\
&= 6.755\,\mathrm{kJ/mole} + 6.022\,\mathrm{kJ/mole} \\
&= 12.78\,\mathrm{kJ/mole}\,, \qquad (9.235)
\end{aligned}
$$

and the passage of each potassium ion into the cell requires

$$
\begin{aligned}
\Delta g'_{K} &= zF\Delta V + RT \ln \frac{[K^+_{in}]}{[K^+_{out}]} \\
&= (+1)(96.5\,\mathrm{kJ/mole}\cdot \mathrm{volt})(-70\,\mathrm{mV}) \\
&\qquad +(2.5786\,\mathrm{kJ/mole}) \ln \frac{0.150}{0.005} \\
&= -6.755\,\mathrm{kJ/mole} + 8.770\,\mathrm{kJ/mole} \\
&= 2.015\,\mathrm{kJ/mole}\,. \qquad (9.236)
\end{aligned}
$$

As a check of signs, the membrane electric field pulls both sodium and potassium ions inward, while thermal diffusion acting through an open channel would transport sodium into the cell and potassium out. In the case of pumping sodium ions out of the cell, work is required to act against both the electric force and the back diffusion, while for pumping potassium ions into the cell, work is expended to counteract diffusion, but the work required is reduced by the presence of an inwardly directed electric field. As a caution: The application of arguments from statistical thermodynamics to diffusive effects of individual particles in membrane channels makes sense only when averages over many channels is implied.

For the set of reactions above, we have

$$
\begin{aligned}
\Delta g'_{total} &= \Delta g'_{ATP} + 3\cdot \Delta g'_{Na+} + 2\cdot \Delta g'_{K+} \\
&= -48.3\,\mathrm{kJ/mole} + 3\cdot 12.78\,\mathrm{kJ/mole} + 2\cdot 1.884\,\mathrm{kJ/mole} \\
&= -6.19\,\mathrm{kJ/mole}\,. \qquad (9.237)
\end{aligned}
$$

As a result, we can deduce that the conversion of a single ATP into ADP plus a phosphate releases sufficient free energy to carry off the $\mathrm{Na}^+/\mathrm{K}^+$ pump action, with a little energy to spare (which goes to kinetic motion and then heat).

9.22 Metabolic Reactions

Advanced life systems have evolved to optimize the rate and increase the efficiency of advantageous chemical reactions. Reaction rate is increased through catalytic proteins (enzymes) and efficiency through the use of a sequence of reactions, each of which is as close to reversible as possible, rather than through single reaction. Cellular metabolic efficiencies average about 20%. The production by the Kreb's cycle of 36 ATP molecules (ATP storing 0.326 eV/molecule of usable energy) for each glucose molecule (glucose which can make 29.8 eV/molecule of usable energy) means this process is about 40% efficient.

In order for an energy-storage molecule to be useful in the chemical environment in which it exists, it must be metastable, i.e. exist at least over a time period to be useful. A system will be metastable if its spontaneous disintegration lifetime is large compared to the times of other needed activities. Metastable molecules holding energy in the tension of binding may be forced to release this energy by first adding 'activation energy' to open the molecule. (See Fig. 10.7.) Enzymes act by reducing the required activation energy, using their shape to distort and hold one or more reactants during a reaction. Enzymes can increase reaction rates by factors of billions, some catalyzing up to 500,000 molecules per second in a cell.

ATP is produced and continuously used in cellular reactions. The entire supply of ATP is recycled once per minute in a healthy cell.

Those energy-producing ("catabolic") reactions which produce ATP allow energy-consuming ("anabolic") reactions and generate activity involving work, such as the coherent motion of molecules.

9.22.1 Process Efficiencies

All processes occurring in a closed system with a large number of entities cannot cause a decrease in entropy in the system. If there are any sources of dissipation in the system, i.e. where energy initially in one state ('coherent') is spread out into many states ('incoherent'), then the entropy in that closed system must increase. Frictional forces doing work is an example in which coherent motion is partly converted into incoherent motion (heat, in this case). In contrast, a reversible process incurs no dissipation. There is no cyclic process that can recover all the energy lost to dissipation. This is equivalent to the statement that there is no 100% efficient heat engine, and equivalent to the statement that incoherent motion cannot spontaneously convert to coherent motion. The 'free energies' introduced earlier give a measure, under given conditions, of the energy in a system that can do useful work.

If a system is open to the transfer of heat to its environment, it can easily lower in entropy (at the expense of increasing the entropy of its environment). We will see that if an open system contains a sufficient variety of interacting entities, it will tend to spontaneously order itself, and expel heat (increasing the entropy of its environment). If an energy flow through the system is maintained, the system may be able to continue its ordering of subsystems, producing complexity in contained structures.

An adiabatic process that involves the transfer of energy from one form to another without a redistribution of the particles among possible energy states, is non-dissipative and non-entropy producing. For example, an electric motor need not produce heat, transferring 100% of electric field energy into coherent mechanical motion. As the flagella motors on bacterial cell walls are partly electric motors, running on proton currents, their efficiency is only partly restricted by thermodynamics.

9.23 Non-equilibrium Thermodynamics

9.23.1 Near-Equilibrium

Near equilibrium, thermodynamic variables for a system with a large number of particles, such as temperature, pressure, and chemical potential, are still measurable, with relatively small fluctuations. Those changes in a closed system which increase its entropy are called dissipative.[67]

Now consider how the entropy changes when a system is near but not at equilibrium. We start with non-dissipative processes in a closed system, for which

$$TdS = dE + pdV - \sum_i \mu_i dN_i \, , \tag{9.238}$$

showing how the internal entropy changes when any of the extensive variables $E, V, \{N_i\}$ change. The relation for the entropy has the form

$$dS = \frac{1}{T}dE + \frac{p}{T}dV - \sum_i \frac{\mu_i}{T}dN_i \, . \tag{9.239}$$

[67]Dissipative changes allow us to deduce the direction of time. People sometimes laugh when viewing a film run backward. We do not expect to see balls bounce higher by themselves, or warm water turn to steam and ice. But the electromagnetic interactions which accurately describe the microscopic behavior of atoms and molecules are symmetric under time reversal. (Experimentally, the decays of kaon particles show a non-time reversal behavior, but this decay is not governed by Maxwellian electromagnetic interactions.) The loss of time reversal symmetry we observe in macroscopic processes comes from energy being dispersed among the available multitude of microstates, with little chance of this process reversing. Discrete models of nature could allow the information to be cryptically stored, so that the entropy produced is a measure of our lack of practical access to that information. In quantum language, wave functions would not 'collapse' during an observation, but rather the wave function for a macrosystem and a macroscopic observer would become entangled and in many circumstances the changes are practically irreversible.

If we know how the temperature, pressure, and chemical potentials depend on the internal energy, volume, and concentration of chemical type, we will know a fundamental equation for the system expressed as $S = S(U, V, \{N_i\})$.

In general form, the entropy change in a closed system due to non-dissipative processes has the general form

$$dS = \sum_i \frac{\partial S}{\partial Y_i} dY_i \equiv -k_B \sum_i y_i dY_i \ , \tag{9.240}$$

where

$$y_i \equiv -\frac{1}{k_B}\frac{\partial S}{\partial Y_i} \tag{9.241}$$

and Y_i is one of the independent variables for the subsystem. For simple systems, $\partial S/\partial E = 1/T$, $\partial S/\partial V = p/T$ and $\partial S/\partial N_i = -\mu_i/T$.

Compare with conservative mechanical forces satisfying $\mathbf{F} = -\nabla\Phi$, so the work done by such a force while a body 'displaces' $d\mathbf{Y}$ is given by $dW = -\nabla\Phi \cdot d\mathbf{Y} = -d\Phi$. In the present case, we have an expression for the entropy change which formally appears as dE does, namely, a sum over gradients times the corresponding changes in the independent variables. For this reason, the y_i are called entropic thermodynamic 'forces', and the Y_i 'entropic thermodynamic displacements'. In addition, each of the terms $y_i dY_i$ must be integrable, because they are perfect differentials when one holds all the other Y_k fixed.

If we take the Y_i to be the extensive variables, We can find the result of this integration using the fact that $S(\lambda Y) = \lambda S(Y)$, which makes

$$S = -k_B \sum_i y_i Y_i \ . \tag{9.242}$$

The form of this expression has led to the nomenclature that the y_i and Y_i are 'conjugate thermodynamic variables'. As S has the units of k_B, each y_i has the inverse units of Y_i.

Let s be the local entropy in a small open subsystem per unit total particle number in the subsystem. If there are no dissipative processes, the disorder of a system with fixed extensive variables will not change. If the system is open, then there could be a flow of disorder (entropy) across the boundaries of the system. Such a flow will be denoted by $\mathbf{J}_s$ giving the entropy per unit area per unit time carried across an area whose normal point in flow direction.[68]

Let $\dot{\sigma}$ be the excess rate of entropy production due to dissipative processes within a small subsystem. Then the local rate of entropy change is given by

[68] A good example of such a flow of disorder is that from a chicken egg as the embryo develops. In this case, the entropy flow is carried by heat flowing out of the egg.

$$\frac{d}{dt}\int s dV = -\oint \mathbf{J_s} \cdot \mathbf{dA} + \int \dot{\sigma} dV \tag{9.243}$$

where the first term on the right is the net flux of entropy through the surface of the volume and the second measures the production of internal entropy due to irreversible processes. In local form, this reads

$$\frac{\partial s}{\partial t} = -\nabla \cdot \mathbf{J_s} + \dot{\sigma} \,. \tag{9.244}$$

Substitution from the expression for S in Eq. (9.239) gives

$$\dot{\sigma} = \frac{1}{T}\frac{\partial E}{\partial t} + \frac{p}{T}\frac{\partial \rho^{-1}}{\partial t} - \sum_k \frac{\mu_k}{T}\frac{\partial n_k}{\partial t} + \nabla \cdot \mathbf{J_s} \geq 0 \tag{9.245}$$

which more generally can be put into the form (from Eq. (9.240)

$$\dot{\sigma} = -k_B \sum_k y_k \frac{\partial (Y_k/N)}{\partial t} + \nabla \cdot \mathbf{J_s} \geq 0 \,. \tag{9.246}$$

Gradients across space of the intensive thermodynamic variables such as gradients of the local temperature, the density, or the chemical potential, will cause currents J_k, such as heat flow, material flow, or chemical type flow. Expanding J_k in a Taylor series in the y_k, the zeroth-order term vanishes, since there will be no current if the gradients vanish. Assuming the quantities y_k do not change too rapidly, quadratic and higher order terms can be negligible. This means there will be a linear relationship between each flux and a corresponding gradient. The table below show examples of such relationships, when each flow acts independently of the other flows:

Flux		Conjugate	Relation	Name
	J	y		
Heat	J_Q	$-(1/T)\nabla T$	$J_Q = -\kappa \nabla T$	Fourier's law
Charge	J_q	$\nabla \phi$	$J_q = -\rho_q \nabla \phi$	Ohm's law
Material	J_{Di}	$-\nabla \mu_i$	$J_{Di} = M_i c_i \nabla \mu_i$	Fick's law

As we should now expect, a gradient of temperature will cause heat flow, a gradient of electric potential will cause a charge current, and a gradient of chemical potential will cause a diffusive current. These observations are incorporated into Fourier's, Ohm's, and Fick's Law. We have already seen that each of these flux relationships help us in understanding and logically describing a number of dynamical processes in biological systems.

9.23.2 Flows Near Equilibrium

A large set of interacting particles, acting classically, such as atoms and molecules in your body, if allowed to relax, will tend to distribute in energy states which follow the Boltzmann distribution, simply because this distribution contains a large number of microstates for a given macrostate and therefore this distribution is the most likely.[69]

In moving toward equilibrium, or under continuous external influences, flows of energy, entropy, material and charges can occur. Internally, these flows are caused by gradients of intrinsic quantities, such as temperature, concentrations, and electric potentials, or by chemical imbalances.

The time rate of change of any one of the extensive variable densities $Y_i/N \equiv \rho_i$ in a small subsystem with fixed total particle number N and located near $\{x\}$, will be denoted $\dot{\rho}_i$. These rates can depend on position and time through the variations in the local values of the Y_k: $\dot{\rho}_i = F_i(Y(x,t))$. At equilibrium, as well as steady-state conditions, there should be no more changes in time. It follows that a Taylor series of $\dot{\rho}_i$ to first order in $\delta Y_j \equiv Y_j - \langle Y_j \rangle$ about equilibrium values should be of the form

$$\dot{\rho}_i = \sum_k \frac{\partial F_i}{\partial Y_k} \delta Y_k \equiv - \sum_k \alpha_{ik} \delta Y_k \ , \tag{9.247}$$

where the α_{ik} are constants. A minus sign is introduced anticipating that the changes with time will exponentially drop with time.

Similarly, the changes in the conjugate thermodynamic variables $y_l = -(1/k_B)\partial S/\partial Y_l$ can be expanded around the equilibrium values of Y_k to give

$$\delta y_l = \sum_m \gamma_{lm} \delta Y_m \ . \tag{9.248}$$

By defining $\lambda_{ij} = \sum_l \alpha_{il} \gamma_{lj}^{-1}$, the last two equations combine to give

$$\dot{\rho}_i = - \sum_j \lambda_{ij} \delta y_j \ . \tag{9.249}$$

For simple systems $y_\varepsilon = -(1/k_B T)$, $y_v = -(p/k_B T)$, and $y_n = (\mu/k_B T)$, and we let $\varepsilon \equiv E/N$. The rates of change with time of the density of extensive variables in each small subsystem of fixed total particle number are given in terms of changes from equilibrium in the intensive quantities, such as temperature, pressure, and chemical potentials.

[69]At or near equilibrium, bosons and fermions densely populating the lowest energy states will deviate from the Boltzmann distribution.

In 1931, Las Onsager[70] showed that if the microscopic equations governing the interactions of the particles have time-reflection symmetry (particle trajectories reversed in time are also solutions to the dynamics), then the $\lambda_{ij} = \lambda_{ji}$, i.e. they form a symmetric matrix.[71] This statement draws profound connections in the behavior of systems near equilibrium. For example, since it is possible for gradients in temperature to cause gradients of density, this process can be used to predict the 'reciprocal process', how gradients in density may cause gradients in temperature. Thus, as another example, when both material and energy flows occur in an otherwise isotropic system as is common in biological solutions, Fourier's and Fick's law are coupled, becoming

$$\mathbf{J_u} = L_{uu}\nabla(1/T) + L_{u\rho}\nabla(-\mu/T) \tag{9.250}$$

$$\mathbf{J}_\rho = L_{\rho u}\nabla(1/T) + L_{\rho\rho}\nabla(-\mu/T)\,, \tag{9.251}$$

in which the current $\mathbf{J}_\varepsilon$ (*or* $\mathbf{J}_\rho$) is the density of internal energy (*or* material) times its speed across the space of the small subsystem volume, and the Onsager relations become $L_{u\rho} = L_{\rho u}$, connecting the two types of flow.

The coupling of diffusion of material to gradients in temperature implied by the first term on the right-hand side of Eq. (9.251) is called the Soret effect, and is a phenomenon of 'thermophoresis'. Thermophoresis is used in the separation of large molecules from smaller ones in a solvent, through a process named 'thermal field flow fractionalization'. The generation of heat flow due to a gradient in a chemical potential is the Dufour effect, represented by the second term on the right-hand side of Eq. (9.250). This behavior can be significant in a mixture of gases.

Gradients in temperature can also produce gradients in the electric potential, as in the Seebeck effect. Conversely, gradients in electric potential can generate cooling, as in the Peltier effect. These two effects are also connected by 'Onsager reciprocal relations'.

We know that gradients in the electric potential (i.e. electric fields) can produce electric currents, but also can simultaneously produce neutral particle flows. Inversely, by the Onsager relations, gradients in particle densities in these same systems can produce an electric field. Many of these types of effects have been discovered in materials, including within biological cells and across membranes.

The addition of an external magnetic field expands further the possible couplings of effects.[72] Here are some which occur:

---

[70] L. Onsager, *Reciprocal relations in irreversible processes, I & II*, Phys Rev **37**, 405–426, 2265–2279 (1931).

[71] The $\{\lambda_{ij}\}$ make up what is now called the Onsager matrix.

[72] If a magnetic field $\mathbf{B}$ is present, then the Onsager symmetry relations are expressed as $\lambda_{ij}(\mathbf{B}) = \lambda_{ji}(-\mathbf{B})$, reflecting that the microscopic laws under time reversal require changing the direction of the magnetic field, since such fields are produced by moving charges.

- Hall Effect: An electric current in a material in the presence of a magnetic field can induce a transverse electric field: $\nabla \phi_E = \kappa_H \mathbf{B} \times \mathbf{J}_q$.
- Nernst Effect: A flow of heat in a material in the presence of a magnetic field can induce a transverse electric field: $\nabla \Phi_E = \kappa_N \mathbf{B} \times \mathbf{J}_u$.
- Ettinghausen Effect: An electric current in a material in the presence of a magnetic field can induce a transverse gradient in temperature: $\nabla(1/T) = \kappa_E \mathbf{B} \times \mathbf{J}_q$.
- Righi-Leduc Effect: A flow of heat in a material in the presence of a magnetic field can cause a transverse gradient of temperature: $\nabla(1/T) = \kappa_R \mathbf{B} \times \mathbf{J}_u$.

9.24 Irreversible Thermodynamics

The study of biological systems using principles of thermodynamics is justified by the fact that thermodynamic variables such as temperature and pressure can often be defined for local but still macroscopic parts of an organism.

Investigating the properties of open systems which are globally far from equilibrium, Prigogine[73] observed that in certain cases order can be spontaneously created within the system as the environment becomes more disordered. Examples include crystallization, hurricanes, some chemical reactions with autocatalysis which run in cycles instead of reaching fixed concentrations, and life itself. Thermal systems can self-generate material flows. Systems which evolve to have subsystems which have greater order than initially, and which accelerate the ordering process, are called "self-organizing." A necessary requirement for continued ordering is a flow of energy into and out of the system. During the process of ordering for cyclic processes in an open system, entropy production in the environment implies that the processes are 'irreversible'. This is both a statistical statement (i.e. it should read, "highly improbable") and a statement about information loss caused by the practical inability to track the dynamics of microstates. We will return to this topic in the chapter on ordering theory (Chap. 15.)

9.25 On the Evolution of Driven Systems

Biological systems are not in thermodynamic equilibrium. They contain ordered structures holding energy. If there is no flow of energy into the system from the environment, those ordered structures will eventually disassociate. After all, having been created by interactions, the ordered structures can decay through interactions,

[73] I. Prigogine, **Thermodynamics of Irreversible Processes**, Second Edition, [Wiley] (1961); Gregoire Nicolis and Ilya Prigogine, **Self-Organization in Non-Equilibrium Systems**, [Wiley] (1977).

producing an internal release of the stored energies, and eventually becoming one temperature and consisting of constituent molecules with long-time stabilities and distributed in energy states according to Boltzmann.

In general no matter how far from equilibrium a system might be, the concept of energy conservation still applies, and the entropy of any macro subsystem within the open system can still be calculated, in principle, through its statistical measure. The temperature of a subsystem, i.e. the inverse of rate at which the internal energy of that subsystem gains entropy, makes sense if its fluctuations are significantly smaller than its value. Such fluctuations tend to be small if the macro subsystems are solutions which can easily dissipate heat, such as subsystems in proximity or containing liquid water (a solvent with an especially high specific heat). The evolution of any such open subsystem follows microscopic laws of nature. Moreover, typically after the thermodynamic relaxation time for the solvent, τ_s, the subsystem will be most likely found in those quasi-stable states which have longer lifetimes than τ_s and that have been built from prior natural and statistical sampling of the possible structures.

A special case of an open system is one through which energy, perhaps in a variety of forms, consistently or periodically flows through the system. These are called 'driven systems'.[74]

If the energy flow is steady, then the phrase, 'steadily driven system' applies. The system accepts energy from an external source and dissipates the energy into the environment. The system can be realized by a mixture of particles capable of taking up energy and transforming that energy into kinetic motion, excited internal states, internal transport processes, and possibly into external work. Now imagine that among those possible excited states are some that are quasi-stable. The concentration of these states, if initially low, can build up. Moreover, some of these quasi-stable states may, by chance, be capable of absorbing the incoming energy more effectively than a system without such states. For example, a molecule might form which resonates at an incoming photon frequency. These more capable molecules will then be more likely to form, and their concentration will increase with time, provided there is sufficient concentration of their constituents.

On top of the ordering process that produces molecules which are efficient at absorbing the incoming energy, some of these molecules may act as intermediaries to other secondary ordered structures. In effect, they act to lower the energy barrier which had made the secondary structures unlikely, but are now more easily accessible. This is the behavior of a catalyst. Again, the most likely secondary structures will be those that are both more easily produced and that benefit from being relatively more stable. The system will undergo a macroscopic change when catalysts drive a large quantity of reactants toward their resultants.

[74]Harold J. Morowitz emphasized that spontaneous order production occurs when a sufficiently complex open system has an energy flow through it. See his book, **Energy Flow in Biology**, [Acad Press] (1968).

We see that driven systems naturally develop into ones with more and more ordered quasi-stable states, with the most likely primary states being those that are effective in utilizing the incoming energy and then those states that can be reached through catalytic or tertiary channels facilitated by the primary states.

This is a form of evolution, applied to any open macrosystem with a flow of energy from an external source to an external sink. The system's 'exploration', 'discovery', and utilization of all those quasi-stable substates and reactions that enhance the energy flow is an ongoing process applicable to both inorganic as well as life systems. This process may be called the 'Principle of Natural Adaptability', and was formalized mathematically by Jeremy England.[75] Steadily driven systems will evolve to utilize the available incoming energy more effectively, and become more structured in this evolution, adapting to the nature of the energy flow. England and his research group explored the theory and practice of this principle, both in computer simulations, in inorganic solutions, and in biological systems. It is clear that driven systems that make up life can display adaptation in addition to Darwin's 'Principle of Natural Selection'. Darwinian evolution occurs when lifeforms with genetic variability are put under stress, thereby giving preference to the genetic variants that can survive, perform, and reproduce more effectively, under new conditions.

These ideas also suggest that biological evolution occurs along a sequence of possible and successful paths. Bacterium flagellar motors will not suddenly pop into existence on the bacterium's cell wall where there was none. They are too complex to be constructed by a single or even a few mutations of a DNA molecule. But they can evolve in small steps if, at each stage, there was a viable purpose to each precursor structure. This kind of evolutionary by re-adaption is now called 'exaptation'. Good examples include the evolution of bird feathers, the eye, and the behavioral trait of laughing.

9.26 Some Applications in Medical Treatment

Thermodynamics can be used to understand much of the microscopic dynamics of life systems, including biochemistry and order building. Given the importance of these processes, any breakdown is serious. Medical applications involve the full spectrum of monitoring, evaluation, diagnostics, and therapy.

Here are a few of the many applications not yet introduced:

- Cryotherapy: The use of cold to kill cancer cells by freezing, such as melanoma of the skin. Liquid nitrogen (which boils at $-195.8\,^{\circ}$C) is often used in a probe design. Damage to cells occurs principally by water crystal growth severing cell membranes. Slow freezing also causes denaturalization of lipo-proteins.

[75] See, for example, Jeremy L. England, *Statistical physics of self-replication*, J Chem Phys **139**, 121923 (2013).

- Cryogenics: Rapid freezing of live tissue in which water crystals do not have time to grow and rupture cells has successfully held fish (cold-blooded) in suspended animation. They can be brought back to life by warming.[76]
- Hypothermia: Uses low temperatures to slow the metabolic processes, such as during surgery. Body temperatures below 35 °C are considered hypothermic. Temperatures below 28 °C lead to unconsciousness, while below 20 °C is lethal. Immersion in water at 10 °C for over 1 h leads to death. At 0 °C, death occurs in less than 15 min. The cerebral metabolic activity decreases by 7% for each degree Celsius drop in temperature.
- Hyperthermia: Uses higher temperatures to increase blood circulation to reduce stiffness, inflammation, edema; therapy for rheumatoid arthritis, relieve neck and upper back pain from muscle tightness, relieve headaches and migraines. Vasodilator IR, microwaves, ultrasonic diathermy uses electromagnetic currents for muscle relaxation (including Ultrasonic therapy, shortwave diathermy, and microwave diathermy). Hyperthermia can destroy warts, neoplasms, and infected tissue. Higher temperatures are used to cauterize blood to stop bleeding. However, whole body temperatures above 41 °C (106 °F) can cause delirium, brain lesions, and stroke.
- Refrigeration: As a method to hold tissue in quasi-stasis. Bacterial growth continues above the freezing point, but at a reduced rate.
- Freeze drying: Preserves the micro-structural elements of tissue. A dehydration process is employed to remove water from perishable food. The material is first cooled, and then placed in a partial vacuum to cause evaporation of water. In this condition, nutrients in food can last indefinitely.
- Incubators: Devices to hold new born requiring neonatal care in a thermally advantageous environment.

Problems

9.1 If your internal temperature is 37°C, your skin temperature is 35.5°C, the average effective thickness of your skin is 5 mm your exposed body area is 1.2 square meters, and the heat conductivity of your skin is 18 kcal-cm/(m^2h-K), estimate the heat loss (in watts) by conduction through your skin.

9.2 A resting adult generates about 100 W of power in the form of heat put into the environment. If the temperature of the environmental air is 22°C, how much disorder (entropy) does this adult generate every day.

[76] A professor of the faculty of The George Washington University Chemistry Department would perform this 'trick' on a goldfish every year in freshman chemistry in the 1960s through immersion in liquid nitrogen.

9.3 If your brain is currently using 20 W of work energy per unit time to make new neuronal structures and populate higher molecular energy states, what is the rate of entropy production in the air surrounding you due to this activity?

9.4 Suppose the muscle attached to your Achilles tendon exerts a force of 705 newtons lifting your weight a distance of 2.6 cm as you rise up onto your toes, and then the muscle relaxes again, returning you to your original height. In the process of raising you up, 12 calories of heat is released from the muscle. What was the efficiency of this muscle? If the environment was at 22°C, how much entropy was produced? Ideally, what was the change in the internal energy of the muscle for the whole process? What was the change in the entropy of the muscle for the whole process?

9.5 A certain thermodynamic system is described by the equilibrium energy relation

$$U(S, V, N) = \frac{KN^{5/3}}{(V - Nb)^{2/3}} \exp\left(\frac{2S}{3N}\right) - \frac{aN^2}{V}$$

where K, a, and b are constants. Find the internal pressure p as a function of temperature T and volume V for this system.

9.6 Suppose at some moment the concentration of sodium ions on the inside of a cell membrane of thickness 5 nm is 304 mM and outside the membrane it is 302 mM (1 nm is one nanometer, 1 mM is one millimolar), with ions passing through at the rate of 4.3×10^{14} s/cm^2. What is the drift velocity of an ion through the membrane? What is the diffusion constant for sodium through this wall?

9.7 The rate of diffusion of small particles in the vitreous humor of the eye is of physiological and medical interest. Estimate the diffusion time for a concentration of protein molecules to spread from an injection point in the center of the eye to a RMS distance of 10 mm across the vitreous if the diffusion constant for these molecules is measured to be $D = 0.47\,\mathrm{mm}^2/\mathrm{h}$.

9.8 If a set of macromolecules each of diameter 812 nm are observed to spread in water a root-mean-square distance of 10.51 μm after 30 s, estimate Avogadro's number. The viscosity of water at room temperature (27°C) is 0.852 centipoise.

9.9 A small vacuole containing a concentrated solution of salt, $NaCl$, is opened on the interior of one side of a cell of diameter 12 μm. The diffusion constant for salt in water is about $1.0 \times 10^3\,(\mu\mathrm{m})^2/\mathrm{s}$. Estimate the time for this salt to reach the opposite side of the cell by diffusion.

9.10 Two membranes, #1 and #2, permeable to a small organic molecule, are found at either end of an impermeable cylinder of length L within a cell. The cell maintains a concentration C_1 on the outside of membrane #1, and C_2 outside membrane #2. The diffusion constant for the molecules in the fluid contained in the cylinder is D. What is the steady-state concentration of the small organic molecules at a distance x from the membrane #1?

9.11 Suppose a mammalian cell wall is blocked from transporting any ions except sodium. As a result (after some time), an equilibrium voltage difference of 60 mV is established. What is the ratio of the concentrations of sodium ions on one side compared to the other side of the cell wall?

9.12 The adult human brain consumes about 20 W of power while resting. About a fifth of that power is used by the cerebral cortex. Assume about 70% of the power is devoted to maintaining the membrane potential. On average, how much ATP (in molecules) is consumed by the cerebral cortex per minute for this purpose?

9.13 Explain how a biochemical reaction with the disassociation of a molecule into two atoms can be spontaneous even if the energy released in the reaction is negative.

9.14 Overall, glycolysis is a conversion symbolized by:
Glucose + 2 ADP + 2 Pi + 2 NAD^+ $\rightarrow$ 2 Pyruvate + 2 ATP + 2 NADH + 2 H^+ + 2 H_2O

This process is performed in cells through the reactions

1. Alpha-D-Glucose to alpha-D-Glucose-6-phosphate via Hexokinase
2. Alpha-D-Glucose to Beta-D-Frutose-6-phosphate via Phosphoglucoisomerase
3. Beta-D-Fructose-6-phosphate to Beta-D-Fructose-1,6-biphosphate via Phospho-fructo-kinase
4. Beta-D-Fructose-1,6-biphosphate to Glyceraldehide-3-phosphate and Dihydroxy-acetone phosphate
5. Dihydroxyacetone phosphate to Glyceraldehide-3-phosphate via Triose phosphate isomerase
6. Glyceraldehide-3-phosphate to 1,3-Biphosphoglycerate via glyceraldehyde-3-phosphate dihydrogenase
7. 1,3-Biphosphoglycerate to 3 Phosphoglycerate via Phosphoglycerate kinase
8. 3-Phosphoglycerate to 2-Phosphoglycerate via phosphoglycerate mutase
9. 2-Phosphoglycerate to Phosphoenolpyruvate via Enolase
10. Phosphoenolpyruvate to Pyuvate via Pyruvate kinase.

Enzyme for reaction 1–10 above	$\Delta G'^{o}$ (kJ/mole)	ΔG (kJ/mole)
1. Hexokinase	−16.7	−33.5
2. Phosphoglucoisomerase	+1.7	−2.5
3. Phosphofructokinase	−14.2	−22.2
4. Aldolase	+23.9	−1.3
5. Triose phosphate isomerase	+7.6	+2.5
6. Glyceraldehyde-3-phosphate dehydrogenase	+12.6	−3.4
7. Phosphoglycerate kinase	−37.6	+2.6
8. Phosphoglycerate mutase	+8.8	+1.6
9. Enolase	+3.4	−6.6
10. Pyruvate kinase	−62.8	−33.4

(a) Which of these reactions are endergonic?
(b) In your cells, what is the equilibrium constant for p-kinase reaction?
(c) How much free energy is available in reaction 2?
(d) After sufficient time, why do not the enzymes in these reactions affect the resultant product concentrations?

9.15 Suppose a developing egg has a net outflow of heat of 280 cal/h into the environment at an incubation temperature of 39° C. What is the average rate of entropy drop in the egg?

9.16 If a cell with a salinity concentration of 1.5 mmolar is placed in pure water at 37 °C, estimate the osmotic pressure that builds in the cell?

9.17 The water vapor in our lung alveoli is close to saturation. On expiration, the air from our lung expands, and its pressure drops. It may also be cooled by the outside air. To see the effect of cooling, take an expiration volume of 0.5 L. If this volume is cooled from 37 to 0 °C, how much water condenses (to tiny droplets)?

Chapter 10
The Statistical Basis of Bioenergetics

If you are out to describe natural laws, leave elegance to the tailor.

— Albert Einstein

Summary This chapter gives the statistical arguments behind thermodynamics, with applications to cellular systems and biochemical reactions. Included is a discussion of the Debye–Hückel model for ionic reactions, a description of fluctuations in thermodynamic variables, and an introduction to kinetic theories for reactions.

10.1 Concepts

The subject of statistical mechanics combines microscopic laws with statistics applied to systems with a large number of particles. A major result is a derivation of the classical laws of thermodynamics using probabilities rather than absolutes in the description. Classical thermodynamics was originally formalized by Clausius[1] as an axiomatic system based on the observations of macroscopic systems at or near equilibrium. Such an axiomatic view has a beauty all its own in the breadth of its applicability. But we know that the behavior of thermodynamic systems comes from microscopic behavior. It is satisfying to see this connection, and to realized how emergent principles can arise from the evolution of microscopic interacting systems. When new statistical relationships spontaneously come forth among macroscopic variables describing dominant structures or processes, we say we have reached a new level in the hierarchy of ordered states.

[1] R. Clausius, **The Mechanical Theory of Heat with its Applications to the Steam Engine and to Physical Properties of Bodies** [John van Vorst Pub., London] (1867).

W. C. Parke, *Biophysics*, https://doi.org/10.1007/978-3-030-44146-3_10

10.2 Statistical Basis of Thermodynamics

With the ideas of Ludwig Boltzmann in 1877, we can now understand the second law of thermodynamics from the statistics of a large number of interacting particles, and the observation that exact information about most systems containing many interacting subsystems is practically lost after a characteristic relaxation time. The classical statements of the second law of thermodynamics using the imperatives "must" and "never" become statistical statements with the phrases from probability language, "very likely" and "very unlikely".[2]

The idea that systems become microscopically intractable introduces a disorder based on our reduced knowledge about such a system. We replace certainty by ascribing probability to the possible configurations.[3] If nature followed Newton's laws exactly, then with big enough computers we should be able to track the throw of dice, given an exact measure of their initial position and motion when launched. Newtonian theory is 'deterministic'. In practice, when we compute physical values with numbers assumed to have values over a continuum (rather than discrete or purely rational), we must truncate those numbers after a certain number of digits. In following many complex dynamical systems, the error introduced by truncation can grow exponentially with each iteration. But for macroscopic systems which have 'settled down', quantities such as temperature and pressure, which are averages over microscopic properties, should be measurable, and relationships between these quantities may be simple, even when the microscopic behavior might be computationally intractable and impractical.

Statistics is applied to macroscopic systems by assigning probabilities to configurations of possible microstates. Thermal equilibrium occurs when the system reaches a quasi-static state, spending most of the time in its most probable states. If we use quantum theory to describe systems, then the properties of that system can be formulated in terms of the wave function for each possible state of the system. The wave function, often called $\Psi(x)$, is a complex number defined throughout space within small volumes surrounding possible locations of each particle for the system. The wave function is used to calculate the probability that each particle will be found in its given small volume by finding its magnitude squared and then multiplying by the small volumes. Quantum theory shows how, in principle, wave functions can be calculated. Quantum theory introduces a new aspect of nature: If a particle can be found in several distinct quantum states, the wave functions for those states add to find the overall wave function, and, when calculating probabilities, the

[2]For good descriptions of the statistical basis of thermodynamics, see W.J. Moore, **Physical Chemistry, 5th Ed.**, [Longman, London] (1960) and N. Davidson, **Statistical Mechanics**, [McGraw Hill, New York] (1962).

[3]In fact, standard probability theory requires the concept of random numbers. But the very tests for random numbers require an infinite number of trials, so that the application of probabilities also has a practical difficulty. We can make an increasing larger number of tests, and thereby become more and more confident in the predictions of a probability theory.

waves for the possible alternative states can constructively or destructively interfere with each other, unless the phases of the component waves have a random character. In the classical limit, averages over each distinct quantum state become averages over positions and momenta.[4]

With our probabilistic interpretation of thermodynamics, a macrostate in thermodynamic equilibrium has a distribution of its particles amongst the most likely distributions. The more likely distributions are those that can be created in many ways microscopically, just as the distribution of heads and tails of 100 tossed pennies will be near half-and-half because there are lots of ways for the coins to land that way, and only a few if most of the coins end up heads (or tails). At a finite temperature, the occupation of many of the microstates in the system will change, but if there are lots of microstates available which make a given macroscopic state, then the system will spend more time among these states than those which have only a few ways to be generated.

To see which distributions are more likely, we imagine all particles in a system occupying possible quantized energy states, with n_i particles in the energy state ϵ_i. The range of the index i may be finite or extend to infinity. As we expressed before, the total energy of the system is then

$$E = \sum_i n_i \epsilon_i \,. \tag{10.1}$$

In the following argument, we take the system as isolated, so that E is fixed. In addition, we assume that the total number of particles N is fixed,[5] with

$$N = \sum_i n_i \,. \tag{10.2}$$

The number of ways W of putting all N of the particles of a system into the possible quantum energy levels ϵ_i will now be considered. The fact that particles such as electrons, atoms, and molecules are indistinguishable is a critical realization in finding W.[6] Unlike billiard balls, according to quantum theory, we cannot tell one

---

[4]Nicely, quantum descriptions over discrete values are simpler to express than over a continuum. An even more extreme extension of discreteness would propose that all natural processes involve a finite number of logical relationships among integers. In such a universe, no continuum would exist. Evolution of this universe would be definite, and all information initially present would be preserved. We mortals, however, may lose the ability to keep track of that information, and in these circumstances introduce entropy as the measure of information which has become practically inaccessible.

[5]If the system is a 'box of photons' in equilibrium with the walls of the box, then N is no longer fixed, so this case is handled separately.

[6]Boltzmann's calculation of W assumed distinguishable particles. With this assumption, the entropy change when two ideal gases are allowed to mix is not zero. But if both are the same gas, there should be no entropy of mixing. This is referred to as the Willard Gibbs 'paradox', which was also resolved by Gibbs. In 1875, Gibbs, considering the entropy of mixing of gases,

electron from another, or one hydrogen atom from another, by any intrinsic property. When the quantum wave for one electron overlaps that of another, the two electrons can exchange places while we are not looking, without any way to find out that this happened.

Quantum theory also tells us the spin of particles must be quantized, of magnitude $\sqrt{s(s+1)}\hbar$, and that $2s$ must be an integer. If $2s$ is even, the particle is a boson. Quantum theory allows any number of bosons in one quantum state. If $2s$ is odd, the particle is a fermion, and quantum theory allows only one fermion in a quantum state (the 'Pauli Exclusion Principle). Electrons have $s = 1/2$, and are fermions. Lucky for us. Otherwise, atoms would collapse into tight little balls. Life and even matter as we know it would not exist.

A third quantum-theory property is important to our counting of W: Microsystems with sufficient symmetry will have some energy levels with many quantum states at or very close to a given energy. For example, isolated hydrogen atoms have eight different electron quantum states at the second energy level. We will use the symbol g_i to represent the number of quantum states hiding in the energy level ϵ_i. The g_i is referred to as the 'degeneracy' of the energy level ϵ_i.

10.3 Independent and Indistinguishable Particles

Let's do a bit of counting, first for classical particles, such as those contemplated by Ludwig Boltzmann. The number of ways of distributing N distinguishable particles into r distinct states, with n_i in the state with label i, is given by[7]

$$W_{Boltzmann} = \frac{N!}{\prod_{i=0}^{r-1} n_i!} . \tag{10.3}$$

This result follows by counting the ways to fill r boxes with N balls, with a given number n_i in the box labeled by 'i'. There are $N(N-1)\cdots(N-n_0+1)$ ways to select n_0 balls to fill the first box. We divide this number by $n_0!$ since the order of selection is immaterial. For the next box, we have only $(N-n_0)$ balls left, so the number of ways to select n_1 balls, independent of the order selected, is $(N-n_0)\cdots(N-n_0-n_1+1)/n_1!$. It follows that the number of ways to fill all r boxes with the given number of balls in each box will be the product given for $W_{\text{Boltzmann}}$ above.[8]

realized that in order for the entropy to be an extensive variable of degree one, the constant of integration in $\int \delta Q/T$ must depend on the number of particles in the system.

[7]Boltzmann used W to stand for *Wahrscheinlichkeit*, which is a measure of the probability of finding the system in a given state.

[8]If the index i is used to label energy levels which have a degeneracy g_i, then another factor of $\prod_i g_i^{n_i}$ is needed to account for the number of ways of re-arranging the particles in each energy state.

Fig. 10.1 Counting boson states. Boson partitioning: The energy level ϵ_o is represented by a horizontal line. Example values of particle number $n_o = 9$ and degeneracy $g_o = 5$ are assumed. Depicted is one of the partitions of nine balls within five bins. Four partitions were placed, one at a time, to left or right of any ball or previous partition, to create separate bins for the particles. Each bin represents a different quantum state, even though all the states have the same energy ϵ_o

Let's now see how the counting goes in quantum theory. Each quantum state will be given an equal chance, a priori, of being occupied by a particle. The ground state energy level with energy ϵ_o has a degeneracy of g_o, with n_o particles. For bosons, imagine laying out all n_o particles as balls along a line. Now add $g_o - 1$ partitions one at a time, anywhere on either side of any ball or previous partition. The first can be placed in $n_o + 1$ ways; the second in $n_o + 2$ ways; the last in $n_o + g_o - 1$ ways. There will be $(n_o + g_o - 1)(n_o + g_o - 2) \cdots (n_o + 1) = (n_o + g_o - 1)!/n_o!$ ways to place all the partitions. Now the balls are segregated into bins. Figure 10.1 shows the partitioning with nine balls (particles) and four partitions. Each bin represents one of the quantum states with the energy ϵ_o, and each ball represents a particle occupying that quantum state. The degeneracy of this state is the number of bins, g_o. As all the partitions are equivalent, and there are $(g_o - 1)!$ rearrangements of these partitions, there will be $(n_o + g_o - 1)!/((g_o - 1)!n_o!)$ independent ways of distributing the bosons in the ground energy level. The same counting method works for any energy state ϵ_i. Thus, for bosons, the number of ways to distribute bosons in energy levels with n_o in the zeroth level, n_1 in the first level, etc., will be

$$W_{\text{bosons}} = \prod_i \frac{(n_i + g_i - 1)!}{(g_i - 1)!n_i!} \,. \tag{10.4}$$

For fermions, we must have $n_i \leq g_i$ to satisfy the Pauli exclusion principle. For a given energy ϵ_i, we can put the first fermion in any one of the possible g_i quantum states. With this quantum state occupied, it becomes unavailable for any more fermions. The next fermion has only $g_i - 1$ ways to be inserted. The last of the fermions will have $g_i - n_i + 1$ ways to be inserted into a quantum state at the given energy level. An example of this partitioning is shown in Fig. 10.2. Since the order in which we inserted fermions results in equivalent configurations, and there are $n_i!$ ways we could have selected from the original set of fermions, we have

$$W_{\text{fermions}} = \prod_i \frac{g_i!}{(g_i - n_i)!n_i!} \,. \tag{10.5}$$

We have found the number of equivalent microstates with the same particle numbers in each energy state. The total number of particles N in a person is

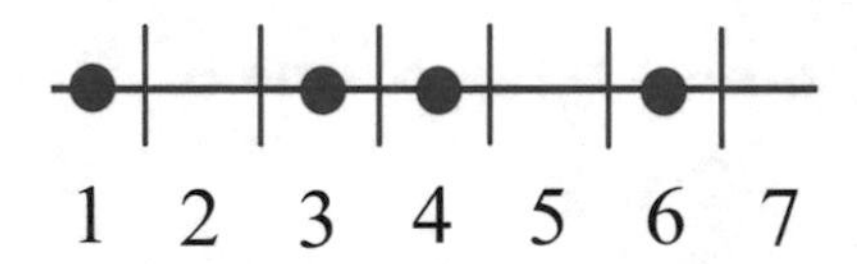

Fig. 10.2 Counting fermion states. Fermion partitioning: The energy level ϵ_o is represented by a horizontal line. Example values of particle number $n_o = 4$ and degeneracy $g_o = 7$ are assumed. Depicted is one of the partitions of four balls within seven bins. At first, six partitions were placed on the line, before any ball is placed. Then, one at a time, a ball is placed in an empty bin. (Since $g_o \geq n_o$, there will be no balls left over.) Each bin represents a different quantum state, even though all the states have the same energy ϵ_o. The Pauli exclusion principle restricts occupancy of a quantum state to no more than one particle

astronomic, of the order of 10^{28}. If the particles are spread out among many levels, W will be even much bigger. At a finite temperature, the distribution of the n_i is constantly changing through interactions. With a system near equilibrium, the chance of finding the system in one of the states with small W will be small. We also should expect that redistributions of the n_i near equilibrium tend to involve small changes in the n_i, i.e. $\delta n_i / N << 1$.

For example, consider a closed room. The air is evacuated from the room. Then the air is re-introduce by injection from one wall, with all the molecules moving in one direction and with one speed, and fixed separation. There is only one way to have such molecules distributed as they are being injected. But as soon as they enter the room, collide with the walls, and interact with each other, they will gain a variety of speeds and directions. With the air restored and thermal equilibrium reached, the molecular speeds and directions will be widely distributed. Is it possible that we might later observe the air all heading with one speed back to one wall in the same way it entered the room? No one has ever seen this happen while sitting in a room. They would be left in a vacuum, so it would likely make the news. But with the microscopic laws time reversible, it could happen! For big systems like the air in a room, we can be assured that the chances are astronomically small. So small that the time to wait for the event to occur with greater that a 50% chance would be much greater than the age of the Universe.

To find the most likely states, we will maximize W as we vary the occupation numbers n_i, while holding the total energy and total particle number fixed. These conditions are called constraints on the variations. The total energy is given by Eq. (10.1) and the total particle number is given by Eq. (10.2). We could first solve these two constraint equations for two of the n_i and eliminate those two from W, then maximize W varying the remaining n_i. This is awkward and asymmetric, as it forces us to pick two of the n_i out of many. For such problems, Lagrange proved that if you wish to maximize $f(\{x_i\})$ with a number L of constraint relations among the x_i, in the form $\kappa_j(\{x_i\}) = 0$, then you can first maximize the function $F(\{x_i\}) + \sum_j^L \alpha_j \kappa_j(\{x_i\})$ without any constraints, and then use the constraint equations to eliminate the constants α_j, called "Lagrange multipliers." (See Appendix Sect. I.1.) In other words, we expand the space of variables so that in the new space, the function we wish to maximize becomes unconstrained. After finding the maximum, the new variables are then eliminated by re-imposing the constraint relations.

As we mentioned, the n_i are typically large numbers, and W even larger. Since the logarithm of a number (i.e. the powers of the number with some base) increase gently and monotonically with the original number, it is more convenient to maximize the logarithm of W under the constraints rather than W. In addition, since the n_i are expected to be large near equilibrium, we can use Stirling's approximation, $N! \sim N^N e^{-N}\sqrt{2\pi N}$ for both N and the n_i. The approximation is good even for N as small as 10. We usually have N's many powers of 10 greater. For the logarithm of factorials, we have

$$\log(N!) \sim N \log N - N\,,$$

dropping the last factor in Stirling's formula because $\log\sqrt{2\pi N} = (1/2)\log N + (1/2)\log 2\pi << N$ for large N.

So, we maximize the function $\overline{W} \equiv \log W + \alpha \sum n_i - \beta \sum n_i \epsilon_i$ by varying the n_i without constraint. At maximum, the slopes of this function along each of the directions of the axes n_i must vanish.

The slopes are first derivatives of $\overline{W}$ with respect to the n_i and give

$$\log \frac{g_i \pm n_i}{n_i} + \alpha - \beta\epsilon_i = 0\,,$$

with the upper sign for bosons, and the lower sign for fermions. Solving for n_i gives

$$n_i = \frac{g_i}{e^{-\alpha} e^{\beta\epsilon_i} \mp 1}\,. \quad \begin{matrix}\text{bosons}\\ \text{fermions}\end{matrix} \tag{10.6}$$

The total particle number and total energy constraints Eqs. (10.2) and (10.1) determine the Lagrange multipliers α and β in terms of the thermodynamic properties of the system.

We have now found how the particles are distributed at or near equilibrium when the total energy and total particle number are fixed.

Photons make a special case, in that there is no conservation law preserving the number of photons.[9] If we drop the constraint $\sum_i n_i = N$, we will find that photons distribute among the possible quantum states according to

$$n_i = \frac{g_i}{e^{\beta\epsilon_i} - 1} \quad \text{photons}\,. \tag{10.7}$$

[9]Photons transfer no intrinsically conserved quantity, such as mass or charge. They are their own antiparticle.

Appendix F shows how this expression leads to the famous Planck's distribution for light from a hot body, like ourselves or the Sun.

10.4 Classical Limit of the Particle Distributions

From a macroscopic view, the discrete nature of energy levels is usually not evident. Classical macrosystems have particles widely spread out over a set of available microscopic energy levels with many quantum states near each level. (So $g_i >> n_i >> 1$.) A simple example is an ideal gas at a finite temperature, wherein there are many more kinetic energy quantum states available in the range of energies carried by the molecules than there are molecules occupying them.

The particles in such sparse (dilute) systems do not come close to filling the available quantum states. The Pauli exclusion principle has little impact on the thermodynamics of these systems. The distinction between bosons and fermions no longer affects the macroscopic measurements of such systems near equilibrium.

For sparse systems, i.e. systems for which the degeneracy, g_i, is much greater than the occupation number n_i (so $g_i >> n_i >> 1$),[10] we will have the approximation, for both W_{bosons} and W_{fermions},

$$W_{\text{boltzons}} \simeq \prod_{i=1}^{r} \frac{g_i^{n_i}}{n_i!} \,. \tag{10.8}$$

The number W_{boltzons} differs from $W_{\text{Boltzmann}}$ because Boltzmann naturally assumed that the particles in the distribution were distinguishable. His work came before the full development of quantum theory. Quantum theory makes particles of the same type indistinguishable.

At or near equilibrium, Eq. (10.8) gives

$$n_i \simeq e^{\alpha} g_i e^{-\beta \epsilon_i} \,. \tag{10.9}$$

This is the distribution of particles found by Boltzmann in 1871, using classical arguments, and derived even earlier for the special case of an ideal gas by Maxwell.[11]

Biosystems, because they take advantage of complexity, tend to have a dense array of energy states with occupations of higher energy states beyond the ultimate

---

[10]This condition works for a wide variety of biological systems, including dilute solutions of macromolecules. An important exception is the system of valence electrons in metals, which do not satisfy the sparsity condition, and so must be treated as Fermi particles.

[11]In 1866, Maxwell showed that the probability of finding a molecule in an ideal gas with a speed within the range $dv_x dv_y dv_z$ is equal to $((m/(2\pi k_B T))^{3/2} \exp(-\frac{1}{2}mv^2/(k_B T)) d^3v)$. This is the 'Maxwellian distribution'.

equilibrium distribution. Even so, we can apply thermodynamic considerations to many biosystems, since

- Life systems tend to maintain the needed biological important macromolecules;
- The time scale for the transformation and degradation of highly energetic structural macromolecules is long compared to biochemical reaction times;
- Many biochemical reactions occur in a 'sea' of water, which acts to distribute heat by thermal diffusion and to reduce concentration gradients by material diffusion;
- Actions which occur locally with only a few molecules taking part are often repeated throughout a cell, so that averages over the large set of actions may have manageable fluctuations.

Now, let's find the Lagrange multipliers α and β for a system of boltzons. Equation (10.2) imposes

$$e^{-\alpha} = \frac{1}{N}\sum_i g_i e^{-\beta\epsilon_i} . \tag{10.10}$$

The 'number constraint' has fixed the first Lagrange multiplier, α.

The second Lagrange multiplier β could be found from the energy constraint Eq. (10.1),

$$E = N\frac{\sum_i \epsilon_i g_i e^{-\beta\epsilon_i}}{\sum_i g_i e^{-\beta\epsilon_i}} = -N\frac{\partial}{\partial\beta}\ln\left(\sum_i g_i e^{-\beta\epsilon_i}\right) . \tag{10.11}$$

However, solving for β from this relation may be difficult. Instead, we will show that all systems in thermal contact have the same β (Sect. 10.5.3) and that

$$\beta = 1/(k_B T)$$

for an ideal gas (Sect. 10.6.1). Since we may use an ideal gas as a thermometer, we can take $\beta = 1/(k_B T)$ for all systems in equilibrium.

From Eq. (10.11), the function

$$Z(\beta, V) \equiv \sum_i g_i e^{-\beta\epsilon_i} \tag{10.12}$$

turns out to be central to the prediction of thermal properties from statistics.[12] We will see that Z encrypts all thermodynamic information. For this reason, Z is given

---

[12]The index i labels energy states. If we had summed over quantum states, then the degeneracy factor g_i would not appear.

its own name, the 'partition function'.[13] The partition function characterizes how the total energy of the system is *partitioned* among the available microscopic energy states possible for the system. If we take the energy to be zero for the lowest ('ground') quantum state, then, roughly, the partition function is a count of the number of quantum states which have a significant probability of being occupied under the given thermal conditions. As $T \to 0$, this number is the degeneracy of the ground state. As the temperature rises, more and more states become populated, corresponding to an increase in Z, i.e. an increase in the number of 'accessible' states.

In the Boltzon partition function of Eq. (10.12), each individual term $g_i e^{-\beta\epsilon_i}$ is proportional to the probability p_i of finding a particle in the energy state ϵ_i:

$$p_i = \frac{n_i}{N} = \frac{g_i e^{-\beta\epsilon_i}}{\sum_k g_k e^{-\beta\epsilon_k}}, \tag{10.13}$$

consistent with the ideas behind Boltzmann's distribution. For example, if one asks what is the chance of finding a particle in an energy level above ϵ_H, the answer would be

$$p_{>\epsilon_H} = \frac{\sum_{k=l}^{\infty} g_l e^{-\beta\epsilon_k}}{\sum_{k=0}^{\infty} g_k e^{-\beta\epsilon_k}}, \tag{10.14}$$

where l is selected so that ϵ_l is the lowest energy level satisfying $\epsilon_l > \epsilon_H$.

10.5 Statistical Definition of Entropy

10.5.1 Statistical Entropy

Let us see how entropy is related to statistical ideas. We know that heat transfer, microscopically, comes from changing the occupation numbers in the microstates available:

$$\delta\!\!\!{}^- Q = \sum_i \epsilon_i \delta n_i\,. \tag{10.15}$$

Varying the occupation numbers n_i is exactly what we did to find where the value of W is maximum. So, near equilibrium,

---

[13]The partition function was introduced by Ludwig Boltzmann and explored by J. Willard Gibbs. The 'Z', used by Max Planck, comes from the first letter of *Zustandssumme*, state sum. The name partition function itself was popularized later in a joint paper by Charles George Darwin (son of the author of **Origin of Species**) and Sir Ralph Howard Fowler entitled *On the partition of energy*, **Philos Mag 44**, 450–479 (1922).

$$\delta \log W = \sum_i (\alpha - \beta\epsilon_i)\,\delta n_i\,. \tag{10.16}$$

The $\sum_i \delta n_i = \delta N$ vanishes, since the total number of particles N is assumed constant. Remembering that $\beta = 1/(k_B T)$, we have

$$\delta \log W = \frac{1}{k_B T}\sum_i \epsilon_i \delta n_i = \frac{1}{k_B}\frac{\delta\!\!\!{}^{-} Q}{T} = \frac{1}{k_B}\delta S\,. \tag{10.17}$$

This relation lets us connect the thermodynamic entropy with our count of microstates at equilibrium:

$$S = k_B\,\ln W + S_o\,. \tag{10.18}$$

If we take the entropy of a state of the system which has $W = 1$ to be zero, then $S_o = 0$. This is Boltzmann's choice, so

$$S = k_B\,\ln W. \tag{10.19}$$

Boltzmann was proud of to have found this connection between entropy and statistics,[14] and rightly so.

In nature, there are systems which have a degeneracy $g_o > 1$ in the lowest energy level ϵ_o. If all the energy which can be extracted is extracted, while preserving the presumed basic entities in the system, there still would be a remaining disorder in a bosonic system of size

$$S(T \to 0) = k_B \ln\left(\prod_{k=1}^{g_o-1}\frac{N+k}{k}\right), \tag{10.20}$$

where we have used Eq. (10.4).[15]

It may also happen that the particles of the system have states not yet observed, perhaps accessible only with higher-energy probes. The count W includes only accessible states, and purposely neglects any 'frozen' degrees of freedom. For example, the entropy of a gas that contains a mixture of atoms whose nuclei can be excited by gamma rays will not include the entropy associated with those excited states unless there is sufficient energy available to excite a significant number of the nuclear states of the atoms.

---

[14] Boltzmann asked that the relation $S = k \log W$ be inscribed on his tombstone. It was.

[15] If the upper limit is below the lower limit on the $\prod$ symbol, the product in taken to be unity, 1. This is consistent with the definition that if the upper limit on a sum is less than the lower limit, the sum vanishes.

Because $\ln W$ is greater for more disordered systems, zero for perfectly ordered ones, and is additive for two independent systems, we use $\ln W$ to DEFINE disorder,[16] whether a system is near equilibrium or not:

> The disorder of a system is proportional to the logarithm of the number of ways of distributing the basic entities amongst the accessible states in the system, holding the number per state fixed.

$$S_D \propto \ln W\,. \tag{10.21}$$

Our generalization of entropy (Eq. (10.21)) to a measure of disorder will be particularly useful in the theory of information storage, information transfer, biological analyzers, and of the dynamics of ordering, to be introduced later in Chap. 12. The expression works even far from equilibrium, where the thermodynamic entropy has little meaning. That information loss is directly connected to the production of entropy was established by Leo Szilárd (see Sect. 12.1).

Near equilibrium, Boltzmann's expression for the entropy becomes (with Eqs. (10.8), (10.9), and (10.12)):

$$\begin{aligned} S &= k_B \ln W \\ &= k_B \left(\sum_i n_i \ln (g_i/n_i) + N \right) \\ &= k_B \sum_i n_i(-\alpha + \beta\epsilon_i) + k_B N \\ &= k_B N \ln (Z/N) + k\beta E + k_B N \end{aligned} \tag{10.22}$$

$$TS = Nk_BT \ln (Z/N) + E + Nk_BT\,. \tag{10.23}$$

In this form, we can see that $\ln W$ near equilibrium is an extensive quantity of degree one.[17]

---

[16]The idea that the disorder of a closed system increases through irreversible processes has far-reaching implications. One is that a wide class of dynamical systems can lose information during their evolution. Another is that the sense of the direction of time can be associated with the direction of increase in entropy. Ergodic Theory as well as Chaos Theory encompasses these ideas.

[17]The thermodynamic entropy contains an arbitrary constant not determined by experiment. The so-called 'third law of thermodynamics' poses that the entropy goes to zero as the temperature goes to zero. However, as we showed in Eq. (10.20), the statistical definition $S = k_B \ln W$ may not give zero in the limit of zero temperature, since the lowest quantum state at energy ϵ_o may have a degeneracy g_o. Also, the classical statement of the third law which includes the impossibility of reaching absolute zero in temperature can, in principle, be overcome by removing the last quantum of available energy from the system.

10.5.2 Thermodynamics from Statistics

The Helmholtz free energy, $A \equiv E - TS$, is simply related to the partition function Z by

$$\begin{aligned} A &= E - TS \\ &= -Nk_BT \ln(Z/N) - Nk_BT \\ &= -Nk_BT \ln(eZ/N) \,. \end{aligned} \tag{10.24}$$

From Eq. (9.110), we know that the entropy, pressure, and chemical potentials in such a system are determined by the Helmholtz free energy function according to

$$S = -\left.\frac{\partial A}{\partial T}\right|_{V,N's} , \tag{10.25}$$

$$p = -\left.\frac{\partial A}{\partial V}\right|_{T,N's} , \tag{10.26}$$

$$\mu_k = \left.\frac{\partial A}{\partial N_k}\right|_{T,V} . \tag{10.27}$$

It follows from Eq. (10.24) that all the thermodynamic variables can be expressed in terms of the partition function. For examples,

$$E = -N\frac{\partial}{\partial \beta} \ln(eZ/N) , \tag{10.28}$$

$$c_v = \left.\frac{\partial E}{\partial T}\right|_{V,N's} = Nk_B\beta^2\frac{\partial^2}{\partial \beta^2} \ln(eZ/N) \tag{10.29}$$

$$\kappa_T = \left.\frac{\partial V}{\partial p}\right|_{T,N's} = \left(\frac{\partial p}{\partial V}\right)^{-1} = \left(Nk_BT\frac{\partial^2}{\partial V^2} \ln(eZ/N)\right)^{-1} . \tag{10.30}$$

where c_v is the heat capacity at constant volume and κ_T is the compressibility at constant temperature. The 'e' in the above three equations is the base of the natural logs.

10.5.3 Balance of Intensive Quantities

Suppose we isolate a pair of systems, allowing the exchange of heat, work, and particles between them. We will assume that each of the systems contains a large number of particles. As the combined systems are isolated, their combined energy, volume, and particle number will be fixed, and we can write

$$
\begin{aligned}
E &= E_A + E_B \\
V &= V_A + V_B \\
N &= N_A + N_B\,.
\end{aligned}
\tag{10.31}
$$

where the subscripts $\{A, B\}$ refer to the selected system. If the two systems are not in thermodynamic equilibrium, then exchanges of heat, work, and particles will make the combined entropy of the two systems increase until it reaches a maximum. If these exchanges occur quasi-statically, so that each system remains close to equilibrium, then the changes in the entropy of each system can be tracked through the small changes in the ways of distributing particles among the quantum states of each, and the small changes in the energy values for these states due to volume changes. We will arrange for the systems in the ensemble to have only small surface interactions, so that the quantum states of any one member of the ensemble in its volume will be little affected by small surface interactions, and we can count distributions of particles in each system independently, expressing the total entropy as

$$S = \ln (W_A W_B) = \ln W_A + \ln W_B = S_A + S_B\,. \tag{10.32}$$

When equilibrium is reached,

$$S(E, V, N) = S_A(E_A, V_A, N_A) + S_B(E_B, V_B, N_B)$$

has reached its maximum, so its slopes vanish there. Small variations in S due to variations in E_A, V_A, or N_A must vanish. But from the first and second law of thermodynamics,

$$\delta S_A = \frac{1}{T_A}\delta E_A + \frac{p_A}{T_A}\delta V_A - \frac{\mu_A}{T_A}\delta N_A\,, \tag{10.33}$$

$$\delta S_B = \frac{1}{T_B}\delta E_B + \frac{p_B}{T_B}\delta V_B - \frac{\mu_B}{T_B}\delta N_B\,, \tag{10.34}$$

so that, with Eqs. (10.31),

$$0 = \delta S = \left(\frac{1}{T_A} - \frac{1}{T_B}\right)\delta E_A + \left(\frac{p_A}{T_A} - \frac{p_B}{T_B}\right)\delta V_A + \left(\frac{\mu_A}{T_A} - \frac{\mu_B}{T_B}\right)\delta N_B\,. \tag{10.35}$$

We conclude that at equilibrium,

$$T_A = T_B \tag{10.36}$$

$$p_A = p_B \tag{10.37}$$

$$\mu_A = \mu_B \; . \tag{10.38}$$

We expect the temperature and pressure to balance with walls between the two systems that can conduct heat and can move, and the chemical potentials will balance at one temperature and pressure for the two systems when the particle diffusion rates from system A to B matches that from B to A.

The identification of β with the inverse of $k_B T$ for an ideal gas will be shown in Sect. 10.6.1. If one of the two systems A or B is an ideal gas, it can act as a thermometer, since the other system must also have the same $\beta = 1/(k_B T)$.

10.5.4 Equilibrium for a Mixture of Independent Particles

Now suppose we have two systems of particles, A and B, put in thermal contact, and so reaching the same temperature, but without any change in the volumes taken up by each set of particles. These systems may be a mixture of two types of particles, or two boxes, one of each type, put into thermal contact, but without the transfer of work or particles. If we consider the combined system, then $W = W_A W_B$, because the distribution of the particles in each system is independent of the other system. The maximization of W with constraints can be found by maximizing

$$\ln (W_A W_B) + \alpha_A \sum_i N_i^A + \alpha_B \sum_j N_j^B - \beta \left(\sum_i N_i^B \epsilon_i^B + \sum_j N_j^B \epsilon_j^B \right) .$$

As before, there are two number constraints (A does not exchange particles with B), but only one energy constraint (A may exchange energy with B). We conclude that two systems in thermal equilibrium have the same β. The maximization of $ln\, W$ with Boltzmann statistics gives

$$N_i^A = g_i^A e^{\alpha_A} e^{-\beta \epsilon_i^A}$$

and

$$N_j^B = g_i^B e^{\alpha_B} e^{-\beta \epsilon_j^B} .$$

Since the entropy of the combined system is

$$S = k \log W = k \log (W_A W_B) = S_A + S_B ,$$

the entropy of the independent subsystems add to give the entropy of the total system. Energy conservation makes

$$E = E_A + E_B .$$

It follows that the enthalpy and the free energies, being extensive thermodynamic variables, add. In the case of the Helmholtz free energy, $A = E - TS$, we have, using Eq. (10.24),

$$A = A_A + A_B = -N_A k_B T \ln(eZ_A/N_A) - N_B k_B T \ln(eZ_B/N_B)$$

and so

$$\mu_A = -k_B T \ln(Z_A/N_A) , \quad \mu_B = -k_B T \ln(Z_B/N_B) . \tag{10.39}$$

10.6 Partition Function for a Gas

The energies accessible to the molecules in an ordinary gas can be separated into a sum of terms corresponding to the kinetic energy and various potential energies each molecule can have. The translational kinetic energy for non-relativistic motion of a molecule takes the usual form of $(1/2)p^2/(2m)$, where p is the momentum of the molecule and m its mass. If the molecule spins, it will also have rotational kinetic energy. Molecules can also be excited into vibrational motion. If the vibrational amplitude is small, the vibrational energy will be harmonic. If molecular stretching occurs due to the molecule having a spin, then rotational and vibrational energies associated with vibration during rotation will not be separable. However, the energies needed to cause rotational stretching by collisions are so high compared to room temperature energies (1/40 eV) that we can neglect their contribution.

During collisions of sufficient impact, the electrons in molecules might be sent to excitation states of their electrons. The energies in this case are called 'electronic'. At the centers of the atoms of molecules are nuclei, which can change state by interactions with the s-shell of the surrounding electrons. Energies associated with nuclear excitations are labeled 'nuclear'. Finally, any intermolecular interactions have potential energies which add to the total energy. For the moment, we will ignore the intermolecular interactions, so that the molecules can be considered acting independently. Each of the possible quantum numbers associated with the total energy of a molecule will be grouped into the index i. In general, we will have

$$\epsilon_i = \epsilon_i^{\text{trans}} + \epsilon_i^{\text{rot}} + \epsilon_i^{\text{vib}} + \epsilon_i^{\text{elec}} + \epsilon_i^{\text{nucl}} + \epsilon_i^{\text{coupling}} . \tag{10.40}$$

The term $\epsilon_i^{\text{coupling}}$ accounts for any added molecular energy due to coupling between vibrations, rotations, electron excitations, and nuclear states. Correspondingly, the partition function of Eq. (10.12), when applied to a single molecule ($Z \to z$) factorizes into

$$z = z_{\text{trans}} \cdot z_{\text{rot}} \cdot z_{\text{vib}} \cdot z_{\text{elec}} \cdot z_{\text{coupling}} \cdot z_{\text{nucl}} \quad . \tag{10.41}$$

Here, z is called the molecular partition function.

10.6.1 The Translational Part

We will work out z_{trans} for an ideal gas left in a closed box for a long enough time that the particles have come to a steady state. The quantum energy levels for the free particles in a box can be found by setting up waves in the box with nodes at the boundaries. Suppose the box has a length L_x along the x direction. The longest wave which fits the box has half a wavelength across the box. The next has two halves, and so forth. So $n_x \lambda_x/2 = L_x$, with $n_x = 1, 2, 3, \cdots$. The momentum of such a particle wave will be $p_x = h/\lambda_x$. The kinetic energy will be $p^2/2m$. So the energy of waves across any direction of the box will be $\epsilon_{n_x,n_y,n_z} = h^2 \left(n_x^2/L_x^2 + n_y^2/L_y^2 + n_z^2/L_z^2\right)/(8m)$. The lowest values for any one of the integers $\{n_x, n_y, n_z\}$ is not zero, but rather one, because of the Heisenberg uncertainty principle: The uncertainty in momentum for a particle limited to a region of space cannot be zero. Thus, the lowest translational energy in a cubicle box of volume V will not be zero, but rather $3h^2/(8mV^{2/3})$. Ordinarily, this is a very small energy compared to k_BT. The largest values for the $\{n_x, n_y, n_z\}$ are determined by the total energy available in the system. If E is the total energy in a system of particles confined to a cubicle box of volume V, then the largest of any of the $\{n_x, n_y, n_z\}$ will be $\sqrt{(8mV^{2/3}/h^2)E - 2}$.

The translational-energy partition function becomes a sum of Gaussians. We can well-approximate this sum by an integral, since the integers n_x, n_y, n_z will be very large for typical states of the particle,[18] and the integer differences in the sum will then be infinitesimal compared to the n's.[19] Therefore, using quantum-state indices for the partition function z sum for a single molecule,

$$z_{\text{trans}} = \sum_{n_x,n_y,n_z} \exp\left(-b\left(n_x^2/L_x^2 + n_y^2/L_y^2 + n_z^2/L_z^2\right)\right)$$
$$\approx \int_0^\infty \exp\left(-b\,n_x^2/L_x^2\right) dn_x \int_0^\infty \exp\left(-b\,n_y^2/L_y^2\right) dn_y \int_0^\infty \exp\left(-b\,n_z^2/L_z^2\right) dn_z \ ,$$

[18]For oxygen in a volume of a cubic micron at room temperature, the n's have to reach above 10^5 before the Gaussian terms in the sum start to decrease in size.

[19]For the partition function $z_{\text{trans}} \approx (\pi/2)\sum_{n=1}^{\infty} n^2 \exp\left(-\beta h^2/(8L^2m)n^2\right)$, the terms in the sum do not reach a peak until $n \approx 1.6 \times 10^9$ for a liter of hydrogen gas at room temperature. However, if the container is only a nanometer cube and the temperature is 10 K, then the peak n is about 3. Note: If small quantum numbers are important, we can use $1 + 2\sum_{n=1}^{\infty} q^{n^2} = \vartheta_3(0, q)$, the obscure but extensively-studied third Jacobi theta function.

$$= \left(\frac{2\pi m}{h^2}\right)^{3/2} \frac{L_x L_y L_z}{\beta^{3/2}} , \tag{10.42}$$

where, for convenience, we put $b \equiv \beta h^2/(8m)$, and have used the Gaussian integral $\int_0^\infty \exp(-x^2)dx = (1/2)\sqrt{\pi}$. The result is

$$z_{\text{trans}} = \left(\frac{2\pi m}{h^2}\right)^{3/2} \frac{V}{\beta^{3/2}} , \tag{10.43}$$

with $V = L_x L_y L_z$ being the volume of the box.

But for an ideal gas whose molecules cannot be excited internally, we know from thermodynamics that its system energy $E = (3/2)N k_B T$. Thus, using Eq. (10.11) in the form $E = -\partial(\ln z)/\partial\beta$, applied to an ideal gas with no internal molecular energy states, we will have

$$\beta = \frac{1}{k_B T} .$$

We now invoke the fact that any system in thermal contact with this ideal gas will have the same temperature, so that this ideal gas will define, as a thermometer, the temperature of any other system.

The quantity in Eq. (10.43)

$$\Lambda_T \equiv \sqrt{\frac{h^2}{2\pi m k_B T}} \tag{10.44}$$

is called the 'thermal de Broglie wavelength'. At a temperature of 293.15 K (i.e. a 'room temperature' of 20 °C), the thermal de Broglie wavelengths of common gases are listed in the table:

Gas		Λ_T (nm)	I(spin)	$\Delta\epsilon$ (eV)
Hydrogen	H_2	0 .072102	0, 1	0.015089
Helium	He	0.050984	0	–
Nitrogen	N_2	0.019270	0, 1, 2	0.004964
Oxygen	O_2	0.018025	0	–
Carbon dioxide	CO_2	0.015372	0	–
Carbon monoxide	CO	0.019270	0	–
Water vapor	H_2O	0.024034	0, 1	0.002951
Argon	Ar	0.016132	0	–

The de Broglie wavelengths Λ_T are calculated at temperature 293.15 K. The numbers in the column labeled I shows total nuclear spins for the common atomic isotopes. The numbers in the column labeled $\Delta\epsilon$ give the difference in energy between the possible two isomers of the molecule in its lowest states

In terms of the thermal de Broglie wavelength,

$$z_{\text{trans}} = \frac{V}{\Lambda_T^3} . \tag{10.45}$$

10.6.2 The Internal Parts

Because we have separated the energies of rotation, vibration and electron orbitals, and nuclear states, we can focus on the partition function for each type of energy carried by the molecules.

Consider the contributions to the partition function in a dilute gas of molecules.[20] The zero of energy is usually taken so that when the participating particles are far separated, they have zero energy. 'Far separated' is an idealized concept, since even separated particles in the laboratory usually encounter containment boundaries, in which case they have a finite lowest translational energy. This is the Heisenberg uncertainty principle at play. But if the boundaries are macroscopically separated by a distance L, the lowest translational energy will be $\epsilon_o = h^2/(8mL^2)$. Even the energy for a hydrogen atom in a cubic millimeter container will be about 2×10^{-19} eV, a negligible amount in comparison to other energies involved.

10.6.3 The Rotational Part

Because electrons are so light compared to nuclei, single isolated atoms do not show atomic rotational excitations at ordinary temperatures. But thermal excitation of rotational states of molecules (with two or more atoms) in a gas at room temperature is likely, as the energy required is typically of the order of a thousandth of an electron volt, while at room temperature, the average kinetic energies are about 0.025 eV.

If we neglect centrifugal distortion,[21] the energy of rotation will be that of a 'rigid' molecule, for which the energy in Hamiltonian form is

[20] If the gas becomes too dense, or changes into a liquid or a solid, the intermolecular interactions inhibit molecular rotations, and cause the vibrations of molecules to couple together into vibrational states for clusters of molecules, or even coherent vibrations of a whole crystal of molecules.

[21] If we needed to included centrifugal distortion, then to first order there would be an additive term in the energy proportional to L^4, and, if there is vibration along with rotation, there would be another 'vibro-rotation' energy term proportional to $L^2(n + 1/2)$, where n is the vibrational excitation quantum number.

Table 10.1 Molecular rotation quantum numbers

Rotor type	Conditions	Quantum #'s	Degeneracy[a]
Linear	$I_A \approx 0, I_B = I_C$	l, m	$(2l+1)$
Symmetric oblate	$I_A = I_B < I_C$	j, k	$(2-\delta_{k,0})(2k+1)$
Symmetric prolate	$I_A < I_B = I_C$	j, k	$(2-\delta_{k,0})(2j+1)$
Spherical	$I_A = I_B = I_C$	j, m, m'	$(2J+1)^2$
Asymmetric	$I_A < I_B < I_C$	J_τ	Varies with J

[a]In table, $\delta_{k,0}$ is one if $k = 0$ and zero otherwise. The index τ ranges from $-J$ to J

Table 10.2 Measured geometric constants of common molecules as a gas

Molecule		Bond lengths (Å)		Angles (deg)	
Hydrogen	H–H	0.741	(HH)	–	–
Water	H–O–H	0.958	(OH)	104.48	(HO)
Ammonia	NH_3	1.012	(NH)	106.67	(HNH)
Methane	CH_4	1.087	(CH)	109.47	(HCH)
Nitrogen	N≡N	1.098	(NN)	–	–
Carbon monoxide	C–O	1.128	(CO)	–	–
Oxygen	O–O	1.208	(OO)	–	–
Carbon dioxide	O=C=O	1.162	(CO)	180.00	(OCO)
Sulfur dioxide	O=S=O	1.432	(SO)	119.50	(OSO)
Carbon tetrachloride	CCl_4	1.767	(CCl)	109.47	(ClCCl)

Data from NIST CCCB Data Base, release 19 (Apr 2008)

$$\widehat{\mathbf{H}}_{\text{rot}} = \sum_{i=1}^{3} \frac{\widehat{\mathbf{L}}_i^2}{2I_i}\,, \tag{10.46}$$

where $i = 1, 2, 3$ labels the x, y, z components. Here, $\widehat{\mathbf{L}}_i$ is the angular momenta of the molecule along a principle axis, and I_i the corresponding moment of inertia. Because the $\widehat{\mathbf{L}}_i$ do not commute in quantum theory, but rather $[\widehat{\mathbf{L}}_x, \widehat{\mathbf{L}}_y] = i\hbar\widehat{\mathbf{L}}_z$, one has to select, before calculating the partition function $Z = \sum_q \langle q|\, e^{-\beta\widehat{\mathbf{H}}}\, |q\rangle$, which set of operators will be observed, explicitly or implicitly. These are usually taken as the complete set of commuting operators which are conserved by the dynamics (Table 10.1).

The best choice of rotation operators, and the consequent selection of quantum numbers to enumerate quantum states, depends on the symmetry of the molecule. Table 10.2 shows the geometry of some simple molecules determined by microwave adsorption spectra. Table 10.3 shows the corresponding moments of inertia. By selecting the moments of inertia around the principle axes in the order $I_A \leq I_B \leq I_C$, we can distinguish five cases, shown in Table 10.1.

10.6.4 Linear Molecule

If the molecule is linear, it has no effective moment of inertia around its axial symmetry axis. (The electrons are too light.) Examples include oxygen O=O, nitrogen N≡N, carbon monoxide C≡O, carbon dioxide O=C=O, nitrogen oxide N=O, hydrogen cyanide HC≡N, and acetylene HC≡CH.

Moments of inertia for an axial symmetric molecule about the two perpendicular axes are equal. The energies become

$$\epsilon_{l,m}^{\text{rot}} = Bl(l+1) \tag{10.47}$$

for $l = 0, 1, 2, \cdots$ and m an integer satisfying $-l \leq m \leq l$. The constant $B = \hbar^2/(8\pi^2 I_B)$. As there are $2l+1$ values of m all giving the same energy, the energy degeneracy is $2l+1$. If the partition function sum is carried over energy levels (instead of quantum states), then

$$z_{\text{rot}} = \frac{1}{\sigma}\sum_l (2l+1)\exp\left(-bl(l+1)\right)) , \tag{10.48}$$

in which we have used $b \equiv \beta B$. The constant σ is called the 'symmetry constant', described below.

If the molecule is reflection symmetric, then there are symmetry restrictions imposed on the angular sums. A molecule after a rotation of 180° that carries the atoms into their counterparts will change the overall wave function for the molecule by a factor $(-1)^J$, where J is the overall spin of the molecule. For example, oxygen-16 nuclei have spin zero, while the electronic wave function for the molecule in its ground state has spin one, and so is odd under the rotation. Since oxygen-16 atoms are bosons, oxygen molecules are bosons, and the total wave function must be even under the rotation that exchanges the atomic position. This restricts l to be odd.

For temperatures significantly higher than $\theta_{rot,B} = \hbar^2/(2k_B I_B)$ (as shown in Table 10.3) we can account (approximately) for the symmetry restriction by including a symmetry number σ in the partition function, as shown in Eq. (10.48). For methane, CH_4, there are four symmetry axes, and three rotations about each which returns the molecule into an indistinguishable state, so $\sigma_{CH_4} = 12$. Room and body temperature is much bigger than $\theta_{rot,B}$. The approximation using σ in the rotational partition function sum gets better with higher temperatures.

10.6.5 Oblate and Prolate Molecule

Oblate molecules have $I_A = I_B < I_C$. A pancake is an oblate object. Examples include ammonia NH_3, cyclobutadiene C_4H_4 and benzene C_6H_6.

Table 10.3 Measured rotational constants of common molecules as a gas

Molecule	$\theta_{rot,B}$	A	B	C	I_A	I_B	I_C
H_2	87.6	–	60.85	60.85	0.0	0.2770	0.2770
H_2O	20.9	27.88	14.51	9.285	0.6047	1.1616	1.8156
NH_3	13.6	9.443	9.443	6.196	1.785	1.785	2.721
CH_4	7.54	5.241	5.241	5.241	3.216	3.216	3.216
N_2	2.88	–	1.998	1.998	0.0	8.436	8.436
CO	2.78	–	1.931	1.931	0.0	8.729	8.729
O_2	2.08	–	1.4377	1.4377	0.0	11.726	11.726
CO_2	0.561	–	0.3902	0.3902	0.0	43.20	43.20
SO_2	0.342	2.027	0.3442	0.2935	8.315	48.981	57.429
CCl_4	0.082	0.0571	0.0571	0.0571	295.08	295.08	295.08

The temperature $\theta_{rot,B}$ is in Kelvin. The terms A, B, and C are in units of inverse centimeter, i.e. energy/hc. The moments of inertia are in units of atomic mass unit times ångström squared. Data from NIST CCCB Data Base, release 19 (Apr 2008), and calculated

Prolate molecules have $I_A < I_B = I_C$. A cigar is a prolate object. Examples include methyl fluoride CH_3F, chloroform $CHCl_3$ and methyl acetylene $CH_3{\equiv}CH$.

The rotation energy states for a molecule whose atoms do not line up (and whose nuclei do not change in relative position) but with symmetry around one axis (also called a 'symmetric top') is given by

$$\epsilon_{j,k} = Bj(j+1) \mp (C - A)k^2 \tag{10.49}$$

where $A = h^2/(8\pi^2 I_A)$, $B = h^2/(8\pi^2 I_B)$, and $C = h^2/(8\pi^2 I_C)$. The quantum number j determines the square of the total angular momentum of the molecule through $\widehat{\mathbf{J}}^2 = j(j+1)\hbar^2$ while k determines the value of an angular momentum projection $\widehat{J}_z = k\hbar$ along a measurement axis. The ranges of these quantum numbers are j non-negative integer and $-j \le k \le j$. An oblate molecule produces the minus sign in front of $(A - C)$, while for a prolate molecule, we use the positive sign. With our ordering of the moments of inertia, $(A - C)$ is always positive, so the shift in energy levels due to the k^2 term is positive for prolate, and negative for oblate shaped molecules, a property evident in microwave/IR spectrometry.

The molecular partition function for rotation becomes

$$q_{\mathrm{rot}} = \sum_{j=0}^{\infty}(2j+1)\sum_{k=-j}^{j} e^{-\beta Bj(j+1)\pm\beta(A-C)k^2} \tag{10.50}$$

$$= \sum_{j=0}^{\infty}(2j+1)e^{-bj(j+1)} + 2\sum_{k=1}^{\infty} e^{\pm(a-c)k^2} \sum_{j=k}^{\infty}(2j+1)e^{-bj(j+1)}$$

$$= q_{\text{rot},0} + 2\sum_{k=1}^{\infty} e^{\pm(a-c)k^2} q_{rot,k} \tag{10.51}$$

where $a \equiv \beta A$, $b \equiv \beta B$, $c = \beta C$, and

$$q_{\text{rot},k} \equiv \sum_{j=k}^{\infty} (2j+1)e^{-bj(j+1)}$$
$$= e^{-bk(k+1)} \sum_{l=0}^{\infty} (2k+2l+1)e^{-bl(2k+l+1)} .$$

For $k >> 1/\sqrt{b}$, $q_{rot,k}$ diminishes rapidly with increasing k.

10.6.6 Spherical Molecules

Spherical molecules have equal moments of inertia around all three principle axes. Examples include methane CH_4 and carbon tetrachloride CCl_4. The allowed energy values have the form

$$\epsilon_j = Aj(j+1) . \tag{10.52}$$

However, now the quantum angular momentum operator acts on functions of the three angles that determine the orientation of the molecule in space. These functions[22] not only depend on the possible measured values of the square of the total angular momentum, $j(j+1)\hbar^2$, but also on two projections of angular momentum, with possible values $m\hbar$ and $m'\hbar$, with both m's ranging by integers from $-j$ to j. For this reason, the energy levels have degeneracy $(2j+1)^2$.

The rotational molecular partition function for a spherical molecule becomes

$$z_{\text{rot}} = \frac{1}{\sigma}\sum_{j=0}^{\infty} (2j+1)^2 \exp\left(-bj(j+1)\right)) , \tag{10.53}$$

where σ is the symmetry number, the number of orientations of the molecule that have identical atomic positions.

[22] The functions are often taken as the Wigner rotation functions $D^{*j}_{m\,m'}(\alpha, \beta, \gamma)$, where α, β, γ are the Euler angles that specify the orientation of a three-dimensional body in space.

10.6.7 Asymmetric Molecules

Asymmetric molecules have different moments of inertia around all three principle axes. A prominent example is water H–O–H. Another is nitrogen dioxide O–N=O, a noxious red gas and an air pollutant from coal and hydrocarbon burning.

Energy levels and their degeneracies are found by solving the Schrödinger equation for the rotor. The solutions are expressible in terms of functions of the Euler angles $\alpha, \beta, \gamma)$ which determine the orientation of the molecule in space. The square of the total angular momentum $\widehat{\mathbf{J}}^2 \equiv \widehat{\mathbf{J}}_a^2 + \widehat{\mathbf{J}}_b^2 + \widehat{\mathbf{J}}_c^2$ will be conserved, so its possible values are fixed to be $J(J+1)\hbar^2$, and we can use J as one of the labels of the energy states. For each J, there will be a set of $2J+1$ distinct energy states, whose degeneracy depends on the shape of the molecule. The solutions can be usefully expanded in terms of the Wigner $D^{*j}_{m,m'}(\alpha, \beta, \gamma)$ functions, or Schwinger rotation states.[23]

With a little algebra, the rotational Hamiltonian (energy operator),

$$\widehat{H}_{\text{rot}} = A\widehat{\mathbf{J}}_a^2 + B\widehat{\mathbf{J}}_b^2 + C\widehat{\mathbf{J}}_c^2$$

can be put into the form

$$\widehat{H}_{\text{rot}} = \frac{1}{2}(A+C)\widehat{\mathbf{J}}^2 + \frac{1}{2}(A-C)h_r(\kappa, J, \tau) \tag{10.54}$$

where

$$h_r(\kappa, J, \tau) \equiv \widehat{\mathbf{J}}_a^2 + \kappa\widehat{\mathbf{J}}_b^2 - \widehat{\mathbf{J}}_c^2,$$

is called the reduced rotational energy which depends on an asymmetry parameter $\kappa \equiv (2B - A - C)/(A - C)$. Note: $\kappa = -1$ for a prolate rotor ($A = B > C$), 1 for an oblate rotor ($A > B = C$), and in the limiting case of a linear molecule, $I_A \to 0$ so $A \to \infty$ and $\kappa \to -1$. The index τ runs from $-J$ to J labels the possible energies for a given J. The measured values taken by h_r called ϵ_r, are the roots of a secular equation[24] of degree $2J+1$. Solving the secular equation for $J \geq 2$ can be a formidable task, so we have no general expression for the molecular partition function in this case. Applying the symmetry group for the molecule helps reduce

[23] See P.D. Jarvis and L.A. Yates, *The molecular asymmetric rigid rotor Hamiltonian as an exactly solvable model*, arXiv:0803.256v1 (2008).

[24] A secular equation is a matrix determinant relation of the form $\det|H - \epsilon I| = 0$, where H is a hermitian matrix, I a unit matrix, and ϵ is one of the unknown 'eigenvalues'. Observables in quantum theory correspond to hermitian matrices, i.e. matrices satisfying $H^{T*} = H$, where T is the transpose operation and $*$ is the complex conjugation.

the secular determinant into block-diagonal form. In any case, numerical methods can be successfully used.[25]

10.6.8 *Molecular Rotation-Partition Functions for 'High' Temperatures*

If the temperatures are 'high' (i.e. $T >> \theta_{rot,B}$, which is satisfied by room and body temperatures), then the summation expressions for the molecular partition functions can be approximated by an integral (since then many large j values contribute). In fact, because room temperatures are sufficient to cause many molecules in a gas to rotate with j significantly bigger than 1, room temperature is 'high'. These characteristic temperatures $\theta_{rot,B}$ are shown in Table 10.3 for various common gases, together with their measured rotational coefficients A, B, and C, as well as their three moments of inertia.

By approximating the rotational partition function sum by an integral, the molecular partition function for rotation for linear molecules is given by

$$z_{\rm rot}^{2D} \approx \frac{1}{\sigma}\left(\frac{8\pi^2 I_B k_B T}{h^2}\right) , \tag{10.55}$$

while for spherical and asymmetric molecules,

$$z_{\rm rot}^{3D} \approx \frac{1}{\sigma}\sqrt{\pi}\left(\frac{8\pi^2 I_A k_B T}{h^2}\right)^{1/2}\left(\frac{8\pi^2 I_B k_B T}{h^2}\right)^{1/2}\left(\frac{8\pi^2 I_C k_B T}{h^2}\right)^{1/2} . \tag{10.56}$$

Since system energies can be found by the negative derivative of the log of the partition function with respect to $\beta = 1/(k_B T)$, we see that at high temperatures, the thermodynamic energy will have contributions of $Nk_BT/2$ from each rotational degree of freedom, a reflection of the equipartition of energy theorem.

10.6.9 *The Vibrational Part*

The vibrational excitation energies of molecules generally are at least an order greater in magnitude than rotational ones. Most of the molecules in a gas at room temperature will not be in vibrational excitation. In quantum theory, the vibrational energy for low amplitude motion is harmonic, of the form $(1/2)\mu\dot{x}^2+(1/2k(x-x_o)^2$

[25] Values for $\epsilon_r(\kappa, J, \tau)$ up to $J = 10$ were first given by G.W. King, R.M. Hainer, and P.C. Cross, *The asymmetric rotor I. Calculation and symmetry classification of energy levels*, J Chem Phys **11**, 27 (1943).

for each mode of vibration of the molecule, where μ is the reduced mass for a pair of masses in relative motion and $\dot{x}$ is the rate of change in the relative coordinate x. The constant x_o is the equilibrium distance between the two masses. By solving the quantum dynamics, one finds that the discrete quantized energy levels for the steady states are given by

$$\epsilon_n^{\text{vib}} = \left(n + \frac{1}{2}\right) hf \,, \tag{10.57}$$

where $f = (1/(2\pi))\sqrt{k/\mu}$ and μ is the reduced mass of the vibrating system. The lowest energy, $(1/2)hf$, is the 'zero-point energy' of the oscillator for a particle with quantum wave-like behavior. In effect, the particle jiggles even in its lowest energy state, and satisfies the Heisenberg uncertainty $\Delta x \Delta p \geq \hbar/2$ which indicates that the particle cannot settle to a fixed position.[26]

Calculations are often performed using observed vibrational frequencies found from IR spectroscopy.

A polyatomic molecule will have a number of distinct modes of vibration. By selecting the appropriate coordinates to describe the positions of the nuclei in the molecule, the energy of vibration can be made to become a sum of kinetic and potential energies in these coordinates. Each vibration in these coordinates then may occur independent of the others. These separated vibrational states are called 'normal modes'.

The position of the atoms in a molecule with n_a atoms require $3n_a$ coordinates. But vibrational coordinates would not involve the center-of-mass coordinates, nor the angles that give the orientation of the molecule in space. Only two angles are needed for a linear molecule, and three to fix the orientation of a non-linear molecule. So, the number of normal modes is $3n_a - 5$ for a linear molecule and $3n - 6$ for a non-linear one.[27]

Because of the separation of the energies into a sum of normal mode vibrational energies, the vibrational partition function becomes a product over the normal modes present.

With explicit expression for the energies, the vibrational partition function becomes

[26]Harmonic vibration assumes that the inter-atomic interaction energy V as a function of separation can be approximated by a parabola. If the vibration energy is near to that needed to break the atomic bonding, this harmonic approximation is no longer valid. The oscillation is said to be 'anharmonic'. The next term in the series expansion of $V(x)$ beyond $V_o + (1/2)\ (\partial^2 V/\partial x^2)\big|_o (x - x_o)^2$ introduces a cubic in the distance $x - x_o$. To first order, this anharmonic term reduces the energy values by a term proportional to $(n + 1/2)^2$. Moreover, if the potential energy approaches zero at large separation r faster than $1/r$, the number of possible energy states is finite.

[27]These numbers are referred to as 'degrees of freedom'. Besides the internal degrees of freedom, The kinetic energies associated with the rate at which the two (linear case) or three (non-linear case) angles change generate the rotational degrees of freedom, and the three kinetic energies going with rate of change of the center of mass coordinates make the translational degrees of freedom.

$$z_{\text{vib}} = \prod_i \sum_{n_i} \exp(-\beta h f_i (n_i + 1/2)) \,. \tag{10.58}$$

The sum over the quantum numbers n_i for fixed i can be done analytically, using $1/(1-x) = \sum_n x^n$, giving

$$z_{\text{vib}} = \prod_i \frac{\exp(-\beta h f_i/2)}{1 - \exp(-\beta h f_i)} = \prod_i \frac{1}{2\sinh(\beta h f_i/2)} \,. \tag{10.59}$$

The product is over distinct normal modes of vibration of the molecule. Generally, the molecule can vibrate in a combination of its normal modes. However, at room temperatures, $\beta h f_0$ is typically much less than one, so that thermally excited vibrational states are not expected.

10.6.10 Electronic Part

The electrons bound in a molecule, if left undisturbed, 'settle' into steady-state molecular orbitals. Excitation of these electrons within a molecule tends to require even more than an electron volt of energy, so the electrons in molecules at room temperature are not likely to be out of their unexcited state. (If many were excited, the gas would glow!) Assuming no excitation contributions, this leads to an electronic partition function of

$$z_{\text{elec}} = g_0^{\text{elec}} \exp(+\beta D_e) \,, \tag{10.60}$$

where g_0^{elec} is the degeneracy of the molecule's ground state, and D_e is the binding energy of the molecule, taken here as a positive number, and gives the energy released by the molecule when disassociated into its atoms.

10.6.11 The Nuclear Part

The energies need to excite an internal state in a nucleus are in the keV to MeV range, far above the average energy of collision of molecules in a gas at room temperature, about $1/25\,\text{eV}$. We should not expect any such excitation. However, because there is magnetic coupling between the nucleus and nearby electrons, the nuclear spin can be re-oriented. This makes the nuclear spin degeneracy contribute to the partition function and thereby to the entropy of the gas. There will be $2I + 1$ possible states of projection of the nuclear spin along a measurement axis, where the integer $I \geq 0$ is the quantum number used to enumerate the possible measured values of the nuclear total spin squared, i.e. $\hat{\mathbf{I}}^2$ takes on one of the values $I(I+1)\hbar$.

The nuclear partition function is just the degeneracy of the spin-orientation states for each of the nuclei in the molecule,

$$z_{\mathrm{nuc}} = \prod_i (2I_i + 1) , \tag{10.61}$$

unless we wish to include the coupling between the molecular spin and the nuclear spin.

10.6.12 Higher-Order Energies in Molecules

Since molecules are not rigid, spinning them may cause a shift in the atomic positions. This means that the momentum of inertia may shift when the molecule rapidly spins. The effect is called 'centrifugal distortion'. The corresponding molecular energy can be approximated by an added term proportional to the angular momentum to the fourth power. (This can be seen classically by expressing the balance of the centrifugal forces with the molecular binding forces in terms of L^2, and then expanding their atomic separations about their non-rotational values.) The proportionality constant can be calculated theoretically, at least for simple molecules, from given potential energies, or measured with photon elastic (Rayleigh) and inelastic (Raman) spectra when dipoles are present or can be induced.

Similarly, if higher states of vibration occur, the first-order 'anharmonicity' can be represented by a term proportional to the harmonic vibrational energy squared.

At high enough temperatures, a coupling between excited states of vibration and rotation may also occur. Recognizing that the vibrational frequencies are much higher than the rotational frequencies, usually by two orders of magnitude, the rotational couplings can be taken as an average over the period of vibration. The result is again an expansion in powers of $L(L + 1)$ and a separate expansion for vibration in terms of powers of $(n + 1/2)hf_o$. Electromagnetic spectroscopy can give the coefficients in the expansions.

10.6.13 Other Factors

If the entities in the system act, for the most part, as independent particles, then the energies of interaction between molecules can be ignored, and the factor in the partition function due to such interaction $z_{inter} \approx 1$. We should expect this to be the case for dilute gases and for particles that have negligible long-range interactions and are spread out in a solution. Calculation of the interaction partition function for dense gases, liquids, and solids are possible knowing some details about the energies associated with intermolecular interactions.[28] An example in which intermolecular interactions cannot be ignored is ionic solutions, an important topic in biological cells. We will present a simplified model for this case in Sect. 10.9.5.

10.6.14 Combining the Parts

Putting together the factors of z for a polyatomic-molecule in a gas or as dilute particles in a solution, acting independently of other molecules, we have

$$z_{\text{mol}} = z_{\text{trans}} z_{\text{rot}} z_{\text{vib}} z_{\text{elec}} z_{\text{nucl}} = \left(\frac{V}{\Lambda_T^3}\right) \times$$
$$\left(\frac{1}{\sigma} \sum_{l=0,rot}^{\infty} (2l+1) e^{-Bl(l+1)}\right) \prod_{i,vib} \frac{e^{-\beta h f_i/2}}{e^{-\beta h f_i} - 1} \left(g_0^{\text{elec}} e^{\beta D_e}\right) \prod_{i,nucl} (2I_i + 1) \,. \tag{10.62}$$

Note that the only factor that depends on the volume of the system is z_{trans}. We will group the factors for molecular internal states into the symbol $z_{internal}$, so that

$$z_{\text{mol}} = \left(\frac{V}{\Lambda_T^3}\right) z_{\text{internal}} \,.$$

10.6.15 Examples for an Ideal Gas

Hydrogen Isomers

Hydrogen gas is a mixture of para- and ortho-hydrogen, two isomers of the hydrogen molecule, H_2.[29] In para-hydrogen, the two proton spins in the hydrogen molecule are anti-aligned, while in ortho-hydrogen, they are aligned. The magnetic coupling between the electrons and the protons is small, so that, without a catalyst, if the two kinds of isomers are mixed, establishing equilibrium concentrations of the two kinds may take hours.

[28]See W.J. Moore, **Physical Chemistry, 5th Ed.**, [Longman, London] (1960) and Donald A. McQuarrie, **Statistical Mechanics**, [Harper & Row, New York] (1976) for strategies and further references.

[29]In chemistry, ortho, meta, and para, from the Greek, meaning aligned, between, and opposite. The designation is used not only for isomers, but for the position of side groups in molecules, such the methyl group in xylene.

For ortho-hydrogen, the two proton spins are parallel, so they add to give a total spin of $I = 1$, with a degeneracy of $2I + 1 = 3$ possible orientations along a measurement axis. Since the spin one state is symmetric under the exchange of the two protons, a rotation of 180° that carries one proton into the prior position of the other must make the orbital part of the hydrogen molecule wave function antisymmetric under such a rotation, since the protons are fermions. These orbital parts with a total orbital angular momentum quantum number L will change by a factor $(-1)^L$ by the 180° rotation, so L must be odd: $L = 1, 3, 5, \cdots$.

For para-hydrogen, when the proton spins are anti-parallel, the nuclear spin state give a total spin of $I = 0$, with just one possible such state, and therefore a degeneracy of $2I + 1 = 1$. This state is antisymmetric on exchange of the two protons, so the orbital parts of the total wave function must have only even L: $0, 2, 4, \cdots$.

Because the protons have only a very weak (magnetic) interaction with the orbiting electrons, their spin states do not often change when the hydrogen molecule gets bounced around by thermal motion at room temperature. Under these conditions, the rotational energy states are dominated by the orbital motion of the two protons revolving around an axis perpendicular to the line joining the protons. Thus,

$$\epsilon_{\text{rot},L} = \frac{\hbar^2}{2\,I_B} L(L+1) \equiv B\,L(L+1)\,. \tag{10.63}$$

From Table 10.3, $I_B = 0.2770\,u\text{Å}$, $B = 60.85/\text{cm} = 0.007544\,\text{eV}$, and $\theta_{rot,B} = 87.6\,\text{K}$. The lowest rotational energy level has $L = 0$, which applies to the para-hydrogen isomer. The next lowest energy level has $L = 1$, which applies to the ortho-hydrogen isomer. The ortho isomer cannot exist without some rotation, but it can convert to the para isomer with no rotation, releasing an energy $\Delta\epsilon_{\text{rot}} = 2\,B = 0.015089\,\text{eV}$.

Thus, we expect that at room temperatures ($k_B T$ about 0.025 eV), there will be sufficient collisional kinetic energy to make some ortho-hydrogen from para-hydrogen. The molecular rotation partition function reads

$$z_{\text{rot}} = \sum_{L,even} (2L+1)\exp\left(-bL(L+1)\right) + 3\sum_{L,odd} (2L+1)\exp\left(-bL(L+1)\right)\,, \tag{10.64}$$

wherein $b = B/k_B T$, and we include the nuclear spin degeneracy $(2S + 1)$, which is one for nuclear spin 0 states and three for nuclear spin 1 states. Since the terms in the partition function are proportional to the probability of finding a particle in the energy for that term, the ratio of the equilibrium number of ortho to para-hydrogen particles will be

$$\frac{N_o}{N_p} = \frac{3\sum_{L,odd}(2L+1)\exp\left(-bL(L+1)\right)}{\sum_{L,even}(2L+1)\exp\left(-bL(L+1)\right)}\,. \tag{10.65}$$

For temperatures $T >> \theta_{rot,B}$, $b << 1$ and large L contribute in the sums, so the sums can be well approximated by integrals. Since only every other L appears in each sum, we use

$$\sum_{L,odd} (2L+1)\exp(-bL(L+1)) \approx \sum_{L,even} (2L+1)\exp(-bL(L+1))$$
$$\approx \frac{1}{2}\int_0^\infty \exp(-bx)dx = \frac{1}{2b}\,. \quad (10.66)$$

So, for high temperatures, $N_o/N_p \to 3$. At room temperature (293.15 K), we find $N_o/N_p = 2.985$. This means that if the two isomers are in equilibrium at room temperature, 74.9% of hydrogen gas will be in the ortho state. Compare with the gas at 80 K, which has $N_o/N_p = 1.00$, so the ortho state is then 50.0% of the gas.

For low temperatures ($T << \theta_{rot,B} = 87.6$ K), only the lowest terms in the partition function contribute, so that

$$\frac{N_o}{N_p} \approx \frac{3 \cdot 3\exp(-2b)}{1} = 9e^{-2B/(k_BT)} \to 0\,, \quad (10.67)$$

making the gas almost all para-hydrogen.

Nitrogen Isomers

Nitrogen gas also comes in different isomer forms, because the nucleus of the atom ^{14}N has spin one, so the nuclei of the molecule N_2 can be in a spin state with a spin quantum number I, with molecules in three possible isomeric states para ($I = 0$), meta ($I = 1$), and ortho ($I = 2$). The spin wave functions under nucleus exchange change sign by $(-1)^I$ Since N_2 is a boson, the overall wave function must be symmetric. The rotation which exchanges the two nitrogen nuclei will change the orbital wave function by $(-1)^L$, so the orbital angular momentum quantum number L must be even if I is even, and odd if I is odd.

The nuclear degeneracies of the nitrogen molecule is $2I + 1$, and so takes values of 1, 3, 5 for $I = 0, 1, 2$. The even I, even L states will be degenerate (in the absence of any strong external magnetic field). The $I = 1, L = odd$ states start in energy slightly higher than the $I = 0, 2, L = even$ states because the orbital rotational energies are

$$\epsilon_L = \frac{\hbar^2}{2I_B}L(L+1) = BL(L+1) \quad (10.68)$$

so the lowest energy for odd L is $2B$ while for even L the lowest rotational energy is zero.

In all, then, the rotational molecular partition function will be

$$z_{\text{rot}} = (1+5) \sum_{L\,even} (2L+1)e^{-bL(L+1)} + 3 \sum_{L\,odd} (2L+1)e^{-bL(L+1)} , \tag{10.69}$$

where $b = \beta B = B/(k_B T) = \theta_B/T$. Table 10.3 gives $\theta_B = 2.88$ K for nitrogen, so at room and body temperature, many nitrogen rotational levels are excited.

The ratio of ortho and para nitrogen to meta nitrogen as a gas becomes

$$\frac{N_{o,p}}{N_m} = \frac{6 \sum_{L,even} (2L+1) \exp(-bL(L+1))}{3 \sum_{L,odd} (2L+1) \exp(-bL(L+1))} . \tag{10.70}$$

At temperatures much above $\theta_B = 2.88$ K, the sums are well approximated by integrals that approach each other (see Eq. (10.66)), so that

$$\frac{N_{o,p}}{N_m} \xrightarrow[T>>\theta_B]{} 2 , \tag{10.71}$$

(67% in the ortho-para state) while at lower temperatures, nitrogen turns to a liquid at 77.347 K at one atmosphere of pressure and then a solid at 63.15 K. The partition function for nitrogen at these temperatures is not so simple as that for the gas.

10.6.16 Partition Function for Many Molecules

When we neglect the intermolecular interactions, the energy for N molecules will be the sum of the energies for the individual ones. Since $Z = \sum_i e^{-\beta\epsilon_i}$ (with i labeling a quantum state), this separation of the energies, one for each molecule, makes the partition function for N molecules proportional to the product of the $z's$ for each molecule. If the molecules are indistinguishable, there will be $N!$ rearrangements of the molecules that produce the same state of the system, so if, in Z, we sum over all possible quantum states for all the molecules, we will over count by a factor $N!$. This over counting can be corrected by dividing the sum by $N!$. As a result,

$$Z_{\text{gas}} = \frac{1}{N!} (z_{\text{mol}})^N . \tag{10.72}$$

As we have seen (Sect. 10.5.2), thermodynamic properties of the system are all expressible in terms of $\ln Z$. For our gas, with large N,

$$\ln Z_{\text{gas}} = N \ln z_{\text{mol}} - N \ln N + N = N \ln(e z_{\text{mol}}/N) . \tag{10.73}$$

We can now find the Helmholtz free energy, in terms of the partition function, and from it the other thermodynamic functions of interest, following Eqs. (10.25) to (10.30). The expression for A given in Eq. (10.25) was derived from counts using indistinguishable particles, so the $N!$ in Eq. (10.72) is, in effect, already included.

Thus,

$$A = -Nk_BT \ln (ez_{\text{mol}}/N) \tag{10.74}$$

for our set of independent and indistinguishable molecules.

10.6.17 Ideal Gas Chemical Potential

Later, in considering air in the lungs, we will employ the chemical potential for an ideal gas,

$$\mu = \left.\frac{\partial A}{\partial N}\right|_{T,V} = -k_BT \ln (z_{\text{gas}}/N) \, , \tag{10.75}$$

$$\mu = k_BT \ln \left((N/V)\Lambda_T^3\right) - k_BT \ln z_{internal} \, . \tag{10.76}$$

If there are several types of molecules in the system, each given a distinct k index, then

$$\mu_k = k_BT \ln \left((N_k/V)\Lambda_{T,k}^3\right) - k_BT \ln z_{internal} \, . \tag{10.77}$$

Alternative to N_k/V, we can use the ideal gas relations to find the chemical potential in terms of concentration or pressure:

$$N_k/V = \mathcal{N}_A n_k/V = \mathcal{N}_A[C_k] = p_k/(k_BT),$$

where $\mathcal{N}_A$ is Avogadro's number, $[C_k] \equiv n_k/V$ is the molar concentration of molecules of type k, and p_k is the partial pressure for molecules of type with index k. So the chemical potential for the k-th component can be expressed in terms of that component's partial pressure as

$$\mu_k = \mu_k^o + k_BT \ln (p_k/p^o) \, , \tag{10.78}$$

where μ_k^o depends on the temperature of the gas (but not its pressure), or, in terms of the molar concentration $[C_k] = n_k/V$ of the k-th component, as

$$\mu_k = \mu_k^c + k_BT \ln ([C_k]/[C_k^o]) \, . \tag{10.79}$$

10.7 Disorder Expressed by Probabilities

Statistical averages are typically written in terms of the probabilities for the occurrence of certain possibilities. There is a profound connection between the entropy S for systems of distinguishable entities and the probability that a given distribution will be found after many trials. If we make N trials, selecting unbiasedly among a set n of entities, each labeled by 'i', then the entropy in the N trials is determined by

$$S/k_B = \ln W \tag{10.80}$$

with

$$W = \frac{N!}{\prod_{i=1}^{n} n_i!}\ .$$

For large n, $n! \approx n \ln n - n$, so

$$\begin{aligned} S/k_B &\approx N \ln N - N - \sum n_i \ln n_i + \sum n_i = N \ln N - \sum n_i \ln n_i \\ &= N \ln N - \sum n_i \ln (n_i/N) - \sum n_i \ln N \\ &= -N \sum_{i=1}^{n} P_i \ln P_i \end{aligned} \tag{10.81}$$

where $P_i = n_i/N$ approaches the probability for the occurrence of the distribution n_i as $N \rightarrow \infty$. The expression Eq. (10.81) was developed by Boltzmann for statistical mechanics and later by Shannon in his definition of information (see Sect. 12.1) in terms of the probability of the occurrence of each character in a string making up a message.

10.8 The Partition Function for Ensembles

I would never join a group that would have me as a member.
— Groucho Marx

10.8.1 Gibbs' Ensemble

In 1902, Willard Gibbs published his analysis over many years[30] on the statistics of systems near thermodynamic equilibrium, establishing the subject of statistical mechanics on a firm theoretical basis.[31] Gibbs generalized Boltzmann's ideas to any macroscopic systems whose energies for all the contained subsystems are known, including interaction energies. This is in contrast to our previous discussion applying statistics to systems whose individual particles act independently.

To describe the statistics of large systems, Gibbs imagined $\mathcal{N}$ replicas of a system, each copy initially having the same complete set of macroscopic variables, such as composition, energy, and volume. Also, all possible microstates consistent with the macroscopic variables are equally represented in the initial collection, before each system in the ensemble is allowed to come to equilibrium. To insure a probabilistic interpretation for the chance of any particular state occurring, the number of systems in the ensemble, $\mathcal{N}$, is imagined taken to infinity. Such a collection is called a 'Gibbs ensemble'.

10.8.2 Systems Described by Classical Theory

Classically, the microscopic dynamical laws of a system of N particles can be expressed in terms of the time behavior of the full set of coordinates $q \equiv \{q_i; i = 1, 3N\}$ and their 'conjugate' momenta $p \equiv \{p_i; i = 1, 3N\}$ for the contained particles. Generalized coordinates and conjugate momenta arise in the Lagrange formulation of mechanics, wherein Newtonian dynamics is determined by the minimization the 'action' $S_A = \int_{t_1}^{t_2} L dt$, a time integral over a function L of the coordinates q and velocities $\dot{q}$ of the particles. The function $L(q, \dot{q}, t)$ is called the 'Lagrangian'. The minimum of the action is found by varying the possible paths $q_i'(t) = q_i(t) + \delta q(t)$ (holding the initial and final values unchanged). Newton's laws result when this variation of the action vanishes. They take the form $(d/dt)\partial L/\partial \dot{q}_i = \partial L/\partial q_i$. The momentum conjugate to q_i is defined to be $p_i \equiv \partial L/\partial \dot{q}_i$. The energy of the system, expressed in terms of the q and p, is called the 'Hamiltonian', which turns out to be $H = \sum_i (\partial L/\partial \dot{q})\,\dot{q} - L$. With the Hamiltonian, Newton's laws come from the pair of equations $dq_i/dt = \partial H/\partial p_i$ and $dp_i/dt = -\partial H/\partial q_i$.

Therefore, classically, the time-evolution of each member of a Gibbs ensemble is determined by the motion of a point in the space of all $6N$ variables $\{q_i, p_i\}$,

[30] J. Willard Gibbs, **Elementary Principles in Statistical Mechanics, Developed with especial reference to the rational foundations of thermodynamics**, Yale University, [Scribners Sons] (1902).

[31] Gibbs' work was praised by leading physicists of the time for its logic and depth, and is still applicable even after the advent of quantum theory.

called 'phase space', and the full ensemble can be described as a fluid in the phase space for such a system. The dynamics of particles can be used to show that the ensemble fluid moves in phase space without changing the density of the systems per unit phase-space volume. The phase-space volume taken up by a given number of adjacent systems of the ensemble does not change in time. (This is called Liouville's Theorem). This volume preserving property allowed Gibbs to give the same unchanging probability for any ensemble state found within each infinitesimal fixed phase-space volume for all those completely isolated systems with the same macroscopically measurable values, such as the energy and volume. This is referred to as the 'assumption of equal a priori probability'. Quantum theory later showed that in the semi-classical limit, the smallest phase-space volume is of size h^{6N}, where h is Planck's constant, a result anticipated by Sackur and by Tetrode in 1912.

10.8.3 Systems Described by Quantum Theory

Quantum theory incorporates the Heisenberg uncertainty principle

$$\Delta q_i \Delta p_i \geq \hbar/2\,,$$

and so does not allow coordinates and momenta to be simultaneously known. Rather, the results of measurement can be described in terms of a set of complex wave functions $\psi_k(\{q_i\}, t)$ for the system, and a set of probabilities $\{P_k\}$ for each possible micro-states of the system. If only one of those probabilities, say P_l, is non-zero (and therefore has a value of 1) then the system is known (by measurement) to be in a 'pure' quantum state labeled by 'l'. If more than one such probability is non-zero, then we say the system is in a mixed state. The index k labels a full set of simultaneously measurable quantities for the system. When we have incomplete knowledge of what microstate the system is in, but only have probabilities that the system will be found in one of the allowed states, it is then convenient to describe the system in terms of the density operator[32]

$$\hat{\rho} = \sum_k P_k \psi_k \psi_k^* \,. \tag{10.82}$$

(For a pure quantum state, and only for a system in a pure state, $\hat{\rho}^2 = \hat{\rho}$.)

We should note that probabilities for the result of measurement of a quantum system decompose into two distinctly different kinds. The first kind comes from the fact that quantum theory, even for systems in pure states, only predicts the

[32]The operator $\hat{\rho}$ is also called the 'density matrix', introduced by John von Neumann in his **Mathematical Foundations of Quantum Mechanics** (1932).

probabilities that certain values will be found for given observables. This probability is intrinsic to quantum theory, and is not due to a lack of information about a system before a measurement.[33] For example, if an atom is held in a volume by finite forces and later its position is measured, there is a chance that the atom will not be found in that volume. After many measurements for the same initial setup, one can estimate the probability of finding the atom in the volume. Quantum theory lets us predict this number. Moreover, because quantum probabilities are constructed from interfering wave alternatives, constructive and destructive interference can manifest themselves, making quantum probabilities for a particle being found at points across space sometime display wave-like character.

The second kind of probability, P_k, comes from our presumed lack of knowledge about what state the system might be found in. The P_k are determined by repeated measurements on copies of the system prepared with the same initial conditions. The P_k cannot manifest interference effects between alternative system wave states.

If each copy of a large isolated system is given sufficient time to 'relax', and become a steady state, then we are likely to lose the information contained in the relative phases of waves between the different states described by quantum waves.[34]

In quantum theory, the values one measures for observables, O_{obs}, such as the total energy of a system in a pure state, or its total angular momentum, can always be expressed as

$$O_{\text{obs}} \equiv \left\langle \hat{O}_{\text{obs}} \right\rangle \equiv \int \psi_k^*(q) \hat{O}_{\text{obs}} \psi_k(q) dq \,, \tag{10.83}$$

where $\hat{O}_{\text{obs}}$ is a linear operator associated with the observable O_{obs}, and q stands for the coordinates in the wave function.

By expanding into a complete set of eigenstates of a selected operator, wave functions are equivalent to a vector of the components in the expansion. Operators are then equivalent to a matrix operating on wave vectors.

For both mixed and pure state systems, the average result of many measurements of an observable can be expressed in terms of the density operator. For example, the average energy becomes,

[33]The entropy of a pure quantum state is zero. If a measurement process involves acquiring information about an observed system, then necessarily the entropy of the measuring system must increase. Conventional quantum theory uses the language that the measurement has 'collapsed' the wave function describing that system. But if one includes the observed system and the measuring system in one larger system, quantum theory, in principle, can follow the transfer of information between the two, and the observed system wave function does not collapse, but rather becomes mixed with the measuring system states, with the observed state dominating the mixture.

[34]The ergodic hypothesis includes the 'randomization' of the relative phases between the quantum waves describing members of a Gibbs ensemble as the systems approach equilibrium. This kind of loss of information is an element of Chaos Theory, wherein exact predictions become intractable when initial conditions have any uncertainty.

$$\langle E \rangle \equiv \overline{E} = \mathrm{Tr}\,[\hat{H}\hat{\rho}\,]\,, \tag{10.84}$$

where $\hat{H}$ is the Hamiltonian operator for the system.[35] The ‘Tr’ performs a trace, i.e., the sum of the diagonal elements of the matrix on which Tr acts.

From Eq. (10.82),

$$\mathrm{Tr}[\rho\,] = \sum_k P_k = 1\,. \tag{10.85}$$

The quantum expression for the entropy of a system as a member of an ensemble is given by[36]

$$S = -k_B \mathrm{Tr}[\hat{\rho}\ln\hat{\rho}\,] = -k_B \sum P_k \ln P_k \tag{10.86}$$

the latter being of the same form as the expression Eq. (10.81). However, Eq. (10.81) gives the entropy of N occurrences of the independent particles in a system, while Eq. (10.86) applies to any system whose possible microstates are known, and the probabilities for each microstate have been discovered.

Underlying Assumption Made by Gibbs

Gibbs argued that for most systems at or near equilibrium, averages of system variables over the ensemble well-approximates averages for a single system if taken over sufficiently long times. This is a form of the ergodic hypothesis.

Choices for Ensemble Conditions

Various conditions can be imposed on a Gibbs ensemble, with the choice depending on how the ensemble will be compared to experimental conditions. Here is a table of useful choices and the name of the ensemble given for each choice:

Fixed			Systems	Ensemble	
N	V	E	Completely isolated	Micro-canonical	z
N	V	T	In thermal contact	Canonical	Z
μ	V	T	” & can exchange particles	Grand canonical	Ξ
N	p	T	” & under constant pressure	Grand NPT	Ƶ
N	p	H	Insulated and under constant pressure	Grand NPH	$\not{Z}$

[35] For ease in writing, we will use an overline to indicate an average, rather than left and right carets.

[36] This form of the entropy was first described by John von Neumann, ibid.

Here, N refers to the average number of particles across the ensemble and μ its conjugate thermodynamic variable called the chemical potential. The quantity $H = E + pV$ is the enthalpy, which you may recall changes by the amount of heat into the system under constant pressure.

For systems with a large number of particles, predicted thermodynamic behavior (i.e. at equilibrium) described by any of the grand canonical ensembles usually is the same as the canonical ensemble predictions. As we will see, this follows from the fact that for large N, fluctuations in extensive macroscopic variables are usually very small, the exception being first-order phase transitions. However, a particular ensemble may be easier to handle than another choice when applied to a system that permits heat, work, or particles to be exchanged with the environment.

10.8.4 Equilibrium Averages with Gibbs Ensemble

In deriving ensemble averages for systems at or near equilibrium, we will start with a quantum description, partly because of its generality (after all, it contains classical behavior), partly because of the simplicity of notation (which arises because of the discrete nature of state labels), and partly because the classical assumption of equal a priori probabilities in phase space simply derives from the assumption of equal a priori probabilities for each distinct microscopic quantum state.

We will often be dealing with systems with a large number N of particles, perhaps of the order of 10^{23}. In that case, a Gibbs ensemble can be well approximated by taking just one representative system for each possible microstate, i.e $\mathcal{N} \approx \overline{N}$, the average across the ensemble of the particle number.

Let the $\mathcal{N}$ members of a Gibbs ensemble be labeled by the full set of quantum numbers for the observables of that system. We map these quantum numbers into a single index k. Each member of the ensemble will have an index k and be in a state ψ_k, with energy E_k. Let the probability for the occurrence of such a state be called P_k. As described in Sect. 10.7, the entropy in the ensemble is given by

$$S = -k_B \sum_k P_k \ln P_k \tag{10.87}$$

in both the quantum and the classical setting.

The Canonical Ensemble

Now suppose that the Gibbs ensemble of $\mathcal{N}$ systems is isolated from the environment, but each member of the ensemble is allowed to exchange heat energy with other members. Let $\mathcal{N}\overline{E}$ be the energy of the full ensemble. As the full ensemble is isolated, $\mathcal{N}\overline{E}$ is fixed, making the average energy $\overline{E}$ in a system also fixed.

In applying statistical ideas to the Gibbs ensemble, we will wait for all members of the ensemble to relax. As the systems within evolve, they will eventually spend most of their time in those states which maximize the entropy of the ensemble. These systems are more numerous in the ensemble as equilibrium is established.

To find the values of the P_k for the canonical ensemble at or near equilibrium, we will maximize the entropy S under the constraints

$$\sum_{k=1}^{\mathcal{N}} P_k = 1 \tag{10.88}$$

and

$$\sum_{k-1}^{\mathcal{N}} P_k E_k = \overline{E} \,. \tag{10.89}$$

The constrained maximization of $S = -k_B \sum_k P_k \ln P_k$ gives the same set of $\{P_k\}$ as the unconstrained maximization of

$$\mathscr{S}_F(\{P_k\}) = S/k_B + \alpha \left(\sum_k P_k - 1 \right) - \beta \left(\sum_k P_k E_k - \overline{E} \right) , \tag{10.90}$$

where α and β are 'Lagrange multipliers', determined by the two constraints.

The maximum of $\mathscr{S}_F$ occurs where all the slopes $\partial \mathscr{S}_F / \partial P_k$ vanish. Thus, at or near equilibrium,

$$-\ln P_k - 1 + \alpha - \beta E_k = 0 \,,$$

i.e.

$$P_k = \exp(-1+\alpha) \exp(-\beta E_k) \,. \tag{10.91}$$

The first constraint, $\sum_k P_k = 1$ tells us that α is fixed by

$$\exp(-1+\alpha) = \frac{1}{\sum_k \exp(-\beta E_k)} \,,$$

so that

$$P_k = \frac{1}{\sum_k \exp(-\beta E_k)} \exp(-\beta E_k) \,. \tag{10.92}$$

The expression

$$Z(\beta, V, N) \equiv \sum_k \exp(-\beta E_k) \tag{10.93}$$

is called the 'canonical partition function'. The volume dependence of Z comes solely from dependence of the energies E_k on the fixed volume V of any one of the systems.

By putting the ensemble of systems into a large bath of an ideal gas (creating a new Gibbs ensemble consisting of the original systems plus the bath), one can show, as we did in Sect. 10.5.3, that

$$\beta = \frac{1}{k_B T}, \tag{10.94}$$

where T is the temperature of the ideal gas in thermal contact with each system of the ensemble. The temperature dependence of the partition function $Z(\beta, V, N)$ resides solely in β. The E_k will vary with the number of particles, N, in a system.

With an equilibrium distribution across the ensemble, the ensemble entropy becomes

$$\begin{aligned} S &= -k_B \sum_k P_k \ln P_k = -k_B \sum_k P_k \left(-\beta E_k - \ln \sum_l \exp(-\beta E_l) \right) \\ &= k_B \beta \overline{E} + k_B \ln \sum_l \exp(-\beta E_l) \\ &= k_B \beta \overline{E} + k_B \ln Z, \end{aligned} \tag{10.95}$$

which we write as

$$\overline{E} - TS = -k_B T \ln Z. \tag{10.96}$$

where the average energy across the ensemble is given by

$$\overline{E} = \frac{1}{Z} \sum_k E_k \exp(-\beta E_k) = -\left.\frac{\partial \ln Z}{\partial \beta}\right|_V. \tag{10.97}$$

As we saw before (Eq. (10.24), the Helmholtz free energy $A \equiv \overline{E} - TS$ can be found from the canonical partition function with

$$A(T, V, N) = -k_B T \ln Z(\beta, V, N), \tag{10.98}$$

and, inversely,

$$Z(\beta, V, N) = \exp(-\beta A). \tag{10.99}$$

Knowing that in thermodynamics $dA = -SdT - pdV + \mu dN$,[37] we have

$$S = -\left.\frac{\partial A}{\partial T}\right|_{V,N} = k_B \ln Z + \frac{1}{T}\overline{E} . \tag{10.100}$$

$$p = -\left.\frac{\partial A}{\partial V}\right|_{T,N} = -k_B T \left.\frac{\partial \ln Z}{\partial V}\right|_{T,N} . \tag{10.101}$$

and

$$\mu = -\left.\frac{\partial A}{\partial N}\right|_{T,V} = \frac{1}{Z}\sum_k \frac{\partial E_k}{\partial N} \exp\left(-\beta E_k\right) \equiv \left\langle \frac{\partial E}{\partial N} \right\rangle . \tag{10.102}$$

The Zero of Energy

It is straightforward to show that our statistical thermodynamic expressions preserve the property of classical and quantum theory (but not Einstein's gravitational theory) that the energy is defined only up to a constant linear shift. Suppose that the zero of energy is shifted to make

$$E_j{}' = E_j + E_o .$$

Then the canonical partition function is changed to

$$Z' = e^{-\beta E_o} Z .$$

Even so, the probabilities for the occurrence of any particular ensemble configuration is unchanged:

$$P_k = \frac{1}{\sum_j e^{-\beta E_j}} e^{(-\beta E_k)} = \frac{1}{\sum_j e^{-\beta E_j'}} e^{-\beta E_k'} . \tag{10.103}$$

Only those properties of the ensemble members that already depend on the choice of the zero of energy, such as the total energy and the free energies, will be affected by the zero-of-energy choice. Other thermodynamic variables, such as the entropy, pressure, temperature, and chemical potential, will not be affected by a new choice.

Less obvious is the answer to the following question: What energy states should be used in equilibrium considerations of a thermal system? After all, there may be a bound state for the system which has a lower energy than the energies of states taking part in the conditions for thermal equilibrium. For example, a crystal

[37] Here, N is the fixed number of particles in each of the members of the Gibbs ensemble.

might be compressible into a denser and more tightly bound set of molecules than a given state. A diamond as a form of carbon is an example. The diamond phase of carbon may not be needed in determining the thermodynamics of carbon if the region being studied has pressures far below those needed to make diamonds. The ultimate compression of any matter would make a black hole, far more tightly bound (by gravity) than ordinary matter. But the thermodynamic description of ordinary matter in ordinary ranges of temperature and pressure does not include black hole states.

The answer to the posed question is: The quantum states that should be included must have the following properties

1. The quantum state must be accessible from other quantum states by heat transfer;
2. The system being considered must have at least one quasi-static state with a lifetime much longer than the expected observation time. Such a state will be called a quasi-equilibrium state.
3. The time for the system to 'fall' into or be excited out of the accessible quantum states must be smaller than the time needed for the system to reach a quasi-equilibrium.

In statistical mechanical considerations, among those states that satisfy these conditions, the one with the lowest energy will have the highest probability of being occupied when the system approaches equilibrium.

Fluctuations in the Internal Energy

The energy in any one system in the ensemble is not fixed, but rather may fluctuate as heat in and out fluctuates. We can calculate this fluctuation using

$$\begin{aligned}
(\Delta E)^2 &\equiv \left\langle (E-\overline{E})^2\right\rangle = \left\langle E^2\right\rangle - \overline{E}^2 \\
&= \frac{1}{Z}\frac{\partial Z}{\partial \beta^2} - \left(-\frac{1}{Z}\frac{\partial Z}{\partial \beta}\right)^2 \\
&= \frac{\partial}{\partial \beta}\left(\frac{1}{Z}\frac{\partial Z}{\partial \beta}\right) = -\frac{\partial E}{\partial \beta} = k_B T^2 \left.\frac{\partial E}{\partial T}\right|_V \\
&= k_B T^2 C_v\,, \qquad (10.104)
\end{aligned}$$

where C_v is the heat capacity of the system. Now, C_v, being an extensive variable proportional to N, we have that the relative fluctuation of energy, $\Delta E/\overline{E} \sim 1/\sqrt{N}$, usually a very small number. However, fluctuations will not be small when a system approaches a phase change, and latent heat is absorbed or released. In that case, the specific heat c_v grows sharply with temperature, limited only by the energy that the total volume of the system can take or release. Even so, the canonical partition function $Z(T, V, N)$ contains the information to completely determine the thermodynamic properties of the given system and fluctuations of the energy near equilibrium.

The Grand Canonical Ensemble

Now let us consider a system that allows the exchange of particles with the environment as well as the exchange of heat. A Gibbs ensemble with $\mathcal{N}$ such systems can be created by allowing each system copy to exchange heat and particles between them. The complete Gibbs ensemble is isolated. Each ensemble member has been prepared in a quantum state with the full set of observables labeled by 'k'. The system labeled by 'k' will have its own particle number N_k. The energy of the system k, called E_k, is determined by its microdynamics. The E_k will be a function of the volume V of any one of the systems and of its particle number N_k: $E_k = E_k(V, N_k)$. Now $\mathcal{N}\overline{E}$ and $\mathcal{N}\overline{N}$ are kept fixed during the evolution of the systems, so that $\overline{E}$ and $\overline{N}$ are also fixed. All $\mathcal{N} - 1$ systems outside the system k act as a heat and particle bath for the selected system k.

The constraints on the ensemble become

$$\sum_k P_k = 1 \,. \tag{10.105}$$

$$\sum_k P_k E_k = \overline{E} \,. \tag{10.106}$$

$$\sum_k P_k N_k = \overline{N} \,. \tag{10.107}$$

At equilibrium, the entropy

$$S = -k_B \sum_k P_k \ln P_k$$

is maximized under the constraints, and

$$\mathcal{S}_A(\{P_k\}) = S/k_B + \alpha\left(\sum_k P_k\right) - \beta\left(\sum_k P_k E_k\right) + \gamma\left(\sum_k P_k N_k\right) \tag{10.108}$$

becomes a maximum without any constraints. The function $\mathcal{S}_A(\{P_k\})$ reaches a maximum when

$$-\ln P_k - 1 + \alpha - \beta E_k + \gamma N_k = 0\,,$$

or

$$P_k = \exp(-1+\alpha)\exp(-\beta E_k + \gamma N_k) . \tag{10.109}$$

As before for the canonical ensemble, the constraint $\sum_k P_k = 1$ determines α, and we write

$$P_k = \frac{1}{\Xi}\exp(-\beta E_k + \gamma N_k) , \tag{10.110}$$

where the 'grand canonical partition function' is defined as

$$\Xi(\beta, V, \gamma) \equiv \sum_k \exp(\gamma N_k - \beta E_k) \tag{10.111}$$

$$= \sum_k \lambda^{N_k} \exp(-\beta E_k) , \tag{10.112}$$

with $\lambda \equiv \exp(\gamma)$ and the sum is taken over all the ensembles, recognizing that the ensemble number k has a distinct value for each of the possible quantum states of the system being considered.

If we group all the ensemble members that have the same number of particles, and label these groups by the index l, then

$$\Xi(\beta, V, \gamma) = \sum_{N=0}^{\infty} \lambda^N \sum_l \exp(-\beta E_{N,l}) = \sum_N \lambda^N Z(\beta, V, N) \tag{10.113}$$

with $Z(\beta, V, N)$ being the canonical partition function for a system with a fixed number of particles N. Using an ideal gas as a heat bath, as in Sect. 10.5.3, we will have again that $\beta = 1/(k_B T)$.

A note of caution: The expression for the grand canonical partition function in Eq. (10.113) starts with $N = 0$. The canonical partition function $Z(\beta, V, 0)$ need not be unity. As the canonical partition function contains a factor $g_0 \exp(-\beta E_0)$ from the selected choice for the zero of energy for all systems,

$$Z(\beta, V, 0) = g_0 \exp(-\beta E_0) .$$

Not taking this zero-point energy into account will make the sizes of the terms in Eq. (10.113) for $N > 0$ incorrect relative to the $N = 0$ term.

With our new grand canonical partition function Ξ, the entropy of the system becomes

$$\overline{S} = -k_B \sum_k P_k \ln P_k = -k_B \sum_k P_k((-\beta E_k + \gamma N_k) - \ln \Xi)$$

$$= k_B \beta \overline{E} - k_B \gamma \overline{N} + k_B \ln \Xi , \tag{10.114}$$

and the Helmholtz free energy $\overline{A} = \overline{E} - T\overline{S}$ in terms of the grand canonical partition function is

$$\overline{A} = \gamma k_B T \overline{N} - k_B T \ln \Xi \,. \tag{10.115}$$

We now wish, as we did for β, to connect the Lagrange multiplier γ to the measurable properties of the system at equilibrium. First, we use the fact that a Lagrange multiplier is determined by the corresponding constraint. In this case, the constraint is in Eq. (10.107), which, near equilibrium, reads

$$\overline{N} = \frac{\sum_k N_k \exp(\gamma N_k - \beta E_k)}{\sum_k \exp(\gamma N_k - \beta E_k)} = \frac{\partial \ln \Xi}{\partial \gamma} \,. \tag{10.116}$$

As with β constraint, an explicit solution of this constraint relation is possible in terms of standard functions only for special cases, such as an ideal gas. However, we know that a functional relationship,

$$\gamma = \gamma(\beta, V, \overline{N}) \,, \tag{10.117}$$

exists, i.e. the constraint makes γ a function of the system temperature, its volume, and its average number of particles. The quantity $\lambda(\beta, V, \overline{N}) \equiv \exp(\gamma)$ is called the 'absolute activity' a.[38]

When all three constraints are applied, the grand canonical partition function becomes a function of T, V, and $\overline{N}$: $\Xi = \Xi(\beta, V, \gamma(\beta, V, \overline{N}))$. Note that the only dependence of the grand canonical partition function on the average particle number is through γ, and the only dependence on the ensemble member volume is through the energies.

Now consider the chemical potential for the system, and apply Eqs. (9.95), (10.115), and (10.116):

$$\mu = \left.\frac{\partial \overline{A}}{\partial \overline{N}}\right|_{T,V} = k_B T \left.\frac{\partial}{\partial \overline{N}} \left(\overline{N}\gamma - \ln \Xi\right)\right|_{T,V} \tag{10.118}$$

$$= k_B T \gamma + k_B T \overline{N} \frac{\partial \gamma}{\partial \overline{N}} - k_B T \frac{\partial \ln \Xi}{\partial \gamma} \frac{\partial \gamma}{\partial \overline{N}} \tag{10.119}$$

$$= k_B T \gamma \,. \tag{10.120}$$

Thus, γ determines the chemical potential for the system:

[38] We introduced the idea of activity in connection to chemical reactions in Sect. 9.21.2.

$$\gamma = \frac{\mu}{k_B T} \; . \tag{10.121}$$

Note that $\mu = k_B T \ln \lambda$. Compare this with the chemical potential for an ideal gas: $\mu = k_B T \ln (cp)$, where c is a number independent of pressure. This comparison motivated the chemist G.N. Lewis to define[39] 'fugacity' f ("fleeing tendency") by $f \equiv p_o \lambda$, with p_o a constant pressure (such as one atmosphere). Thus, the fugacity is a 'generalized pressure', since $f \rightarrow p$ for a material which has an ideal-gas limit. If we let $f = \varphi p$, where p is the pressure in the system, then φ will be the fugacity coefficient introduced in Sect. 9.19.

We can now express the grand canonical partition function as

$$\Xi(\beta, V, \gamma) \equiv \sum_k \exp\left(\beta(\mu N_k - E_k)\right) , \tag{10.122}$$

with the sum taken over the ensemble of systems. This will include all possible quantum states.

Equation (10.115) becomes

$$\overline{A} = \mu \overline{N} - k_B T \ln \Xi \; . \tag{10.123}$$

This can be written in an even simpler form using the thermodynamic identity

$$\overline{A} = -pV + \mu \overline{N} \; , \tag{10.124}$$

giving the average pressure times volume as

$$p\,V = k_B T \ln \Xi \; , \tag{10.125}$$

and, inversely, the grand canonical partition function is expressible as

$$\Xi = \exp\left(p\,V/k_B T\right) . \tag{10.126}$$

From $dE = TdS - pdV + \mu dN$ and $E = TS - pV + \mu N$, we can derive $d(pV) = SdT + pdV + Nd\mu$. This tells us that the 'natural variables' for the function pV are the temperature T, the volume V, and the chemical potential μ, and that, with Eq. (10.125),

$$\overline{S} = k_B \left[1 - \beta \frac{\partial}{\partial \beta}\right] \ln \Xi \; , \tag{10.127}$$

[39] Gilbert Newton Lewis, *The law of physico-chemical change*, Proc Am Acad Arts and Sciences **37 (3)**, 49–69 (June 1901).

$$p = \frac{1}{\beta}\frac{\partial}{\partial V}\ln \Xi \,, \tag{10.128}$$

$$\overline{N} = \frac{\partial}{\partial \gamma}\ln \Xi \,, \tag{10.129}$$

where Ξ is treated as a function of β, V, and γ.

From Eq. (10.122), we can add to our connections between the grand canonical partition function and thermodynamic variables the average internal energy for a member of the ensemble:

$$\overline{E} = -\frac{\partial \ln \Xi}{\partial \beta} \,, \tag{10.130}$$

holding γ and V fixed.

The overall scheme to apply the grand canonical ensemble to calculate the thermodynamics of a system (at one temperature, one volume, but variable number of particles) is the following:

1. From the microscopic dynamics, find all the possible accessible microscopic energies. In a quantum model, that usually means solving the applicable wave equation $H\psi_i = E_{N,i}\psi_i$ for the steady-state energy values $E_{N,i}$ in the N particle system.
2. Calculate the Gibbs grand canonical partition function

$$\Xi = \sum_N \lambda^N \sum_i \exp\left(-\beta E_{N,i}\right).$$

(The second sum is over all possible quantum states for a given N.)
3. From $\ln \Xi$, calculate the thermodynamic variables of interest, such as the average internal energy, entropy, pressure, the average particle number, the specific heat, the compressibility, etc.

Even though, for systems with a large number of particles, the fluctuations in N will usually be small,[40] relaxing the constraint on the ensemble member particle numbers in generating the grand canonical ensemble has a distinct advantage over the canonical ensemble method: We will see that one does not have to restrict summations over occupation numbers for states of the system to a fixed number of particles, thus making these sums easier to handle.

[40]But not near first-order phase transitions.

Application to Independent Particles

Now let's see how the grand canonical ensemble might describe a system of particles that, for the most part, do not interact. Examples will include atoms in an ideal gas, atoms in a solid crystal held by individual binding forces, or 'free' conduction electrons in a metal. Some interaction is necessary to allow the particles in a given system to share their energy, but interactions between particles should be so weak that their energies are dominated by energies internal to each particle. As usual, we assume that after a relaxation time, the particles reach a quasi-steady equilibrium state.

With these assumptions, E_k will be a sum of possible energies for each particle,

$$E(N, V) = E_0 + \sum_i n_j \epsilon_j(N, V) ,$$

where j labels the possible energy states for each particle, n_j is the occupation number for the particle within the k-th ensemble in the j-th quantum energy state, and ϵ_j is the energy of that j-th state. We have included an arbitrary zero of energy term E_0.

Taking $E_0 = 0$, the grand canonical ensemble partition function becomes

$$\Xi(\beta, V, \gamma) = \sum_k \exp(\gamma N_k - \beta E_k) \tag{10.131}$$

$$= \sum_{N=0}^{\infty} \sum_{\{n_j\}}{}' \exp\left(\sum_j (\gamma - \beta\epsilon_j) n_j\right) \tag{10.132}$$

$$= \sum_{N=0}^{\infty} \sum_{\{n_j\}}{}' \prod_j \left(\exp(\gamma - \beta\epsilon_j)\right)^{n_j} \tag{10.133}$$

$$= \sum_{n_o, n_1, \cdots} \prod_j \left(\exp(\gamma - \beta\epsilon_j)\right)^{n_j} \tag{10.134}$$

$$\Xi(\beta, V, \gamma) = \prod_j \sum_{n_j} \left(\exp(\gamma - \beta\epsilon_j)\right)^{n_j} , \tag{10.135}$$

where, in the second line, we form sets of all members of the ensemble with the same $N \equiv N_k$, sum over the occupation numbers $\{n_j\}$ in those members with a given N, and then sum over the sets of ensemble member for various N. The prime on the inner sum indicates that the sum is performed while preserving $\sum_j n_j = N$. The index j labels the individual particle energy levels, in ascending order. In the fourth line we observe that the two sums in the third line become one sum over unrestricted values for the n_j. Line five separates the products of line four into products for individual n_j. (None of these four steps is trivial!)

Shift of the Zero of Energy per Particle

As an aside, we observe that if we had included the arbitrary zero of energy term E_0 in the overall energy for every member of the ensemble, we would have gotten

$$\Xi(\beta, V, \gamma) = e^{-\beta E_0} \prod_j \sum_{n_j} (\exp(\gamma - \beta\epsilon_j))^{n_j} . \tag{10.136}$$

As with the canonical partition function, the effect is to shift all energies, but leave other thermodynamics quantities unchanged. However, for independent particles, we have split each grand canonical ensemble member for a given number of particles into sub-ensembles, one for each particle, sifting the single-particle energies per particle in these sub-ensembles by ϵ_o (with $\epsilon'_j = \epsilon_j + \epsilon_o$) will give

$$\Xi(\beta, V, \gamma) = = \sum_{N=0}^{\infty} \sum_{\{n_j\}}' \left(e^{\beta(\mu+\epsilon_o)}\right)^N \prod_j \left(\exp(\gamma - \beta\epsilon'_j)\right)^{n_j} , \tag{10.137}$$

which has the effect of changing the chemical potential of the grand canonical system to

$$\mu' = \mu + \epsilon_o . \tag{10.138}$$

As we implied in Sect. 10.8.4, the choice of the lowest Boltzmann energy is subtle. Only those energy levels which can be brought into thermal equilibrium in times characteristic for the processes of interest should be included.

Independent Bosons and Fermions in a Grand Canonical Ensemble

For bosons, the quantum-state occupation numbers, $\{n_j\}$, are unrestricted, so we can use the identity $1/(1-x) = \sum_{k=0}^{\infty} x^k$ for $x < 1$ to express

$$\sum_{n_j=0}^{\infty} (\exp(\gamma - \beta\epsilon_j)^{n_j} \equiv \frac{1}{1 - \exp(\gamma - \beta\epsilon_j)} , \tag{10.139}$$

with the condition $\epsilon_o > \mu$.

For fermions, the occupation numbers $\{n_j\}$ can only be zero or one, so that

$$\sum_{n_j=0}^{1} (\exp(\gamma - \beta\epsilon_j))^{n_j} \equiv 1 + \exp(\gamma - \beta\epsilon_j) . \tag{10.140}$$

As a result, the grand partition function becomes

$$\Xi = \prod_j \begin{cases} 1/(1 - \exp(\gamma - \beta\epsilon_j)) & \text{bosons} \\ 1 + \exp(\gamma - \beta\epsilon_j) & \text{fermions} \end{cases} . \tag{10.141}$$

With this result, we can now find the average values of thermodynamic variables. From Eq. (10.123), we have

$$\begin{aligned} \overline{A} &= \mu\overline{N} - k_B T \ln \Xi \\ &= \mu\overline{N} - k_B T \sum_j \begin{cases} -\ln(1 - \exp(\gamma - \beta\epsilon_j)) & \text{bosons} \\ \ln(1 + \exp(\gamma - \beta\epsilon_j)) & \text{fermions} \end{cases} \end{aligned} \tag{10.142}$$

in which

$$\overline{N} = \frac{1}{\Xi}\sum_k N_k \exp(\gamma N_k - \beta E_k) = \frac{\partial \ln \Xi}{\partial \gamma} \tag{10.143}$$

$$= \sum_j \begin{cases} (\exp(-\gamma + \beta\epsilon_j) - 1)^{-1} & \text{bosons} \\ (\exp(-\gamma + \beta\epsilon_j) + 1)^{-1} & \text{fermions} \end{cases} . \tag{10.144}$$

Similarly, the average energy of the system becomes

$$\overline{E} = \frac{1}{\Xi}\sum_k E_k \exp(\gamma N_k - \beta E_k) = -\frac{\partial \ln \Xi}{\partial \beta} \tag{10.145}$$

$$= \sum_j \begin{cases} \epsilon_j(\exp(-\gamma + \beta\epsilon_j) - 1)^{-1} & \text{bosons} \\ \epsilon_j(\exp(-\gamma + \beta\epsilon_j) + 1)^{-1} & \text{fermions} \end{cases} . \tag{10.146}$$

Using Eq. (10.132), we can express the average occupation number of energy level indexed by l as

$$\langle n_l \rangle = -\frac{1}{\beta}\frac{\partial \ln \Xi}{\partial \epsilon_l} \tag{10.147}$$

$$= \begin{cases} (\exp(-\gamma + \beta\epsilon_l) - 1)^{-1} & \text{bosons} \\ (\exp(-\gamma + \beta\epsilon_l) + 1)^{-1} & \text{fermions} \end{cases} . \tag{10.148}$$

Note that, as we should expect, $\overline{N} = \sum_j \langle n_j \rangle$ and $\overline{E} = \sum_j \langle n_j \rangle \epsilon_j$.

When the average occupation numbers become much less than one, the particles are spread thinly over the energy levels. In this case, the distinction between bosons and fermions becomes less dramatic. The condition $\langle n_l \rangle << 1$ will apply if $\epsilon_o >> \mu$. A large negative chemical potential will satisfy this so-called classical statistics limit.[41] In that case, the exponential dominates in Eq. (10.148), and

$$\langle n_l \rangle \approx \exp(\gamma - \beta\epsilon_l) \tag{10.149}$$

i.e. the particles are spread in energy levels according to the Boltzmann distribution. Here, the index i numbers the quantum states.

Application to a Simple Ideal Gas

As a special case, consider a dilute ideal gas in a macroscopic box. From the results of Sect. 10.6, the number of possible available kinetic energy quantum states will be of the order $V/(\Lambda_T^3)$, where $\Lambda_T = \sqrt{h^2/(2\pi mkT)}$ is the thermal de Broglie wavelength, typically a fraction of an ångström in size. Under these circumstances, in applying Eq. (10.132), there will be far more terms in the product of sums that have different quantum states than those with two or more with the same quantum state. We can make the sums unrestricted, neglect the terms with two or more identical states, and then divide the result by $N!$, which is the number of terms that have all different quantum states in the product of the sums. This makes

$$\Xi(\beta, V, \gamma) \approx \sum_N \frac{1}{N!} \sum_i \left(e^{\gamma - \beta\epsilon_i}\right)^N \tag{10.150}$$

$$= \sum_N \frac{1}{N!} (\lambda Z_{\text{ideal-gas}})^N \tag{10.151}$$

$$= \exp(\lambda Z_{\text{ideal-gas}}) \,. \tag{10.152}$$

Note that the zero of energy in the expression for the canonical partition function $Z(\beta, V, N)$ has been taken to be zero when there are no particles in the ensemble member, i.e. $Z(\beta, V, 0) = 1$.

If we include only the kinetic energy of the particles, then $Z_{\text{ideal-gas}} = Z_{\text{trans}}$ and we can use Eq. (10.45) to find

$$\ln \Xi(\beta, V, \gamma) = \frac{\lambda V}{\Lambda_T^3} = \left(\frac{2\pi m}{h^2}\right)^{3/2} \exp(\gamma)\, \beta^{-3/2} V \,. \tag{10.153}$$

It follows that, for the ideal gas whose molecules have energies due just to kinetic motion,

[41] The chemical potential μ of a gas of quasi-independent bosons must be negative, since $\langle n_l \rangle$ must be finite. There is no such restriction for fermions.

$$\overline{N} = \frac{\partial \ln \Xi}{\partial \gamma} = \ln \Xi \,, \tag{10.154}$$

$$\overline{E} = -\frac{\partial \ln \Xi}{\partial \beta} = \frac{3}{2\beta} \ln \Xi \,. \tag{10.155}$$

Now, since generally,

$$\beta p V = \ln \Xi \,,$$

we have

$$pV = \overline{N} k_B T \,, \tag{10.156}$$

$$\overline{E} = \frac{3}{2}\, pV = \frac{3}{2}\overline{N} k_B T \,, \tag{10.157}$$

well known for a monatomic ideal gas.[42]

We can solve Eq. (10.154) for $\gamma = \beta\mu$ to get the chemical potential for a monatomic ideal gas as

$$\mu = \beta^{-1} \ln \left[\Lambda_T^3 \frac{\overline{N}}{V} \right] = k_B T \ln \left[\left(\frac{h^2}{2\pi m k_B T} \right)^{3/2} \frac{\overline{N}}{V} \right] . \tag{10.158}$$

With a little algebra, this agrees with the thermodynamic result Eq. (9.78) and with the statistical mechanics result Eq. (10.76), both applied to a monatomic ideal gas. We can now display the monatomic ideal gas absolute activity as

$$\begin{aligned} \lambda &= \exp(\gamma) = \exp(\beta\mu) \\ &= \Lambda_T^3 \frac{\overline{N}}{V} \\ &= \left(\frac{h^2}{2\pi m k_B T} \right)^{3/2} \frac{\overline{N}}{V} \\ &= \left(\frac{h^2}{2\pi m} \right)^{3/2} \frac{p}{(k_B T)^{5/2}} \,. \end{aligned} \tag{10.159}$$

[42] Looking back at the origin of the factor $3/2$, one can see that any dynamic variable for a particle which appears as a square within the Hamiltonian H for the system will produce, in the classical limit, a term $(1/2)k_B T$ in the average energy per particle. This is the equipartition of energy theorem.

For a dilute gas, we expect $V/\overline{N} >> \Lambda_T^3$, i.e. the volume taken up per particle to be much greater than the volume of the thermal de Broglie wavelength for the particles in the gas. This makes the chemical potential less than zero for an ideal gas. This property is a reflection of the fact that adding a particle to an ideal gas at constant volume and entropy can only be done by reducing the energy of the system.

Substituting in the thermodynamic relation $\overline{S} = (1/T)(\overline{E} + pV - \mu\overline{N})$, we find the entropy in a monatomic ideal gas to be expressible as

$$\overline{S} = \overline{N} k_B \ln \left[e^{5/2} \left(\frac{2\pi m k_B T}{h^2} \right)^{3/2} \frac{V}{\overline{N}} \right]. \tag{10.160}$$

Applications to Biological Systems

To exercise the formalism for a biological system, we will apply the grand canonical formalism to several such cases.[43]

Hemoglobin Uptake of Oxygen

Normally, hemoglobin in red-blood cells flowing within our blood carries needed oxygen to body cells. In the lungs, oxygen diffuses into the blood cells and attaches to the hemoglobin, one O_2 to each of the four iron groups in the molecule. The attachment is not strong, so is sensitive to the concentration of oxygen in the red-blood cell. In fact, the first attached O_2 slightly changes the configuration of the hemoglobin molecule, so that the next O_2 is a little easier to attach. This is called 'cooperative' binding. Gilbert Adair[44] and Linus Pauling[45] derived the degree of saturation of hemoglobin for a given concentration of oxygen, showing that a cooperative effect is present as each additional oxygen binding occurs. In addition, attachment of O_2 is slightly weakened by a lowering of pH, which occurs when carbon dioxide is dissolved in the nearby fluid (a property discovered by Christian Bohr, Niels Bohr's father).

To generate a statistical model, consider the system consisting of a single hemoglobin molecule and a bath of oxygen molecules, which also acts as a thermal bath. There are four sites for attachment of an O_2 per hemoglobin molecule. Each site is either unoccupied or occupied. Let σ_i be zero or one according to occupancy

---

[43] Some examples have been inspired by textbook problems in Charles Kittel, **Introduction to Solid State Physics**, [Wiley, NY] (1st edition, (1953), 8th edition (2005)).

[44] G.S. Adair, *The hemoglobin system. IV. The oxygen dissociation curve of hemoglobin*, J Biol Chem **63**, 529–545 (1925).

[45] Pauling, L., *The Oxygen Equilibrium of Hemoglobin and Its Structural Interpretation*, PNAS **21** **(4)**, 186–191 (Apr 1935).

of site with index i. Suppose the first oxygen molecule attaches to a hemoglobin molecule binds with an energy of ϵ_1 and the second one attaches with a binding energy of ϵ_2. As binding energies, both these values are negative. To make the second oxygen bind more easily than the first, ϵ_2 is more negative than ϵ_1. The Pauling model uses just these two energies. The Adair model effectively takes into account four different energies of binding. As our aim is didactic, we will elaborate on the underlying statistics of the Pauling case, as this model is a little simpler, and does almost as well as the Adair model compared to experiment.

We can write the energy associated with various attachments of the O_2 molecules as

$$\epsilon(\sigma_1, \sigma_2, \sigma_3, \sigma_4) = \epsilon_1 \sum_i \sigma_i + \Delta \sum_{i>j} \sigma_i \sigma_j , \tag{10.161}$$

where $\epsilon_1 < 0$ and $\Delta = \epsilon_2 - \epsilon_1 < 0$.

The grand partition function for this model that describes these sixteen possible states of oxygen molecules attached to the hemoglobin becomes

$$\begin{aligned} \Xi &= \sum_{\sigma_1=0}^{1} \sum_{\sigma_2=0}^{1} \sum_{\sigma_3=0}^{1} \sum_{\sigma_4=0}^{1} \exp\left(\beta\,(\mu - \epsilon_1) \sum_{i=1}^{4} \sigma_i + \beta(-\Delta) \sum_{i>j} \sigma_i \sigma_j \right) \\ &= 1 + 4\lambda e^{-\beta\epsilon_1} + 6\lambda^2 e^{-2\beta\epsilon_1 - \beta\Delta} + 4\lambda^3 e^{-3\beta\epsilon_1 - 3\beta\Delta} + \lambda^4 e^{-4\beta\epsilon_1 - 6\beta\Delta} , \end{aligned} \tag{10.162}$$

where $\lambda = \exp(\beta\mu)$.

Aside. Here we see the importance of how the lowest Boltzmann energy is selected, in that the λ terms relative to unity depends on that choice. We will be using the balance of chemical potential for oxygen as a gas in the lung with the chemical potential $\mu = k_B T \ln \lambda$ in the blood. The lowest Boltzmann energy for the gas (taken to be ideal here) comes to play. At body temperatures, the accessible quantum states for the oxygen molecule as a gas in the lung are those with energies clustered within a few $kT \sim 1/40\,\mathrm{eV}$ just above the atomic ground state.

From the partition function, we can find the average number of oxygen molecules per hemoglobin molecule with

$$\begin{aligned} \langle N \rangle &= \lambda \frac{\partial}{\partial \lambda} \ln \Xi \\ &= 4\lambda\kappa \frac{1 + 3\lambda\kappa\omega + 3\lambda^2\kappa^2\omega^3 + \lambda^3\kappa^3\omega^6}{1 + 4\lambda\kappa + 6\lambda^2\kappa^2\omega + 4\lambda^3\kappa^3\omega^3 + \lambda^4\kappa^4\omega^6} , \end{aligned} \tag{10.163}$$

wherein, to improve the aesthetics, we have introduced $\kappa \equiv \exp(-\beta\epsilon_1)$ and $\omega \equiv \exp(-\beta\Delta)$.

The saturation level, f of oxygen attached to the hemoglobin will be

$$f = \frac{1}{4} \langle N \rangle .$$

At this stage, there remains just one unknown, the chemical potential $\mu = k_B T \gamma$ for the oxygen. As the blood flows through the lungs, the oxygen in the plasma and red-blood corpuscles come into close equilibrium with the oxygen in the air of each lung. This makes the chemical potential of the oxygen in the blood stream close to that of the air in the alveoli. For this model, we will use an ideal gas approximation for the air. Then the chemical potential for the oxygen in the air gas will be (see Eq. (10.77))

$$\mu_{O_2} = k_B T \ln \left(\frac{N_{O_2}}{V\, z_{rot} z_{elec}} \Lambda_T^3 \right) = k_B T \ln \left(\frac{p_{O_2}}{k_B T\, z_{rot} z_{elec}} \left(\frac{h^2}{8\pi^2 m_{O_2} k_B T} \right)^{3/2} \right) . \tag{10.164}$$

where, from Eq. (10.55),

$$z_{rot} = \frac{1}{2} \left(\frac{8\pi^2 I_{O_2} k_B T}{h^2} \right) . \tag{10.165}$$

The electronic ground-state of oxygen is a total electron spin triple, and so has a degeneracy factor $2S+1 = 3$. The first electronic excited state is a total electron spin singlet and occurs at $f/c = 79181\,\text{cm}^{-1}$, for which $\exp(-hf/k_B T) = 0.00292$ at body temperature, and a degeneracy of $2(2S+1) = 2$. The oxygen molecule can disassociate into two atoms if given the binding energy of $D_e = 5.15\,\text{eV}$. Thus, body temperatures are unlikely to excite the electrons in the oxygen molecule, and have an astronomically small chance of separating the oxygen into two atoms. (See Sect. 10.8.4 for the conditions which set the lowest Boltzmann energy level.)

We deduce that the oxygen molecule's electronic partition function can be taken as

$$z_{elec} \approx 3$$

to within about 0.2%. There results

$$\mu_{O_2} = k_B T \ln \left(\frac{2}{3} \left(\frac{h^2}{8\pi^2 m_{O_2} k_B T} \right)^{3/2} \left(\frac{h^2}{8\pi^2 I_{O_2} k_B T} \right) \frac{p_{O_2}}{k_B T} \right) . \tag{10.166}$$

Let

$$\lambda_o = \frac{2}{3} \left(\frac{h^2}{8\pi^2 m_{O_2} k_B T_o} \right)^{3/2} \left(\frac{h^2}{8\pi^2 I_{O_2} k_B T_o} \right) \frac{p_o}{k_B T_o} , \tag{10.167}$$

so that the activity of the oxygen molecule in the lung can be written in a simple form which explicitly shows the temperature and pressure dependence:

$$\lambda = e^{\mu/k_B T} = \lambda_o \left(\frac{T_o}{T}\right)^{7/2} \left(\frac{p}{p_o}\right) . \tag{10.168}$$

We will take $T_o = 298.15\,\text{K}$ and $p_o = 760\,\text{mmHg}$.

Now, in these terms, a balance of chemical potentials gives

$$\nu \equiv \lambda \kappa = \exp\left(\beta(\mu - \epsilon_1)\right) = \lambda_o \left(\frac{T_o}{T}\right)^{7/2} \left(\frac{p}{p_o}\right) \exp\left(-\frac{\epsilon_1}{k_B T}\right) . \tag{10.169}$$

The fractional oxygen saturation becomes

$$f = \frac{\nu + 3\omega\nu^2 + 3\omega^3\nu^3 + \omega^6\nu^4}{1 + 4\nu + 6\omega\nu^2 + 4\omega^3\nu^3 + \omega^6\nu^4} . \tag{10.170}$$

The absorption of oxygen by hemoglobin is highly sensitive to the binding energy of oxygen to hemoglobin, a fact reflected by Eq. (10.170) because this energy is in an exponent in the ν terms. The partial pressure of oxygen in air at sea level is 159 mmHg. Your nose humidifies very cold air passing through, and raises its temperature. Even so, the air reaching your lungs may be as low as 12 °C below body temperature when the outside air temperature is much below freezing. In the lung alveoli, added water vapor (47 mmHg partial pressure) and carbon dioxide (35 mmHg partial pressure) reduces the oxygen partial pressure to about 104 mmHg. Under resting conditions, the de-oxygenated blood entering a pulmonary capillary has an equivalent oxygen partial pressure of about 40 mmHg. After traveling only about a third of the distance through lung alveoli, the partial pressure has reached about 104 mmHg. After leaving the lung, the blood from the lungs then gets mixed with blood that was shunted into the lung tissue, dropping the partial pressure of oxygen passing through the aorta to about 95 mmHg. The concentration of oxygen drops further as the blood flows through the capillaries of the body, while the concentration of carbon dioxide (and blood acidity) grows. Moreover, because carbon dioxide has about 24 times greater solubility than oxygen in water at body temperature, and because carbon dioxide diffuses in water about 20 times faster than oxygen, we have little need for special expulsatory carriers of carbon dioxide in the blood, even though hemoglobin does carry up to about 10% of the carbon dioxide expelled from the lungs.

In the capillaries within exercising muscle, the temperature may be half a degree Celsius above normal body temperature, and during a mild fever, a 2° rise is common. During an infection, the body makes more oxygen available to tissue, and mildly higher temperatures stimulate a greater immune response.

The oxygen saturation curve plotted in Fig. 10.3 demonstrates how nature, through evolution over a sufficient time, selects structures and behavior which optimize survivability. In this case, as we have seen, the behavior of the hemoglobin molecule in response to oxygen produces a 'sigmoid' curve (flat 'S') for the uptake of oxygen as the pressure increases. This curve shows that the response to pressure

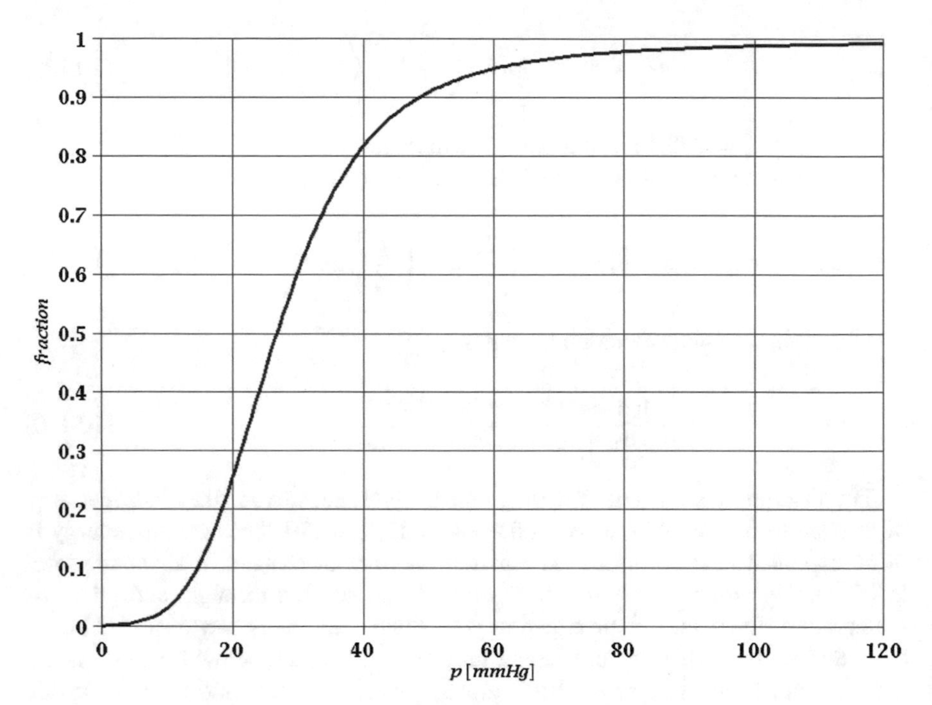

Fig. 10.3 Hemoglobin saturation curve. Fraction of oxygen carried by hemoglobin as a function of oxygen partial pressure. The curve prediction is from Eq. (10.170) using $T = 310.15$ K, $\epsilon_1 = -0.655$ eV, and $\Delta = -0.070$ eV

is similar to a switch, in that below a certain oxygen partial pressure level such as in a capillary, hemoglobin will give up oxygen, and above that level, such as in the lungs, it will take up oxygen. (The 50% saturation level is at about an oxygen partial pressure of 26.6 mmHg.) The sigmoid curve is sharpened by having the first oxygen attachment change the shape of the hemoglobin so that the next oxygen molecules are more easily attached.

For real hemoglobin, there is an enhanced release of oxygen when a greater concentration of carbon dioxide is dissolved in the plasma, when the acidity is raised, when 2,3-bisphosphoglycerate, a product of cellular glycolysis, is present, and when the temperature goes up (such as in active muscle). Each hemoglobin molecule can carry up to four carbon dioxide molecules at molecular sites which do not compete with oxygen affinity. However, 85% of the carbon dioxide expelled from the lung is carried there by hydrogen bicarbonate produced when carbon dioxide reacts with water in the capillary blood plasma, and about 5% is carried by CO_2 in solution. The temperature behavior of the oxygen saturation curve is correctly predicted by Eq. (10.170).

Carbon Monoxide Poisoning

Carbon monoxide is a colorless and odorless gas produced during incomplete burning of hydrocarbons and wood, burning cigarettes, by gasoline engines, and by gas/oil furnaces with insufficient air. The density of carbon monoxide gas is only slightly less than that of air (about 3% less), so, unlike carbon dioxide (with a density1.52 times greater than air), carbon monoxide tends to diffuse uniformly throughout a room. Carbon monoxide molecules have about 80% of the solubility of oxygen in blood. The carbon monoxide molecule can attach to the same sites on the hemoglobin molecule as oxygen, and more tenaciously, so carbon monoxide competes and then blocks oxygen binding. When we humans breath too much carbon monoxide, suffocation will result. (The symptoms are similar to those of cyanide poisoning, but the mechanism in the latter case is a blockage of the uptake of oxygen within a cell due to a binding of cyanide to the ferric ion on cytochrome oxidase within mitochondria, preventing oxygen binding and thereby stopping ATP production.)

Let's look at a simplified statistical model for the process (just to get a rough idea of the equilibrium behavior). Let the fixed (negative) binding energies of attachment be noted as ϵ_{O_2} for O_2 and ϵ_{CO} for CO_2. It is known that CO_2 attaches more strongly to the hemoglobin iron loci than does O_2, so $|\epsilon_{CO}| > \left|\epsilon_{O_2}\right|$. We will neglect any cooperative effect in the binding of either O_2 or CO. Then there are just three possibilities for the states of any one of the ferric attachment sites on hemoglobin molecule: (1) Free of any attachment; (2) Oxygen attachment; and (3) Carbon monoxide attachment. The grand partition function (now for a system containing two types of molecules competing for adsorption onto a distribution of attachment sites) will be

$$\Xi = 1 + \exp\left(\beta(\mu_{O_2} - \epsilon_{O_2})\right) + \exp\left(\beta(\mu_{CO} - \epsilon_{CO})\right). \tag{10.171}$$

From the hemoglobin analysis above,

$$\mu_{O_2} = -0.609\,\text{eV}$$

and

$$\epsilon_{O_2} = -0.655\,\text{eV}$$

for $T = 37\,°\text{C}$, body temperature (and we used $k_B = 8.6173 \times 10^{-5}\,\text{eV/K}$).

If there were no carbon monoxide present, then the probability that any one of the hemoglobin iron sites is occupied by oxygen in an alveolar capillary would be

$$\begin{aligned} f_{O_2} &= \frac{\exp\left(\beta(\mu_{O_2} - \epsilon_{O_2})\right)}{1 + \exp\left(\beta(\mu_{O_2} - \epsilon_{O_2})\right)} \\ &= 85\%\,. \end{aligned} \tag{10.172}$$

To get the chemical potential for carbon monoxide (in the lungs), we note that CO is not reflection symmetric, that the electrons in a CO molecule produce a spin singlet, and that both the carbon-12 and the oxygen-16 nuclei have no spin. This means that μ_{CO} will not have the rotational symmetry number 2 nor the electronic degeneracy factor $1/3$ appearing in Eq. (10.166). As a result, we can write

$$\begin{aligned} \lambda_{CO} &= \frac{3}{2}\left(\frac{m_{O_2}}{m_{CO}}\right)^{3/2}\frac{I_{O_2}}{I_{CO}}\frac{p_{CO}}{p_{O_2}}\lambda_{O_2} \\ &= \frac{3}{2}\left(\frac{32}{28}\right)^{3/2}\frac{11.726}{8.729}\frac{p_{CO}}{p_{O_2}}\lambda_{O_2}\,. \end{aligned} \tag{10.173}$$

For this example, we will take a binding energy of CO to hemoglobin to be

$$\epsilon_{CO} = -0.81\,\text{eV}.$$

For demonstration purposes, suppose the partial pressure of carbon monoxide were 100 times lower than the oxygen. Then $\mu_{CO} = k_B T \ln \lambda_{CO} = -0.708\,\text{eV}$. Now, the probability that any one of the hemoglobin iron sites is occupied by oxygen becomes

$$\begin{aligned} f_{O_2}^{[CO]} &= \frac{\exp\left(\beta(\mu_{O_2} - \epsilon_{O_2})\right)}{1 + \exp\left(\beta(\mu_{O_2} - \epsilon_{O2})\right) + \exp\left(\beta(\mu_{CO} - \epsilon_{CO})\right)} \\ &= 10.8\%\,. \end{aligned} \tag{10.174}$$

Evidently, even very low concentrations of carbon monoxide in the air can devastate the ability of hemoglobin to supply oxygen to the body, simply because carbon monoxide has a significantly stronger attachment to the hemoglobin molecule than oxygen.

OSHA (Occupational Safety and Health Administration) has recommended an upper limit of 50 ppm= 0.038 mmHg over an 8-h period[46] which is less than $1/4000$ of the oxygen partial pressure. In our model above, this would lower the fraction of hemoglobin oxygenation in the lung from 85 to 73%. Cigarette smokers take in 3–100 ppm of CO. Carbon monoxide is also produced in the body during hemoglobin breakdown, and some is taken up by hemoglobin, producing small amounts of carboxyhemoglobin in the blood. Carbon monoxide is slowly released from hemoglobin, with a half-life of four to six hours. The result is a steady-state minimum amount of CO-Hem in the blood, at the 0.5–1% level of the total hemoglobin concentration.[47]

[46]Having one part per million (1 ppm) of a component ideal gas mixed in a volume of another ideal gas is the same as having the component's partial pressure be a millionth of the overall gas pressure.

[47]R.A. Johnson, F. Kozma, E. Colombari, *Carbon monoxide: from toxin to endogenous modulator of cardiovascular functions*, Brazilian J Med & Biol Research **32**, 1–14 (1999).

Tertiary Protein Structure

We now face insurmountable opportunities.

— Walt Kelly, cartoonist

In cells, proteins are constructed by ribosomes that attach amino acids together into a sequence according to a template originating from a code on our DNA. As the peptide chain is created, it begins to form into a secondary configuration and then into its biological tertiary structure. In some cases, the terminal or side chains of the protein are modified after synthesis. Amino acids contain a carbon atom (denoted the alpha carbon) with its four valence electrons bond with (1) hydrogen, (2) an amine (3) a carboxylate and (4) a side chain (shown as '$\mathcal{R}$' in figure below):

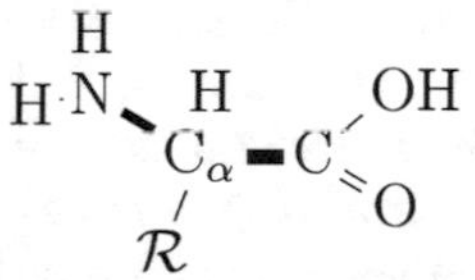

H
H·N H OH
C_α — C
O
$\mathcal{R}$

A peptide bond is created by a ribosome when the hydroxyl of one amino acid and an amine hydrogen from the second amino acid are removed as the two amino acid molecules form a bond linking the exposed nitrogen and carbon atoms, while releasing a water molecule.[48] The peptide bond formed is planar and relatively rigid due to the fact that the bonding valence electrons in the bond between the carbon of one amino acid and the nitrogen of the next amino acid resonate with those of the carbon and the attached oxygen in the first amino acid. However, folding of a protein chain can still occur because, within each amino acid, the single bonds between the two carbons and between the alpha carbon and the nitrogen atom can each be separately rotated about the axis of each bond, although the angles become strongly limited by steric collisions with adjacent parts of the peptide chain.

Natural protein chains can fold together to form a globular structure. In some cases, the globular structure can be stretched out to form long fibrous chains, which group together into the structural fibers such as myosin in muscle tissue.

As the protein settles into its natural structure, the peptide chain initially bends and twists under thermal collisions while ionic and polar forces of the chain with itself and with the media, act. Secondary structure with regions of alpha helicies, beta sheets, and other kinds of bends form.[49] The alpha helix is held together by hydrogen bonding between the N–H group of one amino acid with the C=O group of the amino acid four additional units away. The result is a tight chain holding 3.6 residues per turn, with a turn having 5.4 Å translation along coil's axis. The hydrogen bonds orient along the direction of the axis of the helix. Almost all natural chains are right-handed helices when the axis is directed from the terminus with

[48]After linking, a specific amino acid creates a 'monomer' within a peptide backbone. This monomer is sometimes referred to as the 'residue' of the amino acid since it remains after the peptide bonds materialize and water is extracted.

[49]The alpha helical substructure was deduced by Linus Pauling in 1948.

N–H groups tilted downward to the terminus with the C=O tilted upward. The side chains are on the outside of the helix, slightly tilted downward. There are typically about three or four turns of an alpha coil before an alteration of chain angles.

The beta 'pleated sheets'[50] are also formed by hydrogen bonds, but this time perpendicular to segments of the protein backbone which lie next to each other, with the direction of the chain segments either parallel or anti-parallel. The side-chains point alternately up and down perpendicular to the sheet, with the distance between every second chain about 7 Å.

In turn, the proteins with coil and sheet segments fold into globular tertiary structures. Loop segments of the protein chain (called β turns) form in between the coils and sheets.

The tertiary structure is determined by van der Waals forces, hydrogen bonds, dipole forces, Coulombic interactions of charged side-chains and ions, and sometimes by cross-linking chemical bonds, such as a disulfide bond. Particularly important for the secondary to tertiary transformation of proteins are the side-chain interactions due to water affinity (hydrophobic or hydrophilic).

Quaternary structures are formed when several proteins are held together, such as hemoglobin, which has four subunits, two of alpha and two of beta structures. DNA polymerase has ten subunits.

In thermodynamic terms, the folding process releases energy of binding (but less than 0.43 eV/Da), and also causes a drop in entropy as the protein becomes more ordered. As in all chemical processes at constant temperature and pressure, the Gibbs free energy, $G = E + pV - TS = H - TS$, must drop in value if the reaction is to go forward. Since $\Delta G = \Delta H - T\Delta S$, the heat released is the drop in the enthalpy. This means $-\Delta H$ per degree, must exceed the entropy drop $-\Delta S$, i.e. the energy releases across the protein chain per unit temperature as the chain folds cannot be less than the value of $|\Delta S|$ representing the order produced. As folding occurs in a water solution, hydrophobic regions are likely to be forced inside the protein tertiary structure. The solution pH and various salts also can affect the dynamics of chain folding. In addition, for complex proteins, there are so-called chaperone proteins that stabilize various configurations of the folding protein until the folding of a given region is complete. Also, there are catalysts that can speed the process and facilitate disulfide bonds, when they are required.

However, beginning with the work of Christian Anfinsen,[51] we know that proteins can fold into their natural configurations without the help of chaperones or catalysts, although time scales will lengthen. Most small proteins (less than 100 amino acids) can even be denatured and re-natured by raising and lowering the temperature. Thus, the final configuration is determined by valleys in an energy landscape taken as a function of the multidimensional space of the two angles of

---

[50] The beta sheet structure was first properly described by Linus Pauling and Robert Cory in 1951.

[51] C.B. Anfinsen, E. Haber, M. Sela, F.H. White, *Kinetics of formation of native ribonuclease during oxidation of reduced polypeptide chain*. PNAS **47**, 1309–1314 (1961).

twist in each amino acid, with steric restrictions creating large forbidden zones in the landscape.[52]

The existence of a unique natural protein fold does NOT mean that the protein 'finds' the valley with the lowest energy in an optimization function, since the sequence of folding locks in subfolds before the final fold forms. An example of such subfolds are the alpha helicies and the beta sheets. Other more complex subfolds are common to many proteins. In any case, the final configuration must have some stability under small changes in temperature and chemical environment, and so the energy landscape must have a relatively strong local minimum at the natural fold. In addition, evolution would favor proteins which fold into useful structures even when a subclass of mutations cause variations. Evolution might even make selection of a certain variability in amino-acid sequencing not to protect against mutation, but to permit mutations to produce variability.

Understanding a protein tertiary structure helps us determine its biochemical action, its stability, the variability and pathology when DNA mutation of its code is present, its ancestry in the evolution of proteins, how variations in the amino acids transform its purpose, and how imitators, inhibitors, and enhancers can be constructed for pharmacological applications. Protein folding we recognize as a central topic in describing life on Earth.

As an introduction to the topic of protein folding, we examine if biological protein folding into natural forms occur spontaneously in the cell protoplasm, or if the process requires ancillary structures? An answer comes from a statistical analysis of the folding probabilities as a function of temperature in comparison with experiment.

As indicated, proteins smaller than about 100 amino acids 'spontaneously' fold and unfold according to the temperature. They do so in times on the order of microseconds. This rapidity indicates that the protein folds by a sequence of quick subfolds, creating local substructures first. However, the process is well represented by a two-state system, folded or unfolded, rather than a sequence of quasi-stable intermediates. In addition, the proportion of unfolded chains as a function of temperature has a sharp sigmoid shape, indicating that there are cooperative effects, i.e. once the folded protein starts to unfold, the unfolded segments increase the chance that the remaining folded chain will unfold. This effect can be explained by the increase in thermal motion of the released chain segments and their new exposure to water and other ions in the environment.

For natural protein structures, the particular identity of local sequence of amino acids is not as important as the conformation after global folding. Biological polypeptide chains have a unique fold. Nature has already picked an amino acid sequence that will fold into a single stable form. Sequences that fold into several distinctly different structures with about the same binding energy in the natural

[52]For an early study of the angle restrictions, see G.N. Ramachandran, C. Ramakrishnan, V. Sasisekharan, *Sterochemistry of the polypeptide chain configurations*, J Mol Bio **7**, 95–99 (1963).

environment are selected out in evolution, unless all the produced forms have a biological purpose. The final configuration has important flex joints, surface-charge distributions, binding loci, and crevasses and pockets with specific weak adhesion surfaces, all as needed for the proper functioning of the protein.[53]

Protein Conformational Transitions

As proteins are formed in natural environments, they first tend to form segments of coils and sheets, which then fold into final configurations. Zimm and Bragg[54] first showed how one can establish, by comparing theoretical models to experiment, answers to how protein tertiary structure forms from its secondary structure. Their work and the efforts that followed showed that proteins with residue lengths up to about 100 (1) can fold without the need for templates or auxiliary mechanisms, although the required time needed can be drastically different with templates and/or catalysts; (2) will make the folding transition with a high degree of 'cooperation', in that the process acts positively on itself, speeding the condensation of the whole protein chain once it starts; (3) pass rapidly through any intermediates; (4) may be reversible by environmental changes; and (5) lead to a final configuration that has a certain uniqueness, suggesting it has a minimum energy, at least among various accessible and nearby conformational states.

As a simple sample model, consider a protein chain consisting of N identical smallest segments that can either be twisted into a helical segment (h) or be left in an open configuration (o). (In a real protein, the smallest segment will have five amino acids, with the first forming a hydrogen bond with the fifth, making part of a helix.) A given chain can be characterized by the locations of the helical segments relative to the open stretches, such as 'oohohhhooo'. The hydrogen bonding will make the 'h' regions lower in binding energy than the 'o' regions. We will let $\Delta G = \Delta H - T\Delta S$ be the Gibbs free energy released by the transition of one 'o' to one 'h' region. The energy released in the form of heat will be ΔH and the entropy gained will be ΔS. For our case, as helical segments are being formed, ΔH will be negative since the formation of hydrogen bonding releases energy, and ΔS will be negative since the formation of a helical segment makes the protein chain more ordered. Thus, there will be a 'melting' temperature given by

$$T_m = |\Delta H/\Delta S|$$

[53]There are now many elaborate statistical and kinetic theories to describe protein chain folding. One goal is to have a computer program predict the final folding, given an initial amino acid sequence. Another is to have a program that will design a set of possible sequences for a given function. For an overview, see Ken Dill et al., *The Protein-Folding Problem, 50 Years On*, Science **338, 6110**, 1042–1046 (23 Nov 2012).

[54]B.H. Zimm, J.K. Bragg, *Theory of the Phase Transition between Helix and Random Coils in Polypetide Chains*, J Chem Phys **31**, 526–531 (1959).

above which the transition to an open chain configuration is favored, while below that temperature, the formation of helical segments will be favored.

Suppose there are k 'h' regions on the chain. The number of ways of distributing 'h' segments in a chain with N smallest segments (which can form a helical segment) will be $N(N-1)(N-2)\cdots(N-k+1)/k! = \binom{N}{k}$. Assuming the transitions are independent, we can express the partition function for the system with a distribution of possible 'h' segments as

$$Z_{indep} = \sum_{k=0}^{N} \binom{N}{k} e^{-k\beta\Delta G} = (1+x)^N . \tag{10.175}$$

where we let $x = e^{-\beta\Delta G}$. On the other hand, if the initiation of the transition from 'o' to 'h' triggers the whole chain to fold (from 'ooo...o' to 'hhh...h'), then none of the intermediate states taken in the partition function sum above will contribute (their probabilities are at or close to zero). This is called a 'fully cooperative' transition, for which

$$Z_{coop} = 1 + e^{-\beta N\Delta G} = 1 + x^N . \tag{10.176}$$

We can compare these two cases with the behavior of actual proteins. One of the measurable predictions is the average helical length in a chain, calculated by

$$f^H \equiv \sum_k k\, p(k) \tag{10.177}$$

where $p(k)$ is the probability of finding a chain with k helicies. For the 'independent' model, $p_{indep}(k) = \binom{N}{k} x^k/(1+x)^N$, while for the 'fully cooperative' model, $p_{coop}(0) = 1/(1+x^N)$, $p_{coop}(k) = 0$ for $k = 1, \cdots, N-1$ and $p_{coop}(N) = x^N/(1+x^N)$. Thus,

$$\begin{aligned} f^H_{indep} &= \frac{1}{Z_{indep}} \sum_k k \binom{N}{k} x^k \\ &= x \frac{\partial(\ln Z_{indep})}{\partial x} \\ &= N \frac{1}{1+x^{-1}} \end{aligned} \tag{10.178}$$

while

$$f^H_{coop} = N \frac{1}{1+x^{-N}} . \tag{10.179}$$

In terms of the melting temperature ($T_m = |\Delta H/\Delta S|$),

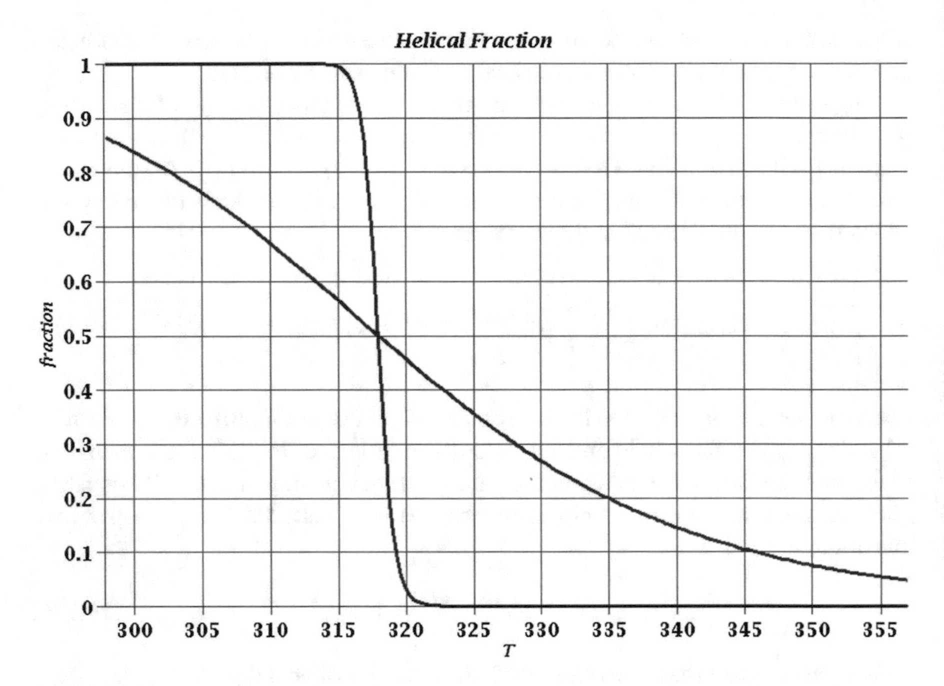

Fig. 10.4 Average helical fraction. The 'independent' folding model (slowly dropping line) compared to the fully 'cooperative' folding model (quickly dropping line). Here, we took $N = 20$ helical-turn segments (72 residues), a melting temperature of 318 K, i.e. body temperature plus 8 °C, and the ratio of the heat released per turn per $k_B T_m$ to be 27.5

$$
\begin{aligned}
x &= \exp\left(-\frac{\Delta H - T\Delta S}{k_B T}\right) \\
&= \exp\left(-\frac{Q}{k_B T_m}\frac{T - T_m}{T}\right)
\end{aligned}
\tag{10.180}
$$

where Q is the (positive) energy released on forming the hydrogen bonds in one helical segment.

Typical values are $T_m \sim 318$ K (8 °C above normal body temperature) and $Q \approx 0.754$ eV per helical segment as a full turn (that is 0.20936 eV for each hydrogen bond times 3.6 residues per turn).

We should expect little or no temperature dependence of ΔS, as it depends only on a count of the reduction in disorder as a helical segment is produced. For high temperatures, f^H approaches a small number, and for low temperatures, f^H approaches one.

The two cases, 'independent' and 'cooperative' helix formation, are shown in Fig. 10.4, plotting the average helical fraction defined by f^H/N as a function of temperature. The cooperative case shows a strong sigmoid shape characteristic of a sharp transition from all helical to all open as the temperature increases. This behavior is preferred biologically, so that structure is preserved over a range of temperatures below the melting point.

Observations of the helical fraction are possible using ultra-violet absorption spectra. Experimental fits favor a cooperative model with 'zipper' behavior, i.e. the helical transition occurs in one direction and completes once started.[55]

Protein Design

The protein folding problem has an associated inverse problem: What sequences of amino acids produce a reliable protein of given structure and function? Reliability for proteins has several aspects:

- Energetic stability: Secondary and tertiary structure that are energetically favorable over competing alternative structures;
- Mutation insensitivity: A variety of single nucleotide mutations produce proteins which retain functionality;
- Structural flexibility: Some alterations of amino acids in backbone and side chains are allowed;
- Environmental durability: Structure should maintain integrity under natural conditions and their expected variations.
- Reproduciblity: A shortened time needed to regenerate a protein adds to its viability;
- Lockout resistivity: Protected and alternate structures with alternate functional pathways ('pathway redundancy') adds resistivity to various biochemical or viral degradations;
- Intergrationality: As the structure of the protein affects its position in the network of functionalities in a cell, good de novo design must include ramifications on this network.

Nature has found solutions to the design problem through the trial and error that follows genetic code mutation. But with 20 amino acids, there are $20^{100} > 10^{126}$ possible proteins with just 100 amino acids in sequence. A human cell produces perhaps 100,000 of these. There is a vast landscape of other possible de novo proteins waiting exploration.

With computer programming, we are now near to having some answers to protein design problems with computer time scales much smaller than evolution has taken![56]

The designed proteins will have good application as

[55]When comparing to experiment, the enthalpy change is usually taken as a fitting parameter, since, in vivo, the change in enthalpy due to the desorption of water is difficult to calculate.

[56]A functioning cell took a billion years of evolution. Future quantum computers together with some organizing principles may make the analysis of even long amino acid chains tractable in hours.

- Small active proteins for pharmaceutical medicines;
- Injectable proteins to replace or supplement natural proteins, such as hormones;
- Injectable designed antibodies, such as against cancer cells;
- Synthetic membranes, both passive and active, for selective diffusion, such as for water purification;
- Proteins that aid medical diagnosis, such as ones that fluoresce on specific attachment;
- Proteins that react to the presence of very small quantities of a gas or particles in air;
- Catalysts in the production of useful biochemical substances.

Fluctuations in the Particle Number

Systems in a Gibbs ensemble at or near equilibrium that are described by the grand canonical ensemble have a fixed temperature and volume. But the number of particles in any one system can vary, since particles can diffuse through permeable walls between the systems. We now can ask what fluctuations in particle numbers can we expect in one of these systems when it is at or near equilibrium? To answer, we will calculate the root-mean-square of the N, called ΔN. We have

$$(\Delta N)^2 = \left\langle (N - \langle N \rangle)^2 \right\rangle = \left\langle N^2 \right\rangle - \langle N \rangle^2 \tag{10.181}$$

$$\begin{aligned} &= \frac{1}{\Xi}\frac{\partial^2 \Xi}{\partial \gamma^2} - \left(\frac{1}{\Xi}\frac{\partial \Xi}{\partial \gamma}\right)^2 \\ &= \frac{\partial}{\partial \gamma}\left(\frac{1}{\Xi}\frac{\partial \Xi}{\partial \gamma}\right) = \left.\frac{\partial \overline{N}}{\partial \gamma}\right|_T . \end{aligned} \tag{10.182}$$

To express the result in terms of more commonly measured thermodynamic variables, we first notice that changes of $\overline{N}$ cause changes of the particle density, $\overline{\rho} \equiv \overline{N}/V$. In these terms (and using $\mu = k_B T\, \gamma$),

$$\left.\frac{\partial \overline{N}}{\partial \gamma}\right|_T = k_B T V \left.\frac{\partial \overline{\rho}}{\partial \mu}\right|_T .$$

The Gibbs–Duhem relation Eq. (9.82) connects changes of μ with changes of pressure and temperature:

$$\overline{N} d\mu = V\, dp - \overline{S}\, dT .$$

In the present case, the temperature is constant, so $d\mu = (1/\overline{\rho})dp$, which makes

$$\frac{\overline{(\Delta N)^2}}{\overline{N}^2} = \frac{1}{\overline{N}^2}\frac{\partial \overline{N}}{\partial \gamma}\bigg|_T = \frac{k_B T}{V}\frac{1}{\overline{\rho}}\frac{\partial \overline{\rho}}{\partial p}\bigg|_T = \frac{k_B T}{\overline{N}}\overline{\rho}\,\kappa_T\,, \tag{10.183}$$

where $\kappa_T = \partial(\ln \overline{\rho})/\partial p\,|_T$ is the isothermal compressibility. Systems with a finite compressibility[57] will have increasingly small relative-number fluctuations as the particle numbers become large. Near equilibrium, the number of particles in a system, N_k, will not deviate significantly from the average $\overline{N}$ over all such systems in the Gibbs ensemble.

If the fluctuations in the number of particles in each system of the ensemble are small, then we can approximate the grand canonical partition function by replacing N in the sum that appears in Ξ by the average $\overline{N}$. The activity $\exp(\gamma \overline{N})$ will then factor out of the sum, making

$$pV = k_B T \ln \Xi \tag{10.184}$$

$$\approx \beta\mu\overline{N} + k_B T \ln Z(T, V, \overline{N}) \tag{10.185}$$

$$= \mu\overline{N} - \overline{A} \tag{10.186}$$

$$= \mu\overline{N} - \overline{E} + T\overline{S}\,, \tag{10.187}$$

as we should expect. Again, neglecting small fluctuations when the particle number is large, the thermodynamics of the system will be determined by the canonical partition function.

10.8.5 Partition Function for Multicomponent System

Suppose each member of a Gibbs ensemble contains $\mathscr{M}$ different types of particles, with some capable of transforming into others, perhaps by chemical reactions. The systems will be allowed to exchange heat and particles with other members of the ensemble.

The statistical analysis for the ensemble as it approaches equilibrium follows that for one-component systems, as given in Sect. 10.8.4, but now, if each type of particle is preserved (e.g. not chemically transformed), there are $\mathscr{M}$ particle number constraints. The grand canonical partition function becomes

$$\Xi(\beta, V, \mu_1, \mu_2, \cdots \mu_{\mathscr{M}})$$

[57] The compressibility may diverge for systems going through a phase transition.

$$= \sum_{N_1,N_2,\cdots N_{\mathcal{M}}} \sum_{i=0}^{\infty} \exp\left(\beta\left(\sum_{J=1}^{\mathcal{M}} \mu_J N_J - E_{N_1,N_2,\cdots N_{\mathcal{M}},i}\right)\right), \tag{10.188}$$

where the i-sum is over the quantum states for the ensemble member with a given number of particles of each type.

The partial derivatives of $\ln \Xi$ with respect to its natural variables $(T, V, \{\mu_J/T\})$ give the average energy, the pressure, and the average number of particles of type J in the system, according to[58]

$$\frac{\overline{E}}{kT^2} = \left.\frac{\partial \ln \Xi}{\partial T}\right|_{V,\{\mu_J/T\}} \tag{10.189}$$

$$\frac{p}{kT} = \left.\frac{\partial \ln \Xi}{\partial V}\right|_{T,\{\mu_J/T\}} \tag{10.190}$$

$$\frac{\overline{N}_J}{kT} = \left.\frac{\partial \ln \Xi}{\partial \mu_J}\right|_{T,V,\{\mu_{K\neq J}/T\}} . \tag{10.191}$$

As before,

$$pV = kT \ln \Xi . \tag{10.192}$$

This also follows from Eq. (10.190) and from the fact that $\ln \Xi(T, V, \{\mu_J/T\})$ is an extensive variable of degree one. so that

$$\frac{\partial \ln \Xi}{\partial V} V = \ln \Xi . \tag{10.193}$$

10.9 Statistics Applied to Biochemical Reactions

Dynamical processes occurring in life systems often happen slow enough that thermodynamic descriptions are appropriate. Biochemical reactions within cells can be described using the ideas behind statistical mechanics, particularly since relative temperature and pressure gradients are not large, and since many of the reactions are duplicated either in the cell fluid, on its walls, or within special organelles. Statistics can be applied to the ensemble of independently acting duplicates, even though individual nanosystems involve only a few molecules at a time.

[58]The partial differentiation of $\ln \Xi$ with respect to T and with respect to V which appears in the calculation of $\overline{E}$ and p is carried out holding the variable μ_J/T constant, i.e. the dependence of the μ_J/T on temperature and volume is inserted *after* differentiation.

10.9.1 Molecular Numbers

Suppose there are $\mathscr{A}$ kinds of indistinguishable atoms within a macrosystem, making up $\mathscr{M}$ species of molecules. Each of the molecules of type J can exist in certain energy states ϵ_{Ji} labeled by the index 'i'. Let the number of molecules of type J in the i-th possible energy state of these molecules be N_{Ji}, with $N_J = \sum_i N_{Ji}$, and the total number of atoms of kind k be n_k. If molecule J has n_{Jk} bound atoms of kind k, then the total of these kinds of atoms reads

$$\sum_{J=1}^{\mathscr{M}} \sum_i n_{Jk} N_{Ji} = n_k \,, \tag{10.194}$$

which we intend to hold fixed. There will be $\mathscr{A}$ of these atom-conservation conditions. In addition, if the molecules are involved in a chemical reaction, we have the stoichiometric balancing condition for each kind of atom:

$$\sum_J \nu_J n_{Jk} = 0 \,. \,, \tag{10.195}$$

a generalization of Eq. (9.210).

10.9.2 The Case of Molecules Acting Independently

We say that molecules in a system act independently if the partition function for the system can be factorized into a product of partition functions for each species of molecule.

Near equilibrium, the entropy of such a mixture of independent reactants and products is

$$\begin{aligned}
S &= k_B \log W \\
&= k_B \sum_J \left(\sum_i N_{Ji} \log\left(g_{Ji}/N_{Ji}\right) + N_J \right) \\
&= k_B \sum_J \left(\sum_{Ni} N_{Ni} \left(-\sum_k n_{Jk}\alpha_k + \beta\epsilon_{Ji}\right) + N_J \right) \\
&= -k_B \sum_J N_J \sum_k n_{Jk}\alpha_k + k_B\beta \sum_J \sum_i N_{Ni}\epsilon_{Ji} + k_B N \\
&= k_B \sum_J N_J \ln\left(eZ_J/N_J\right) + k_B\beta E \,.
\end{aligned} \tag{10.196}$$

This gives a simple form for the Helmholtz free energy for the mixture:

$$A = E - TS = -k_B T \sum_J N_J \ln (eZ_J/N_J) . \tag{10.197}$$

The chemical potentials and the Gibbs free energy follow:

$$\mu_J = \left. \frac{\partial A}{\partial N_J} \right|_{T,V} = -k_B T \ln (Z_J/N_J) . \tag{10.198}$$

$$G = \sum_J \mu_J N_J = -k_B T \sum_J N_J \ln (Z_J/N_J) \tag{10.199}$$

so

$$e^{-\mu_J/k_B T} = Z_J/N_J . \tag{10.200}$$

Note the product

$$\prod_J (Z_J/N_J)^{\nu_J} = \exp\left(-\sum_J \nu_J \mu_J / k_B T \right) = 1 , \tag{10.201}$$

at equilibrium, in which we have used the Gibbs equilibrium condition Eq. (9.216):

$$\sum_J \nu_J \mu_J = 0 . \tag{10.202}$$

The product of the partition functions per unit molecular particle number to powers of the molecular mole numbers gives

$$\prod_J (Z_J/N_J)^{(N_J/\mathcal{N}_A)} = e^{-G/RT} , \tag{10.203}$$

in which Avogadro's number $\mathcal{N}_A$ has been included on each side of the relation to make the powers that appear more manageable.

During a reaction at constant temperature and pressure, changes in Gibbs free energy near equilibrium will be

$$\delta G = \sum_J \mu_J \nu_J \delta\lambda \tag{10.204}$$

$$\approx -k_B T \sum_J \ln \left(\frac{Z_J}{N_J} \right)^{\nu_J} \delta\lambda \tag{10.205}$$

$$= -k_B T \ln \left(\prod_J \left(\frac{Z_J}{N_J} \right)^{\nu_J} \right) \delta\lambda \tag{10.206}$$

which become zero at equilibrium (consistent with Eq. (10.201)).

If the molecules of one species do not affect the internal energy states of another species, then each of the molecules of species J in the system will have its own set of energy levels ϵ_{Ji} over which the molecules of that species can distribute. The number of ways to make this distribution for one set of molecules will be independent of that for any other species, making the total number of ways for all the molecules $W = \prod_{J=1}^{\mathscr{M}} W_J$. To find the equilibrium distributions for the N_{Ji}, we maximize $\ln W$ under the $\mathscr{A}$ constraints given by Eqs. (10.194) plus the constraint of overall energy conservation. The result is

$$N_{Ji} = g_{Ji} \exp \left(\sum_k n_{Jk} \alpha_k \right) \exp\left(-\beta \epsilon_{Ji}\right). \tag{10.207}$$

The so-called Lagrange multipliers α_k are determined by the particle-number constraints Eq. (10.194), through the expression

$$n_k = \sum_J n_{Jk} Z_J \prod_l \left(e^{\alpha_l}\right)^{n_{Jl}}. \tag{10.208}$$

wherein we have defined the partition function for molecules of species J as

$$Z_J \equiv \sum_i g_{Ji} \exp\left(-\beta \epsilon_{Ji}\right). \tag{10.209}$$

Summing the N_{Ji} just over the energy states for molecules of type J gives

$$N_J = \prod_k \left(e^{\alpha_k}\right)^{n_{Jk}} Z_J\,, \tag{10.210}$$

We can use the stoichiometric balance condition Eq, (10.195) to eliminate the Lagrange multipliers α_k. Raise the N_J to the power ν_J and multiply them. Then the product of the exponentials in α_k will be one, as follows:

$$\prod_J (N_J)^{\nu_J} = \prod_k \left(e^{\alpha_k}\right)^{\sum_J \nu_J n_{kJ}} \prod_J (Z_J)^{\nu_J} \tag{10.211}$$

$$= \prod_J (Z_J)^{\nu_J}. \tag{10.212}$$

This is the basis of 'the law of mass action' in chemical reactions. It determines the equilibrium concentration of reactants relative to products. More conventionally, the left-hand side is rewritten in terms of the concentrations of the molecules. The right-hand side is then called the equilibrium constant or the chemical reaction constant K. For example,

$$\prod_J (N_J/V)^{\nu_J} = \prod_J (Z_J/V)^{\nu_J} \tag{10.213}$$

$$= K(T, V)\,. \tag{10.214}$$

For molecules having kinetic energy together with internal degrees of freedom, such as available vibration, rotation, or electron excitation states, the molecular partition function can be factored because the molecular energies are a sum of kinetic and potential terms:

$$Z_J \equiv \sum_i g_{Ji} \exp\left(-\beta(\left[p_{iJ}^2/(2m_J) + V_{Ji}\right]\right) = Z_{Jo} Z_{J\,\mathrm{int}}, \tag{10.215}$$

where the kinetic energy factor Z_{Jo} is the same as the partition function for an ideal gas of the molecule J:

$$Z_{Jo} = \left(\frac{2\pi m_J}{h^2}\right)(k_B T)^{3/2} V\,, \tag{10.216}$$

giving

$$Z_J/V = \left(\frac{2\pi m_J}{h^2}\right)(k_B T)^{3/2} Z_{J\,\mathrm{int}}\,. \tag{10.217}$$

If all the molecules taking part in a chemical reaction are in a gaseous phase, with no long-range forces or internal degrees of freedom, then $Z_{J\mathrm{int}} = 1$, and the equilibrium constant K is a function of temperature alone.

The equilibrium constant K is often re-expressed in terms of the molecular partition functions $\overline{Z}_J$ with the zero of energy taken for the configuration in which all the atoms making up the molecule are separated and unbound from each other, with no kinetic energy. If the atoms in the molecule M_J are bound by ϵ_{JB}, then we take $\epsilon_{Ji} = \overline{\epsilon}_{Ji} + \epsilon_{JB}$. This makes

$$K(T, V) = \exp\left(-\Delta E/k_B T\right) \prod_J \left(\overline{Z}_J/V\right)^{\nu_J} \tag{10.218}$$

where $\Delta E = \sum_J \nu_J \epsilon_{JB}$.

Alternatively, in terms of concentrations,

$$\prod_i [C_i]^{\nu_i} = e^{-\Delta G^o/RT} \prod_i \left(\frac{Z_i^o}{N_i}\right)^{\nu_i} \equiv K_c(T, V) \tag{10.219}$$

where the Z_i^o/N_i are calculated at a standard state for which the concentration of type labeled by 'i' is well-defined. Remember that the stoichiometric coefficients ν_i give the number of molecules of reactants (taken here with negative values) or products (taken with positive values). So

$$\frac{\prod_{products} [C_i]^{\nu_i}}{\prod_{reactants} [C_i]^{-\nu_i}} = K_c(T, V) . \tag{10.220}$$

The reaction constant, K_c, is strongly temperature dependent. If K_c is large (e.g. larger than 1000), then most of the reactants are converted to products. If K_c is less than about 0.001, then little of the reactants make products.

For dilute solutions of reactants and products, with molecules having no long-range interactions, reactions act like those in a gas. The partition function for the system factorizes into those for each type of particle, making the Gibbs free energy additive for each type, and we can apply the law of mass action Eq. (10.220) to determine a relationship between equilibrium concentrations.

10.9.3 The Case of Molecules Not Acting Independently

As we have shown in this Chapter, the statistical mechanics formalism contains the complete thermodynamic description of macroscopic systems at or near equilibrium. All thermodynamic quantities can be calculated from the partition function Z for the system, which in turn requires knowledge of the energies available to the particles in the system, including the energy's dependence on particle momenta and on interparticle distances. In order to establish conditions for equilibrium for reactions in solutions at constant temperature and pressure, we would want to calculate the Gibbs free energy

$$G = A + pV = -k_B T \ln Z + k_B T V \left. \frac{\partial (\ln Z)}{\partial V} \right|_{T,N's}$$

before and after the reaction using the known partition function. This is a formidable task if the particles have important long-range interactions.

Many biological reactions take place in ionic conditions in water solution within cells. Since charges have long-range interactions, one cannot assume that ions act independently. Given the long-range of the Coulomb force, the presence of ions will have an effect on biochemical reactions.

Mitigating these factors in cell solutions is the partial shielding effect of the 'cloud' of oppositely charged ions and the polar nature of water molecules, which tend to accumulate around ions (see Fig. 10.5). This is called a 'screening' of charges. The electric field around a screened charge drops with distance r from the charge as an exponential $\exp(-r/\lambda)/r^2$ instead of the bare Coulombic $1/r^2$

dependence. In addition, higher temperatures will tend to reduce the clustering tendencies of oppositely charged ions and of polar molecules because of thermal diffusion.

Among the complications for biochemical reactions involving ions are:

- Ions of opposite charge will tend to accumulate and also surround other larger charged molecules;
- Orientation of nearby water molecules near charged ions and molecules due to the strong electric dipole moments of water molecules;
- Accumulation of water molecules near small ions because of an induced attractive force on water molecules arising from the strong Coulombic gradients acting on the water molecule dipoles;
- A tendency for the first layer of water molecules accumulated around a small ion to 'crystallize' into quasi-ordered arrays;
- The electric permittivity of water (about eighty times that of a vacuum because of the polar nature of the water molecule) will vary with distance from an ion, dropping nearby because the water molecules are already quasi-oriented and being so are restricted in their reaction to additional external electric fields;
- Changes in the electric dipole moment of a water molecule in nearby region to small ions, i.e. distortion of the position of the hydrogens relative to the oxygen;
- At high concentrations of ions, dynamical clusters of oppositely charged ions may form;
- In an external electric field, ions dissolved in water will move, but with a dynamical 'cloud' of accumulated ions and nearby water, increasing its effective size and mass;
- When a small ion is dragged, its 'cloud' of water and ions changes from quasi-spherical to quasi-elliptical.

Even with these complications, since the 1920s there have been developed a number of strategies for finding equilibrium conditions for the concentration of the molecules involved and reaction rates.

10.9.4 Modeling Reactions Involving Ions in Solution

As with all reactions at constant temperature and pressure, the Gibbs free energy before and after the reaction can be used to describe reaction rates and, in principle, determine equilibrium conditions.

In the case of the formation of ions in solution, charge conservation requires that the reaction occur holding the total charge zero if it started that way. The fact that the electric force is large between ions and electrons compared to the typical intermolecular forces between molecules in a solution, the overall total charge in

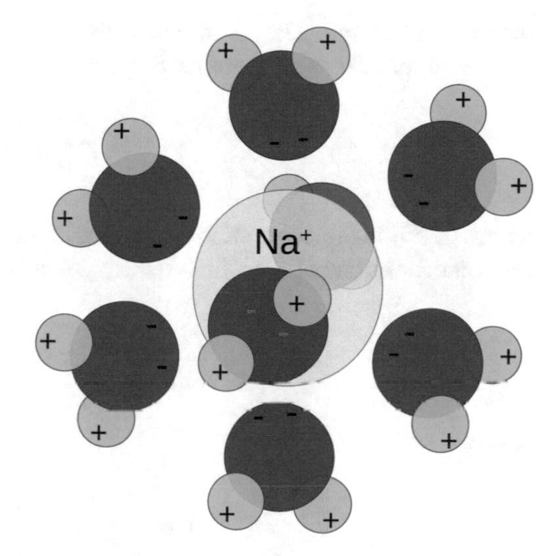

Fig. 10.5 Sodium ion in water

volumes large compared to molecular sizes will be very close to zero. This is the condition of 'net neutrality'[59]:

$$\int_{\delta V} \rho_q \, dV \approx 0 \, ,$$

where ρ_q is the total charge density and dV is an element of volume in the fluid. The sum (integral) is taken over volumes δV containing at least several hundred molecules.

Because of long-range interactions, equilibrium for reactions involving ions will deviate from the law of mass action expressed by Eq. (9.224) even for dilute solutions. As a way to show what is involved, we will describe a simple model developed in the early 1920s for how to handle such reactions. The model actually gives reasonable predictions for sufficiently dilute solutions.

10.9.5 The Debye–Hückel Model

In 1923, Debye and Hückel published an analysis of ionic reactions by looking at how the internal energy of the system depended on the Coulomb force between the ions and background charged particles. They modeled the effect by 'screening' ions with a cloud of surrounding opposite charges. At a finite temperature, thermal motion makes this cloud have a Boltzmann distribution next to the ion, rather

[59] A phrase introduced in Sect. 9.20.1. It was also prominent in public discourse after 2003–2017, but in a completely different context!

than a solid clump. They found how the activity of each ion can be related to its concentration. The activity of participant i is defined by

$$a_i \equiv \gamma_i \frac{[C_i]}{[C_i^{\ominus}]},$$

where $[C_i^{\ominus}]$ is the concentration at some standard state (such as at 1 mmol/L). For low concentrations of ions, their theory predicts the value of γ_i to be

$$\gamma_i = \exp\left(-\frac{kq_i^2/\lambda_D}{2k_bT}\right) = \exp\left(-Az_i^2\sqrt{I}\right), \tag{10.221}$$

where

$$A \equiv \frac{z_i^2\,|e|^3\,\mathcal{N}_A}{2^{5/2}\pi(\epsilon k_B T)^{3/2}};$$

$$I = \frac{1}{2}\sum_i z_i^2 \frac{[C_i]}{[1\,M]};$$

ϵ is the electric permittivity of the solution; $k = 1/(4\pi\epsilon)$; each $q_i = |e|\,z_i$ is the charge on the i-th ion, with z_i being the charge number for that ion (including its sign); and λ_D is the so-called Debye length, which gives the screening distance for the surrounding cloud around an ion. Net neutrality in the solution imposes $\sum_i z_i = 0$.

From Eq. (10.221), one can see that $\gamma_i \leq 1$,[60] which means the cloud reduces the effect of that ion's concentration in reaction equilibrium conditions. As the temperature in the solution rises, the Debye screening length increases in proportion to the square-root of the temperature, and $\gamma_i \to 1$, so that the activities approach the concentrations per unit molar.

The Debye–Hückel model works reasonably well when compared to experiment for concentrations $[C_i]$ less than about one millimolar.

10.10 Fluctuations of Thermal Quantities à la Einstein

10.10.1 Fluctuation Averages

Entropy (disorder) can be calculated even for systems far from equilibrium, using Eq. (10.21), i.e. Boltzmann's famous $S = k_B \ln W$, where W is the number of

[60]For sufficiently high concentration in more sophisticated models, γ_i can be bigger than one.

ways of distributing the energy in the system among the possible energy states. When a system is far from equilibrium, the calculation of S is more difficult, particularly for systems which change rapidly in time, such as one which explodes. But as we have seen, systems such as tissue and fluids in live biological cells, in which small subsystems are not far from equilibrium, can be tractable. The small subsystems must be large enough that fluctuations in thermodynamic variables are small compared to their values.

In investigating thermodynamic fluctuations, Einstein recognized that Boltzmann's formula could be put in a form appropriate for calculating the average of fluctuations in the thermodynamic variables, provided the system is not too far from equilibrium.[61]

Einstein expressed Boltzmann's relation as

$$W = \exp\left(S(Y_1, Y_2, \cdots, Y_n)/k_B\right) , \tag{10.222}$$

where the set $\{Y_1, Y_2, \cdots, Y_n\}$ represent the full set of n extrinsic thermodynamic variables, such as local values of internal energy, volume, and number of various particle species. The W, which we identified as the number of ways particles can be distributed in given energy states, is also proportional to the probability that the system would have the given entropy S for that particle distribution.[62]

Each of the thermodynamic variables Y_i could depend on position in the system and on the time of measurement. If so, the regions around a given position $\mathbf{r}$ will be considered a thermodynamic subsystem when it contains enough particles and energy states to make fluctuations in the thermodynamic variables in the given region small compared to their values after a characteristic relaxation time for the system, i.e. the time for the system to reach equilibrium or a steady state if there were no external influences.

Note that, as we observed with Eq. (9.76), because the entropy is an extensive quantity, we have

$$S(\lambda Y) = \lambda S(Y) \tag{10.223}$$

so that the Euler relation

$$S = \sum_i \frac{\partial S}{\partial Y_i} Y_i \tag{10.224}$$

---

[61] A. Einstein, *Über die Gültigkeitsgrenze des Satzes vom thermodynamischen Gleichgewicht und über die Möglichkeit einer neuen Bestimmung der Elementarquanta*, Annalen der Physik **22** 569–572 (1907). These days, fluctuations in thermodynamic variables are commonly calculated using a partition function, but Einstein's method is an elegant and simple alternative.

[62] Einstein remarked that if he had seen Willard Gibbs' work on statistical mechanics, he would have not written his 1902–1905 papers on the subject. Our discussion here is based on Einstein's development, as it reflects Einstein's intuition, simplicity and his ability to focus on the essence of a problem.

applies.

The quantities

$$y_i \equiv -\frac{1}{k_B}\frac{\partial S}{\partial Y_i} \tag{10.225}$$

are intensive thermodynamic variables, usually simply related to the temperature, pressure, and chemical potential.[63] The y_i are the conjugate thermodynamic variables to the Y_i, introduced in Eq. (9.240). As intensive variables, each will satisfy

$$0 = \sum_k \frac{\partial y_i}{\partial Y_k} Y_k = -\frac{1}{k_B}\sum_k \frac{\partial^2 S}{\partial Y_i \partial Y_k} Y_k\,. \tag{10.226}$$

Except at discontinuities, such as the regions of phase transitions, we expect

$$\frac{\partial^2 S}{\partial Y_i \partial Y_j} = \frac{\partial^2 S}{\partial Y_j \partial Y_i} \tag{10.227}$$

which implies that

$$\frac{\partial y_i}{\partial Y_j} = \frac{\partial y_j}{\partial Y_i}\,. \tag{10.228}$$

These are generalized Maxwell relations (see Sect. 9.16).

To investigate the size of fluctuations in thermodynamic variables, Einstein expanded the entropy appearing in Eq. (10.222) into a Taylor series around its equilibrium value. Since S reaches a maximum (S_e) at equilibrium, the first order derivatives vanish, giving

$$S = S_e + \frac{1}{2}\frac{\partial^2 S}{\partial Y_i \partial Y_j}\delta Y_i \delta Y_j + \cdots\,, \tag{10.229}$$

with $\delta Y_i \equiv Y_i - Y_{ei}$, and we use the convention to sum over repeated indices, unless otherwise stated. The average value of the thermodynamic variable Y_i near equilibrium defines its equilibrium value: $Y_{ei} = \langle Y_i \rangle$. Since S reaches a peak at S_e, the second derivative terms must all add to a negative number. We simplify notation in what follows by defining

$$\beta_{ij} \equiv -\frac{1}{k_B}\frac{\partial^2 S}{\partial Y_i \partial Y_j}\,. \tag{10.230}$$

[63] For simple systems, $dS = (1/T)dE + (p/T)dV - \sum_k(\mu_k/T)dN_k$, with $\sum_l N_l = N$.

These second derivatives are evaluated at equilibrium. Close to equilibrium, the higher order terms in the Taylor expansion beyond the quadratic will be small. As they enter as a small term in an exponential factor, their effect near equilibrium is dramatically reduced compared to the quadratic term.

Also, close to equilibrium, the conjugate variables y_i will be linearly related to the δY_j: From Eq. (10.229),[64]

$$y_i = -\frac{1}{k_B}\frac{\partial S_e}{\partial Y_i} + \frac{1}{k_B}\frac{\partial S}{\partial(\delta Y_i)} = y_{ei} + \beta_{ij}\delta Y_j \tag{10.231}$$

so that fluctuations in Y_i are related to fluctuations in its conjugate by

$$\delta y_i = \beta_{ij}\delta Y_j \ . \tag{10.232}$$

With Einstein's insight, averages of fluctuations for any thermodynamic variable can be simply calculated. For example, the mean square of the fluctuations in Y_i, $(\delta Y_i)^2$, will have the expression

$$\left\langle(\delta Y_k)^2\right\rangle = A\int \exp\left(-\frac{1}{2}\beta_{ij}\delta Y_i\delta Y_j\right)(\delta Y_k)^2 dY_1 dY_2\cdots\ , \tag{10.233}$$

with A a normalization factor.

The integrand is significant only for small δY_i, so the range of the integral can be extended to $\{-\infty, \infty\}$, making Gaussian integrals. These can be simplified by diagonalizing the matrix[65] with a linear transformation of the integration variables. Let

$$Y_i = \langle Y_i\rangle + a_{ij}Z_i \tag{10.234}$$

making

$$\widehat{\beta_{kl}} = a_{ki}\beta_{ij}a_{lj} \equiv \delta_{kl}. \tag{10.235}$$

The Jacobian of this transformation is det(a). We see from Eq. (10.235) in matrix form: $a\beta a^T = I$ that the Jacobian satisfies

$$\det(a) = (\det(\beta))^{-1/2}. \tag{10.236}$$

[64]Care must be taken to distinguish the values of the entropy as a function of the extensive variables in quasi-equilibrium states and the entropy in its dependence on fluctuations in these same variables. The fluctuations are not restricted to be on the thermodynamic surface $S = S(U, V, N, \cdots)$.

[65]Any real symmetric matrix B can be diagonalized by a specific similarity transformation A. $ABA^{-1} = \widehat{B} = \mathrm{diag}(b_1, b_2, \cdots)$. In our case, $AA^T = I$: If this transformation is followed by a scale transformation $S = \mathrm{diag}(b_1^{-1/2}, b_2^{-1/2}, \cdots)$, the result is a unit matrix: $SABA^TS^T = I$.

Since the transformation coefficients a_{ij} are constants (independent of the Y_i) the Jacobian can be factored out of the integral.

After the diagonalizing, we will have

$$\left\langle (\delta Y_k)^2 \right\rangle = A \det(a) \int \exp\left(-\frac{1}{2}\sum_k Z_k^2\right)\left(\sum_l a_{kl} Z_l\right)^2 dZ_1 dZ_2 \cdots \quad (10.237)$$

$$= A \det(a) \sum_l a_{kl}^2 \int \exp\left(-\frac{1}{2}\sum_k Z_k^2\right) Z_l^2 dZ_1 dZ_2 \cdots \quad (10.238)$$

$$= (2\pi)^{n/2} A \det(a) \sum_l a_{kl}^2 \,. \quad (10.239)$$

In the second line, we have dropped all the integrand terms of the form $a_{kl} a_{km} Z_l Z_m$ for $m \neq l$, since these integrate to zero. In the third line, we have utilized

$$\int_{-\infty}^{\infty} \exp(-cx^2) dx = \sqrt{(\pi/c)} \quad (10.240)$$

and

$$\int_{-\infty}^{\infty} \exp(-cx^2) x^2 dx = (1/2)\sqrt{(\pi/c^3)} \,. \quad (10.241)$$

The constant A comes from making the probability of any set of values for the thermodynamic variables equal to one: $\langle 1 \rangle = 1$, i.e.

$$A^{-1} = \det a \int \exp\left(-\frac{1}{2}\sum_k Z_k^2\right) dZ_1 dZ_2 \cdots \,, \quad (10.242)$$

giving

$$A^{-1} = (2\pi)^{n/2} \det a \quad (10.243)$$

so that

$$\left\langle (\delta Y_k)^2 \right\rangle = \sum_l a_{kl}^2 \,. \quad (10.244)$$

To get our feet wet in this formalism, let's see what the results look like for the fluctuation of the internal energy in an ideal gas held in a chamber with a piston, for which

$$S = \frac{1}{T} E + \frac{p}{T} V \quad (10.245)$$

$$dS = \frac{1}{T}dE + \frac{p}{T}dV . \tag{10.246}$$

The coefficients of the differentials should be considered functions of the extrinsic variables E and V. The usual form of the equations of state appear as

$$E = Nc_V T \tag{10.247}$$

$$pV = Nk_B T \tag{10.248}$$

so

$$\frac{\partial S}{\partial E} = \frac{1}{T} = \frac{Nc_V}{E} \tag{10.249}$$

$$\frac{\partial S}{\partial V} = \frac{p}{T} = \frac{Nk_B}{V} \tag{10.250}$$

and

$$\frac{\partial^2 S}{\partial E^2} = -\frac{Nc_V}{E^2} \tag{10.251}$$

$$\frac{\partial^2 S}{\partial E \partial V} = 0 \tag{10.252}$$

$$\frac{\partial^2 S}{\partial V^2} = -\frac{Nk_B}{V^2} . \tag{10.253}$$

From these we have

$$\beta = \begin{pmatrix} \frac{Nc_V}{k_B} E^{-2} & 0 \\ 0 & NV^{-2} \end{pmatrix} . \tag{10.254}$$

It follows from Eq. (10.235) that the simplest transformation of the variables to bring β to a unit matrix is achieved by

$$a = \begin{pmatrix} \sqrt{\frac{k_B}{Nc_V}} E & 0 \\ 0 & \sqrt{\frac{1}{N}} V \end{pmatrix} . \tag{10.255}$$

The root-mean-square fluctuations in E will be the square root of

$$\left\langle (\delta E)^2 \right\rangle = \sum_l a_{1l}^2 \tag{10.256}$$

$$= \frac{k_B}{Nc_V} E^2 . \tag{10.257}$$

We see that the root-mean-square fluctuations in the internal energy per particle decreases with N as the inverse square root of N.[66] The behavior $\Delta E \propto E/\sqrt{N}$ is typical of RMS fluctuations in a large system of particles near equilibrium, except when collective effects become important, such as during condensation of a gas or the freezing of a liquid.

10.10.2 Correlations Between Fluctuations at One Time

We should expect that in a system near thermodynamic equilibrium, fluctuations in temperature will be associated with fluctuations in pressure. The two fluctuations are correlated. Statistically, if the fluctuations in the variables have no connection, then the average of their product should vanish. If not, we say the variables are correlated.

Consider the average of the product of the fluctuations in the variable Y_i and fluctuations in one of the conjugate variables, $y_i = (1/k_B)\partial S/\partial Y_i$. Using Eq. (10.232),

$$\left\langle \delta y_i \delta Y_j \right\rangle = A \int \exp\left(-\frac{1}{2}\beta_{kl}\delta Y_k \delta Y_l\right)\beta_{im}\delta Y_m \delta Y_j dY_1 dY_2 \cdots \tag{10.258}$$

$$= \delta_{ij} \; . \tag{10.259}$$

(Reminder: $\{\delta_{ij}\}$ is the identity matrix.) So the fluctuations in a thermodynamic variable and the fluctuations in a different conjugate variables are not correlated, while fluctuations in a thermodynamic variable and fluctuations in its own conjugate variable are perfectly correlated.

Using Eq. (10.232), we can also give

$$\left\langle \delta Y_i \delta Y_j \right\rangle = \beta_{ij}^{-1} \tag{10.260}$$

and

$$\left\langle \delta y_i \delta y_j \right\rangle = \beta_{ij} \tag{10.261}$$

which show that fluctuation in different densities can be correlated, and fluctuations in the gradients of various intensive quantities can be correlated.

[66]Note that for an ideal gas (with $E = Nc_vT$), $\left\langle (\delta E)^2 \right\rangle = Nk_Bc_vT^2$, in agreement with our result using the canonical partition function, Eq. (10.104).

10.10.3 Correlations Between Fluctuations at Two Times

Fluctuations extend over time, so that if we know that at one time the fluctuation in some thermodynamic variable, such as temperature, in some small subsystem, were above the average, then if we pick a time short enough later, that thermodynamic variable should still be high. After a relaxation time, this kind of correlation in time should die out. These ideas are supported by the following.

The correlation of the fluctuations of the variable Y_i at two different times is measured by

$$\phi_{ij}(t'-t) \equiv \langle \delta Y_i(t')\delta Y_j(t)\rangle \tag{10.262}$$

where we have assumed that a shift of both times t and t' by the same amount will not change the correlation. Note that $\phi_{ij}(\tau) = \phi_{ji}(-\tau)$.

If the microscopic laws governing the time evolution of the Y_i are invariant under time reversal, i.e. $t \rightarrow -t$, then $\phi_{ij}(\tau) = +\phi_{ij}(-\tau)$, i.e. the $\phi_{ij}(\tau)$ are even functions of τ. We can write this symmetry as the condition

$$\langle \delta Y_i(0)\delta Y_j(t)\rangle = \langle \delta Y_j(0)\delta Y_i(t)\rangle \tag{10.263}$$

For small excursions from equilibrium, we expect that the time changes of $\delta Y_i(t)$ to be expandable in terms of the small deviations in the Y_k and in terms of small deviations in the conjugate variables $\delta y_i(t)$:

$$\frac{d}{dt}\delta Y_i(t) = -\alpha_{ij}\delta Y_j(t) = -\lambda_{ij}\delta y_j(t)\ . \tag{10.264}$$

Since $\delta Y_i(t)$ is a fluctuation, its change in time should be calculated over a long enough time Δt that the δY_i averaged over this time is a reasonably smooth function of time.

Solutions to the first of these equations are exponential in the time. The exponential will diminish in time if the coefficients α_{ij} form a positive definite matrix. This is the case under physical conditions, i.e. for the entropy to tend toward a maximum at equilibrium. We see that time-correlations of fluctuations will exponentially die out as the time interval between those fluctuations increases.

If we differentiate the left and right sides of Eq. (10.263) with respect to t, and then let t approach zero from the positive side, we will have, after substituting Eq. (10.264),

$$\lambda_{ij}\langle \delta Y_j \delta y_i\rangle = \lambda_{ji}\langle \delta Y_i \delta y_j\rangle\ . \tag{10.265}$$

But from Eq. (10.259), the averages are Kronecker deltas, so that

$$\lambda_{ij} = \lambda_{ji} \,, \tag{10.266}$$

which constitutes Onsager's proof of his reciprocity relation.[67]

10.11 Kinetic Theories for Chemical Reactions

Thermodynamics and statistical mechanics focus on equilibrium states of thermal systems. Time considerations for a process enters principally in discriminating between reactions that actually reach quasi-equilibrium in the system in times comparable to or shorter than the typical times needed for given processes. Catalysts and other mechanisms can dramatically change these time scales.

Kinetic theory attempts to develop models and understandings for the detailed time evolution of a system, including the effect of catalysts and intermediate states. The models may be empirical or based on the known kinetics and interactions of the molecules and structures involved. A detailed biomolecular kinetic theory of reactions follows the time evolution of biomolecular interactions in solutions and with surfaces, including the dispersive effects of thermal motion. Quantum theory, described in the next chapter, provides a successful framework for these details, all within the subject of 'chemical kinematics'.

10.11.1 Conversion Reactions with Reverse Rates

Because of atom conservation, the rate of change of the concentration of molecules of any given molecule will depend on the change in concentration of other molecules taking part in a reaction. If the proximity of reactants is statically based, then rates depend on the participating concentrations raised to a power. These powers need not be the stoichiometric coefficients, but rather depend on details on how the molecules come together in a reaction. For the reaction rates in one direction, the sum of the powers of the concentrations is called the 'order' of the reaction. The order need not be an integer.

To see what's involved in an empirical model, take an example of the time evolution of a reaction that converts a molecule from type A to type B:

$$A \rightleftharpoons B \,.$$

A phase change is of this category, as is an isomeric transformation.

[67]For more on Onsager's reciprocity relations, see Sect. 9.23.2.

Start (at time $t = 0$) with only type A particles in the system. As time progresses, some type B are created. If the reaction occurs for each particle A independent of the other particles in the system, then we might expect:

$$\frac{dA}{dt} = -\kappa A + \lambda B \,, \tag{10.267}$$

where A is the number of A-type particles at time t, B is the number of B-type particles at time t, starting at zero. The first term accounts for the loss of A-type particles in the forward reaction, which is proportional to the time interval and to how many of the A type particles are present.[68] Similarly, the second term accounts for the rate of conversion of B-type particles back to A-type in the back reaction. The change of B-type follows

$$\frac{dB}{dt} = +\kappa A - \lambda B \,. \tag{10.268}$$

The rate coefficients with signs reverse match those for dA/dt in order to preserve the total particle number $A + B$. The solutions to the pair of first-order differential equations, satisfying $A(0) = A_0$, $B(0) = 0$ and $dA/dt = -dB/dt$, are

$$A = \left(\frac{K}{1+K} e^{-t/\tau} + \frac{1}{1+K}\right) A_0 \tag{10.269}$$

$$B = \frac{K}{1+K}\left(1 - e^{-t/\tau}\right) A_0 \,, \tag{10.270}$$

where $K \equiv \kappa/\lambda$ and $\tau \equiv 1/(\kappa+\lambda)$. The time τ is called the lifetime for the reaction. After a long time,

$$A \underset{t\to\infty}{\longrightarrow} \frac{1}{1+K} A_0 \tag{10.271}$$

$$B \underset{t\to\infty}{\longrightarrow} \frac{K}{1+K} A_0 \,. \tag{10.272}$$

These imply that at equilibrium,

$$\frac{[B]}{[A]} = K \,, \tag{10.273}$$

which is the equilibrium constant

$$K = \exp\left(-\Delta g/(RT)\right)$$

for the reaction.

[68] Our assumption that the rate of loss of A type particles is independent of other particles present makes the forward reaction a 'first-order' process. If the forward reaction requires other A type particles nearby, then the reaction would not be first-order.

Note that the reaction 'goes to near completion' if the relative rate of loss of A-type particles (κ) is much larger than the relative rate of conversion of B-type particles (λ) back to A-type. Also, the lifetime for the conversion is proportional to the inverse of the sum of the relative rates: $(\kappa + \lambda)^{-1}$. These observations are expected.

10.11.2 Pair Reactions with Reverse Rates

Kinetic analytics for reactions of the form

$$|\nu_A|\, A + |\nu_B|\, B \leftrightharpoons |\nu_C|\, C + |\nu_D|\, D \tag{10.274}$$

become quite involved in the general case, as the differential equations are likely to be non-linear. However, with given initial conditions, even these equations can be handled, if not analytically, at least numerically.

Perhaps the simplest example is

$$A + B \leftrightharpoons C + D \tag{10.275}$$

with rates proportional to concentrations to first order:

$$\frac{dA}{dt} = -\kappa\, A\, B + \lambda\, C\, D \tag{10.276}$$

$$\frac{dB}{dt} = -\kappa\, A\, B + \lambda\, C\, D \tag{10.277}$$

$$\frac{dC}{dt} = +\kappa\, A\, B - \lambda\, C\, D \tag{10.278}$$

$$\frac{dD}{dt} = +\kappa\, A\, B - \lambda\, C\, D\, . \tag{10.279}$$

Let the initial conditions be

$$\left.\frac{dA}{dt}\right|_{t=0} = A_o \tag{10.280}$$

$$\left.\frac{dB}{dt}\right|_{t=0} = B_o \tag{10.281}$$

$$\left.\frac{dC}{dt}\right|_{t=0} = 0 \tag{10.282}$$

$$\left.\frac{dD}{dt}\right|_{t=0} = 0\, . \tag{10.283}$$

For simplicity in this example, we will take the initial concentration of B to match that of A. The first two equations for the change of A and B with time insures that A remains equal to B. Similarly, the concentration of D will match that of C. We also have $dC/dt = -dA/dt$, so that $C = A_o - A$. In the following, we will take $\kappa > \lambda$.

With these assumptions, the solution for A, B, C, and D can be found from the solution of

$$\frac{dA}{dt} = -\kappa A^2 + \lambda(A_o - A)^2 . \tag{10.284}$$

This equation can be expressed in terms of a known (logarithmic) integral:

$$t = \int_{A_o}^{A} \frac{dx}{-\kappa x^2 + \lambda(A_o - x)^2} . \tag{10.285}$$

Solving for A gives

$$A = A_o \frac{\sqrt{\lambda\kappa}\left(1 + e^{-2A_o\sqrt{\lambda\kappa}t}\right)}{\left(\kappa + \sqrt{\lambda\kappa}\right) - \left(\kappa - \sqrt{\lambda\kappa}\right)e^{-2A_o\sqrt{\lambda\kappa}t}} . \tag{10.286}$$

A plot of the concentrations of A and C as a function of time is given in Fig. 10.6.

After a time much longer than $1/(A_o\sqrt{\lambda\kappa})$, the concentration of A molecules reaches

$$A(\infty) = A_o \frac{1}{1 + \sqrt{\kappa/\lambda}} , \tag{10.287}$$

as does the concentration of B molecules, while C and D follow $C(\infty) = A_o - A(\infty)$

$$C(\infty) = A_o \frac{\sqrt{\kappa/\lambda}}{1 + \sqrt{\kappa/\lambda}} . \tag{10.288}$$

The equilibrium constant for the reaction is $[C][D]/([A][B]) = \kappa/\lambda$.

In biological reactions, evolution has optimized the production of useful products, so that many biochemical reactions can be taken as 'one directional', proceeding to completion of the products. In these cases, the kinematical differential equations for the reactions become far easier to solve. For the reaction $A + B \leftrightharpoons C + D$ treated above, the reverse reaction will be negligible as $\lambda/\kappa \to 0$. In this case,

$$A(t) = A_o \frac{1}{1 + A_o\kappa t} \to 0$$

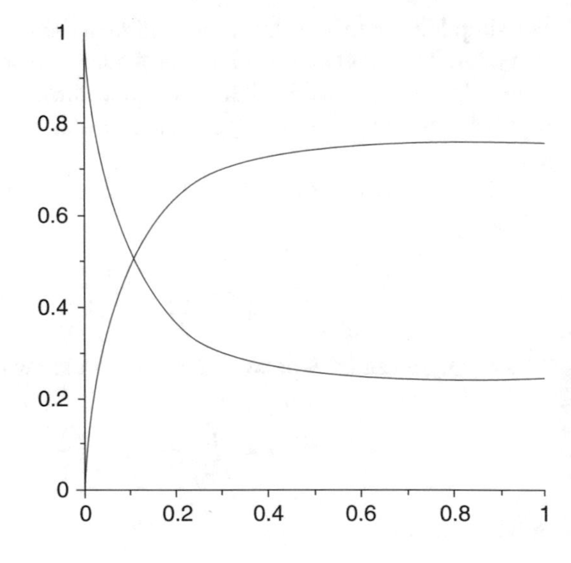

Fig. 10.6 Concentrations of reactants and products. Here, A (dropping curve) and C(rising cure) show concentrations with time for the reaction $A + B \rightleftharpoons C + D$ (with $A_o = 1,\ \kappa/\lambda = 10$)

and

$$C(t) = A_o \frac{\kappa t}{1 + A_o \kappa t} \rightarrow A_o$$

as $t \rightarrow \infty$.

10.12 Catalysts in Reactions

A catalyst is any substance that increases a reaction rate, but does not get consumed in the process. This may involve lowering the energy threshold ('activation energy') needed for a reaction to occur, or enhancing alternate reaction pathways leading to the same products.

Catalysts can speed reaction rates by many orders of magnitude.

Figure 10.7 shows a plot of a possible potential energy U for various separation distances x between two molecules. If the molecules are initially bound at total energy E_1, they can become unbound spontaneously by quantum tunneling[69] 'through the potential energy wall'. However, the probability for tunneling is strongly dependent on the 'height of the wall', which is the activation energy $E_a \equiv E_2 - E_1$, and its width (shown as 'w'). Of course, zapping the bound molecules with an energy greater than E_a can give them sufficient energy to separate. But a catalyst can reduce the height of the 'wall', or carry the molecules through a

[69]Quantum tunneling plays an important role in biology, Tunneling of protons can cause mutations in DNA by altering the base pairing. Electron tunneling occurs in photosynthesis.

sequence of states that effectively skirts around the 'wall'. In quantum theory and in reality, the molecules are not at rest, even in the lowest energy level allowed, The distance x varies between the potential energy curve and the line E_1. The distance between the curve and the line is the kinetic energy of the molecule shown at x when the molecule at the origin is held fixed. Thus, when bound, the second molecule oscillates between the two 'turning points'.

A catalyst may be a molecule in solution, such as an enzyme, but also can be a segment of a macromolecular surface or a chain molecule which holds two or more molecules in close enough proximity to allow them to bind together, and subsequently become free from the surface.

The presence of a 'catalyst' in the conversion of A into C may be considered an example of the reaction $A+M \leftrightharpoons C+M$, where M represents the catalyst. The rate of change of A can depend on the concentration of M. In effect, the concentration of the catalyst changes the rate constants. This can be seen in Eqs. (10.276)–(10.279) after replacing B, D by M, M.

Problems

10.1 If two types of ideal gases are mixed and become in thermal equilibrium at a temperature 25 °C, with the mass of molecules in the first type being five times those in the second, what is the ratio of the RMS speeds for each type of gas?

10.2 In modeling oxygen saturation, instead of hemoglobin, consider the simpler case of myoglobin, which accepts just one oxygen molecule. Find the fraction of oxygen saturation in myoglobin in terms of the oxygen chemical potential and the energy of attachment ϵ.

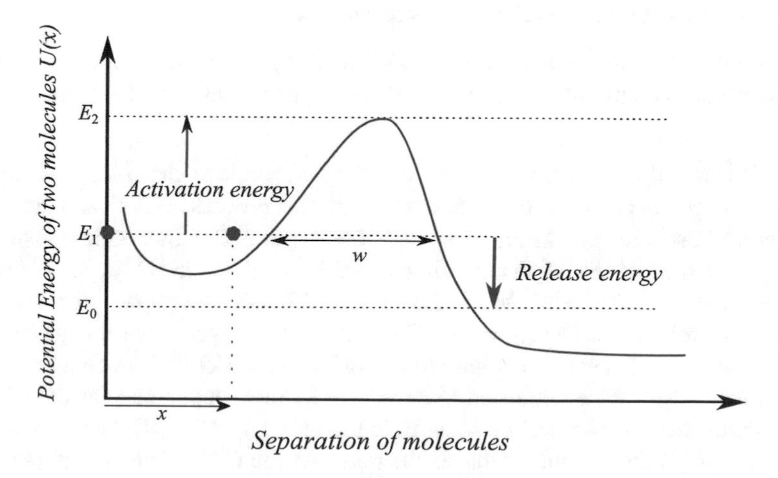

Fig. 10.7 Potential energy stored by two molecules, shown as dots initially bound

10.3 Speculate on how the possible protein foldings in present life forms has been restricted by evolution.

10.4 Describe how the attachment of a small molecule to the surface of a larger molecule can increase the rate of the reaction of the smaller molecule with another smaller molecule.

10.5 Outline a computer program that finds possible foldings of a given short sequence of amino acids forming a protein in a biological cell.

10.6 Consider a molecule in an NMR machine that has four separated hydrogen protons whose spins are effectively uncoupled to neighboring atoms. Being spin 1/2 particles, they are allowed by quantum theory to have only two measurable orientations with the external magnetic field, aligned or anti-aligned. With four arrows indicating spin, show the possible configurations of the proton spins. In an NMR magnetic field of $B = 1.2\,\mathrm{T}$, the energy of the anti-aligned spins are higher than the aligned. If the environmental temperature is $T = 37\,°\mathrm{C}$, what is the fraction of the aligned protons at equilibrium.

10.7 On the International Space Station, astronauts use an airlock to enter and exit at the beginning and end of spacewalks. The pressure in the airlock is first lowered from 1 to 0.694 atm, and the air enriched with oxygen, to reduce the nitrogen dissolved in the astronaut's blood. The astronaut spends the 'night' in this chamber. After the astronaut puts on his spacesuit (containing pure oxygen), the airlock drops pressure to 0.293 atm, the pressure that the spacesuit can sustain in the vacuum of outer space. Finally, the airlock is evacuated and the astronaut can exit to space. If the volume of the airlock chamber is 34 cubic meters, how much does the temperature of the air in the chamber drop during the drop of pressure from 0.694 to 0.293 atm, assuming no heat input, and starting $T_i = 25\,°\mathrm{C}$?

10.8 Estimate the reduction in oxygenated hemoglobin while smoking a cigarette which produces 100 ppm of carbon monoxide gas.

10.9 What is the expected relative fluctuation $(\Delta N/N)$ in water molecule number in a nanoscale volume of water of size $V = 10(\mathrm{nm})^3$ and at body temperature of $37\,°\mathrm{C}$.

10.10 The result of hydrolysis of the ATP in our cells is the release of energy, which is used to promote other reactions. In the process, ADT and inorganic hydrogen phosphate H_3PO_4 are produced. The symbol P_i ('inorganic phosphate') will be used for H_3PO_4. Under standard conditions ($T_s = 25\,°\mathrm{C}$, $p_s = 10^5\,\mathrm{N/m^2}$, $pH = 7.0$, $[ATP] = 1\,\mathrm{M}$, $[ADP] = 1\,\mathrm{M}$, $[P_i] = 1\,\mathrm{M}$), the available energy per mole is the change in the Gibbs free energy per mole for the reaction $ATP \rightarrow ADP + P_i$, which is found to be $\Delta g^o = -30.5\,\mathrm{kJ/mol}$. In an intracellular environment, Δg^o varies between -28 to $-34\,\mathrm{kJ/mol}$, depending on the pH and the concentration of Magnesium ions. If, in the cell, $[ATP] = 1.4\,\mathrm{mM}$, $[ADP] = 0.5\,\mathrm{mM}$, and $[P_i] = 1.2\,\mathrm{mM}$, what is the range of the Gibbs free energy per mole released?

Chapter 11
Biomolecular Structure and Interactions

Those who are not shocked when they first come across quantum theory cannot possibly have understood it.

— Niels Bohr

Summary At a molecular level, organic structures come about from electromagnetic interactions and from nature's quantum behavior. Humans now have a successful, accurate, and predictive description of atoms and molecules, applying the theory of quantum electrodynamics. Included in the theory is all of chemistry, which includes the biochemistry of life. Our instruments for probing the micro-world have recently undergone revolutionary advancements, even to the point of being able to observe the progression of biological actions at the molecular level. We can see viral structure and actions, and the damage of radiation on our DNA.

11.1 Nature in the Small World

Humans get their vision of the world by observing macrosystems with light of wavelength much larger than atoms. Our so-called intuition is a set of ideas distilled and derived from these everyday observations, becoming our 'gut reaction', and percolating from our deeply subconscious analysis of experiences. We are surprised by the behavior of the universe in the nanometer-scale. Even if that behavior is strange, non-intuitional, or paradoxical, our task is to try to understand nature through the formulation of logical connections between what we can determine at all scales. If we are clever enough, these logical relations can be synthesized into a model of the small world. This synthesis occurred between the years 1900 and 1931 in the creation of quantum theory.

Quantum theory, incorporating Einstein's Special Theory of Relativity, is one of the most widely tested and most accurate theories yet formulated by humans. As students in an ordinary laboratory, we are often happy to measure a quantity to a precision of three decimal places. Some predictions of quantum electrodynamics have been checked with experimental measurements to twelve decimal places. Such

W. C. Parke, *Biophysics*, https://doi.org/10.1007/978-3-030-44146-3_11

precise experiments may take years to construct and analyze. Before computers, our theory also took years of calculation to determine a prediction to twelve decimal places.

Quantum theory has been successfully applied to the smallest scale measured, about one billionth of a nanometer (10^{-18} m). Einstein's General Relativity suggests that at scales of 10^{-35} m (the 'Planck length'), quantum theory must be revised, but such small separations cannot be observed with current instruments. Quantum theory can be used to predict the motion of baseballs and planets, but the formalism is more esoteric and less direct than Newton's second law, and the wave interference properties which are within quantum theory are practically unobservable for objects the size of baseballs. When the wave diffraction and interference part of quantum theory are reduced out, Newton's laws emerge, making Newton's laws a macroscopic simplification of the more fundamental quantum theory, just like what occurs when Maxwell's electrodynamics are reduced to ray optics by taking a small wavelength approximation.

Quantum theory explains atomic structure, chemical bonding, chemical reactions, spectral colors in absorption and emission, excitation states, and all other properties and interactions of bodies at the nanoscale, including those in biological systems. For example, the primary, secondary, tertiary, and quaternary structure of proteins, critical to behavior of enzymes, can be predicted by quantum theory. The various kinds of interactions between molecules and ions within a cell are well described by quantum theory.

Quantum effects are evident in the nature of atomic and molecular structure and interactions. The field of Quantum Biology specializes in how quantum mechanics is involved in chemical reactions, light absorption, electron excitations, and proton transfers. Within quantum theory, such examples as the microscopic nature of photosynthesis, vision, and DNA mutation by quantum tunneling can be understood.

11.2 Quantum Behavior of Matter

A new scientific truth does not triumph by convincing its opponents and making them see the light, but rather because its opponents eventually die.

— Max Planck

Here are a number of important aspects of quantum theory:

1. In order to know about the outside world, we must interact with it. We see by gathering light from the environment. This light is scattered or emitted from objects at which we are looking. Atomic particles are so small in mass that the act of observation disturbs the particle, making subsequent observation dependent on the previous observation. This interaction between the observed and the observer is built into quantum theory. Our thinking, however, may be counter to this behavior. We think of bodies as having a motion through

space unaffected by who is looking, or how. Nature does not allow us this luxury. If there is no instrument or system interacting with an electron, we cannot properly think of the electron proceeding along any one path between observations. With quantum theory, cartoons of electrons orbiting nuclei are fundamentally wrong. Also, drawings of individual electrons spread out like a classical wave are wrong.

2. In quantum theory, certain individual processes cannot be predicted. Rather, only the probability of a given outcome can be predicted. For example, if an atom is left unobserved for a second, then the position of one of its electrons within one of the possible electron orbitals cannot be predicted. Rather, we can predict the probability of finding the electron in a certain position. We first have to measure that electron's position many times for the same initially unobserved state to get an average measured position. Then this average, or 'mean position', can be compared with a prediction of quantum theory.
3. Each particle can be represented by a complex valued wave function, denoted $\Psi(t, x)$. In space, the probability density of finding that particle is proportional to the magnitude of the wave function squared, $\Psi^*(t, x)\Psi(t, x)$. The wave function is a solution to a wave equation linear in Ψ, which means that the principle of linear superposition applies to such waves: If $\Psi_1(t, x)$ and $\Psi_2(t, x)$ are two solutions, then so is $\Psi(t, x) = \Psi_1(t, x) + \Psi_2(t, x)$. This principle of superposition also means that any wave function can be 'decomposed' into a linear combination of simpler waves which 'span' the possible waves in the given space.[1]

Quantum theory predicts that the probability of finding a particle somewhere in space can show a pattern of constructive and destructive interference. Interference is a consequence of the superposition principle. In quantum theory, the interference of a particle wave with itself can occur. For example, a sequence of electrons sent (unobserved) toward a double slit will arrive at points on a phosphor screen beyond the slits, but the accumulation of points will have a distribution which matches a classical interference pattern, with nodes and antinodes. In quantum theory language, the electron wave which arrives at the screen is a superposition of the waves which came through each slit. No classical theory of particles is capable of showing such interference. To observe this interference effect for the two-slit experiment, the electron wave must be allowed to pass through both slits. If any instrument is used to determine which slit the electron passes through, the interference pattern is lost.

In a localized bound system left undisturbed for a known relaxation time, particles settle into 'quantum orbitals', i.e. stationary solutions to the appropriate wave equation.

The expression of atomic and molecular orbitals as a linear combination of only a few of the dominant components in an expansion in terms of simpler waves is often used in chemistry to describe valence electrons in atoms and

[1] A Fourier series and its generalizations is an example of such a decomposition.

molecules. For example, the s-p 'hybrid orbitals' describe electron waves participating in chemical bonds in terms of simpler hydrogenic waves. By the superposition principle, expansion of a valence electron wave into hydrogenic waves is always possible. Whether only a few terms in the expansion dominate a given molecular bond depends on the full atomic interactions, and how the electrons behave once they settle into the lowest energy states among those available stationary molecular waves.

4. The evolution of a particle wave (between observations) is determined by a linear diffusive-like differential equation called a quantum wave equation. This means that if an electron is left unobserved, its wave (as described by the theory) 'diffuses' until it reaches spatial boundaries. With non-absorbing boundaries confining the electron in some finite region, the wave evolves into a standing wave. No further change in the probability distribution for these standing waves occurs.

   For electrons making transitions between bound atomic or molecular orbitals, the time to reach a steady state is usually within nanoseconds, but some transitions may take hours (such as within long-lifetime phosphors).
5. An electron initially confined by thin walls with finite confining forces will not be completely confined, but rather the wave function for the unobserved electron will have an exponential tail within the walls, and a small oscillating part in the region outside the wall. This means there is a finite probability of finding the electron outside the confining region, even though classically, the electron would not have had enough energy to overcome the forces of the walls. This effect, called 'quantum tunneling', has not only been observed, but is common in the small world. Radioactive substances would not decay if not for quantum tunneling,[2] nor would electrons be able to conduct through copper wires twisted together with an oxide layer between them. The hydrogen atoms in the ammonia molecule tunnel back and forth past the nitrogen at a microwave frequency. Quantum tunneling of protons can also occur between base pairs in your DNA, possibly leading to a gene mutation.
6. Quantum theory and Relativity predict that a stable localizable particle carries a fixed mass and a quantized spin. The magnitude of the spin of a particle must be $\sqrt{s(s+1)}\hbar$, where s is a quantum number taking one of the possible values $0, 1/2, 1, 3/2, 2, \cdots$. For integer spin, the particle is called a boson. For half-odd-integer spin, the particle is called a fermion. In 1939, Pauli and Fierz proved that within relativistic quantum theory, any number of integer-spin particles can be placed in a single quantum state, but no more than one half-odd-integer-spin particle can be placed in a single quantum state. Photons are bosons (with $s = 1$) and electrons are fermions (with $s = 1/2$). Light from a LASER will be found with lots of photons in one quantum state. We cannot construct an

[2]George Gamow, in 1928, was the first to apply quantum tunneling to explain radioactivity. In 1948, while at the George Washington University, he and his student Ralph Alpher, used quantum theory and General Relativity to devise the hot 'big bang' theory of the Universe.

electron beam with lots of electrons in one quantum state. However, in some metals and semiconductors, electron-lattice interactions allow electrons to have a very weak attraction, and can pair up at sufficiently low temperatures. These pairs act as bosons, and make the material a superconductor.

7. Observation may disturb the state of a system. Each observation of a system may (depending on the nature of the measurement) 'collapse' the wave function into one of the possible observable states of the system. However, for quantum theory consistency, the act of measurement is an event which opens a previously isolated system to the measuring instrument (or the environment), possibly making entropy flow into the original system. The wave function for both the observed system and measuring system becomes entangled[3] when the measurement occurs. There may be a dramatic change in the wave function when the interaction begins, but there is no 'wave-function collapse' of the combined system that includes the measuring device.
8. In quantum theory, for each observable, such as energy, linear momentum, angular momentum, and position, there is an associated operator $\widehat{A}$ which acts on states Ψ of the observed system.[4] The average of repeated measurement of an observable $\widehat{A}$ of a prepared system is predicted by quantum theory to be $\langle A \rangle = \int \Psi^* \widehat{A} \Psi dx$. Each observable $\widehat{A}$ has a spectrum of possible values which can be found by a measurement of $\widehat{A}$, called the eigenvalues $\{a_i\}$ of A, $i = 1, 2, \cdots$. Solutions to the relation $\widehat{A}\psi_{a_i} = a_i \psi_{a_i}$ determine the possible 'eigenstates', ψ_{a_i}, and 'eigenvalues' a_i. Since $\widehat{A}$ is an observable, the a_i must be real numbers. The eigenstates for distinct eigenvalues will be orthogonal: $\int \psi^*_{a_i} \psi_{a_j} dx = \delta_{ij}$. The eigenstates ψ_{a_i} for all possible eigenvalues a_i can be shown to form a complete set, so that any state Ψ can be written as a linear superposition of such states: $\Psi = \sum_i c_i \psi_{a_i}$. After a measurement of $\widehat{A}$ which gives an eigenvalue a_i, the system wave 'collapses' to ψ_{a_i}.[5]
9. If the relation $\widehat{A}\widehat{B} = \widehat{B}\widehat{A}$ (i.e. $\widehat{A}$ and $\widehat{B}$ commute) is true, then the observables $\widehat{A}$ and $\widehat{B}$ can be measured in either order without disturbing the state formed by the first measurement. The system is left in an eigenstate of both $\widehat{A}$ and $\widehat{B}$.
10. If the observables $\widehat{A}$ and $\widehat{B}$ do not commute, then the measurement of one followed by the second necessarily changes the state of the system. Define the uncertainty of $\widehat{A}$ after a set of measurements of $\widehat{A}$ by the root-mean-square

[3]An entangled state is a multiparticle quantum wave which cannot be factored into component particle waves.

[4]A 'hat' on a symbol indicates that the symbol represents an operator, rather than a number. Operators act on quantum states, changing them to a different quantum state.

[5]This phrase 'collapse of the wave function' has been used since Bohr's original thoughts on quantum theory. But as we have noted the phrase cannot be taken too literally. A consistent quantum theory requires the following: If one includes the observer as well as the observed in an isolated system, then the interaction of the observer with the observed in the process of a measurement may change the state of the observed part to be predominantly in one eigenstate of an observable, but in order for the sum of all probabilities to remain one, all the other possible states of the observed system must still be present, now 'entangled' with the observer's states.

$\sqrt{\left\langle\left(\widehat{A}-\langle A\rangle\right)^2\right\rangle} \equiv \Delta A$. Then one can show that after repeated measurements of a system identically initialized before measuring $\widehat{A}$ or $\widehat{B}$, we will have $\Delta A \Delta B \geq (1/2)\left|\langle\widehat{A}\widehat{B}-\widehat{B}\widehat{A}\rangle\right|$. The relation implies that if $\widehat{A}$ does not commute with $\widehat{B}$, then the measurements of $\widehat{A}$ and $\widehat{B}$ will have uncertainties inversely related.

11. A 'complete set of commuting observables' is the maximal set of operators in a quantum theory that obey $\widehat{A}\left(\widehat{B}\Psi\right) = \widehat{B}\left(\widehat{A}\Psi\right)$ for any pair in the set. For example, for the electrons in hydrogen, the energy $\widehat{E}$, the square of the angular momentum $\widehat{L^2}$, the angular momentum projection $\widehat{L_z}$, and the electron spin projection $\widehat{S_z}$, form such a set. The electron wave can be indexed by giving the eigenvalues of these operators, i.e. $\{E_n, l(l+1), m, m_s\}$. Each are called a hydrogen atomic orbital.

P.A.M. Dirac found the appropriate relativistic wave equation for the electron in 1928.[6] In its simplest form, it reads[7]

$$(\not{p}-m)\,\psi = q\not{A}\psi\ . \tag{11.1}$$

Here, $\not{p} \equiv \gamma^\mu p_\mu$, with $p_\mu = i\partial_\mu$ the four-momentum operator of the electron and the γ^μ are a set of four 4×4 matrices satisfying $\gamma^\mu\gamma^\nu + \gamma^\nu\gamma^\mu = 2g^{\mu\nu}$; $g^{\mu\nu}$ is the metric tensor $diag\,[1,-1,-1,-1]$; q is the charge of the electron (including its negative sign); $\not{A} \equiv \gamma^\mu A_\mu$; and A^μ is the external electromagnetic four-vector field due to other nearby charges.

Valence electrons in biological context are non-relativistic. The non-relativistic form of the Dirac equation is the Schrödinger equation for one electron:

$$\left(-\frac{\hbar^2}{2m}\nabla^2 + V(\mathbf{r})\right)\Psi = i\hbar\frac{\partial\Psi}{\partial t}\ . \tag{11.2}$$

The V is the potential energy due to the interactions of the electron with other particles. A stationary solution of the Schrödinger equation is a solution whose wave function square magnitude $(\psi^*\psi)$ does not change with time. For these,

$$\Psi(t,\mathbf{r}) = \exp\left(-iEt/\hbar\right)\psi(\mathbf{r}), \tag{11.3}$$

[6] At the same time, his equation predicted the existence of antimatter, later confirmed. This is an example of the power of logic: Develop a logical theory that closely matches nature. From that theory, draw conclusions from the logic. Some of those conclusions may refer to observations not yet made.

[7] We write the equation not to anticipate using it explicitly, but rather to show how simple it appears in the hands of Dirac and Nature!

where E is a constant with the units of energy. (Note that the time dependence goes away in $\psi^*\psi$.)

The Schrödinger equation for a free electron ($V = 0$) has a simple solution

$$\Psi(t, \mathbf{r}) = N \exp\left(-iEt/\hbar + i\mathbf{p}\cdot\mathbf{r}/\hbar\right) \tag{11.4}$$

with $E = p^2/2m$. This is called a 'plane wave' solution, with the angular frequency $\omega = E/\hbar$, and wave number is $\mathbf{k} = \mathbf{p}/\hbar$. These relations correspond to the well-established connection between wave and particle properties: $f = E/h$ and $1/\lambda = p/h$. (These are the DeBroglie relations.)

A more general solution for a free electron as a superposition of these plane waves is:

$$\Psi(t, \mathbf{r}) = \int \phi(\mathbf{p}) \exp\left((-ip^2t/(2m) + i\mathbf{p}\cdot\mathbf{r})/\hbar\right) d^3p\,. \tag{11.5}$$

The coefficients $\phi(\mathbf{p})$ of the plane waves are called the momentum components of the general wave. We recognize them as the Fourier components of the original wave.

The linear momentum of a particle can be defined experimentally by the impulse (the sum of the forces on the particle times time) needed to bring the particle from zero to its present velocity. The most general theoretical definition of the linear momentum and energy of an isolated system is through the 'generator of spatial and time shifts', described as follows: If the system is in a state $\Psi(t, x)$, we can perform a spatial shift to $x + a$ by a Taylor series:

$$\Psi(t, x + a) = \Psi(t, x) + \frac{\partial\Psi(t, x)}{\partial x}a + \frac{1}{2}\frac{\partial^2\Psi(t, x)}{\partial x^2}a^2 + \cdots\,. \tag{11.6}$$

This relation can be expressed in operator form as

$$\Psi(t, x + a) = \exp\left(a\frac{\partial}{\partial x}\right)\Psi(t, x). \tag{11.7}$$

Time translations can be written as

$$\Psi(t + t_0, x) = \exp\left(t_0\frac{\partial}{\partial t}\right)\Psi(t, x). \tag{11.8}$$

Comparing with the plane wave solution of the Schrödinger equation, and recognizing, according to Einstein, that energy and momentum should appear symmetrically in a theory which satisfies the Principle of Relativity, we identify operators with the total energy and momentum of a system as

$$\widehat{E} = i\hbar\frac{\partial}{\partial t} \tag{11.9}$$

and

$$\widehat{p}_x = -i\hbar\frac{\partial}{\partial x} . \tag{11.10}$$

(The minus sign for the momentum operator is required by Relativity if a plus sign is used in the energy operator.) Similarly, the angular momentum operator

$$\widehat{\mathbf{L}} = \mathbf{r} \times (-i\hbar\nabla) , \tag{11.11}$$

is the generator of rotations.

In operator form, the energy operator that appears in the Schrödinger equation for a non-relativistic electron is

$$\widehat{E} = -(\hbar^2/(2m))\nabla^2 + \widehat{V} = (\widehat{\mathbf{p}}^2/(2m) + \widehat{V} . \tag{11.12}$$

This energy operator, when expressed in terms of the operators for position and momentum, $\widehat{\mathbf{r}}$ and $\widehat{\mathbf{p}}$, is given the name the 'Hamiltonian', $\widehat{H}$, after William Rowan Hamilton.

If an electron is put into a box (an enclosure with impenetrable walls), the potential energy V goes to infinity at the walls, and is zero inside. This makes the electron wave function vanish at the walls, and the electron acts as a free particle inside. The boundary condition forces an integer times a half wavelength in each of the three directions to be the length of the box in that direction, just like the standing waves of a guitar string. For the electron, the constraint on possible values of the wavelength is a quantization condition. In the case of a square box (with side of length L), the energy becomes

$$E = \frac{\pi^2\hbar^2}{2mL^2}(n_x^2 + n_y^2 + n_z^2) \tag{11.13}$$

with integers n_x, n_y, n_z determining the possible energies states.

Typical of stationary solutions to a wave equation is that boundary conditions on the solutions make the particles associated with the wave quantized in energy. Pure plane wave solutions in free space do not have quantization of energy or momentum because the waves are unbounded. Electrons held to nuclei are in states bound by the Coulomb force. This binding allows only certain discrete steady waves, found by solving the Schrödinger equation. For example, for hydrogenic atoms, we use $\widehat{V} = -Zke^2/\widehat{r}$, and find the resulting waves can be enumerated with labels given by a radial quantum number n_r, taking values $0, 1, 2, \cdots$, an angular momentum quantum number l, with values $0, 1, 2, \cdots n_r$, and an angular momentum projection quantum integer number m, with values $-l \leq m \leq l$. Having three such quantum numbers for the spatial wave function is a reflection of binding in three dimensions. Electron spin adds an additional quantum projection quantum number m_s taking the values $\pm 1/2$.

11.3 Atomic Structure

Common sense is the layer of prejudices put down before the age of eighteen.

— Albert Einstein

We can predict the structure of atoms and molecules from electrical interactions and quantum theory. (Magnetic interactions have only a minor role in structure.) Nuclei of atoms have a mass much larger than the surrounding electrons. The electrons are electrically attracted to the nuclei, but are prevented from radiating away their energy and being 'sucked' into the nucleus because of their quantum wave properties: The nucleus is just too small to contain the wave for such a light particle.[8]

The simplest atomic structure is that of the hydrogen atom. The electron waves satisfy the Schrödinger equation. The electron in a hydrogen atom is acted on by the central proton through an electric (Coulombic) field produced by that proton. Since the proton is almost 2000 times the mass of the electron, the proton hardly wiggles as the electron moves around it. A good approximation is then to assume the proton is at rest. The potential energy of the electron in the electric field of the proton is then $V(r) = -ke^2/r$. (In the mks system of units, $k = 1/(4\pi\epsilon_o) \approx 9 \times 10^9\,\mathrm{J/(mC^2)}$.) The wave function for the electron satisfies the Schrödinger equation in the form

$$i\hbar\frac{\partial}{\partial t}\Psi(t,r,\theta,\phi) = \left(\frac{\widehat{p_r}^2}{2m} + \frac{\widehat{L}^2}{2mr^2} - \frac{ke^2}{r}\right)\Psi(t,r,\theta,\phi)\,, \tag{11.14}$$

where

$$\widehat{p_r} \equiv -i\hbar\frac{1}{r}\frac{\partial}{\partial r}r \tag{11.15}$$

is the radial momentum operator and $\widehat{\mathbf{L}} = \widehat{\mathbf{r}} \times \widehat{\mathbf{p}}$ is the angular momentum operator. One can show that in spherical polar coordinates,

$$\widehat{\mathbf{L}}^2 = -\hbar^2\left(\frac{1}{\sin\theta}\frac{\partial}{\partial\theta}\sin\theta\frac{\partial}{\partial\theta} + \frac{1}{\sin^2\theta}\frac{\partial^2}{\partial\phi^2}\right). \tag{11.16}$$

The angular momentum operator along a measurement axis, usually taken as the z-axis, is

$$\widehat{L}_z = -i\hbar\partial/\partial\phi\,, \tag{11.17}$$

[8] However, for some proton-rich nuclei, an orbiting electron can be sucked in by converting a proton to a neutron, while releasing a neutrino that flies out practically undisturbed by any matter nearby.

and its eigenvalue will be $m\hbar$. Now in order for the wave function of the electron to return to its original value after a full cycle of the azimuthal angle ϕ, the value of m must be an integer. This is an example of a cyclic boundary condition on a wave function leading to a quantum property. The quantization of the square of the angular momentum of the electron, $\widehat{\mathbf{L}}^2$ to be $l(l+1)\hbar^2$, with l a non-negative integer, is equivalent to a cyclic condition on θ, and also limits the magnitude of m to be not greater than l. Another quantum number for steady-state waves comes from the boundary condition that the electron wave must be finite at the proton, and must drop sufficiently fast at infinity that the total probability of finding the electron anywhere is one. Energy quantization follows. The result is that the steady-state electron waves can be labeled by their number of radial nodes, 'n_r'. The energy of the system is proportional to the inverse square of the number $n \equiv n_r + l + 1$, called the 'principle quantum number in hydrogen': $E_n = -13.6\,\text{eV}/n^2$.

Because the potential energy is a function of the distance of the electron from the nucleus, but not the angles where the electron might be found, the Schrödinger equation solutions for bound electrons in a steady state can be expressed as a product of functions of time, radius, and spherical polar angles, separately. Thus, hydrogenic electron orbitals take the form

$$\Psi(t, x) = \exp(-iE_n t/\hbar)R_{nl}(r)Y_{lm}(\theta, \phi)\,. \tag{11.18}$$

The angular functions Y_{lm} are called 'spherical harmonics'. They separate into functions of θ and ϕ as $Y_{lm}(\theta, \phi) = N_Y P^l_m(\theta)\exp(im\phi)$, where the constant N_Y is selected to make $\int((Y_{lm})^* Y_{lm} d(\cos\theta)d\phi = 1$.

For an isolated hydrogen atom, its energy and vector angular momentum are conserved. Moreover, the energy $\widehat{H}$, the square of angular momentum $\widehat{\mathbf{L}}^2$, and one component of the angular momentum (taken as $\widehat{L_z}$) can be measured without disturbing the others. (The operators for these three observables commute.) The electron spin along one axis is also quantized to have values $\langle S_z\rangle = m_s\hbar$ with $m_z = -1/2$ or $1/2$. So, we can label the hydrogenic waves by the quantum numbers associated with the quantization of the energy, the magnitude of the angular momentum, the z-component of the angular momentum, and the z-component of the electron spin: n, l, m, m_s. When there are a number of possible quantum states for a given energy, the set of states is called 'degenerate'. Sets of quantum states separated from others by a relatively large energy gap are said to be within one 'shell'. For hydrogen, the number of allowed states for the electron in a given shell labeled by 'n' add up to $2\sum_0^{n-1}(2l+1) = 2n^2$. Transitions from one shell to another, from a bound state to an unbound state, or from an unbound state to a bound one, requires an energy exchange with an external agent, which can be a photon or a nearby atom.

Historically, transitions between hydrogenic atomic states in alkaline metals were seen in spectroscopes and labeled with the letters s, p, d, and f from the appearance on film of the corresponding spectral lines, being: sharp, principal, diffuse, and fundamental. Later, these observations were found associated with electron transitions from states with $l = 0, 1, 2, 3$ to other states. We still use the letters s, p, d, f instead of the quantum numbers $0, 1, 2, 3$ to label the angular momentum quantum number for particle quantum states. For $l = 4, 5 \ldots$ we use $g, h, \ldots$ in the sequence of the alphabet.

Now consider the electrons in an element heavier than hydrogen. Suppose the nucleus has a charge Ze. The electron orbitals can be found by solving the appropriate wave equation with Z electrons bound around the nucleus: $H\Psi(\mathbf{r}_1, \mathbf{r}_2, \ldots, \mathbf{r}_Z) = E\Psi(\mathbf{r}_1, \mathbf{r}_2, \ldots, \mathbf{r}_Z)$. One technique to express the orbitals takes advantage of the fact that the eigenstates of any observable forms a complete set of quantum states. An arbitrary wave can be expanded into a complete set of orthonormal waves.

The valence electrons of atoms are the electrons in orbitals the least bound to the atom, and able to take part in chemical bonding. If the orbital for a valence electron is expressed in terms of hydrogenic orbitals with an effective 'screened' central charge, then only a few such hydrogenic orbitals may be needed to describe the chemical bonding in which this electron takes part. If several valence electrons take part, then linear combinations of their orbitals may be sufficient to loosely characterize their participation in a bond. The combination of orbitals which occurs after the system settles is that which minimizes the energy of the system.

11.4 Molecular Interactions

As we have indicated, interactions between molecules of biological interest are dominated by electrical interactions. The electrical interaction comes about because atoms are made from charged particles: relatively heavy positive nuclei and relatively light electrons. So far (since 1928), the application of wave mechanics to molecular systems, when we can handle the mathematics, successfully predicts atomic bonding and molecular interactions. Of particular interest is the nature of the chemical bond, the structure of macromolecules, the behavior of proteins in the cell, and the sequence of interactions and reactions within a cell that allow for life.

In principle, 'all we have to do' is solve the wave equation for the system: $H\Psi = i\hbar\partial\Psi/\partial t$, with all the Coulombic interactions between each electron i and each nuclei N, $V_{Ni} = -kZe^2/r_{Ni}$, and every pair of electrons, $V_{ij} = +ke^2/r_{ij}$. For the 'ground state' of each molecule, the lowest energy eigenvalue and eigenstate solution of $H\Psi = E\Psi$ would be taken.

In practice, there are no 'closed form' solutions for atoms or molecules with more than two charges. To find such solutions numerically by 'brute force' may require excessive computer time, depending on the number of interacting particles present and on the ingenuity of the programmer. Investigators have developed many clever schemes to approximate and to simplify the calculation. For example, since the nuclei are more than 1836 times the mass of the electrons, their inertia keeps them close to stationary and near the center-of-mass of the atom of which they are a part. One can start a calculation assuming the nuclei are fixed but with an as-yet undetermined separation. For the electrons, one can start by taking the inner shells as undisturbed by the valence shells.

The valence electrons dominate atomic interactions which create bonding. In chemical interactions, atoms have change of energies less than about 20 eV. Under these circumstances, the 'core' of atoms remain largely inert.

11.5 Pauli Exclusion Effects

In quantum theory, the wave function for a set of identical fermions must be antisymmetric in the exchange of a pair of those fermions[9]:

$$\Psi(1,2) = -\Psi(2,1) . \tag{11.19}$$

This exchange antisymmetry means the wave function will vanish if two of the fermions are in identical quantum states. In this way, no two electrons can occupy the same quantum state. The Pauli exclusion principle is a result of quantum behavior. As the fermions approach a single quantum state, their combined wave function approaches zero. Since electrons are fermions (being 'spin 1/2 particles'), if two valence electrons from two atoms are in similar quantum states which approach each other as the atoms are pushed together, the probability density for these electrons will diminish in the region between the atoms.

Because of Pauli exclusion, and because of the relatively small electron magnetic interactions between the electrons and the two protons, the associated energy of a confined two-electron system with anti-aligned spins will be lower than the aligned system. If the electrons are in distinctly different quantum states, the electron probability density may grow in the region between the atoms as the atoms approach, leading to covalent bonding. For example, the dominant part of the wave function for the two electrons bound to two protons in the hydrogen molecule has the electrons in a total spin singlet state, antisymmetric in electron exchange, and a total electron orbital state with $L = 0$, symmetric in the exchange of the electrons.

One might ask how the combination of an even number of fermions to make a boson seems to obviate the Pauli exclusion principle for the combined system. 'Cooper pairs' of electrons in superconductors and atoms in boson condensates are examples. The answer is that when such systems of fermions pair up to act as bosons, the wave function for the boson system is spread over distances much larger than the Compton wavelengths ($h/(mc)$) of the internal fermions. The internal fermions in these boson states are not being forced to be very close together.

[9]More generally Eugene Wigner showed in 1939 that if two identical particles with spin quantum number s, then, under particle exchange, $\Psi(1,2) = (-1)^{2s}\Psi(2,1)$, a relation which follows from how particle waves transform in Special Relativity.

11.6 Binding Interactions Between Atoms

In quantum theory, the interaction energy, also called the potential energy, plays a more fundamental role than force. The potential energy of a particle enters directly into the wave equation for the particle. We solve the wave equation to get the probability of finding the particle at various locations at selected times.

Force in quantum theory is a derived concept coming from the negative gradient of the interaction energy $V(\mathbf{r})$ that a particle has at position $\mathbf{r}$ because of nearby particles: $\mathbf{F} = -\nabla V$. This statement is just the inverse of the definition of work done by the force: $V = -\int \mathbf{F} \cdot d\mathbf{r}$. If you plot the potential energy of one atom relative to a second as a function of the separation between the atoms, the negative slope of the curve for that function is the force on the first atom due to the second atom.

Quantum effects and electromagnetic interactions determine the nanoscale behavior of bound and unbound atoms in biological systems. The valence electrons, i.e. those found in the outer region of atoms, largely mold chemical binding of atoms into molecules. The 'core' electrons prevent atoms from collapsing, because the Pauli exclusion principle allows only one electron per quantum state, and in an atom, there are only a few such states near the nucleus. The electron state most tightly held to the nucleus is distributed over a size comparable to the Bohr radius $(\hbar^2/(m_e c^2)$, about half an ångström, divided by Z, the charge on the nucleus. This size is far greater than the diameter of the atomic nucleus, which is about 1.75×10^{-5} of an ångström for a proton in hydrogen and about 15×10^{-5} of an ångströms for the Uranium nucleus with 239 protons and neutrons.

For simplicity in some quantum calculations for bound atoms and pairs of molecules, the electron Pauli exclusion effect as well as short-range electromagnetic repulsion between electrons and the long-range van der Waals interaction is represented by the behavior of a single effective interaction. The radial dependence of interaction energy is taken to be a simple analytic function which facilitates solving the wave equation for the pair of atoms or molecules, assuming they each act rigidly. Two popular choices are the Lennard-Jones potential energy and the Morse potential energy.

The Lennard-Jones potential energy is given by[10]

$$U_{LJ} = \varepsilon \left((r_e/r)^{12} - 2(r_e/r)^6 \right) , \quad (11.20)$$

where ε is the depth of the potential energy and r_e is the radius to the minimum of U_{LJ} (Fig. 11.1). Small vibrations between two atoms with reduced mass of μ and whose interaction is described by the Lennard-Jones potential energy will have a frequency of $f = (3/2\pi r_e)\sqrt{2\varepsilon/\mu}$.

---

[10] J.E. Lennard-Jones, *On the Determination of Molecular Fields*, Proc R. Soc Lond A **106:738**, 463–477 (1924).

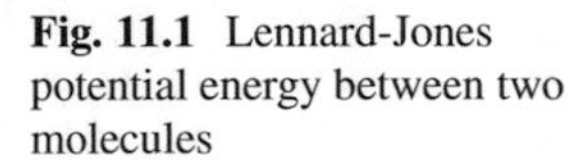
Fig. 11.1 Lennard-Jones potential energy between two molecules

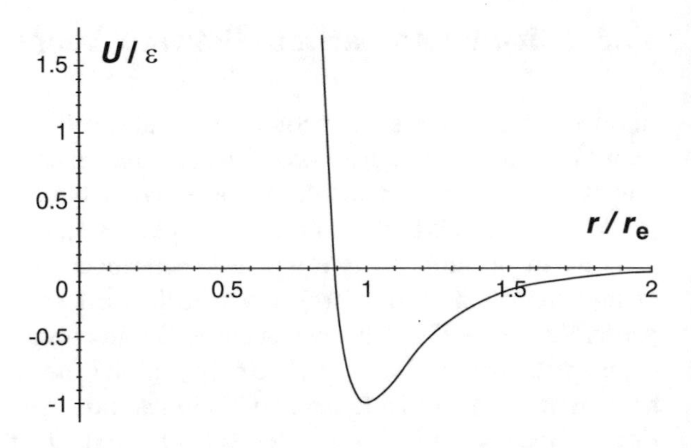

The Morse potential energy is taken to have the form[11]

$$U_M = D_e \left((1 - \exp(-a(r - r_e)))^2 - 1 \right) , \tag{11.21}$$

where $D_e > 0$ is the disassociation energy, and r_e the position of the minimum of U_M. An advantage of the Morse potential energy is that the quantum eigenfunctions and energy eigenvalues for radial vibrations can be found in analytic form. In fact, the energy eigenvalues are simply

$$\epsilon_n = hf \left(n + \frac{1}{2} \right) - \frac{1}{4D_e} (hf)^2 \left(n + \frac{1}{2} \right)^2 , \tag{11.22}$$

where $f = \frac{a}{2\pi} \sqrt{2D_e/\mu}$, with μ the reduced mass for the two atoms.

The gradient of the Lennard-Jones and Morse potential energies at short distances of separation is negative, exhibiting an outwardly directed force, while the long-range force is attractive. The resulting 'well' in the potential energy may be deep enough to give rise to one or more bound-state wave functions for the pair of atoms or molecules.

All the charges in atoms and molecules can affect the long-range electromagnetic interaction between unbound atoms and molecules, and the effects of electromagnetic radiation on atoms. The interactions between atoms and molecules are often listed into categories that loosely follow their strength, as shown in Table 11.1.

[11] P.M. Morse, *Diatomic molecules according to the wave mechanics. II. Vibrational levels* Phys Rev **34**, 57–56 (1929).

Table 11.1 Radial behavior of molecular interactions

Interaction	Radial (r) behavior
Coulomb	$\pm 1/r$
Screened Coulomb	$\pm \exp(-ar)/r$
Charge and dipole	$\pm 1/r^2$
Charge and induced-dipole	$-1/r^4$
Dipole and dipole	$\pm 1/r^6$
Hydrogen bonding	$-1/r^6$
Direct London	$-1/r^6$
Spin-spin (magnetic)	$\pm 1/r^6$
Retarded London	$-1/r^7$
'Shadow' effect	$-1/r$
Pauli exchange	$+1/r^{12}$ (approx)

11.6.1 Chemical Binding

Since the development of quantum electrodynamics (after 1948), electromagnetic interactions and quantum theory have been sufficient to describe all the known properties of chemical binding of atoms to form molecules, including those of life.

As atoms are brought together, the wave functions for outer atomic electrons ('valence electron') are disturbed first. The valence electrons which are strongly affected in light atoms are typically in combinations of $l = 0$ (s) and $l = 1$ (p) orbitals. Above atomic number $Z \approx 18$, the $l = 2$ (d) orbitals can also be strongly affected. If the result of the interaction between the atoms is an attractive force, the atoms can bind together. As the atoms are squeezed together even closer than bond separations, electron Pauli repulsion opposes electric and quantum attraction. Further compression and the Coulombic repulsion between nuclei becomes just as important as the electron Pauli repulsion.

11.6.2 Molecular Orbitals

If a pair of electrons is shared between two atoms, the bonding is called 'covalent'. If the orbital for those electrons has a higher probability near one of the atoms, the bond is called 'polar covalent'. Multiple such bonds between pairs of atoms may occur. For example, acetylene has a triple covalent bond between its two carbon atoms (Figs. 11.2 and 11.3).

A 'sigma bond' is a covalent bond with axial symmetry about the line joining the two atoms. A sigma bond allows one atom to rotate around the symmetry axis relative to the second atom without any energy change. Sigma bonds can be formed by pairs of atomic s-wave or axially aligned p-wave atomic orbitals which have the two atomic waves in phase in the region between the atoms. 'In phase' waves have the same sign in a given region at a given time. Having the atomic waves in phase

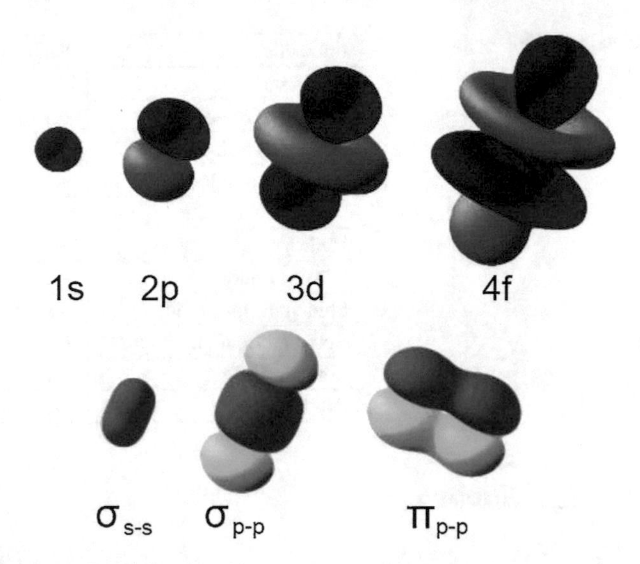

Fig. 11.2 Formation of covalent bonds: The top figures are depictions of some atomic orbitals of hydrogen, with their common labels. The delineated regions encompass volumes in which a given electron has a probability of being found greater than 10%. Light and dark shades indicate where the electron wave function might be positive and negative during half of its sinusoidal oscillation period. The bottom figures show how atomic orbitals combine to form some common molecular orbitals

is needed for the Pauli principle to hold when the molecular orbital has the electron spins anti-aligned.

A 'pi-bond' is a covalent bond with reflection symmetry about a symmetry plane of the electron-pair orbital which passes through the center of the atoms. Pi bonds can be formed by pairs of p-wave atomic orbitals aligned with their symmetry axes parallel, and with their overlapping wave lobes in phase.

A 'delta bond' has two perpendicular symmetry planes with the intersection line being the axis of the bond. Four pairs of overlapping in-phase electron lobes from atomic electron d orbitals in two nearby atoms can form such delta bonds.

A molecular orbital may be delocalized over more than two atoms. Two atoms which share two pairs of electrons with more than two atoms may have neither a single nor a double covalent bond between them. Being, in effect, a combination of single and double bonds, such attachments are sometimes referred to as resonant bonding structures. Benzene has such bonds between pairs of carbon atoms in the ring of six. An electron from one of the p-orbitals of each carbon atom is shared over all six carbons, with each of the original p-orbitals of the carbon atoms aligned perpendicular to the plane of the six carbons, and in phase in their overlap regions. In addition, there is a pair of electrons in sp^2 orbitals, one from each adjacent carbon, forming a sigma bond between the two carbons. A remaining sp^2 orbital of each carbon projects away from the center of the hexagonal carbon structure. These six electrons form a sigma bond with six hydrogen atoms to make benzene.

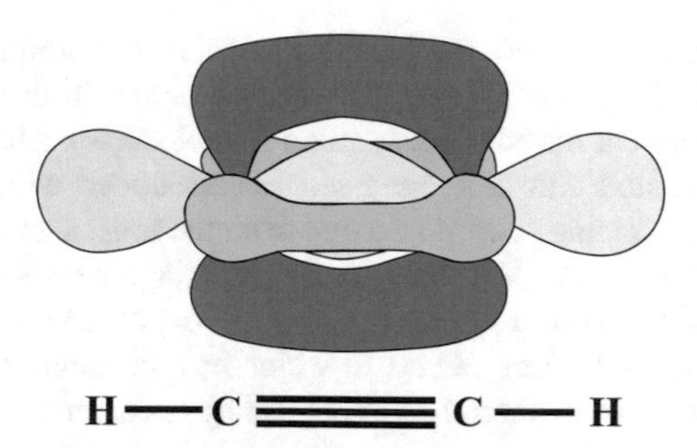

Fig. 11.3 Acetylene covalent bond orbitals: The surfaces of the 'bubbles' enclose regions where the electrons of those orbital bubbles have a probability of being found greater than some fixed number. The triple bond in acetylene is made from two overlapping p atomic orbits from each carbon atom, aligned perpendicular to the axis of the bond and perpendicular to each other, with each atomic p orbital forming one π molecular orbital holding two electrons with oppositely aligned spins. (In the figure, one of these π orbitals is labeled with a darker shade, the other with a lighter shade.) One sp atomic orbital from each carbon atom, aligned along the axis of the bond, combine to form a molecular σ orbital, again holding two electrons. The remaining sp atomic orbital in each carbon stick out to form a combined state with a hydrogen s atomic orbit to form another kind of σ molecular orbit

11.6.3 Ionic Interactions

An electron weakly attached to one atom may jump to a nearby second atom (in an unoccupied valence orbital), releasing energy, and creating an electric force between the resulting ions. If the ions are left free, this Coulomb force will cause the ions to move together until the repulsive electrical and exchange forces keep the ions from getting any closer. In the process, more energy is released. When the ions settle down, we say they are in a bound state. The amount of energy required to get them apart again is called their binding energy. Typical ionic binding energies are $1000\,\text{kJ/mol} = 239\,\text{kcal/mol} = 10.36\,\text{eV/molecule}$. (As comparison, the thermal energy at body temperature (37 °C) available to break bonds is about $k_B T = 0.0267\,\text{eV/molecule}= 2.6\,\text{kJ/mol}$.)

In the realm of chemistry, the strongest of the forces between atoms occurs when the atoms carry a net charge (positive or negative), i.e. they are ions. In this case, the long-range part of the force is given by Coulomb's law. When sodium and chlorine atoms are brought near to each other, an electron is likely to jump from the sodium atom to the chlorine atom. The products, Na^+ and Cl^-, are favored over the reactants Na and Cl by being lower in energy, and the reaction is rapid because the valance electron is only loosely attached to the sodium atom (with the other electrons more tightly held in completely filled atomic shells). Moreover, the electron taken by the chlorine atom fills the chlorine $3p$ atomic level, resulting in a tighter configuration for all the chlorine electrons in the ion than if the subshell were not filled and the spare electron were separated from the atom. For separations of the ions over distances comparable to or smaller than the ion diameters, the force between the ions will show a deviation from Coulomb's law, since the orbitals from one ion to the next will be partially overlapping. A valence electron in the chlorine will now sometimes be found near the sodium, forming a 'molecular orbital'.

In water, sodium and chloride ions become surrounded with water molecules, with their polar sides facing toward or away from the ions, according to the charge on the ions. This tends to keep the ions apart. But with increasing concentrations of sodium and chloride ions in a water solution, salt crystals will form, with greater binding than the separate ions bound to water in a concentrated solution. (This crystallization process is exothermic. Others can be endothermic).

11.7 Weak Molecular Interactions

11.7.1 Hydrogen Bonding

A hydrogen bond can be formed from an atom with a number of electrons exposed in outer orbitals ('electronegative' atoms), such as in oxygen or nitrogen, extending to a hydrogen proton already bound to another molecule. A hydrogen bond is a relatively weak bond, typically less than a quarter of an eV binding each, but this is larger than room-temperature thermal energy, $\sim 1/25\,\mathrm{eV}$. Water molecules in solution form hydrogen bonds with nearby molecules.

Intramolecular hydrogen bonds can be formed, such as in proteins, in their secondary and tertiary configurations. Along the DNA molecule, two hydrogen bonds weakly hold adenine to thymine, and three hydrogen bonds weakly hold guanine to cytosine, with the series of base-pair hydrogen bonds keeping the two strands of the DNA helix together until they are locally spread for transcription of the DNA code or unzipped during replication.

11.7.2 Charge-Dipole Interactions

If an ion of charge q approaches a polar molecule, such as water, with electric dipole moment $\boldsymbol{\delta}$, it will have an electric interaction energy which varies with its distance from the molecule's center according to

$$V_{q\delta}(\mathbf{r}) = -(kq/r^3)\boldsymbol{\delta} \cdot \mathbf{r}.$$

The interaction produces a long-range force on the ion of size $F_x \approx -(kq\delta_x)/r^3$.

11.7.3 Charge-Induced Dipole Interactions

If an ion of charge q approaches a non-polar molecule, it will cause a change in the separation of centers for positive and negative charges in the molecule. The induced dipole moment μ' should be proportional to the external field of the ion: $d' = \alpha E$. The energy stored in the process of making the dipole, starting with a weak field, is then $-\int \alpha E dE$ or

$$V_{q\alpha}(r) = -(1/2)\alpha E^2 = -(1/2)\alpha k^2 q^2/r^4.$$

It follows that the force between the ion and the molecule will vary with r as r^{-5}.

11.7.4 Dipole-Dipole Interactions

Two molecules with fixed electric dipoles δ_1 and δ_2 separated by a distance r will have an energy depending on r and the orientation of the dipoles given by

$$V_{\delta\delta}(r) = k\,(\boldsymbol{\delta}_1 \cdot \boldsymbol{\delta}_2 - 3(\boldsymbol{\delta}_1 \cdot \widehat{\mathbf{r}})(\boldsymbol{\delta}_2 \cdot \widehat{\mathbf{r}}))\,\frac{1}{r^3} \tag{11.23}$$

Putting $\widehat{\mathbf{r}}$ along the $z-$axis gives

$$V_{\delta\delta}(r) = -\frac{k\delta_1\delta_2}{r^3}\,(2\cos\theta_1\cos\theta_2 - \sin\theta_1\sin\theta_2\cos(\phi_1 - \phi_2))\ . \tag{11.24}$$

In a solution at temperature T, the dipoles will rapidly oscillate in orientation. To find the average interaction energy over the possible solid angles for each dipole, we use Boltzmann's distribution for energy levels populated at a given temperature:

$$\langle V_{\delta\delta}\rangle = \frac{\int e^{-\beta V_{\delta\delta}}\, V_{\delta\delta}\,\sin\theta_1\sin\theta_2 d\theta_1 d\theta_2 d\phi_1 d\phi_2}{\int e^{-\beta V_{\delta\delta}}\sin\theta_1\sin\theta_2 d\theta_1 d\theta_2 d\phi_1 d\phi_2}\,, \tag{11.25}$$

or

$$\langle V_{\delta\delta}\rangle = -\frac{\partial}{\partial\beta}\ln\left(\int e^{-\beta V_{\delta\delta}}\sin\theta_1\sin\theta_2 d\theta_1 d\theta_2 d\phi_1 d\phi_2\right), \tag{11.26}$$

where $\beta \equiv 1/k_B T$. The integral is messy in general, but if the thermal energy $k_B T$ is larger than the interaction energy, the integrand can be expanded in powers of β, giving

$$\langle V_{\delta\delta}\rangle = -\frac{\partial}{\partial\beta}\ln\left(1 + \frac{1}{3}\left(\beta\frac{k\delta_1\delta_2}{r^3}\right)^2 + \cdots\right) \tag{11.27}$$

The molecules will have a net average attraction with interaction energy given by

$$\langle V_{\delta\delta} \rangle \simeq -\frac{2}{3k_B T}\frac{k^2\delta_1^2\,\delta_2^2}{r^6}\,. \tag{11.28}$$

11.7.5 Higher Multipole Interactions

Besides dipole fields, the charge distributions within a molecule can create quadrupole, octupole, and higher moments in the external fields. A quadrupole can be thought of as two separated but oppositely directed dipoles on the sides of a square. An octupole can be represented as four dipoles alternately orientated on four opposing edges of a cube.

The multipole effects can be expressed in terms of the electric potential energy V of a charge q at distance $\mathbf{r}$ from the center of the molecule, with total charge Q, net electric dipole moment $\boldsymbol{\delta}$, etc.:

$$V(x, y, z) = kq\frac{Q}{r} + kq\frac{\boldsymbol{\delta}\cdot\widehat{\mathbf{r}}}{r^2} + kq\frac{\widehat{\mathbf{r}}\cdot\mathbf{Q}\cdot\widehat{\mathbf{r}}}{r^3} + \cdots\,, \tag{11.29}$$

in which $\mathbf{Q} \equiv \int \rho(x', y', z')(3\,\mathbf{r}'\,\mathbf{r}' - r'^2\mathbf{1})d^3x'$ expresses the quadrupole moment of the molecule in terms of the charge density ρ within the molecule.

In spherical coordinates, r, θ, ϕ, this reads

$$V(r, \theta, \phi) = kq\sum_{l=0}^{\infty}\sqrt{\frac{4\pi}{2l+1}}\sum_{m=-l}^{l} Q_{lm}Y_{lm}(\theta, \phi)\frac{1}{r^{l+1}}\,, \tag{11.30}$$

where Y_{lm} are spherical harmonics.[12]

The terms

$$Q_{lm} \equiv \sqrt{\frac{4\pi}{2l+1}}\int r'^{l+2}\rho(r', \theta', \phi')Y^*_{lm}(\theta', \phi')dr'd(\cos\theta')d\phi' \tag{11.31}$$

define the multipole moments of the molecule. As can be seen from Eq. (11.30), their contribution to the electric energy at some distance from the molecule diminishes with increasing r, with higher moments interactions dropping faster with distance as the multipole index l increases.

[12] For a general method to convert combinations of spherical harmonics into vector products, see D. Lehman and W.C. Parke, *Angular reduction in multiparticle matrix elements*, J Math Phys **30**, 2797 (1989).

11.7.6 Dipole Induced-Dipole Interactions

A molecule with an electric dipole moment δ can induce another nearby molecule with no natural dipole to get one. The electric field of the first acts on the electrons and nuclei of the second, pushing opposite charges apart. The interaction energy V for the pair is given by

$$V = -k\alpha\delta^2/r^6 . \tag{11.32}$$

The factor α is called the 'polarizability' of the second molecule.

11.7.7 Induced-Dipole Induced-Dipole Interaction

Quantum theory requires all bound particles to have some uncertainty in their motion.[13] A bound particle cannot be at rest. Electrons bound in space must jiggle, even in their lowest quantum state.[14] As a result, fluctuating electric fields must also be created around molecules. These fluctuating fields act on charges in nearby molecules, forcing them to jiggle in unison. In turn, the coherent jiggles of these charges act back on the original charges. The result is force, always attractive, that depends on how easily the charges can be accelerated in their bound states. Electrons can be forced to accelerate by a given field far more easily than the nuclei of atoms, so the charge fluctuation forces are dominated by the jiggling of electrons. The ease with which electrons in a given atom or molecule (labeled by an index 'i') can be so jiggled is measured by its induced polarizability, α_i. The interaction energy $V_{\alpha\alpha}$ due to charge-fluctuations in each of two interacting molecules with polarizabilities of α_i, α_j is found to be

$$V_{\alpha\alpha}(r) = -K(T)\frac{\alpha_i\alpha_j}{r^6} , \tag{11.33}$$

where the coefficient $K(T)$ is always positive and may depend on the temperature T of the surrounding solution and on the excitation energies available to the molecules. The resulting force, is called the 'London dispersion force', a pure quantum effect.[15] There are a number of ways to which the dispersion force

[13]Quantum theory gives the Heisenberg uncertainty principle, $\Delta p_x \Delta x \geq \hbar/2$, where Δp_x is the uncertainty in a particle's momentum, and Δx the uncertainty in its position, both along any direction, such as the x axis. Thus, neither can be zero.

[14]The potential energy for a confined electron can be expanded as a power series about its minimum: $V = V_o + (1/2)\sum_1^3 k_j(x_j - x_{j0})^2 + \cdots$. The lowest electron energy state will be well approximated by that of a harmonic oscillator, for which $E_0 = (1/2)hf$, where $f = \sum_1^3 \sqrt{k_j/m_e}$.

[15]F. London, *Zur Theorie und Systematik der Molekularkräfte*, Zeitschrift für Physik **63:3–4**, 245 (1930).

is referred: Dispersion force; Induced-dipole-induced-dipole force; London force; London-van der Walls force; Quantum fluctuation force; and Charge-fluctuation force. The term dispersion force, invented by London, is used because the same frequency-dependent polarizabilities appearing in the force also affect the dispersion of light in dielectric materials.

Molecular forces may be significantly reduced by the presence of screening counter-ions surrounding macromolecules, forming molecular complexes, and leaving only fluctuation forces between the molecules.

For identical or similar large molecules, there will be a large number of electrons taking part in the attraction of these molecules. A resonance occurs between each fluctuating charge in one molecular group of atoms with its corresponding fluctuating charge in the second molecule, since each has the same 'zero-point fluctuation frequency'. As a result, there is an enhancement of the attractive force between the pair of molecules.

The London dispersion effect can account for the accumulation and clumping in a solution of one kind of macromolecule.[16]

The forces between uncharged molecules, permanent multipolar, induced polar, and dispersion forces, are collectively called the van der Waals forces.

11.7.8 Depletion and Other Shadow Forces

A macromolecule in a medium containing particles moving in all directions will experience diffusive effects when the media particles collide with it. The impact of these collisions produces a 'random walk' (Brownian motion) of the molecule.

Now consider two molecules near to each other. If the gap between these molecules is comparable to the size of the medium particles, then each of these molecules will 'shade' the other from collisions with the medium particles. The same shading will occur if a molecule comes close to a boundary. Because of such shading, there will be a net force due to collisions with the medium particles from the unshaded side toward the shaded region. Near a boundary, there will be a slight increase in the density of molecules, and a slight increase in the pressure toward the boundary. Nearby macromolecules will experience a net attractive force from medium particle collisions. These forces are often referred to as 'depletion forces'. An additional shading effect will occur if the effective surface temperature of a

[16]H. Jehle, W.C. Parke, and A. Salyers, *Intermolecular Charge Fluctuation Interactions in Molecular Biology*, **Electronic Aspects of Biochemistry** (B. Pullman, ed.), 313, [Academic Press, NY], (1964); H. Jehle, W.C. Parke, and A. Salyers, *Charge fluctuation interactions and molecular biophysics*, Biophysics IX, 401–412, [Moscow] (1964), translation to English, **Biophysics 9:4** 433–447, [Pergamon Press, Oxford] (1965); H. Jehle, W.C. Parke, M. Shirven, D. Aein, *Biological Significance of Charge Fluctuation Forces (London-van der Waals) between Sterically Complementary and Sterically Similar Molecules*, Biopolymers Symposia **13**, 209–233 (1964).

macromolecule differs from the medium and the mean-free path of the medium particles is not much shorter than the distance between the macromolecules.

For random impulses coming from all directions, the impulses stopped by the first molecule which would have hit the second will be in proportion to the area of that first molecule over a sphere surrounding the second molecule. If the molecules are separated by a distance r, that area will diminish as the square of r. Thus, for a fixed impulse given by the media particles, the unbalanced force on the second molecule must be in proportion to the inverse square of r. The force will drop even faster than r^{-2} if the media particles have a mean-free path shorter than r. In addition, because of fewer collisions in the shadow regions, the macromolecules may cool on their shadow surfaces. This increases the net attraction due to the shadow force.

This shading effect has a long history. In 1748, Le Sage attempted to explain gravity by postulating a background random flux of particles absorbed by masses. Because the shading depends on the relative size of the area of one mass on a sphere centered on the second mass, such a flux does produce a force inversely proportional to the square of the separation between the masses. However, the requirement of absorption creates far too much heating.

Shadow forces do come to play in the agglomeration of dust in hot plasma near stars, and in the Casimir effect, i.e. the attraction of uncharged conducting plates in a vacuum, which can be explained by the 'quantum shading' of the possible virtual photons between the plates. The drifting together of two boats in rough waters has a similar explanation: There are missing wave modes between the boats.

11.8 Studying Biological Systems at the Nanoscale

11.8.1 Spectral Identification of Biomolecular Structures

The structure of molecules can be investigated by studying how the molecule traps and releases light. Photons, from microwaves to far ultraviolet, are absorbed and emitted within molecules by electron transitions and by molecular vibrations, twists and rotations. As these states are quantized, emitted and absorbed photon energies hf are correspondingly quantized. Moreover, since the photon is a spin-1 particle, the photon-induced transitions must change the angular momentum of the molecule by one unit of Planck's constant in the direction of the photon's momentum.

As indicated in Sects. 7.17.1 and 7.22, the spectral absorption for molecular rotations and vibrations are typically found in the infrared (IR), while valence electron transitions produce spectra in the ultraviolet and visible. These observations we understand by considering the energetics of molecular quantum excitations and the corresponding photons energies.

In attempting to identify unknown organic substances, an IR spectra can serve as a 'fingerprint' of each organic molecule present. In absorption spectra, the internal vibrations of a set of identical molecules are stimulated by light which arrives over a range of frequencies. The light absorption frequencies observed come from the resonant absorption at or near the normal modes of vibration for the system. Molecules have a set of normal modes of vibration. The most general internal vibratory motion of the molecule will be a sum of normal vibrational modes. Vibrational modes of molecules are determined by the structure of the molecules and, to a lesser degree, by the interactions with nearby molecules. Vibrational modes involving stretching of the interatomic distance tend to have higher frequencies than distortions which leave the interatomic distance fixed, and so the spectral wave-numbers for IR resonant absorption due to stretching modes are above those for modes with lateral shifts of atoms and side chains (with those above about 1600/cm likely to be stretching mode absorptions). In the IR wave-number range 480–4000/cm, the resonant absorption is mostly from the ground state to the first excited state if either state has an electric dipole moment. The stretching mode vibration of the first excitation is usually at an energy far below the the bond-breaking energy, so the vibration is close to that of a harmonic oscillator.

The maximum number of distinct normal modes is determined by the number of vibrational degrees of freedom available to the molecule, i.e. the number of independent coordinates needed to specify the internal motions of the system. For a system with N particles in a molecule capable of motion, there will be $3N$ possible motions. However, of these, 3 will be simple uniform translations of the whole molecule through space and there are 2 or 3 more to specify the way the molecule is rotating as a whole. If the molecule has axial symmetry, rotation around the symmetry axis may not be detectable.[17] In this case, only two angular velocities are needed in determining the energy absorbed by rotation of the molecule.

If the three coordinates of the center of mass of the molecule, X, Y, Z are used to track the translational motion, we can eliminate the translational motion from the possible motions by selecting a center-of-mass coordinate system. In this system of coordinates, there will no longer be a translational motion of the molecule and so this motion will no longer contribute to the kinetic energy of the atoms.

We are free to select the direction of the x, y, z axes. We are also free to select three angles which determine the orientation of a molecule, as this choice should not affect how the molecule executes internal vibrations. If the molecule has a symmetry axis, rotations with small angular velocities around that axis are not observable, so that one cannot eliminate a degree of freedom by fixing the angle of rotation around this symmetry axis. This leaves $3N - 6$ independent internal motions for a non-axial molecule, and $3N - 5$ if there is axial symmetry. Distinct vibrational modes correspond to the symmetry group of the molecule.

A water molecule, considered as a three particle system, has no rotational axis of symmetry, so $3N - 6 = 3$. It can vibrate in three different normal modes: bending, with the lowest wave number $1595\,\mathrm{cm}^{-1}$, symmetric stretch, with the lowest wave number $3657\,\mathrm{cm}^{-1}$, and an antisymmetric, with the lowest wave number $3756\,\mathrm{cm}^{-1}$. Carbon dioxide, having three atoms all in a line and so an axis

[17]For fast rotations, the molecule may expand out from the axis, making such motion detectable.

of symmetry, has $3N - 5 = 4$ normal modes of vibration. If we take the z-axis along the symmetry axis, there is a bending mode in the z-x plane at $667\,\mathrm{cm}^{-1}$, a second bending mode in the $z - y$ plane also at $667\,\mathrm{cm}^{-1}$, a symmetric stretch at $1333\,\mathrm{cm}^{-1}$, and an antisymmetric stretch at: $2349\,\mathrm{cm}^{-1}$.

Normal mode vibrations will be excited by resonant absorption of a photon which happens to have a frequency at or near a vibrational frequency of the molecule, provided the vibration generates a charge separation of some multipolar character in the molecule, since photons are absorbed only by acceleration of charges within the molecule. Moreover, in the photon absorption process, the total angular momentum of the whole molecule, made up from orbital angular momentum and intrinsic spins of the electrons and nuclei in the molecule, $\mathbf{J} = \mathbf{L} + \mathbf{S}$, must change by at least one unit of angular momentum, representing the intrinsic spin of the photon. Larger changes in J are possible if the photon carries into the reaction orbital angular momentum as well as spin. From the quantum theory of light, the photon spin is either aligned with its direction of motion, or anti-aligned, corresponding to the two possible helicity states of a photon.

11.8.2 Scanning Probe Microscopy

Scanning Tunneling Microscopy Pictures of individual molecules and atoms have been generated by using quantum tunneling. A sharp conducting needle is carried back and forth over the surface of an atomically smooth surface on which the molecules have been laid. The needle is charge relative to the surface. When the needle comes close to a molecule, electrons tunnel across the gap. This small current is used to control the height of the needle, and to monitor the topology of the surface being explored. The system forms a scanning tunneling microscope. Lateral resolutions the diameter of hydrogen atoms have been achieved. Depth resolutions can be one tenth of a hydrogen atom diameter.

Scanning Atomic Force Microscope An atomic force microscopy uses a needle placed on the end of a flexible small-mass cantilever. The tip of the needle (the 'probe') can be sharpened to about 100 nm. The force between the needle and a surface deflects the cantilever. The displacement of the cantilever is often determined by the deflection of a laser beam from a mirror on the end of the cantilever. To control the needle distance from the surface being scanned, there is a feedback system to adjust this distance according to the measured force. Scanning is usually done by mounting the sample on a piezoelectric material which can be moved in the plane of the sample surface by a change in the electric field in these directions.

This kind of microscope can show individual strands of DNA lying on the surface of a selected substrate. (See Fig. 11.4.)

A variant is the *Near-Field Scanning Microwave Microscope*, which can probe the electric properties of flat materials at the nanoscale. The microwave signal

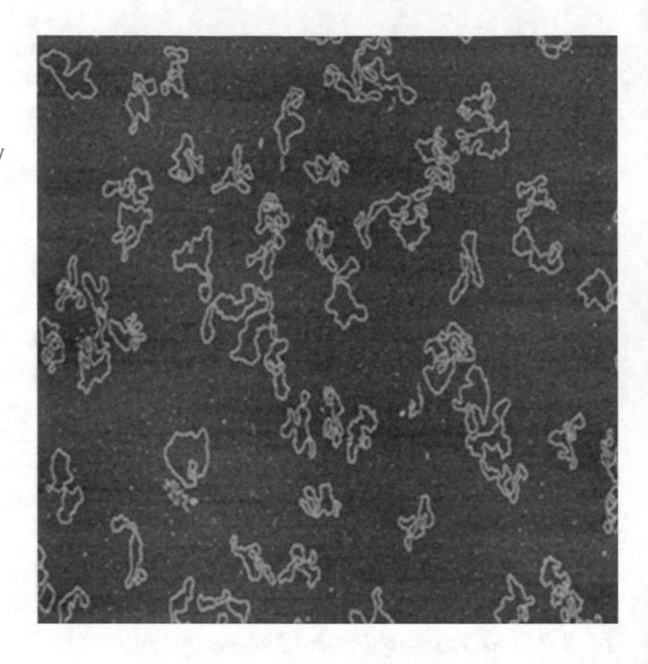

Fig. 11.4 Atomic-Force-Microscope Image of a plasmid DNA (pUC19). The image is from Dalong Pang, Alain R. Thierry, and Anatoly Dritschilol, *DNA studies using atomic force microscopy: capabilities for measurement of short DNA fragments*, Front. Mol. Biosci. (29 January 2015)

extends over tens of nanometers. The Rayleigh limit for the microwaves is avoided by generating and detecting the electric field in the needle's microwave near-zone. Resolutions on the order of a few nanometers are possible.

11.9 Biomolecular Motor

In order for life systems to be able to generate order, those systems have developed the ability to do work. This includes molecular motors in viruses, the transport of molecules within cells, the movement of celia, flagella, pseudopodia, pressure pumps, and muscle contraction. For higher-level organisms, movement helps gather food, interact with other organisms, and explore and exploit the environment. At the nanoscale, such organisms use biomolecular motors to effect movement.

Molecular motors have become an essential mechanism in life systems on Earth, from the smallest organisms, viruses, to us. The activity of life necessarily involves the use of force over a distance to create order. Force times distance is work. That means the molecular motor transforms stored energy into work.

Physically, a motor is any device for transforming some form of energy into mechanical motion. In principle, this transformation can be 100% efficient. Electric motors with magnetic bearings and superconducting currents can transform energy stored in electrical or magnetic pressure into the motion with no loss into heat. However, the motors used in biological systems use energy stored in the molecular bonds, typically the energy in ATP, as a source within cells. The rotation of flagella

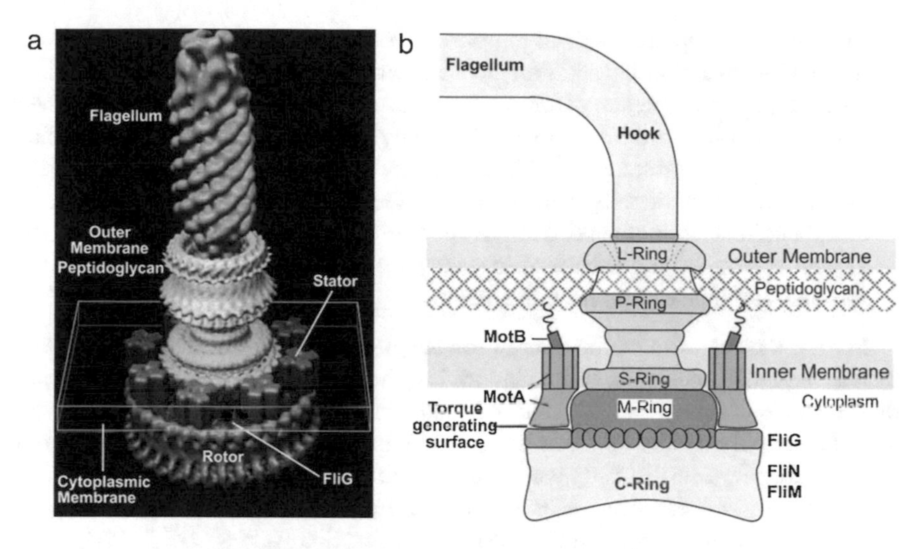

Fig. 11.5 Structures in a Flagellar Motor: Bacteria, such as E-coli in our gut, have nanoscale electric motors built into their cell wall. Each motor rotates a flagellum, that in turn propels the bacterium. (Drawing from paper by David J. DeRosier.)

is powered by a proton current through each flagellum motor embedded in the cell membrane (See Fig. 11.5). The cell ion concentration gradient and electrical gradient across cell membranes is maintained by membrane ion pumps. In turn, these are powered by ATP.

In the conversion of chemical energy to mechanical energy, there is likely to be some disorder as the energy is released. This make the release of energy entropy producing. The process involves dissipative effects. Heat is a likely by-product, and the resultant cyclical conversion of molecular energy to work is not 100% efficient.

Here are some examples:

- The T4 bacteriophage virus steals energy from its E-coli bacterial host to tightly pack its DNA into its head capsule.
- The membrane of cells have pumps and transporters to carry certain ions and molecules across.
- Bacteria use a rotary molecular motor to operate their flagella.
- Inside the nucleus of cells, molecular motors act to manipulate, transcribe, and repair DNA.
- In the cytoplasm of cells, molecular motors transport materials out of the nucleus ('kinesin motors') along microtubules, between organelles ('myosin motors'), and into the nucleus ('dynein motors').
- Pseudopodia generate motion by adding to cell membranes in one direction (by 'exocytosis') and deleting cell membrane in the opposite direction (by 'endocytosis'). The pull is effected by moycin moving along actin fibers within the cell. Our immune cells use this type of amoeboid locomotion.

- Myosin motors operate skeletal muscles. Myosin molecules have swivel heads at their end to ‘walk’ along actin molecular fibers, contracting muscles.
- Large protein complexes within cells are often too large to diffuse from their point of creation to the locations where they are needed. In these cases, small molecular motor proteins called kinesins attach to the complex at one end of the kinesin, and then literally walk with two ‘feet’ at the other end of the kinesin along a cellular microtubule, using ATP to power conformational changes to the kinesin’s feet, causing each foot to sequentially attach, drag the kinesin’s cargo forward, and then detach from the tubule.

For us, the majority of the energy for our molecular motors derives from that stored in a small molecule, adenosine triphosphate, ATP, which can easily release that energy by hydrolyzing to ADP. We also use guanosine triphosphate, GTP, for microtubule polymerization. Even though there is only about 250 g of ATP in all our cells, we recycle our own body weight of ATP in 1 day.

The energy we use from ATP is released from the bond that holds the third phosphate group to its adjacent phosphate in the reaction $ATP \rightarrow ADP + HPO_3$. ATP has a relatively low activation energy, so release of its dissociation energy is not difficult. In the reaction, 30.5 kJ/mol = 0.315 eV/molecule is made available. (This is the change in Gibbs free energy.) Some biological molecular motors can have high efficiency, and so utilize a good fraction of the released energy.

Bacterial motors that drive the rotation of flagella by the flow of a current of protons through the motor can be of high efficiency in the power used to pull charges through the motor compared to energy delivered to the flagellum from ATP-charge pumping. The motor power is expended against viscous drag on the body of the bacterium to propel it through fluids.[18]

Problems

11.1 Give an example of a molecule involving a valence d-orbital contributing to binding. How are these d-orbitals like hydrogenic d-orbitals, and how are they different?

11.2 Suppose an electron in a valence orbital in methane is excited by a photon. What makes this excited state unstable, and why is the state not a steady solution of the wave equation?

11.3 Show that

$$\psi_{1s}(r, \theta, \varphi) = N \exp(-r/a_0)$$

[18]Edward M. Purcell, *The efficiency of propulsion of rotating flagellum*, PNAS **94:21**, 11307–11311 (1997).

is a solution to the Schrödinger equation

$$\left(-\frac{\hbar^2}{2m}\left(\frac{1}{r^2}\left(\frac{\partial}{\partial r}r^2\frac{\partial}{\partial r}\right)\right)-\frac{e^2}{r}\right)\psi=E\psi$$

for an electron which has a spherically symmetric probability around the proton, provided

$$a_0=\frac{\hbar^2}{me^2}$$

and

$$E=-\frac{me^4}{2\hbar^2}\,.$$

Find the value of N in terms of a_0 in order for the solution to give a probability amplitude.

11.4 The a_0 in the previous problem is called the Bohr radius. Using the known values of the mass of the electron, the charge on the electron and proton, and Planck's constant, show that a_0 is about $0.5\times10^{-10}\,\mathrm{m}=0.5\,\text{Å}$. Verify that the energy of the electron in the above state is $-13.6\,\mathrm{eV}$.

11.5 Make a plot of the $3p_z$ electron wave in hydrogen,

$$\psi_{3p_z}=\frac{2}{27}\sqrt{\frac{2}{\pi}}\frac{1}{(a_0)^{3/2}}\exp\left(-r/\left(3a_0\right)\right)\left(1-\frac{r}{6a_0}\right)\frac{r}{a_0}\cos\theta\,,$$

verses radius out to $r=12a_0$, taking $\theta=0$ (i.e. the z axis).

11.6 Estimate the chance of finding the $3p_z$ electron wave in hydrogen in a volume of size $(0.1a_0)^3$ centered on the point $r=9\,a_0$, $\theta=0$, $\varphi=0$. (Take the wave function as constant within that volume.)

11.7 Show that $r=r_e$ at the minimum of the Lennard-Jones potential energy.

11.8 Find the Taylor series of the Lennard-Jones potential energy about its minimum point up to square powers of $(r-r_e)$. Compare with the harmonic oscillator potential energy, $E_H=E_0+\frac{1}{2}k(r-r_e)^2$ to get the effective spring constant k in terms of ε and r_e. In these terms, find the frequency of small vibrations.

11.9 The binding energy of the oxygen molecule is 116 kcal/mol. The infrared light spectral excitation of the first vibrational level in oxygen is found at a frequency/c of 1580/cm. From these data and the mass m of the oxygen atom, find the values of ε (expressed in eV) and r_e (expressed in ångströms) for the Lennard-Jones potential energy approximation applied to the pair of oxygen atoms.

11.10 Estimate the lowest vibrational frequency of a pair of atoms, each of mass 40.2×10^{-27} kg, for the Morse potential energy for the two atoms being $V_e \left(1 - e^{-a(r-r_e)}\right)^2$ with $V_e = 10.67\,\text{eV}$, $a = 0.24 \times 10^{10}\text{m}^{-1}$ and $r_e = 3.1 \times 10^{-10}\text{m}$.

11.11 Why does the dipole-dipole interaction between molecules in a solution depend on the temperature of the solution?

11.12 Why is the exchange force between two atoms negligible when the atoms are several times their diameter away from each other?

11.13 Starting with a given string of amino acids, what determines the tertiary structure of a protein?

11.14 Within cells, important small proteins (kinesins) are used to carry larger structures. These protein motors 'walk' along an intracellular filament (a microtubule). The carrier has two 'feet' that alternately attach to the filament and then let go. To test whether kinesins can make false steps backward, perhaps due to being kicked by the thermal motion of the surrounding water, let's see the slowing effect of such stuttering steps. If we have measured the step size and the time it takes to make a successful step, we can compare the top speed possible along the filament with the average speed measured. Let p be the probability that the carrier protein makes a forward step. Find the average distance the carrier goes after n steps, with each step one unit length.

11.15 Take a diatomic molecule with each atom carrying a mass of 20 times the mass of a proton, and a separation of the atoms of 2.5 Å. If light absorption is able to cause this molecule to rotate to its first rotational excited state, what will the frequency of this light be at resonance? Is this frequency in the microwave, IR, visible or UV part of the spectrum?

11.16 Nitrogen and oxygen are the principle gases in our atmosphere. However, carbon dioxide and water dominate the greenhouse effect. What keeps N_2 and O_2 from taking a significant part in the greenhouse effect?

11.17 The 2p hydrogenic orbital with $l = 1$ and $m = 0$ has a coordinate wave function given by

$$\psi(r, \theta, \phi) = \frac{1}{\sqrt{32\pi}} \left(\frac{Z}{a}\right)^{5/2} r \exp\left(-Zr/2a\right) \cos\theta$$

For the hydrogen atom, find the probability that the electron will be found within five Bohr radii from the origin.

Chapter 12
Entropy and Information in Biology

"Animals have genes for altruism, and those genes have been selected in the evolution of many creatures because of the advantage they confer for the continuing survival of the species."

— Lewis Thomas

Summary With the work of Leo Szilárd and Claude Shannon, we now realize that the information stored in a life system can be measured, and that the loss of that information produces entropy. In fact, there is a universality to open systems having an energy flow through them, making internal ordering a natural process, including evolution of new structures and principles derived from optimization.

12.1 Defining Information

Most people have a good sense of the what information is. But few know that information can be measured by a number.

From our sense about information, we recognize that life systems act to preserve and develop information. At the molecular level, the information necessary for life activity and replication is stored in macromolecules. For life on Earth, this information is coded in the arrangement of the base-pair sequences in deoxyribonucleic acid (DNA) and in ribonucleic acid (RNA).

We know from the statistical basis of thermodynamics that an organism is an open system which takes in energy with a degree of order, maintains and builds structures (which necessarily increases local order), and expels heat energy and waste material to the environment, resulting in an increase in the disorder of the environment. Part of the work of higher level organisms is to use the flow of energy to build information, such as new DNA or memories.

The story of how information can be measured began when James Clerk Maxwell was thinking about the entropy of an ideal gas. He imagined a small door in a partition put half-way through the volume of gas. The door would be operated by

W. C. Parke, *Biophysics*, https://doi.org/10.1007/978-3-030-44146-3_12

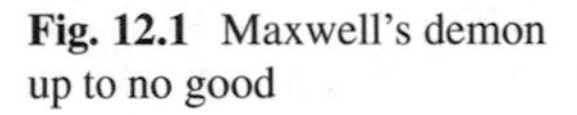

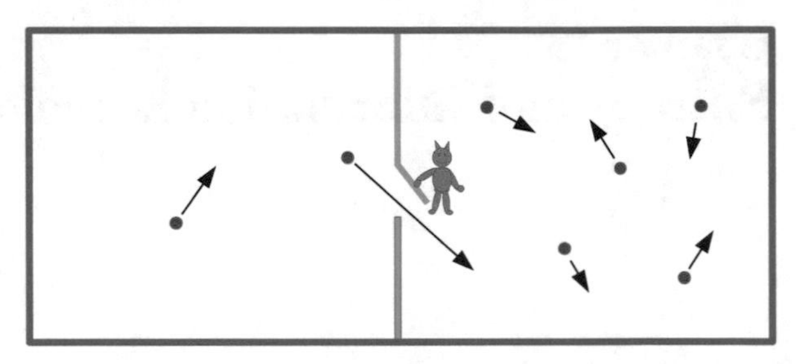

Fig. 12.1 Maxwell's demon up to no good

a little "finite being," who would watch when an atom of the gas heads toward the door from side 'A'. When he sees such an atom moving slowly, he opens the door to side 'B'. But he keeps the door closed for atoms heading fast the other direction. In this way, he can make the gas on side 'B' colder than side 'A', seemingly without doing any work. (The work needed to open and close the door is proportional to the mass of the door, which can be made as small as one wants.) A similar operation by another Maxwell demon could let only atoms through which were going one way, but not the other. Then the gas would end up all on one side. This would decrease the entropy of the gas by $Nk_B \ln 2$. Figure 12.1 shows this situation.

Leo Szilárd was interested in solving the puzzle posed by 'Maxwell's demon', who appeared to cause a violation of the second law of thermodynamics. Szilárd showed[1] that the demon must be a physical agent while acquiring the information to see the speed and direction of the atoms. The demon must produce at least $Nk_B \ln 2$ entropy in dividing the gas while working in a cycle. As this act creates information, Szilárd suggested an equivalence between the entropy created and the information gained about where the molecules in the gas resided, and their velocity direction.

Independently, in 1948, Claude Shannon developed the mathematical theory of communication,[2] defining a measure for information, and how much information can be transferred by a message under the conditions of noise.

Shannon required the following for any definition of information:

1. The information gained depends on the number of ways a given message could have been sent;
2. If two independent messages are received, the information from both is the sum of the information of each;
3. If a message has only two possibilities, receiving one of these messages has transmitted one 'bit' of information.

The first property indicates that we gain information on receiving a message if we initially knew little about what an incoming message might say. If you ask a

[1]L. Szilárd *Über die entropieverminderung in einem thermodynamischen system bei eingriffen intelligenter wesen*, Z Phys **53**, 840–856 (1929).

[2]C. E. Shannon, *A mathematical theory of communication*, Bell System Technical Journal **27**, 379–423 and 623–656 (July and October, 1948). A good exposition can be found in L. Brillouin, **Science and Information Theory**, [Acad Press] (1962).

question which has only two possible answers, you get less information from the answer than if there were a thousand possible answers. We will use the count W for the number of ways a message could be clearly received, and I for the information gained when that message is clearly received.

The second property means that I is logarithmically dependent on W: $I = k \ln W$. The argument is this: Two independent messages have $W = W_1 W_2$ ways of being clearly received, where W_i is the number of ways of clearly receiving message i. We require $I = I_1 + I_2$, i.e. $I(W_1 W_2) = I(W_1) + I(W_2)$. Only logarithms satisfy this condition for any W_1 and W_2.

The third property makes k in $I = k \ln W$ take the value $1/\ln 2$, so that $I = \log_2 W$. This choice of k is quite convenient in quantifying the information sent and received by electronic digital devices, because only two symbols are used, represented in wires by on or off electric pressure. In the wire, a message is then a series of timed pulses, on or off. Within a given time interval, no pulse is a '0' symbol, and a positive pulse is a '1' symbol. These messages can be used to code letters of an alphabet, sounds, or positions, intensities, colors within images, images themselves, and so forth.

If a message is scrambled by noise, we lose information. Let W_a represent the number of ways a message, once received, can be misinterpreted. With ambiguity, we have a reduced number of different messages we can read. If W_c is the number of ways the message could have been composed, then the number of different messages which could be received and clearly interpreted is $W = W_c/W_a$, making the information received: $I = \ln_2 W_c - \ln_2 W_a$, as should be expected, in that ambiguity has reduced the information in a message.

Now suppose a message can use L symbols from the variety $V = \{a_1, a_2, \ldots, a_L\}$ in a sequence. Let the probability of receiving the symbol a_l be p_l, and assume there are no correlations between the received symbols. The average information gathered when the symbol a_l is detected will be $\log_2(1/p_l)$. After a large number N of symbols within the variety V are received, we can detect about Np_l occurrences of the symbol a_l, so that the information received will be $I = \sum Np_l \log_2(1/p_l)$. The average information per symbol is then

$$I/N = -\sum p_l \log_2 p_l \,. \tag{12.1}$$

A simple application of this result is to the game of 21 questions, in which you try to guess what object in a room a person has imagined by asking no more than 21 "yes-no" questions. The best strategy, the experienced player knows, is to ask only questions which have about a 50–50 chance of being answered yes. You will then get the most information from the answer to the question. This observation follows from Shannon's information Eq. (12.1), which is maximum, for varying p_i under the constraint $\sum p_l = 1$, when all the p_l are equal.[3] With only two possible answers, $p_l = 1/2$.

[3]To prove this statement, use the Lagrange-multiplier method (described in Appendix I.1) to maximize $-p_i \log_2 p_i$ under the constraint $\sum p_i = 1$, i.e., $\partial/\partial p_i \left(-p_i \log_2 p_i + \alpha \left(\sum p_i - 1\right)\right) = 0$, which become p_l is a constant, which must then be $1/N$.

12.2 Entropy and Information

As Szilárd showed, there is a direct connection between entropy and information. Shannon formally introduced the mathematical concept of information received in a coded message as proportional to the number of ways that message could have been received divided by the number of independent ways the message could be interpreted after being received, the result being called W.

As we saw above, for long messages, the information per symbol received is

$$i = -\sum p_i \log_2 p_i \ . \tag{12.2}$$

Compare this Shannon information with the Boltzmann entropy for a system which has W ways of being disordered:

$$S = k_B \ln W \simeq k_B \left(N \ln N - \sum n_i \ln n_i \right) = -k_B N \sum p_i \ln p_i \tag{12.3}$$

where $p_i = n_i/N$ approximates the probability of finding a particle in the state with label i for large N. (This expression was presented in the Chapter on The Statistical Basis of Bioenergetics, Eq. (10.81).)

Thus, the information I gained by reading the complete state of a system is related to the entropy S of that system by

$$I = \frac{1}{k_B \ln_e 2} S \ . \tag{12.4}$$

We recall that the statistical entropy of a macrosystem is Boltzmann's constant times the logarithm of the number of ways that the microsystem can be distributed for the given macrosystem. We showed that this definition is the same as thermodynamic entropy when the system was at or near thermal equilibrium. But the measure of disorder, $S_D = k_B \ln W$, applies to any system, including a system of physical objects which can be arranged in a variety of restricted ways. If we know the possible ways your desk can be arranged, we can calculate its disorder. When you put your desk into a certain order, the work you do in creating that arrangement must produce entropy elsewhere, such as in your environment through the heat produced. The total change, a negative quantity for your desk materials plus a positive quantity for the environment, cannot be negative. A perfectly ordered system contains no information, since you already know what you will get by observing the system. In contrast, if the system has disorder, there are a variety of possible 'readings' of that system, so it contains more information than one with lesser entropy.

We now see that Boltzmann's entropy and Shannon information are connected by the following: If we determine how the particles are distributed in the microstates of a system, we gain information about the system as if we received a message which told us how the particles are distributed. The probability of a particle being found in a given energy state is equivalent to the probability of a letter appearing in a message.

With Szilárd's work on Maxwell's demon,[4] we know that if a computer, while operating in a cycle, produces order in a system, such as putting a number of words in alphabetical order, the computer must expel heat. By analyzing a storage device which stores a single bit of information, Landauer proved[5] that each time a bit of information is erased to make that memory available for the next bit, there is an entropy increase of $k_B \ln 2$. Thus, the minimal amount of entropy produced by any cyclic message-generating device which creates I bits of information is

$$\Delta S = (k_B \ln_e 2) I .$$

The cyclic processes which create information necessarily generate entropy. This occurs in our cells as amino acids are assembled into proteins, as DNA is replicated, and in our brains as we learn.

Receiving written messages supplies information. For the English language, messages are coded in letters, space, and punctuation. Depending on the author, the frequency of various letters differ, but over a library of literary works, the probabilities for some 'popular' letters are close to the table below:

Letter	Probability	ASCII (binary)
e	0.127	01100101
t	0.091	01110100
a	0.082	01100001
o	0.075	01101111
i	0.070	01101001
n	0.067	01101110
s	0.063	01110011
h	0.061	01101000
r	0.060	01110010

In a computer, letters, symbols, and other typewriter controls are coded in 8-bits, according to a convention called the American Standard Code for Information Interchange, or 'ASCII'.

Now, because the probabilities of the letters are not all equal, less information is transmitted sending messages by such letters than if the letters were of equal probability. We can save message length (and money, if we are paying for our

[4] L. Szilárd, *On the decrease of entropy in a thermodynamic system by the intervention of intelligent beings*, Z Phys **53**, 840 (1929).

[5] R. Landauer, *Irreversibility and heat generation in the computing process* IBM J Res. Dev **5 (3)**, 183 (1961); see also C.H. Bennett, *Demons, engines and the Second Law*, Sci Am **257 (5)**, 108–116 (1987).

transmissions) by coding the most frequent letters and symbols with fewer bits than eight, and the less frequent with more bits than eight. This strategy is behind compression schemes, such as the Huffman coding.[6] Shannon realized that the information we humans get from English text of length N is significantly less than that calculated from $-N \sum p_i \log_2 p_i$ because, from experience, we anticipate letters, words, and sentences by their context and patterns, thus getting less information when such contextual letters, words, and patterns are received. Shannon estimated that only about one bit of information per character is gained on receipt of a long English text, rather than 4.7 bits per letter that would be gained if the letters occurred with the probabilities implied by frequency tables, such as that given above.

The minimum number of bits to code a given message is a measure of the information content of that message. For example, the number of bits of a 'zip' file is typically less than the original file. If there is no compression possible, then the message, before being read, has no correlations between letters and words, and has maximal entropy (sometimes the letters are then said to be 'pseudo-randomly distributed').

12.3 Information in DNA

Biological information in all life forms on Earth is predominantly stored on deoxyribonucleic acid molecules, a double right-handed helix made from two oppositely-directed chains of linked nucleotides each with a five-carbon sugar called deoxyribose and a phosphate molecule together with one of four nitrogenous bases: adenine (A), thymine (T), cytosine (C), and guanine (G). The bases hold together across the two backbone chains by two or three hydrogen bonds, A with T, C with G. The DNA molecule is 2 nm in diameter with 3.4 nm per helix turn. Proteins are made from twenty biologically active amino acids, with a sequence of amino acids in a given protein, each amino acid coded on the DNA by a triplet of bases. The bases form the "letters" of the code, and a triplet forms a "word" of the code, also called a 'codon'. Thus, the DNA code has 'words' made from three sequential letters. As there are four possible letters, so the number of possible words is: $4^3 = 2^6 = 64$. This means there could be six bits per word. Including code redundancy, there would be 4.2 bits of information per amino acid, assuming no correlations between natural proteins. But through an analysis of the probabilities of given sequences of amino acids in proteins, the information per amino acid is about 2.5 bits[7] rather than 4.2 bits per amino acid.

[6]D.A. Huffman, "*A Method for the Construction of Minimum-Redundancy Codes*," Proceedings of the I.R.E., September 1952, pp 1098–1102.

[7]B. Strait and T. Dewey, *The Shannon Information Entropy of Protein Sequences*, Biophysical Journal **71**, 148–155 (1986).

Table 12.1 Protein coding on RNA

Amino acid	Symbols		RNA codon
Alanine	Ala	A	GCU GCC GCA GCG
Arginine	Arg	R	CGU CGC CGA CGG AGA AGG
Asparagine	Asn	N	AAU AAC
Aspartic acid	Asp	D	GAU GAC
Cystein	Cys	C	UGU UGC
Glutamine	Gln	Q	CAA CAG
Glutamic acid	Glu	E	GAA GAG
Glycine	Gly	G	GGU GGC GGA GGG
Histidine	His	H	CAU CAC
Isoleucine	Ile	I	AUU AUC AUA
Leucine	Leu	L	UUA UUG CUU CUC CUA CUG
Lysine	Lys	K	AAA AAG
Methionine	Met	M	AUG
Phenylalanine	Phe	F	UUU UUC
Proline	Pro	P	CCU CCC CCA CCG
Serine	Ser	S	UCU UCC UCA UCG AGU AGC
Threonine	Thr	T	ACU ACC ACA ACG
Tryptophan	Trp	W	UGG
Tyrosine	Tyr	Y	UAU UAC
Valine	Val	V	GUU GUC GUA GUG
START			AUG
STOP			UAA UGA UAG

Table 12.1 shows the DNA coding for the amino acids, results largely developed by Marshall Nirenberg at NIH in the 1960s.[8]

The human has 23 chromosomes. Each chromosome is a tightly bundled DNA molecule. Human DNA contains over 3 billion base pairs, coding over 18,000 proteins. (See Table 12.2.) As there are four possible bases at each location along the DNA, storing 2 bits each ($\log_2 4 = 2$), the most information our DNA could hold would be 6 billion bits. Redundancy and correlations in the coding makes the actual amount of information far less. Just as in the English language, there are correlations in the sequences of bases. These correlations came about by the requirements of how best to make catalytic and structural proteins from sequences

[8]The first to propose how the DNA base-pair sequence might code protein structures was The George Washington University Professor George Gamow, three months after Watson and Crick published, in February of 1953, a description of the helical structure of DNA.

Table 12.2 Human chromosome genes and bases

Chromosome	Genes	Bases
1	2968	245,203,898
2	2288	243,315,028
3	2032	199,411,731
4	1297	191,610,523
5	1643	180,967,295
6	1963	170,740,541
7	1443	158,431,299
8	1127	145,908,738
9	1299	134,505,819
10	1440	135,480,874
11	2093	134,978,784
12	1652	133,464,434
13	748	114,151,656
14	1098	105,311,216
15	1122	100,114,055
16	1098	89,995,999
17	1576	81,691,216
18	766	77,753,510
19	1454	63,790,860
20	927	63,644,868
21	303	46,976,537
22	288	49,476,972
X (sex chromosome)	1184	152,634,166
Y (sex chromosome)	231	50,961,097
Unplaced bases		25,263,157
DNA total	32,040	3,095,784,273
Mitochondrial-RNA	37	16,569

of amino acids. The folding of the protein is crucial. Certain sequences of amino acids will be preferred to make successful foldings. Moreover, one should expect redundancies for safeguarding essential life processes, and long-range correlations in the coding of proteins which depend on one another. An indication of redundancy and correlation is the ability to compress the Human Genome Data by factors of about 200 using Huffman coding. We should expect we need no more than about 30 million bits of information on our DNA to make one fetus.[9] (The mother's mitochondrial DNA coding also is used, but this coded information is far less than on the DNA, with about 17,000 base pairs holding no more than about 34,000 bits of information.)

[9]There are indications that the information needed is low because the developing embryo uses a strategy of repeating basic themes and then adding variations.

12.4 Dynamics of Information Storage and Transfer

Information storage and transfer is central to life, both in reproduction and in the activity of life. Figure 12.2 shows what is involved in the processing, storing and utilization of information. The diagram applies to life forms, to computers, to the expression of DNA, and to very general systems.[10]

The essential elements are: (1) An energy flow through the system if information is gained from outside the system; (2) Entities that hold the information needed to process incoming information and material; (3) A decoder and transmitter to transport information to (4) A receiver and interpreter that links to (5) an analyzer; (6) An actuator that acts on the information; (7) for cyclic processes that store information or do work, a sink to dump heat. Also, some systems will have one or more sensory inputs together with discriminators to filter the sensory input. During the sequence of information processing, background and system noise will be present to some degree, although digital systems are designed to be relatively immune to degradation by noise. On the other hand, life systems which have progressed by evolution take advantage of variations due to the noise that causes a low level of mutation. The analyzer has alternate names: 'Decider', 'Comparer', 'Weigher', 'Computer', and 'Brain'. An intelligent analyzer will have 'Anticipation

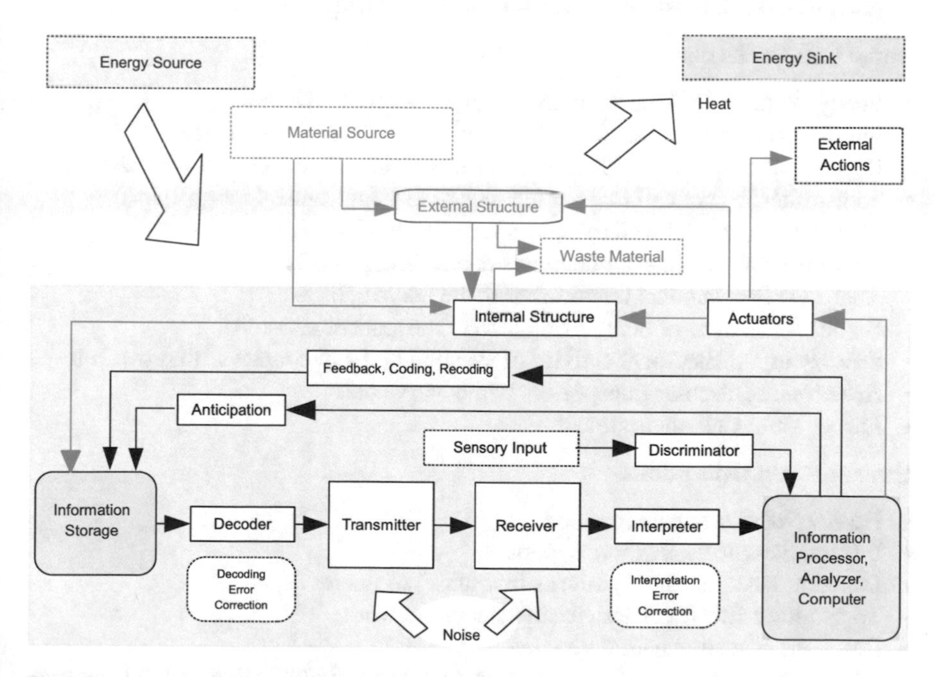

Fig. 12.2 Framework for information-transfer systems

[10]In fact, information and storage systems are a class of ordering processes described in Chap. 15.

Elements', i.e. ways to forecast future events so that preparatory actions can be taken before those events occur. Each of the elements benefit by having repair mechanisms.

To make the diagram in Fig. 12.2 more explicit, consider the following examples.

Cell Operations

- Energy Source: Cell mechanisms for production of ATP
- Information Storage: Principally the DNA molecule in the nucleus of a cell
- Decoder: The transcription of the DNA code to mRNA in the nucleus
- Transmitter: mRNA transport mechanism from nucleus to ribosomes
- Receiver: Ribosomes
- Interpreter: tRNA to proteins
- Analyzer: Biochemical processes which use catalytic and structural proteins
- Feedback: Effectors, inhibitors, and protease production controls
- Internal structure: Built by action of proteins
- Sensory input: Chemical environment of the cell
- Discriminator: G-protein receptors in cell walls
- Actuators: Molecular transporters and molecular motors
- External actions: Cell wall pore and vacuole releases
- Anticipation: Response using the molecular memory storage of past attacks
- Energy Sink: Cell exterior environment accepts heat

Nerve Cell Excitation

- Energy Source: Cell mechanisms for production of ATP
- Information Storage: Dendritic innervation and synaptic plasticity
- Decoder: Pre-synaptic activation mechanisms based on cell wall polarization
- Transmitter: Nerve cell loci for the release of neurotransmitter proteins
- Receiver: Nerve cell wall receptors for neurotransmitters
- Interpreter: Nerve cell ion-channels-controlling proteins
- Analyzer: Biochemical reactions with the nerve cell
- Feedback: History of nerve firing affects subsequent nerve firing
- Sensory input: Electrical activity of synaptic bulbs in contact with nerve cell
- Actuators: Molecular transporters within nerve cell
- Energy sink: Cell environment

Human Brain Operations

- Energy Source: Sugars in blood
- Information Storage: Nerve synapses
- Decoder: The imprinted patterns in a subset of nerves
- Transmitter: Electrical activity in a neural network
- Receiver: A second neural network
- Interpreter: A subset of nerves used to filter activity in the second network including cerebellum
- Analyzer: Pre-programmed and learned coding
- Feedback: Nerve synaptic re-enforcement

- Sensory input: Eyes, ears, nose, and other sensory systems
- Actuators: Muscles attached to nerves and other nerve-stimulated body actions
- Energy sink: Blood and surrounding tissue takes up heat

Social Interactions

- Energy Source: Activated group of humans
- Information Storage: Accumulated knowledge (in people, books, hard drives, etc.)
- Decoder: Experts and translators
- Transmitter: Speakers and communicators
- Receiver: Physical receivers and/or listeners
- Interpreter: Experts and translators
- Analyzer: Combined analysis by a group of people and/or a computer
- Feedback: Saving of analysis and learned experiences
- Sensory input: Observations from others and from the environment
- Actuators: Actions of groups
- Energy Sink: Environment takes heat produced by each individual, including radiation

12.5 Information Channel Capacity

A 'channel' of information is a conduit used to transmit messages. The greatest amount of information which can be transmitted per unit time by a given channel is called the channel capacity C. For example, a glass fiber using a sequence of light pulses to code information has a maximum channel capacity equal to the frequency of the timing signal, since each timing interval contains no more than a single bit of information. With visible light, $C \approx 10^{14}$ per second. Physical systems which transmit information always have background noise[11] which may cause pseudo-random flips of individual bits in the signal carrying the information. The unambiguous interpretation of a message which may be influenced by noise may still succeed if error-correcting code is added to the message by the transmitter. This adds redundancy to the message, and also reduces the channel's capacity. Shannon[12] showed that if the channel capacity is greater than or equal to the rate at which information is transmitted, then error-correction can reduce the information loss to an arbitrarily small amount. Conversely, if the channel capacity is less than the required rate of transmission, then messages will contain unrecoverable errors. Taking noise as a Gaussian uncertainty independent of the signal carrying information, he calculated if the channel has bandwidth B (in Hz), the channel

[11]Noise can arise from many sources, including thermal fluctuations and interference from spurious uncontrollable external interactions.

[12]C.E. Shannon, *Communication in the presence of noise*, Proc IRE **37** 10–21 (Jan 1949).

capacity in the presence of noise will be $C_n = B \log_2(1 + P/N)$, where P is the power of the signal sending the information and N is the power in the noise acting on the signal: The channel capacity increases logarithmically with one plus the signal-to-noise power ratio.

Suppose a channel carries information by binary encoding. If noise causes a binary bit to flip (1–0 or 0–1) with probability p_e, then error-correction code will be needed, reducing the channel capacity by the entropy in the noise, giving a capacity[13] $C_n = C(1 - p_e \ln_2(1/p_e) - (1 - p_e) \ln_2(1/(1 - p_e))$. Note that if noise makes a 50–50 chance of a bit flip, then no information can be transmitted.

12.6 Nature of Dynamical Feedback and Regulatory Systems

A dynamical feedback system is a system that changes over time and has an output that affects its own input. Many people have heard the auditory feedback screech when the sound output of an amplifier is picked up by a microphone of that same amplifier. This so-called positive feedback need not have a run-away effect if the feedback itself is controlled. The control of house temperature can be helped by a negative feedback system, wherein an increase temperature of the house causes the furnace to shut off. Even better is an anticipatory system which monitors the *outside* temperature which changes earlier than the interior temperature. In a crowd, if one person looks up to the sky, many may be similarly inclined. The speed of a motor can be controlled by a 'governor', a device which varies in its moment-of-inertia with the rotational speed. The melting of the polar ice caps exposes more dark ground, increasing the temperature, which increases the melting. An increase in atmospheric carbon dioxide production increases plant vitality and enumeration, reducing the rate of atmospheric carbon dioxide production.

Our DNA has feedback to control the production of proteins, including activating and silencing histones to which genes are bound. The human body has a sophisticated feedback system to keep the internal body temperature at 37 °C. (See Sect. 9.7.) The nervous system includes rapid feedback from sensors, and produces slower feedback by the action of hormones. Our pancreas β-cells and our hypothalamus regulates insulin in the blood, which in turn controls how body cells take up glucose. All these and more are finely-tuned regulatory systems in our bodies to keep us in 'homeostasis' and to respond to distress.

As we have seen, the general scheme of information storage and transfer included a feedback loop that acts back on the original storage unit. (See Sect. 12.4.) Thus, regulatory systems (described by 'control theory') are examples of information transfer with positive and negative feedback. 'Cybernetics' is the study of regulatory systems which involve the use of technology to effect the control.

All dynamical systems with non-linear terms in the amplitude relations have an effective feedback behavior. The strength of some action influences the strength of a

[13] G. Raisbeck, **Information Theory**, [MIT] (1963).

response, but the response can act back on itself. Feedback may be instantaneous or delayed. When action strength is physically limited, positive feedback systems will saturate. Feedback systems can have quasi-stable states (configurations which are centered around steady conditions). For example, there may be two different quasi-stable conditions ('bistable'), which is the basis for digital computer memory units ('flip-flop' devices) for temporarily storing a single bit of information.

A number of questions arise in the study of regulatory systems:

1. What are the probabilities for each possible quasi-stable state?
2. What is the response or delay time for various loops in the regulation, compared to associated activity times?
3. Do oscillations between quasi-stable states occur for the expected range of parameters being controlled by the system? If not, is the system close to such oscillations?
4. Will sporadic solutions[14] occur in the expected range of the parameters being controlled?
5. If the system is perturbed, what is the lifetime of the perturbation before quasi-stability is restored, and how does this lifetime compare with associated activity times?
6. To what degree might the system be disturbed before regulation fails?
7. How robust is the regulation? If parts of the network are blocked or mutated, will the system still be capable of regulation? Are there alternate pathways in the scheme?
8. Are there repair and/or recovery mechanisms?
9. Is the regulatory system transformable under changes in the environment? If so, how rapid is this response compared to times typical to environmental changes?
10. Are there any other nearby or distantly related optimal schemes for the same regulation?

Mathematical models of regulatory systems permitting at least numerical and approximate solutions have been successful in discovering subtle properties. Sometimes, even very simple models carry much of the system's behavior. Mathematical models based on the dynamical behavior of emergent agents in the physical system may be difficult to solve, but are worthy of exploration, helping us use the model for predictions and for alternate designs.

Problems

12.1 If your brain during heavy thinking releases 90 W of heat into air at temperature 22 °C, while during little thinking it releases 70 W, estimate the maximum order (measured by drop in entropy) your brain could have produced each second?

[14]Chaos may develop in non-linear systems. See Chap. 15.

12.2 If you store 10^5 bits of information per second in your brain, what is the minimum rate of entropy production in the environment around you?

12.3 Pick two biological systems which involve information transfer and storage. Draw and identify each of the elements in the generic information transfer diagram as applies to your examples.

12.4 Give examples of prior information about the English language, dialects, and author usages. Show by example that such prior information about a message reduces the information content of that message.

12.5 Show how protein DNA inhibitors fit the general picture of information transfer systems.

12.6 Identify as many sources of noise in the replication of DNA that you think might play a role. Why is the optimal noise reduction in DNA replication not zero noise?

12.7 What conditions would make the proposition that continued ordering of life systems ultimately leads to a catastrophe (the 'singularity postulate') not true?

Chapter 13
Modeling Biological Systems

Science is what we understand well enough to explain to a computer. Art is all the rest.

— Donald E. Knuth

Summary Modeling of biological systems and ecology is a useful exercise, as it allows one to discover and focus on the important aspects of life's system dynamics. Population dynamics is a good example. Modeling also helps us understand system optimization. With a long span of time, evolution picks out optimal systems. Models of optimization can be revealing by testing the importance of various aspects to be optimized, and their constraints. Metabolic scaling laws come about from such optimization.

13.1 Mathematical Modeling in Biology

When a system can be described mathematically, at least to some approximation, we are better able to understand logical relationships in that system, to reduce ambiguities, to deduce consequences, to make definite predictions, and to have a firmer basis for our imaginings and our search for deeper truths.

We have already seen examples of how logic expressed through the shorthand of mathematics helps us understand many of our observations of biological systems. Logical quantitative descriptions have also been applied to such diverse fields as optimization in organ evolution, emergent laws in complex systems, neurological dynamics, robustness in protein design, and the computer simulations of ecosystems, among many others.

In what follows, we will give further examples of the power of logical thinking.

W. C. Parke, *Biophysics*, https://doi.org/10.1007/978-3-030-44146-3_13

13.2 Growth in a Population and Chaos Theory

Anyone who thinks exponential growth continues in a limited world is either a madman or an economist.

— Kenneth Ewart Boulding

13.2.1 *Modeling Population Numbers*

Let's make a simple model of how the number of individuals (rabbits, foxes, or people) varies with time. Of course, individuals are discrete! But when counts get large, the change in the count over a short time is relatively small, so we might expect that a smooth function will fit the variation of the count over time. Smooth functions can also be approached by testing the behavior of many independent samples of populations starting with identical conditions, and then averaging over the samples.

Increases in the count N come about because of births, and decreases by deaths. In some populations, the birth rate per capita (births per year/population) is constant, as may be the death rate per capita. In these cases, changes of N, called ΔN, over a short time Δt should be proportional to N and to the time Δt, giving

$$\Delta N = (\beta N - \gamma N)\Delta t \; , \tag{13.1}$$

where β is the per capita birth rate and γ is the per capita death rate. If $\beta > \gamma$, then the population will grow exponentially.

Realistically, the birth rate is not constant, but depends on factors such as the amount of food available. The amount of food and resources available, in turn, eventually drops as the population increases. A simple model of this behavior is to make $\beta = \beta_o - \beta_1 N$ for N less than β_o/β_1. Alternatively, one may think of this expression for β as the first two terms in a Taylor series of β as a function of N. In consequence, our population number satisfies an equation of the form

$$\frac{dN}{dt} = \alpha(1 - N/\kappa)N \; . \tag{13.2}$$

The value of κ determines at what level the population reaches saturation. One of the advantages of this model is that it has an analytic solution:

$$N = \frac{\kappa}{1 + \exp\left(-\alpha(t - t_o)\right)} \; . \tag{13.3}$$

The number N behaves in time as one might expect (see Fig. 13.1). This model for population growth was invented by Pierre Verhuist in 1844, who named the solution

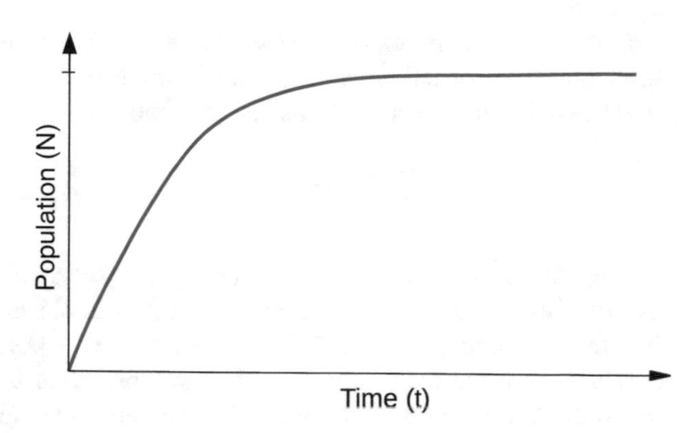

Fig. 13.1 Population growth over time according to Eq. (13.3), starting with the maximum rate. (Taken from Wikipedia, Logistic map)

the 'logistic function'. He was after an answer to Malthus, who worried about an exponential growth of the human population. Note that a shift and rescaling of the time with $t = t_0 + \alpha\tau$, and a rescaling of the population number with $N = \kappa\mathcal{N}$, produces a universal logistic expression[1]:

$$\mathcal{N} = \frac{1}{1 + \exp(-\tau)} . \tag{13.4}$$

In discrete form, $dN \rightarrow N_{i+1} - N_i$. We will use N_0 as the initial population at time $t = t_0$, N_1 at time t_1, etc. Without loss of generality, we can take unit time intervals. Also, to compare with a standard form, we will set $N_i = (1 + \alpha^{-1})\kappa\, x_i$, and define $r = (1+\alpha)$. The logistic equation (13.2) then becomes the 'logistic map'

$$x_{i+1} = r(1 - x_i)x_i . \tag{13.5}$$

Remarkably, this simple nonlinear recursion relation can generate 'chaos', defined to be the behavior of solutions which diverge exponentially when the initial conditions are only infinitesimally far apart.

13.2.2 Chaos

As we have noted in the Sect. 4.3.6, chaotic behavior in the dynamics of the celestial three-body problem was described by Poincaré in the 1890s. In the 1960s, Edward Lorenz discovered chaotic behavior in certain solutions to the Navier–Stokes equation used to make weather predictions, starting with the current atmospheric conditions.

[1]The logistic function in Eq. (13.4) is called a 'sigmoid' because it looks like a tilted 'S' when plotted for both positive and negative τ.

Chaos can arise in non-linear systems, e.g. differential equations expressing the time derivatives of unknown functions in three or more other variables, in which products of the unknown functions appear, such as

$$\frac{df(x, y, z, t)}{dt} = a[f(x, y, z, t)]^3 + \cdots ,$$

Having products of the entities affecting the rates of those same entities is a characteristic of feedback systems. A good example is our population model in the previous section, in which the presence of that population causes its growth to slow. There are many more good examples. Fluid dynamics becomes chaotic due to the non-linear term $\mathbf{v} \cdot \nabla \mathbf{v}$ in the Navier–Stokes Equation. Neuron network models can exhibit chaos when there is feedback. The Hodgkin–Huxley neuron (see Sect. 14.2.2) can show chaotic behavior, as can simple electronic circuits with just one non-linear element.

Our population model above was designed so that we could express an analytic solution. But, in general, non-linear differential equations are notorious for being difficult to solve. Einstein's gravitational field equations are non-linear because localized gravitational energy has an equivalent mass, which acts back on the strength of the gravitational field. This property alone makes Einstein's field equations difficult to solve, unless the system has particular simplicity or strong symmetry.

Even though a system of non-linear differential equations may be difficult to solve analytically, i.e. the solution in terms of known analytic functions is difficult to find, we can 'discretize' the problem[2] by taking small increments for the independent variables. We then have a system of recurrence relations for the unknowns, indexed by the number of time increments. For example,

$$\begin{aligned} f_{i+1} &= \phi f_i + F(f_i, g_i) , \\ g_{i+1} &= \gamma g_i + G(f_i, g_i) \end{aligned} \tag{13.6}$$

where ϕ and γ are constants and F and G are non-linear functions in f_i and g_i.

Now, you say: "Alright. Now we have the equations to solve the problem, at least numerically." Yes, but there is still the 'butterfly effect'. You start with an initial choice for the values f_0, g_0 and use the Eqs. (13.6) to step through time. Great. However, if you change the initial conditions by just a 'little bit', you may find that your new solution differs widely from your first one. If you are trying to predict the weather, your initial data will be uncertain by small amounts. That makes detailed weather prediction over more than about a week practically impossible.

There's another complication: Computers cannot be infinitely precise about the value of an irrational number, or even a very large rational one. This means that

[2]Actually, the problem may have originally been discrete, and had been 'smoothed' by a calculus limit.

even if you have perfect initial data, the initial data stored in a computer will have an uncertainty at the limits of the computer's precision, where the number has been truncated to fit the allocated memory. Again, the uncertainty in the solutions can grow exponentially.

Systems that exhibit chaos will correspondingly generate entropy in the course of time through the information they lose.

Now, back to the 'logistic map' produced by Eq. (13.5). If $r \leq 4$ and $0 \leq x_i \leq 1$, the new value x_{i+1} will also range from 0 to $r/4 \leq 1$. In effect, the range of x_i has been 'folded' onto itself by the logistic map, preparing x_{i+1} to be 'acted' on again. This folding is shown geometrically in Fig. 13.2 by the horizontal lines drawn from the 'y' axis, representing x_{i+1} extending to the diagonal line $y = x$, and then straight down to the x axis to start a new mapping. Solutions having $0 < x_i < 1$ occur as i increases for each value of r. If r is selected less than one, iteration of

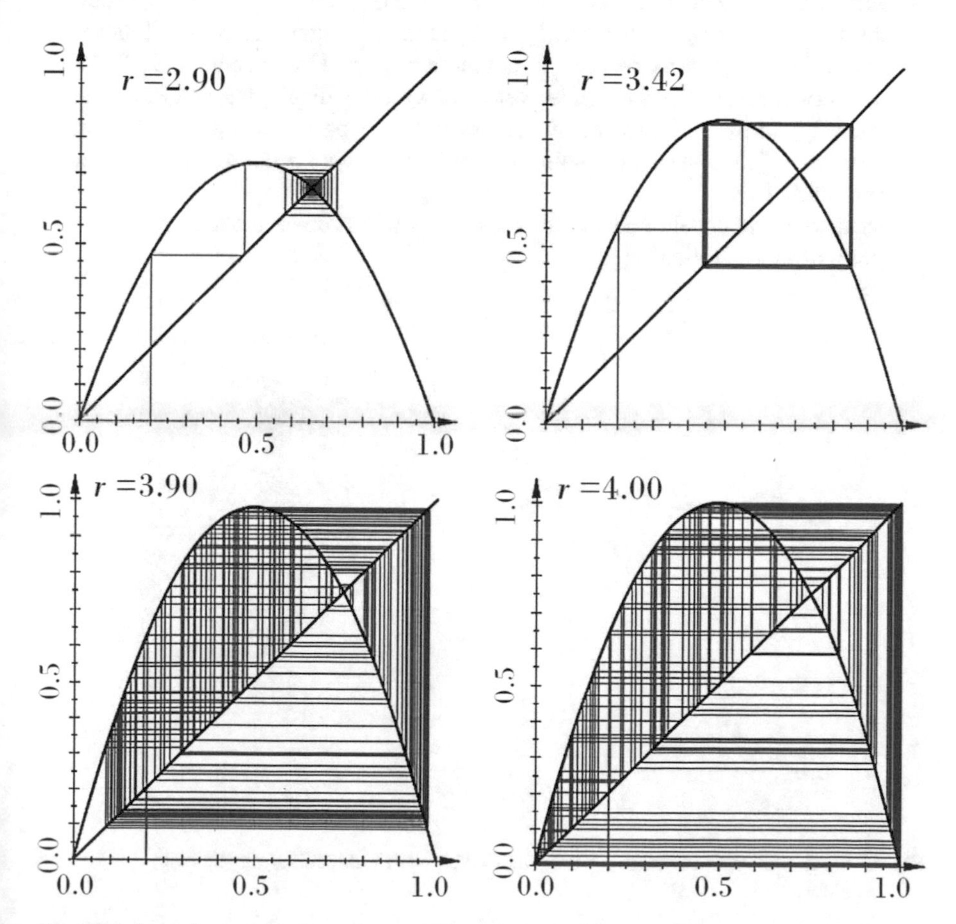

Fig. 13.2 Logistic mapping for various r. Where the horizontal lines in the figures become dense and clustered, there is an attractor value within that cluster. (Taken from Wikipedia, Logistic map)

the map gives solutions which fall to zero. This should not surprise us, because $r < 1$ corresponds to $\alpha < 0$, which is a negative birth rate (which really becomes part of the death rate). If $1 < r < 3$, as i increases, x_i approaches $1 - r^{-1}$. If $3 < r < 1+\sqrt{6} = 3.44949\cdots$, for almost any initial value x_0, the x_i, as i increases, will approach a state of oscillation near one of two values. These values are called 'attractors'. Since the solutions x_i as i is incremented do not stay settled at or near one attractor, but rather flip between the neighborhood of either one of them, we say that the possible solutions has 'bifurcated'. For $1 + \sqrt{6} < r < 3.54409\cdots$, the x_i will approach a state of oscillation near four fixed values. This 'doubling' of the solution states continues as r increases up to $3.56995\cdots$. By then, the doubling has gone to infinity. Successive mappings for three values of r are shown in Fig. 13.2.

Let $r(k)$ be the value of r where a new doubling occurs. The increments of $r(k+1) - r(k)$ between doubling decreases toward zero. The inverse ratio between successive increments, $(r(k) - r(k-1))/(r(k+1 - r(k))$ approaches a constant, a property discovered by Feigenbaum.[3] Figure 13.3 shows how the location of the attractors behave as a function of r. Feigenbaum (1978) found that this constant applies to a wide variety of systems displaying chaos, and so is considered a universal constant of mathematics, on par with π and e. For most values of $r > 3.56995\cdots$, the solutions for large i jump around between 0 and 1 with no regularity.

Because of the onset of chaos, the population of rabbits in the presence of foxes can jump around erratically!

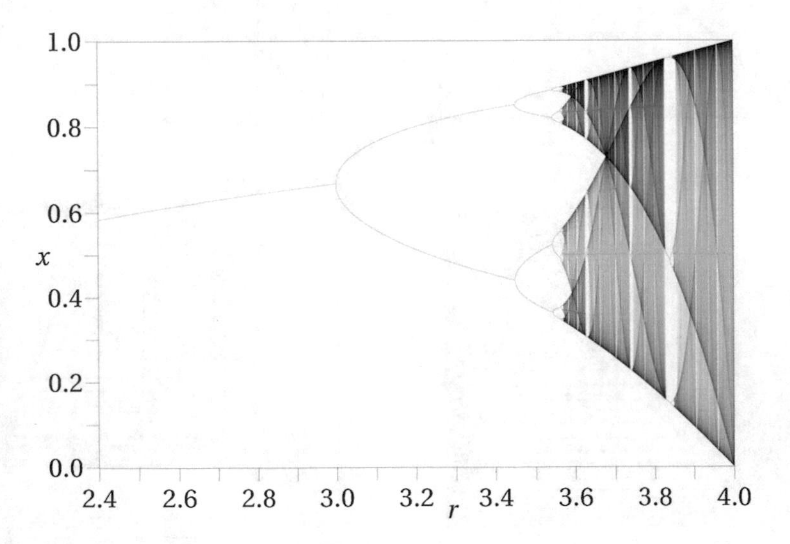

Fig. 13.3 Attractor values versus r parameter (Taken from Wikipedia, Logistic map)

[3]For large k, $r(k+1) - r(k) \approx A \exp(-k\, ln\delta)$, where $\delta = 4.6692\cdots$ is called Feigenbaum's constant.

13.2.3 Connections with Critical Phenomena

In the previous paragraphs, we described how, for non-linear systems, as a parameter associated with the dynamics of that system changes, the system can reach a 'critical point' where the system undergoes a transition increasing its disorder. We also saw that as one increases a parameter, say r, the dynamics of certain non-linear systems show chaos when r reaches a critical value, say r_c.

Phase changes, such as water turning from a liquid to a gas, can have disorder changing drastically upward when there is only a small change in temperature. Order in this case means some regularity in the material structure. One sign of approaching chaos is the development of long-range correlations in space as the system becomes more ordered. Subsystems are formed with dimensions much larger than the typical separation distances between the fundamental entities.

A sudden increase in disorder means a jump in the entropy while passing through the phase-change temperature. The process is easily described: At a given temperature, a solid has more order than a liquid (and liquid more than a gas) for a given number of atoms. When a solid melts, or a liquid vaporizes, each atom must take up energy to release it from local attractive forces, including 'long-range' forces between neighboring atoms. Environmental thermal energy is one source for this melting or vaporizing energy, called latent heat. Latent heat is released or absorbed during a so-called 'discontinuous phase transition'. The first derivatives of the Gibbs free energy appear discontinuous as the temperature or pressure passes through the phase change point.

It is also possible to have a phase change without any exchange of heat. Such 'continuous phase transitions' occur at the so-called critical point. For materials, the critical point is that state where the liquid and gas phases are in equilibrium and have the same specific density and entropy, each measured per particle. A characteristic of continuous phase transitions is the apparent singularities in the susceptibilities, i.e. the second derivatives of the Gibbs free energy. During a continuous phase transition, the correlation in the behavior of distant clusters of particles approaches infinity according to a power law. This can be seen as the onset of a milky appearance as a clear liquid approaches its critical point. The effect is called 'critical obsolescence', and is caused by large-scale variations in clustering of particles.

Measurable quantities such as the heat capacity, when measured near the critical point with temperature T_c, behave close to the power law: $C \propto |T_c - T|^{-\alpha}$. Just as in the case of the logistic map, the exponent $\alpha \approx 0.59$ applies to a diverse set of systems, and so is thought to be a universal constant called the 'critical exponent' for that set of systems.

13.3 Optimality Conditions Applied to Organized Systems

13.3.1 Specifying Optimality Conditions

In dynamical systems undergoing ordering, those subsystems that optimize the efficient utilization of available energy and that optimize the production of critical structures often have tested a wide variety of competing dynamical pathways. Those pathways which optimize conditions for faster and more robust reactions will be the most successful, i.e. they will become dominant. In addition to the level of activity, the conditions of quasi-stability, the degree of adaptability, the extent of viability, and the ease of reproducibility come into the optimization process of life systems.

A variety of approximation schemes have been applied to find the state of relevant variables which minimize functions, such as power consumption, cost, or the so-called 'fitness function'. Finding the extremes of functions and of data is an important challenge in many fields of study, including mathematical biology, biochemical pathway studies, network analysis, ecology, neuronal connection studies, statistics, and complex systems. (See Appendix I.)

Optimization conditions can often be expressed as the minimization of a 'cost' function, $f(x)$, under a set of n constraints $g_l(x) = 0$. Here, the $\{x\} \equiv \{x_1, x_2, \cdots, x_N\}$ is a set of N independent variables which affect the cost, with $1 \leq l \leq n < N$. Analytic techniques exist to minimize known functions under constraints. One of the most useful is the 'Lagrange Multiplier Method' (see Appendix Sect. I.1).

If the function has a large number of variables with many minima, finding those with function values below a certain level can be a daunting task. For example, a long protein chain of amino acids will have many possible flexation and rotation loci along the protein chain, making an astronomically large number of distinct final folds, all with a particular folded energy. Finding what nature picks, or what might be a useful medicinal protein, requires optimization strategies. Some methods used are described in Appendix Sect. I.2.

To show the utility of optimization methods for relatively simple examples, we will examine the questions: Are there optimal arterial radii and bifurcation angles in animals? and: Can one explain the known scaling relationship between an animal's metabolism and its mass?

13.3.2 Optimized Arterial Radius and Bifurcation Angles

The purpose of an artery is to act as a conduit of blood for the delivery of oxygen, nutrients, hormones and other compounds to cells, as well as to transport immune cells and to carry away cell waste material. The heart must pump blood with a flow sufficient to supply each cell of the animal with the needed ingredients to maintain its life and functions. To be efficient, the body has evolved to minimize its power

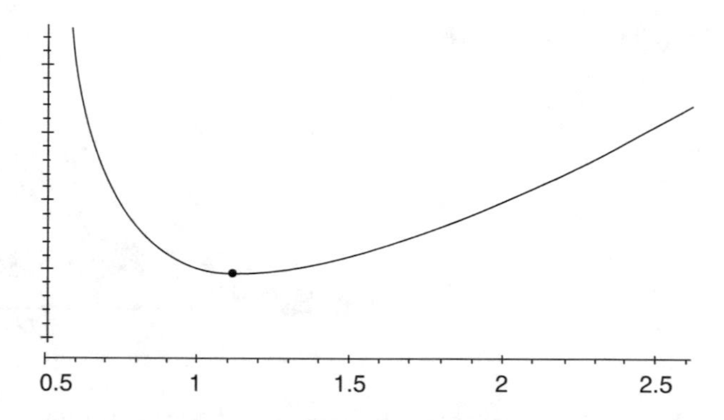

Fig. 13.4 Generic form for the power needed to pump blood through an artery of varying radii

requirements. Pumping blood requires power because of blood viscosity, and blood cells and blood vessels need a flow of energy to sustain their functionality.

Consider a vessel with the heart pumping blood through it.[4] The power expended by the heart in doing so has at least two terms:

$$\mathcal{P} = i^2\mathcal{R} + K\pi r^2 L \, . \tag{13.7}$$

The first term on the right is the power expended in making the blood flow through a vessel of radius r and length L, with i being the fluid current, and $R = 8\eta L/(\pi r^4)$ being the fluid resistance due to its viscosity η. The second term is the power needed to maintain thc associated blood cells and vessel tissue, which we take as proportional to the volume of the vessel, with proportionality constant K. Additional terms should be added if the heart has to act against turbulence, particularly in the aorta.

Note that because of the inverse r^4 behavior of the first term and the r^2 behavior of the second, there will be a value of r which minimizes the power requirements for the given length of vessel and given flow. (See Fig. 13.4.)

This radius we can find by making the slopc of the power as a function of r vanish, which occurs at the lowest value of $\mathcal{P}$ as a function of r. The result is that

$$r^3_{min\mathcal{P}} = \left(\frac{4}{\pi}\sqrt{\frac{\eta}{K}}\right) i \, . \tag{13.8}$$

So the optimal radius does not depend on the length of the vessel, and the flow is proportional to the vessel radius to the third power. The minimum power requirement per unit length is proportional to the vessel radius to the second power.

Now consider the optimal bifurcation angle of an artery, as depicted in Fig. 13.5. Blood flows from A to B and C. At the distal end of B and C are organs, taken at a specific location. At the origin of the supply stands the heart. We will take the points

[4]See Robert Rosen, **Optimality Principles in Biology** [Springer] (1967).

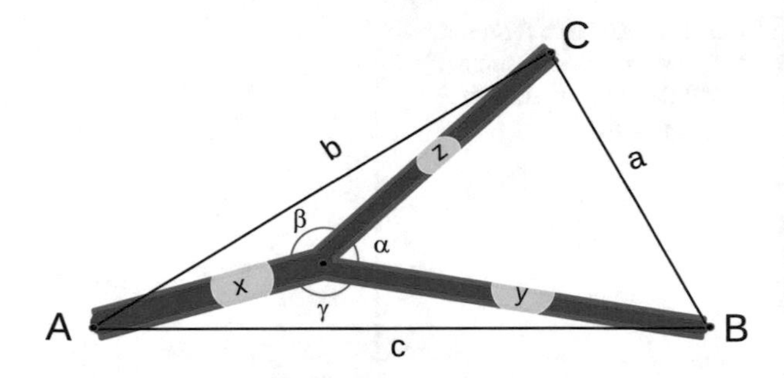

Fig. 13.5 Artery branch

A, B, and C as fixed. The flows i_B and i_C are determined by the organs down the line. Current conservation,

$$i_A = i_B + i_C \ , \tag{13.9}$$

implies that the flow at A is also fixed. For arteries of optimal radius, given by Eq. (13.8), this current conservation law makes

$$r_A^3 = r_B^3 + r_C^3 \ . \tag{13.10}$$

We will vary the position of the bifurcation point to minimize the cost to the body's metabolism.

The cost function applicable to the flow effort and material support for the vessels shown in Fig. 13.5, we take as

$$f = i_A^2 \mathcal{R}_A + i_B^2 \mathcal{R}_B + i_C^2 \mathcal{R}_C + K r_A^2 L_A + K r_B^2 L_B + K r_C^2 L_C \ . \tag{13.11}$$

Since both the power loss due to the resistances and due to the support requirements are both proportional to the length of the vessels involved, the cost function is a weighted sum of the lengths of the vessels:

$$f = \sum_{k=1}^{3} w_k x_k \ , \tag{13.12}$$

where $x_1 = x = L_A$, $x_2 = y = L_B$, $x_3 = z = L_C$, $r_1 = r_A$, $r_2 = r_B$, $r_3 = r_C$, and $w_k = 8 i_k^2 \eta / (\pi r_k^4) + K r_k^2$. Finding the minimum of the sum of lengths from the vertices of a triangle to an inner point is a challenge problem posed by Pierre de Fermat to his friend Evangelista Torricelli way back in about 1642.

Our challenge is similar: We wish to minimize the weighted sum of the lengths. In 1643, Torricelli displayed several solutions to the equal-weight problem by a number of geometric techniques, one of which was to construct three equilateral triangles on the sides a, b, c. Circles centered on these three equilateral triangles passing through the corresponding pair of vertices from A, B, C will all intersect at the Fermat point. It turns out that lines from the distal third vertices of the equilateral triangles to the opposite vertex of the Fermat triangle will also intersect at the Fermat point. In 1750, Thomas Simpson showed how to generalize this geometric construction in the case of weighted lengths.

If we allow the distances x, y, z and the angles α, β, γ to be free in the minimization procedure, then the Lagrange constraints are the following: The end points A, B, C will be held fixed. This is assured by making the lengths a, b, c fixed, where, using the law of cosines, the constraint equations for these fixations can be taken as

$$g_1 = y^2 + z^2 - 2yz\cos\alpha - a^2 = 0, \tag{13.13}$$

$$g_2 = z^2 + x^2 - 2zx\cos\beta - b^2 = 0, \tag{13.14}$$

$$g_3 = x^2 + y^2 - 2xy\cos\gamma - c^2 = 0. \tag{13.15}$$

The three angles, being at one vertex, satisfy the constraint

$$\alpha + \beta + \gamma = 360^\circ = 2\pi \; radians. \tag{13.16}$$

The Lagrange cost function will be

$$F(x, y, z, \alpha, \beta, \gamma, \lambda_1, \lambda_2, \lambda_3, \mu) = \sum_{k=1}^{3} w_k x_k + \sum_{k=1}^{3} \lambda_k g_k + \mu \left(\sum_{k=1}^{3} \alpha_k - 2\pi \right), \tag{13.17}$$

where $\alpha_1 = \alpha$, $\alpha_2 = \beta$, $\alpha_3 = \gamma$. When this cost function is minimized varying all ten of its variables, the function value is at the bottom of a curve or has hit a boundary value. Plotting this function shows that indeed it has only one minimum.

Setting to zero the derivatives of F with respect to its first six variables gives

$$w_2 + \lambda_2(x - z\cos\beta) + \lambda_3(x - y\cos\gamma) = 0 \tag{13.18}$$

$$w_3 + \lambda_1(y - z\cos\alpha) + \lambda_3(y - x\cos\gamma) = 0 \tag{13.19}$$

$$w_1 + \lambda_1(z - y\cos\alpha) + \lambda_2(z - x\cos\beta) = 0 \tag{13.20}$$

$$\lambda_1 yz \sin\alpha + \mu = 0 \tag{13.21}$$

$$\lambda_2 zx \sin\beta + \mu = 0 \tag{13.22}$$

$$\lambda_3 xy \sin\gamma + \mu = 0\,. \tag{13.23}$$

Vanishing derivatives of F with respect to the Lagrange multipliers give back the four constraint Eqs. (13.13)–(13.16).

This system of ten equations has an analytic solution, but the algebra is complicated, and the solution was not published until 2012![5] In our present case, we are focused on finding the bifurcation angles. For this, we can use the argument of Murray.[6] First, take the vessel radii to be optimal according to Eq. (13.8). Then the cost function for the vessels in Fig. 13.5 will be proportional to the three-term sum: $f \propto \sum_i r_i^2 x_i$. Let P be the generalized Fermat Point. Now add a point P' a small distance δ along the line AP, and let θ_{AB} be the angle between AP and PB, and let θ_{AC} be the angle between AP and PC. Since P minimized the cost function, the variation in the cost for the total length as the sum of lengths from the new point P' to each vertex should vanish, to first order in δ. This reads

$$r_A^2 \delta = r_B^2 \delta \cos\theta_{AB} + r_C^2 \delta \cos\theta_{AC} \,. \tag{13.24}$$

Similarly, making variation in cost under infinitesimal extensions of the other two line segments BP and CP will give

$$r_B^2 \delta = r_A^2 \delta \cos\theta_{AB} - r_C^2 \delta \cos(\theta_{AB} + \theta_{AC}) \tag{13.25}$$

and

$$r_C^2 \delta = r_A^2 \delta \cos\theta_{AB} - r_B^2 \delta \cos(\theta_{AB} + \theta_{AC}) \,. \tag{13.26}$$

Solving for the cosine of the two angles gives

$$\cos\theta_{AB} = \frac{r_A^4 + r_B^4 - r_C^4}{2 r_A^2 r_B^2} \,, \tag{13.27}$$

$$\cos\theta_{AC} = \frac{r_A^4 + r_C^4 - r_B^4}{2 r_A^2 r_C^2} \,, \tag{13.28}$$

and the cosine of the angle between the vessels B and C coming from vessel A becomes

$$\cos(\theta_{AB} + \theta_{AC}) = \frac{r_A^4 - r_B^4 - r_C^4}{2 r_B^2 r_C^2} \,. \tag{13.29}$$

With Eq. (13.8), this becomes

$$\cos(\theta_{BC}) = \frac{(1 + (r_C/r_B)^3)^{4/3} - 1 - (r_C/r_B)^4}{2 (r_C/r_B)^2} \,. \tag{13.30}$$

---

[5] Alexei Yu Uteshev, *Analytic Solution for the Generalized Fermat-Torricelli Problem*, Comp Geometry **121:4**, 318–331 (2014).

[6] C.D. Murray, *The Physiological Principle of Minimum Work Applied to the Angle of Branching of Arteries*, Jour Gen Physiol **9**, 835–841 (1926).

Two Interesting Cases If the two arteries B and C have the same radii, then they split off from A at an angle of $\arccos(2^{2/3} - 1) \approx 75°$. If instead $r_C << r_B$, then the angle between vessel A and B is well approximated by $(2/3)(r_C/r_B)$, and that between A and C can be approximated by $90° - (1/\sqrt{2})(r_C/r_B)^2$. So a tiny artery leaves the supply artery at an angle close to 90°, and the larger diameter exiting artery leaves the supply artery at close to 0°.

Murray found that a comparison of the optimal angles predicted are in close agreement with actual bifurcation angles measured in cat tissue. This shows that the cost to supply blood to tissue has been optimized, and that a simple model for minimizing of power needs predicts the radii and bifurcation angles of arteries. Darwin would explain that individuals with less than optimal systems under given environmental pressures have a lower chance of survival than those with optimal systems. After a number of generations, the DNA coding for optimal operations dominate the living population.

13.3.3 *Optimized Aortic Diameter*

The formula for the optimal radius for an artery, Eq. (13.8), works for most arteries, but fails for the aorta, giving a value of 0.4 cm instead of about 1.5 cm.[7] In the flow of blood from the heart, the possibility of turbulence in our aorta plays a significant role. Turbulence produces a back pressure, forcing the heart to work harder. Laminar flow will turn to turbulent flow in a straight pipe when the Reynolds number, $\mathscr{R}_e = \rho v r/\eta$ exceed about 2100, but in the strongly curved upper aorta, turbulence starts for $\mathscr{R}_e$ above about 1000. The speed of flow, v, is inversely proportional to the square of the aorta radius r: $v = i/(\pi r^2)$, where the current i is the cardiac output. With $\rho/\eta \approx 37\,\text{cm}^2/\text{s}$, $i \approx 100\,\text{cm}^3/\text{s}$, and $\mathscr{R}_e \approx 1000$, the minimum radius to have no turbulence is $r \geq 1.3\,\text{cm}$. As the speed in the aorta is actually varying, one should consider the predicted value for the radius in surprisingly good agreement with typical measured values.

13.3.4 *The Pattern of Seeds on a Sunflower*

The arrangement of seeds on a sunflower are in a spiral pattern which minimizes the area taken up by the seeds. The minimization of space makes the number of spirals on a given flower be a Fibonacci number. Alan Newell and Matt Pennybacker (2013) have shown[8] that the growth of seeds on a sunflower can be predicted by a dynamical

---

[7] See Rosen, *ibid.*

[8] Alan C. Newell and Matthew Pennybacker, *Fibonacci patterns: common or rare?* Symposium on Understanding Common Aspects of Extreme Events in Fluids, Procedia IUTAM **9**, 86–109 (2013).

model based on the biochemistry and biomechanics of the process. In the model, the rate of growth has non-linear terms coming from the forces of seeds on each other affecting the production of the growth hormone called auxin.

13.4 Scaling Laws in Life Systems

As complex as life systems are, some very simple relationships among the general properties of such systems follow from universal properties, such as mass conservation, energy conservation, and dimensional limitations. Many such relationships were discovered by attempting to connect observables empirically. For example, there is a remarkable scaling law first described by Max Kleiber in 1932.[9] He observed that the resting metabolism rate[10] of mammals is proportional to the mammals mass to the power of 3/4. (See Fig. 13.6.)

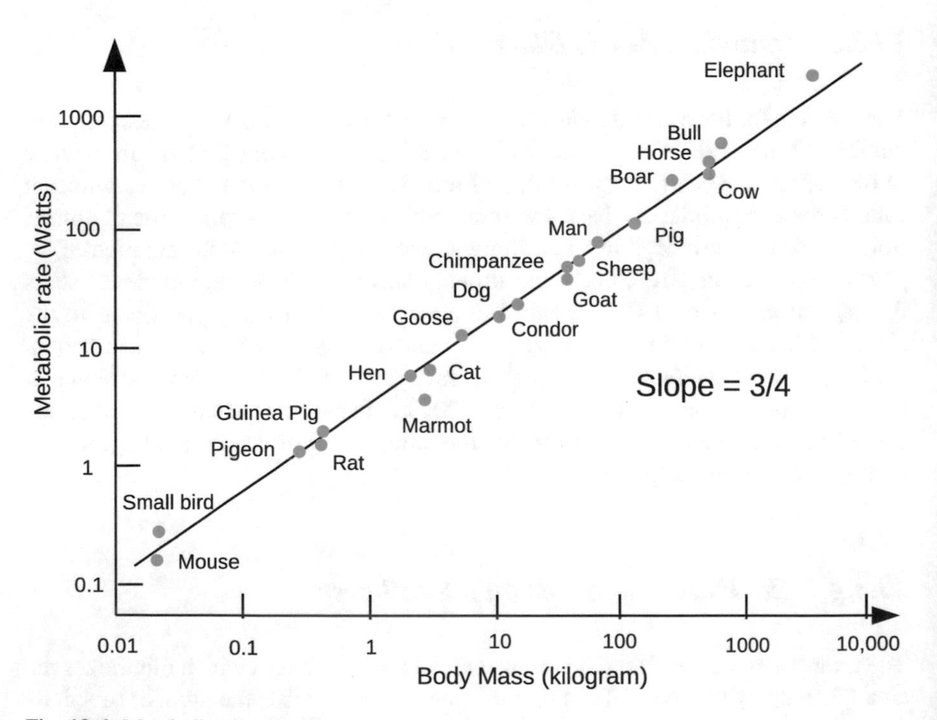

Fig. 13.6 Metabolism scaling law

---

[9]M. Kleiber, *Body size and metabolic rate*, Physiol Rev **27:4**, 511–541 (1947).

[10]The resting metabolism rate (RMR) is close to the basal metabolism rate (BMR) introduced in Sect. 9.8.

It took about 65 years before a reasonable theoretical explanation of this power law was presented by Geoffrey West, James Brown, and Brian Enquist.[11] Their model extends Kleiber's scaling law to a plethora of plants and animals, over 15 decades in range of masses! Moreover, their model describes other universal properties of life subsystems, such as connections between body mass and heart rates and growth rates.

Here is the essence of the West, et al., argument as formulated by Etienne, Apol, and Olff[12] applied to animals:

Assumption 13.1 The nutrient transport system of multicellular animals has a fractal-like splitting of the arteries. With fractal splitting, there is a massive saving in the gene system, in that genetic coding for individual arterial branching, sizes and lengths is not necessary.

Let N_k be the number of pipes after k branching ($1 \leq k \leq C$, the number of capillaries), and $N_1 = 1$, at the aorta. At level k, each pipe has radius r_k and length l_k. Each vessel at level k splits at a node into ν_k at level $k+1$, so $N_k = \sum_{i=1}^{k-1} \nu_i$. Also, we have

$$\nu_k = \frac{N_{k+1}}{N_k} .$$

Let

$$\rho_k = \frac{r_{k+1}}{r_k}$$

$$\lambda_k = \frac{l_{k+1}}{l_k} .$$

These capillary values have a special significance. In terms of these,

$$r_k = \frac{r_C}{\prod_{i=k}^{C-1} \rho_i}$$

$$l_k = \frac{l_C}{\prod_{i=k}^{C-1} \lambda_i}$$

$$N_k = \frac{N_C}{\prod_{i=k}^{C-1} \nu_i} .$$

[11] G.B. West, J.H. Brown, B.J. Enquist, *A general model for the origin of allometric scaling laws in biology*, Science **276:5309**, 122–126 (1997-04-04).

[12] R.S. Etienne, M.E. Apol, H. Olff, *Demystifying the West, Brown & Enquist model of the allometry of metabolism*, Functional Ecology **20**, 394–399 (2006).

Let

$$\alpha_k = \frac{N_{k+1}\pi r_{k+1}^2}{N_k \pi r_k^2} = \nu_k \rho_k^2$$

$$\varphi_k = \frac{\frac{4}{3}\pi N_{k+1} l_{k+1}^3}{\frac{4}{3}\pi N_k l_k^3} = \nu_k \lambda_k^3$$

and define the sums

$$S_1 = \sum_{k=1}^{C} \frac{1}{\prod_{i=k}^{C-1} \nu_i \rho_i^2 \lambda_i}$$

$$S_2 = \sum_{k=1}^{C} \frac{1}{N_k^{1/3} \prod_{i=k}^{C-1} \alpha_i \varphi_i^{1/3}}$$

$$S_3 = \sum_{k=1}^{C} \frac{1}{N_k^{1/3}}$$

The sums S_1 and S_2 are related by

$$S_1 = N_c^{1/3} S_2 \,.$$

Now the total volume of blood contained at branching level k is

$$\begin{aligned} V_{b,k} &= N_k \pi r_k^2 l_k \\ &= \frac{N_C \pi r_C^2 l_C}{\prod_{i=k}^{C-1} \nu_i \rho_i^2 \lambda_i} \end{aligned}$$

so that the total blood volume is

$$\begin{aligned} V_b &= \frac{N_C \pi r_C^2 l_C}{\prod_{i=1}^{C-1} \nu_i \rho_i^2 \lambda_i} \\ &= N_C \pi r_C^2 l_C S_1 \\ &= N_C^{4/3} \pi r_C^2 l_C S_2 \,. \end{aligned}$$

Thus, the total number of capillaries can be expressed as

$$N_C = \left[\frac{V_b}{\pi r_C^2 l_C S_2} \right]^{3/4} .$$

Let the volume flow through one of the vessels at level k be called Q_k and u_k be the average blood speed at level k (averaged over the cross-sectional area). Then

$$Q_k = A_k u_k$$

so that the total flow at level k will be

$$N_k Q_k = N_k A_k u_k \ .$$

Assumption 13.2 The metabolic rate and blood volume flow are proportional to body size. (As we noted in Sect. 9.7 describing the Basal Metabolism Rate, this direct proportionality is called isometric scaling. If the body size in a scaling relationship has an exponent different than one, then the scaling is allometric.) A linear dependence should be expected. Individual animal cell sizes do not vary much. Each cell requires a certain supply of nutrients and oxygen, and a system to expel waste and carbon dioxide. The body mass is proportional to the number of cells.

The basal metabolic rate B is usually calculated from the rate of oxygen consumption. This rate is proportional to the total volume flow:

$$B = f_0 N_1 Q_1 \ .$$

where f_0 is independent of body size.

Assumption 13.3 The blood is incompressible, noting that water has a very large bulk modulus and that blood is made principally of water. Conservation of mass then gives

$$N_k Q_k = N_{k+1} Q_{k+1}$$

We will have

$$\begin{aligned} B &= f_C N_C A_C u_C \\ &= f_C A_C u_C \left[\frac{V_b}{\pi r_C^2 l_C S_2} \right]^{3/4} \ . \end{aligned}$$

Assumption 13.4 The blood volume in an animal using blood oxygenation is proportional to the body size (mass M). Since all cells require a nearby blood flow, this proportionality is expected. The proportionality has been checked by empirical studies of mammals and birds. We have

$$V_v = V_{b0} M \ .$$

Assumption 13.5 Capillary radii and length do not depend on body size. This is an important, but not obvious, property. In essence, the size of a capillary is determined by the need for diffusion of nutrients and oxygen to nearby cells. Diffusion would be impeded if the capillary were larger in radius. If it were longer, the concentration of nutrients and oxygen would be further diminished at its far end while it supplies near-end cells. These constraints are strictly local, depending on a small region of cells near the capillary, and not on the total mass of all the cells.

Assumption 13.6

(a) The network preserves area:

$$N_k \pi r_k^2 = N_{k+1} \pi r_{k+1}^2$$

i.e.

$$\alpha_k = 1$$

This is an impedance-matching assumption. If the total area of the arteries before a node does not match the total area after that node, the blood flow would show pressure reflection at the node, and more heart power would be needed to pump the blood against the increased resistance.

(b) The network is 'space-filling', in the sense that the volume around all the arteries before a given node matches the volume after that node. This reads

$$\frac{4}{3}\pi N_k l_k^3 = \frac{4}{3}\pi N_{k+1} l_{k+1}^3$$

i.e.

$$\varphi_k = 1 \, .$$

The argument for this assumption is one of economy in the utilization of the available volume. Alternative models now exist which involve the thickness of the arterial walls, the pulsation character of the blood in the initial few nodes, and the metabolic requirements needed for sudden increases in activity.[13]

(c) The total number of levels, C, is much greater than one. This makes the quantity S_3 asymptotically reach a constant value not dependent on C, and so not dependent on the body size.

For example, the asymptotic value of S_3 when $\nu_k = n$, so $N_k = N_C/n^{C-k}$, becomes

[13] M.E.F. Apol et al., *Revisiting the evolutionary origin of allometric metabolic scaling in biology*, Functional Ecology **22**, 1070–1080 (2008).

$$S_3 = \sum_{k=1}^{C} \frac{1}{N_k^{1/3}}$$
$$= \left(\frac{n^C}{N_C}\right)^{1/3} \sum_{k=1}^{C} n^{-k/3}$$
$$= \frac{1 - n^{-C/3}}{1 - n^{-1/3}}$$

Now, with Assumption (a) and (b), $S_2 = S_3$, making $B \propto M^{3/4}$. □

Problems

13.1 Argue that the neuronal stimulation of our heart muscle is a non-linear dynamical system, and therefore open to chaotic behavior, manifested as ventricular fibrillation.

13.2 When should you expect mathematical optimization techniques to work in modeling nature?

13.3 Living systems under new environmental pressures naturally undergo Darwinian evolution. Moreover, systems can evolve even without environmental pressure if one or more mutations, reductions or elaborations give an advantage. Some have argued that humans have a number of vestigial subsystems which no longer have significant function. Proposed examples in the past include 'junk' DNA, the appendix, the coccyx, tonsils, wisdom teeth, male nipples, the pulmaris longus muscles, arrector pili muscles, and the hiccup reflex. Argue that if these subsystems continue to be supported by genes after many generations, they must add to our survivability. List proposed or possible function of each system just mentioned. Devise a counter argument that some systems are in a 'transitional' stage of evolution.

13.4 Are scaling laws an 'emergent principle', or do they follow directly from basic principles followed by molecules?

13.5 How can one tell if a biological subsystem has been fully optimized? What determined the time scale for optimization?

Chapter 14
Neural Networks and Brains

. . . of all the fathomless questions that confront and confuse men, the most baffling is the human mind."

—Clarence Darrow

Summary All cells in life systems are capable of communication between cells by the use of signaling molecules such as hormones. The speed of this form of communication is limited by diffusion rates. Multicellular animals early in evolution took advantage of the much faster internal electric signaling to coordinate their activities. Neurons are the specialized cells for this purpose. For quick analysis of inputs, using short connections, a concentrated network of neurons became ganglions, and then brains. A brain consists of a set of interconnected neurons, each acting to process a set of inputs to produce one output. As such, a set of interconnected electronic gates is capable of acting as a brain (Turing's lemma), and is the basis of modern-day computers. We can envision even more powerful computers using quantum coherent states, but our own brain seems to be acting classically, with 'free-will' coming from the complexity in the brain and the almost chaotic nature of its inputs.

14.1 Neural Networks as Logic Systems

A "finite classical computer" is a device which takes a given finite input, acts on that input with Boolean logic using internal memory (which includes programs), and produces a finite output after running for a number N of clock cycles. That output is a result based solely on the input and memory. If the device has not concluded its analysis of the input in N cycles, the result is a failure signal (also called a "halt" state). The output must have a unique correspondence to the input and memory, although many inputs may produce the same output. Often, the output is made to be an "answer" to the input "question" or proposition.

Alan Turing has argued that a finite classical computer can be constructed that is capable of reproducing, for a given input, the output of any other given finite

W. C. Parke, *Biophysics*, https://doi.org/10.1007/978-3-030-44146-3_14

computer.[1] Thus, our electronic computers, made from interconnected 'logic gates', with sufficient capacity, can perform arbitrarily close to any other constructed brain based on logic elements. To the extent that our 'wet brains' act deterministically, we expect our deterministic electronic brains can act as our biological brains do. This is Turing's theorem: Any deterministic computer is capable of producing the same output as any other deterministic computer, irrespective of how it is constructed.

Our brains do not contain neurons acting as digital logic gates. But to a good approximation, the input and output of a biological neuron can be digitized, in that the neurotransmitters are made from individual molecules, and their effect on the receiving neuron can be quantized. Our brains can be well approximated by a set of idealized neurons.

However, brains that are constructed to act like human brains can be made from electronic components. We can then use the "Turing Test" to see how good an approximation we have. The Turing Test uses interrogation of the simulation computer to see if its answers match that of our brain sufficiently that a human cannot tell the difference.

Brains give life forms a sophisticated ability to coordinate actions, analyze and respond to sensory inputs, control internal functions, balance objectives, anticipate based on experience, and make plans. Besides analysis, there are two other major advancements for life forms with brains: Learning and memory capacity. A brain can be flexible, in that it is capable of learning. Learning is also a property of individual protozoan, but brains add elaboration to the capacity to learning. Learning strategies for survival, from experience and from communication with parents and others, relieves the DNA from having to store such information, which often is ephemeral, and therefore better stored outside the DNA. In fact, for mammals, learning accounts for billions of times more information than that stored on their DNA.

The fundamental element of a biological brain is a neuron, abstractly represented in (Fig. 14.1). Biological networks of neurons learn by having activity affect their response thresholds and their innervations.[2]

Neuron inputs are carried on conduction lines called dendrites; the output on a line called an axon, which divides into branches called axon terminals and ending in terminal bulbs. The neuron can be either excitatory or inhibitory in action on another neuron, muscle, or gland. The strengths of the inputs, $\{E_i, I_j\}$ and the output threshold T are 'plastic', in that conditions and history can affect these numbers. The connection locus between terminal bulbs and dendrites of another neuron or muscle are called 'synapses'. The synapse and the state of the neural endings determine the strengths $\{E_i\}$ and $\{I_j\}$. Neurons can also sprout new dendrites and axon terminals.

Neurons require a recovery time before they are able to fire again, called the 'refractory period'. At the end an absolute refractory period (about 2 ms), the neuron

[1]Alan Turing, *Computing Machinery and Intelligence*, Mind LIX (236): 433–460, (October 1950), doi:10.1093/mind/LIX.236.433

[2]Donald O. Hebb, **Organization of Behavior**, [Wiley & Sons] (1949).

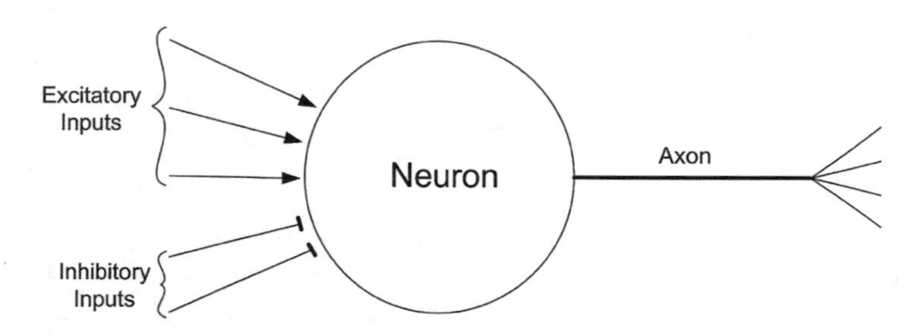

Fig. 14.1 Essence of a neuron

enters into a relative refractory period (about 3 ms), during which the firing threshold is higher than during the neuron's resting state.

Neurons principally transfer information by the timing of their impulses, and not by impulse strength. The speed of an impulse along an axon varies from 0.6 to 120 m/s, depending on whether the axon is surrounded by myelin sheath insulating cells, which allows faster transmission.

Models of Neuronal Networks

An idealized 'artificial' neuron was proposed by McCulloch and Pitts[3] with the intention of exploring how far model brains can be used to understand the behavior of real brains. Their model had fixed strengths for the synaptic thresholds and used a fixed time τ to represent the action and refractory period of each neuron. A clock with period τ signaled when any of the neurons could fire. With a few simple McCulloch–Pitts neurons, one can construct the Boolean operations of AND, OR, NAND, NOR, NOT, XOR, and NXOR. Thus, all logical operations can be constructed.

Each McCulloch–Pitts neuron accepts a set of excitatory and inhibitory inputs as timed pulses and, based on an internal threshold, produces an impulse output. Let E_i represent the strength of the excitor labeled by 'i', and I_j the strength of inhibitor labeled by 'j'. The condition that a given neuron "fires", i.e. produces a pulse, when a subset of the inputs have a pulse, is expressed by

$$\sum_i \theta_i E_i - \sum_j \theta_j I_j \geq T ,$$

where θ is 0 or 1 depending on whether an input has a impulse delivered to the neuron in a given time interval, and T is the threshold level for the neuron.

[3]W.S. McCulloch, and W.H. Pitts in *A logical calculus of the ideas immanent in nervous activity*. Bulletin of Math Biophys **5**, 115–133 (1943).

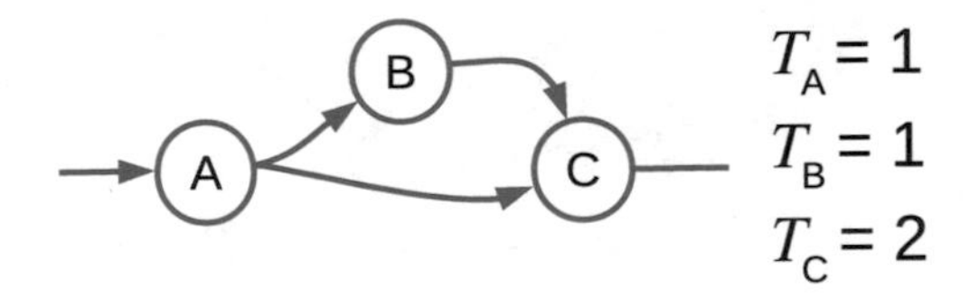

Fig. 14.2 McCulloch–Pitts example network I

Table 14.1 Network I: time versus neuron input-output

t/τ	i_A	o_A	i_B	o_B	i_{C1}	i_{C2}	o_C
0	0	0	0	0	0	0	0
1	0	0	0	0	0	0	0
2	1	0	0	0	0	0	0
3	0	1	1	0	0	1	0
4	0	0	0	1	1	0	0
5	1	0	0	0	0	0	0
6	1	1	1	0	0	1	0
7	0	1	1	1	1	1	0
8	0	0	0	1	1	0	1
9	1	0	0	0	0	0	0
10	0	1	1	0	0	1	0
11	1	0	0	1	1	0	0
12	0	1	1	0	0	1	0
13	0	0	0	1	1	0	0

Within this model, learning networks can be designed by allowing the synaptic strengths (i_e, i_i) and thresholds θ to change through positive and negative feedback based on the results of a given analysis. In this way, computer programs can be made to learn to play chess without initially knowing the rules.

Even some of the simplest networks have interesting properties. Consider the network shown in Fig. 14.2 with only three McCulloch–Pitts (M-P) neurons. All inputs i's and outputs o's are taken either zero or one. We construct Table 14.1 showing in each row the state of the input and output pulses of the involved neurons at one particular time. Time is indicated in the left (first) column. The first row carries the initial state of the neurons (at clock time zero). The second column labeled i_A has the input stream of pulses. These we select with various choices for the impulse timings to test how the network reacts. Once the input stream is selected, the intermediate states of the neurons and their output are completely determined.[4] The threshold of neurons A and B are taken as 1, while the threshold for firing of C is taken as 2. As can be seen in the table, this simple network requires two excitatory pulses in sequence to give an output pulse. (This is the property of the trigger hairs in a Venus flytrap!)

A second example is shown in Fig. 14.3, which will continue to send a sequence of pulses out after only one excitation by "i_A". The pulses are a regular train of on

[4]For example, $o_C(k) = (i_{C1}(k-1) + i_{C2}(k-1))$ *modulo* 2, where $k = t/\tau$.

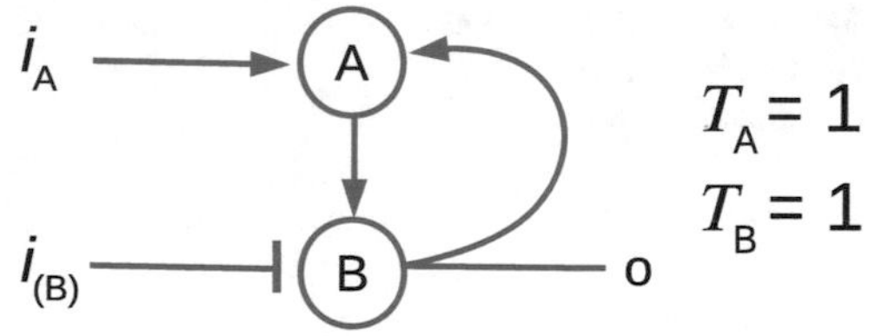

Fig. 14.3 McCulloch–Pitts example network II

Table 14.2 Network II: time versus neuron input-output

t/τ	i_A	$i_{(B)}$	o_{AB}	i_{BA}	o_{BA}	i_{AB}	o
0	0	0	0	0	0	0	0
1	0	0	0	0	0	0	0
2	1	0	0	0	0	0	0
3	0	0	1	1	0	0	0
4	0	0	0	0	1	1	1
5	0	0	1	1	0	0	0
6	0	0	0	0	1	1	1
7	0	0	1	1	0	0	0
8	0	0	0	0	1	1	1
9	0	0	1	1	0	0	0
10	0	0	0	0	1	1	1
11	0	1	1	1	0	0	0
12	0	0	0	0	0	0	0
13	0	0	0	0	0	0	0
14	0	0	0	0	0	0	0

and off signals (1's and 0's) shown in the column "o" of Table 14.2 at a frequency of $1/(2\tau)$. The pulses are turned off by a single activation of the input inhibitory line into neuron B (called "$i_{(B)}$" in the Table), acting just after an output pulse. (An inhibitory line in these M-P diagrams has a bar next to neuron on which it terminates.)

Every logical propositional formula can be represented by a network of binary-threshold units. This statement follows from the fact that such formulae can be translated into logical Boolean operations, and that each of the logical operations can be performed by a small number of McCulloch–Pitts neurons with binary-thresholds. The Fig. 14.4 gives a McCulloch–Pitts 'NOT' unit example. Both neurons in the drawing have a threshold of '1', and nerve impulses are taken to have unit strength. The neuron labeled 'C' performs as a local clock, producing a pulse at the beginning of each clock period. The local clock can be turned on or off by the 'start' and 'stop' lines. A logic circuit NOT gate is shown to the right in the figure, together with its 'truth' table. (A unit value is taken to be a 'true' indicator, and a zero value to be a 'false' indicator.) The McCulloch–Pitts unit that acts as an AND logic gate is shown in Fig. 14.5. A McCulloch–Pitts NAND gate can be made by just adding a NOT to a AND unit.

'Logic gates' act on pulse signals whose smallest pulse width is the clock period, and whose amplitude is one unit. As the gate devices are made to react to the

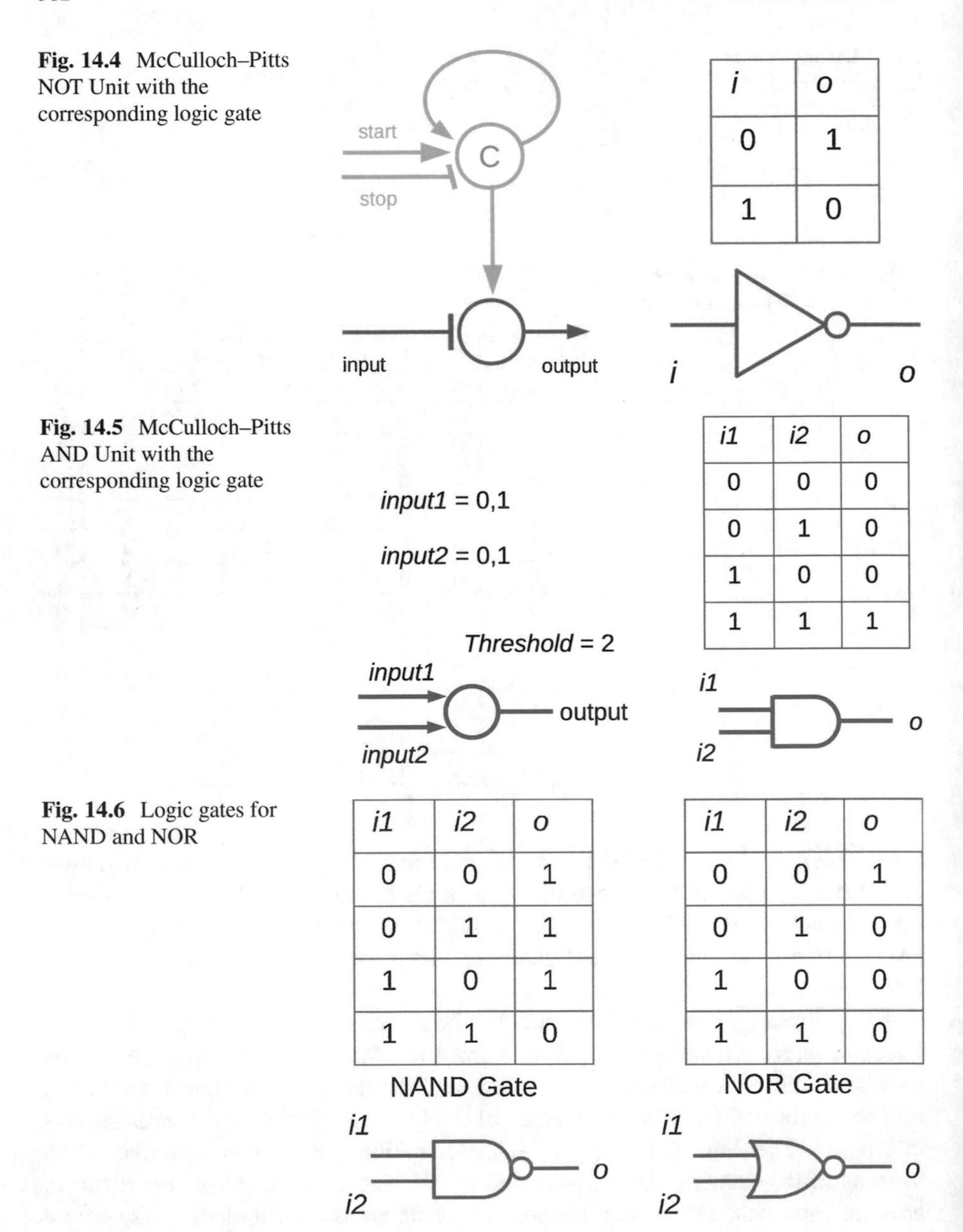

i	o
0	1
1	0

i1	i2	o
0	0	0
0	1	0
1	0	0
1	1	1

i1	i2	o
0	0	1
0	1	1
1	0	1
1	1	0

i1	i2	o
0	0	1
0	1	0
1	0	0
1	1	0

Fig. 14.4 McCulloch–Pitts NOT Unit with the corresponding logic gate

Fig. 14.5 McCulloch–Pitts AND Unit with the corresponding logic gate

Fig. 14.6 Logic gates for NAND and NOR

presence of a pulse of amplitude above a fixed threshold, they are digital devices as opposed to analog devices, making them far less susceptible to noise. One can show that any of the logical units can be made by a combination of NAND units alone. The same is true of NOR units. (The logic gates for these Boolean operations are shown in Fig. 14.6.)

A finite "automaton" is a device which accepts a finite set of input symbols, acts on this input with a finite internal state, including programming operations, and produces a new internal state and an output. Arbib[5] has shown that the behavior of every finite automaton corresponds to the behavior of a McCulloch–Pitts network, and visa versa. We also should expect that the behavior of our brain can at least be approximated by a McCulloch–Pitts network. The seemingly continuous elements, such as nerve impulse frequencies and axon delay-time lengths, we propose are all discretizable.

Although the brain, like all other natural systems, is successfully described by quantum theory, we should not expect that the effects of non-classical quantum behavior to be important in the operation of our brain, except to the extent that quantum uncertainty can cause uninitiated neuronal firings. Evolution would favor mechanisms which would minimize these effects. Now, you may ask what about free thought? If our brains act mechanistically, you argue, then our thoughts are determined by input and memory.

We can define 'free thought' as neuronal activity which was not initiated by a mechanistic reaction to input and memory. Classical noise, chaotic nerve firings, and even quantum decoherence could be used to explain such effects. However, given the need and advantage for determined brain function, evolution likely suppressed such undetermined thoughts. Besides, our close interaction with our environment through our senses is sufficient to cause many fluctuations in our thoughts associated with external inputs. These external inputs also contain noise from various sources and the consequences of any chaotic behavior in non-linear external systems. In fact, external noise can be very difficult to eliminate. We do not yet know how to eliminate the passage of cosmic neutrinos through our brain! Quantum theory allows for the coherence of wave functions for particles at far distances with the wave functions of our particles. (Einstein's 'spooky action at a distance'.) We can build such couplings, but decoherence is rampant for particles open to the environment, as ours are.

14.2 More Realistic Neuron Models

14.2.1 Electrical Activity of a Neuron Membrane

Cell membranes have special channels made from protein molecules embedded in their lipid bilayer. These channels transport ions and other small biologically important molecules across the membrane. Neurons use these channels to build a resting potential of about $-70\,\mathrm{mV}$, making the inside of the cell membrane negatively charged, and the outside positively charged. This is achieved by allowing potassium ions into the cell, while pumping sodium ions out. A single sodium-potassium pump

[5]Michael A. Arbib, **Brains, Machines and Mathematics** [Springer-Verlag] (1964).

channel will use the energy of one ATP ($\rightarrow$ADP) to transport two potassium ions into the cell while carrying three sodium ions out.[6] The concentration of potassium ions inside the cell can be twenty times that on the outside. A resting neuron will build up a large reservoir of potassium ions within the cell, and expel sodium ions. Only a very small fraction of these pass through channels in the membrane during any one firing of the neuron. A neuron can fire hundreds of thousands of times without a replenishment of the resting concentrations of these ions.

In addition to the slow-acting sodium-potassium pumps, there are voltage-gated ion channels for the rapid diffusion of sodium and potassium ions during a nerve impulse. There are membrane channels, as well, for the transport of calcium, chlorine, and magnesium ions. (In smooth muscle cells, the calcium voltage-gated channels, which act 10–20 times slower than the nerve cell sodium voltage-gated channels, dominate the nerve pulse initiation. Calcium ion pumps keep a lower concentration of calcium ions within the cell.)

The gates open and close depending on the electric field across the membrane. In switching, gates draw energy available in the electric field across the membrane working to affect conformational changes in channel proteins. The energy available in the membrane electric field comes from the slow but continual work done by the ion pumps, which operate using ATP as their source of energy.

Activation (inactivation) voltage-gated channels open (close) with a probability that increases with the depolarization of the cell membrane. Gates can also differ in how quickly they respond to voltage changes.

Present also are so-called leakage channels. They let ions pass by the combined passive effects of diffusion and electric force, although most let certain ions pass preferentially in one direction. Some can be closed by the binding of a ligand molecule, such as a neurotransmitter. Other nerve-membrane channels are controlled by changes in the cell environment, such as temperature or pressure.

If a region of a nerve cell membrane experiences a depolarizing electric field carrying its membrane potential from its resting value of $-70\,\mathrm{mV}$ to a value above about $-55\,\mathrm{mV}$, sodium diffusion channel gates open, letting positive charge into the cell in this region of the membrane, and producing an 'action potential'. The drop in potential difference, in turn, polarizes the next region along the nerve membrane. When the potential goes positive, the sodium diffusion channel gates begin to close, and the potassium diffusion channel gates open, reversing the potential toward negative values below $-70\,\mathrm{mV}$. This desensitizes the membrane for the refractory period, so that a propagating nerve impulse traveling down a nerve will not cause another impulse traveling backward.

[6] The transport of both sodium and potassium ions is against the direction of their unforced diffusion, but moving the potassium ions in the direction of the already established electric field helps get back a little of the ATP$\rightarrow$ ADP energy.

14.2.2 *The Hodgkin–Huxley Model*

Through their extensive and detailed studies of the electrical properties of squid neuron membranes (squid axons are particularly big), Alan Hodgkin and Andrew Huxley described[7] (1952) an electric circuit model of a real neuron's electrical activity. Unlike the discrete time sequences used to describe the input and output of a McCulloch-Pitts artificial neuron, the dynamics of the Hodgkin–Huxley model neuron are taken continuous in time. Because they used insightful and consistent mathematical formulations of how ion channels might operate, and used measurement to pin down what description and parameters to use in the formulation, their model was impressively successful in matching how a real neuron behaves when stimulated, earning them a Nobel prize.

Hodgkin and Huxley found that in producing a nerve impulse, nerve cell axon membranes act as an electric cable with an interior wall insulated from an exterior wall, except for a sequence of bridging elements: A voltage-gated sodium ion conduit, a voltage-gated potassium ion conduit, a leakage conduit, and a capacitor[8] to represent the behavior of a cell membrane that is capable of storing energy by separating charge on its outer surface relative to its inner surface. Figure 14.7 shows this Hodgkin and Huxley electric circuit. Batteries are inserted in the conduits to account for the local electric fields generated by diffusive ion charge separation.[9] The energy to make the nerve impulse comes from borrowing a little of the energy of the separated charges, i.e. the electric field produced by the separated charges does work in operating the gates and in draining a little of the charge back across the membrane.

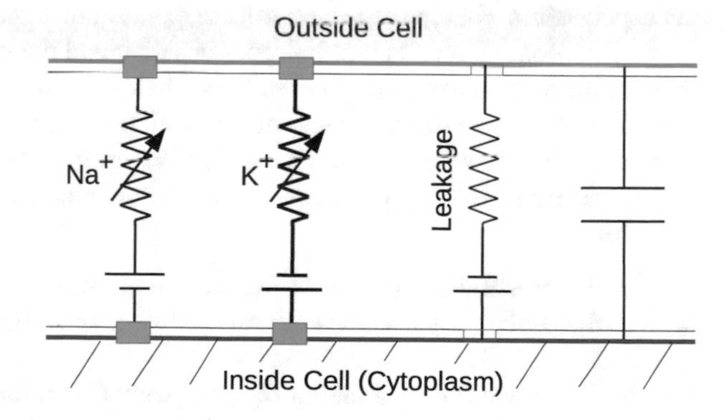

Fig. 14.7 Hodgkin–Huxley model circuit for a neural membrane

[7] A.L. Hodgkin and A. Huxley, *A quantitative description of membrane current and its application to conduction and excitation in nerve*, J Physioy **117:4**, 500–544 (August 1952).

[8] Charges separated between the inside and outside of the lipid bilayer, about 5 nanometers thick, act as a fixed capacitor (of about 1 $\mu F/cm^2$).

[9] The statistical ideas behind a description of diffusion would not apply to an individual channel, as they accommodate only a few ions each. But the ideas would apply to a large number of channels. For a discussion of the diffusion of ions across membranes, see Sect. 9.20.1.

The ion active-transport channels are not included in the Hodgkin–Huxley model, as only a small discharge of the membrane occurs on each pulse. Moreover, the impulse drains the small charge in a millisecond or so, while the active-transport channels take about 8 ms to pump the same amount of charge back.

Let's see how the Hodgkin–Huxley neuron membrane operates. Including an external source of current I into the membrane, charge conservation makes

$$I = g_{Na}(V - V_{Na}) + g_K(V - V_K) + g_L(V - V_L) + C\frac{dV}{dt} , \tag{14.1}$$

where V is the potential difference across the membrane and each g_i, called the conductivity, is the inverse of the resistance in the respective channel. The constants V_i are set so that if the potential across the membrane matches V_i, then the current in that channel is zero. The passive nature of the leakage channel makes g_L a constant. However, the voltage-gated sodium and potassium channels will have conductivities which depend on V itself. Here is where the non-linearity of the system arises.

By fitting the measured conductivity to particular models of the voltage-gated channels, Hodgkin and Huxley discovered that the sodium channels have three separate fast gates which must open before the sodium ions can pass, and another slower-acting gate that closes to inactivate the channel. They found that the potassium channels had four slower-acting gates to activate the potassium channel. An external stimulus that increases the membrane potential from −70 mV to a certain threshold value (about −45 mV), triggers a nerve impulse. Over the next half a millisecond, the three sodium channel activation gates open, letting sodium ions into the cell, raising the potential to about +30 mV. The slower-acting inactivating sodium gate then close the sodium channel, while the four potassium gates in each potassium channel open to release potassium from the cell. After another half millisecond, the cell potential goes negative and the potassium gates close, but acting relatively slowly, the potential goes below −70 mV, leading to a refractory period of about two milliseconds, during which the membrane does not generate a new impulse even when encountering the triggering stimulus.

Hodgkin and Huxley proceeded in their mathematical formulation as follows. They let

m be the probability that one of the sodium activation gates is open,
h be the probability that the sodium channel inactivation gate is open, and
n be the probability that one of the potassium activation gates is open.

The use of probabilities is justified by noting that even though any one channel handles only a few ions at a time, there are many channels on the membrane to average over.

Each of these probabilities varies as the voltage V varies. Combining the probabilities gives

$$g_{Na} = m^3 h g_{Na}^{max} \tag{14.2}$$

$$g_K = n^4 g_K^{max} \,, \tag{14.3}$$

where g^{max} stands for the maximum value of the corresponding conductivity g. When the gates are all open, $m = 1, \ h = 1$, and $n = 1$. The values for the constants in the Hodgkin–Huxley model are given in the Table below.

Const		value	unit
C	=	1.0	μ F/cm^2
g_{Na}^{max}	=	120.	1/(Kilo ohm·cm^2)
g_K^{max}	=	36.	1/(Kilo ohm·cm^2)
g_L	=	0.3	1/(Kilo ohm·cm^2)
V_{Na}	=	55.	mV
V_K	=	−72.	mV
V_L	=	−49.	mV
T	=	6.3	$^\circ$C

For a given gate in a voltage-gated channel, the electric field across the membrane determines the chance that the gate is open. Thermal noise and collisions within the channels cause fluctuations in the gate open and close condition. The gates respond to changes in the electric field, but not instantaneously. Let

p be the probability that a particular gate is open.

If the electric field changes over the time interval dt, then any increase in p will be proportional to how many were closed. Any decrease in p will be proportional to number of gates that were open. This observation is expressed as

$$dp/dt = \alpha(1 - p) - \beta p \,, \tag{14.4}$$

where α is the rate at which a shut gate opens, and β is the rate at which an open gate closes. These rates α and β will differ when the closed configuration of a gate has a different energy than the open configuration.

We expect that the rates α and β will depend on the membrane electric field strength, but not on time,[10] so we can integrate in Eq. (14.4) to get

$$p = p_\infty + (p_o - p_\infty) \exp\left(-(\alpha + \beta)t\right) . \tag{14.5}$$

After a change in the electric field, the probability of a gate being open goes from an initial value p_o to a value near

[10] A change in the electric field over the thickness of the nerve membrane acts on the molecules in that field 'instantaneously', which here means in times less than a femtosecond.

$$p_\infty = \frac{\alpha}{\alpha + \beta}\,, \tag{14.6}$$

with a time constant of

$$\tau \equiv 1/(\alpha + \beta)\,. \tag{14.7}$$

The gate-open probability expression Eq. (14.5) applies to all three of the Hodgkin–Huxley gate probabilities m, h, and n, with different rates for each, except that for h, which gives the probability of the sodium inactivation gate being open, the rates α and β reverse roles.

The next task for Hodgkin and Huxley was to find the electric field dependence of the rates α and β for each of the gates. To do this, they measured the externally injected current I through the membrane that would hold voltage V across the membrane fixed at selected values ('voltage clamped' measurements). They could also vary the exterior concentration of potassium and sodium, affecting the corresponding gated channels. In this way, they could measure the values of m, n, and h as well as their associated time constants τ, and then solve for the αs and βs. By plotting these rates as a function of V, they could see that the α and β showed what appeared to be simple exponential behavior. At the time, they had insufficient knowledge to support an explanation based on physical principles. To proceed, they fitted the six functions, producing:

$$\alpha_m(V) = -0.1\,\frac{V+45}{\exp\left(-(V+45)/10\right) - 1}\,, \tag{14.8}$$

$$\beta_m(V) = 4.0\,\exp\left(-(V+70)/18\right)\,, \tag{14.9}$$

$$\alpha_n(V) = -0.01\,\frac{V+60}{\exp\left(-(V+60)/10\right) - 1}\,, \tag{14.10}$$

$$\beta_n(V) = 0.125\,\exp\left(-(V+70)/80\right)\,, \tag{14.11}$$

$$\alpha_h(V) = 0.07\,\exp\left(-(V+70)/20\right)\,, \tag{14.12}$$

$$\beta_h(V) = 1.0\,\frac{1}{\exp\left(-(V+40)/10\right) + 1}\,. \tag{14.13}$$

Here, V is the potential measured across the neuron membrane, $(V_{outside} - V_{inside})$, in units of millivolts, with a resting potential of -70 mV. The units of α and β are inverse milliseconds.[11]

Figure 14.8 shows how their model for the action potential of a neuron compares with the measured potential across a squid neuron. The model not only works well, but it also gives insight into how a nerve membrane can generate and propagate

[11] To account for temperature differences in their measurements, Hodgkin and Huxley appended a factor $\phi = 3^{(T-6.3C)/10}$ to each of the α and β.

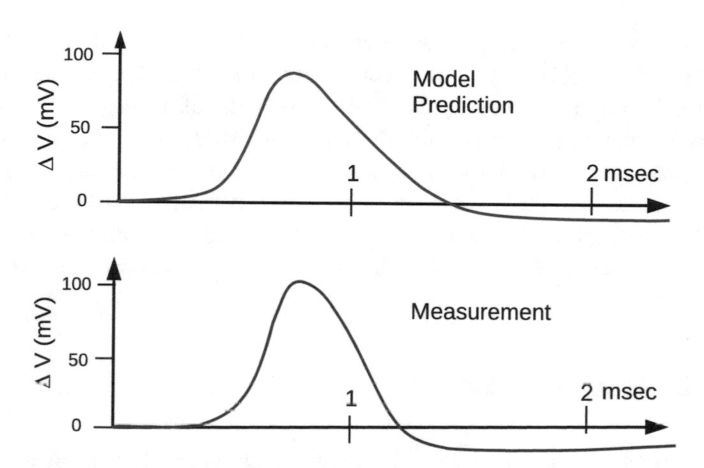

Fig. 14.8 Nerve impulse (action potential) for Hodgkin–Huxley model (top drawing) compared to measurement (bottom drawing) (Figure adapted from 1952 paper by Hodgkin and Huxley)

a nerve impulse. It shows threshold behavior, in that a minimum depolarization must happen before the action potential is generated. It indicates how fluxes of sodium and potassium ions, with short and longer activation times, can make the action impulse occur, and then allow the neuron to recover. A sustained external stimulus of sufficient magnitude will cause a sustained sequence of pulses, with a certain minimum frequency which increases with increased external potential across the membrane. This all was found by Hodgkin and Huxley, who made the right measurements, constructed the right mathematics, and numerically solved their non-linear system of equations with the use of just a calculator.

14.3 Brain Models

The human brain, with a myriad of interconnected neural networks, is capable of learning by having the connections 'plastic' and by having the number of connections variable. Modeling the whole brain based on the properties of neurons requires large-memory exascale computers, i.e. computers which can perform more than a quintillion floating point operations per second, and can hold of the order of a quadrillion bytes of directly accessible memory.

Models of a system based on the deeper physical properties are called 'mechanistic'. The discovery of mathematical rules for the observational correlations is sometimes referred to as the 'descriptive' model. These require no need for reference to underlying structures. Models which introduce relationships between the subsystems, such as neural networks, are called 'interpretive'. All three techniques have had limited success in describing the real brain.

In the 1940s, Alan Turing, and independently John von Neumann, pioneered mechanistic descriptions of the computers and the brain, starting with programmable logic elements. Mechanistic models of biological brains ('wet computers') often start with simplified Hodgkin–Huxley neurons, and then build computer programs to model how networks engage in specific tasks. Interpretive models of biological brains can use the ideas behind degrees and levels of complexity (see Sect. 15.4). Neural networks become agents, with their own rules of engagement.[12] With agents and rules, the behavior of the brain becomes fathomable.

14.4 The Human Brain

A healthy human brain has about 10^{11} neurons and about 10^{14} synapses. (Single neurons are stimulated or inhibited by about 1000–15,000 axon terminal bulbs forming synapses with the neuron.) By using a number of different neurotransmitters, each synapse stores 10 to 100 bits of information. Thus the whole brain can store about 10^{15}–10^{16} bits. (For more data about the brain, see Appendix J.) To support the activity of the brain, the brain uses up to 20–25% of the body's oxygen supply, even though it weighs only 2.5% of the total body weight. The brain must dissipate about 25 W in heat (5.5 g of glucose per hour). Up to 90% of the free energy used by the brain is consumed by the work done in the nerve membrane sodium pumps. In turn, the pumps are run by ATP.

The human auditory nerve has about 30,000 nerve fibers in the basilar membrane, each capable of a few hundred action potentials per second, making an upper limit of about 30 million pulses per second. But there is significant redundancy, with strong correlations between nearby fibers, reducing the information to about 3 million bits per second.

The human retina transmits data at the rate of about 10 million bits per second. Some of this information, the kind considered critical, such as quick motions, is sent through faster ganglions at about 875,000 bits per second.

As individual neurons can die, redundancy in the brain is an important strategy for protection of important information and programming. One effective strategy for memory redundancy would be holographic storage, i.e. delocalized wave interference patterns stored over a sub-network of neurons. (See Appendix C for a further description of holography and holographic memory.) If there were a tenfold redundancy, the brain would still be able to hold about 10^{14} to 10^{15} bits of information (about 10–100 TB).

Our brain has an information capacity far greater than the DNA which created it. (Human DNA, with 3 billion base pairs, can hold no more than about 6×10^9 bits in its coding.) We know from the development of a fetus that a significant amount

[12] Magnasco et al., *Self-Tuned Critical Anti-Hebbian Networks*, Phys Rev Let **102:25**, 258102 (2009).

of coding of the brain is contained within our DNA. Even the general structure of language (coded in the Broca area) is on our DNA, such as the subject, verb, and object elements of a simple thought made into a sentence. Much of this information is in the programming of the neurons, i.e. how they are initially connected. But we are endowed with a cerebral cortex capable of storing life experience, stories, and principles, and of creating new ideas through the analysis of what we have learned.[13] By being able to logically think about our condition and environment, we survive better. Curiosity comes from the advantage learning gives us.

14.5 Consciousness

Humans and other animals with adaptive brains are aware of themselves and their environment in a form called consciousness. It is clear that our conscious brain activity is only a small part of our total brain activity. The human brain is set up to perform automated tasks, to subliminally analyze deferred problems and anxieties, and to process a large amount of input data, without conscious thought. Some fraction of this processing is done by primitive and pre-programmed neural networks. Some of these networks are responsible for the activity of our bodily functions, communicating with organs through the parasympathetic nervous system. The basic structure of the autonomous networks are largely genetically coded.

Others subconscious networks contain learned programs, such as the melodies of songs. Chewing, walking, and even singing, once initiated, require little higher-level computation. Rather, a programmed sequence of neural firings is initiated in the cerebellum, and then normally left to 'run' on its own. Pattern recognition and some more sophisticated analyses are done in designated areas of the cerebrum, but also largely subconsciously.[14]

However, when analysis of urgent new information from sensors is required, or a problem has been set up but not yet solved,[15] then each of these inputs and programs are allowed to alert the conscious network. They are weighed according to whether they urgently require attention. Those with the highest weight are sent to our conscious network, and we become aware of them.[16] Thus, 'consciousness' is the activity of a particular network within the brain that makes decisions on new actions

[13]It is humbling to realize that the human genome information plus the information in our brain can be transmitted over a visible light beam in about 10 s, and that all of human knowledge transmits in about 20 min.

[14]For example, our ability to recognize human faces of particular individuals we know is done subconsciously. In fact, each distinctive feature of a face triggers a particular cell in the neuronal complex called the fusiform-face area in the ventral visual processing stream of the temporal lobe.

[15]Pointers to pending programs are stored in an 'anxiety' center within the subconscious network. These anxiety-center problems vie for conscious attention.

[16]The 'ah ha' moment is the time when the subconscious network devoted to solving a problem has succeeded, and then informed the conscious network. People who cannot describe how they

based on the input of various subconscious processes whose urgencies are high. The actual problem solving is done by-and-large subconsciously.[17] The activity of the conscious network reaches to every memory associated with each thought. Most of these memory stimulations and their permeations occur subconsciously, except for the strongest, or the ones that match a pattern search, or the ones which have a lower threshold of activation due to associated anxieties imprinted with the memory.

During sleep or unconsciousness, many of the thresholds for activating the conscious network are set higher than during wakefulness. These thresholds are maintained by neural suppression actions. The purpose of sleep appears to be not only to allow cells and organs to recover from any debilitating activity, but also to review recent memories, and to select those with consequential impact to be allocated for permanent memory. While in the sleeping review condition, memories send signals to the parts of the brain which ordinarily cause hormone production, muscles movement, and other stimulations corresponding to those memories. However, the thresholds for these inter-brain stimulations are also set higher in sleep. Unconsciousness and semi-consciousness can also be self-induced, or induced by chemical changes in the blood, and by trauma to the brain. Hypnosis, memory avoidance, pain suppression, and various form of trances are forms of self-induced altered consciousness, wherein some neural networks are partially blocked while others may be allowed to have increased activity. Post-trauma stress disorder (PTSD) is an example. Autism also affects cognition, but is associated with a variety of more permanent physiological pathologies. Certain networks of neurons of the brain may become hyperactive due to external stimulation, such as light flashing at an alpha-wave frequency, to physiologically unusual conditions, as in epilepsy, or to pharmacological stimulation. Hyperactive neurons may be an advantage in brain function, but has the disadvantage of being close to uncontrolled neuronal firing, producing seizure.

14.6 Artificial Intelligence

A sidelight to brain investigations is the exploration of neuronal systems that show intelligence. 'Intelligence' may be defined as the ability to successfully respond to new problems, and to be able to solve (or, if deemed too difficult, to dissolve) a new problem in a reasonable length of time. The shorter the time, the more intelligent the response. A problem is a challenge to find a series of actions which allows one to proceed from an initial state to a desired final state. Some problems are close to

---

solved a problem are simply not aware of the subconscious program working 'in the background' while they concentrated on other matters.

[17]There is good evidence that the reticular activating system, a neural network which has pathways to the cerebrum, cerebellum, and spinal cord, is that part of the brain's network that monitors, evaluates, and allocates brain resources. This network, therefore, satisfies our definition of the conscious network.

already solved ones, so that memory of those helps. Other problems require a leap of imagination, i.e. the ability to use wide searches for patterns which might lead to a solution, and then to be able to calculate the consequences of each possible pattern. Evidently, computational strength, memory, and good search algorithms are key elements to intelligence in a given area. Crows, dolphins, and often humans show high intelligence.

'Artificial intelligence' is a constructive logic network which has the characteristics of intelligence (either restricted or general). Computers have already surpassed humans in solving chess problems, and in playing the game of "Jeopardy" that involves the subtle interpretation of language and culture.

14.7 Quantum Computers

Although biological brains operate largely through mechanisms which can be associated with classical algorithms, it is possible to design a computer which takes advantage of coherent quantum entangled states to handle internal configurations which represent many possibilities before one is realized. Acting on such quantum states rather than their realizations can exponentially speed the solution to well-posed problems in comparison to known classical calculations. In practice, keeping many entities in quantum coherence is difficult due to the action of the environment on the system. No examples of a brain using quantum coherence has been found in nature except for the ones that humans have constructed.[18]

Problems

14.1 Give arguments for the proposition that electrons in delocalized orbits is important in the neuronal activity of our brains.

14.2 Support the proposition that quantum uncertainty is important in our so-called free will.

14.3 Explain in your own words what must be done to generalize the Goldman–Hodgkin–Katz voltage equation to include the calcium ion concentrations and permeability.

14.4 Suppose the concentration of sodium ions inside a neuron in your body (at 37 °C) is found to be 12 millimolar, while just outside the neuron it is 145

[18]For a further discussion of the principles behind quantum computers, see William C. Parke, *The Essence of Quantum Theory for Computers*, **Logic and Algebraic Structures in Quantum Computing**, eds. J. Chubb, A. Eskandarian, & V. Harizanov, [Cambridge Univ. Press] (2016).

millimolar. Use the Nernst–Planck equation to estimate the electric potential (in millivolts) that would be created by this concentration difference acting alone.

14.5 Now consider the cell membrane for the neuron in the previous problem as also permeable to potassium, chloride, and calcium ions, with equilibrium concentrations of $C(K, out) = 2.2\,\text{mM}$, $C(K, in) = 125\,\text{mM}$, $C(Cl, out) = 78\,\text{mM}$, $C(Cl, in) = 1.6\,\text{mM}$, $C(Ca, out) = 10\,\text{mM}$, $C(Ca, in) = 10^{-4}\,\text{mM}$, with effective permeabilities of the membrane to each ion found to be: $p(Na) = 5 \times 10^{-10}$ m/s; $p(K) = 5 \times 10^{-9}$ m/s, $p(Cl) = 1 \times 10^{-10}$ m/s, and $p(Ca) = 1 \times 10^{-12}$ m/s. Use the Goldman–Hodgkin–Katz voltage equation to predict the resulting electric potential which the neuron must maintain. Because the permeability of the calcium ions is small compared to the other ions, you do not need to include their effect to get a reasonable prediction.

14.6 Argue either for or against the proposition that awake dogs are normally conscious.

14.7 Show that if a brain has more than four neurons, they cannot all be simply connected to each other in less than a three-dimensional space. However, with a more complicated design for connection crossings, complex networks can be formed.

Chapter 15
Ordering Theory

We have seen that the formation and maintenance of self-organizing systems are compatible with the laws of physical chemistry.

—Ilya Prigogine

Summary Order can be measured in a system, whether it be the order on your desk or the order within your body. For open systems with activity, ordering is a natural process, such as the simple crystallization of a salt initially dissolved in water, or as the convoluted evolution of life systems. In general, the necessary elements for ordering can be spelled out, and examples given, including DNA replication, protein synthesis, and brain operations. Throughout the Universe's development, thresholds for rapid changes in the rates of ordering have occurred. Life formation is one, and so is the invention of the printing press. For any open subsystem, ordering can lead to complexity, with new 'rules' governing optimized behavior.

15.1 Defining Order

We have already introduced the idea of entropy in a system. The statistical, rather than the thermodynamic, definition of entropy is better because it is defined even when thermodynamic quantities are ill-defined. For systems with a fixed total energy and given microstates, the entropy of that system is proportional to the logarithm of the number of ways that system might be found by a redistribution of its fundamental particles within its microstates. A microstate of the system is a possible quantum state for the system's particles, but included are only those quantum states whose occupation can be varied by interactions within the system and with the particle's local environment. A fundamental particle of a system is a localizable entity that has identifiable and unchanging properties, at least over observation times. By implication, interactions with the particle will provide insufficient energy to change the internal properties of the particle. Ordering occurs when the entropy of a system drops.

W. C. Parke, *Biophysics*, https://doi.org/10.1007/978-3-030-44146-3_15

We can define the order of a system by its information content. As Landauer has shown, the erasing of information necessarily produces a corresponding amount of entropy, or more. A perfectly ordered system has no entropy. A diamond near absolute zero temperature has nearly perfect order for the given carbon atoms. If you receive a message that has no ambiguity, that message carried maximal information (for the given set of letters, symbols words and their correlations). Before the message is read, the possible such messages had maximal disorder. The logarithm (base two) of the possible ways that the base pairs in DNA could be arranged is information that the DNA can transmit.

A system becomes more ordered when it can be completely characterized by fewer rules. The transcendental number π, although it has quasi-random digits, can be predicted by a simple formula. The orbit of a structureless mass around a much heavier one, with only gravity keeping them together, can be interesting, but the orbit is predictable by a few of differential equations (e.g. Newton's laws) once the initial state of the two is known. Newton's laws 'compress' a large amount of data on the positions of planets over time. However, if the minimum number of algorithms and conditions needed to describe a system approaches the number of entities within that system, then the system has little order.

A test of the order of a system comes from attempts to compress the data needed to characterize a system. On computers, 'zip' files[1] often can be much smaller than the 'unzipped' versions. Let N be the number of entities in the system. Then if the data can be compressed by factors $N \ln N$ or greater,[2] the system is strongly ordered.[3]

For open systems with a flow of energy through them, order evolves. As we have discussed, ordering in an open system is a natural process. Examples include crystallization and the creation of water droplets in rain clouds. Another is the clustering of similar macro-molecules in solution.[4] In the production of order in

[1]Phil Katz introduced this term in the 1980s to characterize the bundling of files, each compressed using his own algorithms. The name he picked to not infringe on the copyrighted term 'arc', for archived files. Those algorithms involved searching for patterns, and then using shorter code symbols for those discovered patterns. In the case of encoded text, many patterns and symbol correlations are already known from the structure of language. Zip pattern searches do not yet search for other word and phrase correlations, such as culturally preferred linguistic patterns. So they, thus far, do not necessarily produce the smallest compression. Never-the-less, current searches will discover some patterns.

[2]For large N, the factor $N \ln N$ is the largest part of $\ln(N!)$, which is how the entropy of the system grows with the number of 'fundamental entities' within. A fundamental entity is a localizable subsystem with distinct properties that do not change with the available local interaction energies and in observation times. Such entities might be the digits in a number, the characters in a message, or the particles in a physical system.

[3]Kolmogorov defined a form of information content of a sequence of symbols as the bit length of the shortest program which can reproduce the systems data: Kolmogorov, A.N.,"*Three Approaches to the Quantitative Definition of Information*," Problems Inform. Transmission **1:1**, 1–7 (1965).

[4]Herbert Jehle, *Specificity of Intermolecular Forces due to Quantum-Mechanical and Thermal Charge Fluctuations*, Proc Natl Acad Sci **43:9**, 847–855 (1957); see also Chap. 11, footnote 16.

one system through cyclic processes, entropy must be transferred to another. Most commonly, this entropy is transferred by heating the environment.

15.2 Factors Needed for Ordering to Occur

A description of the natural laws and emergent principles that lead to ordering and self-organization of an open system we will call 'Ordering Theory'.

The important factors and relations can be displayed as in Fig. 15.1. Two examples are shown in Figs. 15.2 and 15.3. Clearly, several conditions are necessary and sufficient for ordering to occur in an open subsystem of a larger system:

1. **Energy Flow**: Energy must be flowing through the subsystem;
2. **Variability**: Many complexions with quasi-stable configurations must be dynamically possible;
3. **Rules and Specificity**: A subset of the varieties must have selectable character based on rules stored and perhaps even previously created within the subsystem.

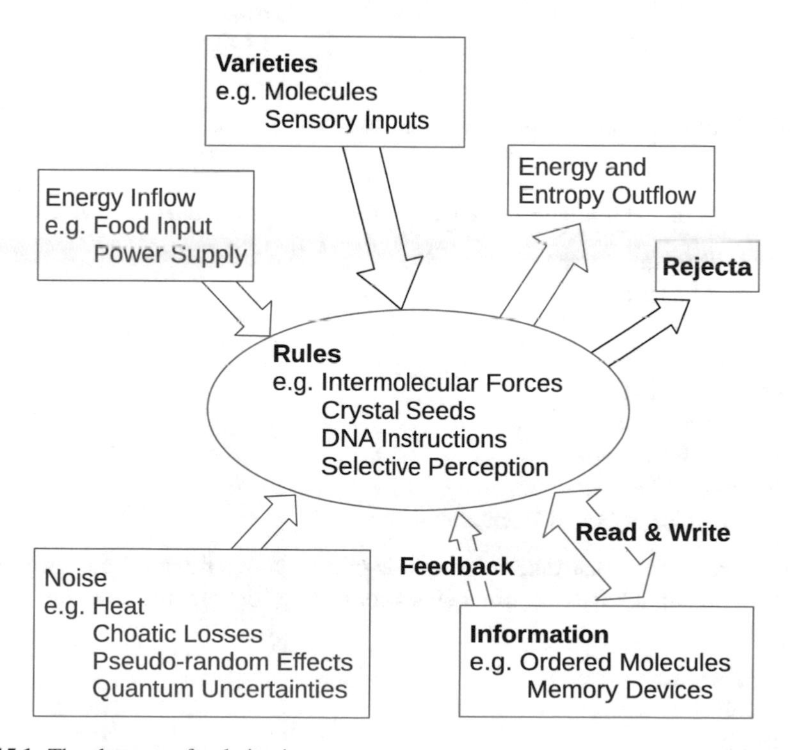

Fig. 15.1 The elements of ordering in an open system

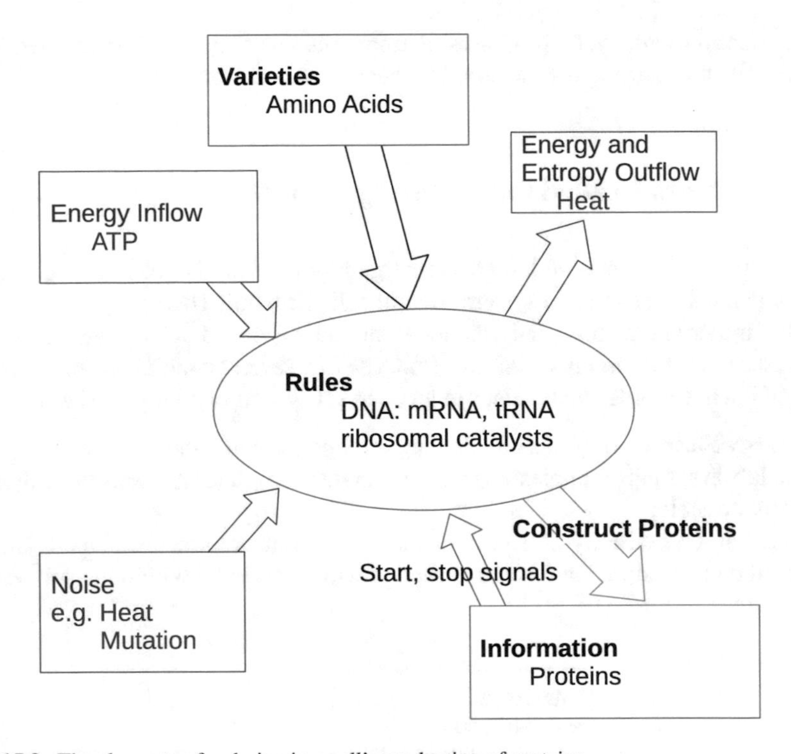

Fig. 15.2 The elements of ordering in a cell's production of proteins

The following additional examples are all describable in the above terms:

- Particle Transformations
- Chemical Reactions
- Crystallization
- Formation of Life
- DNA Replication
- Evolution of Species
- Learning Mechanisms
- Evolution of Consciousness
- The Dynamics of Social Interactions

Universal factors that determine the degree of competitive advantage a given ordered state might have over other states competing for resources are:

- Stability
- Maintainability
- Adaptability
- Viability
- Activity
- Reproducibility

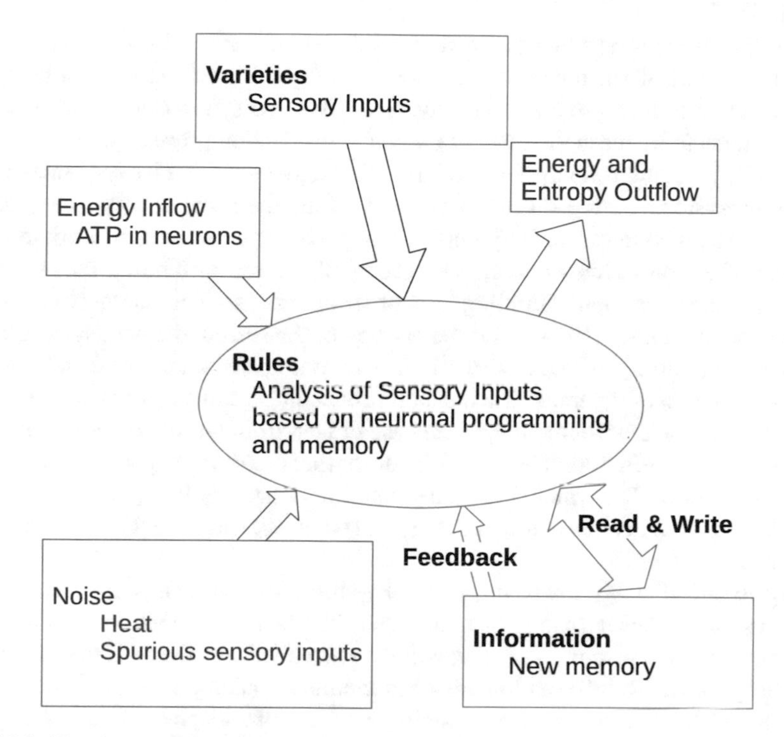

Fig. 15.3 The elements of ordering in a brain's recording of a sensation

Stability includes structural integrity. Ordered systems encompass lifetimes over the full range of time intervals available to subsystems. A crystal in outer space may have a lifetime of billions of years. A metastable atomic state typically has lifetimes measured in microseconds. Nuclear states which last over about 10^{-22} s are often considered metastable. For life systems, certain structures may last years. For more ephemeral structures, repair and renewal mechanisms have evolved.

'Reliability theory' concentrates on those aspects of a system that insure the system's survival under fault and adverse conditions, and also under the natural degradation of active components. Strategies include redundancy, built-in alternative pathways, error testing and correction, protection schemes against adverse environments, external influence filtering, self correction and repair systems.

Maintainability is a property of a system that gauges the ease and speed of restorative actions, renewals, and repairs. With degradation due to the wear-and-tear of activity and to environmental attacks, having the ability to effectively and quickly replace broken parts is a distinct advantage.

Adaptability involves the ability to change the rules of specificity at a rate comparable or faster than the rate of critical environmental change. In evolution, adaptability requires a variety of order variations. Mutation produces such a variety, so variability and mutability become strategies under adaptability. Learning systems and relearning systems also incorporate a form of adaptability.

Viability is measured by how easily the subsystem is created and sustained. Survivability is part of sustainability. Synergism is the ability of different subsystems to act together to increase their viability. A catalyst acts synergistically with certain other systems to increase the rates of particular production processes.

Activity is required for systems to have life. Higher activity gives advantage over slower subsystems acting on the same constraints, the same energy flow and the same material resources. Activity implies work is being done. In turn, work requires an input of energy. This work may be eventually stored as binding energy in the building of the subsystem. Building is an ordering process. This can only take place for subsystems running in cycles if the entropy of the surroundings increases. Thus, the subsystem must produce disorder in its environment equal to or greater than the interior reduction of disorder. The most often encountered mechanism that increases the entropy of the environment is the release of heat from the subsystem, usually by conduction and radiation. There will be an optimal activity depending on energy and material flow. Thus, included in any measure of activity is the effectiveness for a system to gather needed external energy and supplies, and to vent heat and remove spent waste material.

Reproducibility gives advantage to subsystems, in that many copies can utilize resources to a greater extent than one can. Those subsystems which can limit reproduction when resources are low will have an added advantage. If the subsystem is a life system with information stored in memory, making a new individual with fresh structure and new memory capacity may be more adaptable than redoing the parents, particularly when the environment is changing.[5] If the progeny also have slightly different structural coding, and suffer changes in the environment, those with greater survivability have an advantage. This also means a finite lifetime for individuals may also have an advantage. The span of life should be sufficient to protect and train the next generation, and to take advantage of learned skills.

The universal factors, stability, adaptability, viability, and activity, cannot be simultaneously maximized. For example, stability and mutability are counteracting traits. With time, nature finds a system approaching an optimal combination of the relevant factors, given the natural constraints and the environmental conditions.

15.3 Major Thresholds in the Development of Order

Ordering is a process of accumulating information. Life systems first used information stored on molecules to be able to recreate new molecules, and then organisms. Further information added to the molecules made the activity of life further capable of survival. New flexible storage through proteins regulators allowed for variable responses to variable environments. Information for producing chemical transmitter systems, hydraulic systems and ganglions permitted coordination among many

[5]This is an argument against immortality!

cells. The development of flexible neuronal systems was a threshold in memory storage in life systems, as information was no longer held on individual molecules determined by genes, but in a network which could immediately adapt and learn. As brains developed, the total amount of information stored and available for survival under varying conditions also took a leap. Training could be passed from generation to generation from the parent's life experience. Another threshold in information utilization came when human stories were recorded outside the brain. Then, libraries of information could be accessed.[6]

In the evolution of the Universe, ordering has been a natural process in subsystems. The following list shows the epochs and thresholds in such ordering. A threshold in ordering occurs when nature creates a way to dramatically speed up the ordering of a subsystem while maintaining or increasing viability.

1. Space-time creation; Universe expands, temperature drops
2. Particle condensation (quarks and leptons) from gravitational energy
3. Formation of light elements ($H, He, {}^2H, Li$), creating a Universe with a variety of entities and interactions
4. The Universe becomes neutral and dark
5. Star formation; light returns; stellar production of heavier elements
6. Secondary star formation with planets (from supernovae debris)
7. Formation of macromolecules on terrestrial planets
8. Ability of macromolecules to self-replicate: Life genesis; life evolution
9. Molecular coding of auxiliary structures
10. Development of encapsulated systems
11. Catalytic metabolism
12. Facility for reaction to environmental changes
13. Coding for the coordination of cells
14. Development of molecular anticipatory schemes
15. Life systems with some actions based on internal memory of experiences
16. Coordination and control through neuronal systems
17. Reservation of neuronal memory
18. Development of organs for analysis and learning
19. Use of external tools
20. Ability to form models of environment
21. Development of a conscious network
22. Creation of language to relate and pass on life experience
23. Memory storage outside an organism (external memory)
24. Success of science in generating logical models of observations
25. Ability to cause mass extinction
26. Rational control over primitive drives (such as the 'we-they' syndrome)
27. Understanding genetic coding (genotype to phenotype)
28. Management of diseases and environmental stability

[6]The amount of information in the Library of Congress is in excess of about 10^{18} bits. Some computer databases have more.

29. Actions to alter self: Genetic engineering
30. Spread of conscious systems across galaxies
31. Artificial life created by biological life systems
32. Ability to influence the evolution of our Universe
33. Ability to create new Universes

15.4 Complexity

Complexity in an open system is distinguished from disorder. We will introduce a calculable measure of complexity: The 'degree of complexity' of a system is the minimum number of 'emergent' relationships to describe the dynamics of 'agents' within that system. An 'emergent principle or relationship' ('Specificity Rule') is a rule or propensity that is not simply an elaboration of microscopic dynamics, but rather the result of a statistical likelihood for how a large number of entities interact within the system or a subsystem, with that likelihood resulting from the prior adaptive abilities of the system. For statistics to apply, the subsystems must be capable of 'passing through' many possible alternative microscopic pathways.[7] In the evolution of a complex system, quasi-stable substructures can be created that have the ability to make the system better adaptive to external pressures and to better optimize the utilization of power flowing through the system. These substructures are referred to as 'agents'. As time evolves, agents can change as new pathways for more rapid or more efficient activity are spontaneously 'discovered'. Also, new agents at a larger scale may develop. This makes the system hierarchical, with agents at each level of complexity.

As external pressures and power flow conditions change, the agents can change in response. As these responsive reconfigurations make a feedback system, such complex systems have non-linear behavior, and can exhibit and even take advantage of chaotic states. They also are subject to 'cascading failure', wherein conditions are reached that make all entities at a given system level unstable.

It has been shown that as extended open systems evolve, they naturally approach a phase change, producing a dramatic increase of order within.[8] Natural systems spread over a region of space evolve complexity all by themselves, with adaptive changes occurring simply from the advantages adaption gives to effectiveness and

[7] On the speculative side, it is even possible that the 'laws of physics' are emergent properties of our current Universe. This view arose from a physics discussion group organized by Prof. Ali Eskandarian in the late 1990s. A cogent proponent in the group was Congressional Science and Technology Committee consultant Anthony Scoville, who had become enamored with the related ideas of John Archibald Wheeler, and invoked Gödel's Theorem to say physical relationships can be discovered without having antecedents. As new substructures are created, we should expect that new emergent relationships also evolve.

[8] P. Bak, C. Tang and K. Wiesenfeld, *Self-organized criticality: an explanation of 1/f noise*, Phys Rev Lett **59**, 381 (1987) and *Self-organized criticality*, Phys Rev A **38**, 364–374 (1988).

therefore survivability of dominant pathways. The threshold for this kind of change in ordering is sometimes called 'self-organized criticality'.[9]

Biological systems have a large number of structured subsystems, and therefore are more ordered than that of the thermodynamic equilibrium state of the system's entities. But in addition, biological systems have evolved dynamical pathways that are far more likely than those that might be physically realizable but are largely unexpressed. Catalytic proteins are examples of evolved agents. People are another. Biological systems have self-organized complexity. The theory of 'steadily directed systems' (see Sect. 9.25) describes the evolution of self-organized complexity using the language of statistical mechanics.

The 'rules' for how DNA produces proteins is an example of an emergent relationship among agents in a complex system. Rules that describe the network of effective biochemical reactions in cells is an emergent set of relationships. The behavior of a brain in reaction to a stimulus is yet another example.

As the complexity of a system evolves, the hierarchical 'levels of complexity' may become distinguishable. For example, for humans, atoms are the fundamental particles in the 'zeroth' layer, macromolecules are the agents at the first level of complexity, organelles in cells are within a second level, cells themselves in a third, organs in a fourth, individuals in a fifth, and societies in a sixth. Each level has its own emergent 'rules' for its dynamics. Thus, complexity has a measurable degree and may have observable levels.

15.5 Catastrophes in Complex Systems

If you pick up a starving dog and make him prosperous, he will not bite you. This is the principal difference between a dog and a man.

— Mark Twain

Open systems with non-linear dynamics can suddenly change state. We have seen that chaotic dynamical systems show such behavior. Population dynamics (Sect. 13.2) was our first example. A system as simple as an anharmonic oscillator with a driving force that slowly increases in frequency can suddenly change in amplitude near resonance. This behavior occurs in the solutions to $d^2y/dt^2 + ady/dt + by + cy^3 = A \sin(\Omega t)$, which is Newton's law for the damped motion of a mass held by a non-linear spring, and forced to oscillate. Figure 15.4 shows how the

[9]The creation of reproductive systems is not confined to molecular systems. In June, 1959, Lionel Penrose and son Roger showed how to build wooden entities with hooks and springs, which, when placed in a box and shaken, could link together and then, after reaching a certain configuration, release subassemblies. Also, the evolution of new entities in a complex system can be demonstrated within a seemingly simple solitary game devised by mathematician John Horton Conway and published in Martin Gardner's column in the October 1970 edition of **Scientific American**. The computer program called 'The Game of Life' is based on Conway's scheme.

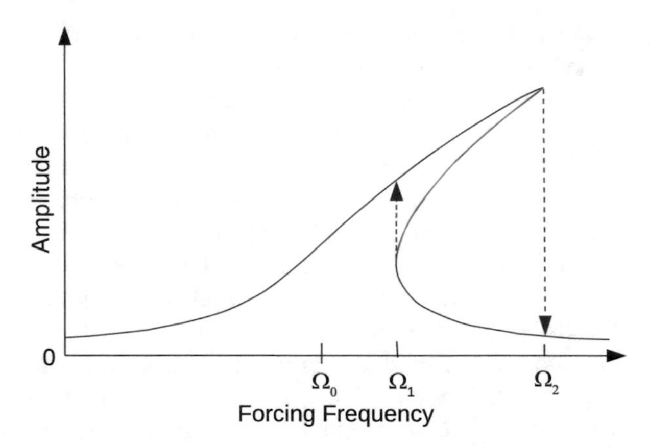

Fig. 15.4 Maximum amplitude versus forcing frequency for a forced anharmonic oscillator described by $d^2y/dt^2 + ady/dt + by + cy^3 = A \sin(\Omega t)$. Here, the non-linear term has a positive coefficient: $c > 0$. Ω_o is the natural frequency of the unforced oscillator without the non-linear term. Ω_2 is the frequency of the forcing term where the amplitude discontinuously drops to a lower value as the frequency increases from Ω_o, and Ω_1 is the frequency of the forcing term where the maximum amplitude discontinuously increases as the frequency is lowered back down. The lighter-shaded part of the plotted curve contains physically unstable points

maximum amplitude of motion varies with the forcing frequency. (Compare with the linear case, represented in Fig. 5.4.)

Complex systems have non-linear dynamics due to feedback of emergent subsystems affecting the future dynamics. Even the control network for setting a healthy heartbeat rate has chaotic fluctuation. A lack of this fluctuation is a sign of disease.

Complex systems can also show large changes due to small ones after applying emergent dynamics: A single base mutation in a DNA molecule can cause dramatic changes in the phenotype. The brain's activity and hormonal state can suddenly change from 'fight' to 'flight' by a small change in an external stimulus. If that sudden large change produces a drastic increase in disorder, or negatively affects the prospects for that system to prosper, then we call the event a 'catastrophe'.[10]

Being close to a chaotic condition may actually give advantage to subsystem agents, in that many configurations of the subsystem are being explored near a chaotic state. In human societies, this condition is sometimes referred to as 'living near the edge'. This might even explain why a brain near an epileptic fit tends to operate with high intelligence.

Whether all civilized societies are likely to reach a self-inflicted extinction catastrophe is an open question. Asteroid hits are also a form of catastrophe in our

[10]René Thom's definition of catastrophe in 1968 included both positive and negative discontinuous events as the parameters in a system varied over an optimization surface. See R. Thom, *A Dynamical Theory of Morphogenesis*, Theoretical Bio Symp, [Bellagio, Italy] (1968).

planetary system.[11] Spreading to different planets greatly reduces the probability of complete loss and waste of the life experiment that Nature is performing on Earth. Besides our vile lust for domination and our decent level of curiosity, consideration of the unpredictability in the timing of catastrophes argues for populating the stars and galaxies.

Problems

15.1 What is the entropy of a crystal at equilibrium near absolute zero if there are only seven units of energy available for thermal vibration? Take the vibration of the atoms in only one direction, and with a separation of vibrational energy levels of one unit.

15.2 If two liquids with the same densities are thoroughly mixed, each liquid having molecules which have no dipole moments, what would cause those liquids to spontaneously separate? While separating, will the liquids release or take in energy from the environment? Does the separated liquids have more or less entropy than the mixture had? If less, how does the process satisfy the second law of thermodynamics?

15.3 Make a 'catastrophe plot' appropriate to the 'fight or flight' syndrome. Devise a model with the production of adrenalin and other biosystem changes in response to danger that might explain 'fight or flight'.

15.4 Argue that biological scaling laws are an example of an 'emergent' principle in the ordering of systems.

15.5 Humans have reached the 'threshold of capability for mass extinction'. Is this an inevitable fate of evolving intelligent systems?

15.6 What factors are there to impede a life system from over saturating its environment, producing a catastrophic collapse.

15.7 Outline what might be involved in a theory of punctuated evolution, i.e. the cause for sudden spurts in the evolution of life forms. One avenue: Think of life system evolution as adding structures and configurations that enhance purposeful activity. Even though there are an astronomically large number of possible organic structures, up to now, evolution must proceed by exploring only the effects of small changes (some with large consequences!). The various possible configurations can be associated with a number giving its viability. They also have a measure of distance apart by how much the structures differ. Thus, a viability plot ('surface') can be made, looking like a very jagged landscape. A given life form starts at or

[11] Extending Newtonian dynamics to predict asteroid orbits to longer and longer times becomes more and more difficult due to the chaotic behavior of the dynamics.

near one of the peaks of that landscape. As the form evolves, it makes short trips to nearby peaks (or falls into a valley of reduced viability.)

15.8 What do you think might be the minimal set of chemical elements and interactions ('rules') to sustain an organic life system? Try to justify your answer.

15.9 Think of how to write a primitive computer program that would take a minimal set of entities ('varieties') and one or two types of interaction between those entities, set the entities in random encounters, and see if life evolves.

15.10 In the previous problem, how would punctuated evolution show itself?

Chapter 16
Energy Flow in the Production of Order

The flow of energy through a system acts to organize that system.

— Harold Morowitz

Summary The Earth itself is an open system, with specific sources and sinks of energy, and a number of continuing ordering processes, including geological differentiation and life.

16.1 The Flow of Energy Through the Earth System

Much of what the Earth receives as energy comes from sunlight. In addition, the Earth receives ocean tidal energy from the Moon and Sun, moonlight, starlight, cosmic rays, extraterrestrial neutrinos, relic microwaves, and even gravitational waves from exploding supernovae in space. It also has a geothermal heat energy flow arising predominantly from radioactive material decay in its crust and from tidal heat. Table 16.1 shows the energy flow (in terawatts) into the Earth from various sources. In addition, there is energy stored in chemical and nuclear form contained in deposits, ores, and other materials of the Earth.[1]

[1]Referring to Table 16.1, the Earth has a mantle and crust containing mostly lithophiles (rock-lovers) and a core containing mostly siderophiles ('iron lovers'). The lithophiles are the oxides, silicates, and other such compounds, the 'scum' and slag that was buoyed toward the surface of the Earth by being less dense than iron and other unoxidized metals. Uranium and thorium easily form oxides, and so are mostly found in the lithosphere. Therefore, heat from radioactive decay comes mostly from the crust and mantle: about 8 TW from Uranium-238, 8 TW from thorium-232, and 4 TW from Potassium-40. The absorption cross section of neutrinos by neutrons rises linearly with the neutrino energy to at least 350 GeV: $\sigma_\nu = 0.68 \times 10^{-45}\,\mathrm{m}^2(E_\nu/\mathrm{MeV})$, giving the rate of neutrinos absorbed by the Earth as $dN_\nu/dt = 4\pi I_\nu(d\sigma_\nu/dE_\nu)N_N$. Here, N_N is the number of neutrons in the volume of the Earth. For relic neutrinos, $I_\nu = (7/16)\sigma T_\nu^4$ is the neutrino total energy flux. Stefan's constant $\sigma = 5.67 \times 10^{-8}\,\mathrm{W/m^2/K^4}$ and $T_\nu = 1.95\,\mathrm{K}$. The peak of the relic gravitational-wave Planck's distribution occurs at $f \sim 10^{11}$ Hz, where the stretching parameter h

W. C. Parke, *Biophysics*, https://doi.org/10.1007/978-3-030-44146-3_16

Table 16.1 Earth: sources of power (in 10^{12} W)

Source	Power (TW)
Solar and lunar	
Sunlight (174,000 input)	128,000
Water condensation	40,000
Hurricane heat (↑ strato.)	$\lesssim 600$
Hurricane wind	$\lesssim 3$
Geohydropeftos (water falls)	~40
Wind (from thermal gradients)	~16
Water waves (from wind)	~2
Tornado	$\sim 1 \times 10^{-5}$
Ocean thermals	~5
Tidal from Moon and Sun	3.7
Ocean height variation	1.1
Seabed frictional heat	2.4
Earth viscoflexing heat	0.2
Moonlight	0.12
Solar wind (p, e^-, He^{++}, etc.)	0.0016
Solar neutrinos	$\sim 2 \times 10^{-27}$
Geologic	
Geothermal	47
Earth radioactivity	20
^{238}U decay	7.7
^{235}U decay	0.3
^{232}Th decay	8
^{40}K decay	4
Core relic heat flow	25
Iron core heat of solidification	1.9
Mantle viscous heat	~0.1
Other cosmic sources	
Extrasolar cosmic rays	0.01
Starlight	0.0035
Relic microwaves	0.0016
Cosmic dust heating	$\lesssim 0.001$
Asteroid and comet collisions	$\lesssim 3 \times 10^{-4}$
Relic gravitational waves	$\lesssim 2 \times 10^{-22}$
Relic neutrinos	$\sim 1 \times 10^{-46}$
Radiation to space	174,050
Photosynthesis storage rate	~1800
Human usage (2013)	18

is about 3×10^{-32}. The estimate for relic ggravitational wave density uses $\rho_{GW} \leq 0.15\, \rho_c$ and the Dyson estimate of damping in the Earth of $\varepsilon \approx 10^{-21}$.

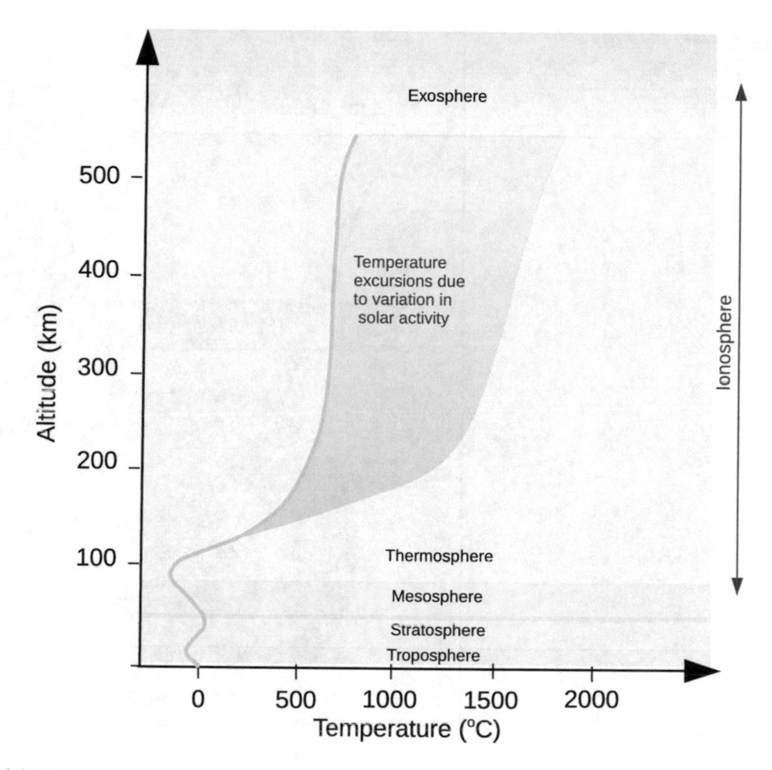

Fig. 16.1 Temperature with height in the Earth's atmosphere

Life forms can use any of these sources, but sustained life requires a 'sink' into which the energy can flow. Such a sink is ordinarily a 'heat bath' at a lower temperature than the temperature that the source produces in some material. In thermodynamic terms, the sink is a reservoir in contact with a heat engine which extracts work from part of the energy in a relatively low entropic source while sending the rest of the energy into the sink, in our case outer space. As a cyclic thermodynamic process, there must always be some heat energy expelled into the sink, since organizing subsystems are being reduced in entropy. The rate of reduction in entropy because of life will always be less than the entropy flowing into the system. We order ourselves at the expense of disordering outer space.

Sunlight is dominated by Planckian radiation from a hot body with a surface temperature of $T_S = 5778\,\mathrm{K}$. The intensity of light radiating from a hot 'black' body (energy per unit time per unit area) is given by Stefan's law (Eq. (7.18)) as

$$I_\gamma = \sigma T^4 = \frac{2}{15}\frac{\pi^5 k^4}{h^3 c^2} T^4 = (5.67 \times 10^{-8}\,\mathrm{W/m^2})(T/\,\mathrm{K})^4\,. \qquad (16.1)$$

As the Earth has maintained a relatively constant radiation temperature for millions of years, a close balancing of energy flow must occur for energy into and

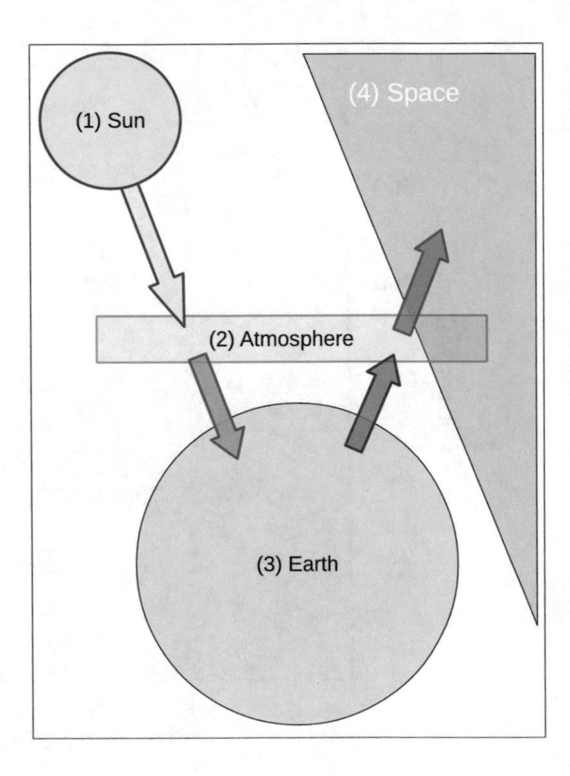

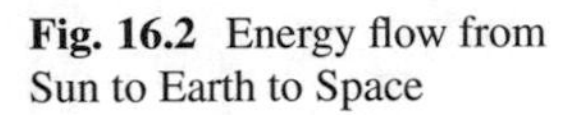
Fig. 16.2 Energy flow from Sun to Earth to Space

out of the Earth (Fig. 16.2). If the radiation absorbed by the Sun caused the Earth's surface to heat to a higher temperature, the Earth's radiation back to space would increase, eventually rebalancing the flow of energy in and out. Measurement shows that the Earth's surface and atmosphere have had an average temperature which has varied over hundreds of millions of years from 248 to 266 K. We are assured that whatever rate of radiant energy absorbed by the Earth, that same rate of energy flow must be radiating back to outer space. Outer space has an effective temperature of 2.76 K through the presence of the relic cosmic microwave radiation left over from the hot big bang 13.82 billion years ago. (This background radiation arriving at the Earth contributes little to the Earth's heat balance.)

Using the energy balancing condition, lets us calculate the effective temperature of the Earth. To do so, we need to know how much of sunlight is reflected back to space without being absorbed. The reflectivity of a surface is measured by its albedo, α, ranging from 0 to 1. An albedo value of 0 means that all the light is absorbed. The Earth's albedo has been measured to be about 0.3. Another factor is the emissivity ε of the Earth, which accounts for the fact that the Earth does not act as a perfect blackbody: $\varepsilon \approx 0.77$. So the balancing of radiation energy flow reads

$$P_{sun} = (R_E^2/(4R_{E-S}^2))\sigma T_{sun}^4(1-\alpha)\pi R_E^2 \approx P_{Earth} = 4\pi R_E^2 \varepsilon\sigma T_{Earth}^4 \,. \quad (16.2)$$

Here, we take $R_E = 6.371 \times 10^6$ m; $R_{E-S} = 150 \times 10^9$ m; $\alpha = 0.3$; $\varepsilon = 0.77$; $T_{sun} = 5778$ K. The above gives an effective radiation temperature of the Earth as $T_{Earth} = 18\,°C$. A better calculation would include the radiative energy flow from the Earth's atmosphere and surface, taking account the atmospheric spectral absorption and emission in each layer above the Earth, and give an effective temperature of 15 °C.

16.2 The Negentropy of Life

The Earth not only has a flux of energy from the Sun and a nearly equal amount of energy expelled into outer space, it also has a flow of solar entropy in and a flow of entropy out. Although energy is conserved, entropy is not. The rate of entropy out is much larger than that coming in from the Sun, while the open geological and life systems become more ordered. The reduction of entropy of open subsystems is measured by 'negentropy', i.e. its loss of entropy, which Shannon has shown is a form of information storage in that subsystem.

The entropy flow from the Sun to the Earth is given by $(4/3)P_{sun}/T_{sun}$. The entropy flow into space from the Earth is $(4/3)P_{Earth}/T_{space}$. (The factors $4/3 = 1 + 1/3$ include $1/3$ to account for the entropy generated by the dissipation of the work done by the pressure of light.) Thus the net entropy flow is

$$\frac{dS}{dt} \geq \frac{4}{3}\left(\frac{P_{Earth}}{T_{space}} - \frac{P_{sun}}{T_{sun}}\right) = \frac{4}{3}P_{Earth}\left(\frac{1}{T_{space}} - \frac{1}{T_{sun}}\right) > 0\,. \tag{16.3}$$

In the above, the radiation temperature of outer space is $T_{space} = 2.76$ K. Since $T_{sun} >> T_{space}$, the entropy flow into outer space from the Earth must exceed the flow in from the Sun. The production of entropy in the Universe by the Earth allows subsystems, including life systems on Earth, to reduce in entropy, up to the rate given in Eq. (16.3). This ordering process can be in the formation of complex molecules, chemical energy storage, and the utilization of work for transport processes.

Problems

16.1 Under what conditions would the activity of life benefit from periods of stasis, suspended animation, hibernation, and stagnation?

16.2 If energy flow can be controlled by an evolving system, what factors might determine an optimal flow?

16.3 What strategies are available to life systems on Earth to overcome the limitations of order production by the limiting flux of entropy to outer space?

16.4 Light from the Sun arrives above the Earth's atmosphere with an almost random distribution of polarizations. If sunlight arrived perfectly polarized, how much less entropy would it carry per unit second per square meter?

16.5 Some of the heat keeping us warm comes from the heat released by solidification of iron on the surface of the inner (solid) core of the Earth. This causes the core to grow in radius by about 1 mm a year. (The radius of the inner core is 1220 km, where the temperature is near 6000 K and pressure is about 330 GPa. The heat of fusion of iron is 13.8 kJ/mol, and the density of iron at the inner core is 13 gm/cm^3.) From this data, calculate the Joule heat released per second, and compare with the number in the energy-flow table.

Chapter 17
Life in the Universe

Chemists have been inspired by Nature for hundreds of years, not only trying to understand the chemistry that occurs in living systems, but also trying to extend Nature based on the learned facts.

— Kirsten Zeitler

Summary From an estimate of the number of planets in our galaxy that orbit in the habitable zone of their star and that have the right range of size to hold a light atmosphere, astrobiologists expect there to be millions of examples of life throughout the galaxy. Whether these life forms have evolved to the point of being able to destroy all life on their planet is a more difficult calculation. In this chapter, we explore the requirements for life and the conditions needed for life's initiation and progression.

17.1 Minimum Requirements for Life

Life can develop where there was no life before. We have at least one example on the Earth. Clearly, (1) the materials, (2) the energy flow, and (3) the environmental conditions must be just right.

(1) Certain materials are necessary: Those that combine in a number of ways to allow for complexity in the system, and that are capable of transformations both into and out of higher energy states in the given environment.
(2) There must be a source of relatively high energy states that can do work, and a sink to deposit energy, such as in the form of heat.
(3) The environment must allow the materials of life to be accessible and capable of intimate mixing.

W. C. Parke, *Biophysics*, https://doi.org/10.1007/978-3-030-44146-3_17

The above three conditions are sufficient. This can be shown by computer modeling. Physical modeling in an experimental laboratory is more difficult, as the critical step of evolving code, within an encapsulated life form, to hold the information need for replication, takes a long time in human lifespans.

Modeling shows that the steps to life occur with time sequencing of ordering and the formation of complexity in spurts and jumps.[1] Spurts occur when a 'door' opens with the right combination of catalysts and synthesizing pathways, so that the production of organized subsystems dramatically increases. Starting with liquid water and basic elements present on the Earth, the organic building blocks naturally form in the soup. As we have seen, the ordering of subsystems in an open system (the soup) is a natural and even expected process as heat energy is released to the environment. Experiments of the type of Miller-Urey in 1952[2] using mixtures of carbon dioxide, methane, ammonia, and liquid water, together with electric sparks, amply demonstrate how one can get organic molecules in short order, including all the amino acids. About 3.6 billion years ago, favorable conditions would have existed within certain water pools on the surface of the Earth, and near sea vents at the bottom in the ocean. The next step on the Earth is the linking of monomers, such as amino acids, to make polymers, such as proteins, and the linking of nucleotides to make ribonucleic acid chains. The linking process of nucleotides is catalyzed naturally on some rock crystalline surfaces.

The following step probably took eons, but at some point in time, the first RNA molecule formed which contained the coding to make replication catalysts. This was a watershed event, since such a molecule could then reproduce copiously in a soup of nucleotides. Variability and mutability in structure made competition possible, and evolution followed.

We still have open questions regarding the minimal requirements for life to form. The life-forming materials must have some entities which can form bonds with others. Those bonded systems must be able to bond with a multitude of entities in order to contain the instructions for replication. They also must be able to synthesize their own structure and any auxiliary catalysts and other functional units that have evolved, using the surrounding environmental material ('food stuff'). There must be a source of energy capable of breaking typical bonds perhaps with the help of a catalyst. The energy released as kinetic motion must have a way of being dissipated. Can life exist with only carbon, hydrogen, and oxygen as the material elements, and no others?

For life on Earth, 'CHOPS-N-things' (Carbon, Hydrogen, Oxygen, Phosphorus, Sulfur, Nitrogen) started us off.[3] The flow of solar energy that amounts to 240 W/m^2

[1]Spurts also characterize 'thresholds' in ordering theory (see Chap. 15). For life systems, the sequence of spurts and lulls is called 'punctuated evolution'.

[2] Stanley L. Miller, *Production of Amino Acids Under Possible Primitive Earth Conditions*, Science **117:3046**, 528–529 (1953); Stanlely L. Miller and Harold C. Urey, *Organic Compound Synthesis on the Primitive Earth*, Science **130:3370**, 245–251 (1959).

[3]More than half of the dry weight of a biological cell is carbon.

into and out of the planet, comes in with significant UV and visible light, and leaves largely as infrared radiation from the upper atmosphere. Overall, this transformation is entropy producing, as it must be. But some inflow of energy does work in making complex molecules and performing necessary processes. The production of complex molecules is entropy reducing in an open subsystem.

On the ocean floor, geothermal vents sustain life. There has been plenty of room for order to form and for complexity to evolve in the open system which is the Earth. That ordering selectively picks out pathways that allow further ordering and that speed the process. This occurs simply from the fact that statistical interactions between a variety of dispersed and diffusing molecules will encounter those pathways if they exist. Once the system encounters a catalytic ordering pathway, it undergoes a kind of phase change in the production of more ordered configurations. The existence of the catalytic 'shortcuts' will be strongly dependent on the available materials and the nature of the incoming energy.

When the system by these processes encounters molecular reproduction, Darwinian selection becomes operative, including the preservation of mutability balancing stability. Many of the molecules and processes used by life forms early in evolution were so successful that we find them common in all analyzable past and present life, including prokaryotes. This means that alternatives, if they existed, died off early on. It also means that certain features of present life is strongly preserved. To be adaptive, other attributes must be flexible and mutable.

We should not be too anthropomorphic by assuming life must be carbon based. Silicon can also form long chains, although not quite as easily as carbon. Chains, sheets, and three-dimensional structures of silicon and oxygen formed in the Earth's crust, with various mineral varieties created when an occasional silicon atom is replaced by a metal atom. Boron also can form chain molecules with the boron-nitrogen bond acting similarly to the carbon-carbon bond. Both water and ammonia are highly polar molecules, and therefore can dissolve many salts and other polar molecules. Perhaps a silicon-based life exists, with liquid ammonia as the solute. Even more exotic life systems may exist, such as those formed secondarily by the primary intelligent beings. Self-replicating robots would be an example.

In analogy and in parallel, an Asimov robot might run amok after a conflict with 'The First Law' directing no harm to humans, or robots may even evolve away from strict rules.

A threshold in the evolution of order was passed when humans altered and made organisms through genetic engineering. Humans will be tempted to improve their own genes, to cure diseases and to enhance what might be considered favorable characteristics. But if changed or created organisms supplant the natural ones, the slow (millions of years) natural testing of viability of naturally evolved subsystems is short-circuited. An engineered corn resistant to insects might replace all the wild varieties, but later show fatal vulnerability to a rare virus, due to the changed DNA. And if 'super humans' are created, new conflicts are bound to occur.

17.2 Life Elsewhere

Astrobiologists study the conditions on other heavenly bodies and space to see if life has or can form there. There are several objectives. One is to learn, by comparison, more about how life formed on Earth. A second is to see if there are any other ways to evolve life. A third is to prepare for the discovery of other intelligent life. NASA's Kepler project has found hundreds of extrasolar nearby planets, many that would support life. The SETI project (Search for Extraterrestrial Intelligence) looks for signals from other beings, at frequencies cleverly selected and bearing a 'hello' artfully coded, perhaps with a preface of enticing math identities. Likely originating hundreds or thousands of light-years away, we could not, as individuals, respond to any signal sent by such intelligent beings. But we might learn something from a long message in a radio wave directed toward us.

If we humans apply our intelligence and reasoning, our drive to explore and expand will inevitably lead us to form colonies on bodies in our solar system and those in extrasolar systems. With relativistic spaceships, individuals will be able to travel, in their own lifetime, to locations where other intelligent life may have been detected, although centuries or millennia may pass before Earthlings will hear of the progress of such explorations.

17.3 Self-stabilizing Life Systems

In the evolution of the Earth, there was a time, over three billion years ago, when the atmosphere contained little oxygen. Molecular life forms had developed near hot vents on the ocean floor, metabolizing sulfur, and taking advantage of the plentiful organic molecules and chemical gradients. Encapsulated organisms, the archaea and bacteria, evolved. Viral particles soon followed. After the Earth cooled to have a surface temperature below about 70 °C (higher temperatures prohibit photosynthetic processes) 2.3 billion years ago, cyanobacteria came about, utilizing the energy of sunlight, together with water and dissolved carbon dioxide, to produce carbohydrates and other organic material needed for the structure, growth, reproduction, and the activity in life. Fungi evolved with a symbiotic relationship with bacteria. As the basic photosynthetic process is $6\,CO_2 + 6\,H_2O \rightarrow (C_6H_{12}O_6) + 6O_2$, excess oxygen was released by the plants while carbon dioxide was captured.

Life systems, being mutagenic, respond across generations to changes in the environment, provided that the time for environmental changes causing a measurable effect on survivability is longer than the time needed to improve life in the new environment. This time needed for improvement can be linked to the time required for genetic changes, to the production of memory (records) containing survival strategies passed across generations, or to intelligent redesigns for life systems capable of making purposeful self-correcting changes.

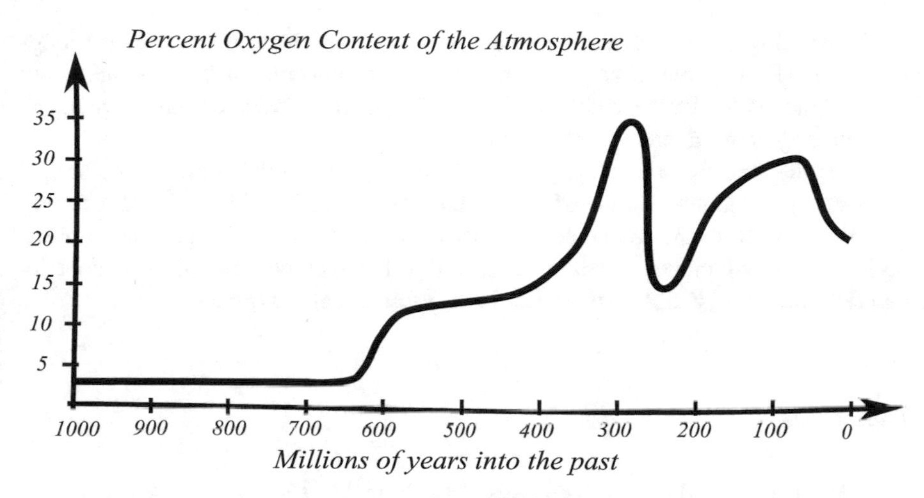

Fig. 17.1 Percent of oxygen in the Earth's atmosphere over time

The asteroid hit that killed off the land and sea dinosaurs caused environment changes too quick and too disastrous for such large animals to adapt, either as individuals or in a selection processes.

There is room for thoughtful and scientific debate whether humans can survive possible climatic changes. If all life on Earth is killed, there is not enough remaining solar time for Nature to 'try again'. Life on Earth probably took at least a half billion years to evolve to just one-celled organisms. Also, there are changes occurring in the Solar System. The Sun is getting hotter by 10% every billion years, so in about a billion years, the Earth will no long be in a habitable zone if it remains in its present orbit. The oceans of the Earth will evaporate, just as those on Venus did.

But there is another factor in the issue of response to environmental change. Life forms can, either by evolution or by intelligence, initiate actions which mitigate and relieve environmental pressures. Animals can insulate, heat, and cool their immediate habitat. But even more, they can produce stabilizing factors tending to improve their natural world. An increase in the greenhouse gas carbon-dioxide causes plants to increase their metabolism. With sufficient mass, the plants can stop further increase in global CO_2, and even reverse it. The production of oxygen by plants adds to the ozone layer in the stratosphere that, in turn, protects plants from damaging UV radiation (Fig. 17.1).

Humans are now capable of controlling and changing the global climate. Active measures to lower the average global temperature are easily contrived (but most are hard to implement), such as changing the water vapor level in the stratosphere, or building solar collectors in space, or making heat pumps between the ground and the upper clouds ('controlled hurricanes!'), or painting the poles white, or dusting the deserts with white calcium silicate, or changing how the Sahara sends dust to South America, or . . .

Self-stabilizing life systems not only evolve themselves, but can cause long-term and advantageous changes in their own environment. Such life systems are an example of 'Complex Adaptive Systems', which includes climatic dynamics, economic systems, and computer networks.

In another hundred thousand years, we can hope that humankind has colonized Mars and the suitable moons of the Jovian planets, and perhaps planets orbiting nearby stars. It is also conceivable that we will have engineered our worlds, including moving mother Earth to a larger orbit. For example, this might be done by carefully directing Kuiper-belt bodies toward Earth-tugging orbits.

Problems

17.1 Hydrogen bond breaking releases from about 2 to 150 kJ/mole. How does this energy per mole compare with $k_B T$ per molecule at body temperature? Why is this comparison significant?

17.2 Suggest further strategies for controlling the global temperature.

17.3 Make a list of all those factors you can think of that might affect the percent oxygen content of the Earth's atmosphere in any particular epoch.

17.4 Describe how a given life system in a given ecology might become 'too optimized' ('too specialized').

Chapter 18
Future Developments

Our preferences do not determine what's true.

— Carl Sagan

Summary Our intelligence allows us to dream up new technologies and new techniques for bettering our lives or for enhancing our individual and collective survivability. We can anticipate, in the not-too distant future, real Star Trek 'tricorders', and computer diagnostics which looks at all possible diseases based on symptoms, recommending tests and treatments. Drug testing and treatments will be done within computer simulators, not on humans or animals. We need not be passive in the direction we think desirable, nor succumb to our primitive instincts. The social polarization and aggression due to our 'We-They' syndrome should not be taken as inevitable, since we can see the follies and pathologies of our destructive tendencies.

18.1 DNA Sequencing

There are now machines which can take a given human's chromosomes and produce a complete sequencing of base-pairs within the DNA.

The efforts continue for very rapid and inexpensive DNA sequencing. Two new possibilities are (1) Laying the DNA down on a substrate groove, which may even have the ability through molecular interaction with the sugar-phosphate groups to unwind the helix. Then a scanning tunneling microscope could read the bases sequentially; or (2) The DNA can be pulled through a nanopore, and the cross tunneling current will vary depending on the base-pair passing through.

Human history, planetary history, living organism's habitat and behavior, and the RNA/DNA sequencing of all life forms should be stored several places, including away from the Earth, as a safeguard against natural or man-made catastrophes, and to study and possibly regenerate lost species.

W. C. Parke, *Biophysics*, https://doi.org/10.1007/978-3-030-44146-3_18

18.2 The Distal Multi-Corder

The Star Trek tricorder has elements which we know how to construct. Just outside your neck the following technology can be employed: Visual monitory of your carotid artery to get your pulse rate; IR reflection to get your blood oxygenation; electronic nose to get any pheromones, steroids, volatiles from drugs, and other skin expellants as well as emissions from skin microbiota; a microphone to get your throat and lung noises; an ultrasonic reflector to get your skin blood vessel velocities; a CCD camera to get your skin tone, signs of age, and possible moles, sunspots, or melanoma spots; an infrasound emitter and detector to get tissue elasticity, and, for the more invasive device, a laser blaster to knock off a microspeck of skin and vacuum to capture it, then analyze DNA, etc. There is no reason in principle that the detection systems just mentioned could not be miniaturized and placed into a battery-operated hand-held device.

18.3 The Doc-Tour Code

Someday real soon, we will have a database of all reviewed documents on disease medical studies, of all known diseases, of all verified symptoms of diseases, of all known efficacious medications for given diseases, of all medication side effects, and of all substantiated techniques for diagnosis.

A parallel code, 'The Cur-ator', finds all current methods for treatment and/or cure, with probabilities for effectiveness and for downsides. The program would also make suggestions for more clinical observations and tests, and patient data gathering. Also retrievable are demoted, failed or fake techniques, and reasons for ineffectiveness. Each element of the code's information must have been thoroughly reviewed by panels of accredited physicians.

Eventually, people will not die from disease, becoming virtually immortal, incurring social and biological challenges.

18.4 The Imi-tator Code

Someday soon, we will have a computer code that can perform predictions of what a given drug of known molecular structure will do if introduced into a human bloodstream, all from modeling of the human system, with real examples in the literature to check and fine-tune the model.

18.5 The Holo-reality Code

Even today, computers can be used to create holographic scenes in space of objects which may never have existed physically, and even make them sequentially act over time. Filmmakers could resurrect the complete 3-D images of past actors via computer animation, and have them act in 3-D with new scripts, advancing on current digital animation techniques.

However, one should note that animated digital holograms are encumbered by the large amount of information that needs to be stored. Compare this information with that required for ordinary video. The hologram has the equivalent of many views, as if a large number of cameras were being used. In contrast, a 3-D viewer requires only two times the information storage of an ordinary video. Of course, since one expects a large amount of redundancy and correlation from one to the next view, the opportunity for significant data compression is available.

18.6 The X-pression Code

Someday, a computer code will take as input a human DNA sequence, and from it generate as output the consequent adult 3-D tomographic image, including the facial features, internal organs, and list any serious genetic diseases. Of course, there may be many accidents, mutations, deprivations, and other mayhem along the way during any real development from embryo to adult. But the outcome of DNA expression is strongly determined. Look how similar we are one to another, and how much our biochemistry is so close to other animals and even plants. Reliability mechanisms have become part of the DNA code.

The Xpression code will have all the molecular dynamics needed to know how natural mutations, viral mutagenesis, or intentional changes in the DNA sequence will affect the adult living system. The dynamical coupling of genes and feedback systems for controlling protein production must necessarily be a part of the code, as well as the rules for protein folding and interactions.

Xpression will not start with the quantum theory of atoms, as this approach would make the computer memory requirements far larger than necessary. Rather, emergent principles in nano- and mesoscopic biophysics and biochemistry derived from the dynamics of cells would be a more feasible beginning point.

18.7 The Simu-later Code

Someday, far into the future, we will have a code which can closely simulate the evolution of life, including its genesis from organic compounds, the thresholds produced by catalysts, by the utilization of high-energy molecules and external energy sources, by the ability to learn and store present experiences, by the advantages of colonization, by the use of hormones for distal communications, by the application of electrical pulses for rapid intracellular communication, by the development of specialized organs, by the evolution of sensory organs, by the processing of information in a centralized ganglion and then brain, by the passing of life survival and enhancing information by teaching, by the external storage of information, by self-awareness, and by anticipation through conceptual model building. Each threshold introduces an acceleration in evolution with slower periods between ('punctuated' evolution).

18.8 The Im-Mortality Code

Someday, far far into the future, we will have a code which can closely approximate the DNA structure of our ancestors, extrapolating backward and forward from known progenitors and descendants. Let's call the original ancestors 'orgs'. Planets could be given inhabitants with individuals, let's call them 'propes', containing these approximate ancestral DNA. Each such prope could be given the opportunity to be educated in ways similar to the original ancestor, so that the productivity of the original ancestor would have a chance to become active again. If the propes grow up in environments similar to the orgs, and if they study the lives of the orgs, they can become close approximates to the full personality of their originals. In this way, the orgs, for all intent and purposes, live again. But we should realize that such continued living can happen now when an individual decides to take on the ideas and philosophy and even the experience of another. Having the same DNA helps the process, but is not essential, since individual humans are so similar in capacity, aspirations, and drives. The prope's existence is a form of unexpected immortality. Implementing the immorality code is not cloning. Cloning will likely be banned in any democracy.

18.9 The Physics of the Light-Beam Transporter

The Star Trek transporter is difficult to make, but not impossible. First: Retrieve the record of your complete genome. Second (the hard part): Record the current state of your brain (i.e. every neuron configuration, every synapse, their environment, connections, and plasticity). You might have to drink an illumination cocktail, to make your synapses light up in a MRI machine. Third: Test the transporter by having it send the information about you to the next room, where there is a human replicator/Turing machine. To be reassured, you can ask your clone about your mother. Fourth: Encode this information and send it over a laser beam to a distant location. Sending should take only a few seconds, since the total information content is less than 10^{15} bits. Reception is delayed by limitations on signals traveling not greater than the speed of light. The signal would take about four and a half years to get to the nearest star system (Alpha Centuri). Fifth (the really hard part): At the receiving end, have the sentient beings there reconstruct a clone of you from the information received.[1] Advantage can be taken in the transmission process and the reconstruction process of our prior knowledge of the general structures common to all humans. (Note that the reconstructed version of you will need a number of essential fungi, bacteria, and microscopic epidermal arthropods that we have come to rely on in symbiotic relationships).

Now there are two of you in the Universe. Your clone[2] will not have your blemishes, or your healed broken bones, but that, we suppose, is good! Some societies will require you to sign a 'vaporization agreement' to limit the number of your twins in the Universe.

18.10 The Future of Life

It appears that one of the essential elements of evolution is the presence of competition between individual organisms and groups of organisms. If so, life forms necessarily carry aggressive tendencies, either explicitly, or behind a veil covering the primitive parts of their brain. The result is often a 'we-they' syndrome: A polarization of populations based on perceived distinctions, with the evolutionary purpose of enhancing and spreading those in the same group. The genes of successful groups become more dominant.

If a particular life form evolves high intelligence with the associated increased survivability,[3] technology may give that life form the ability to annihilate not just other groups, but all life on the home planet. I.S. Shklovskii and Carl Sagan,[4] while trying to estimate the number of life forms in the galaxy that might be able to communicate with us, guessed that the lifetime of life forms, once that life form had gained the knowledge and technology for planetary destruction, is about 100 years. They still estimated millions of planets in the Milky Way with life that could send out radio waves to us.

[1]Without parity information, they might get your left hand where you have a right. Solution: Sometime earlier, send an intense beam of 'left-handed' neutrinos to the receivers.

[2]The 'no-clone' theorem in quantum theory does not apply, since we are not trying to clone your complete quantum state. A 'classical' copy will do. Reliability in evolution means it is very unlikely that you need to reproduce quantum entangled states for your clone to operate. This statement would not exclude the possibility that some choices ('free will') come from the collapse of entangled states in their interaction with the environment.

[3]This is not the only strategy for survivability: Sharks have evolved into perfectly marvelous carnivores, with only subtle changes in hundreds of millions of years.

[4]I.S. Shklovskii & Carl Sagan, **Intelligent Life in the Universe**, [Holden Day, Inc., San Francisco] (1966).

A saving grace is our ability to reason even in the presence of prejudice and sterotypes. Recognizing that primitive emotions and tendencies may cloud our reasoning power, we can decide not to act on the darker pressures. Such an awakening may let us see our way to a future. With our knowledge and abilities in the field of genetics, we do not need the primitive 'we-they' syndrome to determine the direction of our species.

Freeman Dyson[5] considered the far future of life, even after all the stars have extinguished, in a Universe which continues to expand (but not exponentially), and described ways in which intelligent life could continue to exist.

It is even conceivable that life spread over the galaxies can control how the Universe evolves. After all, we already know how to keep stars from imploding: 'Simply' blow them up and then keep the debris from collapsing back! This reserves the hydrogen fuel, extending the time of light. We no longer need be bystanders in this Universe.

Problems

18.1 Think of ways we might 'blow up a star', and then hold its hydrogen in reserve. (Even now, we can think of ways of changing the evolution of the whole universe!)

18.2 Give your arguments for or against the proposition that intelligent life systems that have a leader with unchallengeable control over a large population is more suitable for extended survival than a democratic structure.

18.3 What would prevent or curtail the creation of super humans by genetic engineering?

18.4 How would a species of robots with intelligence greater than humans nevertheless be less fit than humans?

18.5 What would mitigate against the creation of immortal humans?

18.6 Once an intelligent and technologically advanced society develops the ability to annihilate themselves, what factors might drastically reduce this possibility?

[5]F.J. Dyson, *Time without end*, Rev Mod Phys **51**, 447–460 (1979).

Appendices

The following appendices contain supplemental and auxiliary materials for the topics in this book. Some appendices hold esoteric topics and sophisticated mathematical arguments. These are not intended for student imprintation, but rather to show that the logic behind the conclusions and tools of our subject can be displayed without mystery.

W. C. Parke, *Biophysics*, https://doi.org/10.1007/978-3-030-44146-3

Appendix A
Optical Limit of Maxwell's Equations

From a long view of the history of mankind the most significant event of the nineteenth century will be judged as Maxwell's discovery of the laws of electrodynamics.

— Richard Feynman

A.1 The Eikonal

Optics is a study of visible light and the application of visible light in instruments which enhance images and communications. Optical instruments rely on the behavior of electromagnetic waves in the visible range of wavelengths (400–700 nm) passing through regions and instruments which have sizes much larger than the wavelength of the light. Under these circumstances, wave interference and diffraction are too small an effect to notice. However, viewing very small objects invokes wavelength limits on image resolution.

As Maxwell's theory of electromagnetism is complete, the 'laws of optics' are derivable from Maxwell's equations in the short wavelength limit. Start with Maxwell's equations in the form

$$\nabla \cdot (\epsilon \mathbf{E}) = 4\pi \rho_f$$

$$\nabla \times \mathbf{E} = -\frac{1}{c}\frac{\partial \mathbf{B}}{\partial t}$$

$$\nabla \cdot \mathbf{B} = 0$$

$$\nabla \times \left(\frac{1}{\mu}\mathbf{B}\right) = \frac{4\pi}{c}\mathbf{J}_f + \frac{1}{c}\frac{\partial (\epsilon \mathbf{E})}{\partial t} .$$

The subscript f indicates that the corresponding quantities are for free charge and current densities, not bound ones.

W. C. Parke, *Biophysics*, https://doi.org/10.1007/978-3-030-44146-3

Consider a single frequency solution[1]:

$$\mathbf{E} = \mathbf{e}\,(r)\, e^{ik_0 S - i\omega t}$$
$$\mathbf{B} = \mathbf{b}\,(r)\, e^{ik_0 S - i\omega t}$$

where $\omega = k_0 c$. For a plane wave, $S = \widehat{\mathbf{k}} \cdot \mathbf{r}$. More generally, S is a non-linear function of coordinates, and is called 'the eikonal' for the wave. When there are no free charges in the material, such as insulators with no net charge, Maxwell's equations will give

$$\begin{aligned} \nabla \cdot \mathbf{E} + \nabla \ln \epsilon \cdot \mathbf{E} &= 0 \\ \nabla \times \mathbf{E} &= ik_0 \mathbf{B} \\ \nabla \cdot \mathbf{B} &= 0 \\ \nabla \times \mathbf{B} - \nabla \ln \mu \times \mathbf{B} &= -ik_0 \mu\epsilon \mathbf{E} \end{aligned}$$

which become

$$\begin{aligned} \nabla \cdot \mathbf{e} + ik_0 \mathbf{e} \cdot \nabla S + \nabla \ln \epsilon \cdot \mathbf{e} &= 0 \\ \nabla \times \mathbf{e} + ik_0 \mathbf{e} \times \nabla S &= ik_0 \mathbf{b} \\ \nabla \cdot \mathbf{b} + ik_0 \mathbf{b} \cdot \nabla S &= 0 \\ \nabla \times \mathbf{b} + ik_0 \mathbf{b} \times \nabla S - \nabla \ln \mu \times \mathbf{b} &= -ik_0 \mu\epsilon \mathbf{e} \end{aligned}$$

Now for short-wavelengths, k_0 is large, so that

$$\begin{aligned} \mathbf{e} \cdot \nabla S &\simeq 0 \\ \mathbf{e} \times \nabla S &\simeq \mathbf{b} \\ \mathbf{b} \cdot \nabla S &\simeq 0 \\ \mathbf{b} \times \nabla S &\simeq -\mu\epsilon \mathbf{e} \end{aligned}$$

These mean that **e**, **b**, and ∇S are all perpendicular, and that

$$(\mathbf{e} \times \nabla S) \times \nabla S = -\mu\epsilon\, \mathbf{e}$$

or

$$(\nabla S)^2 = \mu\epsilon = n^2$$

---

[1]Fourier's theorem and the linearity of Maxwell's equations makes a general solution expressible in terms of single frequency solutions

where $n = c/v$, i.e. the refractive index in the material, which is the speed of light in a vacuum divided by its speed in the region where n is calculated. This is the so-called 'eikonal equation'. The gradient of S points in the direction of the propagation of the wave fronts.

A.2 The Ray Equation

Define a ray to be a line in the direction that a wave propagates. Then the unit tangent vector to the ray will be given by

$$\frac{d\mathbf{r}}{ds} = \frac{1}{n}\nabla S$$

where s measures the length along the ray:

$$ds^2 = d\mathbf{r} \cdot d\mathbf{r}$$

A differential equation for determining the line for a given ray can be found by considering how $n\,(d\mathbf{r}/ds)$ changes along a given ray:

$$\begin{aligned}
\frac{d}{ds}\left(n\frac{d\mathbf{r}}{ds}\right) &= \frac{d\mathbf{r}}{ds} \cdot \nabla\left(n\frac{d\mathbf{r}}{ds}\right) \\
&= \frac{1}{n}\nabla S \cdot \nabla\left(\nabla S\right) \\
&= \frac{1}{2n}\nabla\left((\nabla S)^2\right) \\
&= \frac{1}{2n}\nabla\left(n^2\right) \\
&= \nabla n \,.
\end{aligned}$$

We see that the ray will deviate from a straight line whenever there is a gradient of the index of refraction, and will point in the direction of increasing index (slower speed of light). The change in the ray's direction per unit change in the distance along the ray will be given by

$$\begin{aligned}
\frac{d^2\mathbf{r}}{ds^2} &= \frac{d}{ds}\left(\frac{1}{n}\left(n\frac{d\mathbf{r}}{ds}\right)\right) \\
&= -\frac{1}{n^2}\frac{dn}{ds}\left(n\frac{d\mathbf{r}}{ds}\right) + \frac{1}{n}\nabla n \,,
\end{aligned}$$

or

$$\frac{d^2\mathbf{r}}{ds^2} + \frac{d\ln n}{ds}\frac{d\mathbf{r}}{ds} = \nabla \ln n \,.$$

If we let $\widehat{\boldsymbol{\rho}}$ be the vector pointing from a position on a ray toward the center of the osculating circle at that point on the ray's path, and ρ be the osculating circle radius, then geometry dictates that

$$\frac{d^2\mathbf{r}}{ds^2} = \frac{1}{\rho}\widehat{\boldsymbol{\rho}} .$$

From the above, the gradient of the index of refraction must lie in the plane of the tangent vector to the ray and the normal to the ray pointing toward the center of the osculating circle.

The radius of the osculating circle is given by

$$\rho = \frac{1}{\widehat{\boldsymbol{\rho}} \cdot \nabla \ln n} .$$

A.3 Fermat's Principle

Fermat's principle states that light rays take that path between two points which minimizes the time of travel: $\Delta t = \int dt = (1/c) \int n\, ds$. The principle is a consequence of the behavior of optical rays. The time is minimized if

$$\int n\, ds = \int n \sqrt{\frac{d\mathbf{r}}{d\tau} \cdot \frac{d\mathbf{r}}{d\tau}}\, d\tau$$

reaches a minimum between two points. The Euler-Lagrange equations to minimize this integral become

$$\frac{d}{ds}\left(n \frac{d\mathbf{r}}{ds}\right) = \nabla n$$

which is the same as the ray equation.

The full set of laws for reflection and refraction can be derived from the ray equation or from Fermat's Principle.

A.4 Snell's Law

Suppose a boundary exists between two regions of constant but distinct materials with different indices of refraction. If the boundary is a plane, and coordinates are selected with the z axis perpendicular to the plane, then the gradient of the index of refraction is along the z axis. If a ray is in the $x - z$ plane, then the projection of the ray equation terms along the x axis reads $\Delta\, (n \sin\theta) = 0$, where θ is the angle between the ray tangent vector and the normal to the surface. This is Snell's law: $n_1 \sin\theta_1 = n_2 \sin\theta_2$.

A.5 Deflection of Sunlight

As an exercise in applying the ray equation, consider the image of the Sun as we watch a sunset. That image is not in the direction of the actual Sun because of refraction of sunlight in the atmosphere.[2]

If we specify points by coordinates from the center of the Earth, then the index of refraction will largely depend only on the distance r from the Earth center. Our ray equation will now read

$$\frac{d}{ds}\left(n\frac{d\mathbf{r}}{ds}\right) = \frac{dn}{dr}\mathbf{r}$$

which, when crossed with $\mathbf{r}$, gives

$$n\mathbf{r} \times \frac{d\mathbf{r}}{ds} = \text{constant} ,$$

i.e.

$$nr \sin\theta = \text{constant} .$$

If the Sun is viewed on the horizon, $\theta_R = 90°$, so, along the ray,

$$r \sin\theta = Rn_R .$$

We also know that the incident ray is a distance $R + h$ from the center of the Earth, where h is the effective height of the atmosphere, so we can replace $r \sin\theta$ by $R + h$. Finally, the angle $\cos\Phi$ between the incident ray and the refracted ray which arrives at the Earth tangent to its surface is also the angle between the corresponding normals, giving

$$\cos\Phi = \frac{R}{R+h} = \frac{1}{n_R} .$$

The value of the index of refraction at the Earth's surface is 1.003, making $\Phi = 4.4°$. Including the effect of light time delay, the Sun is seen $4.1°$ behind where it is physically.

[2] There is also an eight and a half minute time delay for light to get to our eyes from the Sun. The Sun's image moves so slowly across the sky that this effect is small, producing an angle deviation of $8.5\ \text{min}/24 \times 60\ \text{min} = 0.3\,°$.

Appendix B
Huygens' Principle

One may conceive light to spread successively, by spherical waves.

— Christiaan Huygens

In the analysis of how waves move in three dimensions, Christiaan Huygens, in 1690, poised a marvelous and useful idea: Each wavefront of a given wave may be taken as a new source of spherical waves! (Fig. B.1). Consider, for example, a plane wave traveling through a hole. If you are given the wavelength of the wave, the interference and diffraction of the wave past the hole can be constructed as if new little wavelets are being generated over small regions covering the hole, all initially in phase with each other. The resulting wave produced past the hole generates a diffracted wave. For light, the diffraction causes a diffraction pattern on a screen or in a set of detectors.

G. Kirchhoff in 1882 indicated how the descriptions of wave propagation and diffraction by Huygens, Young, and Fresnel, could come from the properties of the solutions to the wave equation. Rayleigh and Sommerfeld later showed how to express the theory of wave diffraction in a form amenable to a consistent mathematical expansion for short wavelengths. If the wave with amplitude $\Psi(\mathbf{r}, t)$ is taken to have one frequency $f = \omega/(2\pi)$, so that $\Psi(\mathbf{r}, t) = \psi(\mathbf{r})e^{-i\omega t}$, then the wave equation requires that the space part satisfy:

$$\nabla^2\psi(\mathbf{r}) + k^2\psi(\mathbf{r}) = 0, \tag{B.1}$$

where $k = \omega/v$, and v is the speed of the wave. The above is Helmholtz' equation. If the wave vanishes on a large surface area $\mathcal{A}''$ except for a finite segment $\mathcal{A}'$, and it propagates from that area into an effectively unbounded region, then an exact solution to Helmholtz' equation is given by

$$\psi(\mathbf{r}) = \int_{\mathcal{A}'} \psi(\mathbf{r}')\nabla G(\mathbf{r} - \mathbf{r}') \cdot d\mathcal{A}' \,. \tag{B.2}$$

W. C. Parke, *Biophysics*, https://doi.org/10.1007/978-3-030-44146-3

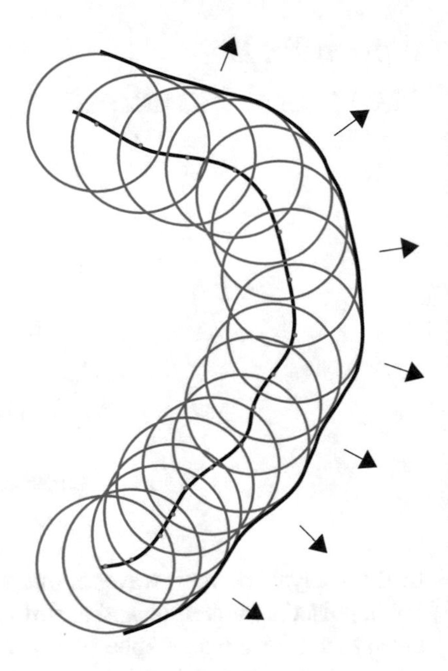

Fig. B.1 Huygens wavelets construction: solid black lines are wave fronts. Arrows show direction of propagation. New wave front to right is constructed by wavelets from prior wave front to left

In this relation, the so-called Green's function, $G(\mathbf{r}-\mathbf{r}')$, is a solution to the wave equation which vanishes when $\mathbf{r}'$ locates points on a surface $\mathcal{A}''$ which includes $\mathcal{A}'$ and the extension of $\mathcal{A}'$ area to large distances:

$$G(\mathbf{r}-\mathbf{r}')\big|_{\mathbf{r}'\in\mathcal{A}''} = 0\,. \tag{B.3}$$

It can be expressed as

$$G(\mathbf{r}-\mathbf{r}') = \frac{1}{|\mathbf{r}-\mathbf{r}'|} + F(\mathbf{r}-\mathbf{r}') \tag{B.4}$$

where $F(\mathbf{r}-\mathbf{r}')$ is a solution of the Laplace equation

$$\nabla^2 F(\mathbf{r}-\mathbf{r}') = 0 \tag{B.5}$$

in the region into which the wave is propagating.

If the area segment $\mathcal{A}'$ is reasonably flat (i.e. its curvature is large compared to the wavelength of the wave) and is of extent much larger than the wavelength, then Eq. (B.2) becomes

$$\psi(\mathbf{r}) \approx \frac{k}{2\pi i}\int_{\mathcal{A}'} \psi(\mathbf{r}')\frac{\exp\left(ik\left|\mathbf{r}-\mathbf{r}'\right|\right)}{|\mathbf{r}-\mathbf{r}'|^3}(\mathbf{r}-\mathbf{r}')\cdot d\mathcal{A}'\,. \tag{B.6}$$

This shows that the propagated wave can be constructed from outgoing spherical waves coming from the wave on the surface $\mathcal{A}'$, a statement of Huygens' Principle.

Appendix C
Holography

Until now, man has been up against Earth's nature. From now on he will be up against his Own nature.

— Dennis Gabor

C.1 Introduction to Holography

Holography was invented by Dennis Gabor in 1947 through his work on improving images produced in electron microscopes. He wanted to use the whole of the information in the electron waves which formed an image, i.e. not just the intensity but also the phase information. (The Greek word for whole is όλος, pronounced holos.) For electromagnetic wave holography, he used a highly-filtered mercury lamp 'point' source to get a reference wave that was close to one frequency and spatially coherent. The later development of lasers in 1960 made optical holography practical.

Holography is the technique of recording and displaying a three-dimensional image using a two-dimensional interference pattern. A hologram is made by splitting coherent waves, with one split beam illuminating an object and the other made to interfere with waves reflected from the object. (See Fig. C.1. The interference pattern is recorded on a two-dimensional surface, such as a photographic film. Holograms are usually produced optically using lasers because laser light can be intense and coherent. However, the principle works for any coherent or partially coherent wave, including sound waves.

One can show that the three-dimensional information determining the surface of bodies is stored on the two-dimensional hologram surface as a Fourier transform of the object. The positions of the object's surface is recovered by a second Fourier transform.

For imaging of surfaces and volumes, holograms require storage of both the intensity and the phase of the wave components. Ordinary photography records only intensity of light onto a two-dimensional surface. Color information is, in effect, recorded separately, one image for each of at least three distinct colors. A single photographic image does not contain information about the distance from the image surface to each of the object points where light was emitted, so

W. C. Parke, *Biophysics*, https://doi.org/10.1007/978-3-030-44146-3

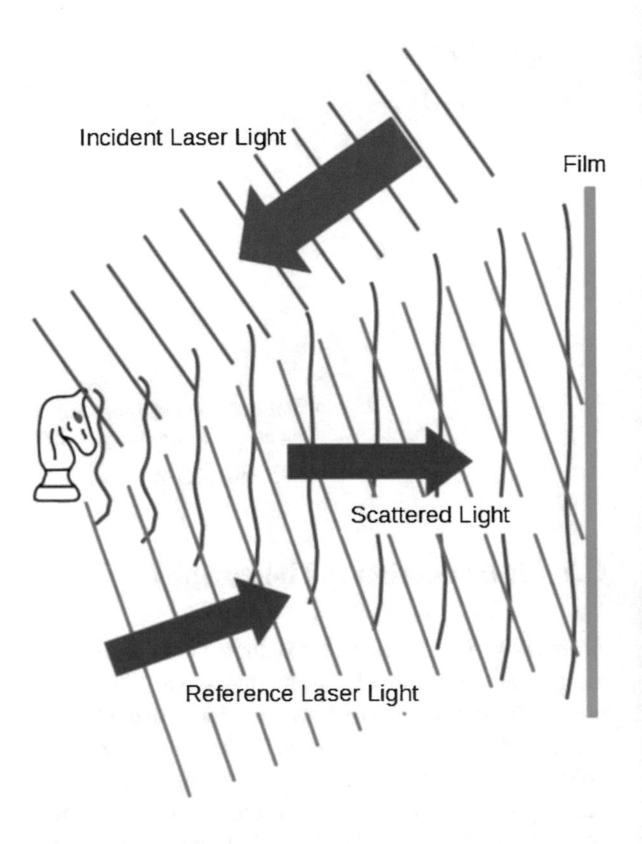

Fig. C.1 Holographic recording: the reference laser light comes from the same source as the incident laser light. The film records the intensity of the interference between the scattered light and the reference laser light

that three-dimensional (3-D) information is discarded. One way to store distance information is to record both intensity and phase of a set of waves scattered from an object onto a recording surface, such as a photographic film or CCD (charge-coupled device). A single light wave of one frequency scattered from one object point will vary in phase when arriving at a recording surface as the distance from the object point is varied.

By using a 'reference wave' originating from the same source of light as the light illuminating the object, but sent directly onto the recording surface, an interference pattern between the direct wave and the wave scattered from the object will be generated. Now, because of constructive and destructive interference, the brightness over the recording surface will vary according to the interference of the object wave with the reference wave. The record will show an interference pattern. For a complicated object, the interference pattern will be so fine that the human eye will hardly see it on the recorded surface. (Larger diffraction patterns caused by dust in the air will be easily visible.)

An image of the object can be reconstructed by passing a plane wave through a two-dimensional holographic recording. Now, the interference pattern on the recording causes the incoming plane wave to produce a diffraction pattern in the wave generated past the recording. If the plane wave has the same frequency as the original reference wave used to make the recording, then the object will appear in the same location it had been physically when the recording was made. The image will be virtual, since the light appears to come from behind the hologram surface where the object was, but no light need come from the image points.

C.2 Principles Behind Holography

Now, to understand holograms more deeply, we will introduce more of the physics details. Consider a hologram made by light coming from a single laser illuminating an object and photographic film.[1]

As the vector character of the electric and magnetic field intensities is not essential to the argument, we will use scalar waves. Also, since the wave travels at the speed of light, we can neglect time delays for waves traveling between our localized objects. At each fixed time, the amplitude of the light wave at the surface of the film will be a superposition of the laser reference wave added to the wavelets of incident laser light scattered from the object:

$$A = A_r + A_o \,. \tag{C.1}$$

The reference wave amplitude A_r has a simple plane-wave behavior on the film. The object wave amplitude varies over the film in a more complicated fashion, as it is formed by the sum of all the scattering wavelets from every illuminated point on the object that arrive at each point on the film. The film exposure is proportional to the resultant intensity of light, which is proportional to

$$|A|^2 = |A_r|^2 + |A_o|^2 + A_r^* A_o + A_r A_o^* \,.$$

Now let's see what happens when the film is developed and then put into a laser beam in the same position as that of the reference wave in forming the hologram. This procedure is called 'holographic reconstruction'. The new reference wave (which, we take also with amplitude A_r) will pass through the clear spots on the film, while diffracting around the dark spots. Since the interference pattern recorded on the film will have opacity variations with dimensions comparable to the wavelength of the laser light, significant diffraction will occur, sending some light into non-colinear directions after passing through the film. Viewing the film from the side opposite to where the object was originally, the light from the reference beam will have passed through the film and enters the eye. Locations on the retina will have a

[1]Negative photographic film can be made by a gel containing silver bromide crystals. When exposed to light, electrons are released in the crystal which combine with the silver ions to make metallic silver. Accumulated silver metal in the gel is opaque. Development of the film stabilizes the remaining silver ions into a thiosulfate.

confluence of interfering wavelets coming from different locations on the film. The eye sees the image produced by a diffraction pattern made by the film, a pattern which itself was produced by diffracted light from an object. The eye should focus not on the film, but rather where the object was originally placed.

For simplicity, let's assume that the film is a positive linear recording, i.e. it has a transmission at each location in proportion to the original light intensity, and that the reference wave has the same amplitude as the original one. In this case, the light transmitted by the film, just past the film's surface, will have an amplitude proportional to the intensity of the wave that occurred just past the film as the recording was being made. We will express this observation in terms of the diffracted wave amplitude, A_D, just past the film:

$$A_D \propto |A_r|^2 A_o + A_r^2 A_o^* + (|A_r|^2 + |A_o|^2) A_r \ . \tag{C.2}$$

The first term makes a virtual image of the object, seen at the same position the object originally sat. The second term produces a 'conjugate real image' of the object, seen in front of the film. The third term is seen as a light originating from the position of the reference beam source and proportional to the intensity of that reference beam. If the original object position and laser beam angles were not colinear, then the images and reference light sources will not be seen in the same direction (Fig. C.2).

C.3 Digital Holography

With present computational power, reconstructing a digitized image of an object recorded on a hologram can be performed directly from the holographic recording on a CCD or CMOS pixelated surface. Advantages to such a digital analysis include the ability to program enhancements such as magnification, aberration correction, noise reduction, low-pass (frequency) filtering, phase contrast enhancement, digital color superposition from images made from various wavelength or by a multitude

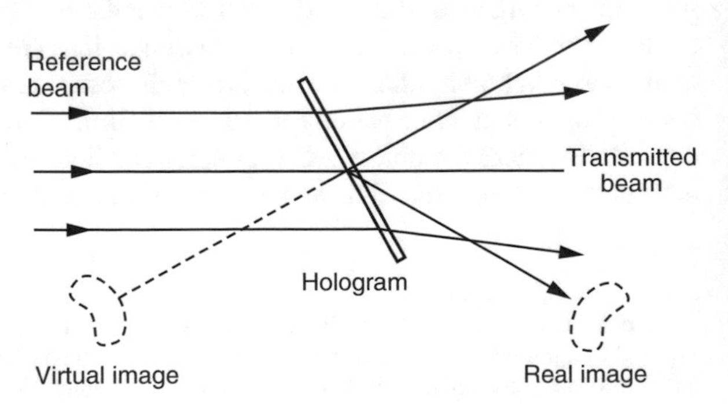

Fig. C.2 Images for off-axis hologram

of digital sensor arrays, and the creation of video sequences to see the motion of objects with no depth of focus problems.

C.4 Some Special Properties of Holograms

In order to make a hologram to form in the first place, the light source must have spatial and temporal coherence, so that the phase of the scattered light from the object is correlated with the phase of the reference beam. Temporal coherence can be achieved by using nearly one frequency of light. The distance over which light from a given source split into two beams keeps a measurable phase relationship between the two beams is called the light 'coherence length'.[2]

Light from an incandescent source at 3000 degrees Kelvin has a coherence length of about 1.2 μm. Laser light can have a coherence length of kilometers, although typical sources have coherence lengths in centimeters to meters. If the object one intends to use for a hologram is more distant from the film than the coherence length, the hologram fails. Also, if any part of the film moves more than a half of the wavelength of the light during exposure, that part will not record an interference pattern. For a flexible film, low level sound in the room can ruin the creation of a hologram. Similarly, the object must be rigidly held to within the same limitations.

The illumination of a hologram after it is made need not be as coherent as the light that produced the hologram. There are various strategies for using point-like light sources (even the Sun) and color discriminators to produce 3-D images from a hologram.

As we noted in the description of optical holograms, they are able to store information about objects with redundancy spread over an area. The storage includes not just illumination of the object, but also relative phases between light scattered from various points on the object, and also, if the object has some transparency, the distribution of density variations and location of scattering point are recorded.

C.5 Holographic Memory Storage

Note that holograms store the information about an object non locally, i.e. the change of one point on the object does not produce one local change in the hologram. Moreover, the information is redundant, in that every finite region of

[2]Specifically, a light source is split into two beams and then brought back together to allow the beams to interfere. One of the beams is now made to travel an extra distance more than the other. The maximum intensity in the resulting interference fringe pattern minus the minimum divided the maximum plus the minimum, called the 'visibility', is measured. At zero extra distance, the visibility is one. The extra distance when the visibility drops to one half is defined as the 'coherence length'.

the hologram has some information of every object point. This non locality and redundancy makes using a holographic model of brain memory attractive. We know we can look through a small segment of a hologram and still see the object, although it may look fuzzy and you can no longer see around the object by moving your head side-to-side. Thus, destruction of a good fraction of a hologram still allows recovery of some complete information. This is similar to our brain memory, in that each part of our memory does not depend on the survival of a single neuron. A pattern of brain waves stimulated by recalling a memory may have been created by an interference pattern among the synapses.

Appendix D
Bubbles

D.1 Biobubbles

The behavior of bubbles in biological systems is of some interest. Cases in point: Cavitation within cells by strong ultrasonic waves, or the formation of nitrogen bubbles in the blood as divers surface too rapidly from ocean depths, or pilots who fly to high altitudes too quickly, producing 'the bends', also called decompression sickness.[1]

The vibrational frequencies of microsized gas bubbles in biological cells falls in the ultrasonic region. Strong ultrasonic waves can cause cavitation bubbles. Correspondingly, weaker ultrasonic waves can increase the rate nitrogen bubbles in the blood dissolves.

Ultrasonic image contrast is facilitated by the prior injection of microbubbles, but with added dangers (See Sect. 5.22.1.)

D.2 Minimum Area Surfaces

Consider the surface of a bubble. Such a surface might exist as a two-dimensional liquid surface on a closed frame in three dimensions. Locally, surface tension in the liquid draws the fluid into a surface with minimal area. If the surface has minimal area, then small variations in that surface for the given frame boundary B must vanish:

$$\delta \int_B dA = \int_B dA \bigg|_{A_2} - \int_B dA \bigg|_{A_1} = 0 \,, \tag{D.1}$$

[1]Pilots can breathe pure oxygen a half hour before rapid decompression to avoid the bends.

W. C. Parke, *Biophysics*, https://doi.org/10.1007/978-3-030-44146-3

where A_1 is a surface which differs from A_2 infinitesimally. Now express

$$\int dA = \int \hat{\mathbf{n}} \cdot d\mathbf{A} \,, \tag{D.2}$$

where $\hat{\mathbf{n}}$ is a normal vector to the surface element $d\mathbf{A}$. This means that Eq. (D.1) can be written

$$\oint_B \hat{\mathbf{n}} \cdot d\mathbf{A} \bigg|_{A+\delta A} = \oint_B \hat{\mathbf{n}} \cdot d\mathbf{A} \bigg|_A \,. \tag{D.3}$$

We will reverse the direction of the normal on the surface A, so that the normal points in the same direction on both surfaces $A + \delta A$ and on A. In this way, $\hat{\mathbf{n}}$ becomes a vector field in the volume between both nearby surfaces. Applying Gauss' law, we will have

$$\oint \nabla \cdot \hat{\mathbf{n}} \, dV \tag{D.4}$$

over the thin volume enclosed by the two areas A and $A+\delta A$. Since the variations in the areas were arbitrary (apart from being constrained at the edges), we must have, near the bubble surface:

$$\nabla \cdot \hat{\mathbf{n}} = 0 \,. \tag{D.5}$$

The divergence of the surface normal can be expressed in terms of the principal radii of curvature by $\nabla \cdot \hat{\mathbf{n}} = 1/R_1 + 1/R_2$.

To show this becomes the usual expression for a minimal surface, let the minimal surface be expressed as $z = f(x, y)$. Then a surface normal can be expressed by

$$\hat{\mathbf{n}} = \left(1 + f_x^2 + f_y^2\right)^{-1/2} \{-f_x, -f_y, 1\} \tag{D.6}$$

Eq. (D.5) becomes

$$(1 + f_x^2) f_{xx} + (1 + f_y^2) f_{yy} - 2 f_x f_y f_{xy} = 0 \,. \tag{D.7}$$

Now compare to conditions which minimize the surface area

$$A = \int dxdy / \cos(\hat{\mathbf{k}} \cdot \hat{\mathbf{n}}) = \int (1 + f_x^2 + f_y^2)^{1/2} dxdy \,. \tag{D.8}$$

The Euler-Lagrange equations which minimize A are given by

$$\partial_x \left(\frac{f_x}{(1 + f_x^2 + f_y^2)^{1/2}} \right) + \partial_y \left(\frac{f_y}{(1 + f_x^2 + f_y^2)^{1/2}} \right) = 0 \,. \tag{D.9}$$

This reduces to the relation given Eq. (D.7).

Because the minimal area relation is a second order partial differential equation, non-linear in the unknown function $f(x, y)$, only a few exact solutions are known, such as a catenoid between two separated rings and a helicoid. Other solutions can be found numerically under conditions where the fluid surface boundary is known. Solving Eq. (D.9) is called Plateau's problem.

The minimal area surface equation when there is a pressure difference across the surface is Eq. (3.14), is sometimes expressed as

$$\sigma \nabla \cdot \hat{\mathbf{n}} = \sigma \left(\frac{1}{R_1} + \frac{1}{R_2} \right) = p_2 - p_1 \,, \tag{D.10}$$

where σ is the surface tension of the interface between the fluids on either side of the interface, and R_1, R_2 are the principle radii of curvature at the location on the surface where $\nabla \cdot \widehat{\mathbf{n}}$ is evaluated. In terms of the surface equation $z = f(x, y)$, the minimum area equation becomes

$$(1 + f_y^2) f_{xx} + (1 + f_y^2) f_{yy} - 2 f_x f_y f_{xy} = \frac{\Delta p}{\sigma} (1 + f_x^2 + f_y^2)^{3/2} \,. \tag{D.11}$$

D.3 Vibration of Bubbles

If a bubble occurs within a cell, it is subject to vibrational modes due to passing sonic waves. Bubble vibrations can take energy from the passing wave, possibly leading to excessive heat in the fluid surround the bubble and destruction of organelle or cell walls.

For a gas or a cavitation bubble, the dominant mode of vibration is expansion and contraction of the bubble radius. This case has been described by Rayleigh and for traveling cavitation bubbles by Plesset.[2]

The dynamics of bubble vibrations follows from the Navier-Stokes equation and the proper boundary conditions, resulting in the Rayleigh-Plesset equation for the radial motion of the bubble surface:

$$p = p_o + \rho \left(r \frac{d^2 r}{dt^2} + \frac{3}{2} \left(\frac{dr}{dt} \right)^2 \right) + \frac{4\eta}{r} \frac{dr}{dt} + \frac{\sigma}{r} \,, \tag{D.12}$$

[2]Lord Rayleigh, *On the pressure developed in a liquid during the collapse of a spherical cavity* Phil. Mag. **34**, 94–98 (1917); M.S. Plesset, *The dynamics of cavitation bubbles*, ASME J Appl Mech **16**, 228–231 (1949).

where p is the pressure in the bubble, p_0 the ambient pressure in the fluid, r the radius of the bubble, ρ the density of the fluid, η the viscosity of the fluid, σ the surface tension of the fluid in contact with the vapor in the bubble, and t is time.

A large bubble (with adiabatic behavior of the enclosed vapor) will naturally vibrate in radial mode near the resonance frequency

$$f_R = \frac{1}{2\pi r}\sqrt{\gamma\frac{3p_o}{\rho_o}}\,, \tag{D.13}$$

where $\gamma = c_p/c_v$ is the ratio of specific heats at constant pressure to constant volume for the vapor in the bubble.

Small bubbles (with isothermal behavior of the enclosed vapor) naturally vibrate in radial mode near

$$f_R = \frac{1}{2\pi r}\sqrt{\frac{3p_o + 4\sigma/r}{\rho_o}}\,. \tag{D.14}$$

For example, if the cavitation bubble in interstitial fluid had a radius 1 μm, then at body temperature (with $\sigma \sim 50 \times 10^{-3}$N/m) f_R is about 3 MHz. A radius of a tenth of a micron gives a resonant cavitation frequency of about 70 MHz. These are ultrasonic frequencies.

Inversely, the most likely cavitation bubble size will be those bubbles near the resonant frequency. When surface tension dominates over ambient pressure times radius, i.e. very small bubbles, we will have, from Eq. (D.14),

$$r \sim \left(\frac{4\sigma}{\rho(2\pi f)^2}\right)^{1/3} \sim 1.7\,\mu\text{m}\left(\frac{1\,\text{MHz}}{f}\right)^{2/3}.$$

Higher frequency modes of vibration (harmonics) are also expected. As an example of such modes, we give the results of an early investigation Sir Horace Lamb,[3] who calculated the vibrational frequencies of a bubble formed by the immersion of one fluid of density ρ_i into another with density ρ_o. For simplicity, he assumed the fluids were incompressible and non-viscous. His result is

$$f_l = \frac{1}{2\pi}\sqrt{\frac{\sigma}{r^3}\frac{(l-1)l(l+1)(l+2)}{(l+1)\rho_i + l\rho_o}}\,. \tag{D.15}$$

The index l takes values 2, 3, $\cdots$ corresponding to harmonics on the surface of the bubble.[4]

---

[3]Horace Lamb, **Hydrodynamics** [Cambridge Univ. Press] (1879).

[4]The radial mode ($l = 0$) does not occur for an incompressible bubble. The $l = 1$ is excluded as the motion corresponds to an oscillation of the center-of-mass.

High-intensity ultrasonic waves can be focused to a small region to produce extreme cavitation. The concentration of sound energy at the focus can be sufficient to vaporize and ionize molecules and atoms in the fluid, with the material in the focal region reaching temperatures in the millions of degrees Celsius, emitting visible, ultraviolet, and even X-rays. The effect is called 'sonoluminescence'.

Appendix E
DNA Splitting by Radiation

The study and practice of medicine could benefit from an enhanced engagement with the new perspectives provided by the emerging areas of complexity science and systems biology.

— Geoffrey B. West

An important observable that a radiation-damage theory can predict is the distribution of fragment lengths created by a given exposure and given environment. As another example of mathematical modeling in biology, we here present a statistical model to predict the splitting of the DNA backbone by radiation.[1]

E.1 Cleavage of Linear Molecules

Consider an initial volume of DNA molecules dispersed in a sample volume of target material, such as a water solution. If this volume is irradiated, causing locally deposited energy sufficient to break both backbones of the double helix (directly or indirectly), then cleavage will have some probability of occurring. (Here, direct break-up refers to cleavage caused by energy deposited onto the DNA from primary and secondary radiation while indirect break-up includes energy transferred from nearby ions and radicals created by the radiation.) To model the radiation-induced cleavage events, assume that the molecule is made of segments of minimum length δ, and that the initial molecules have length $n\,\delta$, where n is an integer. Let N_i be the number of DNA molecule fragments present at a particular time having a length $i\,\delta$. Initially, all N_i are zero except N_n, which starts as the total number of DNA molecules present in the initial volume, which is taken as N. As the molecules are irradiated, the change in the number of molecules having length $i\,\delta$ comes about from two mechanisms: an increase from cleavage of larger molecules; and a decrease due to cleavage of the molecules of length $i\,\delta$. For uniform radiation of randomly distributed molecules all equally exposed to the radiation, the rate of

[1] W.C. Parke, *"Kinetic Model of DNA Cleavage by Radiation,"* **Phys Rev E 56**, 5819–5822 (1997).

W. C. Parke, *Biophysics*, https://doi.org/10.1007/978-3-030-44146-3

cleavage of the molecules of length $i\,\delta$ will be proportional to the number of possible cleavage points, $i-1$. (For simplicity, fixed length monomers and equal probability for site breakage is taken. These assumptions can be relaxed without reformulating the method.) A single cleavage of the molecules of length longer than $i\,\delta$, say $k\,\delta$, will have equal chance to make fragments of length from δ to $(k-1)\,\delta$. From the set of such single-cleavage possibilities of a given molecule, two of this set will produce segments of length $i\,\delta$, adding to the number with this length.

It follows that

$$\frac{dN_i}{dt} = -(i-1)\,r\,N_i + 2r\sum_{k=i+1}^{n} N_k \tag{E.1}$$

where r is the rate at which the given radiation causes a cleavage at a given site of the DNA molecule. In particular

$$\begin{aligned}
\frac{dN_n}{rdt} &= -(n-1)\,N_n\ . \\
\frac{dN_{n-1}}{rdt} &= -(n-2)\,N_{n-1} + 2\,N_n \\
\frac{dN_{n-2}}{rdt} &= -(n-3)\,N_{n-2} + 2\,(N_{n-1}+N_n) \\
&\cdots \\
\frac{dN_1}{rdt} &= 2\,(N_2+N_3+\cdots+N_n).
\end{aligned} \tag{E.2}$$

The solutions for the N_i which satisfy the initial conditions (at $t=0$)

$$\begin{aligned}
N_n(0) &= N \\
N_i(0) &= 0 \quad \text{for} \quad i < n
\end{aligned} \tag{E.3}$$

can be found in a straightforward manner, starting with N_n, and leading to

$$\begin{aligned}
N_n &= N\,e^{-(n-1)\,r\,t} \\
N_{n-1} &= N\left(-2\,e^{-(n-1)\,r\,t} + 2\,e^{-(n-2)\,r\,t}\right) \\
N_{n-2} &= N\,(1\,e^{-(n-1)\,r\,t} - 4\,e^{-(n-2)\,r\,t} + 3\,e^{-(n-3)\,r\,t}) \\
N_{n-3} &= N\,(2\,e^{-(n-2)\,r\,t} - 6\,e^{-(n-3)\,r\,t} + 4\,e^{-(n-4)\,r\,t}) \\
&\cdots \\
N_2 &= N\,((n-1)\,e^{-r\,t} - 2\,(n-2)\,e^{-2r\,t} \\
&\quad +(n-3)\,e^{-3r\,t})
\end{aligned}$$

$$N_1 = N\,(2(n-1)(1-e^{-r\,t}) \\ -(n-2)\,(1-e^{-2r\,t})) \tag{E.4}$$

or, in general,

$$N_l = N\,(\overline{\delta}_{nl}(n-l-1)\,e^{-(l+1)\,r\,t} - 2(n-l)\,e^{-l\,r\,t} \\ +(n-l+1)\,e^{-(l-1)\,r\,t}) \tag{E.5}$$

where $\overline{\delta}_{nl}$ excludes $n = l$, i.e., $\overline{\delta}_{nl} = (1 - \delta_{nl})$, and δ_{nl} is the Kronecker delta. Note that as t approaches ∞,

$$N_l \to 0 \;\; (l > 1), \quad N_1(t) \to n\,N\,, \tag{E.6}$$

so that after sufficient time, all the original N molecules have been divided into n segments of length δ. These solutions also satisfy

$$\sum_{l=1}^{n} l\,N_l(t) = nN \tag{E.7}$$

for all time t, showing that the total length of all the broken segments remains unchanged. Atomic-force microscopy can be used to determine N_l by direct measurement. Techniques which separate fragments according to their molecular weight, such as electrophoresis, determine the distribution $f_l \equiv l N_l$.

From the form of the predicted number of original lengths (having $n-1$ cleavage sites),

$$N_n = N\,\left(e^{-\,r\,t}\right)^{n-1},$$

the expression e^{-rt} may be interpreted as the probability that a given site is not cleaved by the radiation. The probability a given site becomes cleaved is then

$$\alpha \equiv 1 - e^{-rt}\,, \tag{E.8}$$

so that from Eq. (E.5) the number of segments present with length l (for $l < n$) is

$$N_l = N\alpha(1-\alpha)^{l-1}\,[2 + (n-1-l)\alpha] \tag{E.9}$$

while for $l = n$,

$$N_n = N(1-\alpha)^{n-1}. \tag{E.10}$$

Eqs. (E.9) and (E.10) agree precisely with the results of Montroll and Simha[2] who used a combinatorial argument.

E.2 Fragmentation and Dose

Here, a connection is drawn between the radiation dose given to a sample of DNA and the cleavage probability. This will allow us to express the fragmentation numbers in terms of the radiation suffered by the sample. At this stage, a simple but reasonable target model will be taken in order to make the relationships clear.

Consider dividing the sample into M volumes ('sites'), each of size v, small enough so that, if radiation of sufficient energy is deposited into one site surrounding a location on a DNA molecule, it will cause a cleavage of the DNA. Suppose the radiation generates m localized deposition events randomly distributed throughout the sample and of sufficient energy to cause a cleavage. The probability that a given site is hit exactly k times by the first k events will be $(1/M)^k$. The chance that the remaining $(m-k)$ radiation events hit the other sites is $(1-1/M)^{m-k}$. But the k events on a given site may have occurred interspersed in time among the m depositions events. There are $(m-k+1)(m-k+2)\cdots m$ ways in which an ordered set of the k events in the given site could have occurred among the remaining $(m-k)$ events which did not hit the given site. Any ordering of the k events is equivalent, so that the number of ways that an unordered set of the k events can occur is $(m-k+1)(m-k+2)\cdots m/k!$. Therefore, the probability that a given site experiences exactly k hits after m events among M loci is

$$p_k = \binom{m}{k}\left(\frac{1}{M}\right)^k\left(1-\frac{1}{M}\right)^{m-k}, \tag{E.11}$$

where $\binom{m}{k}$ is the binomial coefficient. The probability that a given locus is hit one or more times will be

$$\sum_{k=1}^{m} p_k = 1-\left(1-\frac{1}{M}\right)^m \tag{E.12}$$

and the probability that it is hit two or more times will be

$$\sum_{k=2}^{m} p_k = 1-\left(1-\frac{1}{M}\right)^{m-1}\left(1+\frac{(m-1)}{M}\right). \tag{E.13}$$

[2]E.W. Montroll, and R. Simha, *Theory of depolymerization of long chain molecules*, J. Chem Phys **8**, 721 (1940).

If only one hit in a localized volume is required to cause (directly or indirectly) a cleavage of DNA, then the Montroll-Simha parameter α, i.e., the probability that a given site is broken, becomes

$$\alpha_1 = 1 - \left(1 - \frac{1}{M}\right)^m . \tag{E.14}$$

If two or more hits are required, then α becomes

$$\alpha_2 = 1 - \left(1 - \frac{1}{M}\right)^{m-1} \left(1 + \frac{(m-1)}{M}\right) . \tag{E.15}$$

In terms of the cleavage number, rt, Eq. (E.8) gives

$$r_1 t = m \ln\left(\frac{1}{1 - 1/M}\right) \approx \frac{m}{M} \tag{E.16}$$

and

$$\begin{aligned} r_2 t &= (m-1)\ln\left(\frac{1}{1-1/M}\right) - \ln\left(1 + \frac{m-1}{M}\right) \\ &\approx \frac{m}{M} - \ln\left(1 + \frac{m}{M}\right) \approx \frac{m^2}{2M^2} \end{aligned} \tag{E.17}$$

where the last approximations follow by taking $M >> 1$ and $M >> m >> 1$.

Now the dose, D, of radiation left in the sample is the total energy deposited per unit mass of sample. The number of localized energy deposits left in the sample, m, should be proportional to the dose over a wide range of exposures as

$$m = \frac{\mathcal{M}}{\varepsilon} D \tag{E.18}$$

where $\mathcal{M}$ is the mass of the sample and ε is defined by this relation and measures the energy needed for a double-strand cleavage. From this connection between dose and number of localized energy deposits, the rate of site cleavage will be proportional to the dose if only a single hit is needed. In contrast, if two hits within a given site are needed, then the rate of site cleavage will be proportional to the square of the dose. If the sample has mass density ρ, then

$$M = \frac{\mathcal{M}}{\rho v}, \quad \text{making} \quad \frac{m}{M} = \frac{\rho v D}{\varepsilon} . \tag{E.19}$$

It then follows from Eq. (E.16) that

$$r_1 t = \frac{\rho v D}{\varepsilon_1}, \tag{E.20}$$

while, if $M >> m$, Eq. (E.17) gives

$$r_2 t = \frac{1}{2}\left(\frac{\rho v D}{\varepsilon_2}\right)^2 . \tag{E.21}$$

Since, under the second scenario, two or more energy deposits are needed for cleavage, the average ε_2 should be about half of ε_1. The expressions for rt, given in Eqs. (E.16) and (E.17), determine the dependence of the DNA fragmentation numbers, N_l (Eqs. (E.9)–(E.10)), on radiation dose. In terms of the frequency of cleavage α (even if $\rho v D/\varepsilon$ is not small),

$$\alpha_1 = 1 - e^{-\rho v D/\varepsilon_1}$$

and

$$\alpha_2 = 1 - e^{-\rho v D/\varepsilon_2}(1 + \rho v D/\varepsilon_2). \tag{E.22}$$

If $\rho v D/\varepsilon << 1$ (low doses), then $\alpha_1 \sim \rho v D/\varepsilon_1$ and $\alpha_2 \sim (\rho v D/\varepsilon_2)^2/2$. The ratio α/D will be constant with dose for a given sample only for small doses and only in the case of single-event cleavages.

To appreciate the size of these numbers, let us estimate the magnitude of the expression in Eq. (E.20). Suppose DNA molecules are in a water solution and are given a dose of 100 Gy. The density ρ is approximately that of water, 10^3 kg/m^3. Note that the volume v is not the size of the primary and possible secondary ionization volumes covering an ionization track, but rather the volume surrounding a DNA site which, if sufficient energy is deposited within, causes cleavage. If we take the radius of the interaction volume to be 5 nm, then v will be approximately 10^{-25} m^3. Now take the effective energy ε needed in the volume v to cause a double-strand break to be 25 eV. (The threshold energy has been measured to be about 8 eV for photons, with 20–30 eV needed for electrons.) Then $r_1 t \sim 0.01$. With the cleavage number rt much less than one, rt will be close to the average fraction of cuts α. Even so, the number of possible cleavage sites n along a DNA molecule can be much larger than 10^2, so that the exponents in Eqs. (E.5) must be used for the large fragments ($l \sim n$), rather than their small rt approximation. Measured values for gamma ray irradiation of mammalian DNA give rather small values for α. Measurement gives α/D in the range $(6 \pm 2) \times 10^{-9}$ double-strand breaks per Gray per base-pair. With an effective energy deposit of 8 eV, the radius of the interaction volume if breakage had been dominated by a single-hit would be under 0.2 nm while a double-hit breakage at 100 Gy would give an interaction radius near 2 nm.

Appendix F
The Planck Distribution

We have no right to assume that any physical laws exist.

— Max Planck

In 1900, Max Planck was the first to explain the distribution of light coming from a hot body. (If all the chemical color of an object is removed, making it black, it still radiates, so the hot-body radiation is also called 'black-body radiation'.) Before 1900, classical Maxwell theory and thermodynamics could not explain the distribution of frequencies emitted by a hot body. In fact, classical theory predicts that a hot body radiates at very high frequencies. This was not seen. Also, the body would very quickly radiate all its internal thermal energy, dropping in temperature to absolute zero, far faster than observed cooling rates.

Consider a macroscopic cavity surrounded by a material kept at temperature T. The atomic vibration within ordinary material will generate electromagnetic waves, which will permeate the cavity. The radiation will also be absorbed by the walls of the cavity, so that an equilibrium will be established between the radiation and the vibrating charges. If those vibrations are thermal, then we should expect, according to Boltzmann, that various modes of vibration are excited, with a distribution proportional to $\exp(-\beta\epsilon_i)$, where the possible energies are indexed by i. This means the average energy in the photons for each possible photon frequency is given by

$$\langle\epsilon\rangle = \sum_i \epsilon_i e^{-\beta\epsilon_i} / \sum_i e^{-\beta\epsilon_i} , \tag{F.1}$$

where $\beta = 1/(k_B T)$. The excitations in the electromagnetic field act like harmonic oscillators. For a given frequency f, these excitations have energy $\epsilon_i = (i + 1/2)hf$. From Eq. (F.1), there results[1]

[1] We have shifted the average energy value to make it zero when the frequency is zero. The shift is the so-called quantum zero-point energy fluctuation, $(1/2)hf$, which can be removed from the total energy as long as it remains constant.

W. C. Parke, *Biophysics*, https://doi.org/10.1007/978-3-030-44146-3

$$\langle \epsilon \rangle = \frac{hf}{e^{\beta hf} - 1} \ . \tag{F.2}$$

As the effect of photon-photon interaction is extremely small, even at the center of stars, the total energy in the cavity at all frequencies will be $E = \sum_f \langle \epsilon \rangle = \int \langle \epsilon \rangle \, (d\mathcal{N}/df) df$, where $d\mathcal{N}/df$ is the number of possible photon states per unit frequency interval. To find this number density, consider a cubical cavity of side L. It will sustain an electromagnetic wave of wavelength $\lambda = 2L/n$ in each direction, where n labels the number of antinodes in the wave. In terms of frequencies, $n = 2Lf/c$. This means the number of possible waves in any direction, in small intervals of frequency, is given by $d^3n = (2L/c)^3 d^3 f$. Taking only positive n, and using spherical symmetry, gives $d\mathcal{N} = 2(1/8)(8V/c^3)4\pi f^2 df$. The extra factor of two accounts for the two possible polarization directions (helicities) of the photon. We have[2]

$$E = \frac{8\pi V}{c^3} \int \frac{hf}{e^{\beta hf} - 1} f^2 df \ . \tag{F.3}$$

The energy density, $u \equiv E/V$, per unit frequency interval becomes

$$du/df = \frac{8\pi h}{c^3} \frac{f^3}{e^{\beta hf} - 1} \ . \tag{F.4}$$

If a small hole in the cavity is used to detect this distribution, the intensity per unit frequency at the hole will be

$$dI/df = c \frac{du}{df} \frac{1}{4\pi} \int \int \cos\theta \, d\Omega \tag{F.5}$$

$$= \frac{1}{4\pi} 2\pi \frac{du}{df} \int_0^{\pi/2} \cos\theta \, d\cos\theta \tag{F.6}$$

$$= \frac{1}{2} \frac{du}{df} \int_0^1 x dx \tag{F.7}$$

$$= \frac{c}{4} \frac{du}{df} \tag{F.8}$$

$$= \frac{2\pi h}{c^2} \frac{f^3}{e^{hf/(kT)} - 1} . \tag{F.9}$$

This is Planck's distribution, giving the intensity of light per unit frequency interval. The solar radiation in small intervals of frequency taken above the atmosphere can be fit to a Planck's distribution, resulting in a temperature of 5778 K. The distribution is shown in Fig. F.1.

[2] The result for the average energy also follows from Eq. (10.7) using $\langle \epsilon \rangle = \sum n_i \epsilon_i$.

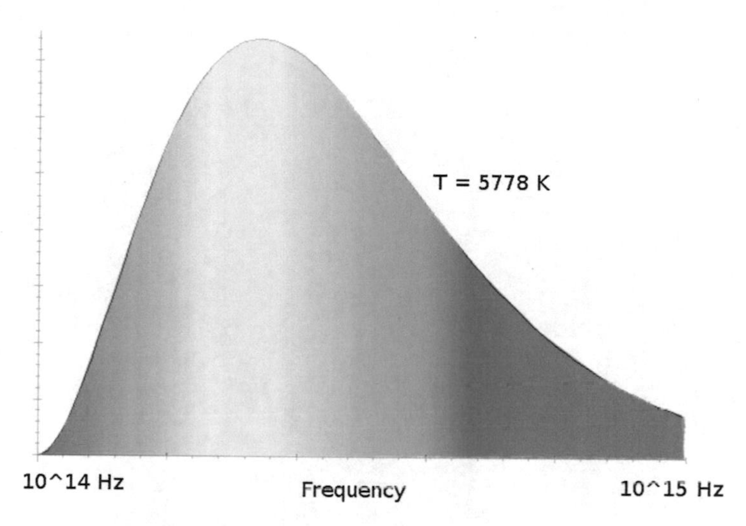

Fig. F.1 Planck's distribution Eq. (F.9) for the Sun

The light intensity over all frequencies will be

$$I = \frac{2\pi}{c^2h^3}k^4T^4\int_0^\infty \frac{x^3}{e^x - 1}dx. \qquad \text{(F.10)}$$

The integral gives $\pi^4/15$, so

$$I = \frac{2\pi^5}{15c^2h^3}k^4T^4 \equiv \sigma T^4. \qquad \text{(F.11)}$$

This is Stefan's law.[3] Stefan's constant[4] $\sigma = 5.6704 \times 10^{-8}\ \mathrm{W/m^2/K^4}$. The strong temperature dependence of the total radiation means that if a body's temperature doubles, the emitted power of its light goes up by a factor of 16. Of course, you would not like to be that body.

The peak of Planck's distribution occurs at the point where the derivative dI/df vanishes. Letting $x = \beta hf$, that point occurs where $xe^x/(e^x - 1) = 3$, which is at $x = 2.821435\cdots$, i.e. where

$$f \approx 2.82\, kT/h\, . \qquad \text{(F.12)}$$

This is Wein's law. As the temperature rises, the peak frequency rises in proportion.[5]

[3]That the radiation intensity is proportional to the fourth power of the temperature can be derived from classical thermodynamics, but classical theory has a diverging energy density with increasing frequency, which implies that the temperature drops far more rapidly toward zero by radiation than is observed.

[4]In the mks system of units, Stefan's constant has a memorable ordering of digits: 5,6,7,-8 !

[5]For the Sun, the color of the light at peak frequency is yellow. The additive mixture of adjacent colors give the Sun a near-white appearance.

Appendix G
Some Tools of Vector Analysis

If I were again beginning my studies, I would follow the advice of Plato and start with mathematics.

— Galileo Galilei

G.1 Displacements and Their Length in General Coordinates

The points in a three-dimensional space can be specified by three numbers (x_1, x_3, x_3). These might be the Cartesian coordinates of a point. But locations in space are sometimes more conveniently specified by intersections of grid lines not parallel to each other, such as those for spherical polar coordinates.[1] Curvilinear coordinates are defined to have grid lines which are locally perpendicular to each other, so that the small vector displacements can be express as

$$d\mathbf{r} = \sum_i h_i dx_i \widehat{\mathbf{e}}_i$$

where the $\widehat{\mathbf{e}}_i$ is a unit vector along one of the grid lines, and

$$\widehat{\mathbf{e}}_i \cdot \widehat{\mathbf{e}}_j = \delta_{ij} \ .$$

Three common choices for curvilinear coordinates are Cartesian, cylindrical, and spherical polar. In these cases, h_i and the usual choices for symbols are given below:

[1]The choice of coordinates is made strategically, according to the underlying symmetry of the system. A good choice can make the solution to the functional relationships for a given problem simpler to find and express than a bad choice.

W. C. Parke, *Biophysics*, https://doi.org/10.1007/978-3-030-44146-3

	h_1	h_2	h_3	dx_1	dx_2	dx_3	$\widehat{\mathbf{e}}_1$	$\widehat{\mathbf{e}}_2$	$\widehat{\mathbf{e}}_3$
Cartesian	1	1	1	dx	dy	dz	$\widehat{\mathbf{i}}$	$\widehat{\mathbf{j}}$	$\widehat{\mathbf{k}}$
Cylindrical	1	ρ	1	$d\rho$	$d\varphi$	dz	$\widehat{\mathbf{u}}_r$	$\widehat{\mathbf{u}}_\varphi$	$\widehat{\mathbf{u}}_z$
Spherical polar	1	$r\sin\theta$	r	dr	$d\varphi$	$d\theta$	$\widehat{\mathbf{u}}_r$	$\widehat{\mathbf{u}}_\varphi$	$\widehat{\mathbf{u}}_\theta$

The length squared between nearby points in curvilinear coordinates becomes

$$ds^2 = d\mathbf{r} \cdot d\mathbf{r} = \sum_i h_i^2 dx_i^2 . \tag{G.1}$$

This is the Pythagorean theorem in curvilinear coordinates. The relation defines what is called the 'metric' of the space, i.e. how one measures the distance between points. For nearby points, that distance is ds. For arbitrary choice of grid lines, not necessarily curvilinear, the metric is determined by

$$ds^2 = \sum_i g_{ij} dx_i dx_j . \tag{G.2}$$

The g_{ij} are called the components of the 'metric tensor'. This expression even works on curved spaces. For curvilinear coordinates, $g_{ij} = h_i^2 \delta_{ij}$.

G.2 Vectors and Tensors

A mathematical vector is a set of ordered numbers. A physical vector is a mathematical vector with the added properties that (1) two physical vectors defined near one physical point and being of the same physical type can be added by adding their components, just the way small displacements add, and (2) physical vectors transform under rotations the same way as small displacements. Vector A will be denoted symbolically as $\mathbf{A}$. In terms of components, $\mathbf{A} = \{A_1, A_2, \cdots\}$. If $\mathbf{A}$ is a physical vector, the 'components' $\{A_i\}$ are generally real and all carry the same physical units. The addition property lets us express a vector as a sum of three, one each along three perpendicular spatial directions, with each of these as a scale times a unit vector in one of the three directions: $i = 1, 2, 3$ for (x, y, z). $\widehat{\mathbf{e}}_i$:

$$\mathbf{A} = \sum_i A_i \widehat{\mathbf{e}}_i . \tag{G.3}$$

The length is defined for the vector $\mathbf{A}$ in a 'flat' (Euclidean) space by

$$|A| = \sqrt{\sum A_i A_i} . \tag{G.4}$$

A rotation by an angle θ around the z axis changes the components of a small displacement according to

$$dx' = dx \cos\theta + dy \sin\theta \tag{G.5}$$

$$dy' = -dx \sin\theta + dy \cos\theta \tag{G.6}$$

$$dz' = dz\ . \tag{G.7}$$

These can be displayed in matrix form:

$$\begin{pmatrix} dx' \\ dy' \\ dz' \end{pmatrix} = \begin{pmatrix} \cos\theta & \sin\theta & 0 \\ -\sin\theta & \cos\theta & 0 \\ 0 & 0 & 1 \end{pmatrix} \begin{pmatrix} dx \\ dy \\ dz \end{pmatrix}$$

or in index form as

$$dx_i' = \sum_{j=1}^{3} R_{ij} dx_j\ .$$

By definition, the components of any vector change under a rotation the same way. In fact, any linear homogeneous transformation of a vector which preserves the length of the vector will be a rotation. If the new components of the vector are denoted by a prime, then a rotation of the vector can be represented by a linear transformation

$$A'_i = \sum R_{ij} A_j\ . \tag{G.8}$$

The length of **A** will be preserved if

$$\sum_i R_{ij} R_{ik} = \delta_{jk}\ , \tag{G.9}$$

where δ_{jk} is called the Kronecker delta, and is unity if $j = k$ and zero otherwise.

Equation (G.9) gives six independent conditions on the nine R_{ij}, leaving three of them free. These three can be taken to be the three angles used to rotate a given Cartesian coordinate frame to any other Cartesian coordinate frame. The conditions on the transformation can be put in matrix form:

$$R^T R = I\ , \tag{G.10}$$

with R^T being the transpose of R. Note that $\det R = \pm 1$. The plus sign is a 'proper' rotation, i.e. a rotation reachable by a continuous transformation from the identity. The negative sign applies to an 'improper' rotation, which is a reflection together with a proper rotation.

To vectors **A** and **B** define a 'scalar' (i.e. a number unchanged by a rotation) through

$$\mathbf{A} \cdot \mathbf{B} = \sum_i A_i B_i \ . \tag{G.11}$$

From the definition, it is easy to see that after a rotation, the value of $\mathbf{A} \cdot \mathbf{B}$ is unchanged. Taking one axis along vector **A**, it is also easy to see that $\mathbf{A} \cdot \mathbf{B} = |\mathbf{A}|\,|\mathbf{B}| \cos\theta_{AB}$, where θ_{AB} is the angle between the two vectors.

The cross product of **A** and **B** defines a vector generated from two different vectors, and can be defined by

$$(\mathbf{A} \times \mathbf{B})_i = \sum_{jk} \epsilon_{ijk} A_j B_k \ . \tag{G.12}$$

In this definition, the three-indexed quantity, ϵ_{ijk}, called the 'alternating tensor',[2] is completely antisymmetric in its three indices (e.g. $\epsilon_{ijk} = -\epsilon_{ikj}$), and $\epsilon_{123} = 1$.

An alternative form of Eq. (G.12), using a determinant, is

$$\mathbf{A} \times \mathbf{B} = \begin{vmatrix} \widehat{\mathbf{e}}_1 & \widehat{\mathbf{e}}_2 & \widehat{\mathbf{e}}_3 \\ A_1 & A_2 & A_3 \\ B_1 & B_2 & B_3 \end{vmatrix} \tag{G.13}$$

By selection of axes, it is straightforward to show that

$$|\mathbf{A} \times \mathbf{B}| = |A|\,|B| \sin\theta_{AB} \tag{G.14}$$

and that the direction of $\mathbf{A} \times \mathbf{B}$ is perpendicular to the plane defined by **A** and **B** according to the 'right-hand rule'.

From the definitions, it follows that

$$\mathbf{A} \cdot (\mathbf{B} \times \mathbf{C}) = (\mathbf{A} \times \mathbf{B}) \cdot \mathbf{C} \ . \tag{G.15}$$

This number represents the 'volume' enclosed by the parallelepiped with edges along **A**, **B**, and **C**.

Also,

$$\mathbf{A} \times (\mathbf{B} \times \mathbf{C}) = (\mathbf{A} \cdot \mathbf{C})\mathbf{B} - (\mathbf{A} \cdot \mathbf{B})\mathbf{C} \ , \tag{G.16}$$

Sets of quantities $T_{ijk\cdots}$ which transform according to

[2]The set of quantities ϵ_{ijk} are also called the Levi-Civita symbol.

$$T'_{i_1 i_2 i_3 \cdots i_n} = \sum_{j_1 j_2 j_3 \cdots} R_{i_1 j_1} R_{i_2 j_2} R_{i_3 j_3} \cdots R_{i_n j_n} T_{j_1 j_2 j_3 \cdots j_n} \tag{G.17}$$

are called the components of tensors of rank n. A vector is an example of a tensor. By multiplying two tensors, one can form new tensors. The range of the indices of the components of a tensor gives the dimension of the space in which a tensor operates. A product of two vectors with components $A_i B_j$ form the components of a factorizable tensor of rank two. If two indices of the components of a tensor are summed, the results are the components of a new tensor. Such a sum is called a 'contraction', as it creates a tensor of rank $n - 2$. A contraction of a rank-two tensor is also called the 'trace' of that tensor. If the components of two tensors A and B are multiplied, and then contracted over indices across the pair, the result will be a tensor with components $\sum_k A_{i_1 i_2 \cdots i_{m-1} k} B_{k j_2 j_3 \cdots j_n}$. This operation is called a 'dot-product' of A and B. Similarly, the construction $\sum_{jk} \epsilon_{ijk} A_{i_1 i_2 \cdots i_{m-1} j} B_{k j_2 j_3 \cdots j_n}$ is called the 'cross-product' of the tensors A and B.

The Kronecker delta, δ_{ij}, and the Levi-Civita symbol, ϵ_{ijk}, are also components of tensors, but of a special kind, since, under rotations, $\delta'_{ij} = \delta_{ij}$ and $\epsilon'_{ijk} = \epsilon_{ijk}$, i.e. they are the components of tensors invariant under rotations.[3] $I = \{\delta_{ij}\}$ is called the 'identity tensor', since contraction with any other tensor gives back that other tensor.

Any tensor can be 'decomposed' into tensors which transform by rotation into tensors with the same symmetry. Consider the case of a rank two tensor. The components of a rank two tensor can always be written in the following way:

$$T_{ij} = \left[\frac{1}{2}(T_{ij} + T_{ji}) - \frac{1}{3}\delta_{ij} \sum_k T_{kk}\right] + \left[\frac{1}{2}(T_{ij} - T_{ji})\right] + \left[\frac{1}{3}\delta_{ij} \sum_k T_{kk}\right] \tag{G.18}$$

The terms in first bracket make the components of a symmetric and traceless tensor, with five independent components. The second bracketed terms make up the components of an antisymmetric tensor (which is automatically traceless and has three independent components), and the third is proportional to the components of identity tensor, and has only one independent component. Rotations turn a symmetric traceless tensor into a symmetric traceless tensor; they turn an antisymmetric tensor into an antisymmetric tensor, and the Kronecker delta into the Kronecker delta. For this reason, the decomposition Eq. (G.18) is a splitting of the original tensor into invariant subspaces.

Note that the antisymmetric part in Eq. (G.18) is equivalent to a vector:

$$T^a{}_i = \sum_{jk} \epsilon_{ijk} \frac{1}{2}(T_{jk} - T_{kj}) = \sum_{jk} \epsilon_{ijk} T_{jk} \ . \tag{G.19}$$

[3]Invariance of δ_{ij} follows from Eq. (G.9). Invariance of the tensor with components ϵ_{ijk} follows from the fact that for a rotation, $\det R = 1$.

We see that there are only three kinds of tensors which can be constructed from a product of two vectors **A** and **B**, and which are not mixed by transformations which rotate the coordinate frame: (1) A scalar $\mathbf{A} \cdot \mathbf{B}$, (2) A vector, $\mathbf{A} \times \mathbf{B}$, and (3) A symmetric traceless tensor $(1/2)(\mathbf{AB} + \mathbf{BA}) - (1/3)\,(\mathbf{A} \cdot \mathbf{B})\,\mathbf{I}$. (Here, **I** is the Kronecker-delta tensor.)

The ϵ_{ijk} tensor is useful in expressing the determinant of a square matrix:

$$\det A = \frac{1}{3!} \sum_{ijk,qrs} \epsilon_{qrs}\epsilon_{ijk} A_{qi} A_{rj} A_{sk} = \sum_{ijk} \epsilon_{ijk} A_{1i} A_{2j} A_{3k} \,, \tag{G.20}$$

from which it is evident that the determinant of a matrix is a scalar. This definition of the determinant can easily be generalized to spaces with more than three dimensions. For example, in four dimensions,

$$\det A = \sum_{ijkl} \epsilon_{ijkl} A_{1i} A_{2j} A_{3k} A_{4l} \,, \tag{G.21}$$

The proof of the identity (G.16) is helped by knowing that

$$\sum_{i} \epsilon_{ijk}\epsilon_{ilm} = \delta_{jl}\delta_{km} - \delta_{jm}\delta_{kl} \,. \tag{G.22}$$

This relation (G.22) also gives

$$\sum_{ij} \epsilon_{ijk}\epsilon_{ijm} = 2\delta_{km} \,. \tag{G.23}$$

Eq. (G.22) follows from a more general identity:

$$\epsilon_{ijk}\epsilon_{pqr} = \delta_{ip}\delta_{jq}\delta_{kr} + \delta_{iq}\delta_{jr}\delta_{kp} + \delta_{ir}\delta_{jp}\delta_{kq} - \delta_{ip}\delta_{jr}\delta_{kq} - \delta_{ir}\delta_{jq}\delta_{kp} - \delta_{iq}\delta_{jp}\delta_{kr} \,. \tag{G.24}$$

A tensor H is hermitian if and only if $H^\dagger = H$, where the 'dagger' represents a transpose and a complex conjugation operation. All finite-dimensional hermitian matrices can be diagonalized by a unitary operation $UHU^{-1} = diag\,[h_1, h_2, \cdots]$, with $U^\dagger U = I$ and the diagonal values h_i are all real. If all the components of a hermitian matrix T are real, then T is a symmetric matrix, and the diagonalization operation S can be real. The operation is then called a similarity transformation: $STS^{-1} = diag\,[t_1, t_2, \cdots]$. Rotations R of a second-rank tensor T is an example of a similarity transformation, since the preservation of lengths requires that $RR^T = I$ and $T' = RTR^T$. If all the eigenvalues of a symmetric matrix are positive, then the matrix is called positive definite.

G.3 Scalar, Vector, and Tensor Fields

A field is any quantity defined at each point in a given space. A scalar is a tensor of rank zero, so its value is left unchanged by a rotation of coordinates. A scalar field is a scalar defined at each point is space. The temperature across your body forms a scalar field.

A vector field is a field of vectors of a particular kind, defined over a space. Vector fields naturally occur in biological systems. The field of velocities in a flowing fluid, such as blood or protoplasm, is a good example. The flux of heat from inside your body to your skin is a vector field. The electric and magnetic fields within and across cells is another. The proportionality coefficients relating stress to strain in your bones forms a tensor field across the bone.

If a field smoothly changes across space, then it is possible to define spatial derivatives of that field.

G.4 Grad Operation on a Scalar Field

The differential of a scalar field $f(x_1, x_2, x_3)$, i.e. the change of this function for a small shift in the coordinates, is given by

$$df = \sum_i \frac{\partial f}{\partial x_i} dx_i \, . \tag{G.25}$$

The 'grad' operation on a scalar field f, denoted by ∇f, is defined through

$$df = \nabla f \cdot d\mathbf{r} \, . \tag{G.26}$$

In curvilinear coordinate, we must have

$$\nabla f = \sum_i \frac{1}{h_i} \frac{\partial f}{\partial x_i} \widehat{\mathbf{e}}_i \, . \tag{G.27}$$

G.5 The Divergence of a Vector Field and Gauss' Theorem

Let $\mathbf{J}$ represent the current density for the flow of some fluid entity Q, such as mass, volume, or charge. Then the amount of that entity passing through a given area A per unit time can be represented as

$$\frac{dQ}{dt} = \int_A \mathbf{J} \cdot d\mathbf{S} \, , \tag{G.28}$$

where $d\mathbf{S}$ is a small surface element of area dS with a direction normal $\widehat{\mathbf{n}}$ to the surface, and with a 'right-hand' sense: If the surface is coordinated with values of x_1, x_2, then $\widehat{\mathbf{e}}_1 \times \widehat{\mathbf{e}}_2 = \widehat{\mathbf{n}}$. If the surface encloses a simply connected volume, then the net amount of the entity flowing out of the volume is

$$\frac{dQ}{dt} = \oint_A \mathbf{J} \cdot d\mathbf{S} \,. \tag{G.29}$$

Divide the volume up into little cubes. Summing the flow through all the cubes will have equal and opposite contributions between adjacent cubes, so the result of the sum will be just the flow out of the surface of the original volume:

$$\begin{aligned}
\frac{dQ}{dt} &= \sum_{cubes} \oint_{small\ cube} \mathbf{J} \cdot d\mathbf{S} \\
&= \sum_{cubes} \left((J_1 h_2 h_3)|_{x_1+dx_1} - (J_1 h_2 h_3)|_{x_1} \right) dx_2 dx_3 + \cdots \\
&= \sum_{cubes} \sum_i \frac{\partial (J_i h_j h_k)}{\partial x_i} dx_1 dx_2 dx_3 \\
&= \iiint \frac{1}{h_1 h_2 h_3} \sum_i \frac{\partial (J_i h_j h_k)}{\partial x_i} h_1 dx_1 h_2 dx_2 h_3 dx_3 \\
&= \int \frac{1}{h_1 h_2 h_3} \sum_i \frac{\partial (J_i h_j h_k)}{\partial x_i} \, dV \,,
\end{aligned} \tag{G.30}$$

where $\{ijk\}$ are taken cyclically.

The quantity inside the volume integral of Eq. (G.30) is called the divergence of $\mathbf{J}$, with the short-hand notation $\nabla \cdot \mathbf{J}$. For any vector field $\mathbf{A}$, we define

$$\nabla \cdot \mathbf{A} = \lim_{\Delta V \to 0} \frac{1}{\Delta V} \oint \mathbf{A} \cdot d\mathbf{S} \,, \tag{G.31}$$

with the integral taken over the surface of the volume. The advantage of this definition over an explicit coordinate one is that this form can be used in any coordination of the space. In the case of curvilinear coordinates,

$$\nabla \cdot \mathbf{A} = \frac{1}{h_1 h_2 h_3} \sum_i \frac{\partial (A_i h_j h_k)}{\partial x_i} \,. \tag{G.32}$$

For any simply connected volume (this excludes doughnuts!), we have proven Gauss' Integral Theorem:

$$\int \nabla \cdot \mathbf{A} \, dV = \oint \mathbf{A} \cdot d\mathbf{S} \,. \tag{G.33}$$

G.6 Laplacian in General Coordinates

If the vector field **A** is the gradient of a scalar field ϕ: $\mathbf{A} \equiv \nabla\phi$, then the divergence of $\mathbf{A} = \nabla \cdot \nabla\phi = \nabla^2\phi$ is called the Laplacian of ϕ. From our general definition of the grad and divergence operations, we have, for any field $\phi(x)$,

$$\nabla^2\phi = \frac{1}{h_1h_2h_3}\sum_i \frac{\partial}{\partial x_i}\left(\frac{h_jh_k}{h_i}\frac{\partial\phi}{\partial x_i}\right) . \tag{G.34}$$

The indices j and k are taken in cyclic order from i.

Even more generally, for spaces with a metric given by Eq. (G.2), the Laplacian becomes

$$\nabla^2\phi = \frac{1}{\|g\|^{1/2}}\sum_i \frac{\partial}{\partial x_i}\left(\|g\|^{1/2}\sum_j g^{ij}\frac{\partial\phi}{\partial x_j}\right) \tag{G.35}$$

where $\|g\|$ is the magnitude of the determinant of the metric tensor $\{g_{ij}\}$ and $\{g^{ij}\}$ is the matrix inverse of $\{g_{ij}\}$.

G.7 The Curl Operation and Stokes' Theorem

Now consider integrating on a closed line the projection of a vector field **F** onto a distant increment on the line: $\oint \mathbf{F} \cdot d\mathbf{r}$.[4] If we divide up the enclosed area into small squares, and add the line integrals around each square, then the pieces of the integrals for common segments in adjacent squares will cancel. (All the integrals are taken in the same sense, conventionally counter clockwise looking down on the closed path.) What is left after the sum is the integral on the boundary. Thus

$$\begin{aligned}
\oint \mathbf{F} \cdot d\mathbf{r} &= \sum_{squares} \oint \mathbf{F} \cdot d\mathbf{r} \\
&= \sum_{squares} \left(h_2F_2|_{x_1+dx_1} - h_2F_2|_{x_1}\right) dx_2 + \cdots \\
&= \sum_{squares} \left(\frac{\partial(h_2F_2)}{\partial x_1} - \frac{\partial(h_1F_1)}{\partial x_2}\right) dx_1dx_2 + \cdots \\
&= \sum_{squares} \left(\frac{\partial(h_2F_2)}{h_2h_1\partial x_1} - \frac{\partial(h_1F_1)}{h_1h_2\partial x_2}\right) h_1dx_1h_2dx_2 + \cdots
\end{aligned} \tag{G.36}$$

[4] A good example is the work done by a force carrying a body around a curve, back to the starting point.

The result is a surface integral bounded by the original path, and is expressed as

$$\oint \mathbf{F} \cdot d\mathbf{r} = \int \nabla \times \mathbf{F} \cdot d\mathbf{S} \,, \tag{G.37}$$

where

$$\nabla \times \mathbf{F}|_i = \left(\frac{\partial (h_j F_j)}{h_j h_k \partial x_k} - \frac{\partial (h_k F_k)}{h_k h_j \partial x_j} \right) \tag{G.38}$$

is called the curl of **F** in the ith direction, with (ijk) taken in cyclic permutation of $(1, 2, 3)$. Equation (G.37) is Stokes' Integral Theorem.

Gauss' and Stokes' integral theorems are special cases of the transformation of integrals of functions over closed boundaries embedded in an n dimensional space into integrals of derivatives of those functions in the interior regions of that boundary.

Differential Forms

In 1899, Élie Cartan introduced differential forms to express a generalization of both Gauss' and Stokes' integral theorems. His generalization includes spaces for which distances have not even been defined. A differential form is a sum of infinitesimal volume elements, each times an arbitrary function of the coordinates. In Cartesian coordinates, a k-dimensional form can be written as

$$\omega = \sum_{i_1 < i_2 \cdots < i_k} \omega_{i_1 \cdots i_k}(x) dx^{i_1} \wedge dx^{i_2} \wedge \cdots \wedge dx^{i_k} \,. \tag{G.39}$$

The 'wedge product' ($\wedge$) creates surface elements that are necessarily antisymmetric in the infinitesimal coordinate tangent vectors on that surface, i.e. $dx^1 \wedge dx^2 = -dx^2 \wedge dx^1$. The derivative of a form, such as $d\omega$, can be defined by

$$d\omega = \sum_{i, i_1 < i_2 \cdots < i_k} \frac{\partial \omega_{i_1 \cdots i_k}(x)}{\partial x^i} dx^i \wedge dx^{i_1} \wedge dx^{i_2} \wedge \cdots \wedge dx^{i_k} \,. \tag{G.40}$$

In an arbitrary $k + 1$-dimensional orientable manifold Ω,[5] Gauss' and Stokes' law generalize to the elegant expression

$$\oint_{\partial \Omega} \omega = \int_{\Omega} d\omega, \tag{G.41}$$

[5]The manifold is orientable if all directed simple loops when moved around the manifold and returned, have the same direction relative to the normals to the manifold. A Möbius surface is not orientable.

where ω is a k-dimensional differential form and $\partial\Omega$ is a closed piecewise-continuous boundary of the manifold Ω whose interior contains no singularities or holes (i.e. 'simply connected'), and is also piecewise continuous.

Maxwell's equations for electromagnetism and Einstein's equations for gravitational curvature also have simple expressions in terms of differential forms. Because the relations are not explicitly dependent on a choice of coordinates, they can aid in finding physical solutions which also should not depend on our choice of coordinates.[6]

G.8 Some Useful Vector-Calculus Identities

$$\nabla \cdot (\mathbf{A} \times \mathbf{B}) = (\nabla \times \mathbf{A}) \cdot \mathbf{B} - (\nabla \times \mathbf{B}) \cdot \mathbf{A} \tag{G.42}$$

$$\nabla \times (\nabla \times \mathbf{A}) = -\nabla^2 \mathbf{A} + \nabla(\nabla \cdot \mathbf{A}) \tag{G.43}$$

$$\nabla \times (\mathbf{A} \times \mathbf{B}) = (\mathbf{B} \cdot \nabla)\mathbf{A} - (\mathbf{A} \cdot \nabla)\mathbf{B} + (\nabla \cdot \mathbf{B})\mathbf{A} - (\nabla \cdot \mathbf{A})\mathbf{B} \tag{G.44}$$

These identities are easily proved in coordinate index form, taking advantage of Eqs. (G.12) for the cross product and (G.22) for a double cross.

G.9 Applications of Vector Calculus

The gradient, divergence, and curl of a vector field finds wide application: In mechanics, the curl of the force vanishes if the action of that force conserves energy. In fluid mechanics, twice the curl of the velocity field of the fluid determines the local angular velocity of the fluid where the curl is non-zero. The curl of the velocity field is also called the local vorticity of the fluid. Maxwell's electrodynamics were originally expressed in term of the divergence and curl of the electric and magnetic fields.

G.10 Poisson's Equation

Poisson's equation arises in many areas of mathematical physics, including the study of gravitational fields, electrodynamics and fluid mechanics. It takes the form of a differential equation for a field ϕ given a 'driving term' f, also sometimes called

[6] A good reference is the book by Harley Flanders, **Differential Forms with Application to the Physical Sciences**, [Acad Press] (1963); [Dover Books] (1989).

the 'effective force':

$$\nabla^2 \phi = f \ , \tag{G.45}$$

This relation is a special case of a linear operator acting on an unknown function y giving a known function f:

$$\mathcal{L} y = f \ . \tag{G.46}$$

In fact, even Newton's 2nd law can be put in this form, with y being the coordinate of a particle (as a function of time) and f being the force on the particle, and $\mathcal{L} = md^2/dt^2$. A formal solution to Eq. (G.46) can be found by the following idea. The 'forcing term' can be thought of as a combinations of impulses, each of short duration. If f does not depend on y, the equation is linear in the unknown y, so the general solution can be expressed as the sum of impulse solutions. Let the impulse be expressed as $f = \delta$, where δ is a function which vanishes outside a small region in the space $\{x\}$ over which $\mathcal{L}$ operates, and δ is taken to have unit total impulse, i.e. $\int \delta(x) dx = 1$. For application to Newton's law, $x = t$. The solution to the impulse problem is simpler than the arbitrary-force case. Let $G(x, x')$ be a solution to the impulse equation:

$$\mathcal{L} G(x, x') = \delta(x - x') \ , \tag{G.47}$$

where the region of non-vanishing impulse can be varied by changing x'.[7] Then the general solution for arbitrary force is

$$y(x) = y_o(x) + \int G(x, x') f(x') dx' \ , \tag{G.48}$$

where $y_0(x)$ is the solution to the homogeneous equation: $\mathcal{L} y_0 = 0$. That this is the solution to the inhomogeneous equation (G.46) can be seen by operating on the left and right by $\mathcal{L}$.

For the Poisson equation (G.45), one can verify that the Green's function in an unbounded region is[8]

$$G_P(x, x') = -\frac{1}{4\pi} \frac{1}{|\mathbf{r} - \mathbf{r}'|} \ , \tag{G.49}$$

[7] G is called the Green's function, or the 'kernel' of the equation. The impulse $\delta(x - x')$ is call the Dirac delta. Note: We do not assume that the impulse time goes to zero, so we do not have to introduce a singular Dirac delta. Our δ is non-zero over a finite region, but whose width is much smaller than the extent of variations of y.

[8] For those who know some electrostatics, you may recognize this Green's function as proportional to the electric potential for a point charge. Such a point charge acts as the 'impulse' for the equation for the electric potential $\nabla^2 V = -4\pi \rho_q$.

i.e., this Green's function satisfies

$$\nabla^2 G_P(x, x') = \delta^3(x - x') . \tag{G.50}$$

For the solution to Eq. (G.45) in three-dimensional space, we have

$$\phi(\mathbf{r}) = \phi_0(\mathbf{r}) - \frac{1}{4\pi} \int \frac{1}{|\mathbf{r} - \mathbf{r}'|} f(\mathbf{r}') dV' . \tag{G.51}$$

The function $\phi_0(\mathbf{r})$ is a solution of the homogeneous equation, and is determined by boundary conditions on ϕ.

G.11 Derivative Decomposition of a Vector Field

Every differentiable vector field, **E**, in a simply connected region, can be expressed in terms of the gradient of a scalar field, ϕ, plus the curl of a divergenceless field, **A**:

$$\mathbf{E} = -\nabla\phi + \nabla \times \mathbf{A} . \tag{G.52}$$

(The minus sign is put in for later convenience.) In fact, we can give explicit forms for how to find the fields ϕ and **A**:

$$\begin{aligned} \phi(\mathbf{r}) &= \frac{1}{4\pi} \int \frac{\nabla' \cdot \mathbf{E}(\mathbf{r}')}{|\mathbf{r}' - \mathbf{r}|} dV', \\ \mathbf{A}(\mathbf{r}) &= \frac{1}{4\pi} \int \frac{\nabla' \times \mathbf{E}(\mathbf{r}')}{|\mathbf{r}' - \mathbf{r}|} dV'. \end{aligned} \tag{G.53}$$

These expressions can be found by taking the divergence and curl of the left and right-hand sides of Eq. (G.52) and applying our solution to Poisson's equation (G.51).

Appendix H
The Fourier Decomposition

A mathematician who is not also something of a poet will never be complete.

— Karl Weierstrass

H.1 The Fourier Series: Harmonic Analysis

In 1822, Jean Fourier showed that complicated waves which repeat in a definite period T can always be decomposed into a superposition of simple sinusoidal waves, each with a frequency which is a multiple of $1/T$. The same kind decomposition works for waves in space, over a length of λ. The wave can be expressed as a sum of sinusoidal waves having wave numbers which are a multiple of $1/\lambda$. Each such decomposition is called a Fourier series. The technique of decomposition is called 'harmonic analysis'. Because of the Fourier theorem, one can analyze the behavior of any complex wave satisfying a linear differential equation by first studying the behavior of simple sinusoidal waves.

For a wave with a period T, i.e. $\xi(x, t+T) = \xi(x, t)$, the Fourier decomposition into simple sine and cosine waves looks like this:

$$\xi(x, t) = a_o(x)/2 + \sum_{j=1}^{\infty} a_j(x) \cos(2\pi\, j\, t/T) + \sum_{j=1}^{\infty} b_j(x) \sin(2\pi\, j\, t/T) \ . \tag{H.1}$$

The coefficients $a_j(x)$ and $b_j(x)$ are found from

$$a_j(x) = \frac{2}{T} \int_0^T dt\ \xi(x, t) \cos(2\pi\, j\, t/T) \ , \tag{H.2}$$

$$b_j(x) = \frac{2}{T} \int_0^T dt\ \xi(x, t) \sin(2\pi\, j\, t/T) \ . \tag{H.3}$$

W. C. Parke, *Biophysics*, https://doi.org/10.1007/978-3-030-44146-3

Using the Euler identity $\exp(i\theta) = \cos\theta + i\sin\theta$, the Fourier series has the more succinct form

$$\xi(x,t) = \sum_{j=-\infty}^{\infty} c_j(x) \exp(-2\pi\, i j\, t/T) \ . \tag{H.4}$$

with the coefficients $c_j(x)$ found from

$$c_j(x) = \frac{1}{T} \int_0^T dt\ \xi(x,t) \exp(2\pi\, i j\, t/T) \ . \tag{H.5}$$

The complex coefficients c_j are related to the real coefficients a_j and b_j through

$$c_j = (1/2)\left(a_j + i b_j\right) . \tag{H.6}$$

Note that the frequencies of the Fourier component waves, $f_j = j/T$, are integer multiples of a lowest frequency, $f_1 = 1/T$, which is called the 'fundamental frequency'. Each of sine or cosine waves with frequencies which are an integer multiple of a fundamental are called 'harmonics'. The fundamental frequency is also called the 'first harmonic', corresponding to $j = 1$. The 'second harmonic', also called the 'first overtone', has a frequency $f = 2f_1$, etc.

Th terminology 'harmonic analysis' comes from the nomenclature of music. But the properties of harmonic analysis can now be found in pure mathematics, with no reference to sound or music, even though many terms, such as harmonic, have survived the abstraction.

H.2 The Fourier Transform

For waves which do not repeat in time, the period of the lowest frequency wave in the Fourier decomposition may be taken as infinity. Analogously, waves which do not repeat in space, or which are not maintained between physical boundaries, will have Fourier components which have wavelengths approaching infinity. In these cases, the Fourier series becomes a sum over a continuous set of frequencies or wavelengths. The series is then referred to as a Fourier transform.[1]

To get an expression for the Fourier transform, first consider the Fourier series for a function with a finite period T. Combining (H.4) and (H.5), this series can expressed as

[1]An approximate transform, called the Fast-Fourier-Transform, or FFT, is used in computers for rapid analysis.

$$f(t) = \sum_{j=-\infty}^{\infty} \frac{1}{2\pi} \int_{-T/2}^{T/2} \exp\left(i\omega_j\,(t'-t)\right) f(t')dt'(2\pi/T)\ . \tag{H.7}$$

where $\omega_j \equiv 2\pi j/T$. As the period T becomes large, the factor $2\pi/T = \omega_{j+1} - \omega_j$ is small, and becomes the incremental interval $d\omega$ in the summation over the angular frequencies ω. Thus, in the limit of an infinite period,

$$f(t) = \frac{1}{2\pi} \int_{-\infty}^{\infty} \int_{-\infty}^{\infty} \exp\left(i\omega\,(t'-t)\right) f(t')dt'd\omega\ . \tag{H.8}$$

Separating the operations in each integral, one can define the so-called 'inverse Fourier transform' as

$$\tilde{f}(\omega) = \sqrt{\frac{1}{2\pi}} \int_{-\infty}^{\infty} \exp\left(i\omega\,t'\right) f(t')dt'\ , \tag{H.9}$$

which makes the Fourier transform of $f(t)$ have the expression

$$f(t) = \sqrt{\frac{1}{2\pi}} \int_{-\infty}^{\infty} \exp\left(-i\omega\,t\right) \tilde{f}(\omega)d\omega\ . \tag{H.10}$$

The Fourier transform of a Fourier transform is the original function (apart from a flip of the sign of the argument). Conventionally, for physical applications, the sign in exponent for the time-transform component wave, $\exp(-i\omega t)$ is taken negative, while the corresponding the component for a spatial Fourier transform, $\exp(ikx)$, is taken positive, with opposite signs in the inverse Fourier transform. This makes a pure plane wave moving forward along the x-axis, in the form $A\exp(ikx - i\omega t)$, the default Fourier component in an expansion over both space and time.

H.3 Generalized Fourier Transform

Expressing a given function as an infinite series of simpler functions is a useful exercise when the given function is a solution to a linear differential equation. This will be the case for sound waves obeying a linear wave equation, for light waves, and for quantum wave functions. If the choice of simpler functions is based on having them match the same boundary conditions as the given function, then the analysis of the given function reduces to knowing the properties of the simpler functions.

The expansion of the given function $f(x)$ into simpler functions $\phi(k, x)$ appears as

$$f(x) = \int g(k)\phi(k, x)dk\ . \tag{H.11}$$

where the $g(k)$ are the coefficients of the expansion. For the convenience which follows, the simpler functions $\phi(k, x)$ are usually taken as orthogonal for different k:

$$\int \phi^*(k', x)\phi(k, x)dx = \delta(k' - k) , \tag{H.12}$$

where $\delta(k' - k)$ is the Dirac delta, defined by the property

$$\int \delta(k' - k)h(k')dk' = h(k) \tag{H.13}$$

for any piecewise continuous function $h(k)$.

One can then solve for the coefficients $g(k)$ as

$$g(k) = \int \phi^*(k, x)f(x)dx . \tag{H.14}$$

So the $g(k)$ are the generalized Fourier transform of the original $f(x)$. If any function $f(x)$ defined over x within the given boundaries can be so expanded, then we say the expansion functions $\phi(k, x)$ are complete. Completeness can be expressed as

$$\int \phi(k, x')\phi^*(k, x)dk = \delta(x' - x) \tag{H.15}$$

since this relation can be placed in

$$f(x) = \int \delta(x' - x)f(x')dx' \tag{H.16}$$

to generate the expansion of $f(x)$.

Each of the above relations can be re-written as a discrete sum to represent the corresponding relations for a generalized Fourier series. For some useful cases, the complete set of expansion functions contains both discrete and continuous function labels k.

Many of the linear differential equations in physics are of the Strum-Liouville type. In these cases, the expansion functions $\phi(k, x)$ can be taken as eigenfunctions of the corresponding Strum-Liouville equation:

$$\left(\frac{d}{dx}p(x)\frac{d}{dx} + q(x)\right)\phi(k, x) = -\lambda(k)w(x)\phi(k, x) . \tag{H.17}$$

With fixed finite boundary conditions on the solutions, only certain 'eigenvalues' $\lambda(k)$ give a solution. Each possible eigenvalue is labeled by 'k'. Then, the arbitrary solution for the same boundary conditions is a linear combination of the eigenfunctions for the various k values.

Appendix I
Optimization Techniques

There is "... a general tendency in driven many-particle systems toward self-organization into states formed through exceptionally reliable absorption and dissipation of work energy from the surrounding environment."

— Jeremy England, Nikolai Peruvov, and Robert Marshland

I.1 The Lagrange Multiplier Method

There are many examples of optimization and minimization related to life systems such as those in mathematical biology, theoretical biochemistry, network analysis, neuronal systems, statistical mechanics, and complex systems.

Joseph Louis Lagrange formulated a beautiful method for such minimization when there are constraints among the variables. The method is particularly useful if the constraints are awkward or difficult to solve for one subset of variables in terms of the others.

Lagrange showed that the minimization of a function $f(x_1, x_2, \cdots, x_N)$ under a number $n < N$ of constraints among the values in $\{x\}$ given by a set of relations $\{\kappa_l(x) = 0 \quad 1 \leq l \leq n\}$ is equivalent to the unconstrained minimization of $F(x, \lambda) \equiv f(x) + \sum_l \lambda_l \kappa_l(x)$, i.e. a minimum of f with constraints occurs at those values of $\{x\}$ which make the slopes of F vanish:

$$\frac{\partial}{\partial x_k} F(x) = \frac{\partial}{\partial x_k} f(x) + \sum_{l=1}^{n} \lambda_l \frac{\partial}{\partial x_k} \kappa_l(x) = 0 \, . \tag{I.1}$$

These N equations, plus the n constraint relations $\kappa_l(x) = 0$ can be solved for the $N + n$ values in the set $\{x, \lambda\}$. Now, n of these solutions can be used to eliminate the so-called Lagrange multipliers $\{\lambda\}$, giving the N solutions in the set $\{x\}$ that may minimize, maximize, or reach a boundary in the range of $\{x\}$ for the cost function f under constraints. The locations where the derivatives of F vanish within the boundaries are called 'critical' points, and may be local minima, maxima, or 'saddle' points of F. A condition for the minima can be generated as follows: At each critical point, solve the 'secular' equation $\det\left|F_{ij} - a_i\delta_{ij}\right| = 0$ for the

W. C. Parke, *Biophysics*, https://doi.org/10.1007/978-3-030-44146-3

'eigenvalues' a_i, where $F_{ij} \equiv \partial^2 F/(\partial x_i \partial x_j)$. If all the a_i are positive, the critical point is a minimum of F.

To get a flavor of the proof, consider the case of just two variables, x and y. We wish to minimize the function $f(x, y)$, with a constraint $\kappa(x, y) = 0$. Suppose we solve the constraint equation to get $y = h(x)$ and substitute in f to get $f(x, h(x))$. We then find where this new function of x reaches a minimum, i.e. where $df/dx = 0$. We can express this derivative by

$$\left(\frac{\partial f(x,y)}{\partial x} + \frac{\partial f(x,y)}{\partial y}\frac{\partial h(x)}{\partial x}\right)\Bigg|_{y=h(x)} = 0\,. \tag{I.2}$$

We can find $\partial h(x)/\partial x$ by differentiating the identity $\kappa(x, h(x)) \equiv 0$ to get

$$\left(\frac{\partial \kappa(x,y)}{\partial x} + \frac{\partial \kappa(x,y)}{\partial y}\frac{\partial h(x)}{\partial x}\right)\Bigg|_{y=h(x)} \equiv 0\,. \tag{I.3}$$

so

$$\frac{\partial h(x)}{\partial x} = -\frac{\partial \kappa/\partial x}{\partial \kappa/\partial y}\,. \tag{I.4}$$

Substituting in Eq. (I.2) gives an expression for minimization in the form

$$\frac{\partial f(x,y)}{\partial x} - \frac{\partial f(x,y)}{\partial y}\frac{\partial \kappa/\partial x}{\partial \kappa/\partial y} = 0\,. \tag{I.5}$$

Now consider the slopes of the function $F(x, y, \lambda) = f(x, y) + \lambda\kappa(x, y)$ without any constraint on x, y except for possible boundary conditions. The three slopes will be

$$\frac{\partial f(x,y)}{\partial x} + \lambda\frac{\partial \kappa(x,y)}{\partial x} = 0\,, \tag{I.6}$$

$$\frac{\partial f(x,y)}{\partial y} + \lambda\frac{\partial \kappa(x,y)}{\partial y} = 0\,, \tag{I.7}$$

$$\kappa(x,y) = 0. \tag{I.8}$$

If we put λ from the second relation into the first, we get the condition Eq. (I.5). So the unconstrained minimization of $F(x, y, \lambda)$ gives the same solution as the constrained minimization of $f(x, y)$.

I.2 Optimization Through Iterative Searches

It often occurs that the optimization of a function $f(x_1, x_2, \cdots, x_N) \equiv f(\mathbf{r})$ leads to a system of equations that cannot be solved analytically. Also, the value of the function may be only given at discrete points, and may exist over wide ranges of numerous variables. In these cases, an iterative technique may be employed to find the minima of the function. (The space spanned by the values of $\mathbf{r}$ will be called the phase space of f.)

If f is known analytically, the process is simpler than the case of discrete data for f. In fact, fitting the data first to a known analytic function can improve the search, since the 'true' extrema of f may occur between points in the data set. The choice of the fitting function should be based on any information known about the nature of the data, or, if there is none, one can always represent the data as an expansion into a series of orthonormal functions with the same boundary conditions as the data. The expansion coefficients become the parameters of the fit.

Fitting is done by adjusting the parameters in the fitting function

$$\varphi(x_1, x_2, \cdots, x_N; \alpha_1, \alpha_2, \cdots, \alpha_n)$$

to minimize the sum of the square of the differences between the fit and the data at the discrete locations where f is known, with each term in the sum weighted by the square of the inverse of the standard deviations of the data.[1] If the number of parameters n in the fitting function is far less than the number of data points, then we say that the fit is a 'good' model of the data. Some caution must be observed in fitting if the data has uncertainty: Forcing the fit to data not on a 'smooth' representation of that same data may create unnatural kinks or even singularities in the fit between the data.

If one wishes the 'raw' data to be searched, rather than first fitting that data, then the derivatives indicated below should be replaced by their discrete expressions across the known locations of the data. For example, $\nabla_i f \approx (f(\mathbf{r} + \boldsymbol{\delta}) - f(\mathbf{r}))/\delta_i$, where $\boldsymbol{\delta}$ is a small displacement connecting two nearby points in the phase space.

The iteration idea is this:

1. Pick a likely point in the phase space to be near a minimum of f, or, if there is no criteria to know of a likely point, use a 'randomly' picked position $\mathbf{r} = [x_1, x_2, \cdots, x_N]$.
2. Find ∇f (either analytically or numerically). If the gradient is zero to within a selected tolerance, an extremum has been found. If not, shift to a new position $\mathbf{r}'$ toward $-\nabla f$, i.e. follow f 'down hill':

[1] This 'method of least squares' was shown by Gauss to be optimal when the errors in the data come from 'random' fluctuations.

$$\mathbf{r}' = \mathbf{r} - \frac{\nabla f}{|\nabla f|}\delta. \qquad \text{(I.9)}$$

The value of δ should be taken small enough that one does not 'overshoot' the minimum, such as $\delta = (1/2)|\nabla f|/(|\nabla^2 f| + \varepsilon^2|$.

3. Repeat the δ shifts until the gradient ∇f is with a specified tolerance of zero or a boundary of the function is reached. If not a boundary point, test that $\nabla^2 f > 0$. If a boundary has been reached, test if f at the boundary point, f_B, is smaller than $f_B + |\nabla f|\delta$. If so, the boundary location is a minimum point of f.
4. Record the minimum position $\mathbf{r}_m$, and also the value of $f_m = f(\mathbf{r}_m)$ there.
5. Repeat this process for other locations in the phase space, either picked with some foreknowledge, such as with a known probability distribution for $\mathbf{r}$ across phase space, or at 'random' positions.[2]
6. When the tests have densely 'peppered' the phase space, stop.
7. Now compare the values of f at its minima to see which one (or ones) is (or are) smaller than the rest.

When the number of possible configurations of a system we want to optimize is astronomically large, we are driven to formulate optimization strategies. The folding of proteins is a good example, because the number of possible bond rotations and bond flexing can be in the googols. In the 'extreme optimization' scheme,[3] searches for the minimum energy configurations are narrowed by knowledge of the energy of backbone configurations and of various rotamer possibilities for segments of the protein. The presence of steric or chemical catalytic agents, such as particular molecules, ions, or surfaces, may strongly limit possibilities. Also, the path a particular natural amino-acid chain follows from secondary to tertiary structure has an evolutionary history. More complex structures may have evolved from simpler ones, or iteration of simpler ones. In searches for these natural cases, the full phase space of possible configurations does not need to be explored, but only those paths near those discovered by natural systems in biological evolution.

An any case, choices for changes in the next configuration to be tested can come from a power-law probability distribution of the largest to the smallest local energies along the protein chain.

---

[2]A random selection produces the so-called Monte Carlo method.

[3]See, for example, Naigong Zhang and Chen Zeng, *Reference Energy Extremal Optimization as a Stochastic Search Algorithm Applied to Computational Protein Design*, Journal of Computational Chemistry **29:11**, 1762–1771 (2008).

Appendix J
Useful Biophysical Data

Time flies like an arrow. Fruit flies like a banana.

— Groucho Marx

J.1 Mathematical Constants

'In Math, and in Physics, size Matters' — wcp

Table J.1 Transcendentals

Circumference/diameter of circle	π	$3.14159265358979323846 26\cdots$
Base of natural logarithms	e	$2.71828182845904523536 02\cdots$
Euler's constant	γ	$0.57721566490153286060 65\cdots$
Feigenbaum's constant	δ	$4.66920160910299067185 32\cdots$

J.2 Physical Constants

'Let's get physical.' — wcp

W. C. Parke, *Biophysics*, https://doi.org/10.1007/978-3-030-44146-3

Table J.2 Number games

Septrillionth	yocto	10^{-24}
Sextrillionth	zepto	10^{-21}
Quintrillionth	atto	10^{-18}
Quadtrillionth	femto	10^{-15}
Trillionth	pico	10^{-12}
Billionth	nano	10^{-9}
Millionth	micro	10^{-6}
Thousandth	milli	10^{-3}
Hundredth	centi	10^{-2}
Tenth	deci	10^{-1}
Ten	deca	10^{1}
Hundred	hecto	10^{2}
Thousand	kilo	10^{3}
Million	mega	10^{6}
Billion (U.S.)	giga	10^{9}
Trillion	tera	10^{12}
Quadtrillion	peta	10^{15}
quintrillion	exa	10^{18}
Sextillion	zetta	10^{21}
Septillion	yotta	10^{24}
Googol	googola	10^{100}
Googolplex	googolaplexi	$10^{10^{100}}$

Table J.3 Physical constants

Speed of light in a vacuum (*defined*)	c	299 792 458 m/s
Gravitational constant	G	$6.6408 \times 10^{-11} \mathrm{m^3/(kg\, s^2)}$
Acceleration due to gravity	g	$9.80665 \cdot \mathrm{m/s^2}$
Boltzmann's constant	k_B	$1.3806485 \times 10^{-23}$ J/K
Coulomb's constant	k	$c^2 \times 10^{-7} \mathrm{N\, m^2/coul^2}$
Elementary charge	e	$1.6021367 \times 10^{-19}$ coul
Electron mass	m_e	$9.1093897 \times 10^{-31}$ kg
Proton mass	m_p	$1.6726231 \times 10^{-27}$ kg
Proton g-factor	g_H	$5.585 \cdot$
Neutron mass	m_n	$1.6749286 \times 10^{-27}$ kg
Atomic Mass Unit m[^{12}C]/12	amu	1.66054×10^{-27} kg
Planck's constant	h	$6.62607004 \times 10^{-34}$ J s
Stefan-Boltzmann constant	σ	$5.67037 \times 10^{-8} \mathrm{W/(m^2\, K^4)}$
Faraday's constant	F	96,485.333 coul/mol
Absolute zero	0 K	−273.15 °C
STP temperature	T_s	$273.15\, K = 0\,°C$
STP pressure	p_s	$1.00 \times 10^5 \mathrm{N/m^2}$

Table J.4 Physical conversions

1 fermi	10^{-15} m
1 barn	100 fermi^2
1 ångström	10^{-10} m
1 curie	3.7×10^7/s
1 rad	100 erg/gm= 0.01 joule/kg= 0.01 Gy
1 röntgen	1 esu/cm^3 of air at STP = 2.58×10^{-4}C/kg
1 kilocalorie (def)	4184 J
1 poise (viscosity)	1 dyne-s/cm^2
1 Astronomical unit (def)	149 597 870 700 m
Energy from 1 kg of glucose	1.59×10^7 joules (Krebs cycle)
Energy in 1 barrel oil (42 gal)	6.5×10^9 joules
Energy from 1 ton of coal	2.8×10^{10} joules
Energy from 1 kiloton TNT	4.184×10^{12} joules
Energy from 1 kg ^{235}U	8.1×10^{13} joules
Energy from 1 kg ^2H	3.28×10^{14} joules
Radiation power from sun	3.8×10^{26} watts
Earthquake energy	6.3×10^4 joules$\times 10^{(3/2)\,M_{Richter}}$
Temperature of outer space	2.725· K
Asteroid hits to Earth	$\sim(10^5\,\text{kg}/Mass)^{11/15}$ per year

Table J.5 Material constants

Gas constant	R	8.31446·J/(mol-K)
Avogadro's number	$\mathcal{N}_A$	$6.02214086 \times 10^{23}$/mol
Atomic number	Z	Number of protons in nucleus
Neutron number	N	Number of neutrons in nucleus
Isotope number	A	$Z + N$
Mass of carbon-12 atom	m_{C-12}	$1.992646547 \times 10^{-23}$ gm
Atomic weight	w_A	$Atomic\,Mass/(m_{C-12}/12)$
One atmosphere pressure	p_o	101,325 N/m^2
Volume of 1 mole Ideal Gas @STP	v_{STP}	22.414·liters/mol
Density of air @ 20 °C	ρ_a	1.204· kg/m^3
Density of water @ 20 ° C	ρ_w	0.9982· gm/cm^3 (0.9933 @ 37 °C)
Dynamic viscosity of air @ 20 ° C	η_a	0.0181· centipoise
Dynamic viscosity water @ 20 °C	η_w	1.01 centipoise (0.692 @ 37 °C)
Speed of sound in dry air @ 20 °C	v_a	343.2· m/s
Speed of sound in water @ 20 °C	v_w	1484· m/s
Specific heat of air @ 20 °C, 1 atm	c_a	1.005 kJ/(kg °C)
Specific heat of water @ 20 °C	c_w	4.182 kJ/(kg °C)
Triple point of water	$T_o,\ p_o$	273.16 K, 611.73 N/m^2
Critical point of water	$T_c,\ p_c$	647 K, 22.1×10^6 N/m^2
Heat of fusion of water	ΔH_{fus}	79.72· cal/gm

(continued)

Table J.5 (continued)

Heat of vaporization of water	ΔH_{vap}	539.2· cal/gm
Thermal conductivity of air @ 20 °C	k_a	25.7 · mW/(m °C)
Thermal conductivity of water	k_w	598.4 · mW/(m °C)
Equivalent weight of ion	EW	Atomic Weight/Valence
Quantity of an electrolyte	Q_{el}	milligrams/(liter-EW) (mEq/L)

J.3 Biophysical Constants

'Here we have Biological Physiques' — wcp

Table J.6 You've got your nerve

Number of synapses for a "typical" neuron	1000–10,000
Resting potential across axon membrane	−70 mV
Threshold firing potential	−55 mV
Maximum action potential	+30 mV
Time to reach max action potential	1 ms
Time to depolarize	0.5 ms
Time to recover resting potential	3 ms
Conduction speed of action potential	0.6–120 m/s
Single Na^+ pump max transport rate	200 Na^+, 130 K^+ ions/s
Typical number of sodium pumps	1000 pumps/μm^2
Total Na^+ pumps for a small neuron	1 million
Density of sodium channels	300 μm^2
Voltage-gated Na^+ channels at each node	1000–2000/μm^2
channels between nodes	25/μm
channels in unmyelinated axon	100–200/μm^2
Diameter of microtubule	20–25 nm
Diameter of microfilament	5 nm
Diameter of neurofilament	7–10 nm
Thickness of neuronal membrane	5 nm (Squid: 10 nm)
Membrane surface area of a typical neuron	250,000 μm^2
Average axon diameter	0.3 μm
Average dendrite diameter	0.9 μm
Typical synaptic cleft distance	20–40 nm across
Slow axoplasmic transport rate	2.4–48 nm/s (actin, tubulin)
Intermediate axoplasmic transport rate	180–600 nm/s (mitochondrial protein)
Fast axoplasmic transport rate	2400–4800 nm/s (peptides, glyolipids)
Molecules of neurotransmitter in a vesicle	5000
Diameter of synaptic vesicle	50 nm–200 nm
Internodal length	150–1500 μm
Composition of myelin	70–80% lipid; 20–30% protein
Neocortex glucose consumption	0.40 μmol/g/min

(continued)

Table J.6 (continued)

Neocortex ATP total consumption	3.5×10^{21} molecules/min
Neocortex ATP ion pumping	1.7×10^{21} molecules/min
Neocortex ATP to keep resting potential	1.3×10^{21} molecules/min
Neocortex ATP for neural spiking	0.39×10^{21} molecules/min

Table J.7 Don't let your Herring Flounder

Ear power consumption	10^{-6} watts
Response delay time	0.18 s
Min. energy for response	5×10^{-11} ergs
Impedance of ear to air	1.5×10^5 g/(cm^{2}s)
Resolution, intensity	8%
Resolution, frequency	0.2%
Resolution, arrival time	55 ms
Resolution, angle from ear	4°
Surface area of the tympanic membrane	85 mm^2
Length of the Eustachian tube	3.5–3.9 cm
Number of hair cells in inner cochlea	10,000
Number of hair cells in outer cochlea	30,000
Number of fibers in auditory nerve	29,000
Length of auditory nerve	2.5 cm
Most sensitive range of human hearing	1000–4000 Hz
Length, diameter of ear canal	2.7 cm, 0.7 cm
Weight, length of malleus	23 mg, 8–9 mm
Weight, size of incus	30 mg, 5 mm×7 mm
Weight of stapes	3–4 mg
Size of stapes (length, width, height)	3 mm, 3.5 mm, 1.4 mm
Length, width of cochlea	35 mm, 10 mm
Number of turns in the cochlea	2.2–2.9
Length basilar membrane	25–35 mm
Width (at base) of basilar membrane	150 microns
Number of neurons in cochlear nuclei	8800
Number of neurons in inferior colliculus	392,000
Number of neurons in medial geniculate	570,000
Number of neurons in auditory cortex	100,000,000
Hearing range (young adult human)	20–20,000 Hz
Hearing range (elderly human)	50–8000 Hz
Hearing range (rat)	1000–50,000 Hz
Hearing range (cat)	100–60,000 Hz
Hearing range (dolphin)	200–150,000 Hz
Hearing range (elephant)	1–20,000 Hz
Hearing range (goldfish)	5–2000 Hz
Hearing range (moth, noctuid)	1000–240,000 Hz
Hearing range (mouse)	1000–100,000 Hz
Hearing range (sea lion)	100–40,000 Hz
Threshold of hearing (3000 Hz)	0 dB = 10^{-12} W/m^2
Auditory pain threshold	120 dB

Table J.8 Balance yourself

Organs	Eye/brain	Visual analysis
Inner ear system	Semicircular canals[a]	Change in rotation
Inner ear system	Utricle	Horizontal acceleration
Inner ear system	Vestibule	Vertical acceleration
Distributed	Proprioceptors	Body position

[a]Specific data are sparse, but see Timothy A. Huller, *Semicircular Canal Geometry,*···, Anat Rec A Discov Mol Cell Evol Biology **288:4**,466–472 (2006). Useful for the prediction of extinct animal behavior using skull measurements.

Table J.9 With taste

Total number of taste buds (tongue, palate, cheeks)	10,000
Number of taste buds on the tongue	9000
Height of taste bud	50–100 microns
Diameter of taste bud	30–60 microns
Number of receptors on each taste bud	50–150
Diameter of taste receptor	10 micron
Diameter of taste fiber	less than 4 micron
Taste threshold for quinine sulfate	3.376 mg/liter water

Table J.10 That Ol'factory smells

Number of human olfactory receptor cells	12 million
Number of rabbit olfactory receptor cells	100 million
Number of dog olfactory receptor cells	1 billion
Number of bloodhound olfactory receptor cells	4 billion
Surface area of olfactory epithelium (humans)	10 cm^2
Surface area of bloodhound olfactory epithelium	760 cm^2
Area of olfactory epithelium in some dogs	170 cm^2
Area of olfactory epithelium in cats	21 cm^2
Thickness of olfactory epithelium mucous layer	20–50 microns
Diameter of olfactory receptor axons	0.1–0.2 micron
Diameter of distal end olfactory receptor cell	1 micron
Diameter of olfactory receptor cell	40–50 micron
Number of cilia per olfactory receptor cell	10–30
Length of cilia on olfactory receptor cell	100–150 micron
Concentration for detection threshold of musk	0.04 μg/liter air

Table J.11 We shall see

Length of eyeball	24.5 mm
Volume of eyeball	5.5 cm^3
Weight of eyeball	7.5 g
Average time between blinks	2.8 s
Average duration of a single blink	0.1–0.4 s
Thickness of cornea at center	~0.5 mm
Thickness of cornea at periphery	~1 mm
Diameter of cornea	11.5 mm
Thickness of lens	4 mm
Diameter of lens	9 mm
Composition of lens	65% water; 35% protein
Indices of refraction (and $1/f$ power):	
Cornea	1.376 (40.)
Aqueous humor	1.336
Lens	1.386–1.406 (19.- 33.)
Vitreous humor	1.337
Number of retinal rod receptor cells	120–140 million
Number of retinal cone receptor cells	5–6 million
Number of retinal ganglion cells	800 thousand to 1 million
Number of fibers in optic nerve	1,200,000
Neurons in lateral geniculate body	570,000
Number of cells in visual cortex	538,000,000
Wavelength of visible light (human)	400–700 nm
Amount of light necessary to excite a rod	1 photon
Amount of light necessary to excite a cone	100 photons
Location of the greatest density of rods	20 ° from fovea
Highest density of rods	160,000/mm^2
Peak density of rods (cat)	400,000/mm^2
Density of cones in fovea	200,000/mm^2
Diameter of fovea	1.5 mm
Intraocular pressure (above atmospheric)	10–20 mmHg
Volume of orbit	30 ml
Area of retina	~1000 mm^2
Thickness of retina	100–230 microns
Production rate of aqueous humor	2 microliters/min
Turnover of aqueous humor	15 times/day
Volume of eye occupied by the vitreous	80%
Maximal sensitivity of red cones	$\lambda = 570$ nm
Maximal sensitivity of green cones	$\lambda = 540$ nm
Maximal sensitivity of blue cones	$\lambda = 440$ nm

Table J.12 I feel your pain

Nociceptor locations:	
Skin Types:	Mech., Therm., Chem., Polymodal
Joint Types:	Mech., Polymodal, Silent[a]
Visceral Types:	Pressure, Temp., Chem., Silent
Density of $A\delta$ type[b] in skin	$<2/\text{mm}^2$
Density of C type in skin	$4-8/\text{mm}^2$
Depth of skin nociceptors	$20-500\ \mu\text{m}$
Separation, visceral nociceptors	0.2–2 cm
Speed of signal in $A\delta$ type	2–20 m/s
Speed of signal in C type	<2 m/s
Thermal activation of $A\delta$	>46 °C
Thermal activation of C	>41 °C

[a] Silent nociceptors are activated after a time of stimulation above a certain threshold.
[b] $A\delta$ type cause quick sharp pain; C type are slower; acting after lingering damage.

Table J.13 The human touch

Meissner corpuscle	Fine touch
Merkel cell	Touch pressure
Ruffini corpuscle	Skin shear
Pacinian corpuscle	Skin vibration
Root hair plexus	Proximal movements
Free nerve endings	(See pain table)
Weight of skin (adult human)	4.1 kg
Surface area of skin (adult human)	$\sim 1.8\ \text{m}^2$
Number of tactile receptors in the hand	17,000
Number of nerve endings in hand	$200/\text{cm}^2$
von Frey sensitivity threshold (Face)	4.9 dynes
Two point separation threshold (Finger)	2–3 mm
Length of dermis Meissner's nerve corpuscle	90–120 micron
Density of receptors on finger tips	$2500/\text{cm}^2$
Density of Meissner's corpuscles on finger tips	$1500/\text{cm}^2$
Density of Merkel's cells on finger tips	$750/\text{cm}^2$
Density of Pacinian corpuscles on finger tips	$75/\text{cm}^2$
Density of Ruffini's corpuscles on finger tips	$75/\text{cm}^2$
Thermal pain threshold	45 °C

Index

W. C. Parke, *Biophysics*, https://doi.org/10.1007/978-3-030-44146-3

Q

R

S

Printed in the United States
by Baker & Taylor Publisher Services